Life

The Science of Biology

NINTH EDITION

Sinauer Associates, Inc.

W. H. Freeman and Company

NINTH
EDITION

Life The Science of Biology

DAVID
SADAVA
The Claremont Colleges
Claremont, California

DAVID M.
HILLIS
University of Texas
Austin, Texas

H. CRAIG
HELLER
Stanford University
Stanford, California

MAY R.
BERENBAUM
University of Illinois
Urbana-Champaign, Illinois

About the Cover

The cover of *Life* captures many themes that echo throughout the book. The photograph shows a lesser long-nosed bat pollinating a saguaro cactus. This cactus has evolved large flowers that produce copious quantities of nectar. The nectar attracts many species that pollinate the cactus, including bats. The ability of bats to hover as they feed on the nectar of the cactus is an excellent example of adaptation of body form and physiology. These themes of adaptation, evolution, nutrition, reproduction, species interactions, and integrated form and function are ideas that are repeated throughout the chapters of *Life*. Photograph copyright © Dr. Merlin D. Tuttle/Photo Researchers, Inc.

The Frontispiece

Blue wildebeest and Burchell's zebra migrate together through Serengeti National Park, Tanzania. Copyright © Art Wolfe, www.artwolfe.com.

LIFE: The Science of Biology, Ninth Edition

Address editorial correspondence to:
Sinauer Associates Inc., 23 Plumtree Road, Sunderland, MA 01375 U.S.A.
www.sinauer.com
publish@sinauer.com

Address orders to:
MPS / W. H. Freeman & Co., Order Dept., 16365 James Madison Highway,
U.S. Route 15, Gordonsville, VA 22942 U.S.A.
Examination copy information: 1-800-446-8923
Orders: 1-888-330-8477

Planet Friendly Publishing
✓ Made in the United States
✓ Printed on Recycled Paper
Text: 10%   Cover: 10%
GREEN EDITION
Learn more: www.greenedition.org

SUSTAINABLE FORESTRY INITIATIVE
Certified Chain of Custody
Promoting Sustainable Forest Management
www.sfiprogram.org

Library of Congress Cataloging-in-Publication Data

Life, the science of biology / David Sadava .. [et al.]. — 9th ed.
      p. cm.
   Includes index.
   ISBN 978-1-4292-1962-4 (hardcover) — 978-1-4292-4645-3 (pbk. : v. 1) —
ISBN 978-1-4292-4644-6 (pbk. : v. 2) — ISBN 978-1-4292-4647-7 (pbk. : v. 3)
1. Biology. I. Sadava, David E.
   QH308.2.L565 2011
   570—dc22
                                                                      2009036693

Printed in U.S.A.
Second Printing July 2010
The Courier Companies, Inc.

To Bill Purves and Gordon Orians,
extraordinary colleagues, biologists, and teachers,
and the original authors of LIFE

The Authors

CRAIG HELLER DAVID HILLIS MAY BERENBAUM DAVID SADAVA

DAVID SADAVA is the Pritzker Family Foundation Professor of Biology, Emeritus, at the Keck Science Center of Claremont McKenna, Pitzer, and Scripps, three of The Claremont Colleges. In addition, he is Adjunct Professor of Cancer Cell Biology at the City of Hope Medical Center. Twice winner of the Huntoon Award for superior teaching, Dr. Sadava taught courses on introductory biology, biotechnology, biochemistry, cell biology, molecular biology, plant biology, and cancer biology. In addition to *Life: The Science of Biology*, he is the author or coauthor of books on cell biology and on plants, genes, and crop biotechnology. His research has resulted in many papers coauthored with his students, on topics ranging from plant biochemistry to pharmacology of narcotic analgesics to human genetic diseases. For the past 15 years, he has investigated multi-drug resistance in human small-cell lung carcinoma cells with a view to understanding and overcoming this clinical challenge. At the City of Hope, his current work focuses on new anti-cancer agents from plants.

DAVID HILLIS is the Alfred W. Roark Centennial Professor in Integrative Biology and the Director of the Center for Computational Biology and Bioinformatics at the University of Texas at Austin, where he also has directed the School of Biological Sciences. Dr. Hillis has taught courses in introductory biology, genetics, evolution, systematics, and biodiversity. He has been elected into the membership of the National Academy of Sciences and the American Academy of Arts and Sciences, awarded a John D. and Catherine T. MacArthur Fellowship, and has served as President of the Society for the Study of Evolution and of the Society of Systematic Biologists. His research interests span much of evolutionary biology, including experimental studies of evolving viruses, empirical studies of natural molecular evolution, applications of phylogenetics, analyses of biodiversity, and evolutionary modeling. He is particularly interested in teaching and research about the practical applications of evolutionary biology.

CRAIG HELLER is the Lorry I. Lokey/BusinessWire Professor in Biological Sciences and Human Biology at Stanford University. He earned his Ph.D. from the Department of Biology at Yale University in 1970. Dr. Heller has taught in the core biology courses at Stanford since 1972 and served as Director of the Program in Human Biology, Chairman of the Biological Sciences Department, and Associate Dean of Research. Dr. Heller is a fellow of the American Association for the Advancement of Science and a recipient of the Walter J. Gores Award for excellence in teaching. His research is on the neurobiology of sleep and circadian rhythms, mammalian hibernation, the regulation of body temperature, the physiology of human performance, and the neurobiology of learning. Dr. Heller has done research on a huge variety of animals and physiological problems ranging from sleeping kangaroo rats, diving seals, hibernating bears, photoperiodic hamsters, and exercising athletes. Some of his recent work on the effects of temperature on human performance is featured in the opener to Chapter 40, "Physiology, Homeostasis, and Temperature Regulation."

MAY BERENBAUM is the Swanlund Professor and Head of the Department of Entomology at the University of Illinois at Urbana-Champaign. She has taught courses in introductory animal biology, entomology, insect ecology, and chemical ecology, and has received awards at the regional and national level for distinguished teaching from the Entomological Society of America. A fellow of the National Academy of Sciences, the American Academy of Arts and Sciences, and the American Philosophical Society, she served as President of the American Institute for Biological Sciences in 2009. Her research addresses insect–plant coevolution, from molecular mechanisms of detoxification to impacts of herbivory on community structure. Concerned with the practical application of ecological and evolutionary principles, she has examined impacts of genetic engineering, global climate change, and invasive species on natural and agricultural ecosystems. Devoted to fostering science literacy, she has published numerous articles and five books on insects for the general public.

Contents in Brief

Investigating Life/Tools for Investigating Life

TOOLS FOR INVESTIGATING LIFE

Preface

Biology is a dynamic, exciting, and important subject. It is dynamic because it is constantly changing, with new discoveries about the living world being made every day. (Although it is impossible to pinpoint an exact number, approximately 1 million new research articles in biology are published each year.) The subject is exciting because life in all of its forms has always fascinated people. As active scientists who have spent our careers teaching and doing research in a wide variety of fields, we know this first hand.

Biology has always been important in peoples' daily lives, if only through the effects of achievements in medicine and agriculture. Today more than ever the science of biology is at the forefront of human concerns as we face challenges raised both by recent advances in genome science and by the rapidly changing environment.

Life's new edition brings a fresh approach to the study of biology while retaining the features that have made the book successful in the past. A new coauthor, the distinguished entomologist May R. Berenbaum (University of Illinois at Urbana-Champaign) has joined our team, and the role of evolutionary biologist David Hillis (University of Texas at Austin) is greatly expanded in this edition. The authors hail from large, medium-sized, and small institutions. Our multiple perspectives and areas of expertise, as well as input from many colleagues and students who used previous editions, have informed our approach to this new edition.

Enduring Features

We remain committed to blending the presentation of core ideas with an emphasis on introducing students to the *process of scientific inquiry*. Having pioneered the idea of depicting seminal experiments in specially designed figures, we continue to develop this here, with 79 **INVESTIGATING LIFE** figures. Each of these figures sets the experiment in perspective and relates it to the accompanying text. As in previous editions, these figures employ a structure: Hypothesis, Method, Results, and Conclusion. They often include questions for further research that ask students to conceive an experiment that would explore a related question. Each *Investigating Life* figure has a reference to BioPortal (*yourBioPortal.com*), where citations to the original work as well as additional discussion and references to follow-up research can be found.

A related feature is the **TOOLS FOR INVESTIGATING LIFE** figures, which depict laboratory and field methods used in biology. These, too, have been expanded to provide more useful context for their importance.

Over a decade ago—in *Life's* Fifth Edition—the authors and publishers pioneered the much-praised use of **BALLOON CAPTIONS** in our figures. We recognized then, and it is even truer today, that many students are visual learners. The balloon captions bring explanations of intricate, complex processes directly into the illustration, allowing students to integrate information without repeatedly going back and forth between the figure, its legend, and the text.

Life is the only introductory textbook for biology majors to begin each chapter with a story. These **OPENING STORIES** provide historical, medical, or social context and are intended to intrigue students while helping them see how the chapter's biological subject relates to the world around them. In the new edition, all of the opening stories (some 70 percent of which are new) are revisited in the body of the chapter to drive home their relevance.

We continue to refine our well-received *chapter organization*. The chapter-opening story ends with a brief **IN THIS CHAPTER** preview of the major subjects to follow. A **CHAPTER OUTLINE** asks questions to emphasize scientific inquiry, each of which is answered in a major section of the chapter. A **RECAP** at the end of each section asks the student to pause and answer questions to review and test their mastery of the previous material. The end-of-chapter summary continues this inquiry framework and highlights key figures, bolded terms, and activities and animated tutorials available in BioPortal.

New Features

Probably the most important new feature of this edition is *new authorship*. Like the biological world, the authorship team of *Life* continues to evolve. While two of us (Craig Heller and David Sadava) continue as coauthors, David Hillis has a greatly expanded role, with full responsibility for the units on evolution and diversity. New coauthor May Berenbaum has rewritten the chapters on ecology. The perspectives of these two acclaimed experts have invigorated the entire book (as well as their coauthors).

Even with the enduring features (see above), this edition has a different look and feel from its predecessor. A fresh *new design* is more open and, we hope, more accessible to students. The extensively *revised art program* has a contemporary style and color palette. The information flow of the figures is easier to follow, with numbered balloons as a guide for students. There are new conceptual figures, including a striking visual timeline for the evolution of life on Earth (Figure 25.12) and a single overview figure that summarizes the information in the genome (Figure 17.4).

In response to instructors who asked for more real-world data, we have incorporated a feature introduced online in the Eighth Edition, **WORKING WITH DATA**. There are now 36 of these exercises, most of which relate to an *Investigating Life* figure. Each is referenced at the end of the relevant chapter and is available online via BioPortal (*yourBioPortal.com*). In these exercises, we describe in detail the context and approach of the research paper that forms the basis of the figure. We then ask the student to examine the data, to make calculations, and to draw conclusions.

We are proud that this edition is a *greener Life,* with the goal of reducing our environmental impact. This is the first introductory biology text to be printed on paper earning the Forest Stewardship Council label, the "gold standard" in green paper products, and it is manufactured from wood harvested from sustainable forests. And, of course, we also offer *Life* as an eBook.

The Ten Parts

We have reorganized the book into ten parts. **Part One, The Science of Life and Its Chemical Basis**, sets the stage for the book: the opening chapter focuses on biology as an exciting science. We begin with a startling observation: the recent, dramatic decline of amphibian species throughout the world. We then show how biologists have formed hypotheses for the causes of this environmental problem and are testing them by carefully designed experiments, with a view not only to understanding the decline, but reversing it. This leads to an outline of the basic principles of biology that are the foundation for the rest of the book: the unity of life at the cellular level and how evolution unites the living world. This is followed by chapters on the basic chemical building blocks that underlie life. We have added a new chapter on nucleic acids and the origin of life, introducing the concepts of genes and gene expression early and expanding our coverage of the major ideas on how life began and evolved at its earliest stages.

In **Part Two, Cells**, we describe the view of life as seen through cells, its structural units. In response to comments by users of our previous edition, we have moved the chapter on cell signaling and communication from the genetics section to this part of the book, with a change in emphasis from genes to cells. There is an updated discussion of ideas on the origin of cells and organelles, as well as expanded treatment of water transport across membranes.

Part Three, Cells and Energy, presents an integrated view of biochemistry. For this edition, we have worked to clarify such challenging concepts as energy transfer, allosteric enzymes, and biochemical pathways. There is extensive revision of the discussions of alternate pathways of photosynthetic carbon fixation, as well as a greater emphasis on applications throughout these chapters.

Part Four, Genes and Heredity, is extensively revised and reorganized to improve clarity, link related concepts, and provide updates from recent research results. Separate chapters on prokaryotic genetics and molecular medicine have been removed and their material woven into relevant chapters. For example, our chapter on cell reproduction now includes a discussion of how the basic mechanisms of cell division are altered in cancer cells. The chapter on transmission genetics now includes coverage of this phenomenon in prokaryotes. New chapters on gene expression and gene regulation compare prokaryotic and eukaryotic mechanisms and include a discussion of

epigenetics. A new chapter on mutation describes updated applications of medical genetics.

In **Part Five, Genomes**, we reinforce the concepts of the previous part, beginning with a new chapter on genomes—how they are analyzed and what they tell us about the biology of prokaryotes and eukaryotes, including humans. This leads to a chapter describing how our knowledge of molecular biology and genetics underpins biotechnology (the application of this knowledge to practical problems). We discuss some of the latest uses of biotechnology, including environmental cleanup. Part Five finishes with two chapters on development that explore the themes of molecular biology and evolution, linking these two parts of the book.

Part Six, The Patterns and Processes of Evolution, emphasizes the importance of evolutionary biology as a basis for comparing and understanding all aspects of biology. These chapters have been extensively reorganized and revised, as well as updated with the latest thinking of biologists in this rapidly changing field. This part now begins with the evidence and mechanisms of evolution, moves into a discussion of phylogenetic trees, then covers speciation and molecular evolution, and concludes with the evolutionary history of life on Earth. An integrated timeline of evolutionary history shows the timing of major events of biological evolution, the movements of the continents, floral and faunal reconstructions of major time periods, and depicts some of the fossils that form the basis of the reconstructions.

In **Part Seven, The Evolution of Diversity**, we describe the latest views on biodiversity and evolutionary relationships. Each chapter has been revised to make it easier for the reader to appreciate the major changes that have evolved within the various groups of organisms. We emphasize understanding the big picture of organismal diversity, as opposed to memorizing a taxonomic hierarchy and names (although these are certainly important). Throughout the book, the tree of life is emphasized as a way of understanding and organizing biological information. A *Tree of Life Appendix* allows students to place any group of organisms mentioned in the text of our book into the context of the rest of life. The web-based version of this appendix provides links to photos, keys, species lists, distribution maps, and other information to help students explore biodiversity of specific groups in greater detail.

After modest revisions in the past two editions, **Part Eight, Flowering Plants: Form and Function**, has been extensively reorganized and updated with the help of Sue Wessler, to include both classical and more recent approaches to plant physiology. Our emphasis is not only on the basic findings that led to the elucidation of mechanisms for plant growth and reproduction, but also on the use of genetics of model organisms. There is expanded coverage of the cell signaling events that regulate gene expression in plants, integrating concepts introduced earlier in the book. New material on how plants respond to their environment is included, along with links to both the book's earlier descriptions of plant diversity and later discussions of ecology.

Part Nine, Animals: Form and Function, continues to provide a solid foundation in physiology through comprehensive coverage of basic principles of function of each organ system and then emphasis on mechanisms of control and integration. An important reorganization has been moving the chapter on immunology from earlier in the book, where its emphasis was on molecular genetics, to this part, where it is more closely allied to the information systems of the body. In addition, we have added a number of new experiments and made considerable effort to clarify the sometimes complex phenomena shown in the illustrations.

Part Ten, Ecology, has been significantly revised by our new coauthor, May Berenbaum. A new chapter of biological interactions has been added (a topic formerly covered in the community ecology chapter). Full of interesting anecdotes and discussions of field studies not previously described in biology texts, this new ecology unit offers practical insights into how ecologists acquire, interpret, and apply real data. This brings the book full circle, drawing upon and reinforcing prior topics of energy, evolution, phylogenetics, Earth history, and animal and plant physiology.

Exceptional Value Formats

We again provide *Life* both as the full book and as a cluster of *paperbacks*. Thus, instructors who want to use less than the whole book can choose from these split volumes, each with the book's front matter, appendices, glossary, and index.

Volume I, The Cell and Heredity, includes: Part One, The Science of Life and Its Chemical Basis (Chapters 1–4); Part Two, Cells (Chapters 5–7); Part Three, Cells and Energy (Chapters 8–10); Part Four, Genes and Heredity (Chapters 11–16); and Part Five, Genomes (Chapters 17–20).

Volume II, Evolution, Diversity, and Ecology, includes: Chapter 1, Studying Life; Part Six, The Patterns and Processes of Evolution (Chapters 21–25); Part Seven, The Evolution of Diversity (Chapters 26–33); and Part Ten, Ecology (Chapters 54–59).

Volume III, Plants and Animals, includes: Chapter 1, Studying Life; Part Eight, Flowering Plants: Form and Function (Chapters 34–39); and Part Nine, Animals: Form and Function (Chapters 40–53).

Responding to student concerns, we offer two options of the entire book at a *significantly reduced cost*. After it was so well received in the previous edition, we again provide *Life* as a *loose-leaf version*. This shrink-wrapped, unbound, 3-hole punched version fits into a 3-ring binder. Students take only what they need to class and can easily integrate any instructor handouts or other resources.

Life was the first comprehensive biology text to offer the entire book as a truly robust *eBook*. For this edition, we continue to offer a flexible, interactive ebook that gives students a new way to read the text and learn the material. The ebook integrates the student media resources (animations, quizzes, activities, etc.) and offers instructors a powerful way to customize the textbook with their own text, images, Web links, documents, and more.

Media and Supplements for the Ninth Edition

The wide range of media and supplements that accompany *Life*, Ninth Edition have all been created with the dual goal of helping students learn the material presented in the textbook more efficiently and helping instructors teach their courses more effectively. Students in majors introductory biology are faced with learning a tremendous number of new concepts, facts, and terms, and the more different ways they can study this material, the more efficiently they can master it.

All of the *Life* media and supplemental resources have been developed specifically for this textbook. This provides strong consistency between text and media, which in turn helps students learn more efficiently. For example, the animated tutorials and activities found in BioPortal were built using textbook art, so that the manner in which structures are illustrated, the colors used to identify objects, and the terms and abbreviations used are all consistent.

For the Ninth Edition, a new set of Interactive Tutorials gives students a new way to explore many key topics across the textbook. These new modules allow the student to learn by doing, including solving problem scenarios, working with experimental techniques, and exploring model systems. All new copies of the Ninth Edition include access to the robust new version of BioPortal, which brings together all of *Life's* student and instructor resources, powerful assessment tools, and new integration with Prep-U adaptive quizzing.

The rich collection of visual resources in the Instructor's Media Library provides instructors with a wide range of options for enhancing lectures, course websites, and assignments. Highlights include: layered art PowerPoint® presentations that break down complex figures into detailed, step-by-step presentations; a collection of approximately 200 video segments that can help capture the attention and imagination of students; and PowerPoint slides of textbook art with editable labels and leaders that allow easy customization of the figures.

For a detailed description of all the media and supplements available for the Ninth Edition, please turn to *"Life's* Media and Supplements," on page xvii.

Many People to Thank

"If I have seen farther, it is by standing on the shoulders of giants." The great scientist Isaac Newton wrote these words over 330 years ago and, while we certainly don't put ourselves in his lofty place in science, the words apply to us as coauthors of this text. This is the first edition that does not bear the names of Bill Purves and Gordon Orians. As they enjoy their "retirements," we are humbled by their examples as biologists, educators, and writers.

One of the wisest pieces of advice ever given to a textbook author is to "be passionate about your subject, but don't put your ego on the page." Considering all the people who looked over our shoulders throughout the process of creating this book, this advice could not be more apt. We are indebted to many people who gave invaluable help to make this book what it is. First and foremost are our colleagues, biologists from over 100 institutions. Some were users of the previous edition, who suggested many improvements. Others reviewed our chapter drafts in detail, including advice on how to improve the illustrations. Still others acted as accuracy reviewers when the book was almost completed. All of these biologists are listed in the Reviewer credits.

Of special note is Sue Wessler, a distinguished plant biologist and textbook author from the University of Georgia. Sue looked critically at Part Eight, Flowering Plants: Form and Function, wrote three of the chapters (34–36), and was important in the revision of the other three (37–39). The new approach to plant biology in this edition owes a lot to her.

The pace of change in biology and the complexities of preparing a book as broad as this one necessitated having two developmental editors. James Funston coordinated Parts 1–5, and Carol Pritchard-Martinez coordinated Parts 6–10. We benefitted from the wide experience, knowledge, and wisdom of both of them. As the chapter drafts progressed, we were fortunate to have experienced biologist Laura Green lending her critical eye as in-house editor. Elizabeth Morales, our artist, was on her third edition with us. As we have noted, she extensively revised almost all of the prior art and translated our crude sketches into beautiful new art. We hope you agree that our art program remains superbly clear and elegant. Our copy editors, Norma Roche, Liz Pierson, and Jane Murfett, went far beyond what such people usually do. Their knowledge and encyclopedic recall of our book's chapters made our prose sharper and more accurate. Diane Kelly, Susan McGlew, and Shannon Howard effectively coordinated the hundreds of reviews that we described above. David McIntyre was a terrific photo editor, finding over 550 new photographs, including many new ones of his own, that enrich the book's content and visual statement. Jefferson Johnson is responsible for the design elements that make this edition of *Life* not just clear and easy to learn from, but beautiful as well. Christopher Small headed the production department—Joanne Delphia, Joan Gemme, Janice Holabird, and Jefferson Johnson—who contributed in innumerable ways to bringing *Life* to its final form. Jason Dirks once again coordinated the creation of our array of media and supplements, including our superb new Web resources. Carol Wigg, for the ninth time in nine editions, oversaw the editorial process; her influence pervades the entire book.

W. H. Freeman continues to bring *Life* to a wider audience. Associate Director of Marketing Debbie Clare, the Regional Specialists, Regional Managers, and experienced sales force are effective ambassadors and skillful transmitters of the features and unique strengths of our book. We depend on their expertise and energy to keep us in touch with how *Life* is perceived by its users. And thanks also to the Freeman media group for eBook and BioPortal production.

Finally, we are indebted to Andy Sinauer. Like ours, his name is on the cover of the book, and he truly cares deeply about what goes into it. Combining decades of professionalism, high standards, and kindness to all who work with him, he is truly our mentor and friend.

DAVID SADAVA

DAVID HILLIS

CRAIG HELLER

MAY BERENBAUM

Reviewers for the Ninth Edition

Between-Edition Reviewers

David D. Ackerly, University of California, Berkeley

Amy Bickham Baird, University of Leiden

Jeremy Brown, University of California, Berkeley

John M. Burke, University of Georgia

Ruth E. Buskirk, University of Texas, Austin

Richard E. Duhrkopf, Baylor University

Casey W. Dunn, Brown University

Erika J. Edwards, Brown University

Kevin Folta, University of Florida

Lynda J. Goff, University of California, Santa Cruz

Tracy A. Heath, University of Kansas

Shannon Hedtke, University of Texas, Austin

Richard H. Heineman, University of Texas, Austin

Albert Herrera, University of Southern California

David S. Hibbett, Clark University

Norman A. Johnson, University of Massachusetts

Walter S. Judd, University of Florida

Laura A. Katz, Smith College

Emily Moriarty Lemmon, Florida State University

Sheila McCormick, University of California, Berkeley

Robert McCurdy, Independence Creek Nature Preserve

Jacalyn Newman, University of Pittsburgh

Juliet F. Noor, Duke University

Theresa O'Halloran, University of Texas, Austin

K. Sata Sathasivan, University of Texas, Austin

H. Bradley Shaffer, University of California, Davis

Rebecca Symula, Yale University

Christopher D. Todd, University of Saskatchewan

Elizabeth Willott, University of Arizona

Kenneth Wilson, University of Saskatchewan

Manuscript Reviewers

Tamarah Adair, Baylor University

William Adams, University of Colorado, Boulder

Gladys Alexandre, University of Tennessee, Knoxville

Shivanthi Anandan, Drexel University

Brian Bagatto, University of Akron

Lisa Baird, University of San Diego

Stewart H. Berlocher, University of Illinois, Urbana-Champaign

William Bischoff, University of Toledo

Meredith M. Blackwell, Louisiana State University

David Bos, Purdue University

Jonathan Bossenbroek, University of Toledo

Nicole Bournias-Vardiabasis, California State University, San Bernardino

Nancy Boury, Iowa State University

Sunny K. Boyd, University of Notre Dame

Judith L. Bronstein, University of Arizona

W. Randy Brooks, Florida Atlantic University

James J. Bull, University of Texas, Austin

Darlene Campbell, Cornell University

Domenic Castignetti, Loyola University, Chicago

David T. Champlin, University of Southern Maine

Shu-Mei Chang, University of Georgia

Samantha K. Chapman, Villanova University

Patricia Christie, MIT

Wes Colgan, Pikes Peak Community College

John Cooper, Washington University

Ronald Cooper, University of California, Los Angeles

Elizabeth Cowles, Eastern Connecticut State University

Jerry Coyne, University of Chicago

William Crampton, University of Central Florida

Michael Dalbey, University of California, Santa Cruz

Anne Danielson-Francois, University of Michigan, Dearborn

Grayson S. Davis, Valparaiso University

Kevin Dixon, Florida State University

Zaldy Doyungan, Texas A&M University, Corpus Christi

Ernest F. Dubrul, University of Toledo

Roland Dute, Auburn University

Scott Edwards, Harvard University

William Eldred, Boston University

David Eldridge, Baylor University

Joanne Ellzey, University of Texas, El Paso

Susan H. Erster, State University of New York, Stony Book

Brent E. Ewers, University of Wyoming

Kevin Folta, University of Florida

Brandon Foster, Wake Technical Community College

Richard B. Gardiner, University of Western Ontario

Douglas Gayou, University of Missouri, Columbia

John R. Geiser, Western Michigan University

Arundhati Ghosh, University of Pittsburgh

Alice Gibb, Northern Arizona University

Scott Gilbert, Swarthmore College

Matthew R. Gilg, University of North Florida

Elizabeth Godrick, Boston University

Lynda J. Goff, University of California, Santa Cruz

Elizabeth Blinstrup Good, University of Illinois, Urbana-Champaign

John Nicholas Griffis, University of Southern Mississippi

Cameron Gundersen, University of California, Los Angeles

Kenneth Halanych, Auburn University

E. William Hamilton, Washington and Lee University

Monika Havelka, University of Toronto at Mississauga

Tyson Hedrick, University of North Carolina, Chapel Hill

Susan Hengeveld, Indiana University, Bloomington

Albert Herrera, University of Southern California

Kendra Hill, South Dakota State University

Richard W. Hill, Michigan State University

Erec B. Hillis, University of California, Berkeley

Jonathan D. Hillis, Carleton College

William Huddleston, University of Calgary

Dianne B. Jennings, Virginia Commonwealth University

Norman A. Johnson, University of Massachusetts, Amherst

William H. Karasov, University of Wisconsin, Madison

Susan Keen, University of California, Davis

Cornelis Klok, Arizona State University, Tempe

Olga Ruiz Kopp, Utah Valley University

William Kroll, Loyola University, Chicago

Allen Kurta, Eastern Michigan University

Rebecca Lamb, Ohio State University

Brenda Leady, University of Toledo

Hugh Lefcort, Gonzaga University

Sean C. Lema, University of North Carolina, Wilmington

Nathan Lents, John Jay College, City University of New York

Rachel A. Levin, Amherst College

Donald Levin, University of Texas, Austin

Bernard Lohr, University of Maryland, Baltimore County

Barbara Lom, Davidson College

David J. Longstreth, Louisiana State University

Catherine Loudon, University of California, Irvine

Francois Lutzoni, Duke University

Charles H. Mallery, University of Miami

Kathi Malueg, University of Colorado, Colorado Springs

Richard McCarty, Johns Hopkins University

Sheila McCormick, University of California, Berkeley

Francis Monette, Boston University

Leonie Moyle, Indiana University, Bloomington

Jennifer C. Nauen, University of Delaware

Jacalyn Newman, University of Pittsburgh

Alexey Nikitin, Grand Valley State University

Shawn E. Nordell, Saint Louis University

Tricia Paramore, Hutchinson Community College

Nancy J. Pelaez, Purdue University

Robert T. Pennock, Michigan State University

Roger Persell, Hunter College

Debra Pires, University of California, Los Angeles

Crima Pogge, City College of San Francisco

Jaimie S. Powell, Portland State University

Susan Richardson, Florida Atlantic University

David M. Rizzo, University of California, Davis

Benjamin Rowley, University of Central Arkansas

Brian Rude, Mississippi State University

Ann Rushing, Baylor University

Christina Russin, Northwestern University

Udo Savalli, Arizona State University, West

Frieder Schoeck, McGill University

Paul J. Schulte, University of Nevada, Las Vegas

Stephen Secor, University of Alabama

Vijayasaradhi Setaluri, University of Wisconsin, Madison

H. Bradley Shaffer, University of California, Davis

Robin Sherman, Nova Southeastern University

Richard Shingles, Johns Hopkins University

James Shinkle, Trinity University

Richard M. Showman, University of South Carolina

Felisa A. Smith, University of New Mexico

Ann Berry Somers, University of North Carolina, Greensboro

Ursula Stochaj, McGill University

Ken Sweat, Arizona State University, West

Robin Taylor, Ohio State University

William Taylor, University of Toledo

Mark Thogerson, Grand Valley State University

Sharon Thoma, University of Wisconsin, Madison

Lars Tomanek, California Polytechnic State University

James Traniello, Boston University

Jeffrey Travis, State University of New York, Albany

Terry Trier, Grand Valley State University

John True, State University of New York, Stony Brook

Elizabeth Van Volkenburgh, University of Washington

John Vaughan, St. Petersburg College

Sara Via, University of Maryland

Suzanne Wakim, Butte College (Glenn Community College District)

Randall Walikonis, University of Connecticut

Cindy White, University of Northern Colorado

Elizabeth Willott, University of Arizona

Mark Wilson, Humboldt State University

Stuart Wooley, California State University, Stanislaus

Lan Xu, South Dakota State University

Heping Zhou, Seton Hall University

Accuracy Reviewers

John Alcock, Arizona State University

Gladys Alexandre, University of Tennessee, Knoxville

Lawrence A. Alice, Western Kentucky University

David R. Angelini, American University

Fabia U. Battistuzzi, Arizona State University

Arlene Billock, University of Louisiana, Lafayette

Mary A. Bisson, State University of New York, Buffalo

Meredith M. Blackwell, Louisiana State University

Nancy Boury, Iowa State University

Eldon J. Braun, University of Arizona

Daniel R. Brooks, University of Toronto

Jennifer L. Campbell, North Carolina State University

Peter C. Chabora, Queens College, CUNY

Patricia Christie, MIT

Ethan Clotfelter, Amherst College

Robert Connour, Owens Community College

Peter C. Daniel, Hofstra University

D. Michael Denbow, Virginia Polytechnic Institute

Laura DiCaprio, Ohio University

Zaldy Doyungan, Texas A&M University, Corpus Christi

Moon Draper, University of Texas, Austin

Richard E. Duhrkopf, Baylor University

Susan A. Dunford, University of Cincinnati

Brent E. Ewers, University of Wyoming

James S. Ferraro, Southern Illinois University

Rachel D. Fink, Mount Holyoke College

John R. Geiser, Western Michigan University

Elizabeth Blinstrup Good, University of Illinois, Urbana-Champaign

Melina E. Hale, University of Chicago

Patricia M. Halpin, University of California, Los Angeles

Jean C. Hardwick, Ithaca College

Monika Havelka, University of Toronto at Mississauga

Frank Healy, Trinity University

Marshal Hedin, San Diego State University

Albert Herrera, University of Southern California

David S. Hibbett, Clark University

James F. Holden, University of Massachusetts, Amherst

Margaret L. Horton, University of North Carolina, Greensboro

Helen Hull-Sanders, Canisius College

C. Darrin Hulsey, University of Tennessee, Knoxville

Timothy Y. James, University of Michigan

Dianne B. Jennings, Virginia Commonwealth University

Norman A. Johnson, University of Massachusetts, Amherst

Susan Jorstad, University of Arizona

Ellen S. Lamb, University of North Carolina, Greensboro

Dennis V. Lavrov, Iowa State University

Hugh Lefort, Gonzaga University

Rachel A. Levin, Amherst College

Bernard Lohr, University of Maryland, Baltimore County

Sharon E. Lynn, College of Wooster

Sarah Mathews, Harvard University

Susan L. Meacham, University of Nevada, Las Vegas

Mona C. Mehdy, University of Texas, Austin

Bradley G. Mehrtens, University of Illinois, Urbana-Champaign

James D. Metzger, Ohio State University

Thomas W. Moon, University of Ottowa

Thomas M. Niesen, San Francisco State University

Theresa O'Halloran, University of Texas, Austin

Thomas L. Pannabecker, University of Arizona

Nancy J. Pelaez, Purdue University

Nicola J. R. Plowes, Arizona State University

Gregory S. Pryor, Francis Marion University

Laurel B. Roberts, University of Pittsburgh

Anjana Sharma, Western Carolina University

Richard M. Showman, University of South Carolina

John B. Skillman, California State University, San Bernadino

John J. Stachowicz, University of California, Davis

Brook O. Swanson, Gonzaga University

Robin A. J. Taylor, Ohio State University

William Taylor, University of Toledo

Steven M. Theg, University of California, Davis

Mark Thogerson, Grand Valley State University

Christopher D. Todd, University of Saskatchewan

Jeffrey Travis, State University of New York, Albany

Joseph S. Walsh, Northwestern University

Andrea Ward, Adelphi University

Barry Williams, Michigan State University

Kenneth Wilson, University of Saskatchewan

Carol L. Wymer, Morehead State University

LIFE's Media and Supplements

BIOPORTAL featuring Prep-U

yourBioPortal.com

BioPortal is the new gateway to all of *Life's* state-of-the-art on-line resources for students and instructors. BioPortal includes the breakthrough quizzing engine, Prep-U; a fully interactive eBook; and additional premium learning media. The textbook is tightly integrated with BioPortal via in-text references that connect the printed text and media resources. The result is a powerful, easily-managed online course environment. Bio-Portal includes the following features and resources:

Life, Ninth Edition eBook

- Integration of all activities, animated tutorials, and other media resources.
- Quick, intuitive navigation to any section or subsection, as well as any printed book page number.
- In-text links to all glossary entries.
- Easy text highlighting.
- A bookmarking feature that allows for quick reference to any page.
- A powerful Notes feature that allows students to add notes to any page.
- A full glossary and index.
- Full-text search, including an additional option to search the glossary and index.
- Automatic saving of all notes, highlighting, and bookmarks.

Additional eBook features for instructors:

- Content Customization: Instructors can easily add pages of their own content and/or hide chapters or sections that they do not cover in their course.
- Instructor Notes: Instructors can choose to create an annotated version of the eBook with their own notes on any page. When students in the course log in, they see the instructor's personalized version of the eBook. Instructor notes can include text, Web links, images, links to all Bio-Portal content, and more.

Smarter than the average quiz

Built by educators, Prep-U focuses student study time exactly where it should be, through the use of personalized, adaptive quizzes that move students toward a better grasp of the material—and better grades. For *Life,* Ninth Edition, Prep-U is fully integrated into BioPortal, making it easy for instructors to take advantage of this powerful quizzing engine in their course. Features include:

- Adaptive quizzing
- Automatic results reporting into the BioPortal gradebook

- Misconception index
- Comparison to national data

Student Resources

Diagnostic Quizzing. The diagnostic quiz for each chapter of *Life* assesses student understanding of that chapter, and generates a Personalized Study Plan to effectively focus student study time. The plan includes links to specific textbook sections, animated tutorials, and activities.

Interactive Summaries. For each chapter, these dynamic summaries combine a review of important concepts with links to all of the key figures from the chapter as well as all of the relevant animated tutorials, activities, and key terms.

Animated Tutorials. Over 100 in-depth animated tutorials, in a new format for the Ninth Edition, present complex topics in a clear, easy-to-follow format that combines a detailed animation with an introduction, conclusion, and quiz.

Activities. Over 120 interactive activities help students learn important facts and concepts through a wide range of exercises, such as labeling steps in processes or parts of structures, building diagrams, and identifying different types of organisms.

NEW! Interactive Tutorials. New for the Ninth Edition, these tutorial modules help students master key concepts through hands-on activities that allow them to learn through action. With these tutorials, students can solve problem scenarios by applying concepts from the text, by working with experimental techniques, and by using interactive models to discover how biological mechanisms work. Each tutorial includes a self-assessment quiz that can be assigned.

Interactive Quizzes. Each question includes an image from the textbook, thorough feedback on both correct and incorrect answer choices, references to textbook pages, and links to eBook pages, for quick review.

BioNews from Scientific American. BioNews makes it easy for instructors to bring the dynamic nature of the biological sciences and up-to-the minute currency into their course. Accessible from within BioPortal, BioNews is a continuously updated feed of current news, podcasts, magazine articles, science blog entries, "strange but true" stories, and more.

NEW! BioNavigator. This unique visual resource is an innovative way to access the wide variety of *Life* media resources. Starting from the whole-Earth view, instructors and students can zoom to any level of biological inquiry, encountering links to a wealth of animations, activities, and tutorials on the full range of topics along the way.

Working with Data. Built around some of the original experiments depicted in the Investigating Life figures, these exercises help build quantitative skills and encourage student interest in how scientists do research, by looking at real experimental data and answering questions based on those data.

Flashcards. For each chapter of the book, there is a set of flashcards that allows the student to review all the key terminology from the chapter. Students can review the terms in study mode, and then quiz themselves on a list of terms.

Experiment Links. For each Investigating Life figure in the textbook, BioPortal includes an overview of the experiment featured in the figure and related research or applications that followed, a link to the original paper, and links to additional information related to the experiment.

Key Terms. The key terminology introduced in each chapter is listed, with definitions and audio pronunciations from the glossary.

Suggested Readings. For each chapter of the book, a list of suggested readings is provided as a resource for further study.

Glossary. The language of biology is often difficult for students taking introductory biology to master, so BioPortal includes a full glossary that features audio pronunciations of all terms.

Statistics Primer. This brief introduction to the use of statistics in biological research explains why statistics are integral to biology, and how some of the most common statistical methods and techniques are used by biologists in their work.

Math for Life. A collection of mathematical shortcuts and references to help students with the quantitative skills they need in the laboratory.

Survival Skills. A guide to more effective study habits. Topics include time management, note-taking, effective highlighting, and exam preparation.

Instructor Resources

Assessment

- Diagnostic Quizzing provides instant class comprehension feedback to instructors, along with targeted lecture resources for those areas requiring the most attention.
- Question banks include questions ranked according to Bloom's taxonomy.
- Question filtering: Allows instructors to select questions based on Bloom's category and/or textbook section.
- Easy-to-use customized assessment tools allow instructors to quickly create quizzes and many other types of assignments using any combination of the questions and resources provided along with their own materials.
- Comprehensive question banks include questions from the test bank, study guide, textbook self-quizzes, and diagnostic quizzes.

Media Resources *(see Instructor's Media Library below for details)*

- Videos
- PowerPoint® Presentations (Textbook Figures, Lectures, Layered Art)

- Supplemental Photos
- Clicker Questions
- Instructor's Manual
- Lecture Notes

Course Management

- Complete course customization capabilities
- Custom resources/document posting
- Robust Gradebook
- Communication Tools: Announcements, Calendar, Course Email, Discussion Boards

Note: The printed textbook, the eBook, BioPortal, and Prep-U can all be purchased individually as stand-alone items, in addition to being available in a package with the printed textbook.

Student Supplements

Study Guide (ISBN 978-1-4292-3569-3)

Jacalyn Newman, *University of Pittsburgh;* Edward M. Dzialowski, *University of North Texas;* Betty McGuire, *Cornell University;* Lindsay Goodloe, *Cornell University;* and Nancy Guild, *University of Colorado*

For each chapter of the textbook, the *Life* Study Guide offers a variety of study and review tools. The contents of each chapter are broken down into both a detailed review of the Important Concepts covered and a boiled-down Big Picture snapshot. New for the Ninth Edition, Diagram Exercises help students synthesize what they have learned in the chapter through exercises such as ordering concepts, drawing graphs, linking steps in processes, and labeling diagrams. In addition, Common Problem Areas and Study Strategies are highlighted. A set of study questions (both multiple-choice and short-answer) allows students to test their comprehension. All questions include answers and explanations.

Lecture Notebook (ISBN 978-1-4292-3583-9)

This invaluable printed resource consists of all the artwork from the textbook (more than 1,000 images with labels) presented in the order in which they appear in the text, with ample space for note-taking. Because the Notebook has already done the drawing, students can focus more of their attention on the concepts. They will absorb the material more efficiently during class, and their notes will be clearer, more accurate, and more useful when they study from them later.

Companion Website www.thelifewire.com

(Also available as a CD, which can be optionally packaged with the textbook.)

For those students who do not have access to BioPortal, the *Life*, Ninth Edition Companion Website is available free of charge (no access code required). The site features a variety of resources, including animations, flashcards, activities, study ideas, help with math and statistics, and more.

CatchUp Math & Stats

Michael Harris, Gordon Taylor, and Jacquelyn Taylor (ISBN 978-1-4292-0557-3)

This primer will help your students quickly brush up on the quantitative skills they need to succeed in biology. Presented in brief, accessible units, the book covers topics such as working with powers, logarithms, using and understanding graphs, calculating standard deviation, preparing a dilution series, choosing the right statistical test, analyzing enzyme kinetics, and many more.

Student Handbook for Writing in Biology, Third Edition

Karen Knisely, *Bucknell University* (ISBN 978-1-4292-3491-7)

This book provides practical advice to students who are learning to write according to the conventions in biology. Using the standards of journal publication as a model, the author provides, in a user-friendly format, specific instructions on: using biology databases to locate references; paraphrasing for improved comprehension; preparing lab reports, scientific papers, posters; preparing oral presentations in PowerPoint®, and more.

Bioethics and the New Embryology: Springboards for Debate

Scott F. Gilbert, Anna Tyler, and Emily Zackin (ISBN 978-0-7167-7345-0)

Our ability to alter the course of human development ranks among the most significant changes in modern science and has brought embryology into the public domain. The question that must be asked is: Even if we can do such things, should we?

BioStats Basics: A Student Handbook

James L. Gould and Grant F. Gould (ISBN 978-0-7167-3416-1)

BioStats Basics provides introductory-level biology students with a practical, accessible introduction to statistical research. Engaging and informal, the book avoids excessive theoretical and mathematical detail, and instead focuses on how core statistical methods are put to work in biology.

Instructor Media & Supplements

Instructor's Media Library

The *Life,* Ninth Edition Instructor's Media Library (available both online via BioPortal and on disc) includes a wide range of electronic resources to help instructors plan their course, present engaging lectures, and effectively assess student comprehension. The Media Library includes the following resources:

Textbook Figures and Tables. Every image and table from the textbook is provided in both JPEG (high- and low-resolution) and PDF formats. Each figure is provided both with and without balloon captions, and large, complex figures are provided in both a whole and split version.

Unlabeled Figures. Every figure is provided in an unlabeled format, useful for student quizzing and custom presentation development.

Supplemental Photos. The supplemental photograph collection contains over 1,500 photographs (in addition to those in the text), giving instructors a wealth of additional imagery to draw upon.

Animations. Over 100 detailed animations, revised and enlarged for the Ninth Edition, all created from the textbook's art program, and viewable in either narrated or step-through mode.

Videos. A collection of over 200 video segments that covers topics across the entire textbook and helps demonstrate the complexity and beauty of life. Includes the Cell Visualization Videos.

PowerPoint® Resources. For each chapter of the textbook, several different PowerPoint presentations are available. These give instructors the flexibility to build presentations in the manner that best suits their needs. Included are:

- Textbook Figures and Tables
- Lecture Presentation
- Figures with Editable Labels
- Layered Art Figures
- Supplemental Photos
- Videos
- Animations

Clicker Questions. A set of questions written specifically to be used with classroom personal response systems, such as the iClicker system, is provided for each chapter. These questions are designed to reinforce concepts, gauge student comprehension, and engage students in active participation.

Chapter Outlines, Lecture Notes, and the complete **Test File** are all available in Microsoft Word® format for easy use in lecture and exam preparation.

Intuitive Browser Interface provides a quick and easy way to preview and access all of the content on the Instructor's Media Library.

Instructor's Resource Kit

The *Life,* Ninth Edition Instructor's Resource Kit includes a wealth of information to help instructors in the planning and teaching of their course. The Kit includes:

Instructor's Manual, featuring (by chapter):

- A "What's New" guide to the Ninth Edition
- Brief chapter overview
- Chapter outline
- Key terms section with all of the boldface terms from the text

Lecture Notes. Detailed notes for each chapter, which can serve as the basis for lectures, including references to figures and media resources.

Media Guide. A visual guide to the extensive media resources available with the Ninth Edition of *Life.* The guide includes thumbnails and descriptions of every video, animation, lecture PowerPoint®, and supplemental photo in the Media Library, all organized by chapter.

Overhead Transparencies

This set includes over 1,000 transparencies—including all of the four-color line art and all of the tables from the text—along with convenient binders. All figures have been formatted and color-enhanced for clear projection in a wide range of conditions. Labels and images have been resized for improved readability.

Test File

Catherine Ueckert, *Northern Arizona University;* Norman Johnson, *University of Massachusetts;* Paul Nolan, *The Citadel;* Nicola Plowes, *Arizona State University*

The Test File offers more than 5,000 questions, covering the full range of topics presented in the textbook. All questions are referenced to textbook sections and page numbers, and are ranked according to Bloom's taxonomy. Each chapter includes a wide range of multiple choice and fill-in-the-blank questions. In addition, each chapter features a set of diagram questions that involve the student in working with illustrations of structures, graphs, steps in processes, and more. The electronic versions of the Test File (within BioPortal, the Instructor's Media Library, and the Computerized Test Bank CD) also include all of the textbook end-of-chapter Self-Quiz questions, all of the BioPortal Diagnostic Quiz questions, and all of the Study Guide multiple-choice questions.

Computerized Test Bank

The entire printed Test File, plus the textbook end-of-chapter Self-Quizzes, the BioPortal Diagnostic Quizzes, and the Study Guide multiple-choice questions are all included in Wimba's easy-to-use Diploma® software. Designed for both novice and advanced users, Diploma enables instructors to quickly and easily create or edit questions, create quizzes or exams with a "drag-and-drop" feature, publish to online courses, and print paper-based assignments.

Course Management System Support

As a service for *Life* adopters using WebCT, Blackboard, or ANGEL for their courses, full electronic course packs are available.

www.whfreeman.com/facultylounge/ majorsbio
NEW! The new Faculty Lounge for Majors Biology is the first publisher-provided website for the majors biology community that lets instructors freely communicate and share peer-reviewed lecture and teaching resources. It is continually updated and vetted by majors biology instructors—there is always something new to see. The Faculty Lounge offers convenient access to peer-recommended and vetted resources, including the following categories: Images, News, Videos, Labs, Lecture Resources, and Educational Research.

In addition, the site includes special areas for resources for lab coordinators, resources and updates from the *Scientific Teaching* series of books, and information on biology teaching workshops.

Developed for educators by educators, iclicker is a hassle-free radio-frequency classroom response system that makes it easy for instructors to ask questions, record responses, take attendance, and direct students through lectures as active participants. For more information, visit www.iclicker.com.

LabPartner

www.whfreeman.com/labpartner

NEW! LabPartner is a site designed to facilitate the creation of customized lab manuals. Its database contains a wide selection of experiments published by W. H. Freeman and Hayden-McNeil Publishing. Instructors can preview, choose, and re-order labs, interleave their original experiments, add carbonless graph paper and a pocket folder, and customize the cover both inside and out. LabPartner offers a variety of binding types: paperback, spiral, or loose-leaf. Manuals are printed on-demand once W. H. Freeman receives an order from a campus bookstore or school.

The Scientific Teaching Book Series is a collection of practical guides, intended for all science, technology, engineering and mathematics (STEM) faculty who teach undergraduate and graduate students in these disciplines. The purpose of these books is to help faculty become more successful in all aspects of teaching and learning science, including classroom instruction, mentoring students, and professional development. Authored by well-known science educators, the Series provides concise descriptions of best practices and how to implement them in the classroom, the laboratory, or the department. For readers interested in the research results on which these best practices are based, the books also provide a gateway to the key educational literature.

Scientific Teaching

Jo Handelsman, Sarah Miller, and Christine Pfund, *University of Wisconsin-Madison* (ISBN 978-1-4292-0188-9)

NEW! Transformations: Approaches to College Science Teaching

A Collection of Articles from CBE Life Sciences Education
Deborah Allen, *University of Delaware;* Kimberly Tanner, *San Francisco State University* (ISBN 978-1-4292-5335-2)

Contents

PART TWO

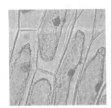

CELLS

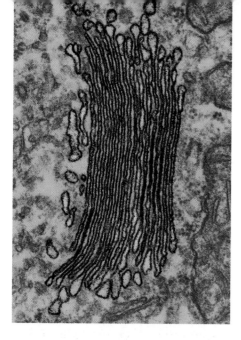

7 Cell Signaling and Communication 128

PART THREE

CELLS AND ENERGY

8 Energy, Enzymes, and Metabolism 148

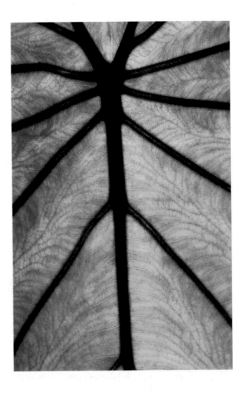

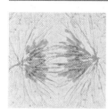

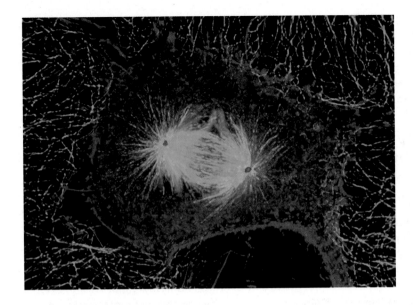

14 From DNA to Protein: Gene Expression 290

15 Gene Mutation and Molecular Medicine 316

PART FIVE
GENOMES

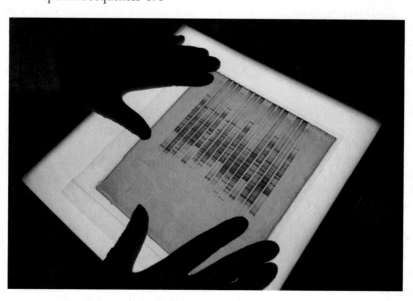

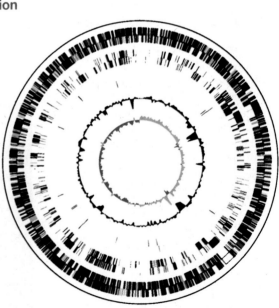

PART SIX

THE PATTERNS AND PROCESSES OF EVOLUTION

PART SEVEN

THE EVOLUTION OF DIVERSITY

30 Fungi: Recyclers, Pathogens, Parasites, and Plant Partners 626

31 Animal Origins and the Evolution of Body Plans 645

32 Protostome Animals 666

PART EIGHT

FLOWERING PLANTS: FORM AND FUNCTION

PART NINE
ANIMALS: FORM AND FUNCTION

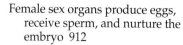

43 Animal Reproduction 899

44 Animal Development 922

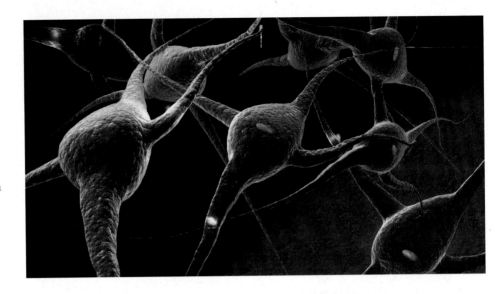

45 Neurons and Nervous Systems 943

46 Sensory Systems 964

53 Animal Behavior 1113

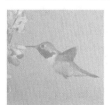

Life

The Science of Biology

NINTH EDITION

1 Studying Life

Why are frogs croaking?

Amphibians—frogs, toads, and salamanders—have been around for a long time. They watched the dinosaurs come and go. But today amphibian populations around the world are in dramatic decline, with more than a third of the world's amphibian species threatened with extinction. Why?

Biologists work to answer this question by making observations and doing experiments. A number of factors may be involved, and one possible cause may be the effects of agricultural pesticides and herbicides. Several studies have shown that many of these chemicals tested at realistic concentrations do not kill amphibians. But Tyrone Hayes, a biologist at the University of California at Berkeley, probed deeper.

Hayes focused on atrazine, the most widely used herbicide in the world and a common contaminant in fresh water. More than 70 million pounds of atrazine are applied to farmland in the United States every year, and it is used in at least 20 countries. Atrazine is usually applied in the spring, when many amphibians are breeding and thousands of tadpoles swim in the ditches, ponds, and streams that receive runoff from farms.

In his laboratory, Hayes and his associates raised frog tadpoles in water containing no atrazine and in water with concentrations ranging from 0.01 parts per billion (ppb) up to 25 ppb. The U.S. Environmental Protection Agency considers environmental levels of atrazine of 10 to 20 ppb of no concern; the level it considers safe in drinking water is 3 ppb. Rainwater in Iowa has been measured to contain 40 ppb. In Switzerland, where the use of atrazine is illegal, the chemical has been measured at approximately 1 ppb in rainwater.

In the Hayes laboratory, concentrations as low as 0.1 ppb had a dramatic effect on tadpole development: it feminized the males. In some of the adult males that developed from these larvae, the vocal structures used in mating calls were smaller than normal, female sex organs developed, and eggs were found growing in the testes. In other studies, normal adult male frogs exposed to 25 ppb had a tenfold reduction in testosterone levels and did not produce sperm. You can imagine the disastrous effects these developmental and hormonal changes could have on the capacity of frogs to breed and reproduce.

But Hayes's experiments were performed in the laboratory, with a species of frog bred for laboratory use. Would his results be the same in nature? To find out, he and his students traveled from Utah to Iowa, sampling water and collecting frogs. They analyzed the water

Frogs Are Having Serious Problems An alarming number of species of frogs, such as this tiny leaf frog (*Agalychnis calcarifer*) from Ecuador, are in danger of becoming extinct. The numerous possible reasons for the decline in global amphibian populations have been a subject of widespread scientific investigation.

A Biologist at Work Tyrone Hayes grew up near the great Congaree Swamp in South Carolina collecting turtles, snakes, frogs, and toads. Now a professor of biology at the University of California at Berkeley, he has more than 3,000 frogs in his laboratory and studies hormonal control of their development.

for atrazine and examined the frogs. In the only site where atrazine was undetectable in the water, the frogs were normal; in all the other sites, male frogs had abnormalities of the sex organs.

Like other biologists, Hayes made observations. He then made predictions based on those observations, and designed and carried out experiments to test his predictions. Some of the conclusions from his experiments, described at the end of this chapter, could have profound implications not only for amphibians but also for other animals, including humans.

IN THIS CHAPTER we identify and examine the most common features of living organisms and put those features into the context of the major principles that underlie all biology. Next we offer a brief outline of how life evolved and how the different organisms on Earth are related. We then turn to the subjects of biological inquiry and the scientific method. Finally we consider how knowledge discovered by biologists influences public policy.

1.1 What Is Biology?

Biology is the scientific study of living things. Biologists define "living things" as all the diverse organisms descended from a single-celled ancestor that evolved almost 4 billion years ago. Because of their common ancestry, living organisms share many characteristics that are not found in the nonliving world. Living organisms:

- consist of one or more cells
- contain genetic information
- use genetic information to reproduce themselves
- are genetically related and have evolved
- can convert molecules obtained from their environment into new biological molecules
- can extract energy from the environment and use it to do biological work
- can regulate their internal environment

This simple list, however, belies the incredible complexity and diversity of life. Some forms of life may not display all of these characteristics all of the time. For example, the seed of a desert plant may go for many years without extracting energy from the environment, converting molecules, regulating its internal environment, or reproducing; yet the seed is alive.

And what about viruses? Viruses do not consist of cells, and they cannot carry out physiological functions on their own; they must parasitize host cells to do those jobs for them. Yet viruses contain genetic information, and they certainly mutate and evolve (as we know, because evolving flu viruses require constant changes in the vaccines we create to combat them). The existence of viruses depends on cells, and it is highly probable that viruses evolved from cellular life forms. So, are viruses alive? What do you think?

This book explores the characteristics of life, how these characteristics vary among organisms, how they evolved, and how they work together to enable organisms to survive and reproduce. *Evolution* is a central theme of biology and therefore of this book. Through differential survival and reproduction, living systems evolve and become adapted to Earth's many environments. The processes of evolution have generated the enormous diversity that we see today as life on Earth.

Cells are the basic unit of life

We lay the chemical foundation for our study of life in the next three chapters, after which we will turn to cells and the processes by which they live, reproduce, age, and die. Some organisms are *unicellular,* consisting of a single cell that carries out

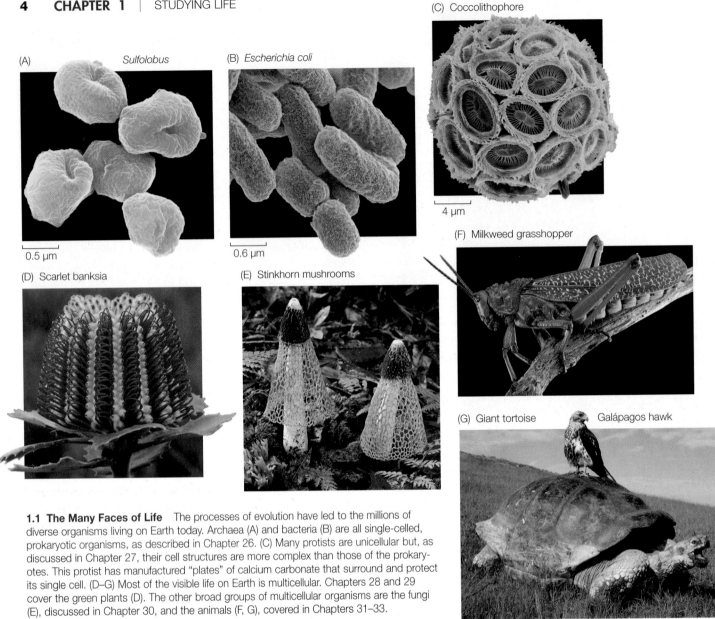

(A) *Sulfolobus*

(B) *Escherichia coli*

(C) Coccolithophore

4 µm

0.5 µm

0.6 µm

(D) Scarlet banksia

(E) Stinkhorn mushrooms

(F) Milkweed grasshopper

(G) Giant tortoise   Galápagos hawk

1.1 The Many Faces of Life   The processes of evolution have led to the millions of diverse organisms living on Earth today. Archaea (A) and bacteria (B) are all single-celled, prokaryotic organisms, as described in Chapter 26. (C) Many protists are unicellular but, as discussed in Chapter 27, their cell structures are more complex than those of the prokaryotes. This protist has manufactured "plates" of calcium carbonate that surround and protect its single cell. (D–G) Most of the visible life on Earth is multicellular. Chapters 28 and 29 cover the green plants (D). The other broad groups of multicellular organisms are the fungi (E), discussed in Chapter 30, and the animals (F, G), covered in Chapters 31–33.

all the functions of life (**Figure 1.1A–C**). Others are *multicellular*, made up of many cells that are specialized for different functions (**Figure 1.1D–G**). Viruses are *acellular*, although they depend on cellular organisms.

The discovery of cells was made possible by the invention of the microscope in the 1590s by the Dutch spectacle makers Hans and Zaccharias Janssen (father and son). In the mid- to late 1600s, Antony van Leeuwenhoek of Holland and Robert Hooke of England both made improvements on the Janssens' technology and used it to study living organisms. Van Leeuwenhoek discovered that drops of pond water teemed with single-celled organisms, and he made many other discoveries as he progressively improved his microscopes over a long lifetime of research. Hooke put pieces of plants under his microscope and observed that they were made up of repeated units he called *cells* (**Figure 1.2**). In 1676, Hooke wrote that van Leeuwenhoek had observed "a vast number of small animals in his Excrements which were most abounding when he was troubled with a Loosenesse and very few or none when he was well." This simple observation

represents the discovery of bacteria—and makes one wonder why scientists do some of the things they do.

More than a hundred years passed before studies of cells advanced significantly. As they were dining together one evening in 1838, Matthias Schleiden, a German biologist, and Theodor Schwann, from Belgium, discussed their work on plant and animal tissues, respectively. They were struck by the similarities in their observations and came to the conclusion that the basic structural elements of plants and animals were essentially the same. They formulated their conclusion as the **cell theory**, which states that:

- Cells are the basic structural and physiological units of all living organisms.
- Cells are both distinct entities and building blocks of more complex organisms.

But Schleiden and Schwann also believed (wrongly) that cells emerged by the self-assembly of nonliving materials, much as crystals form in a solution of salt. This conclusion was in ac-

1.2 Cells Are the Building Blocks of Life The development of microscopes revealed the microbial world to seventeenth-century scientists such as Robert Hooke, who proposed the concept of cells based on his observations. (A) Hooke drew the cells of a slice of plant tissue (cork) as he saw them under his optical microscope. (B) A modern optical, or "light," microscope reveals the intricacies of cells in a leaf. (C) Transmission electron microscopes (TEMs) allow scientists to see even smaller objects. TEMs do not visualize color; here color has been added to a black-and-white micrograph of cells in a duckweed stem.

(A)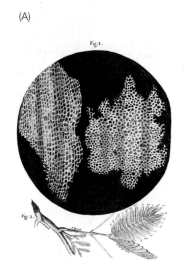

(B)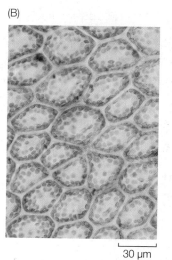

30 μm

(C)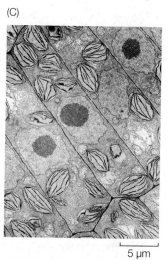

5 μm

cordance with the prevailing view of the day, which was that life can arise from non-life by spontaneous generation—mice from dirty clothes, maggots from dead meat, or insects from pond water.

The debate continued until 1859, when the French Academy of Sciences sponsored a contest for the best experiment to prove or disprove spontaneous generation. The prize was won by the great French scientist Louis Pasteur, who demonstrated that sterile broth directly exposed to the dirt and dust in air developed a culture of microorganisms, but a similar container of broth not directly exposed to air remained sterile (see Figure 4.7). Pasteur's experiment did not prove that it was microorganisms in the air that caused the broth to become infected, but it did uphold the conclusion that life must be present in order for new life to be generated.

Today scientists accept the fact that all cells come from preexisting cells and that the functional properties of organisms derive from the properties of their cells. Since cells of all kinds share both essential mechanisms and a common ancestry that goes back billions of years, modern cell theory has additional elements:

- All cells come from preexisting cells.
- All cells are similar in chemical composition.
- Most of the chemical reactions of life occur in aqueous solution within cells.
- Complete sets of genetic information are replicated and passed on during cell division.
- Viruses lack cellular structure but remain dependent on cellular organisms.

At the same time Schleiden and Schwann were building the foundation for the cell theory, Charles Darwin was beginning to understand how organisms undergo evolutionary change.

All of life shares a common evolutionary history

Evolution—change in the genetic makeup of biological populations through time—is the major unifying principle of biol-

ogy. Charles Darwin compiled factual evidence for evolution in his 1859 book *On the Origin of Species*. Since then, biologists have gathered massive amounts of data supporting Darwin's theory that all living organisms are descended from a common ancestor. Darwin also proposed one of the most important processes that produce evolutionary change. He argued that differential survival and reproduction among individuals in a population, which he termed **natural selection**, could account for much of the evolution of life.

Although Darwin proposed that living organisms are descended from common ancestors and are therefore related to one another, he did not have the advantage of understanding the mechanisms of genetic inheritance. Even so, he observed that offspring resembled their parents; therefore, he surmised, such mechanisms had to exist. That simple fact is the basis for the concept of a **species**. Although the precise definition of a species is complicated, in its most widespread usage it refers to a group of organisms that can produce viable and fertile offspring with one another.

But offspring do differ from their parents. Any population of a plant or animal species displays variation, and if you select breeding pairs on the basis of some particular trait, that trait is more likely to be present in their offspring than in the general population. Darwin himself bred pigeons, and was well aware of how pigeon fanciers selected breeding pairs to produce offspring with unusual feather patterns, beak shapes, or body sizes (see Figure 21.2). He realized that if humans could select for specific traits in domesticated animals, the same process could operate in nature; hence the term *natural selection* as opposed to artificial (human-imposed) selection.

How would natural selection function? Darwin postulated that different probabilities of survival and reproductive success would do the job. He reasoned that the reproductive capacity of plants and animals, if unchecked, would result in unlimited growth of populations, but we do not observe such growth in nature; in most species, only a small percentage of offspring survive to reproduce. Thus any trait that confers even a small increase in the probability that its possessor will survive and reproduce would be spread in the population.

Many leaves are wide and flat, a configuration that presents a maximum of photosynthetic surface to the sun. Some trees, such as this Japanese maple, lose their leaves in response to cold or dry weather.

The leaves of many evergreen conifers, such as spruce trees, are waxy-coated needles that resist water loss and are not shed on a yearly basis.

These water lilies are rooted in the pond bottom; their large leaves are flat "pads" that float on the surface.

The leaves of pitcher plants form a vessel that holds water. The plant receives extra nutrients from the decomposing bodies of insects that drown in the pitcher.

The ability to climb can be advantageous to a plant, enabling it to reach above other plants to obtain more sunlight. Some of the leaves of this climbing cucumber are tightly furled tendrils that wrap around a stake.

1.3 Adaptations to the Environment The leaves of all plants are specialized for photosynthesis—the sunlight-powered transformation of water and carbon dioxide into larger structural molecules called carbohydrates. The leaves of different plants, however, display many different adaptations to their individual environments.

Because organisms with certain traits survive and reproduce best under specific sets of conditions, natural selection leads to **adaptations**: structural, physiological, or behavioral traits that enhance an organism's chances of survival and reproduction in its environment (**Figure 1.3**). In addition to natural selection, evolutionary processes such as sexual selection (selection due to mate choice) and genetic drift (the random fluctuation of gene frequencies in a population due to chance events) contribute to the rise of diverse adaptations. These processes operating over evolutionary history have led to the remarkable array of life on Earth.

If all cells come from preexisting cells, and if all the diverse species of organisms on Earth are related by descent with modification from a common ancestor, then what is the source of information that is passed from parent to daughter cells and from parental organisms to their offspring?

Biological information is contained in a genetic language common to all organisms

Cells are the basic building blocks of organisms, but even a single cell is complex, with many internal structures and many functions that depend on information. The information required for a cell to function and interact with other cells—the "blueprint" for existence—is contained in the cell's **genome**, the sum total of all the DNA molecules it contains. **DNA** (deoxyribonucleic acid) molecules are long sequences of four different subunits called **nucleotides**. The sequence of the nucleotides contains genetic information. **Genes** are specific segments of DNA encoding the information the cell uses to make **proteins** (**Figure 1.4**). Protein molecules govern the chemical reactions within cells and form much of an organism's structure.

By analogy with a book, the nucleotides of DNA are like the letters of an alphabet. Protein molecules are the sentences. Combinations of proteins that form structures and control biochemical processes are the paragraphs. The structures and processes that are organized into different systems with specific tasks (such as digestion or transport) are the chapters of the book, and the complete book is the organism. If you were to write out your own genome using four letters to represent the four nucleotides, you would write more than 3 billion letters. Using the size type you are reading now, your genome would fill about a thousand books the size of this one. The mechanisms of evolution, including natural selection, are the authors and editors of all the books in the library of life.

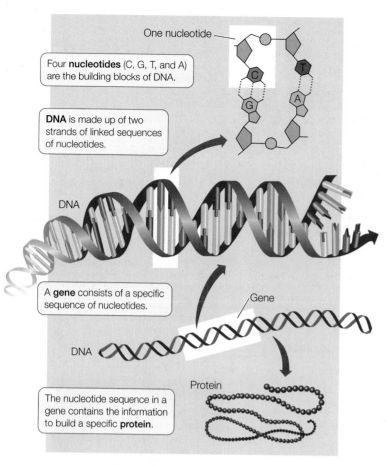

1.4 DNA Is Life's Blueprint The instructions for life are contained in the sequences of nucleotides in DNA molecules. Specific DNA nucleotide sequences comprise genes. The average length of a single human gene is 16,000 nucleotides. The information in each gene provides the cell with the information it needs to manufacture molecules of a specific protein.

of biochemical reactions that occur inside cells. Some of these reactions break down nutrient molecules into smaller chemical units, and in the process some of the energy contained in the chemical bonds of the nutrients is captured by high-energy molecules that can be used to do different kinds of cellular work.

One obvious kind of work cells do is mechanical—moving molecules from one cellular location to another, moving whole cells or tissues, or even moving the organism itself, as muscles do (**Figure 1.5A**). The most basic cellular work is the building, or *synthesis*, of new complex molecules and structures from smaller chemical units. For example, we are all familiar with the fact that carbohydrates eaten today may be deposited in the body as fat tomorrow (**Figure 1.5B**). Still another kind of work is the electrical work that is the essence of information processing in nervous systems. The sum total of all the chemical transformations and other work done in all the cells of an organism is its **metabolism**, or **metabolic rate**.

The myriad of biochemical reactions that go on in cells are integrally linked in that the products of one are the raw materials of the next. These complex networks of reactions must be integrated and precisely controlled; when they are not, the result is disease.

Living organisms regulate their internal environment

Multicellular organisms have an *internal environment* that is not cellular. That is, their individual cells are bathed in extracellular fluids, from which they receive nutrients and into which they excrete waste products of metabolism. The cells of multicellu-

All the cells of a multicellular organism contain the same genome, yet different cells have different functions and form different structures—contractile proteins form in muscle cells, hemoglobin in red blood cells, digestive enzymes in gut cells, and so on. Therefore, different types of cells in an organism must express different parts of the genome. How cells control gene expression in ways that enable a complex organism to develop and function is a major focus of current biological research.

The genome of an organism consists of thousands of genes. If the nucleotide sequence of a gene is altered, it is likely that the protein that gene encodes will be altered. Alterations of the genome are called *mutations*. Mutations occur spontaneously; they can also be induced by outside factors, including chemicals and radiation. Most mutations are either harmful or have no effect, but occasionally a mutation improves the functioning of the organism under the environmental conditions it encounters. Such beneficial mutations are the raw material of evolution and lead to adaptations.

Cells use nutrients to supply energy and to build new structures

Living organisms acquire *nutrients* from the environment. Nutrients supply the organism with energy and raw materials for carrying out biochemical reactions. Life depends on thousands

1.5 Energy Can Be Used Immediately or Stored (A) Animal cells break down and release the energy contained in the chemical bonds of food molecules to do mechanical work—in this kangaroo's case, to jump. (B) The cells of this Arctic ground squirrel have broken down the complex carbohydrates in plants and converted their molecules into fats, which are stored in the animal's body to provide an energy supply for the cold months.

lar organisms are specialized, or *differentiated*, to contribute in some way to the maintenance of the internal environment. With the evolution of specialization, differentiated cells lost many of the functions carried out by single-celled organisms, and must depend on the internal environment for essential services.

To accomplish their specialized tasks, assemblages of differentiated cells are organized into *tissues.* For example, a single muscle cell cannot generate much force, but when many cells combine to form the tissue of a working muscle, considerable force and movement can be generated (see Figure 1.5B). Different tissue types are organized to form *organs* that accomplish specific functions. For example, the heart, brain, and stomach are each constructed of several types of tissues. Organs whose functions are interrelated can be grouped into *organ systems*; the stomach, intestine, and esophagus, for example, are parts of the digestive system. The functions of cells, tissues, organs, and organ systems are all integral to the multicellular *organism*. We cover the biology of organisms in Parts Eight and Nine of this book.

Living organisms interact with one another

The internal hierarchy of the individual organism is matched by the external hierarchy of the biological world (**Figure 1.6**). Organisms do not live in isolation. A group of individuals of the same species that interact with one another is a *population,* and populations of all the species that live and interact in the same area are called a *community.* Communities together with their abiotic environment constitute an *ecosystem.*

Individuals in a population interact in many different ways. Animals eat plants and other animals (usually members of another species) and compete with other species for food and other resources. Some animals will prevent other individuals of their own species from exploiting a resource, whether it be food, nesting sites, or mates. Animals may also *cooperate* with members of their species, forming social units such as a termite colony or a flock of birds. Such interactions have resulted in the evolution of social behaviors such as communication.

Plants also interact with their external environment, which includes other plants, animals, and microorganisms. All terrestrial plants depend on complex partnerships with fungi, bacteria, and animals. Some of these partnerships are necessary to obtain nutrients, some to produce fertile seeds, and still others to disperse seeds. Plants compete with each other

1.6 Biology Is Studied at Many Levels of Organization
Life's properties emerge when DNA and other molecules are organized in cells. Energy flows through all the biological levels shown here.

yourBioPortal.com
GO TO Web Activity 1.1 • The Hierarchy of Life

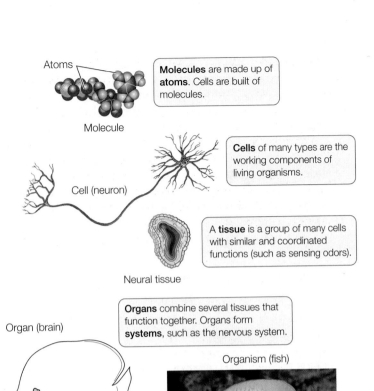

Atoms

Molecule

Molecules are made up of **atoms**. Cells are built of molecules.

Cell (neuron)

Cells of many types are the working components of living organisms.

Neural tissue

A **tissue** is a group of many cells with similar and coordinated functions (such as sensing odors).

Organ (brain)

Organs combine several tissues that function together. Organs form **systems**, such as the nervous system.

Organism (fish)

An **organism** is a recognizable, self-contained individual. Complex multicellular organisms are made up of organs and organ systems.

Population (school of fish)

A **population** is a group of many organisms of the same species.

Communities consist of populations of many different species.

Community (coral reef)

Biological communities in the same geographical location form **ecosystems**. Ecosystems exchange energy and create Earth's **biosphere**.

Biosphere

for light and water, and they have ongoing evolutionary interactions with the animals that eat them, evolving anti-predation adaptations or ways to attract the animals that assist in their reproduction. The interactions of populations of different plant and animal species in a community are major evolutionary forces that produce specialized adaptations.

Communities interacting over a broad geographic area with distinguishing physical features form ecosystems; examples might include an Arctic tundra, a coral reef, or a tropical rainforest. The ways in which species interact with one another and with their environment in communities and in ecosystems is the subject of *ecology* and of Part Ten of this book.

Discoveries in biology can be generalized

Because all life is related by descent from a common ancestor, shares a genetic code, and consists of similar building blocks—cells—knowledge gained from investigations of one type of organism can, with care, be generalized to other organisms. Biologists use **model systems** for research, knowing that they can extend their findings to other organisms, including humans. For example, our basic understanding of the chemical reactions in cells came from research on bacteria but is applicable to all cells, including those of humans. Similarly, the biochemistry of photosynthesis—the process by which plants use sunlight to produce biological molecules—was largely worked out from experiments on *Chlorella*, a unicellular green alga (see Figure 10.13). Much of what we know about the genes that control plant development is the result of work on *Arabidopsis thaliana*, a relative of the mustard plant. Knowledge about how animals develop has come from work on sea urchins, frogs, chickens, roundworms, and fruit flies. And recently, the discovery of a major gene controlling human skin color came from work on zebrafish. Being able to generalize from model systems is a powerful tool in biology.

1.2 How Is All Life on Earth Related?

What do biologists mean when they say that all organisms are *genetically related*? They mean that species on Earth share a *common ancestor*. If two species are similar, as dogs and wolves are, then they probably have a common ancestor in the fairly recent past. The common ancestor of two species that are more different—say, a dog and a deer—probably lived in the more distant past. And if two organisms are very different—such as a dog and a clam—then we must go back to the *very* distant past to find their common ancestor. How can we tell how far back in time the common ancestor of any two organisms lived? In other words, how do we discover the evolutionary relationships among organisms?

For many years, biologists have investigated the history of life by studying the *fossil record*—the preserved remains of organisms that lived in the distant past (**Figure 1.7**). Geologists supplied knowledge about the ages of fossils and the nature of the environments in which they lived. Biologists then inferred the evolutionary relationships among living and fossil organisms by comparing their anatomical similarities and differences. Frequently big gaps existed in the fossil record, forcing biologists to predict the nature of the "missing links" between two lineages of organisms. As the fossil record became more complete, those missing links were filled in.

Molecular methods for comparing genomes, described in Chapter 24, are enabling biologists to more accurately establish the degrees of relationship between living organisms and to use that information to interpret the fossil record. Molecular information can occasionally be gleaned from fossil specimens, such as recently deciphered genetic material from fossil bones of Ne-

---

1.1 RECAP

Living organisms are made of (or depend on) cells, are related by common descent and evolve, contain genetic information and use it to reproduce, extract energy from their environment and use it to do biological work, synthesize complex molecules to construct biological structures, regulate their internal environment, and interact with one another.

- Describe the relationship between evolution by natural selection and the genetic code. See pp. 6–7

- Why can the results of biological research on one species often be generalized to very different species? See p. 9

Now that you have an overview of the major features of life that you will explore in depth in this book, you can ask how and when life first emerged. In the next section we will summarize briefly the history of life from the earliest simple life forms to the complex and diverse organisms that inhabit our planet today.

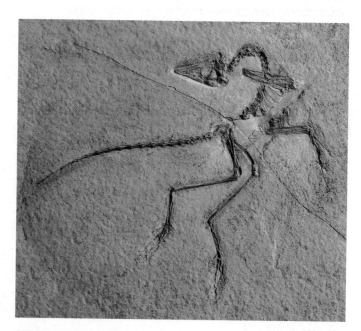

1.7 Fossils Give Us a View of Past Life This fossil, formed some 150 million years ago, is that of an *Archaeopteryx*, the earliest known representative of the birds. Birds evolved from the same group of reptiles as the modern crocodiles.

anderthals that led to the conclusion that even though Neanderthals and modern humans coexisted, they did not interbreed.

In general, the greater the differences between the genomes of two species, the more distant their common ancestor. Using molecular techniques, biologists are exploring fundamental questions about life. What were the earliest forms of life? How did simple organisms give rise to the great diversity of organisms alive today? Can we reconstruct a family tree of life?

Life arose from non-life via chemical evolution

Geologists estimate that Earth formed between 4.6 and 4.5 billion years ago. At first, the planet was not a very hospitable place. It was some 600 million years or more before the earliest life evolved. If we picture the history of Earth as a 30-day month, life first appeared somewhere toward the end of the first week (**Figure 1.8**).

When we consider how life might have arisen from nonliving matter, we must take into account the properties of the young

Earth's atmosphere, oceans, and climate, all of which were very different than they are today. Biologists postulate that complex biological molecules first arose through the random physical association of chemicals in that environment. Experiments simulating the conditions on early Earth have confirmed that the generation of complex molecules under such conditions is possible, even probable. The critical step for the evolution of life, however, had to be the appearance of molecules that could reproduce themselves and also serve as templates for the synthesis of large molecules with complex but stable shapes. The variation of the shapes of these large, stable molecules (described in Chapters 3 and 4) enabled them to participate in increasing numbers and kinds of chemical reactions with other molecules.

Cellular structure evolved in the common ancestor of life

The second critical step in the origin of life was the enclosure of complex biological molecules by *membranes* that contained them in a compact internal environment separate from the surrounding external environment. Fatlike molecules played a critical role because they are not soluble in water and they form membranous films. When agitated, these films can form spherical *vesicles*, which could have enveloped assemblages of biological molecules. The creation of an internal environment that concentrated the reactants and products of chemical reactions opened up the possibility that those reactions could be integrated and controlled. As described in Section 4.4, scientists postulate that this natural process of membrane formation resulted in the first cells with the ability to replicate themselves—the evolution of the first cellular organisms.

For more than 2 billion years after cells originated, all organisms consisted of only one cell. These first unicellular organisms were (and are, as multitudes of their descendants exist in similar form today) **prokaryotes**. Prokaryotic cells consist of DNA and other biochemicals enclosed in a membrane.

These early prokaryotes were confined to the oceans, where there was an abundance of complex molecules they could use as raw materials and sources of energy. The ocean shielded them from the damaging effects of ultraviolet light, which was intense at that time because there was little or no oxygen (O_2) in the atmosphere, and hence no protective ozone (O_3) layer.

Photosynthesis changed the course of evolution

To fuel their cellular metabolism, the earliest prokaryotes took in molecules directly from their environment and broke these small molecules down to release and use the energy contained in their chemical bonds. Many modern species of prokaryotes still function this way, and very successfully. During the early eons of life on Earth, there was no oxygen in the atmosphere. In fact, oxygen was toxic to the life forms that existed then.

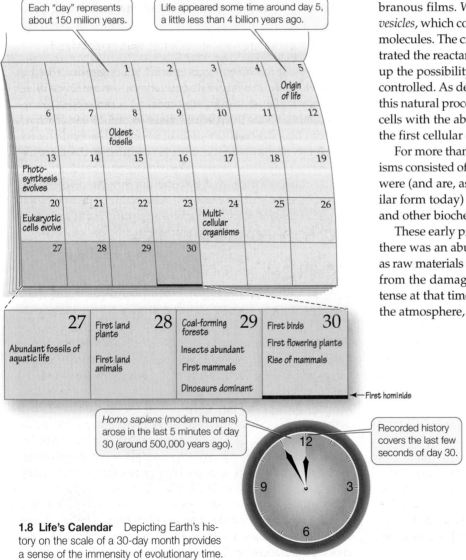

1.8 Life's Calendar   Depicting Earth's history on the scale of a 30-day month provides a sense of the immensity of evolutionary time.

350 μm

1.9 Photosynthetic Organisms Changed Earth's Atmosphere
These strands are composed of many cells of cyanobacteria. This modern species (*Oscillatoria tenuis*) may be very similar to the early photosynthetic prokaryotes responsible for the buildup of oxygen in Earth's atmosphere.

About 2.7 billion years ago, the evolution of **photosynthesis** changed the nature of life on Earth. The chemical reactions of photosynthesis transform the energy of sunlight into a form of biological energy that can power the synthesis of large molecules (see Chapter 10). These large molecules are the building blocks of cells, and they can be broken down to provide metabolic energy. Photosynthesis is the basis of much of life on Earth today because its energy-capturing processes provide food for other organisms.

Early photosynthetic cells were probably similar to present-day prokaryotes called *cyanobacteria* (**Figure 1.9**). Over time, photosynthetic prokaryotes became so abundant that vast quantities of O_2, which is a by-product of photosynthesis, slowly began to accumulate in the atmosphere. Oxygen was poisonous to many of the prokaryotes that lived at that time. Those organisms that did tolerate oxygen, however, were able to proliferate as the presence of oxygen opened up vast new avenues of evolution. *Aerobic metabolism* (energy production based on the conversion of O_2) is more efficient than *anaerobic* (non-O_2-using) *metabolism*, and today it is used by the majority of Earth's organisms. Aerobic metabolism allowed cells to grow larger.

Oxygen in the atmosphere also made it possible for life to move onto land. For most of life's history, ultraviolet (UV) radiation falling on Earth's surface was too intense to allow life to exist outside the shielding water. But the accumulation of photosynthetically generated oxygen in the atmosphere for more than 2 billion years gradually produced a layer of ozone in the upper atmosphere. By about 500 million years ago, the ozone layer was sufficiently dense and absorbed enough UV radiation to make it possible for organisms to leave the protection of the water and live on land.

Eukaryotic cells evolved from prokaryotes

Another important step in the history of life was the evolution of cells with discrete intracellular compartments, called **organelles**, which were capable of taking on specialized cellular functions. This event happened about 3 weeks into our calendar of Earth's history (see Figure 1.8). One of these organelles, the dense-appearing *nucleus* (Latin *nux*, "nut" or "core"), came to contain the cell's genetic information and gives these cells their name: **eukaryotes** (Greek *eu*, "true"; *karyon*, "kernel" or "core"). The eukaryotic cell is completely distinct from the cells of prokaryotes (*pro*, "before"), which lack nuclei and other internal compartments.

Some organelles are hypothesized to have originated by **endosymbiosis** when cells ingested smaller cells. The *mitochondria* that generate a cell's energy probably evolved from engulfed prokaryotic organisms. And *chloroplasts*—organelles specialized to conduct photosynthesis—could have originated when photosynthetic prokaryotes were ingested by larger eukaryotes. If the larger cell failed to break down this intended food object, a partnership could have evolved in which the ingested prokaryote provided the products of photosynthesis and the host cell provided a good environment for its smaller partner.

Multicellularity arose and cells became specialized

Until just over a billion years ago, all the organisms that existed—whether prokaryotic or eukaryotic—were unicellular. An important evolutionary step occurred when some eukaryotes failed to separate after cell division, remaining attached to each other. The permanent association of cells made it possible for some cells to specialize in certain functions, such as reproduction, while other cells specialized in other functions, such as absorbing nutrients and distributing them to neighboring cells. This **cellular specialization** enabled multicellular eukaryotes to increase in size and become more efficient at gathering resources and adapting to specific environments.

Biologists can trace the evolutionary tree of life

If all the species of organisms on Earth today are the descendants of a single kind of unicellular organism that lived almost 4 billion years ago, how have they become so different? A simplified answer is that as long as individuals within a population mate with one another, structural and functional changes can evolve within that population, but the population will remain one species. However, if something happens to isolate some members of a population from the others, the structural and functional differences between the two groups may accumulate over time. The two groups may diverge to the point where their members can no longer reproduce with each other and are thus distinct species. We discuss this evolutionary process, called *speciation*, in Chapter 23.

Biologists give each species a distinctive scientific name formed from two Latinized names (a **binomial**). The first name identifies the species' *genus*—a group of species that share a recent common ancestor. The second is the name of the species. For example, the scientific name for the human species is *Homo sapiens*: *Homo* is our genus and *sapiens* our species. *Homo* is Latin for "man"; *sapiens* is from the Latin for word for "wise" or "rational."

Tens of millions of species exist on Earth today. Many times that number lived in the past but are now extinct. Many millions of speciation events created this vast diversity, and the unfolding of these events can be diagrammed as an evolutionary "tree" whose branches describe the order in which populations split and eventually evolved into new species, as described in Chapter 22. Much of biology is based on comparisons among species, and these comparisons are useful precisely because we can place species in an evolutionary context relative to one another. Our ability to do this has been greatly enhanced in recent decades by our ability to sequence and compare the genomes of different species.

Genome sequencing and other molecular techniques have allowed *systematists*—scientists who study the evolution and classification of life's diverse organisms—to augment evolutionary knowledge based on the fossil record with a vast array of molecular evidence. The result is the ongoing compilation of *phylogenetic trees* that document and diagram evolutionary relationships as part of an overarching tree of life, the broadest categories of which are shown in **Figure 1.10**. (The tree is expanded in this book's Appendix; you can also explore the tree interactively at http://tolweb.org/tree.)

Although many details remain to be clarified, the broad outlines of the tree of life have been determined. Its branching patterns are based on a rich array of evidence from fossils, structures, metabolic processes, behavior, and molecular analyses of genomes. Molecular data in particular have been used to separate the tree into three major **domains**: Archaea, Bacteria, and Eukarya. The organisms of each domain have been evolving separately from those in the other domains for more than a billion years.

Organisms in the domains **Archaea** and **Bacteria** are single-celled prokaryotes. However, members of these two groups differ so fundamentally in their metabolic processes that they are believed to have separated into distinct evolutionary lineages very early. Species belonging to the third domain—**Eukarya**—have eukaryotic cells whose mitochondria and chloroplasts may have originated from the ingestion of prokaryotic cells, as described on page 11.

The three major groups of multicellular eukaryotes—plants, fungi, and animals—each evolved from a different group of the eukaryotes generally referred to as *protists*. The chloroplast-containing, photosynthetic protist that gave rise to plants was completely distinct from the protist that was ancestral to both animals and fungi, as can be seen from the branching pattern of Figure 1.10. Although most protists are unicellular (and thus sometimes called *microbial eukaryotes*), multicellularity has evolved in several protist lineages.

The tree of life is predictive

There are far more species alive on Earth than biologists have discovered and described to date. In fact, most species on Earth

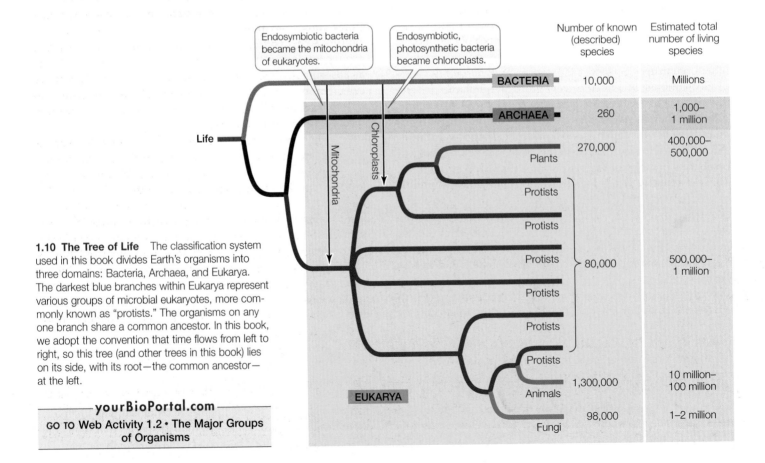

1.10 The Tree of Life The classification system used in this book divides Earth's organisms into three domains: Bacteria, Archaea, and Eukarya. The darkest blue branches within Eukarya represent various groups of microbial eukaryotes, more commonly known as "protists." The organisms on any one branch share a common ancestor. In this book, we adopt the convention that time flows from left to right, so this tree (and other trees in this book) lies on its side, with its root—the common ancestor—at the left.

yourBioPortal.com
GO TO Web Activity 1.2 • The Major Groups of Organisms

	Number of known (described) species	Estimated total number of living species
BACTERIA	10,000	Millions
ARCHAEA	260	1,000–1 million
Plants	270,000	400,000–500,000
Protists		
Protists		
Protists	80,000	500,000–1 million
Protists		
Protists		
Protists		
Animals	1,300,000	10 million–100 million
Fungi	98,000	1–2 million

have yet to be discovered by humans (see Section 32.4 for a discussion of how we know this). When we encounter a new species, its placement on the tree of life immediately tells us a great deal about its biology. In addition, understanding relationships among species allows biologists to make predictions about species that have not yet been studied, based on our knowledge of those that have.

For example, until phylogenetic methods were developed, it took years of investigation to isolate and identify most newly encountered human pathogens, and even longer to discover how these pathogens moved into human populations. Today, pathogens that cause diseases such as the flu are identified quickly on the basis of their evolutionary relationships. Placement in an evolutionary tree also gives us clues about the disease's biology, possible effective treatments, and the origin of the pathogen (see Chapters 21 and 22).

1.2 RECAP

The first cellular life on Earth was prokaryotic and arose about 4 billion years ago. The complexity of the organisms that exist today is the result of several important evolutionary events, including the evolution of photosynthesis, eukaryotic cells, and multicellularity. The genetic relationships of all organisms can be shown as a branching tree of life.

- Discuss the evolutionary significance of photosynthesis. See pp. 10–11

- What do the domains of life represent? What are the major groups of eukaryotes? See p. 12 and Figure 1.10

In February of 1676, Robert Hooke received a letter from the physicist Sir Isaac Newton in which Newton famously re-

marked, "If I have seen a little further, it is by standing on the shoulders of giants." We all stand on the shoulders of giants, building on the research of earlier scientists. By the end of this course, you will know more about evolution than Darwin ever could have, and you will know infinitely more about cells than Schleiden and Schwann did. Let's look at the methods biologists use to expand our knowledge of life.

1.3 How Do Biologists Investigate Life?

Regardless of the many different tools and methods used in research, all scientific investigations are based on *observation* and *experimentation*. In both, scientists are guided by the *scientific method*, one of the most powerful tools of modern science.

Observation is an important skill

Biologists have always observed the world around them, but today our ability to observe is greatly enhanced by technologies such as electron microscopes, DNA chips, magnetic resonance imaging, and global positioning satellites. These technologies have improved our ability to observe at all levels, from the distribution of molecules in the body to the distribution of fish in the oceans. For example, not too long ago marine biologists were only able to observe the movement of fish in the ocean by putting physical tags on the fish, releasing them, and hoping that a fisherman would catch that fish and send back the tag—and even that would reveal only where the fish ended up. Today we can attach electronic recording devices to fish that continuously record not only where the fish is, but also how deep it swims and the temperature and salinity of the water around it (**Figure 1.11**). The tags download this information to a satellite, which relays it back to researchers. Suddenly we are acquiring a great deal of knowledge about the distribution of life in the oceans—information that is relevant to studies of climate change.

Technologies that enable us to *quantify* observations are very important in science. For example, for hundreds of years species were classified by generally qualitative descriptions of the physical differences between them. There was no way of objectively calculating evolutionary distances between organisms, and biologists had to depend on the fossil record for insight. Today our ability to rapidly analyze DNA sequences enables quantitative estimates of evolutionary distances, as described in Parts Five and Six of this book. The ability to gather quantitative observations adds greatly to the biologist's ability to make strong conclusions.

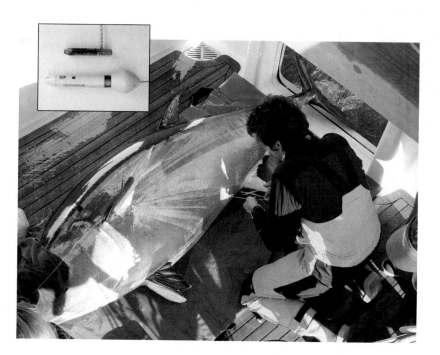

1.11 Tuna Tracking Marine biologist Barbara Block attaches computerized data recording tags (inset) to a live bluefin tuna before returning it to the ocean. Such tags make it possible to track an individual tuna wherever it travels in the world's oceans.

The scientific method combines observation and logic

Observations lead to questions, and scientists make additional observations and do experiments to answer those questions. The conceptual approach that underlies most modern scientific investigations is the **scientific method**. This powerful tool, also called the *hypothesis–prediction (H–P) method*, has five steps: (1) making *observations*; (2) asking *questions*; (3) forming *hypotheses*, or tentative answers to the questions; (4) making *predictions* based on the hypotheses; and (5) *testing* the predictions by making additional observations or conducting experiments (**Figure 1.12**).

After posing a question, a scientist uses *inductive logic* to propose a tentative answer. Inductive logic involves taking observations or facts and creating a new proposition that is compatible with those observations or facts. Such a tentative proposition is called a **hypothesis**. In formulating a hypothesis, scientists put together the facts they already know to formulate one or more possible answers to the question. For example, at the opening of

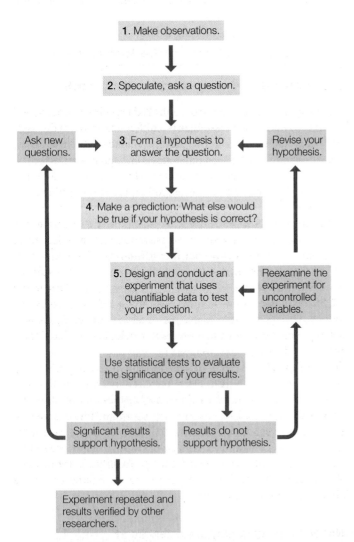

1.12 The Scientific Method   The process of observation, speculation, hypothesis, prediction, and experimentation is the cornerstone of modern science. Answers gleaned through experimentation lead to new questions, more hypotheses, further experiments, and expanding knowledge.

this chapter you learned that scientists have observed the rapid decline of amphibian populations worldwide and are asking why. Some scientists have hypothesized that a fungal disease is a cause; other scientists have hypothesized that increased exposure to ultraviolet radiation is a cause. Tyrone Hayes hypothesized that exposure to agricultural chemicals could be a cause. He knew that the most widely used chemical herbicide is atrazine; that it is mostly applied in the spring, when amphibians are breeding; and that atrazine is a common contaminant in the waters in which amphibians live as they develop into adults.

The next step in the scientific method is to apply a different form of logic—*deductive logic*—to make predictions based on the hypothesis. Deductive logic starts with a statement believed to be true and then goes on to predict what facts would also have to be true to be compatible with that statement. Based on his hypothesis, Tyrone Hayes predicted that frog tadpoles exposed to atrazine would show adverse effects of the chemical once they reached adulthood.

Good experiments have the potential to falsify hypotheses

Once predictions are made from a hypothesis, experiments can be designed to test those predictions. The most informative experiments are those that have the ability to show that the prediction is wrong. If the prediction is wrong, the hypothesis must be questioned, modified, or rejected.

There are two general types of experiments, both of which compare data from different groups or samples. A *controlled* experiment manipulates one or more of the factors being tested; *comparative* experiments compare unmanipulated data gathered from different sources. As described at the opening of this chapter, Tyrone Hayes and his colleagues conducted both types of experiment to test the prediction that the herbicide atrazine, a contaminant in freshwater ponds and streams throughout the world, affects the development of frogs.

In a **controlled experiment**, we start with groups or samples that are as similar as possible. We predict on the basis of our hypothesis that some critical factor, or **variable**, has an effect on the phenomenon we are investigating. We devise some method to manipulate *only that variable* in an "experimental" group and compare the resulting data with data from an unmanipulated "control" group. If the predicted difference occurs, we then apply statistical tests to ascertain the probability that the manipulation created the difference (as opposed to the difference being the result of random chance). **Figure 1.13** describes one of the many controlled experiments performed by the Hayes laboratory to quantify the effects of atrazine on male frogs.

The basis of controlled experiments is that one variable is manipulated while all others are held constant. The variable that is manipulated is called the *independent variable,* and the response that is measured is the *dependent variable*. A good controlled experiment is not easy to design because biological variables are so interrelated that it is difficult to alter just one.

A **comparative experiment** starts with the prediction that there will be a difference between samples or groups based on the hypothesis. In comparative experiments, however, we can-

INVESTIGATING LIFE

1.13 Controlled Experiments Manipulate a Variable

The Hayes laboratory created controlled environments that differed only in the concentrations of atrazine in the water. Eggs from leopard frogs (*Rana pipiens*) raised specifically for laboratory use were allowed to hatch and the tadpoles were separated into experimental tanks containing water with different concentrations of atrazine.

HYPOTHESIS Exposure to atrazine during larval development causes abnormalities in the reproductive system of male frogs.

METHOD

1. Establish 9 tanks in which all attributes are held constant except the water's atrazine concentrations. Establish 3 atrazine conditions (3 replicate tanks per condition): 0 ppb (control condition), 0.1 ppb, and 25 ppb.
2. Place *Rana pipiens* tadpoles from laboratory-reared eggs in the 9 tanks (30 tadpoles per replicate).
3. When tadpoles have transitioned into adults, sacrifice the animals and evaluate their reproductive tissues.
4. Test for correlation of degree of atrazine exposure with the presence of abnormalities in the reproductive systems of male frogs.

RESULTS

Abnormal testes development

Oocytes (eggs) in normal-size testis (sex reversal)

■ Gonadal dysgenesis
■ Testicular oogenesis

In the control condition, only one male had abnormalities.

Male frogs with gonadal abnormalities (%)

Atrazine (ppb)

CONCLUSION Exposure to atrazine at concentrations as low as 0.1 ppb induces abnormalities in the male reproductive systems of frogs. The effect is not proportional to the level of exposure.

Go to **yourBioPortal.com** for original citations, discussions, and relevant links for all INVESTIGATING LIFE figures.

---

not control the variables; often we cannot even identify all the variables that are present. We are simply gathering and comparing data from different sample groups.

When his controlled experiments indicated that atrazine indeed affects reproductive development in frogs, Hayes and his colleagues performed a comparative experiment. They collected frogs and water samples from eight widely separated sites across the United States and compared the incidence of abnormal frogs from environments with very different levels of atrazine (**Figure 1.14**). Of course, the sample sites differed in many ways besides the level of atrazine present.

The results of experiments frequently reveal that the situation is more complex than the hypothesis anticipated, thus raising new questions. In the Hayes experiments, for example, there was no clear direct relationship between the *amount* of atrazine present and the percentage of abnormal frogs: there were fewer abnormal frogs at the highest concentrations of atrazine than at lower concentrations. There are no "final answers" in science. Investigations consistently reveal more complexity than we expect. The scientific method is a tool to identify, assess, and understand that complexity.

────────── **yourBioPortal.com** ──────────
GO TO Animated Tutorial 1.1 • The Scientific Method

Statistical methods are essential scientific tools

Whether we do comparative or controlled experiments, at the end we have to decide whether there is a difference between the samples, individuals, groups, or populations in the study. How do we decide whether a measured difference is enough to support or falsify a hypothesis? In other words, how do we decide in an unbiased, objective way that the measured difference is significant?

Significance can be measured with statistical methods. Scientists use statistics because they recognize that variation is always present in any set of measurements. Statistical tests calculate the probability that the differences observed in an experiment could be due to random variation. The results of statistical tests are therefore probabilities. A statistical test starts with a **null hypothesis**—the premise that no difference exists. When quantified observations, or **data**, are collected, statistical methods are applied to those data to calculate the likelihood that the null hypothesis is correct.

More specifically, statistical methods tell us the probability of obtaining the same results by chance even if the null hypothesis were true. We need to eliminate, insofar as possible, the chance that any differences showing up in the data are merely the result of random variation in the samples tested. Scientists generally conclude that the differences they measure are significant if statistical tests show that the *probability of error* (that is, the probability that the same results can be obtained by mere chance) is 5 percent or lower.

Not all forms of inquiry are scientific

Science is a unique human endeavor that is bounded by certain standards of practice. Other areas of scholarship share with science the practice of making observations and asking ques-

INVESTIGATING LIFE

1.14 Comparative Experiments Look for Differences among Groups

To see whether the presence of atrazine correlates with reproductive system abnormalities in male frogs, the Hayes lab collected frogs and water samples from different locations around the U.S. The analysis that followed was "blind," meaning that the frogs and water samples were coded so that experimenters working with each specimen did not know which site the specimen came from.

HYPOTHESIS Presence of the herbicide atrazine in environmental water correlates with reproductive system abnormalities in frog populations.

METHOD
1. Based on commercial sales of atrazine, select 4 sites (sites 1–4) less likely and 4 sites (sites 5–8) more likely to be contaminated with atrazine.
2. Visit all sites in the spring (i.e., when frogs have transitioned from tadpoles into adults); collect frogs and water samples.
3. In the laboratory, sacrifice frogs and examine their reproductive tissues, documenting abnormalities.
4. Analyze the water samples for atrazine concentration (the sample for site 7 was not tested).
5. Quantify and correlate the incidence of reproductive abnormalities with environmental atrazine concentrations.

RESULTS

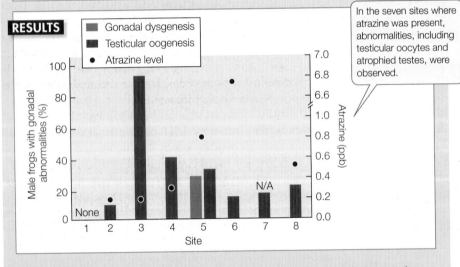

In the seven sites where atrazine was present, abnormalities, including testicular oocytes and atrophied testes, were observed.

CONCLUSION Reproductive abnormalities exist in frogs from environments in which aqueous atrazine concentration is 0.2 ppb or above. The incidence of abnormalities does not appear to be proportional to atrazine concentration at the time of transition to adulthood.

FURTHER INVESTIGATION: The highest proportion of abnormal frogs was found at site 3, located on a wildlife reserve in Wyoming. What kind of data and observations would you need to suggest possible explanations for this extremely high incidence?

Go to **yourBioPortal.com** for original citations, discussions, and relevant links for all INVESTIGATING LIFE figures.

Scientific explanations for natural processes are objective and reliable because the hypotheses proposed *must be testable* and *must have the potential of being rejected* by direct observations and experiments. Scientists must clearly describe the methods they use to test hypotheses so that other scientists can repeat their results. Not all experiments are repeated, but surprising or controversial results are always subjected to independent verification. Scientists worldwide share this process of testing and rejecting hypotheses, contributing to a common body of scientific knowledge.

If you understand the methods of science, you can distinguish science from non-science. Art, music, and literature all contribute to the quality of human life, but they are not science. They do not use the scientific method to establish what is fact. Religion is not science, although religions have historically purported to explain natural events ranging from unusual weather patterns to crop failures to human diseases. Most such phenomena that at one time were mysterious can now be explained in terms of scientific principles.

The power of science derives from the uncompromising objectivity and absolute dependence on evidence that comes from *reproducible and quantifiable observations*. A religious or spiritual explanation of a natural phenomenon may be coherent and satisfying for the person holding that view, but it is not testable, and therefore it is not science. To invoke a supernatural explanation (such as a "creator" or "intelligent designer" with no known bounds) is to depart from the world of science.

Science describes the facts about how the world works, not how it "ought to be." Many scientific advances that have contributed to human welfare have also raised major ethical issues. Recent developments in genetics and developmental biology, for example, enable us to select the sex of our children, to use stem cells to repair our bodies, and to modify the human genome. Although scientific knowledge allows us to do these things, science cannot tell us whether or not we should do them, or, if we choose to do so, how we should regulate them.

To make wise decisions about public policy, we need to employ the best possible ethical reasoning in deciding which outcomes we should strive for.

---

tions, but scientists are distinguished by what they do with their observations and how they answer their questions. Data, subjected to appropriate statistical analysis, are critical in the testing of hypotheses. The scientific method is the most powerful way humans have devised for learning about the world and how it works.

1.3 RECAP

The scientific method of inquiry starts with the formulation of hypotheses based on observations and data. Comparative and controlled experiments are carried out to test hypotheses.

- Explain the relationship between a hypothesis and an experiment. **See p. 14 and Figure 1.12**

- What is controlled in a controlled experiment? **See p. 14 and Figure 1.13**

- What features characterize questions that can be answered only by using a comparative approach? **See pp. 14–15 and Figure 1.14**

- Do you understand why arguments must be supported by quantifiable and reproducible data in order to be considered scientific? **See pp. 15–16**

The vast scientific knowledge accumulated over centuries of human civilization allows us to understand and manipulate aspects of the natural world in ways that no other species can. These abilities present us with challenges, opportunities, and above all, responsibilities.

1.4 How Does Biology Influence Public Policy?

Agriculture and medicine are two important human activities that depend on biological knowledge. Our ancestors unknowingly applied the principles of evolutionary biology when they domesticated plants and animals, and people have speculated about the causes of diseases and searched for methods to combat them since ancient times. Long before the microbial causes of diseases were known, people recognized that infections could be passed from one person to another, and the isolation of infected persons has been practiced as long as written records have been available.

Today, thanks to the deciphering of genomes and our new-found ability to manipulate them, vast new possibilities exist for controlling human diseases and increasing agricultural productivity, but these capabilities raise ethical and policy issues. How much and in what ways should we tinker with the genes of humans and other species? Does it matter whether the genomes of our crop plants and domesticated animals are changed by traditional methods of controlled breeding and crossbreeding or by the biotechnology of gene transfer? What rules should govern the release of genetically modified organisms into the environment? Science alone cannot provide all the answers, but wise policy decisions must be based on accurate scientific information.

Biologists are increasingly called on to advise government agencies concerning the laws, rules, and regulations by which society deals with the increasing number of challenges that have at least a partial biological basis. As an example of the value of scientific knowledge for the assessment and formulation of public policy, let's return to the tracking study of bluefin tuna introduced in Section 1.3. Prior to this study, both scientists and fishermen knew that bluefins had a western breeding ground in the Gulf of Mexico and an eastern breeding ground in the Mediterranean Sea (**Figure 1.15**). Overfishing had led to declining numbers of fish in the western-breeding populations, to the point of these populations being endangered.

1.15 Bluefin Tuna Do Not Recognize Boundaries It was assumed that tuna from western-breeding populations and those from eastern-breeding populations also fed on their respective sides of the Atlantic, so separate fishing quotas were established on either side of 45° W longitude (dashed line) to allow the endangered western population to recover. However, tracking data shows that the two populations *do not* remain separate after spawning, so in fact the established policy does not protect the western population.

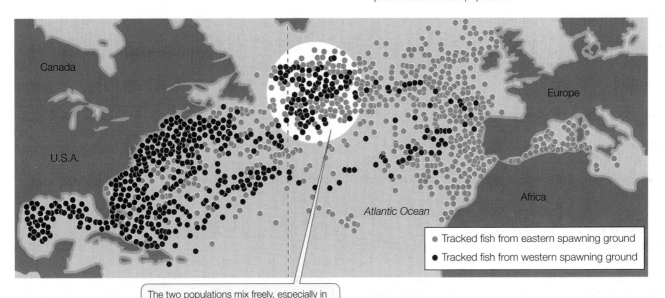

Canada

U.S.A.

Europe

Africa

Atlantic Ocean

● Tracked fish from eastern spawning ground
● Tracked fish from western spawning ground

The two populations mix freely, especially in the heavily fished waters of the North Atlantic.

Initially it was assumed by scientists, fishermen, and policy makers alike that the eastern and western populations had geographically separate feeding grounds as well as separate breeding grounds. Acting on this assumption, an international commission drew a line down the middle of the Atlantic Ocean and established strict fishing quotas on the western side of the line, with the intent of allowing the western population to recover. New tracking data, however, revealed that in fact the eastern and western bluefin populations mix freely on their feeding grounds across the entire North Atlantic—a swath of ocean that includes the most heavily fished waters in the world. Tuna caught on the eastern side of the line could just as likely be from the western breeding population as the eastern; thus the established policy was not achieving its intended goal.

Policy makers take more things into consideration than scientific knowledge and recommendations. For example, studies on the effects of atrazine on amphibians have led one U.S. group, the Natural Resources Defense Council, to take legal action to have atrazine banned on the basis of the Endangered Species Act. The U.S. Environmental Protection Agency, however, must also consider the potential loss to agriculture that such a ban would create and has continued to approve atrazine's use as long as environmental levels do not exceed 30 to 40 ppb—which is 300 to 400 times the levels shown to induce abnormalities in the Hayes studies. Scientific conclusions do not always prevail in the political world.

Another reason for studying biology is to understand the effects of the vastly increased human population on its environment. Our use of natural resources is putting stress on the ability of Earth's ecosystems to continue to produce the goods and services on which our society depends. Human activities are changing global climates, causing the extinctions of a large number of species like the amphibians featured in this chapter, and spreading new diseases while facilitating the resurgence of old ones. The rapid spread of flu viruses has been facilitated by modern modes of transportation, and the recent resurgence of tuberculosis is the result of the evolution of bacteria that are resistant to antibiotics. Biological knowledge is vital for determining the causes of these changes and for devising wise policies to deal with them.

Beyond issues of policy and pragmatism lies the human "need to know." Human beings are fascinated by the richness and diversity of life, and most people want to know more about organisms and how they interact. Human curiosity might even be seen as an adaptive trait—it is possible that such a trait could have been selected for if individuals who were motivated to learn about their surroundings were likely to have survived and reproduced better, on average, than their less curious relatives. Far from ending the process, new discoveries and greater knowledge typically engender questions no one thought to ask before. There are vast numbers of questions for which we do not yet have answers, and the most important motivator of most scientists is curiosity.

CHAPTER SUMMARY

1.1 What Is Biology?

- **Biology** is the scientific study of living organisms, including their characteristics, functions, and interactions. Cells are the basic structural and physiological units of life. The **cell theory** states that all life consists of cells and that all cells come from preexisting cells.

- All living organisms are related to one another through descent with modification. **Evolution** by **natural selection** is responsible for the diversity of **adaptations** found in living organisms.

- The instructions for a cell are contained in its **genome**, which consists of **DNA** molecules made up of sequences of **nucleotides**. Specific segments of DNA called **genes** contain the information the cell uses to make **proteins**. **Review Figure 1.4**

- Living organisms regulate their internal environment. They also interact with other organisms of the same and different species. Biologists study life at all these levels of organization. **Review Figure 1.6, WEB ACTIVITY 1.1**

- Biological knowledge obtained from a **model system** may be generalized to other species.

1.2 How Is All Life on Earth Related?

- Biologists use fossils, anatomical similarities and differences, and molecular comparisons of genomes to reconstruct the history of life. **Review Figure 1.8**

- Life first arose by chemical evolution. Cells arose early in the evolution of life.

- **Photosynthesis** was an important evolutionary step because it changed Earth's atmosphere and provided a means of capturing energy from sunlight.

- The earliest organisms were **prokaryotes**. Organisms called **eukaryotes**, with more complex cells, arose later. Eukaryotic cells have discrete intracellular compartments, called **organelles**, including a nucleus that contains the cell's genetic material.

- The genetic relationships of **species** can be represented as an evolutionary tree. Species are grouped into three **domains**: **Archaea**, **Bacteria**, and **Eukarya**. Archaea and Bacteria are domains of unicellular prokaryotes. Eukarya contains diverse groups of protists (most but not all of which are unicellular) and the multicellular plants, fungi, and animals. **Review Figure 1.10, WEB ACTIVITY 1.2**

1.3 How Do Biologists Investigate Life?

- The **scientific method** used in most biological investigations involves five steps: making observations, asking questions, forming hypotheses, making predictions, and testing those predictions. **Review Figure 1.12**

- **Hypotheses** are tentative answers to questions. Predictions made on the basis of a hypothesis are tested with additional observations and two kinds of **experiments: comparative** and **controlled experiments**. Review Figures 1.13 and 1.14, **ANIMATED TUTORIAL 1.1**

- Statistical methods are applied to **data** to establish whether or not the differences observed are significant or whether they could be the result of chance. These methods start with the **null hypothesis** that there are no differences.

- Science can tell us how the world works, but it cannot tell us what we should or should not do.

1.4 How Does Biology Influence Public Policy?

- Biologists are often called on to advise government agencies on the solution of important problems that have a biological component.

FOR DISCUSSION

1. Even if we knew the sequences of all of the genes of a single-celled organism and could cause those genes to be expressed in a test tube, it would still be incredibly difficult to create a functioning organism. Why do you think this is so? In light of this fact, what do you think of the statement that the genome contains all of the information for a species?

2. Why is it so important in science that we design and perform tests capable of falsifying a hypothesis?

3. What features characterize questions that can be answered only by using a comparative approach?

4. Cite an example of how you apply aspects of the scientific method to solve problems in your daily life.

ADDITIONAL INVESTIGATION

1. The abnormalities of frogs in Tyrone Hayes's studies were associated with the presence of a herbicide in the environment. That herbicide did not kill the frogs, but it feminized the males. How would you investigate whether this effect could lead to decreased reproductive capacity for the frog populations in nature?

2. Just as all cells come from preexisting cells, all mitochondria—the cell organelles that convert energy in food to a form of energy that can do biological work—come from preexisting mitochondria. Cells do not synthesize mitochondria from the genetic information in their nuclei. What investigations would you carry out to understand the nature of mitochondria?

WORKING WITH DATA (GO TO yourBioPortal.com)

Feminization of Frogs   Analogous to the experiment shown in Figure 1.13, this exercise asks you to graph data about the size of the laryngeal (throat) muscles required to produce male mating calls in the frog *Xenopus laevis*. After plotting data from frogs exposed to different levels of the herbicide atrazine during their development, you will formulate conclusions about the effects of the herbicide on this physical attribute and speculate about what these effects might mean.

2 Small Molecules and the Chemistry of Life

A hairy story

"You are what you eat—and that is recorded in your hair." Two scientists at the University of Utah are responsible for adding the last phrase to this famous saying about body chemistry. Ecologist Jim Ehleringer and chemist Thure Cerling showed that the composition of human hair reflects the region where a person lives.

As we pointed out in Chapter 1, living things are made up of the same kinds of atoms that make up the inanimate universe. Two of those atoms are hydrogen (H) and oxygen (O), which combine to form water (H_2O). Both atoms have naturally occurring variants called *isotopes*, which have the same chemical properties but different weights because their nuclei have different numbers of particles called *neutrons*.

When water evaporates from the ocean, it forms clouds that move inland and release rain. Water made up of the heavier H and O isotopes is heavier and tends to fall more readily than water containing the lighter isotopes. Warm rains tend to be heavier than cooler precipitation. People living on the coast or in regions where there are frequent warm rains consume heavier water and foods made from water than people living in cooler, inland areas (assuming, of course, that their beverages and produce come from the same area they live in). And, since you are what you eat, the heavy H and O atoms become part of their bodies.

Our hair contains abundant H and O atoms, many obtained from local water. Ehleringer and Cerling wondered whether the ratios of heavy-to-light H and O in hair reflected the ratio of heavy-to-light H_2O in the local water. To address this question, Ehleringer's wife and Cerling's children and their friends went on a hair-collecting trip across the United States, collecting hair trimmings from barbershop floors while at the same time filling test tubes with local water. Back at the lab, scientists tested the samples and found that the ratios of heavy to light isotopes in the hair did indeed reflect these same ratios in the local water.

While this information is intrinsically fascinating, it is also potentially useful. For example, police could use hair analysis to evaluate a suspect's alibi: "You say you've been in Montana for the past month? Your hair sample indicates that you were in a warm coastal area." Such conflicting evidence could form the basis of further investigation.

Hair Tells a Tale The ratio in hair protein of the heavy isotope ^{18}O to its lighter counterpart ^{16}O reflects the ratios in local water.

Free Samples Need hair samples for a research project? Try the local barber shop.

Or anthropologists might analyze hair samples from graves to work out migration patterns of human groups.

The understanding that life is based on chemistry and obeys universal laws of chemistry and physics is relatively new in human history. Until the nineteenth century, a "vital force" (from the Latin *vitalis*, "of life") was presumed be responsible for life. This vital force was seen as distinct from the mechanistic forces governing physics and chemistry. Many people still assume that a vital force exists, but the physical–chemical view of life has led to great advances in biological science and is the cornerstone of modern medicine and agriculture.

IN THIS CHAPTER we will introduce the constituents of matter: atoms, their variety, their properties, and their capacity to combine with other atoms. We will consider how matter changes, including changes in state (solid to liquid to gas), and changes caused by chemical reactions. We will examine the structure and properties of water and its relationship to chemical acids and bases.

2.1 How Does Atomic Structure Explain the Properties of Matter?

All matter is composed of **atoms**. Atoms are tiny—more than a trillion (10^{12}) of them could fit on top of the period at the end of this sentence. Each atom consists of a dense, positively charged **nucleus**, around which one or more negatively charged **electrons** move (**Figure 2.1**). The nucleus contains one or more positively charged **protons** and may contain one or more **neutrons** with no electrical charge. Atoms and their component particles have volume and mass, which are characteristics of all matter. *Mass* is a measure of the quantity of matter present; the greater the mass, the greater the quantity of matter.

The mass of a proton serves as a standard unit of measure called the *dalton* (named after the English chemist John Dalton) or **atomic mass unit** (**amu**). A single proton or neutron has a mass of about 1 dalton (Da), which is 1.7×10^{-24} grams (0.0000000000000000000000017 g). That's tiny, but an electron is even tinier at 9×10^{-28} g (0.0005 Da). Because the mass of an electron is negligible compared with the mass of a proton or a neutron, the contribution of electrons to the mass of an atom can usually be ignored when measurements and calculations are made. It is electrons, however, that determine how atoms will combine with other atoms to form stable associations.

Each proton has a positive electric charge, defined as +1 unit of charge. An electron has a negative charge equal and opposite to that of a proton (–1). The neutron, as its name suggests, is electrically neutral, so its charge is 0. Charges that are different (+/–) attract each other, whereas charges that are alike (+/+, –/–) repel each other. Atoms are electrically neutral because the number of electrons in an atom equals the number of protons.

An element consists of only one kind of atom

An **element** is a pure substance that contains only one kind of atom. The element hydrogen consists only of hydrogen atoms; the element iron consists only of iron atoms. The atoms of each element have certain characteristics or properties that distinguish them from the atoms of other elements. These properties include their mass and how they interact and associate with other atoms.

The more than 100 elements found in the universe are arranged in the *periodic table* (**Figure 2.2**). Each element has its own one- or two-letter chemical symbol. For example, H stands for hydrogen, C for carbon, and O for oxygen. Some symbols come from other languages: Fe (from the Latin, *ferrum*) stands for iron, Na (Latin, *natrium*) for sodium, and W (German, *wolfram*) for tungsten.

Properties

...in This representation of a helium atom is called a ... it exaggerates the space occupied by the nucleus. In reality, ...ugh the nucleus accounts for virtually all of the atomic mass, it occupies only about 1/10,000 of the atom's volume. The Bohr model is also inaccurate in that it represents the electron as a discrete particle in a defined orbit around the nucleus.

—Nucleus

bon, hydrogen, nitrogen, oxygen, phosphorus, and sulfur. The chemistry of these six elements will be our primary concern in this chapter, but other elements found in living organisms are important as well. Sodium and potassium, for example, are essential for nerve function; calcium can act as a biological signal; iodine is a component of a vital hormone; and magnesium is bound to chlorophyll in plants. The physical and chemical (reactive) properties of atoms depend on the number of subatomic particles they contain.

Each element has a different number of protons

An element differs from other elements by the number of protons in the nucleus of each of its atoms; the number of protons is designated the **atomic number**. This atomic number is unique

The elements of the periodic table are not found in equal amounts. Stars have abundant amounts of hydrogen and helium. Earth's crust, and the surfaces of the neighboring planets, are almost half oxygen, 28 percent silicon, 8 percent aluminum, and between 2 and 5 percent each of sodium, magnesium, potassium, calcium, and iron. They contain much smaller amounts of the other elements.

About 98 percent of the mass of every living organism (bacterium, turnip, or human) is composed of just six elements: car-

Atomic number
(number of protons)

2

He

Chemical symbol
(for helium)

Atomic mass
(number of protons plus
number of neutrons)

4.003

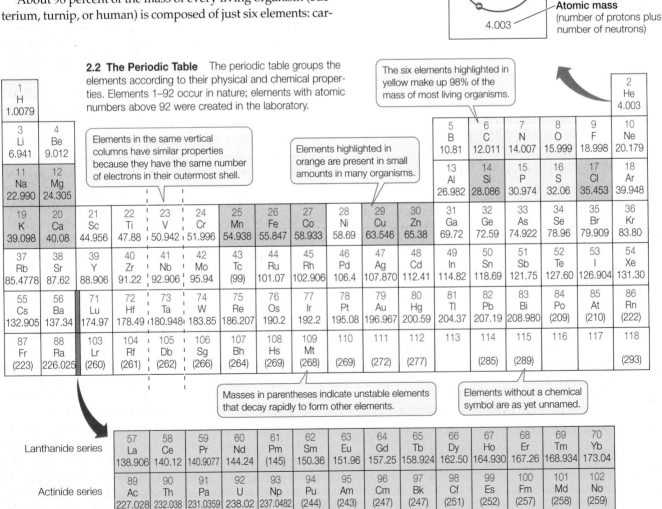

2.2 The Periodic Table The periodic table groups the elements according to their physical and chemical properties. Elements 1–92 occur in nature; elements with atomic numbers above 92 were created in the laboratory.

The six elements highlighted in yellow make up 98% of the mass of most living organisms.

Elements in the same vertical columns have similar properties because they have the same number of electrons in their outermost shell.

Elements highlighted in orange are present in small amounts in many organisms.

Masses in parentheses indicate unstable elements that decay rapidly to form other elements.

Elements without a chemical symbol are as yet unnamed.

1 H 1.0079																	2 He 4.003
3 Li 6.941	4 Be 9.012											5 B 10.81	6 C 12.011	7 N 14.007	8 O 15.999	9 F 18.998	10 Ne 20.179
11 Na 22.990	12 Mg 24.305											13 Al 26.982	14 Si 28.086	15 P 30.974	16 S 32.06	17 Cl 35.453	18 Ar 39.948
19 K 39.098	20 Ca 40.08	21 Sc 44.956	22 Ti 47.88	23 V 50.942	24 Cr 51.996	25 Mn 54.938	26 Fe 55.847	27 Co 58.933	28 Ni 58.69	29 Cu 63.546	30 Zn 65.38	31 Ga 69.72	32 Ge 72.59	33 As 74.922	34 Se 78.96	35 Br 79.909	36 Kr 83.80
37 Rb 85.4778	38 Sr 87.62	39 Y 88.906	40 Zr 91.22	41 Nb 92.906	42 Mo 95.94	43 Tc (99)	44 Ru 101.07	45 Rh 102.906	46 Pd 106.4	47 Ag 107.870	48 Cd 112.41	49 In 114.82	50 Sn 118.69	51 Sb 121.75	52 Te 127.60	53 I 126.904	54 Xe 131.30
55 Cs 132.905	56 Ba 137.34	71 Lu 174.97	72 Hf 178.49	73 Ta 180.948	74 W 183.85	75 Re 186.207	76 Os 190.2	77 Ir 192.2	78 Pt 195.08	79 Au 196.967	80 Hg 200.59	81 Tl 204.37	82 Pb 207.19	83 Bi 208.980	84 Po (209)	85 At (210)	86 Rn (222)
87 Fr (223)	88 Ra 226.025	103 Lr (260)	104 Rf (261)	105 Db (262)	106 Sg (266)	107 Bh (264)	108 Hs (269)	109 Mt (268)	110 (269)	111 (272)	112 (277)	113	114 (285)	115 (289)	116	117	118 (293)

Lanthanide series

57 La 138.906	58 Ce 140.12	59 Pr 140.9077	60 Nd 144.24	61 Pm (145)	62 Sm 150.36	63 Eu 151.96	64 Gd 157.25	65 Tb 158.924	66 Dy 162.50	67 Ho 164.930	68 Er 167.26	69 Tm 168.934	70 Yb 173.04

Actinide series

89 Ac 227.028	90 Th 232.038	91 Pa 231.0359	92 U 238.02	93 Np 237.0482	94 Pu (244)	95 Am (243)	96 Cm (247)	97 Bk (247)	98 Cf (251)	99 Es (252)	100 Fm (257)	101 Md (258)	102 No (259)

to each element and does not change. The atomic number of helium is 2, and an atom of helium always has two protons; the atomic number of oxygen is 8, and an atom of oxygen always has eight protons.

Along with a definitive number of protons, every element except hydrogen has one or more neutrons in its nucleus. The **mass number** of an atom is the total number of protons and neutrons in its nucleus. The nucleus of a carbon atom contains six protons and six neutrons, and has a mass number of 12. Oxygen has eight protons and eight neutrons, and has a mass number of 16. The mass number is essentially the mass of the atom in daltons (see below).

By convention, we often print the symbol for an element with the atomic number at the lower left and the mass number at the upper left, both immediately preceding the symbol. Thus hydrogen, carbon, and oxygen can be written as $^{1}_{1}$H, $^{12}_{6}$C, and $^{16}_{8}$O, respectively.

The number of neutrons differs among isotopes

In some elements, the number of neutrons in the atomic nucleus is not always the same. Different **isotopes** of the same element have the same number of protons, but different numbers of neutrons. Many elements have several isotopes. The isotopes of hydrogen shown below have special names, but the isotopes of most elements do not have distinct names.

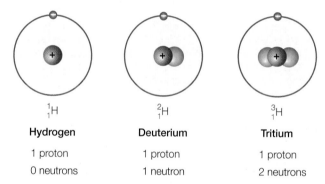

$^{1}_{1}$H	$^{2}_{1}$H	$^{3}_{1}$H
Hydrogen	**Deuterium**	**Tritium**
1 proton	1 proton	1 proton
0 neutrons	1 neutron	2 neutrons

The natural isotopes of carbon, for example, are ^{12}C (six neutrons in the nucleus), ^{13}C (seven neutrons), and ^{14}C (eight neutrons). Note that all three (called "carbon-12," "carbon-13," and "carbon-14") have six protons, so they are all carbon. Most carbon atoms are ^{12}C, about 1.1 percent are ^{13}C, and a tiny fraction are ^{14}C. But all have virtually the same chemical reactivity, which is an important property for their use in experimental biology and medicine. An element's **atomic weight** (or atomic mass) is the average of the mass numbers of a representative sample of atoms of that element, with all the isotopes in their normally occurring proportions. The atomic weight of carbon, taking into account all of its isotopes and their abundances, is thus 12.011. The fractional atomic weight results from averaging the contributing weights of all of the isotopes.

Most isotopes are stable. But some, called **radioisotopes**, are unstable and spontaneously give off energy in the form of α (alpha), β (beta), or γ (gamma) radiation from the atomic nucleus. Known as *radioactive decay*, this release of energy transforms the original atom. The type of transformation varies depending on

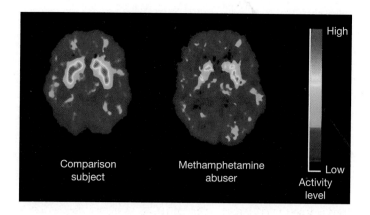

2.3 Tagging the Brain In these images from live people, a radioactively labeled sugar is used to detect differences between the brain activity of a healthy person and that of a person who abuses methamphetamines. The more active a brain region is, the more sugar it takes up. The healthy brain (left) shows more activity in the region involved in memory (the red area) than the drug abuser's brain does.

the radioisotope, but some can change the number of protons, so that the original atom becomes a different element.

With sensitive instruments, scientists can use the released radiation to detect the presence of radioisotopes. For instance, if an earthworm is given food containing a radioisotope, its path through the soil can be followed using a simple detector called a Geiger counter. Most atoms in living organisms are organized into stable associations called **molecules**. If a radioisotope is incorporated into a molecule, it acts as a tag or label, allowing researchers or physicians to trace the molecule in an experiment or in the body (**Figure 2.3**). Radioisotopes are also used to date fossils, an application described in Section 25.1.

Although radioisotopes are useful in research and in medicine, even a low dose of the radiation they emit has the potential to damage molecules and cells. However, these damaging effects are sometimes used to our advantage; for example, the radiation from ^{60}Co (cobalt-60) is used in medicine to kill cancer cells.

The behavior of electrons determines chemical bonding and geometry

The characteristic number of electrons in an atom determines how it will combine with other atoms. Biologists are interested in how chemical changes take place in living cells. When considering atoms, they are concerned primarily with electrons because the behavior of electrons explains how chemical *reactions* occur. Chemical reactions alter the atomic compositions of substances and thus alter their properties. Reactions usually involve changes in the distribution of electrons between atoms.

The location of a given electron in an atom at any given time is impossible to determine. We can only describe a volume of space within the atom where the electron is likely to be. The region of space where the electron is found at least 90 percent of the time is the electron's **orbital**. Orbitals have characteristic shapes and orientations, and a given orbital can be occupied by

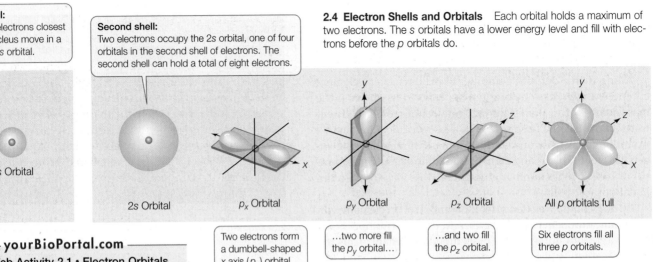

First shell:
The two electrons closest to the nucleus move in a spherical s orbital.

1s Orbital

Second shell:
Two electrons occupy the 2s orbital, one of four orbitals in the second shell of electrons. The second shell can hold a total of eight electrons.

2s Orbital

p_x Orbital

p_y Orbital

p_z Orbital

All p orbitals full

Two electrons form a dumbbell-shaped x axis (p_x) orbital...

...two more fill the p_y orbital...

...and two fill the p_z orbital.

Six electrons fill all three p orbitals.

2.4 Electron Shells and Orbitals Each orbital holds a maximum of two electrons. The s orbitals have a lower energy level and fill with electrons before the p orbitals do.

yourBioPortal.com
GO TO Web Activity 2.1 • Electron Orbitals

a maximum of two electrons (**Figure 2.4**). Thus any atom larger than helium (atomic number 2) must have electrons in two or more orbitals. As we move from lighter to heavier atoms in the periodic chart, the orbitals are filled in a specific sequence, in a series of what are known as **electron shells**, or *energy levels*, around the nucleus.

- *First shell:* The innermost electron shell consists of just one orbital, called an s orbital. A hydrogen atom ($_1$H) has one electron in its first shell; helium ($_2$He) has two. Atoms of all other elements have two or more shells to accommodate orbitals for additional electrons.

- *Second shell:* The second shell contains four orbitals (an s orbital and three p orbitals), and hence holds up to eight electrons. As depicted in Figure 2.4, the s orbitals have the shape of a sphere, while the p orbitals are directed at right angles to one another. The orientations of these orbitals in space contribute to the three-dimensional shapes of molecules when atoms link to other atoms.

2.5 Electron Shells Determine the Reactivity of Atoms Each shell can hold a specific maximum number of electrons. Each shell must be filled before electrons can occupy the next shell. The energy level of an electron is higher in a shell farther from the nucleus. An atom with unpaired electrons in its outermost shell can react (bond) with other atoms.

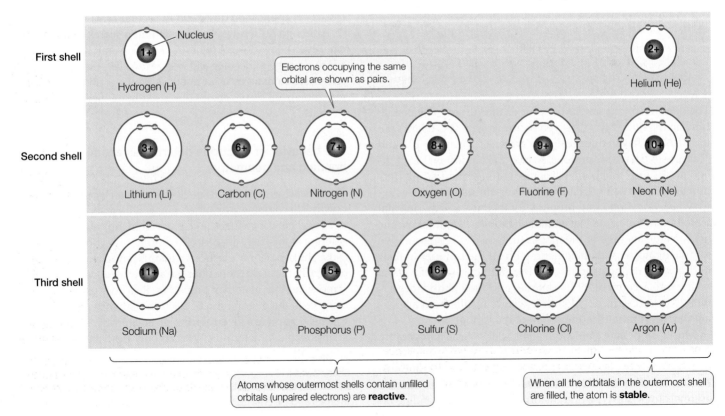

First shell

Nucleus

Hydrogen (H) 1+

Helium (He) 2+

Electrons occupying the same orbital are shown as pairs.

Second shell

Lithium (Li) 3+

Carbon (C) 6+

Nitrogen (N) 7+

Oxygen (O) 8+

Fluorine (F) 9+

Neon (Ne) 10+

Third shell

Sodium (Na) 11+

Phosphorus (P) 15+

Sulfur (S) 16+

Chlorine (Cl) 17+

Argon (Ar) 18+

Atoms whose outermost shells contain unfilled orbitals (unpaired electrons) are **reactive**.

When all the orbitals in the outermost shell are filled, the atom is **stable**.

- *Additional shells*: Elements with more than ten electrons have three or more electron shells. The farther a shell is from the nucleus, the higher the energy level is for an electron occupying that shell.

The s orbitals fill with electrons first, and their electrons have the lowest energy level. Subsequent shells have different numbers of orbitals, but the outermost shells usually hold only eight electrons. In any atom, the outermost electron shell (the *valence shell*) determines how the atom combines with other atoms—that is, how the atom behaves chemically. When a valence shell with four orbitals contains eight electrons, there are no unpaired electrons, and the atom is *stable*—it will not react with other atoms (**Figure 2.5**). Examples of chemically stable elements are helium, neon, and argon. On the other hand, atoms that have one or more unpaired electrons in their outer shells are capable of reacting with other atoms.

Atoms with unpaired electrons (i.e., partially filled orbitals) in their outermost electron shells are unstable, and will undergo reactions in order to fill their outermost shells. Reactive atoms can attain stability either by sharing electrons with other atoms or by losing or gaining one or more electrons. In either case, the atoms involved are *bonded* together into stable associations called molecules. The tendency of atoms to form stable molecules so that they have eight electrons in their outermost shells is known as the *octet rule*. Many atoms in biologically important molecules—for example, carbon (C) and nitrogen (N)—follow this rule. An important exception is hydrogen (H), which attains stability when two electrons occupy its single shell (consisting of just one s orbital).

2.1 RECAP

The living world is composed of the same set of chemical elements as the rest of the universe. An atom consists of a nucleus of protons and neutrons, and a characteristic configuration of electrons in orbitals around the nucleus. This structure determines the atom's chemical properties.

- Describe the arrangement of protons, neutrons, and electrons in an atom. **See Figure 2.1**

- Use the periodic table to identify some of the similarities and differences in atomic structure among different elements (for example, oxygen, carbon, and helium). How does the configuration of the valence shell influence the placement of an element in the periodic table? **See p. 25 and Figures 2.2 and 2.5**

- How does bonding help a reactive atom achieve stability? **See p. 25 and Figure 2.5**

We have introduced the individual players on the biochemical stage—the atoms. We have shown how the energy levels of electrons drive an atomic "quest for stability." Next we will describe the different types of chemical bonds that can lead to stability, joining atoms together into molecular structures with hosts of different properties.

2.2 How Do Atoms Bond to Form Molecules?

A **chemical bond** is an attractive force that links two atoms together in a molecule. There are several kinds of chemical bonds (**Table 2.1**). In this section we will begin with *covalent bonds*, the strong bonds that result from the sharing of electrons. Next we will examine *ionic bonds*, which form when an atom gains or loses one or more electrons to achieve stability. We will then consider other, weaker, kinds of interactions, including hydrogen bonds, which are enormously important to biology.

— **yourBioPortal.com** —
GO TO Animated Tutorial 2.1 • Chemical Bond Formation

Covalent bonds consist of shared pairs of electrons

A **covalent bond** forms when two atoms attain stable electron numbers in their outermost shells by *sharing* one or more pairs of electrons. Consider two hydrogen atoms coming into close proximity, each with an unpaired electron in its single shell (**Figure 2.6**). When the electrons pair up, a stable association is formed, and this links the two hydrogen atoms in a covalent bond, resulting in H_2.

A **compound** is a substance made up of molecules with two or more elements bonded together in a fixed ratio. Methane gas (CH_4), water (H_2O), and table sugar (sucrose, $C_{12}H_{22}O_{11}$) are examples of compounds. The chemical symbols identify the different elements in a compound, and the subscript numbers indicate how many atoms of each element are present. Every compound has a **molecular weight** (molecular mass) that is the

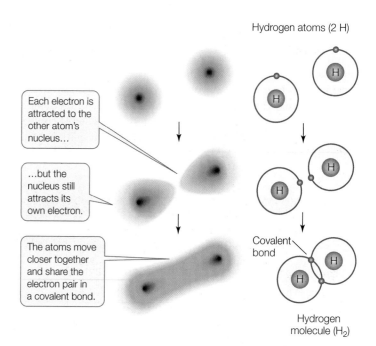

Hydrogen atoms (2 H)

Each electron is attracted to the other atom's nucleus…

…but the nucleus still attracts its own electron.

The atoms move closer together and share the electron pair in a covalent bond.

Covalent bond

Hydrogen molecule (H_2)

2.6 Electrons Are Shared in Covalent Bonds Two hydrogen atoms can combine to form a hydrogen molecule. A covalent bond forms when the electron orbitals of the two atoms overlap in an energetically stable manner.

TABLE 2.1

Chemical Bonds and Interactions

NAME	BASIS OF INTERACTION	STRUCTURE	BOND ENERGY[a] (KCAL/MOL)
Covalent bond	Sharing of electron pairs		50–110
Ionic bond	Attraction of opposite charges		3–7
Hydrogen bond	Sharing of H atom		3–7
Hydrophobic interaction	Interaction of nonpolar substances in the presence of polar substances (especially water)		1–2
van der Waals interaction	Interaction of electrons of nonpolar substances		1

[a]Bond energy is the amount of energy needed to separate two bonded or interacting atoms under physiological conditions.

sum of the atomic weights of all atoms in the molecule. Looking at the periodic table in Figure 2.2, you can calculate the molecular weights of the three compounds listed above to be 16.04, 18.01, and 342.29, respectively. Molecules that make up living organisms range in molecular weight from two to half a billion, and covalent bonds are common to all.

How are covalent bonds formed in a molecule of methane gas (CH_4)? The carbon atom in this compound has six electrons: two electrons fill its inner shell, and four unpaired electrons travel in its outer shell. Because its outer shell can hold up to eight electrons, carbon can share electrons with up to four other atoms—*it can form four covalent bonds* (**Figure 2.7A**). When an atom of carbon reacts with four hydrogen atoms, methane forms. Thanks to electron sharing, the outer shell of methane's carbon atom is now filled with eight electrons, a stable configuration. The outer shell of each of the four hydrogen atoms is also filled. Four covalent bonds—four shared electron pairs—

hold methane together. **Figure 2.7B** shows several different ways to represent the molecular structure of methane. **Table 2.2** shows the covalent bonding capacities of some biologically significant elements.

STRENGTH AND STABILITY Covalent bonds are very strong, meaning that it takes a lot of energy to break them. At temperatures in which life exists, the covalent bonds of biological molecules are quite stable, as are their three-dimensional structures. However, this stability does not preclude change, as we will discover.

ORIENTATION For a given pair of elements—for example, carbon bonded to hydrogen—the length of the covalent bond is always the same. And for a given atom within a molecule, the angle of each covalent bond, with respect to the other bonds, is generally the same. This is true regardless of the type of larger molecule that contains the atom. For example, the four filled orbitals around the carbon atom in methane are always distributed in space so that the bonded hydrogens point to the corners of a regular tetrahedron, with carbon in the center (see Figure 2.7B). Even when carbon is bonded to four atoms other than hydrogen, this three-dimensional orientation is more or less maintained. The orientation of covalent bonds in space gives the molecules their three-dimensional geometry, and the shapes of molecules contribute to their biological functions, as we will see in Section 3.1.

MULTIPLE COVALENT BONDS A covalent bond can be represented by a line between the chemical symbols for the linked atoms:

- A *single bond* involves the sharing of a single pair of electrons (for example, H—H or C—H).

TABLE 2.2

Covalent Bonding Capabilities of Some Biologically Important Elements

ELEMENT	USUAL NUMBER OF COVALENT BONDS
Hydrogen (H)	1
Oxygen (O)	2
Sulfur (S)	2
Nitrogen (N)	3
Carbon (C)	4
Phosphorus (P)	5

(A)

1 C and 4 H   Methane (CH₄)

Covalent bond

Bohr models

Carbon can complete its outer shell by sharing the electrons of four hydrogen atoms, forming methane.

2.7 Covalent Bonding Can Form Compounds (A) Bohr models showing the formation of covalent bonds in methane, whose molecular formula is CH₄. Electrons are shown in shells around the nucleus. (B) Three additional ways of representing the structure of methane. The ball-and-stick model and the space-filling model show the spatial orientations of the bonds. The space-filling model indicates the overall shape and surface of the molecule. In the chapters that follow, different conventions will be used to depict molecules. Bear in mind that these are models to illustrate certain properties, and not the most accurate portrayal of reality.

(B)

Each line or pair of dots represents a shared pair of electrons.

The hydrogen atoms form corners of a regular tetrahedron.

This model shows the shape methane presents to its environment.

Structural formulas   Ball-and-stick model   Space-filling model

- A *double bond* involves the sharing of four electrons (two pairs) (C=C).

- *Triple bonds*—six shared electrons—are rare, but there is one in nitrogen gas (N≡N), which is the major component of the air we breathe.

UNEQUAL SHARING OF ELECTRONS If two atoms of the same element are covalently bonded, there is an equal sharing of the pair(s) of electrons in their outermost shells. However, when the two atoms are of different elements, the sharing is not nec-

essarily equal. One nucleus may exert a greater attractive force on the electron pair than the other nucleus, so that the pair tends to be closer to that atom.

The attractive force that an atomic nucleus exerts on electrons in a covalent bond is called its **electronegativity**. The electronegativity of a nucleus depends on how many positive charges it has (nuclei with more protons are more positive and thus more attractive to electrons) and on the distances between the electrons in the bond and the nucleus (the closer the electrons, the greater the electronegative pull). **Table 2.3** shows the electronegativities (which are calculated to produce dimensionless quantities) of some elements important in biological systems.

If two atoms are close to each other in electronegativity, they will share electrons equally in what is called a *nonpolar covalent bond*. Two oxygen atoms, for example, each with an electronegativity of 3.5, will share electrons equally. So will two hydrogen atoms (each with an electronegativity of 2.1). But when hydrogen bonds with oxygen to form water, the electrons involved are unequally shared: they tend to be nearer to the oxygen nucleus because it is the more electronegative of the two. When electrons are drawn to one nucleus more than to the other, the result is a *polar covalent bond* (**Figure 2.8**).

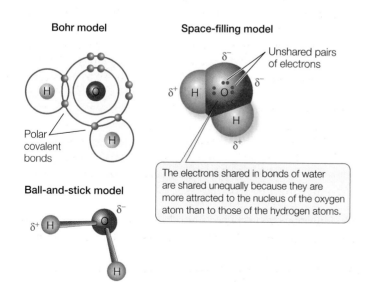

Bohr model   Space-filling model

Unshared pairs of electrons

Polar covalent bonds

Ball-and-stick model

The electrons shared in bonds of water are shared unequally because they are more attracted to the nucleus of the oxygen atom than to those of the hydrogen atoms.

2.8 Water's Covalent Bonds Are Polar These three representations all illustrate polar covalent bonding in water (H₂O). When atoms with different electronegativities, such as oxygen and hydrogen, form a covalent bond, the electrons are drawn to one nucleus more than to the other. A molecule held together by such a polar covalent bond has partial (δ⁺ and δ⁻) charges at different surfaces. In water, the shared electrons are displaced toward the oxygen atom's nucleus.

TABLE 2.3
Some Electronegativities

ELEMENT	ELECTRONEGATIVITY
Oxygen (O)	3.5
Chlorine (Cl)	3.1
Nitrogen (N)	3.0
Carbon (C)	2.5
Phosphorus (P)	2.1
Hydrogen (H)	2.1
Sodium (Na)	0.9
Potassium (K)	0.8

Because of this unequal sharing of electrons, the oxygen end of the hydrogen–oxygen bond has a slightly negative charge (symbolized by δ^- and spoken of as "delta negative," meaning a partial unit of charge), and the hydrogen end has a slightly positive charge (δ^+). The bond is **polar** because these opposite charges are separated at the two ends, or poles, of the bond. The partial charges that result from polar covalent bonds produce polar molecules or polar regions of large molecules. Polar bonds within molecules greatly influence the interactions that they have with other polar molecules. Water (H_2O) is a polar compound, and this polarity has significant effects on its physical properties and chemical reactivity, as we will see in later chapters.

Ionic bonds form by electrical attraction

When one interacting atom is much more electronegative than the other, a complete transfer of one or more electrons may take place. Consider sodium (electronegativity 0.9) and chlorine (3.1). A sodium atom has only one electron in its outermost shell; this condition is unstable. A chlorine atom has seven electrons in its outermost shell—another unstable condition. Since the electronegativity of chlorine is so much greater than that of sodium, any electrons involved in bonding will tend to transfer completely from sodium's outermost shell to that of chlorine (**Figure 2.9**). This reaction between sodium and chlorine makes the resulting atoms more stable because they both have eight fully paired electrons in their outer shells. The result is two *ions*.

Ions are electrically charged particles that form when atoms gain or lose one or more electrons:

- The sodium ion (Na^+) in our example has a +1 unit of charge because it has one less electron than it has protons. The outermost electron shell of the sodium ion is full, with eight electrons, so the ion is stable. Positively charged ions are called **cations**.

- The chloride ion (Cl^-) has a –1 unit of charge because it has one more electron than it has protons. This additional electron gives Cl^- a stable outermost shell with eight electrons. Negatively charged ions are called **anions**.

Some elements can form ions with multiple charges by losing or gaining *more than* one electron. Examples are Ca^{2+} (the cal-

cium ion, a calcium atom that has lost two electrons) and Mg^{2+} (the magnesium ion). Two biologically important elements can each yield more than one stable ion. Iron yields Fe^{2+} (the ferrous ion) and Fe^{3+} (the ferric ion), and copper yields Cu^+ (the cuprous ion) and Cu^{2+} (the cupric ion). Groups of covalently bonded atoms that carry an electric charge are called *complex ions*; examples include NH_4^+ (the ammonium ion), SO_4^{2-} (the sulfate ion), and PO_4^{3-} (the phosphate ion). Once formed, ions are usually stable and no more electrons are lost or gained.

Ionic bonds are bonds formed as a result of the electrical attraction between ions bearing opposite charges. Ions can form bonds that result in stable solid compounds, which are referred to by the general term *salts*. Examples are sodium chloride (NaCl) and potassium phosphate (K_3PO_4). In sodium chloride—familiar to us as table salt—cations and anions are held together by ionic bonds. In solids, the ionic bonds are strong because the ions are close together. However, when ions are dispersed in water, the distance between them can be large; the strength of their attraction is thus greatly reduced. Under the conditions in living cells, an ionic attraction is less strong than a nonpolar covalent bond (see Table 2.1).

Not surprisingly, ions can interact with polar molecules, since they both carry electric charges. Such an interaction results when a solid salt such as NaCl dissolves in water. Water molecules surround the individual ions, separating them (**Figure**

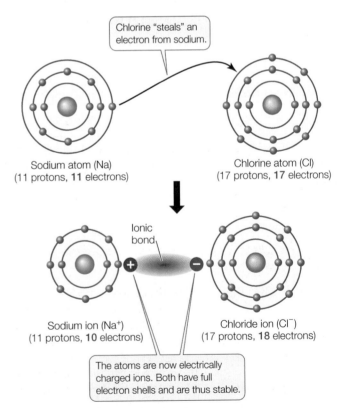

Chlorine "steals" an electron from sodium.

Sodium atom (Na)
(11 protons, **11** electrons)

Chlorine atom (Cl)
(17 protons, **17** electrons)

Ionic bond

Sodium ion (Na^+)
(11 protons, **10** electrons)

Chloride ion (Cl^-)
(17 protons, **18** electrons)

The atoms are now electrically charged ions. Both have full electron shells and are thus stable.

2.9 Formation of Sodium and Chloride Ions When a sodium atom reacts with a chlorine atom, the more electronegative chlorine fills its outermost shell by "stealing" an electron from the sodium. In so doing, the chlorine atom becomes a negatively charged chloride ion (Cl^-). With one less electron, the sodium atom becomes a positively charged sodium ion (Na^+).

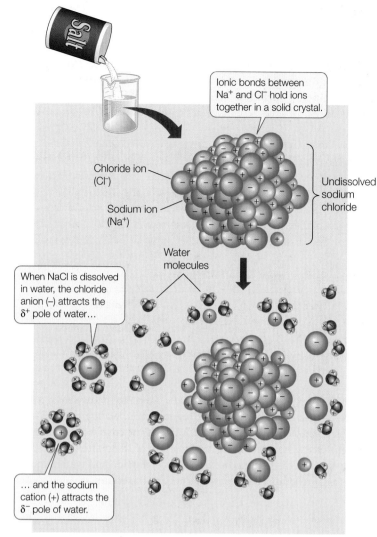

Ionic bonds between Na$^+$ and Cl$^-$ hold ions together in a solid crystal.

Chloride ion (Cl$^-$)

Sodium ion (Na$^+$)

Undissolved sodium chloride

Water molecules

When NaCl is dissolved in water, the chloride anion (–) attracts the δ$^+$ pole of water…

… and the sodium cation (+) attracts the δ$^-$ pole of water.

2.10 Water Molecules Surround Ions When an ionic solid dissolves in water, polar water molecules cluster around the cations and anions, preventing them from re-associating.

2.10). The negatively charged chloride ions attract the positive poles of the water molecules, while the positively charged sodium ions attract the negative poles of the water molecules. This is one of the special properties of water molecules, due to their polarity.

Hydrogen bonds may form within or between molecules with polar covalent bonds

In liquid water, the negatively charged oxygen (δ$^-$) atom of one water molecule is attracted to the positively charged hydrogen (δ$^+$) atoms of another water molecule (**Figure 2.11A**). The bond resulting from this attraction is called a **hydrogen bond**. Hydrogen bonds are not restricted to water molecules; they may also form between a strongly electronegative atom and a hydrogen atom that is covalently bonded to a different electronegative atom, as shown in **Figure 2.11B**.

A hydrogen bond is weaker than most ionic bonds because its formation is due to partial charges (δ$^+$ and δ$^-$). It is much weaker than a covalent bond between a hydrogen atom and an oxygen atom (see Table 2.1). Although individual hydrogen bonds are weak, many of them can form within one molecule or

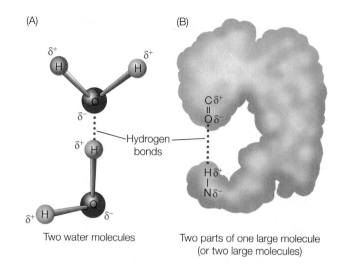

(A)

(B)

Two water molecules

Two parts of one large molecule (or two large molecules)

Hydrogen bonds

2.11 Hydrogen Bonds Can Form Between or Within Molecules (A) A hydrogen bond between two molecules is an attraction between a negative charge on one molecule and the positive charge on a hydrogen atom of the second molecule. (B) Hydrogen bonds can form between different parts of the same large molecule.

between two molecules. In these cases, the hydrogen bonds together have considerable strength, and greatly influence the structure and properties of substances. Later in this chapter we'll see how hydrogen bonding between water molecules contributes to many of the properties that make water so significant for living systems. Hydrogen bonds also play important roles in determining and maintaining the three-dimensional shapes of giant molecules such as DNA and proteins (see Section 3.2).

Polar and nonpolar substances: Each interacts best with its own kind

Just as water molecules can interact with one another through hydrogen bonds, any molecule that is polar can interact with other polar molecules through the weak (δ$^+$ to δ$^-$) attractions of hydrogen bonds. If a polar molecule interacts with water in this way, it is called **hydrophilic** ("water-loving") (**Figure 2.12A**).

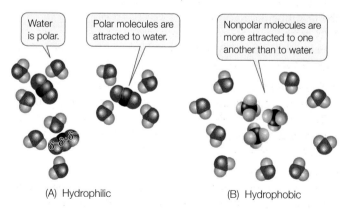

Water is polar.

Polar molecules are attracted to water.

Nonpolar molecules are more attracted to one another than to water.

(A) Hydrophilic

(B) Hydrophobic

2.12 Hydrophilic and Hydrophobic (A) Molecules with polar covalent bonds are attracted to polar water (they are hydrophilic). (B) Molecules with nonpolar covalent bonds show greater attraction to one another than to water (they are hydrophobic).

Nonpolar molecules tend to interact with other nonpolar molecules. For example, carbon (electronegativity 2.5) forms nonpolar bonds with hydrogen (electronegativity 2.1), and molecules containing only hydrogen and carbon atoms—called *hydrocarbon molecules*—are nonpolar. In water these molecules tend to aggregate with one another rather than with the polar water molecules. Therefore, nonpolar molecules are known as **hydrophobic** ("water-hating"), and the interactions between them are called *hydrophobic interactions* (**Figure 2.12B**). Of course, hydrophobic substances do not really "hate" water; they can form weak interactions with it, since the electronegativities of carbon and hydrogen are not exactly the same. But these interactions are far weaker than the hydrogen bonds between the water molecules, so the nonpolar substances tend to aggregate.

The interactions between nonpolar substances are enhanced by **van der Waals forces**, which occur when the atoms of two nonpolar molecules are in close proximity. These brief interactions result from random variations in the electron distribution in one molecule, which create opposite charge distributions in the adjacent molecule. Although a single van der Waals interaction is brief and weak, the sum of many such interactions over the entire span of a large nonpolar molecule can result in substantial attraction. This makes nonpolar molecules stick together in the polar (aqueous) environment inside organisms. We will see this many times, for example in the structure of biological membranes.

2.2 RECAP

Some atoms form strong covalent bonds with other atoms by sharing one or more pairs of electrons. Unequal sharing of electrons produces polarity. Other atoms become ions by losing or gaining electrons, and they interact with other ions or polar molecules.

- Why is a covalent bond stronger than an ionic bond? See pp. 26–28 and Table 2.1

- How do variations in electronegativity result in the unequal sharing of electrons in polar molecules? See pp. 27–28 and Figure 2.8

- What is a hydrogen bond and how is it important in biological systems? See p. 29 and Figure 2.11

The bonding of atoms into molecules is not necessarily a permanent affair. The dynamic of life involves constant change, even at the molecular level. Let's look at how molecules interact with one another—how they break up, how they find new partners, and what the consequences of those changes can be.

2.3 How Do Atoms Change Partners in Chemical Reactions?

A **chemical reaction** occurs when moving atoms collide with sufficient energy to combine or change their bonding partners. Consider the combustion reaction that takes place in the flame of a propane stove. When propane (C_3H_8) reacts with oxygen gas (O_2), the carbon atoms become bonded to oxygen atoms instead

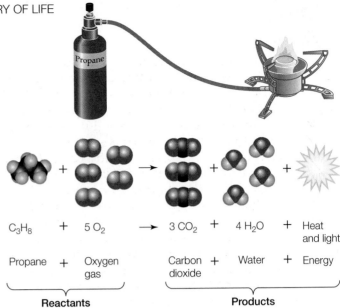

C_3H_8	+	5 O_2	$\longrightarrow$	3 CO_2	+	4 H_2O	+	Heat and light
Propane	+	Oxygen gas		Carbon dioxide	+	Water	+	Energy

Reactants — Products

2.13 Bonding Partners and Energy May Change in a Chemical Reaction   One molecule of propane from this burner reacts with five molecules of oxygen gas to give three molecules of carbon dioxide and four molecules of water. This reaction releases energy in the form of heat and light.

of hydrogen atoms, and the hydrogen atoms become bonded to oxygen instead of carbon (**Figure 2.13**). As the covalently bonded atoms change partners, the composition of the matter changes; propane and oxygen gas become carbon dioxide and water. This chemical reaction can be represented by the equation

$$C_3H_8 + 5 O_2 \rightarrow 3 CO_2 + 4 H_2O + \text{Energy}$$

Reactants $\rightarrow$ Products

In this equation, the propane and oxygen are the **reactants**, and the carbon dioxide and water are the **products**. In fact, this is a special type of reaction called an oxidation–reduction reaction. Electrons and protons are transferred from propane (the reducing agent) to oxygen (the oxidizing agent) to form water. You will see this kind of reaction involving electron/proton transfer many times in later chapters.

The products of a chemical reaction have very different properties from the reactants. In the case shown in Figure 2.13, the reaction is *complete*: all the propane and oxygen are used up in forming the two products. The arrow symbolizes the direction of the chemical reaction. The numbers preceding the molecular formulas indicate how many molecules are used or produced.

Note that in this and all other chemical reactions, *matter is neither created nor destroyed*. The total number of carbon atoms on the left (3) equals the total number of carbon atoms on the right (3). In other words, the equation is *balanced*. However, there is another aspect of this reaction: the heat and light of the stove's flame reveal that the reaction between propane and oxygen releases a great deal of energy.

Energy is defined as the capacity to do work, but in the context of chemical reactions, it can be thought of as the capacity for change. Chemical reactions do not create or destroy energy, but *changes in the form of energy* usually accompany chemical reactions.

In the reaction between propane and oxygen, a large amount of heat energy is released. This energy was present in another form, called *potential chemical energy*, in the covalent bonds within the propane and oxygen gas molecules. Not all reactions release energy; indeed, many chemical reactions require that energy be supplied from the environment. Some of this energy is then stored as potential chemical energy in the bonds formed in the products. We will see in future chapters how reactions that release energy and reactions that require energy can be linked together.

Many chemical reactions take place in living cells, and some of these have a lot in common with the oxidation–reduction reaction that happens in the combustion of propane. In cells, the reactants are different (they may be sugars or fats), and the reactions proceed by many intermediate steps that permit the released energy to be harvested and put to use by the cells. But the products are the same: carbon dioxide and water. We will discuss energy changes, oxidation–reduction reactions, and several other types of chemical reactions that are prevalent in living systems in Part Three of this book.

2.3 RECAP

In a chemical reaction, a set of reactants is converted to a set of products with different chemical compositions. This is accomplished by breaking and making bonds. Reactions may release energy or require its input.

- Explain how a chemical equation is balanced. **See p. 30 and Figure 2.13**

- How can the form of energy change during a chemical reaction? **See p. 30**

We will present and discuss energy changes, oxidation–reduction reactions, and several other types of chemical reactions that are prevalent in living systems in Part Two of this book. First, however, we must understand the unique properties of the substance in which most biochemical reactions take place: water.

2.4 What Makes Water So Important for Life?

Water is an unusual substance with unusual properties. Under conditions on Earth, water exists in solid, liquid, and gas forms, all of which have relevance to living systems. Water allows chemical reactions to occur inside living organisms, and it is necessary for the formation of certain biological structures. In this section we will explore how the structure and interactions of water molecules make water essential to life.

2.14 Hydrogen Bonding and the Properties of Water
Hydrogen bonding exists between the molecules of water in both its liquid and solid states. Ice is more structured but less dense than liquid water, which is why ice floats. Water forms a gas when its hydrogen bonds are broken and the molecules move farther apart.

Water has a unique structure and special properties

The molecule H_2O has unique chemical features. As we have already learned, water is a polar molecule that can form hydrogen bonds. The four pairs of electrons in the outer shell of the oxygen atom repel one another, giving the water molecule a tetrahedral shape:

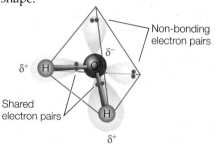

These chemical features explain some of the interesting properties of water, such as the ability of ice to float, the melting and freezing temperatures of water, the ability of water to store heat, the formation of water droplets, and water's ability to dissolve—and not dissolve—many substances.

ICE FLOATS In water's solid state (ice), individual water molecules are held in place by hydrogen bonds. Each molecule is bonded to four other molecules in a rigid, crystalline structure (**Figure 2.14**). Although the molecules are held firmly in place, they are not as tightly packed as they are in liquid water. In other words, *solid water is less dense than liquid water*, which is why ice floats.

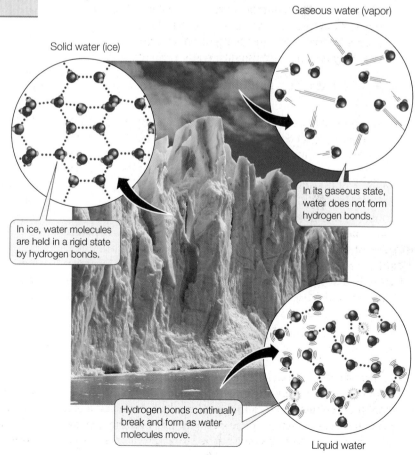

Think of the biological consequences if ice were to sink in water. A pond would freeze from the bottom up, becoming a solid block of ice in winter and killing most of the organisms living there. Once the whole pond is frozen, its temperature could drop well below the freezing point of water. But in fact ice floats, forming an insulating layer on the top of the pond, and reducing heat flow to the cold air above. Thus fish, plants, and other organisms in the pond are not subjected to temperatures lower than 0°C, which is the freezing point of pure water.

MELTING, FREEZING, AND HEAT CAPACITY   Compared with many other substances that have molecules of similar size, ice requires a great deal of heat energy to melt. This is because so many hydrogen bonds must be broken in order for water to change from solid to liquid. In the opposite process—freezing—a great deal of energy is released to the environment.

This property of water contributes to the surprising constancy of the temperatures found in oceans and other large bodies of water throughout the year. The temperature changes of coastal land masses are also moderated by large bodies of water. Indeed, water helps minimize variations in atmospheric temperature across the planet. This moderating ability is a result of the high *heat capacity* of liquid water, which is in turn a result of its high specific heat.

The **specific heat** of a substance is the amount of heat energy required to raise the temperature of 1 gram of that substance by 1°C. Raising the temperature of liquid water takes a relatively large amount of heat because much of the heat energy is used to break the hydrogen bonds that hold the liquid together. Compared with other small molecules that are liquids, water has a high specific heat.

Water also has a high **heat of vaporization**, which means that a lot of heat is required to change water from its liquid to its gaseous state (the process of *evaporation*). Once again, much of the heat energy is used to break the many hydrogen bonds between the water molecules. This heat must be absorbed from the environment in contact with the water. Evaporation thus has a cooling effect on the environment—whether a leaf, a forest, or an entire land mass. This effect explains why sweating cools the human body: as sweat evaporates from the skin, it uses up some of the adjacent body heat.

COHESION AND SURFACE TENSION   In liquid water, individual molecules are able to move about. The hydrogen bonds between the molecules continually form and break (see Figure 2.14). Chemists estimate that this occurs about a trillion times a minute for a single water molecule, making it a truly dynamic structure.

At any given time, a water molecule will form an average of 3.4 hydrogen bonds with other water molecules. These hydrogen bonds explain the *cohesive strength* of liquid water. This cohesive strength, or **cohesion**, is defined as the capacity of water molecules to resist coming apart from one another when placed under tension. Water's cohesive strength permits narrow columns of liquid water to move from the roots to the leaves of tall trees. When water evaporates from the leaves, the entire column moves upward in response to the pull of the molecules at the top.

2.15 Surface Tension   Water droplets form "beads" on the surface of a leaf because hydrogen bonds keep the water molecules together. The leaf is coated in a nonpolar wax that does not interact with the water molecules.

The surface of liquid water exposed to the air is difficult to puncture because the water molecules at the surface are hydrogen-bonded to other water molecules below them (**Figure 2.15**). This *surface tension* of water permits a container to be filled slightly above its rim without overflowing, and it permits insects to walk on the surface of a pond.

Water is an excellent solvent—the medium of life

A human body is over 70 percent water by weight, excluding the minerals contained in bones. Water is the dominant component of virtually all living organisms, and most biochemical reactions take place in this watery, or aqueous, environment.

A **solution** is produced when a substance (the **solute**) is dissolved in a liquid (the **solvent**). If the solvent is water, then the solution is an *aqueous solution*. Many of the important molecules in biological systems are polar, and therefore soluble in water. Many important biochemical reactions occur in aqueous solutions within cells. Biologists study these reactions in order to identify the reactants and products and to determine their amounts:

- *Qualitative analyses* deal with the identification of substances involved in chemical reactions. For example, a qualitative analysis would be used to investigate the steps involved, and the products formed, during the combustion of glucose in living tissues.

- *Quantitative analyses* measure concentrations or amounts of substances. For example, a biochemist would seek to describe *how much* of a certain product is formed during the combustion of a given amount of glucose using a quantitative analysis. What follows is a brief introduction to some of the quantitative chemical terms you will see in this book.

Fundamental to quantitative thinking in chemistry and biology is the concept of the mole. A **mole** is the amount of a substance (in grams) that is numerically equal to its molecular weight.

So a mole of table sugar ($C_{12}H_{22}O_{11}$) weighs about 342 grams; a mole of sodium ion (Na^+) weighs 23 grams; and a mole of hydrogen gas (H_2) weighs 2 grams.

Quantitative analyses do not yield direct counts of molecules. Because the amount of a substance in 1 mole is directly related to its molecular weight, it follows that the number of molecules in 1 mole is constant for all substances. So 1 mole of salt contains the same number of molecules as 1 mole of table sugar. This constant number of molecules in a mole is called **Avogadro's number**, and it is 6.02×10^{23} molecules per mole. Chemists work with moles of substances (which can be weighed out in the laboratory) instead of actual molecules (which are too numerous to be counted). Consider 34.2 grams (just over 1 ounce) of table sugar, $C_{12}H_{22}O_{11}$. This is one-tenth of a mole, or as Avogadro puts it, 6.02×10^{22} molecules.

If you have trouble grasping the concept of a mole, compare it with the concept of a dozen. We buy a dozen eggs or a dozen doughnuts, knowing that we will get 12 of whichever we buy, even though they don't weigh the same or take up the same amount of space.

A chemist can dissolve a mole of sugar (342 g) in water to make 1 liter of solution, knowing that the mole contains 6.02×10^{23} individual sugar molecules. This solution—1 mole of a substance dissolved in water to make 1 liter—is called a 1 molar (1 *M*) solution. When a physician injects a certain molar concentration of a drug into the bloodstream of a patient, a rough calculation can be made of the actual number of drug molecules that will interact with the patient's cells.

The many molecules dissolved in the water of living tissues are not present at concentrations anywhere near 1 molar. Most are in the micromolar (millionths of a mole per liter of solution; μM) to millimolar (thousandths of a mole per liter; m*M*) range. Some, such as hormone molecules, are even less concentrated than that. While these molarities seem to indicate very low concentrations, remember that even a 1 μM solution has 6.02×10^{17} molecules of the solute per liter.

Aqueous solutions may be acidic or basic

When some substances dissolve in water, they release *hydrogen ions* (H^+), which are actually single, positively charged protons. Hydrogen ions can attach to other molecules and change their properties. For example, the protons in "acid rain" can damage plants, and you probably have experienced the excess of hydrogen ions that we call "acid indigestion."

Here we will examine the properties of **acids** (defined as substances that release H^+) and **bases** (defined as substances which accept H^+). We will distinguish between strong and weak acids and bases and provide a quantitative means for stating the concentration of H^+ in solutions: the pH scale.

ACIDS RELEASE H⁺ When hydrochloric acid (HCl) is added to water, it dissolves, releasing the ions H^+ and Cl^-:

$$HCl \rightarrow H^+ + Cl^-$$

Because its H^+ concentration has increased, such a solution is *acidic*.

Acids are substances that *release* H^+ ions in solution. HCl is an acid, as is H_2SO_4 (sulfuric acid). One molecule of sulfuric acid will ionize to yield two H^+ and one SO_4^{2-}. Biological compounds that contain —COOH (the carboxyl group) are also acids because

$$-COOH \rightarrow -COO^- + H^+$$

Acids that fully ionize in solution, such as HCl and H_2SO_4 are called *strong acids*. However, not all acids ionize fully in water. For example, if acetic acid (CH_3COOH) is added to water, some will dissociate into two ions (CH_3COO^- and H^+), but some of the original acetic acid remains as well. Because the reaction is *not complete*, acetic acid is a *weak acid*.

BASES ACCEPT H⁺ Bases are substances that *accept* H^+ in solution. Just as with acids, there are strong and weak bases. If NaOH (sodium hydroxide) is added to water, it dissolves and ionizes, releasing OH^- and Na^+ ions:

$$NaOH \rightarrow Na^+ + OH^-$$

Because the concentration of OH^- increases and OH^- absorbs H^+ to form water ($OH^- + H^+ \rightarrow H_2O$), such a solution is *basic*. Because this reaction is complete, NaOH is a *strong base*.

Weak bases include the bicarbonate ion (HCO_3^-), which can accept a H^+ ion and become carbonic acid (H_2CO_3), and ammonia (NH_3), which can accept a H^+ and become an ammonium ion (NH_4^+). Biological compounds that contain —NH_2 (the amino group) are also bases because

$$-NH_2 + H^+ \rightarrow -NH_3^+$$

ACID–BASE REACTIONS MAY BE REVERSIBLE When acetic acid is dissolved in water, two reactions happen. First, the acetic acid forms its ions:

$$CH_3COOH \rightarrow CH_3COO^- + H^+$$

Then, once the ions are formed, some of them re-form acetic acid:

$$CH_3COO^- + H^+ \rightarrow CH_3COOH$$

This pair of reactions is reversible. A **reversible reaction** can proceed in either direction—left to right or right to left—depending on the relative starting concentrations of the reactants and products. The formula for a reversible reaction can be written using a double arrow:

$$CH_3COOH \rightleftharpoons CH_3COO^- + H^+$$

In terms of acids and bases, there are two types of reactions, depending on the extent of the reversibility:

- The ionization of strong acids and bases in water is virtually irreversible.
- The ionization of weak acids and bases in water is somewhat reversible.

WATER IS A WEAK ACID AND A WEAK BASE The water molecule has a slight but significant tendency to ionize into a hydroxide ion (OH^-) and a hydrogen ion (H^+). Actually, two water molecules participate in this reaction. One of the two molecules "captures" a hydrogen ion from the other, forming a hydroxide ion and a hydronium ion:

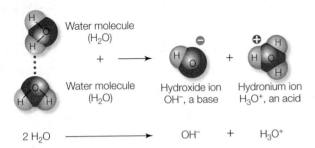

$$2\,H_2O \longrightarrow OH^- + H_3O^+$$

The hydronium ion is, in effect, a hydrogen ion bound to a water molecule. For simplicity, biochemists tend to use a modified representation of the ionization of water:

$$H_2O \rightarrow H^+ + OH^-$$

The ionization of water is important to all living creatures. This fact may seem surprising, since only about one water molecule in 500 million is ionized at any given time. But this is less surprising if we focus on the abundance of water in living systems, and the reactive nature of the H^+ ions produced by ionization.

pH: HYDROGEN ION CONCENTRATION   Compounds or ions can be acids or bases, and thus, solutions can be acidic or basic. We can measure how acidic or basic a solution is by measuring its concentration of H^+ in moles per liter (its *molarity*; see page 33). Here are some examples:

- Pure water has a H^+ concentration of $10^{-7}\ M$.
- A $1\ M$ HCl solution has a H^+ concentration of $1\ M$ (recall that all the HCl dissociates into its ions).
- A $1\ M$ NaOH solution has a H^+ concentration of $10^{-14}\ M$.

This is a very wide range of numbers to work with—think about the decimals! It is easier to work with the *logarithm* of the H^+ concentration, because logarithms compress this range: the $\log_{10}$ of 100, for example is 2, and the $\log_{10}$ of 0.01 is –2. Because most H^+ concentrations in living systems are less than 1M, their $\log_{10}$ values are negative. For convenience, we convert these negative numbers into positive ones, by using the *negative* of the logarithm of the H^+ molar concentration (the molar concentration is designated by square brackets: $[H^+]$). This number is called the **pH** of the solution.

Since the H^+ concentration of pure water is $10^{-7}\ M$, its pH is $-\log(10^{-7}) = -(-7)$, or 7. A smaller negative logarithm means a larger number. In practical terms, a lower pH means a higher H^+ concentration, or greater acidity. In $1\ M$ HCl, the H^+ concentration is $1\ M$, so the pH is the negative logarithm of 1 ($-\log 10^0$), or 0. The pH of $1\ M$ NaOH is the negative logarithm of 10^{-14}, or 14.

A solution with a pH of less than 7 is acidic—it contains more H^+ ions than OH^- ions. A solution with a pH of 7 is *neutral* (without net charge), and a solution with a pH value greater than 7 is basic. **Figure 2.16** shows the pH values of some common substances.

Why is this discussion of pH so important in biology? Many biologically important molecules contain charged groups

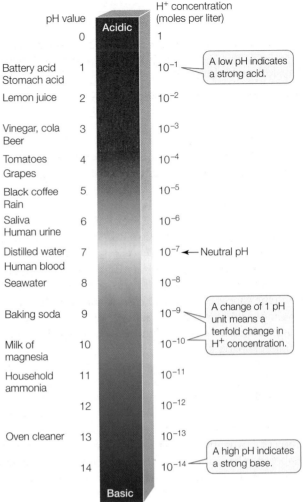

2.16 pH Values of Some Familiar Substances

(e.g., —COO⁻) that can interact with the polar regions of water to form their structures. But these groups can combine with H^+ or other ions in their environment to form uncharged groups (e.g., —COOH, see above). These uncharged groups have much less tendency to interact with water. If such a group is part of a larger molecule, it might now induce the molecule to fold in such a way that it stays away from water because it is hydrophobic. In a more acidic environment, a negatively charged group such as —COO⁻ is more likely to combine with H^+. So the pH of a biological tissue is a key to the three-dimensional structures of many of its constituent molecules. Organisms do all they can to minimize changes in the pH of their watery medium. An important way to do this is with buffers.

BUFFERS   The maintenance of internal constancy—*homeostasis*—is a hallmark of all living things and extends to pH. As we mentioned earlier, if biological molecules lose or gain H^+ ions their properties can change, thus upsetting homeostasis. Internal constancy is achieved with buffers: solutions that maintain a relatively constant pH even when substantial amounts of acid or base are added. How does this work?

A **buffer** is a solution of a weak acid and its corresponding base—for example, carbonic acid (H_2CO_3) and bicarbonate ions (HCO_3^-). If an acid is added to a solution containing this buffer,

2.17 Buffers Minimize Changes in pH   With increasing amounts of added base, the overall slope of a graph of pH is downward. Without a buffer, the slope is steep. Inside the buffering range of an added buffer, however, the slope is shallow. At very high and very low values of pH, where the buffer is ineffective, the slopes are much steeper.

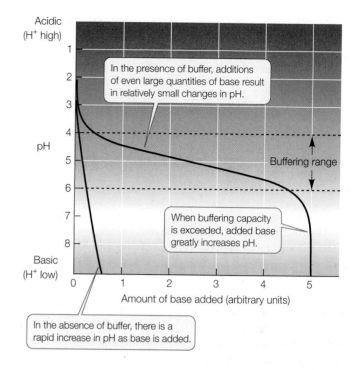

Acidic (H⁺ high)

In the presence of buffer, additions of even large quantities of base result in relatively small changes in pH.

Buffering range

When buffering capacity is exceeded, added base greatly increases pH.

Basic (H⁺ low)

Amount of base added (arbitrary units)

In the absence of buffer, there is a rapid increase in pH as base is added.

not all the H⁺ ions from the acid stay in solution. Instead, many of them combine with the bicarbonate ions to produce more carbonic acid:

$$HCO_3^- + H^+ \rightarrow H_2CO_3$$

This reaction uses up some of the H⁺ ions in the solution and decreases the acidifying effect of the added acid. If a base is added, the reaction essentially reverses. Some of the carbonic acid ionizes to produce bicarbonate ions and more H⁺, which counteracts some of the added base. In this way, the buffer minimizes the effect that an added acid or base has on pH. This buffering system is present in the blood, where it is important for preventing significant changes in pH that could disrupt the ability of the blood to carry vital oxygen to tissues. A given amount of acid or base causes a smaller pH change in a buffered solution than in a non-buffered one (**Figure 2.17**).

Buffers illustrate an important chemical principle of reversible reactions, called the *law of mass action*. Addition of a reactant on one side of a reversible system drives the reaction in the direction that uses up that compound. In the case of buffers, addition of an acid drives the reaction in one direction; addition of a base drives the reaction in the other direction.

We use a buffer to relieve the common problem of indigestion. The lining of the stomach constantly secretes hydrochloric acid, making the stomach contents acidic. Excessive stomach acid inhibits digestion and causes discomfort. We can relieve this discomfort by ingesting a salt such as NaHCO₃ ("bicarbonate of soda"), which acts as a buffer.

2.4 RECAP

Most of the chemistry of life occurs in water, which has molecular properties that make it suitable for its important biochemical roles. A special property of water is its ability to ionize (release hydrogen ions). The presence of hydrogen ions in solution can change the properties of biological molecules.

- Describe some of the biologically important properties of water arising from its molecular structure. **See pp. 31–32 and Figure 2.14**

- What is a solution, and why do we call water "the medium of life"? **See pp. 32–33**

- What is the relationship between hydrogen ions, acids, and bases? Explain what the pH scale measures. **See pp. 33–34 and Figure 2.16**

- How does a buffer work, and why is buffering important to living systems? **See pp. 34–35 and Figure 2.17**

An Overview and a Preview

Now that we have covered the major properties of atoms and molecules, let's review them and see how these properties relate to the major molecules of biological systems.

- *Molecules vary in size.* Some are small, such as those of hydrogen gas (H₂) and methane (CH₄). Others are larger, such as a molecule of table sugar (C₁₂H₂₂O₁₁), which has 45 atoms. Still others, especially proteins and nucleic acids, are gigantic, containing tens of thousands or even millions of atoms.

- *All molecules have a specific three-dimensional shape.* For example, the orientations of the bonding orbitals around the carbon atom give the methane molecule (CH₄) the shape of a regular tetrahedron (see Figure 2.7B). Larger molecules have complex shapes that result from the numbers and kinds of atoms present, and the ways in which they are linked together. Some large molecules, such as the protein hemoglobin (the oxygen carrier in red blood cells), have compact, ball-like shapes. Others, such as the protein called keratin that makes up your hair, have long, thin, ropelike structures. Their shapes relate to the roles these molecules play in living cells.

- *Molecules are characterized by certain chemical properties* that determine their biological roles. Chemists use the characteristics of composition, structure (three-dimensional shape), reactivity, and solubility to distinguish a pure sample of one molecule from a sample of a different molecule. The presence of certain groups of atoms can impart distinctive chemical properties to a molecule.

Between the small molecules discussed in this chapter and the world of the living cell are the macromolecules. These larger molecules—proteins, lipids, carbohydrates, and nucleic acids— will be discussed in the next two chapters.

CHAPTER SUMMARY

2.1 How Does Atomic Structure Explain the Properties of Matter?

- Matter is composed of atoms. Each **atom** consists of a positively charged **nucleus** made up of **protons** and **neutrons**, surrounded by **electrons** bearing negative charges. Review Figure 2.1

- The number of protons in the nucleus defines an **element**. There are many elements in the universe, but only a few of them make up the bulk of living organisms: C, H, O, P, N, and S. **Review Figure 2.2**

- **Isotopes** of an element differ in their numbers of neutrons. **Radioisotopes** are radioactive, emitting radiation as they break down.

- Electrons are distributed in **shells**, which are volumes of space defined by specific numbers of orbitals. Each **orbital** contains a maximum of two electrons. **Review Figures 2.4 and 2.5, WEB ACTIVITY 2.1**

- In losing, gaining, or sharing electrons to become more stable, an atom can combine with other atoms to form a **molecule**.

2.2 How Do Atoms Bond to Form Molecules?

SEE ANIMATED TUTORIAL 2.1

- A **chemical bond** is an attractive force that links two atoms together in a molecule. **Review Table 2.1**

- A **compound** is a substance made up of molecules with two or more elements bonded together in a fixed ratio, such as water (H_2O) or table sugar ($C_6H_{12}O_6$).

- **Covalent bonds** are strong bonds formed when two atoms share one or more pairs of electrons. **Review Figure 2.6**

- When two atoms of unequal electronegativity bond with each other, a **polar** covalent bond is formed. The two ends, or poles, of the bond have partial charges (δ^+ or δ^-). **Review Figure 2.8**

- **Ions** are electrically charged bodies that form when an atom gains or loses one or more electrons in order to form more stable electron configurations. **Anions** and **cations** are negatively and positively charged ions, respectively. Different charges attract, and like charges repel each other.

- **Ionic bonds** are electrical attractions between oppositely charged ions. Ionic bonds are strong in solids (salts), but weaken when the ions are separated from one another in solution. **Review Figure 2.9**

- A **hydrogen bond** is a weak electrical attraction that forms between a δ^+ hydrogen atom in one molecule and a δ^- atom in another molecule (or in another part of a large molecule). Hydrogen bonds are abundant in water.

- Nonpolar molecules interact very little with polar molecules, including water. Nonpolar molecules are attracted to one another by very weak bonds called **van der Waals forces**.

2.3 How Do Atoms Change Partners in Chemical Reactions?

- In **chemical reactions**, atoms combine or change their bonding partners. **Reactants** are converted into **products**.

- Some chemical reactions release **energy** as one of their products; other reactions can occur only if energy is provided to the reactants.

- Neither matter nor energy is created or destroyed in a chemical reaction, but both change form. **Review Figure 2.13**

- Some chemical reactions, especially in biology, are reversible. That is, the products formed may be converted back to the reactants.

- In living cells, chemical reactions take place in multiple steps so that the released energy can be harvested for cellular activities.

2.4 What Makes Water So Important for Life?

- Water's molecular structure and its capacity to form hydrogen bonds give it unique properties that are significant for life. **Review Figure 2.14**

- The high **specific heat** of water means that water gains or loses a great deal of heat when it changes state. Water's high **heat of vaporization** ensures effective cooling when water evaporates.

- The **cohesion** of water molecules refers to their capacity to resist coming apart from one another. Hydrogen bonds between water molecules play an essential role in these properties.

- A **solution** is produced when a solid substance (the **solute**) dissolves in a liquid (the **solvent**). Water is the critically important solvent for life.

- **Acids** are solutes that release hydrogen ions in aqueous solutions. **Bases** accept hydrogen ions.

- The **pH** of a solution is the negative logarithm of its hydrogen ion concentration. Values lower than pH 7 indicate that a solution is acidic; values above pH 7 indicate a basic solution. **Review Figure 2.16**

- A **buffer** is a mixture of a weak acid and a base that limits changes in the pH of a solution when acids or bases are added.

SELF-QUIZ

1. The atomic number of an element
   a. equals the number of neutrons in an atom.
   b. equals the number of protons in an atom.
   c. equals the number of protons minus the number of neutrons.
   d. equals the number of neutrons plus the number of protons.
   e. depends on the isotope.

2. The atomic weight (atomic mass) of an element
   a. equals the number of neutrons in an atom.
   b. equals the number of protons in an atom.
   c. equals the number of electrons in an atom.

   d. equals the number of neutrons plus the number of protons.
   e. depends on the relative abundances of its electrons and neutrons.

3. Which of the following statements about the isotopes of an element is *not* true?
   a. They all have the same atomic number.
   b. They all have the same number of protons.
   c. They all have the same number of neutrons.
   d. They all have the same number of electrons.
   e. They all have identical chemical properties.

4. Which of the following statements about covalent bonds is *not* true?
   a. A covalent bond is stronger than a hydrogen bond.
   b. A covalent bond can form between atoms of the same element.
   c. Only a single covalent bond can form between two atoms.
   d. A covalent bond results from the sharing of electrons by two atoms.
   e. A covalent bond can form between atoms of different elements.

5. Hydrophobic interactions
   a. are stronger than hydrogen bonds.
   b. are stronger than covalent bonds.
   c. can hold two ions together.
   d. can hold two nonpolar molecules together.
   e. are responsible for the surface tension of water.

6. Which of the following statements about water is *not* true?
   a. It releases a large amount of heat when changing from liquid into vapor.
   b. Its solid form is less dense than its liquid form.
   c. It is the most effective solvent for polar molecules.
   d. It is typically the most abundant substance in a living organism.
   e. It takes part in some important chemical reactions.

7. The reaction $HCl \rightarrow H^+ + Cl^-$ in the human stomach is an example of the
   a. cleavage of a hydrophobic bond.
   b. formation of a hydrogen bond.
   c. elevation of the pH of the stomach.
   d. formation of ions by dissolving an acid.
   e. formation of polar covalent bonds.

8. The hydrogen bond between two water molecules arises because water is
   a. polar.
   b. nonpolar.
   c. a liquid.
   d. small.
   e. hydrophobic.

9. When table salt ($NaCl$) is added to water,
   a. a covalent bond is broken.
   b. an acidic solution is formed.
   c. the Na^+ and Cl^- ions are separated.
   d. the Na^+ ions are attracted to the hydrogen atoms of water.
   e. water molecules surround the Na^+ (but not Cl^-) ions.

10. The three most abundant elements in a human skin cell are
    a. calcium, carbon, and oxygen.
    b. carbon, hydrogen, and oxygen.
    c. carbon, hydrogen, and sodium.
    d. carbon, nitrogen, and potassium.
    e. nitrogen, hydrogen, and argon.

FOR DISCUSSION

1. Using the information in the periodic table (Figure 2.2), draw a Bohr model (see Figures 2.5 and 2.7) of silicon dioxide, showing electrons shared in covalent bonds.

2. Compare a covalent bond between two hydrogen atoms with a hydrogen bond between a hydrogen and an oxygen atom, with regard to the electrons involved, the role of polarity, and the strength of the bond.

3. Write an equation describing the combustion of glucose ($C_6H_{12}O_6$) to produce carbon dioxide and water.

4. The pH of the human stomach is about 2.0, while the pH of the small intestine is about 10.0. What are the hydrogen ion concentrations [H^+] inside these two organs?

ADDITIONAL INVESTIGATION

Would you expect the elemental composition of Earth's crust to be the same as that of the human body? How could you find out?

3 Proteins, Carbohydrates, and Lipids

Molecular fossils

About 68 million years ago, a *Tyrannosaurus rex*, the fearsome dinosaur of movie stardom, died in what is now Wyoming in the United States. Over time, the giant carcass became buried 60 feet below the surface of what geologists call the Hell Creek Formation. In 2003, a thigh bone from the long-dead beast was found by the famous dinosaur hunter/biologist, John Horner from the Museum of the Rockies. Mary Schweitzer, a molecular paleontologist, was visiting Horner's Montana lab from North Carolina State University. She cut into the bone and found that it contained the remnants of soft tissues (such as bone marrow). This discovery was remarkable, because up until then scientists had thought that after about a million years, all the soft tissues in bone were replaced with minerals.

Back on the east coast, Lewis Cantley, a biochemist at Harvard University, read about Schweitzer's find in a newspaper and saw the possibility for a unique opportunity: for the first time, a scientist would be able to isolate and study the complex molecules of soft tissues from an extinct organism. He asked Schweitzer to send him a sample, and when he and his colleagues analyzed the dinosaur material, they found fragments of protein molecules.

Protein molecules are composed of long chains of individual molecules called amino acids. The protein fragments extracted from the *T. rex* bone were identified as collagen, a substance found in many modern animals. Moreover, the identity and specific order of the amino acids in the dinosaur collagen fragments closely matched that of collagen from chickens, and the dinosaur collagen folded into shapes very similar to those of bird collagen. This similarity to birds is not surprising, because, based on other evidence, scientists believe that birds are evolutionarily closely related to dinosaurs. Cantley's molecular analysis further confirmed this belief.

Proteins are one of the four major kinds of large molecules that characterize living systems. These *macromolecules*, which also include *carbohydrates*, *lipids*, and *nucleic acids*, differ in several significant ways from the small molecules and ions described in Chapter 2. First—no surprise—they are larger; the molecular weights of some

Molecular Clues   A thigh bone from a *Tyrannosaurus rex* that died 68 million years ago contained fragments of the protein collagen.

Molecular Evolution The sequence of amino acids in collagen dictates the shape the protein folds into. Collagen's amino acid sequence is similar in *T. rex* and in chickens, indicating that the two species share a common evolutionary ancestor.

nucleic acids reach billions of daltons. Second, these molecules all contain carbon atoms, and so belong to a group of what are known as *organic* chemicals. Third, the atoms of individual macromolecules are held together mostly by covalent bonds, which gives them important structural stability and distinctive three-dimensional geometries. These distinctive shapes are the basis of many of the functions of macromolecules, particularly the proteins.

Finally, carbohydrates, proteins, lipids, and nucleic acids are all unique to the living world. None of these molecular classes occurs in inanimate nature. You aren't likely to find protein in a rock—but if you do, you can be sure it came from a living organism.

IN THIS CHAPTER we will describe the chemical and biological properties of proteins, carbohydrates, and lipids. We will identify the components that make up these larger molecules, describe their assembly and geometries, as well as the roles they play in living organisms.

3.1 What Kinds of Molecules Characterize Living Things?

Four kinds of molecules are characteristic of living things: proteins, carbohydrates, lipids, and nucleic acids. With the exception of the lipids, these *biological molecules* are **polymers** (*poly*, "many"; *mer*, "unit") constructed by the covalent bonding of smaller molecules called **monomers**. The monomers that make up each kind of biological molecule have similar chemical structures:

- *Proteins* are formed from different combinations of 20 *amino acids*, all of which share chemical similarities.

- *Carbohydrates* can form giant molecules by linking together chemically similar sugar monomers (*monosaccharides*) to form polysaccharides.

- *Nucleic acids* are formed from four kinds of nucleotide monomers linked together in long chains.

- *Lipids* also form large structures from a limited set of smaller molecules, but in this case noncovalent forces maintain the interactions between the lipid monomers.

Polymers with molecular weights exceeding 1,000 grams per mole are considered to be **macromolecules**. The proteins, carbohydrates, and nucleic acids of living systems certainly fall into this category. Although large lipid structures are not polymers in the strictest sense, it is convenient to treat them as a special type of macromolecule (see Section 3.4).

How the macromolecules function and interact with other molecules depends on the properties of certain chemical groups in their monomers, the *functional groups*.

---
yourBioPortal.com
GO TO Animated Tutorial 3.1 • Macromolecules

Functional groups give specific properties to biological molecules

Certain small groups of atoms, called **functional groups**, are consistently found together in very different biological molecules. You will encounter several functional groups repeatedly in your study of biology (**Figure 3.1**). Each functional group has specific chemical properties and, when it is attached to a larger molecule, it confers those properties on the larger molecule. One of these properties is polarity. Looking at the structures in Figure 3.1, can you determine which functional groups are the most

Functional group	Class of compounds and an example	Properties
Hydroxyl R—OH	**Alcohols** Ethanol	Polar. Hydrogen bonds with water to help dissolve molecules. Enables linkage to other molecules by dehydration.
Aldehyde	**Aldehydes** Acetaldehyde	C=O group is very reactive. Important in building molecules and in energy-releasing reactions.
Keto	**Ketones** Acetone	C=O group is important in carbohydrates and in energy reactions.
Carboxyl	**Carboxylic acids** Acetic acid	Acidic. Ionizes in living tissues to form —COO⁻ and H⁺. Enters into dehydration synthesis by giving up —OH. Some carboxylic acids important in energy-releasing reactions.
Amino	**Amines** Methylamine	Basic. Accepts H⁺ in living tissues to form —NH₃⁺. Enters into dehydration synthesis by giving up H⁺.
Phosphate	**Organic phosphates** 3-Phosphoglycerate	Negatively charged. Enters into dehydration synthesis by giving up —OH. When bonded to another phosphate, hydrolysis releases much energy.
Sulfhydryl R—SH	**Thiols** Mercaptoethanol	By giving up H, two —SH groups can react to form a disulfide bridge, thus stabilizing protein structure.

3.1 Some Functional Groups Important to Living Systems
Highlighted here are the seven functional groups most commonly found in biologically important molecules. "R" is a variable chemical grouping.

— **yourBioPortal.com** —
GO TO Web Activity 3.1 • Functional Groups

different groups interact on the same macromolecule. These diverse groups and their properties help determine the shapes of macromolecules as well as how they interact with other macromolecules and with smaller molecules.

Isomers have different arrangements of the same atoms

Isomers are molecules that have the same chemical formula—the same kinds and numbers of atoms—but the atoms are arranged differently. (The prefix *iso-*, meaning "same," is encountered in many biological terms.) Of the different kinds of isomers, we will consider two: structural isomers and optical isomers.

Structural isomers differ in how their atoms are joined together. Consider two simple molecules, each composed of four carbon and ten hydrogen atoms bonded covalently, both with the formula C_4H_{10}. These atoms can be linked in two different ways, resulting in different molecules:

Butane   Isobutane

The different bonding relationships in butane and isobutane are distinguished by their structural formulas, and the two molecules have different chemical properties.

Optical isomers occur when a carbon atom has four different atoms or groups of atoms attached to it. This pattern allows two different ways of making the attachments, each the mirror image of the other (**Figure 3.2**). Such a carbon atom is called an *asymmetrical carbon*, and the two resulting molecules are optical isomers of each other. You can envision your right and left hands as optical isomers. Just as a glove is specific for a particular hand, some biochemical molecules that can interact with one optical isomer of a carbon compound are unable to "fit" the other.

The structures of macromolecules reflect their functions

The four kinds of biological macromolecules are present in roughly the same proportions in all living organisms (**Figure 3.3**). Furthermore, a protein that has a certain function in an apple tree probably has a similar function in a human being because its chemistry is the same wherever it is found. Such *biochemical unity* reflects the evolution of all life from a common ancestor, by descent with modification. An important advantage of biochemical unity is that some organisms can acquire

polar? (Hint: Look for C—O, N—H, and P—O bonds.) The consistent chemical behavior of functional groups helps us understand the properties of the molecules that contain them.

Because macromolecules are so large, they contain many different functional groups (see Figure 3.1). A single large protein may contain hydrophobic, polar, and charged functional groups, each of which gives different specific properties to local sites on the macromolecule. As we will see, sometimes these

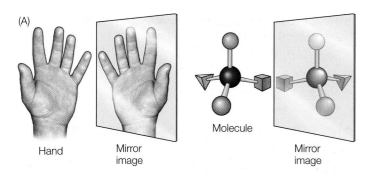

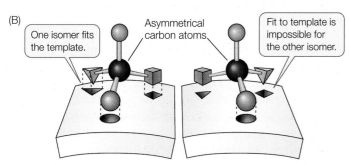

3.2 Optical Isomers (A) Optical isomers are mirror images of each other. (B) Molecular optical isomers result when four different atoms or groups are attached to a single carbon atom. If a template (representing a larger biological molecule in a living system) is laid out to match the groups on one carbon atom, the groups on that carbon's optical isomer cannot be rotated to fit the same template. This is a source of specificity in biological structure and biochemical transformations.

needed raw materials by eating other organisms. When you eat an apple, the molecules you take in include carbohydrates, lipids, and proteins that can be broken down and rebuilt into the varieties of those molecules needed by humans.

Each type of macromolecule performs some combination of functions, such as energy storage, structural support, protection, catalysis (speeding up a chemical reaction), transport, defense, regulation, movement, and information storage. These roles are not necessarily exclusive; for example, both carbohydrates and proteins can play structural roles, supporting and protecting tissues and organs. However, only the nucleic acids specialize in

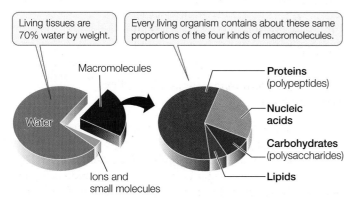

3.3 Substances Found in Living Tissues The substances shown here make up the nonmineral components of living tissues (bone would be an example of a mineral component).

information storage and transmission. These macromolecules function as hereditary material, carrying the traits of both species and individuals from generation to generation.

The functions of macromolecules are directly related to their three-dimensional shapes and to the sequences and chemical properties of their monomers. Some macromolecules fold into compact spherical forms with surface features that make them water-soluble and capable of intimate interaction with other molecules. Some proteins and carbohydrates form long, fibrous systems (such as those found in hair) that provide strength and rigidity to cells and tissues. The long, thin assemblies of proteins such as those in muscles can contract, resulting in movement.

Most macromolecules are formed by condensation and broken down by hydrolysis

Polymers are constructed from monomers by a series of reactions called **condensation reactions** (sometimes called *dehydration* reactions; both terms refer to the loss of water). Condensation reactions result in covalent bonds between monomers. A molecule of water is released with each covalent bond formed (**Figure 3.4A**). The condensation reactions that produce the different kinds of polymers differ in detail, but in

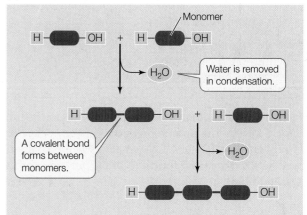

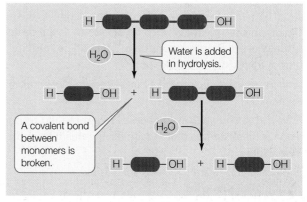

3.4 Condensation and Hydrolysis of Polymers (A) Condensation reactions link monomers into polymers and produce water. (B) Hydrolysis reactions break polymers into individual monomers and consume water.

all cases, polymers form only if water molecules are removed and energy is added to the system. In living systems, specific energy-rich molecules supply the necessary energy.

The reverse of a condensation reaction is a **hydrolysis reaction** (*hydro*, "water"; *lysis*, "break"). Hydrolysis reactions result in the breakdown of polymers into their component monomers. Water reacts with the covalent bonds that link the polymer together. For each covalent bond that is broken, a water molecule splits into two ions (H^+ and OH^-), which each become part of one of the products (**Figure 3.4B**). The linkages between monomers can thus be formed and broken inside living tissues.

3.1 RECAP

The four kinds of large molecules that distinguish living tissues are proteins, lipids, carbohydrates, and nucleic acids. These biological molecules carry out a wide range of life-sustaining functions. Most of them are polymers, made up of linked monomeric subunits. Very large polymers are called macromolecules.

- How do functional groups affect the structure and function of macromolecules? (Keep this question in mind as you read the rest of this chapter.) See pp. 39–40 and Figure 3.1

- Why is biochemical unity, as seen in the proportions of the four types of macromolecules present in all organisms, important for life? See p. 40 and Figure 3.3

- How do monomers link up to make polymers and how do they break down into monomers again? See pp. 41–42 and Figure 3.4

The four types of macromolecules can be seen as the building blocks of life. The unique properties of the nucleic acids will be covered in Chapter 4. The remainder of this chapter describes the structures and functions of the proteins, carbohydrates, and lipids.

3.2 What Are the Chemical Structures and Functions of Proteins?

While all of the kinds of large molecules are essential to the function of organisms, few have such diverse roles as the proteins. In virtually every chapter of this book, you will be studying examples of their extensive functions:

- *Enzymes* are catalytic proteins that speed up biochemical reactions.
- *Defensive proteins* such as antibodies recognize and respond to non-self substances that invade the organism from the environment.
- *Hormonal and regulatory proteins* such as insulin control physiological processes.
- *Receptor proteins* receive and respond to molecular signals from inside and outside the organism.

- *Storage proteins* store chemical building blocks—amino acids—for later use.
- *Structural proteins* such as collagen provide physical stability and movement.
- *Transport proteins* such as hemoglobin carry substances within the organism.
- *Genetic regulatory proteins* regulate when, how, and to what extent a gene is expressed.

Among the functions of macromolecules listed earlier, only two—energy storage and information storage—are not usually performed by proteins.

All **proteins** are polymers made up of different proportions and sequences of 20 amino acids. Proteins range in size from small ones such as insulin, which has a molecular weight of 5,733 daltons and 51 amino acids, to huge molecules such as the muscle protein titin, with a molecular weight of 2,993,451 daltons and 26,926 amino acids. All proteins consist of one or more *polpeptide chains*—unbranched (linear) polymer of covalently linked amino acids. The *composition* of a protein refers to the relative amounts of the different amino acids present in its polypeptide chains. Variation in the *sequence* of the amino acids in polypeptide chains is the source of the diversity in protein structure and function, because each chain folds into specific three-dimensional shape that is defined by the precise sequence of the amino acids present in the chain.

Many proteins are made up of more than one polypeptide chain. For example, the oxygen-carrying protein hemoglobin has four chains that are folded separately and come together to make up the functional protein. Proteins can also associate with one another, forming multi-protein complexes that carry out intricate tasks such as DNA synthesis.

To understand the many functions of proteins, we must first explore protein structure. We begin by examining the properties of amino acids and how they link together to form polypeptide chains. Then we will describe how a linear chain of amino acids is consistently folded into a specific, compact, three-dimensional shape. Finally, we will see how this three-dimensional structure provides a definitive physical and chemical environment that influences how other molecules can interact with the protein.

Amino acids are the building blocks of proteins

The amino acids have both a carboxyl functional group and an amino functional group (see Figure 3.1) attached to the same carbon atom, called the α (alpha) carbon. Also attached to the α carbon atom are a hydrogen atom and a **side chain**, or **R group**, designated by the letter R.

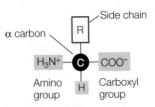

The α carbon is asymmetrical because it is bonded to four different atoms or groups of atoms. Therefore, amino acids exist in two isomeric forms, called D-amino acids and L-amino acids. D and L are abbreviations of the Latin terms for right (*dextro*) and left (*levo*). Only L-amino acids are commonly found in proteins in most organisms, and their presence is an important chemical "signature" of life.

At the pH values commonly found in cells, both the carboxyl and amino groups of amino acids are ionized: the carboxyl group has lost a hydrogen ion, and the amino group has gained one. Thus *amino acids are simultaneously acids and bases.*

The side chains of amino acids contain functional groups that are important in determining the three-dimensional structure and thus the function of the protein. As **Table 3.1** shows, the 20 amino acids found in living organisms are grouped and distinguished by their side chains:

- The five amino acids that have electrically charged side chains (+1, −1) attract water (are hydrophilic) and attract oppositely charged ions of all sorts.

- The five amino acids that have polar side chains (δ^+, δ^-) tend to form hydrogen bonds with water and with other polar or charged substances. These amino acids are also hydrophilic.

- Seven amino acids have side chains that are nonpolar hydrocarbons or very slightly modified hydrocarbons. In the watery environment of the cell, these hydrophobic side chains may cluster together in the interior of the protein. These amino acids are hydrophobic.

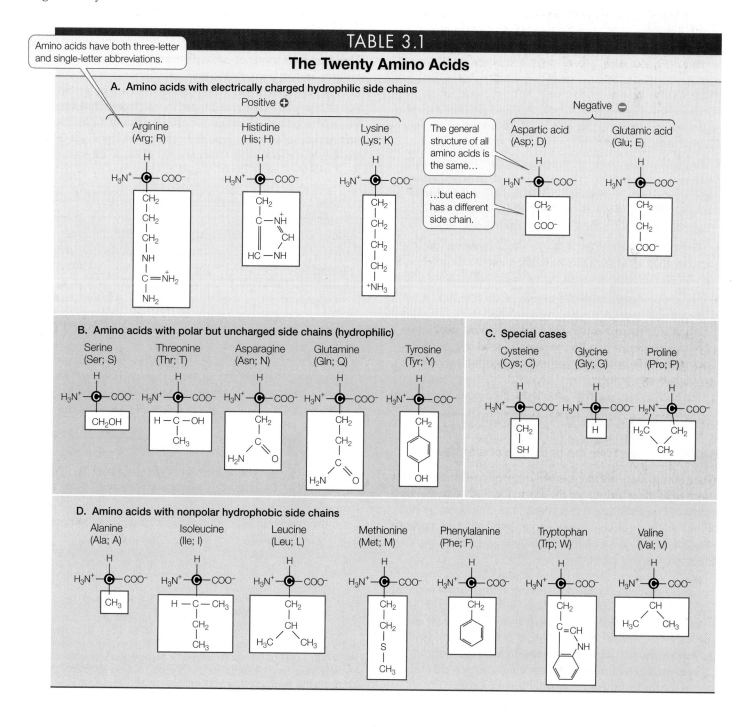

TABLE 3.1
The Twenty Amino Acids

Amino acids have both three-letter and single-letter abbreviations.

A. Amino acids with electrically charged hydrophilic side chains

The general structure of all amino acids is the same...

...but each has a different side chain.

B. Amino acids with polar but uncharged side chains (hydrophilic)

C. Special cases

D. Amino acids with nonpolar hydrophobic side chains

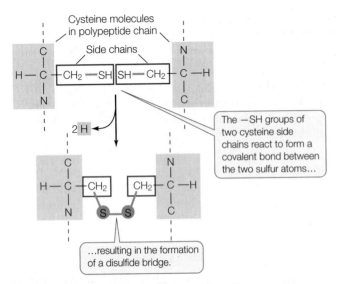

3.5 A Disulfide Bridge Two cysteine molecules in a polypeptide chain can form a disulfide bridge (—S—S—) by oxidation (removal of H atoms).

Three amino acids—cysteine, glycine, and proline—are special cases, although the side chains of the latter two are generally hydrophobic.

- The *cysteine* side chain, which has a terminal —SH group, can react with another cysteine side chain in an oxidation reaction to form a covalent bond called a **disulfide bridge**, or *disulfide bond* (—S—S—) (**Figure 3.5**). Disulfide bridges help determine how a polypeptide chain folds.

- The *glycine* side chain consists of a single hydrogen atom and is small enough to fit into tight corners in the interior of a protein molecule, where a larger side chain could not fit.

- *Proline* possesses a modified amino group that lacks a hydrogen and instead forms a covalent bond with the hydrocarbon side chain, resulting in a ring structure. This limits both its hydrogen-bonding ability and its ability to rotate about the α carbon. Thus proline is often found where a protein bends or loops.

—**yourBioPortal.com**—
GO TO Web Activity 3.2 • Features of Amino Acids

Peptide linkages form the backbone of a protein

When amino acids polymerize, the carboxyl and amino groups attached to the α carbon are the reactive groups. The carboxyl group of one amino acid reacts with the amino group of another, undergoing a condensation reaction that forms a **peptide linkage** (also called a *peptide bond*). **Figure 3.6** gives a simplified description of this reaction.

Just as a sentence begins with a capital letter and ends with a period, polypeptide chains have a beginning and an end. The "capital letter" marking the beginning of a polypeptide is the amino group of the first amino acid added to the chain and is known as the *N terminus*. The "period" is the carboxyl group of the last amino acid added; this is the *C terminus*.

Two characteristics of the peptide bond are especially important in the three-dimensional structure of proteins:

- In the C—N linkage, the adjacent α carbons (αC—C—N—αC) are not free to rotate fully, which limits the folding of the polypeptide chain.

- The oxygen bound to the carbon (C=O) in the carboxyl group carries a slight negative charge (δ^-), whereas the hydrogen bound to the nitrogen (N—H) in the amino group is slightly positive (δ^+). This asymmetry of charge favors hydrogen bonding within the protein molecule itself and with other molecules, contributing to both the structure and the function of many proteins.

Before we explore the significance of these characteristics of the peptide linkage, however, we will describe the significance of the sequence of amino acids in determining a protein's structure.

The primary structure of a protein is its amino acid sequence

There are four levels of protein structure: primary, secondary, tertiary, and quaternary. We will consider each of these in turn over the next few pages. The precise sequence of amino acids in a polypeptide chain held together by peptide linkages constitutes the **primary structure** of a protein (**Figure 3.7A**). The peptide backbone of the polypeptide chain consists of the repeating sequence —N—C—C—made up of the N atom from the

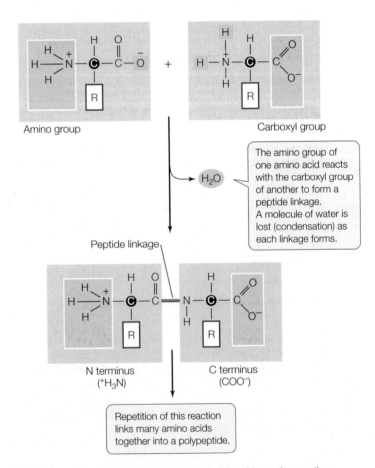

The amino group of one amino acid reacts with the carboxyl group of another to form a peptide linkage. A molecule of water is lost (condensation) as each linkage forms.

Repetition of this reaction links many amino acids together into a polypeptide.

3.6 Formation of Peptide Linkages In living things, the reaction leading to a peptide linkage (also called a peptide bond) has many intermediate steps, but the reactants and products are the same as those shown in this simplified diagram.

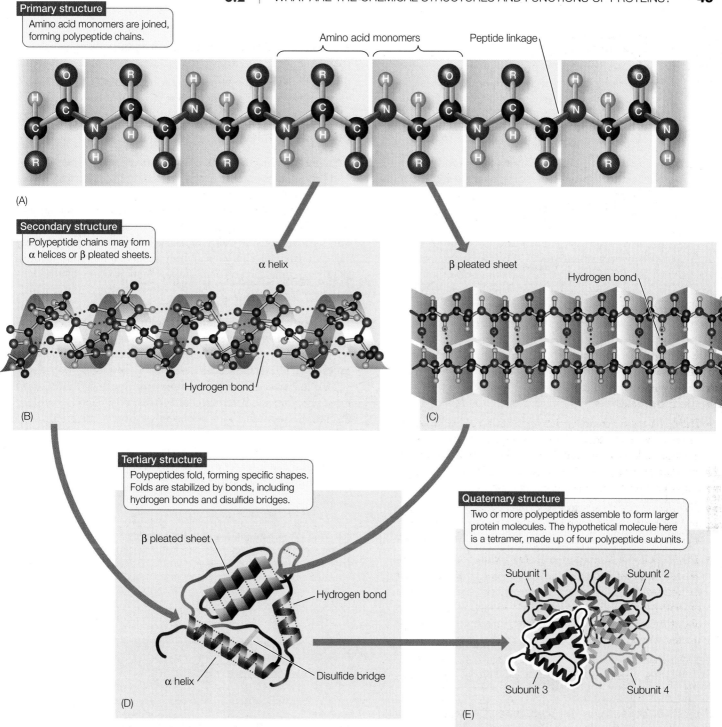

Primary structure

Amino acid monomers are joined, forming polypeptide chains.

Amino acid monomers

Peptide linkage

(A)

Secondary structure

Polypeptide chains may form α helices or β pleated sheets.

α helix

Hydrogen bond

(B)

β pleated sheet

Hydrogen bond

(C)

Tertiary structure

Polypeptides fold, forming specific shapes. Folds are stabilized by bonds, including hydrogen bonds and disulfide bridges.

β pleated sheet

Hydrogen bond

α helix

Disulfide bridge

(D)

Quaternary structure

Two or more polypeptides assemble to form larger protein molecules. The hypothetical molecule here is a tetramer, made up of four polypeptide subunits.

Subunit 1   Subunit 2

Subunit 3   Subunit 4

(E)

3.7 The Four Levels of Protein Structure Secondary, tertiary, and quaternary structure all arise from the primary structure of the protein.

amino group, the α carbon atom, and the C atom from the carboxyl group of each amino acid.

Scientists have determined the primary structure of many proteins. The single-letter abbreviations for amino acids (see Table 3.1) are used to record the amino acid sequence of a protein. Here, for example, are the first 20 amino acids (out of a total of 124) in the protein ribonuclease from a cow:

KETAAAKFERQHMDSSTSAA

The theoretical number of different proteins is enormous. Since there are 20 different amino acids, there could be $20 \times 20 = 400$ distinct dipeptides (two linked amino acids), and $20 \times 20 \times 20 =$ 8,000 different tripeptides (three linked amino acids). Imagine this process of multiplying by 20 extended to a protein made up of 100 amino acids (which would be considered a small protein). There could be 20^{100} (that's approximately 10^{130}) such small proteins, each with its own distinctive primary structure. How large is the number 20^{100}? Physicists tell us that there aren't that many electrons in the entire universe.

At the higher levels of protein structure (secondary, tertiary and quaternary), local coiling and folding of the polypeptide chain(s) give the molecule its final functional shape. All of these levels, however, derive from the protein's primary structure—that is, the precise location of specific amino acids in the polypeptide chain. The properties associated with a precise sequence of amino acids determine how the protein can twist and fold, thus adopting a specific stable structure that distinguishes it from every other protein.

Primary structure is established by covalent bonds. The next level of protein structure makes use of weaker hydrogen bonds.

The secondary structure of a protein requires hydrogen bonding

A protein's **secondary structure** consists of regular, repeated spatial patterns in different regions of a polypeptide chain. There are two basic types of secondary structure, both determined by hydrogen bonding between the amino acids that make up the primary structure, the α helix and the β pleated sheet.

THE α HELIX The α (alpha) **helix** is a right-handed coil that turns in the same direction as a standard wood screw (**Figure 3.7B**). The R groups extend outward from the peptide backbone of the helix. The coiling results from hydrogen bonds that form between the δ^+ hydrogen of the N—H of one amino acid and the δ^- oxygen of the C=O of another. When this pattern of hydrogen bonding is established repeatedly over a segment of the protein, it stabilizes the coil.

THE β PLEATED SHEET A β (beta) **pleated sheet** is formed from two or more polypeptide chains that are almost completely extended and aligned. The sheet is stabilized by hydrogen bonds between the N—H groups on one chain and the C=O groups on the other (**Figure 3.7C**). A β pleated sheet may form between separate polypeptide chains, as in spider silk, or between different regions of a single polypeptide chain that is bent back on itself. Many proteins contain regions of both α helix and β pleated sheet in the same polypeptide chain.

The tertiary structure of a protein is formed by bending and folding

In many proteins, the polypeptide chain is bent at specific sites and then folded back and forth, resulting in the **tertiary structure** of the protein (**Figure 3.7D**). Although α helices and β pleated sheets contribute to the tertiary structure, usually only portions of the macromolecule have these secondary structures, and large regions consist of tertiary structure unique to a particular protein. Tertiary structure results in a macromolecule's definitive three-dimensional shape, often including a buried interior as well as a surface that is exposed to the environment.

The protein's exposed outer surfaces present functional groups capable of interacting with other molecules in the cell. These molecules might be other proteins (as happens in quaternary structure, as we will see below) or smaller chemical reactants (as in enzymes; see Section 7.4).

While hydrogen bonding between the N—H and C=O groups within and between chains is responsible for secondary structure, the interactions between R groups—the amino acid side chains—determine tertiary structure. We described the various strong and weak interactions between atoms in Section 2.2. Many of these interactions are involved in determining and maintaining tertiary structure.

3.8 Three Representations of Lysozyme Different molecular representations of a protein emphasize different aspects of its tertiary structure: surface features, sites of bends and folds, sites where alpha or beta structure predominate. These three representations of lysozyme are similarly oriented.

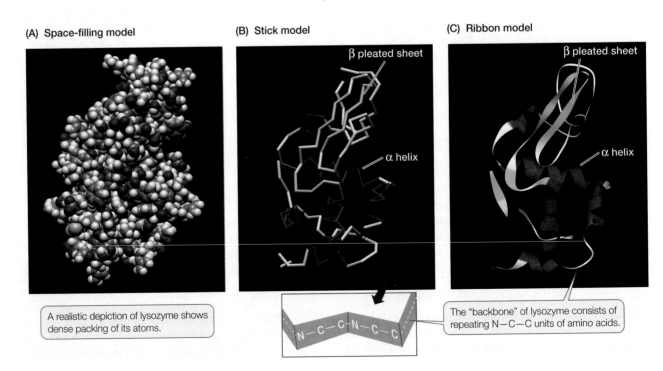

(A) Space-filling model

A realistic depiction of lysozyme shows dense packing of its atoms.

(B) Stick model

β pleated sheet

α helix

N—C—C—N—C—C

(C) Ribbon model

β pleated sheet

α helix

The "backbone" of lysozyme consists of repeating N—C—C units of amino acids.

- Covalent *disulfide bridges* can form between specific cysteine side chains (see Figure 3.5), holding a folded polypeptide in place.
- *Hydrogen bonds* between side chains also stabilize folds in proteins.
- *Hydrophobic* side chains can aggregate together in the interior of the protein, away from water, folding the polypeptide in the process.
- *van der Waals forces* can stabilize the close interactions between hydrophobic side chains.
- *Ionic bonds* can form between positively and negatively charged side chains, forming *salt bridges* between amino acids. Ionic bonds can also be buried deep within a protein, away from water.

A complete description of a protein's tertiary structure would specify the location of every atom in the molecule in three-dimensional space relative to all the other atoms. Such a description is available for the protein lysozyme (**Figure 3.8**).

The different ways of depicting the molecule have their uses. The space-filling model might be used to study how other molecules interact with specific sites and R groups on a protein's surface. The stick model emphasizes the sites where bends occur in order to make the folds of the polypeptide chain. The ribbon model, perhaps the most widely used, shows the different types of secondary structure and how they fold into the tertiary structure.

Remember that both secondary and tertiary structure derive from primary structure. If a protein is heated slowly, the heat energy will disrupt only the weak interactions, causing the secondary and tertiary structure to break down. The protein is then said to be **denatured**. But the protein can return to its normal tertiary structure when it cools, demonstrating that all the information needed to specify the unique shape of a protein is contained in its primary structure. This was first shown (using chemicals instead of heat to denature the protein) by biochemist Christian Anfinsen for the protein ribonuclease (**Figure 3.9**).

The quaternary structure of a protein consists of subunits

Many functional proteins contain two or more polypeptide chains, called *subunits*, each of them folded into its own unique tertiary structure. The protein's **quaternary structure** results from the ways in which these subunits bind together and interact (**Figure 3.7E**).

The models of hemoglobin in **Figure 3.10** illustrate quaternary structure. Hydrophobic interactions, van der Waals forces, hydrogen bonds, and ionic bonds all help hold the four subunits together to form a hemoglobin molecule. However, the weak nature of these forces permits small changes in the quaternary structure to aid the

INVESTIGATING LIFE

3.9 Primary Structure Specifies Tertiary Structure

Using the protein ribonuclease, Christian Anfinsen showed that proteins spontaneously fold into a functionally correct three-dimensional configuration. As long as the primary structure is not disrupted, the information for correct folding under the right conditions is retained.

HYPOTHESIS Under controlled conditions that simulate normal cellular environment in the laboratory, the primary structure of a denatured protein can reestablish the protein's three-dimensional structure.

METHOD Chemically denature functional ribonuclease, disrupting disulfide bridges and other intramolecular interactions that maintain the protein's shape, so that only primary structure (i.e., the amino acid sequence) remains. Once denaturation is complete, remove the disruptive chemicals.

1 Extract and purify a functional protein, ribonuclease, from tissue.

α helix

β pleated sheet

Disulfide bridge

2 Add chemicals that disrupt hydrogen and ionic bonds (urea) and disulfide bridges (mercaptoethanol).

Denatured protein

3 Slowly remove the chemical agents.

RESULTS When the disruptive agents are removed, three-dimensional structure is restored and the protein once again is functional.

CONCLUSION In normal cellular conditions, the primary structure of a protein specifies how it folds into a functional, three-dimensional structure.

Go to **yourBioPortal.com** for original citations, discussions, and relevant links for all INVESTIGATING LIFE figures.

3.10 Quaternary Structure of a Protein Hemoglobin consists of four folded polypeptide subunits that assemble themselves into the quaternary structure shown here. In these two graphic representations, each type of subunit is a different color. The heme groups contain iron and are the oxygen-carrying sites.

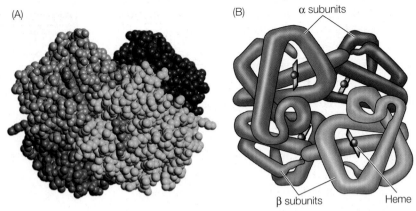

(A)

(B)

α subunits

β subunits

Heme

protein's function—which is to carry oxygen in red blood cells. As hemoglobin binds one O_2 molecule, the four subunits shift their relative positions slightly, changing the quaternary structure. Ionic bonds are broken, exposing buried side chains that enhance the binding of additional O_2 molecules. The quaternary structure changes again when hemoglobin releases its O_2 molecules to the cells of the body.

Shape and surface chemistry contribute to protein function

The shape and structure of a protein allow specific sites on its exposed surface to bind noncovalently to another molecule, which may be large or small. The binding is said to be specific because only certain compatible chemical groups will bind to one another. The specificity of protein binding depends on two general properties of the protein: its shape, and the chemistry of its exposed surface groups.

- *Shape.* When a small molecule collides with and binds to a much larger protein, it is like a baseball being caught by a catcher's mitt: the mitt has a shape that binds to the ball and fits around it. Just as a hockey puck or a ping-pong ball does not fit a baseball catcher's mitt, a given molecule will not bind to a protein unless there is a general "fit" between their three-dimensional shapes.

- *Chemistry.* The exposed amino acid R groups on the surface of a protein permit chemical interactions with other substances (**Figure 3.11**). Three types of interactions may be involved: ionic, hydrophobic, and hydrogen bonding. Many important functions of proteins involve interactions between exposed-surface R groups and other molecules.

Environmental conditions affect protein structure

Because it is determined by weak forces, the three-dimensional structure of proteins is influenced by environmental conditions. Conditions that would not break covalent bonds can disrupt the

weaker, noncovalent interactions that determine secondary and tertiary structure. Such alterations may affect a protein's shape and thus its function. Various conditions can alter the weak, noncovalent interactions:

- *Increases in temperature* cause more rapid molecular movements and thus can break hydrogen bonds and hydrophobic interactions.

- *Alterations in pH* can change the pattern of ionization of exposed carboxyl and amino groups in the R groups of amino acids, thus disrupting the pattern of ionic attractions and repulsions.

- *High concentrations of polar substances* such as urea can disrupt the hydrogen bonding that is crucial to protein structure. This was used in the experiment on reversible protein denaturation shown in Figure 3.9.

- *Nonpolar substances* may also disrupt normal protein structure in cases where hydrophobic groups are essential to maintain the structure.

Denaturation can be irreversible when amino acids that were buried in the interior of the protein become exposed at the surface, and vice versa, causing a new structure to form or different molecules to bind to the protein. Boiling an egg denatures its proteins and is, as you know, not reversible.

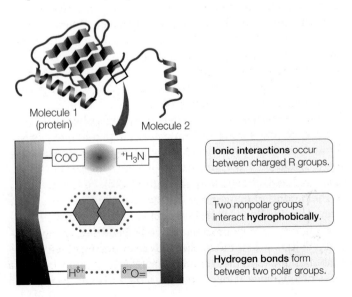

Molecule 1 (protein)

Molecule 2

Ionic interactions occur between charged R groups.

Two nonpolar groups interact **hydrophobically**.

Hydrogen bonds form between two polar groups.

3.11 Noncovalent Interactions Between Proteins and Other Molecules Noncovalent interactions (see pp. 28–30) allow a protein (brown) to bind tightly to another molecule (green) with specific properties. Noncovalent interactions also allow regions within the same protein to interact with one another.

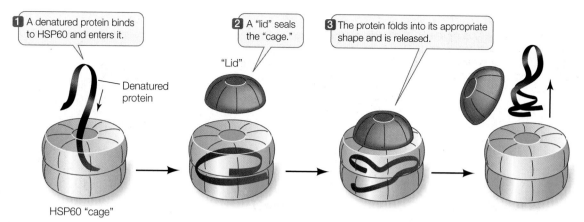

1 A denatured protein binds to HSP60 and enters it.

Denatured protein

2 A "lid" seals the "cage."

"Lid"

3 The protein folds into its appropriate shape and is released.

HSP60 "cage"

3.12 Chaperones Protect Proteins from Inappropriate Binding
Chaperone proteins surround new or denatured proteins and prevent them from binding to the wrong substance. Heat shock proteins such as HSP60, whose actions are illustrated here, are one class of chaperone proteins.

Molecular chaperones help shape proteins

Because of their specific shapes and the exposure of chemical groups on their surfaces, proteins can bind specific substances. Within a living cell, a polypeptide chain is sometimes in danger of binding the wrong substance. Two important examples of such a situation are:

- Following denaturation: Inappropriate environmental conditions in a cell, such as elevated temperature, can cause the denatured protein to re-fold incorrectly.

- Just after a protein is made: When a protein has not yet folded completely, it can present a surface that binds the wrong molecule.

In these cases, change may be irreversible. Eukaryotic cells have a special class of proteins that act to counteract threats to three-dimensional structure. Proteins in this class, called **chaperones**, act as molecular caretakers for other proteins. Like the chaperones at a high school dance, they prevent inappropriate interactions and enhance the appropriate ones.

Molecular chaperones were discovered by accident in 1962, when the temperature of an incubator holding fruit flies was accidentally turned up. Italian geneticist Ferruccio Ritossa noticed that this "heat shock" did not kill the flies. Instead, there was enhanced synthesis of a set of proteins that were later described as chaperones. They bound to many target proteins in the fruit fly cells and kept them from being denatured, and in some cases facilitated the correct refolding of proteins.

The general class of stress-induced chaperone proteins is called the **heat shock proteins (HSPs)**, after this discovery. HSPs are made by most eukaryotic cells, and many enhance protein folding in addition to their protective role during periods of stress. As an example, HSP60 forms a cage that sucks a protein in, causes it to fold into the correct shape, and then releases it (**Figure 3.12**). Tumors make abundant HSPs, possibly to stabilize proteins important in the cancer process, and so HSP-inhibiting drugs are being designed. In some clinical situations, treatment with these inhibitors results in the inappropriate folding of tumor-cell proteins, causing the tumors to stop growing and even disappear.

3.2 RECAP

Proteins are polymers of amino acids. The sequence of amino acids in a protein determines its primary structure. Secondary, tertiary, and quaternary structures arise through interactions between the amino acids. A protein's three-dimensional shape and exposed chemical groups establish binding specificity for other substances.

- What are the attributes of an amino acid's R group that would make it hydrophobic? Hydrophilic? **See pp. 42–43 and Table 3.1**

- Sketch and explain how two amino acids link together to form a peptide linkage. **See p. 44 and Figure 3.6**

- What are the four levels of protein structure and how are they all ultimately determined by the protein's primary structure (i.e., its amino acid sequence)? **See pp. 44–48 and Figure 3.7**

- How do environmental factors such as temperature and pH affect the weak interactions that give a protein its specific shape and function? **See p. 48**

The seemingly infinite number of protein configurations made possible by the biochemical properties of the 20 amino acids has driven the evolution of life's diversity. The linkage configurations of sugar monomers (monosaccharides) drives the structure of the next group of macromolecules, the carbohydrates that provide energy for life.

3.3 What Are the Chemical Structures and Functions of Carbohydrates?

Carbohydrates are a large group of molecules that all have a similar atomic composition but differ greatly in size, chemical properties, and biological functions. Carbohydrates have the general formula $C_n(H_2O)_n$, which makes them appear as hydrates of carbon (association between water molecules and carbon in the ratio $C_1H_2O_1$), hence their name. When their molecular structures are examined, the linked carbon atoms are seen to be bonded with hydrogen atoms (—H) and hydroxyl groups

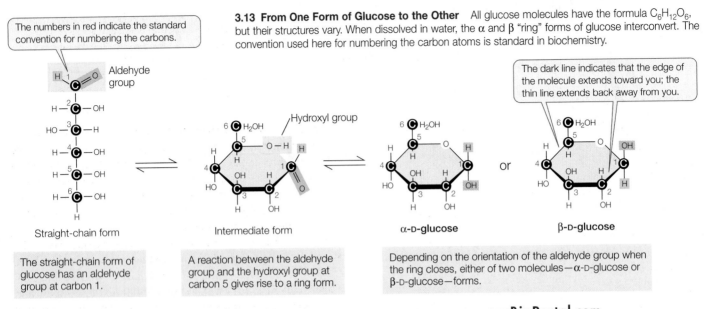

3.13 From One Form of Glucose to the Other   All glucose molecules have the formula $C_6H_{12}O_6$, but their structures vary. When dissolved in water, the α and β "ring" forms of glucose interconvert. The convention used here for numbering the carbon atoms is standard in biochemistry.

The numbers in red indicate the standard convention for numbering the carbons.

The dark line indicates that the edge of the molecule extends toward you; the thin line extends back away from you.

Straight-chain form

Intermediate form

α-D-glucose

or

β-D-glucose

The straight-chain form of glucose has an aldehyde group at carbon 1.

A reaction between the aldehyde group and the hydroxyl group at carbon 5 gives rise to a ring form.

Depending on the orientation of the aldehyde group when the ring closes, either of two molecules—α-D-glucose or β-D-glucose—forms.

─yourBioPortal.com─

GO TO Web Activity 3.3 • Forms of Glucose

(—OH), the components of water. Carbohydrates have three major biochemical roles:

- They are a source of stored energy that can be released in a form usable by organisms.
- They are used to transport stored energy within complex organisms.
- They serve as *carbon skeletons* that can be rearranged to form new molecules.

Some carbohydrates are relatively small, with molecular weights of less than 100 Da. Others are true macromolecules, with molecular weights in the hundreds of thousands.

There are four categories of biologically important carbohydrates:

- **Monosaccharides** (*mono*, "one"; *saccharide*, "sugar"), such as glucose, ribose, and fructose, are *simple sugars*. They are the monomers from which the larger carbohydrates are constructed.
- **Disaccharides** (*di*, "two") consist of two monosaccharides linked together by covalent bonds. The most familiar is sucrose, which is made up of covalently bonded glucose and fructose molecules.
- **Oligosaccharides** (*oligo*, "several") are made up of several (3–20) monosaccharides.
- **Polysaccharides** (*poly*, "many"), such as starch, glycogen, and cellulose, are polymers made up of hundreds or thousands of monosaccharides.

Monosaccharides are simple sugars

All living cells contain the monosaccharide **glucose**; it is the familiar "blood sugar," used to transport energy in humans. Cells use glucose as an energy source, breaking it down through a series of reactions that release stored energy and produce water and carbon dioxide; this is a cellular form of the combustion reaction described in Chapter 2.

Glucose exists in straight chains and in ring forms. The ring forms predominate in virtually all biological circumstances because they are more stable under physiological conditions. There are two versions of glucose ring, called α- and β-glucose, which differ only in the orientation of the —H and —OH attached to carbon 1 (**Figure 3.13**). The α and β forms interconvert and exist in equilibrium when dissolved in water.

Different monosaccharides contain different numbers of carbons. Some monosaccharides are structural isomers, with the same kinds and numbers of atoms, but in different arrangements (**Figure 3.14**). Such seemingly small structural changes can significantly alter properties. Most of the monosaccharides in living systems belong to the D (right-handed) series of isomers.

Pentoses (*pente*, "five") are five-carbon sugars. Two pentoses are of particular biological importance: the backbones of the nucleic acids RNA and DNA contain ribose and deoxyribose, respectively (see Section 4.1). These two pentoses are not isomers of each other; rather, one oxygen atom is missing from carbon 2 in deoxyribose (*de-*, "absent"). The absence of this oxygen atom is an important distinction between RNA and DNA.

The **hexoses** (*hex*, "six"), a group of structural isomers, all have the formula $C_6H_{12}O_6$. Included among the hexoses are glucose, fructose (so named because it was first found in fruits), mannose, and galactose.

Glycosidic linkages bond monosaccharides

The disaccharides, oligosaccharides, and polysaccharides are all constructed from monosaccharides that are covalently bonded together by condensation reactions that form **glycosidic linkages**. A single glycosidic linkage between two monosaccharides forms a disaccharide. For example, sucrose—common table sugar in the human diet and a major disaccharide in plants—is a disaccharide formed from a glucose and a fructose molecule.

Three-carbon sugar

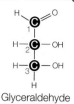

Glyceraldehyde is the smallest monosaccharide and exists only as the straight-chain form.

Glyceraldehyde

Five-carbon sugars (pentoses)

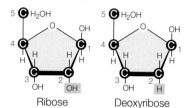

Ribose and deoxyribose each have five carbons, but very different chemical properties and biological roles.

Ribose   Deoxyribose

Six-carbon sugars (hexoses)

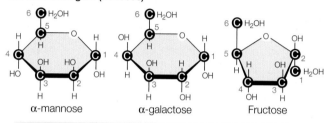

α-mannose   α-galactose   Fructose

These hexoses are structural isomers. All have the formula $C_6H_{12}O_6$, but each has distinct biochemical properties.

3.14 Monosaccharides Are Simple Sugars Monosaccharides are made up of varying numbers of carbons. Some hexoses are structural isomers that have the same kind and number of atoms, but the atoms are arranged differently. Fructose, for example, is a hexose, but forms a five-membered ring like the pentoses.

The disaccharides maltose and cellobiose are made from two glucose molecules (**Figure 3.15**). Maltose and cellobiose are structural isomers, both having the formula $C_{12}H_{22}O_{11}$. However, they have different chemical properties and are recognized by different enzymes in biological tissues. For example, maltose can be hydrolyzed into its monosaccharides in the human body, whereas cellobiose cannot.

Oligosaccharides contain several monosaccharides bound by glycosidic linkages at various sites. Many oligosaccharides have additional functional groups, which give them special properties. Oligosaccharides are often covalently bonded to proteins and lipids on the outer cell surface, where they serve as recognition signals. The different human blood groups (for example, the ABO blood types) get their specificity from oligosaccharide chains.

3.15 Disaccharides Form by Glycosidic Linkages Glycosidic linkages between two monosaccharides can create many different disaccharides. Which disaccharide is formed depends on which monosaccharides are linked; on the site of linkage (i.e., which carbon atoms are involved); and on the form (α or β) of the linkage.

The presence of a carbon atom (C) at a junction such as this is implied.

In sucrose, glucose and fructose are linked by an α-1,2 glycosidic linkage.

Maltose is produced when an α-1,4 glycosidic linkage forms between two glucose molecules. The hydroxyl group on carbon 1 of one D-glucose in the α (down) position reacts with the hydroxyl group on carbon 4 of the other glucose.

In cellobiose, two glucoses are linked by a β-1,4 glycosidic linkage.

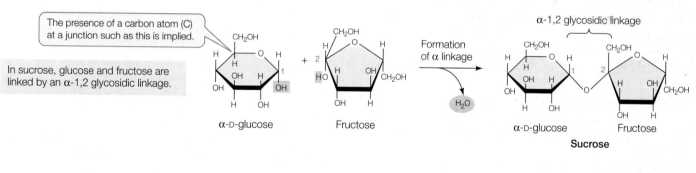

α-D-glucose   Fructose   α-D-glucose   Fructose
Sucrose

α-D-glucose   β-D-glucose   α-D-glucose   β-D-glucose
Maltose

β-D-glucose   β-D-glucose   β-D-glucose   β-D-glucose
Cellobiose

Polysaccharides store energy and provide structural materials

Polysaccharides are large (sometimes gigantic) polymers of monosaccharides connected by glycosidic linkages (**Figure 3.16**). In contrast to proteins, polysaccharides are not necessarily linear chains of monomers. Each monomer unit has several sites that may be capable of forming glycosidic linkages, and thus branched molecules are possible.

STARCH Starches comprise a family of giant molecules of broadly similar structure. While all starches are polysaccharides of glucose with α-glycosidic linkages (α–1,4 and α–1,6 glycosidic

bonds; Figure 3.16A), the different starches can be distinguished by the amount of branching that occurs at carbons 1 and 6 (Figure 3.16B). Starch is the principal energy storage compound of plants. Some plant starches, such as amylose, are unbranched; others are moderately branched (amylopectin, for example). Starch readily binds water. When that water is removed, however, hydrogen bonds tend to form between the unbranched polysaccharide chains, which then aggregate, as in the large starch grains observed in the storage material of plant seeds (see Figure 3.16C).

3.16 Representative Polysaccharides Cellulose, starch, and glycogen have different levels of branching and compaction of the polysaccharides.

(A) Molecular structure

Cellulose

Hydrogen bonding to other cellulose molecules can occur at these points.

Cellulose is an unbranched polymer of glucose with β-1,4 glycosidic linkages that are chemically very stable.

Starch and glycogen

Branching occurs here.

Glycogen and starch are polymers of glucose with α-1,4 glycosidic linkages. α-1,6 glycosidic linkages produce branching at carbon 6.

(B) Macromolecular structure

Linear (cellulose)

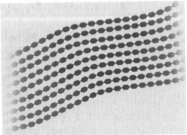

Parallel cellulose molecules form hydrogen bonds, resulting in thin fibrils.

Branched (starch)

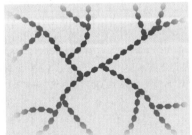

Branching limits the number of hydrogen bonds that can form in starch molecules, making starch less compact than cellulose.

Highly branched (glycogen)

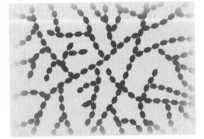

The high amount of branching in glycogen makes its solid deposits more compact than starch.

(C) Polysaccharides in cells

Layers of cellulose fibrils, as seen in this scanning electron micrograph, give plant cell walls great strength.

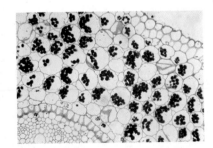

Within these plant cells, starch deposits (dyed purple in this micrograph) have a granular shape.

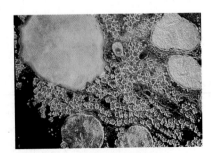

The pink-stained granules in this electron micrograph are glycogen deposits in the human liver.

GLYCOGEN Glycogen is a water-insoluble, highly branched polymer of glucose. It stores glucose in liver and muscle, serving as an energy storage compound for animals as starch does for plants. Both glycogen and starch are readily hydrolyzed into glucose monomers, which in turn can be broken down to liberate their stored energy.

But if it is glucose that is needed for fuel, why store it in the form of glycogen? The reason is that 1,000 glucose molecules would exert 1,000 times the *osmotic pressure* of a single glycogen molecule, causing water to enter the cells (see Section 6.3). If it were not for polysaccharides, many organisms would expend a lot of energy expelling excess water from their cells.

CELLULOSE As the predominant component of plant cell walls, cellulose is by far the most abundant organic compound on Earth. Like starch and glycogen, cellulose is a polysaccharide of glucose, but its individual monosaccharides are connected by β- rather than by α-glycosidic linkages. Starch is easily degraded by the actions of chemicals or enzymes. Cellulose, however, is chemically more stable because of its β-glycosidic linkages. Thus, whereas starch is easily broken down to supply glucose for energy-producing reactions, cellulose is an excellent structural material that can withstand harsh environmental conditions without substantial change.

Chemically modified carbohydrates contain additional functional groups

Some carbohydrates are chemically modified by the addition of functional groups, such as phosphate and amino groups (**Figure 3.17**). For example, carbon 6 in glucose may be oxidized from —CH_2OH to a carboxyl group (—COOH), producing glucuronic acid. Or a phosphate group may be added to one or more of the —OH sites. Some of the resulting *sugar phosphates*, such as fructose 1,6-bisphosphate, are important intermediates in cellular energy reactions, which will be discussed in Chapter 9.

When an amino group is substituted for an —OH group, *amino sugars*, such as glucosamine and galactosamine, are produced. These compounds are important in the extracellular matrix (see Section 5.4), where they form parts of glycoproteins, which are molecules involved in keeping tissues together. Galactosamine is a major component of cartilage, the material that forms caps on the ends of bones and stiffens the ears and nose. A derivative of glucosamine is present in the polymer *chitin*, the principal structural polysaccharide in the external skeletons of insects and many crustaceans (e.g., crabs and lobsters) and a component of the cell walls of fungi. Because these organisms are among the most abundant eukaryotes on Earth, chitin rivals cellulose as one of the most abundant substances in the living world.

(A) Sugar phosphate

Fructose 1,6 bisphosphate is involved in the reactions that liberate energy from glucose. (The numbers in its name refer to the carbon sites of phosphate bonding; *bis-* indicates that two phosphates are present.)

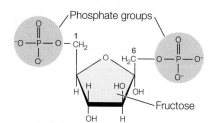

Phosphate groups

Fructose

Fructose 1,6 bisphosphate

3.17 Chemically Modified Carbohydrates Added functional groups can modify the form and properties of a carbohydrate.

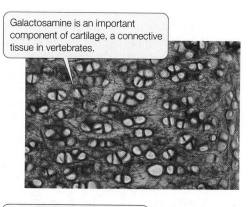

Galactosamine is an important component of cartilage, a connective tissue in vertebrates.

(B) Amino sugars

The monosaccharides glucosamine and galactosamine are amino sugars with an amino group in place of a hydroxyl group.

Amino group

Glucosamine   **Galactosamine**

The external skeletons of insects are made up of chitin.

(C) Chitin

Chitin is a polymer of *N*-acetylglucosamine; *N*-acetyl groups provide additional sites for hydrogen bonding between the polymers.

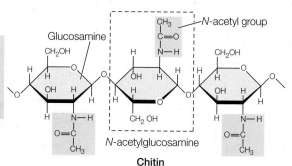

Glucosamine

N-acetyl group

N-acetylglucosamine

Chitin

3.3 RECAP

Carbohydrates are composed of carbon, hydrogen, and oxygen in the general ratio of 1:2:1. They provide energy and structure to cells and are precursors of numerous important biological molecules. Monosaccharide monomers can be connected by glycosidic linkages to form disaccharides, oligosaccharides, and polysaccharides.

- Draw the chemical structure of a disaccharide formed by two monosaccharides. **See Figure 3.15**

- What qualities of the polysaccharides starch and glycogen make them useful for energy storage? **See pp. 52–53 and Figure 3.16**

- From looking at the cellulose molecules in Figure 3.16A, can you see where a large number of hydrogen bonds are present in the linear structure of cellulose shown in Figure 3.16B? Why is this structure so strong?

We have seen how amino acid monomers form protein polymers and how sugar monomers form the polymers of carbohydrates. Now we will look at the lipids, which are unique among the four classes of large biological molecules in that they are not, strictly speaking, polymers.

3.4 What Are the Chemical Structures and Functions of Lipids?

Lipids—colloquially called *fats*—are hydrocarbons that are insoluble in water because of their many nonpolar covalent bonds. As we saw in Section 2.2, nonpolar hydrocarbon molecules are hydrophobic and preferentially aggregate among themselves, away from water (which is polar). When nonpolar hydrocarbons are sufficiently close together, weak but additive van der Waals forces hold them together. The huge macromolecular aggregations that can form are not polymers in a strict chemical sense, because the individual lipid molecules are not covalently bonded. With this understanding, it is still useful to consider aggregations of individual lipids as a different sort of polymer.

There are several different types of lipids, and they play a number of roles in living organisms:

- Fats and oils store energy.

- Phospholipids play important structural roles in cell membranes.

- Carotenoids and chlorophylls help plants capture light energy.

- Steroids and modified fatty acids play regulatory roles as hormones and vitamins.

- Fat in animal bodies serves as thermal insulation.

- A lipid coating around nerves provides electrical insulation.

- Oil or wax on the surfaces of skin, fur, and feathers repels water.

Fats and oils are hydrophobic

Chemically, fats and oils are *triglycerides*, also known as *simple lipids*. Triglycerides that are solid at room temperature (around 20°C) are called **fats**; those that are liquid at room temperature are called **oils**. Triglycerides are composed of two types of building blocks: *fatty acids* and *glycerol*. **Glycerol** is a small molecule with three hydroxyl (—OH) groups (thus it is an alcohol). A **fatty acid** is made up of a long nonpolar hydrocarbon chain and a polar carboxyl group (—COOH). These chains are very hydrophobic, with their abundant C—H and C—C bonds, which have low electronegativity and are nonpolar (see Section 2.2).

A **triglyceride** contains three fatty acid molecules and one molecule of glycerol. Synthesis of a triglyceride involves three condensation (dehydration) reactions. In each reaction, the carboxyl group of a fatty acid bonds with a hydroxyl group of glycerol, resulting in a covalent bond called an **ester linkage** and the release of a water molecule (**Figure 3.18**). The three fatty acids in a triglyceride molecule need not all have the same hydrocarbon chain length or structure; some may be saturated fatty acids, while others may be unsaturated:

- In **saturated fatty acids**, all the bonds between the carbon atoms in the hydrocarbon chain are single bonds—there are no double bonds. That is, all the bonds are saturated with

3.18 Synthesis of a Triglyceride In living things, the reaction that forms a triglyceride is more complex, but the end result is the same as shown here.

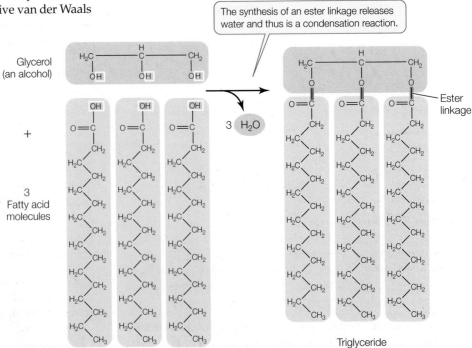

The synthesis of an ester linkage releases water and thus is a condensation reaction.

Glycerol (an alcohol)

3 Fatty acid molecules

Ester linkage

Triglyceride

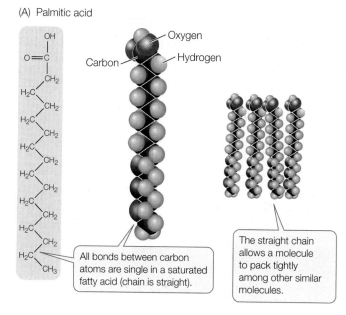

(A) Palmitic acid

All bonds between carbon atoms are single in a saturated fatty acid (chain is straight).

The straight chain allows a molecule to pack tightly among other similar molecules.

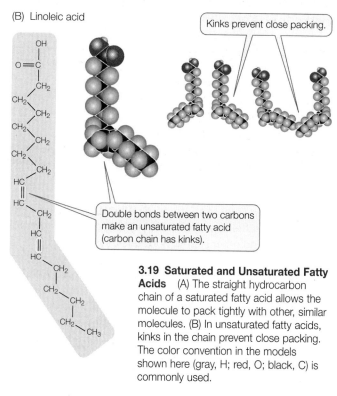

(B) Linoleic acid

Kinks prevent close packing.

Double bonds between two carbons make an unsaturated fatty acid (carbon chain has kinks).

3.19 Saturated and Unsaturated Fatty Acids (A) The straight hydrocarbon chain of a saturated fatty acid allows the molecule to pack tightly with other, similar molecules. (B) In unsaturated fatty acids, kinks in the chain prevent close packing. The color convention in the models shown here (gray, H; red, O; black, C) is commonly used.

of animal fats tend to have many long-chain saturated fatty acids, packed tightly together; these fats are usually solids at room temperature and have a high melting point. The triglycerides of plants, such as corn oil, tend to have short or unsaturated fatty acids. Because of their kinks, these fatty acids pack together poorly and have a low melting point, and these triglycerides are usually liquids at room temperature.

Fats are excellent storehouses for chemical energy. As you will see in Chapter 9, when the C—H bond is broken, it releases significant energy that an organism can use for its own purposes, such as movement or building up complex molecules. On a per weight basis, broken-down fats yield more than twice as much energy as do degraded carbohydrates.

Phospholipids form biological membranes

We have mentioned the hydrophobic nature of the many C—C and C—H bonds in fatty acids. But what about the carboxyl functional group at the end of the molecule? When it ionizes and forms COO^-, it is strongly hydrophilic. So a fatty acid is a molecule with a hydrophilic end and a long hydrophobic tail. It has two opposing chemical properties; the technical term for this is **amphipathic**. This explains what happens when oil (fatty acid) and water mix: the fatty acids orient themselves so that their polar ends face outward (i.e., toward the water) and their nonpolar tails face inward (away from water). Although no covalent bonds link individual lipids in large aggregations, such stable aggregations form readily in aqueous conditions. So these large lipid structures can be considered a different kind of macromolecule.

Like triglycerides, **phospholipids** contain fatty acids bound to glycerol by ester linkages. In phospholipids, however, any one of several phosphate-containing compounds replaces one of the fatty acids, giving these molecules amphipathic properties—that is properties of both water soluble and water insoluble molecules (**Figure 3.20A**). The phosphate functional group has a negative electric charge, so this portion of the molecule is hydrophilic, attracting polar water molecules. But the two fatty acids are hydrophobic, so they tend to avoid water and aggregate together or with other hydrophobic substances.

In an aqueous environment, phospholipids line up in such a way that the nonpolar, hydrophobic "tails" pack tightly together and the phosphate-containing "heads" face outward, where they interact with water. The phospholipids thus form a **bilayer**: a sheet two molecules thick, with water excluded from the core (**Figure 3.20B**). Biological membranes have this kind of **phospholipid bilayer** structure, and we will devote Chapter 6 to their biological functions.

Lipids have roles in energy conversion, regulation, and protection

In the previous section, we focused on lipids involved in energy storage and cell structure, whose molecular structures are variations on the glycerol–fatty acid structure. However, there are other nonpolar and amphipathic lipids that are not based on this structure.

hydrogen atoms (**Figure 3.19A**). These fatty acid molecules are relatively rigid and straight, and they pack together tightly, like pencils in a box.

- In **unsaturated fatty acids**, the hydrocarbon chain contains one or more double bonds. Linoleic acid is an example of a *polyunsaturated* fatty acid that has two double bonds near the middle of the hydrocarbon chain, which causes kinks in the molecule (**Figure 3.19B**). Such kinks prevent the unsaturated fat molecules from packing together tightly.

The kinks in fatty acid molecules are important in determining the fluidity and melting point of a lipid. The triglycerides

(A) Phosphatidylcholine

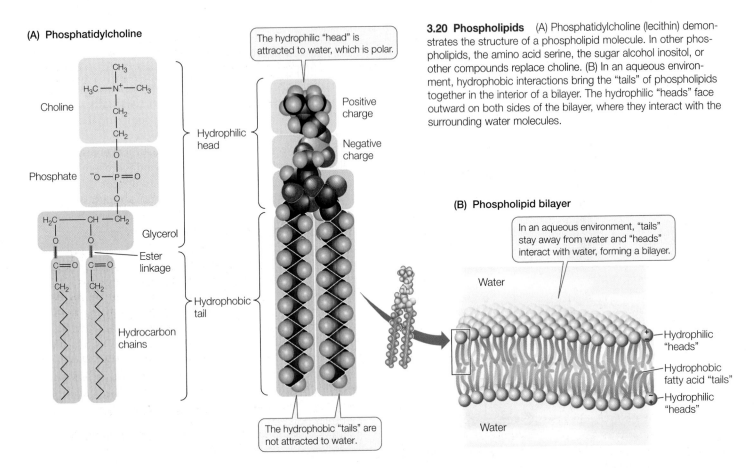

3.20 Phospholipids (A) Phosphatidylcholine (lecithin) demonstrates the structure of a phospholipid molecule. In other phospholipids, the amino acid serine, the sugar alcohol inositol, or other compounds replace choline. (B) In an aqueous environment, hydrophobic interactions bring the "tails" of phospholipids together in the interior of a bilayer. The hydrophilic "heads" face outward on both sides of the bilayer, where they interact with the surrounding water molecules.

CAROTENOIDS The carotenoids are a family of light-absorbing pigments found in plants and animals. Beta-carotene (β-carotene) is one of the pigments that traps light energy in leaves during photosynthesis. In humans, a molecule of β-carotene can be broken down into two vitamin A molecules (**Figure 3.21**), from which we make the pigment *cis*-retinal, which is required for vision. Carotenoids are responsible for the colors of carrots, tomatoes, pumpkins, egg yolks, and butter.

STEROIDS The steroids are a family of organic compounds whose multiple rings share carbons (**Figure 3.22**). The steroid cholesterol is an important constituent of membranes. Other steroids function as hormones, chemical signals that carry messages from one part of the body to another (see Chapter 41). Cholesterol is synthesized in the liver and is the starting material for making testosterone and other steroid hormones, such as estrogen.

VITAMINS *Vitamins* are small molecules that are not synthesized by the human body and so must be acquired from the diet (see Chapter 50). For example, *vitamin A* is formed from the β-carotene found in green and yellow vegetables (see Figure 3.21). In humans, a deficiency of vitamin A leads to dry skin, eyes, and internal body surfaces, retarded growth and development, and night blindness, which is a diagnostic symptom for the deficiency. Vitamins D, E, and K are also lipids.

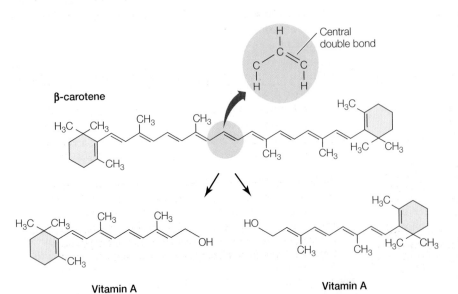

3.21 β-Carotene is the Source of Vitamin A The carotenoid β-carotene is symmetrical around its central double bond. When that bond is broken, two molecules of vitamin A are formed. The structural formula presented here is standard chemical shorthand for large organic molecules with many carbon atoms; it is simplified by omitting the C (indicating a carbon atom) at the intersections representing covalent bonds. The presence of hydrogen atoms (H) to fill all the available bonding sites on each C is assumed.

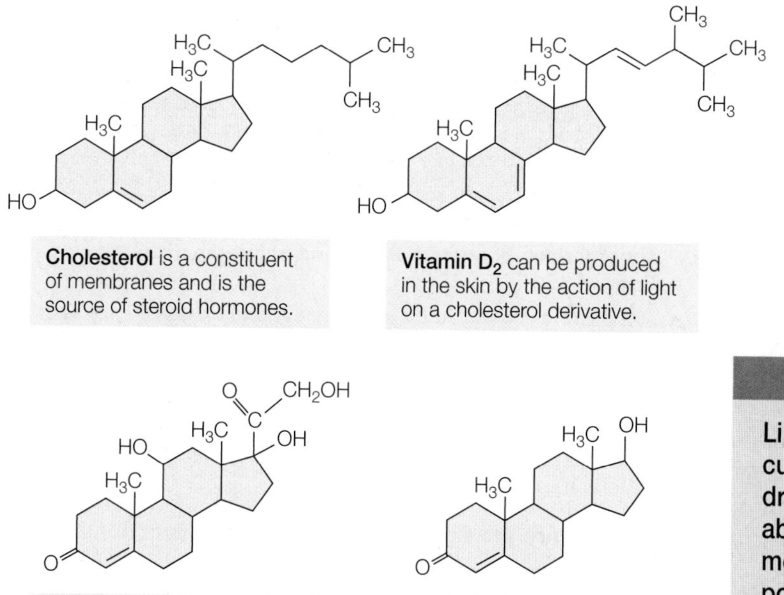

3.22 All Steroids Have the Same Ring Structure The steroids shown here, all important in vertebrates, are composed of carbon and hydrogen and are highly hydrophobic. However, small chemical variations, such as the presence or absence of a hydroxyl group, can produce enormous functional differences among these molecules.

Cholesterol is a constituent of membranes and is the source of steroid hormones.

Vitamin D₂ can be produced in the skin by the action of light on a cholesterol derivative.

Cortisol is a hormone secreted by the adrenal glands.

Testosterone is a male sex hormone.

WAXES The sheen on human hair is more than cosmetic. Glands in the skin secrete a waxy coating that repels water and keeps the hair pliable. Birds that live near water have a similar waxy coating on their feathers. The shiny leaves of plants such as holly, familiar during winter holidays, also have a waxy coating. Finally, bees make their honeycombs out of wax. All waxes have the same basic structure: they are formed by an ester linkage between a saturated, long-chain fatty acid and a saturated, long-chain alcohol. The result is a very long molecule, with 40–60 CH_2 groups. For example, here is the structure of beeswax:

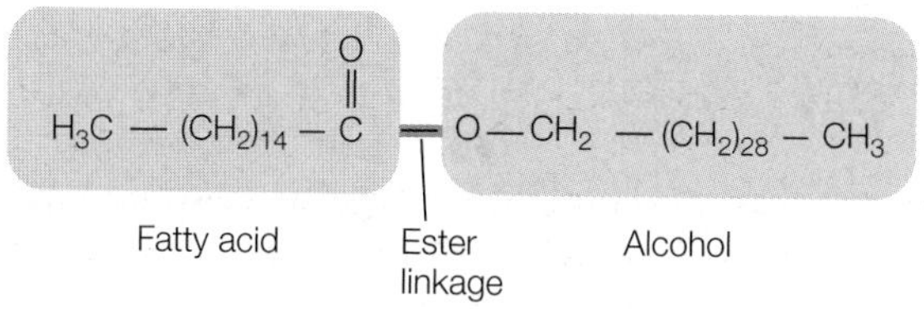

$$H_3C - (CH_2)_{14} - C(=O) - O - CH_2 - (CH_2)_{28} - CH_3$$

Fatty acid   Ester linkage   Alcohol

This highly nonpolar structure accounts for the impermeability of wax to water.

3.4 RECAP

Lipids include both nonpolar and amphipathic molecules that are largely composed of carbon and hydrogen. They are important in energy storage, light absorption, regulation and biological structures. Cell membranes contain phospholipids, which are composed of hydrophobic fatty acids linked to glycerol and a hydrophilic phosphate group.

- Draw the molecular structures of fatty acids and glycerol and show how they are linked to form a triglyceride. See p. 54 and Figure 3.18

- What is the difference between fats and oils? See p. 54

- How does the polar nature of phospholipids result in their forming a bilayer? See p. 55 and Figure 3.20

- Why are steroids and some vitamins classified as lipids? See p. 56

All the types of molecules we have discussed in this chapter are found only in living organisms, but a final class of biological macromolecules has special importance to the living world. The function of the nucleic acids is nothing less than the transmission of life's "blueprint" to each new organism. This chapter showed the wonderful biochemical unity of life, a unity that implies all life has a common origin. Essential to this origin were the monomeric nucleotides and their polymers, nucleic acids. In the next chapter, we turn to the related topics of nucleic acids and the origin of life.

CHAPTER SUMMARY

3.1 What Kinds of Molecules Characterize Living Things?

SEE ANIMATED TUTORIAL 3.1

- **Macromolecules** are **polymers** constructed by the formation of covalent bonds between smaller molecules called **monomers**. Macromolecules in living organisms include polysaccharides, proteins, and nucleic acids. Large lipid structures may also be considered macromolecules.

- **Functional groups** are small groups of atoms that are consistently found together in a variety of different macromolecules. Functional groups have particular chemical properties that they

confer on any larger molecule of which they are a part. **Review Figure 3.1, WEB ACTIVITY 3.1**

- Structural and optical **isomers** have the same kinds and numbers of atoms, but differ in their structures and properties. **Review Figure 3.2**

- The many functions of macromolecules are directly related to their three-dimensional shapes, which in turn result from the sequences and chemical properties of their monomers.

- Monomers are joined by **condensation reactions**, which release a molecule of water for each bond formed. **Hydrolysis reactions** use water to break polymers into monomers. **Review Figure 3.4**

3.2 What Are the Chemical Structures and Functions of Proteins?

- The functions of proteins include support, protection, catalysis, transport, defense, regulation, and movement.

- **Amino acids** are the monomers from which proteins are constructed. Four groups are attached to a central carbon atom: a hydrogen atom, an amino group, a carboxyl group, and a variable R group. The particular properties of each amino acids depend on its **side chain**, or **R group**, which may be charged, polar, or hydrophobic. Review Table 3.1, **WEB ACTIVITY 3.2**

- **Peptide linkages**, also called peptide bonds, covalently link amino acids into polypeptide chains. These bonds form by condensation reactions between the carboxyl and amino groups. Review Figure 3.6

- The **primary structure** of a protein is the sequence of amino acids in the chain. This chain is folded into a **secondary structure**, which in different parts of the protein may form an α **helix** or a β **pleated sheet**. Review Figure 3.7A–C

- **Disulfide bridges** and noncovalent interactions between amino acids cause polypeptide chains to fold into three-dimensional **tertiary structures** and allow multiple chains to interact in a **quaternary structure**. Review Figure 3.7D,E

- Heat, alterations in pH, or certain chemicals can all result in protein **denaturation**, which involves the loss of tertiary and/or secondary structure as well as biological function. Review Figure 3.9

- The specific shape and structure of a protein allows it to bind noncovalently to other molecules. Review Figure 3.11

- **Chaperone proteins** enhance correct protein folding and prevent binding to inappropriate ligands. Review Figure 3.12

3.3 What Are the Chemical Structures and Functions of Carbohydrates?

- **Carbohydrates** contain carbon bonded to hydrogen and oxygen atoms in a ratio of 1:2:1, or $(CH_2O)_n$.

- **Monosaccharides** are the monomers that make up carbohydrates. **Hexoses** such as **glucose** are six-carbon monosaccharides; **pentoses** have five carbons. Review Figure 3.14, **WEB ACTIVITY 3.3**

- **Glycosidic linkages**, which have either an α or a β orientation in space, covalently link monosaccharides into larger units such as **disaccharides**, **oligosaccharides**, and **polysaccharides**. Review Figure 3.15

- **Starch** stores energy in plants. Starch and **glycogen** are formed by α-glycosidic linkages between glucose monomers and are distinguished by the amount of branching they exhibit. They can be easily broken down to release stored energy. Review Figure 3.16

- **Cellulose** is a very stable glucose polymer and is the principal structural component of plant cell walls.

3.4 What Are the Chemical Structures and Functions of Lipids?

- Fats and oils are **triglycerides**, composed of three fatty acids covalently bonded to a molecule of glycerol by ester linkages. Review Figure 3.18

- **Saturated** fatty acids have a hydrocarbon chain with no double bonds. The hydrocarbon chains of **unsaturated** fatty acids have one or more double bonds that bend the chain, making close packing less possible. Review Figure 3.19

- **Phospholipids** have a hydrophobic hydrocarbon "tail" and a hydrophilic phosphate "head"; that is, they are **amphipathic**. In water, the interactions of the tails and heads of phospholipids generate a **phospholipid bilayer**. The heads are directed outward, where they interact with the surrounding water. The tails are packed together in the interior of the bilayer, away from water. Review Figure 3.20

- Other lipids include vitamins A and D, steroids and plant pigments such as carotenoids.

SELF-QUIZ

1. The most abundant molecule in the cell is
   a. a carbohydrate.
   b. a lipid.
   c. a nucleic acid.
   d. a protein.
   e. water.

2. All lipids are
   a. triglycerides.
   b. polar.
   c. hydrophilic.
   d. polymers of fatty acids.
   e. more soluble in nonpolar solvents than in water.

3. All carbohydrates
   a. are polymers.
   b. are simple sugars.
   c. consist of one or more simple sugars.
   d. are found in biological membranes.
   e. are more soluble in nonpolar solvents than in water.

4. Which of the following is *not* a carbohydrate?
   a. Glucose
   b. Starch
   c. Cellulose
   d. Hemoglobin
   e. Deoxyribose

5. All proteins
   a. are enzymes.
   b. consist of one or more polypeptide chains.
   c. are amino acids.
   d. have quaternary structures.
   e. are more soluble in nonpolar solvents than in water.

6. Which of the following statements about the primary structure of a protein is *not* true?
   a. It may be branched.
   b. It is held together by covalent bonds.
   c. It is unique to that protein.
   d. It determines the tertiary structure of the protein.
   e. It is the sequence of amino acids in the protein.

7. The amino acid leucine
   a. is found in all proteins.
   b. cannot form peptide linkages.
   c. has a hydrophobic side chain.
   d. has a hydrophilic side chain.
   e. is identical to the amino acid lysine.
8. The quaternary structure of a protein
   a. consists of four subunits—hence the name quaternary.
   b. is unrelated to the function of the protein.
   c. may be either alpha or beta.
   d. depends on covalent bonding among the subunits.
   e. depends on the primary structures of the subunits.

9. The amphipathic nature of phospholipids is
   a. determined by the fatty acid composition.
   b. important in membrane structure.
   c. polar but not nonpolar.
   d. shown only if the lipid is in a nonpolar solvent.
   e. important in energy storage by lipids.
10. Which of the following statements about condensation reactions is *not* true?
   a. Protein synthesis results from them.
   b. Polysaccharide synthesis results from them.
   c. They involve covalent bonds.
   d. They consume water as a reactant.
   e. Different condensation reactions produce different kinds of macromolecules.

FOR DISCUSSION

1. Suppose that, in a given protein, one lysine is replaced by aspartic acid (see Table 3.1). Does this change occur in the primary structure or in the secondary structure? How might it result in a change in tertiary structure? In quaternary structure?

2. If there are 20 different amino acids commonly found in proteins, how many different dipeptides are there? How many different tripeptides?

ADDITIONAL INVESTIGATION

Human hair is composed of a protein, keratin. At the hair salon, two techniques are used to modify the three-dimensional shape of hair. Styling involves heat, and a perm involves cleaving and reforming disulfide bonds. How would you investigate these phenomena in terms of protein structure?

WORKING WITH DATA (GO TO yourBioPortal.com)

Primary Structure Specifies Tertiary Structure   In this hands-on exercise based on Figure 3.9, you will learn about the methods used to disrupt the chemical interactions that determine the tertiary structure of proteins. You will examine the original data that led Anfinsen to conclude that denaturation of ribonuclease is reversible.

4 Nucleic Acids and the Origin of Life

Looking for life

The trip had lasted a long and anxious ten months when, in the summer of 1976, the first of two visitors from Earth landed on a plain on the Martian surface. A second spacecraft arrived in September. The task of these robotic laboratories, part of NASA's Viking project, was to search for life.

On Earth, life has existed for several billion years and has spread over most of the planet's surface. Determining life's origins is difficult, however, because (with few exceptions) simple organisms leave no fossils. On Mars, scientists thought, things might be different. A primitive form of life might exist there now, or might have left chemical signatures that remain in place, untouched by other organisms.

The two Viking spacecraft that landed on Mars in 1976 analyzed soil samples for the small molecules of life, including simple sugars and amino acids. None were found. The robotic laboratories immersed soil samples in an aqueous solution of sugars, amino acids, and minerals. Living organisms take in and break down such substances from their environment, releasing gases such as CO_2. A small amount of CO_2 was detected in one experiment, but, frustratingly, no gases were released in further experiments.

The results from the Viking landers remain controversial. Why did that one experiment detect a sign of life? The 1976 robotic landers are still on Mars but have long since stopped working. In 2008, more probes were sent from Earth, carrying more sophisticated instruments. One of them, the Phoenix lander, is in a northern region of Mars, at a latitude corresponding to that of Alaska on Earth. Phoenix has a robotic arm like the backhoes used in a construction site. When the arm dug a small trench into the Martian soil, shiny dice-sized beads of what turned out to be ice were exposed, although the beads disappeared in a few days as exposure to the atmosphere caused them to vaporize. Dissolved ions such as sodium, magnesium, potassium and chloride were all present in the frozen water, indicating that at least those requirements for life are present on Mars. Once again, the soil was analyzed for traces of current or past organisms; once again, the results were negative. But even if there probably is no life on Mars today, there might have been in the past.

Lab Seeking Life   Landers such as the robotic space laboratory Phoenix, shown here on Earth, have been sent to look for traces of life on Mars.

Ice on Mars The Phoenix landing site (blue dot) is near the Martian north pole, where chemical traces of life might be preserved in the hypercold environment. When the lander scooped up a patch of soil for analysis, it also took photos that revealed ice crystals just below the surface of the Red Planet.

Ice

As we saw in Chapter 2, water is a key requirement for life. Remote measurements from orbiting spacecraft and chemical measurements using special telescopes have shown that water is present on Mars and, indeed, on some of the moons of other planets in our solar system.

Scientists are using their knowledge of the small and large molecules that are present in living organisms to search for the chemical signatures of life on other planets. Chapters 2 and 3 described molecules that are important for biological structure and function. In Chapter 4, we turn to certain molecules involved in the origin and perpetuation of life itself.

IN THIS CHAPTER we first describe the structure of nucleic acids, the informational macromolecules needed for the perpetuation of life. We then turn to biologists' speculations on the origin of life and describe early experimental evidence that life on Earth today comes from pre-existing life. We present some ideas on the formation of the building blocks of life, including the monomers and polymers that characterize biological systems. Finally, we describe some proposals for the origin of cells.

4.1 What Are the Chemical Structures and Functions of Nucleic Acids?

From medicine to evolution, from agriculture to forensics, the properties of nucleic acids impact our lives every day. It is with nucleic acids that the concept of "information" entered the biological vocabulary. Nucleic acids are uniquely capable of coding for and transmitting biological information.

The **nucleic acids** are polymers specialized for the storage, transmission between generations, and use of genetic information. There are two types of nucleic acids: **DNA** (*d*eoxyribo*nu*cleic *a*cid) and **RNA** (*ri*bo*nu*cleic *a*cid). DNA is a macromolecule that encodes hereditary information and passes it from generation to generation. Through an RNA intermediate, the information encoded in DNA is used to specify the amino acid sequences of proteins. Information flows from DNA to DNA during reproduction. In the non-reproductive activities of the cell, information flows from DNA to RNA to proteins. It is the proteins that ultimately carry out life's functions.

Nucleotides are the building blocks of nucleic acids

Nucleic acids are composed of monomers called **nucleotides**, each of which consists of a pentose sugar, a phosphate group, and a nitrogen-containing **base**. (Molecules consisting of a pentose sugar and a nitrogenous base—but no phosphate group—are called *nucleosides*.) The bases of the nucleic acids take one of two chemical forms: a six-membered single-ring structure called a **pyrimidine**, or a fused double-ring structure called a **purine** (**Figure 4.1**). In DNA, the pentose sugar is **deoxyribose**, which differs from the **ribose** found in RNA by the absence of one oxygen atom (see Figure 3.14).

In both RNA and DNA, the backbone of the macromolecule consists of a chain of alternating pentose sugars and phosphate groups (sugar–phosphate–sugar–phosphate). The bases are attached to the sugars and project from the polynucleotide chain (**Figure 4.2**). The nucleotides are joined by **phosphodiester linkages** between the sugar of one nucleotide and the phosphate of the next (*diester* refers to the two covalent bonds formed by —OH groups reacting with acidic phosphate groups). The phosphate groups link carbon 3 in one pentose sugar to carbon 5 in the adjacent sugar.

Most RNA molecules consist of only one polynucleotide chain. DNA, however, is usually double-stranded; its two polynucleotide chains are held together by hydrogen bonding between their nitrogenous bases. The two strands of DNA run in opposite directions. You can see what this means by drawing an arrow through a phosphate group from carbon 5 to

4.1 Nucleotides Have Three Components   Nucleotide monomers are the building blocks of DNA and RNA polymers.

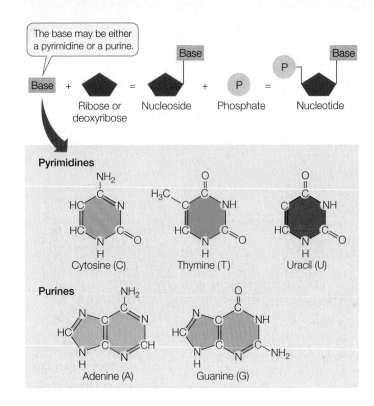

The base may be either a pyrimidine or a purine.

Base + Ribose or deoxyribose = Nucleoside + Phosphate = Nucleotide

Pyrimidines

Cytosine (C)   Thymine (T)   Uracil (U)

Purines

Adenine (A)   Guanine (G)

---

yourBioPortal.com

GO TO Web Activity 4.1 • Nucleic Acid Building Blocks

---

carbon 3 in the next ribose. If you do this for both strands of the DNA in Figure 4.2, the arrows will point in opposite directions. This *antiparallel* orientation allows the strands to fit together in three-dimensional space.

Base pairing occurs in both DNA and RNA

Only four nitrogenous bases—and thus only four nucleotides—are found in DNA. The DNA bases and their abbreviations are **adenine (A)**, **cytosine (C)**, **guanine (G)**, and **thymine (T)**. Adenine and guanine are purines; thymine and cytosine are pyrimidines. RNA is also made up of four different monomers, but its nucleotides differ from those of DNA. In RNA the nucleotides are termed *ribonucleotides* (the ones in DNA are *deoxyribonucleotides*). They contain ribose rather than deoxyribose, and in-

4.2 Distinguishing Characteristics of DNA and RNA Polymers
RNA is usually a single strand. DNA usually consists of two strands running in opposite directions (antiparallel).

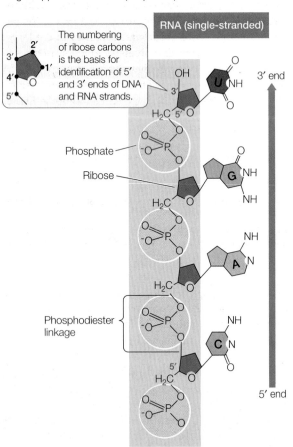

RNA (single-stranded)

The numbering of ribose carbons is the basis for identification of 5′ and 3′ ends of DNA and RNA strands.

Phosphate

Ribose

Phosphodiester linkage

3′ end

5′ end

In RNA, the bases are attached to ribose. The bases in RNA are the purines adenine (A) and guanine (G) and the pyrimidines cytosine (C) and uracil (U).

DNA (double-stranded)

Deoxyribose

Pyrimidine base   Purine base

Phosphate

Hydrogen bond

5′ end

3′ end

5′ end

3′ end

In DNA, the bases are attached to deoxyribose, and the base thymine (T) is found instead of uracil. Hydrogen bonds between purines and pyrimidines hold the two strands of DNA together.

TABLE 4.1
Distinguishing RNA from DNA

NUCLEIC ACID	SUGAR	BASES	STRANDS
RNA	Ribose	Adenine	Single
		Cytosine	
		Guanine	
		Uracil	
DNA	Deoxyribose	Adenine	Double
		Cytosine	
		Guanine	
		Thymine	

stead of the base thymine, RNA uses the base **uracil (U)**. The other three bases are the same in RNA and DNA (**Table 4.1**).

The key to understanding the structure and function of nucleic acids is the principle of **complementary base pairing**. In double-stranded DNA, adenine and thymine always pair (A-T), and cytosine and guanine always pair (C-G).

Three factors make base pairing complementary:

- The sites for hydrogen bonding on each base
- The geometry of the sugar–phosphate backbone, which brings complementary bases near each other

- The molecular sizes of the paired bases; the pairing of a larger purine with a smaller pyrimidine ensures stability and uniformity in the double-stranded molecule of DNA

Although RNA is generally single-stranded, complementary hydrogen bonding between ribonucleotides plays important roles in determining the three-dimensional shapes of some types of RNA molecules, since portions of the single-stranded RNA can fold back and pair with each other (**Figure 4.3**). Complementary base pairing can also take place between ribonucleotides and deoxyribonucleotides. In RNA, guanine and cytosine pair (G-C), as in DNA, but adenine pairs with uracil (A-U). Adenine in an RNA strand can pair either with uracil (in another RNA strand) or with thymine (in a DNA strand).

The three-dimensional physical appearance of DNA is strikingly uniform. The segment shown in **Figure 4.4** could be from any DNA molecule. The variations in DNA—the different sequences of bases—are strictly internal. Through hydrogen bonding, the two complementary polynucleotide strands pair and twist to form a **double helix**. When compared with the complex and varied tertiary structures of proteins, this uniformity is surprising. But this structural contrast makes sense in terms of the functions of these two classes of macromolecules. As we describe in Section 3.2, the different and unique shapes of proteins permit these macromolecules to recognize specific "target" molecules. The area on the surface of a protein that interacts with the target molecule must match the shape of at least part of the target molecule. In other words, structural diversity in the target molecules requires corresponding diversity in the structures of the proteins themselves. Structural diversity is necessary in DNA as well. However, the diversity of DNA is found in its base sequence rather than in the physical shape of the molecule. Different DNA base sequences encode specific information.

---
yourBioPortal.com

GO TO Web Activity 4.2 • DNA Structure

DNA carries information and is expressed through RNA

DNA is a purely *informational* molecule. The information is encoded in the sequence of bases carried in its strands—the infor-

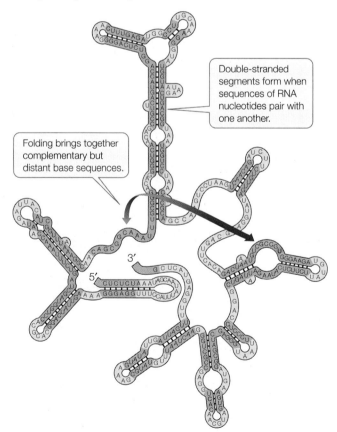

4.3 Hydrogen Bonding in RNA When a single-stranded RNA folds in on itself, hydrogen bonds between complementary sequences can stabilize it into a three-dimensional shape with complicated surface characteristics.

Double-stranded segments form when sequences of RNA nucleotides pair with one another.

Folding brings together complementary but distant base sequences.

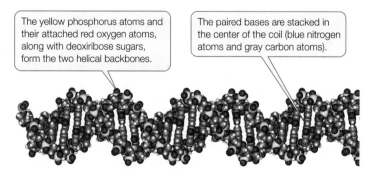

The yellow phosphorus atoms and their attached red oxygen atoms, along with deoxiribose sugars, form the two helical backbones.

The paired bases are stacked in the center of the coil (blue nitrogen atoms and gray carbon atoms).

4.4 The Double Helix of DNA The backbones of the two strands in a DNA molecule are coiled in a double helix that is held together by hydrogen bonds between the purines and pyrimidines in the interior of the structure. In this model, the small white atoms represent hydrogen.

4.5 DNA Stores Information The DNA macromolecule stores information that can either be copied (replicated) or transcribed into RNA. RNA can then be translated into protein.

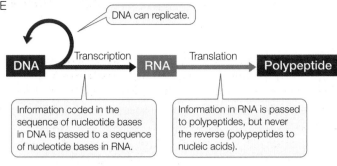

mation encoded in the sequence TCAGCA is different from the information in the sequence CCAGCA. DNA has two functions in terms of information. Taken together, they comprise the *central dogma of molecular biology* (**Figure 4.5**).

- DNA can reproduce itself exactly. This is called *DNA replication*. It is done by polymerization on a template.

- DNA can copy its information into RNA, in a process called *transcription*. The nucleotide sequence in RNA can specify a sequence of amino acids in a polypeptide. This is called *translation*.

While the details of these important processes are described in later chapters, it is important to realize two things at this point:

1. DNA replication and transcription depend on the base pairing properties of nucleic acids. The hydrogen-bonded base pairs are A-T and G-C in DNA and A-U and G-C in RNA (see Figure 4.2). Consider this double-stranded DNA region:

TCAGCA
AGTCGT

Transcription of the lower strand will result in a single strand of RNA with the sequence UCAGCA. Can you figure out what the top strand would produce?

2. DNA replication usually involves the entire DNA molecule, but only relatively small sections of the DNA are transcribed into RNA molecules. Since DNA holds essential information, it must be replicated completely so that each new cell or new organism receives a complete set of DNA from its parent. The complete set of DNA in a living organism is called its **genome**. However, not all of the information in the genome is needed at all times (**Figure 4.6A**).

The sequences of DNA that encode specific proteins are transcribed into RNA and are called **genes** (**Figure 4.6B**). In humans, the genes that encode the subunits of the protein hemoglobin (see Figure 3.10) are expressed only in the precursors of red blood cells. The genetic information in each globin gene is transcribed into RNA and then translated into a globin polypeptide. In other tissues, such as the muscles, the genes that encode the globin subunits are not transcribed, but others are—for example, the genes for the myosin proteins that are the major component of muscle fibers (see Section 48.1).

The DNA base sequence reveals evolutionary relationships

Because DNA carries hereditary information from one generation to the next, a theoretical series of DNA molecules, with changes in base sequences, stretches back through the lineage of every organism to the beginning of biological evolution on Earth, about 4 billion years ago. Therefore, closely related living species should have more similar base sequences than species that are more distantly related. The details of how scientists use this information are covered in Chapter 24.

The elucidation and examination of DNA base sequences has confirmed many of the evolutionary relationships that were inferred from more traditional comparisons of body structures, biochemistry, and physiology. Many studies of anatomy, physiology, and behavior have concluded that the closest living relative of humans (*Homo sapiens*) is the chimpanzee (genus *Pan*). In fact, the chimpanzee genome shares more than 98 percent of its DNA base sequence with the human genome. Increasingly, scientists turn to DNA analyses to elucidate evolutionary relationships when other comparisons are not possible or are not conclusive. For example, DNA studies revealed a close relationship between starlings and mockingbirds that was not expected on the basis of their anatomy or behavior.

Nucleotides have other important roles

Nucleotides are more than just the building blocks of nucleic acids. As we will describe in later chapters, there are several nucleotides with other functions:

- ATP (adenosine triphosphate) acts as an energy transducer in many biochemical reactions (see Section 8.2).

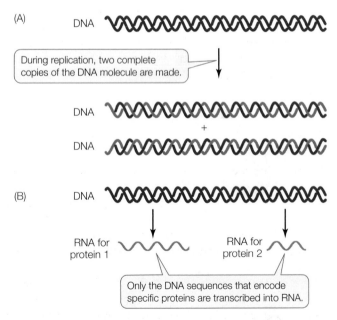

4.6 DNA Replication and Transcription DNA is usually completely replicated (A) but only partially transcribed (B). RNA transcripts encode the genes for specific proteins. Transcription of the many different proteins is activated at different times and, in multicellular organisms, in different cells of the body.

- GTP (guanosine triphosphate) serves as an energy source, especially in protein synthesis. It also plays a role in the transfer of information from the environment to cells (see Section 7.2).

- cAMP (cyclic adenosine monophosphate) is a special nucleotide in which an additional bond forms between the sugar and phosphate group. It is essential in many processes, including the actions of hormones and the transmission of information by the nervous system (see Section 7.3).

4.1 RECAP

The nucleic acids DNA and RNA are polymers made up of nucleotide monomers. The sequence of nucleotides in DNA carries the information that is used by RNA to specify primary protein structure. The genetic information in DNA is passed from generation to generation and can be used to understand evolutionary relationships.

- List the key differences between DNA and RNA. Between purines and pyrimidines. **See p. 61, Table 4.1, and Figure 4.1**

- How do purines and pyrimidines pair up in complementary base pairing? **See pp. 62–63 and Figure 4.2**

- What are the differences between DNA replication and transcription? **See pp. 63–64 and Figures 4.5 and 4.6**

- How can DNA molecules be very diverse, even though they appear to be structurally similar? **See p. 64**

We have seen that the nucleic acids RNA and DNA carry the blueprint of life, and that the inheritance of these macromolecules reaches back to the beginning of evolutionary time. But when, where, and how did nucleic acids arise on Earth? How did the building blocks of life such as amino acids and sugars originally arise?

4.2 How and Where Did the Small Molecules of Life Originate?

Chapter 2 points out that living things are composed of the same atomic elements as the inanimate universe—the 92 naturally occurring elements of the periodic table (see Figure 2.2). But the arrangements of these atoms into molecules are unique in biological systems. You will not find biological molecules in inanimate matter unless they came from a once-living organism.

It is impossible to know for certain how life on Earth began. But one thing is sure: life (or at least life as we know it) is not constantly being re-started. That is, *spontaneous generation* of life from inanimate nature is not happening before our eyes. Now and for many millenia past, all life has come from life that existed before. But people, including scientists, did not always believe this.

Experiments disproved spontaneous generation of life

The idea that life could have originated from nonliving matter is common in many cultures and religions. During the European Renaissance (from about 1450 to 1700, a period that witnessed the birth of modern science), most people thought that at least some forms of life arose repeatedly and directly from inanimate or decaying matter by *spontaneous generation*. Many thought that mice arose from sweaty clothes placed in dim light; that frogs sprang directly from moist soil; and that rotting meat produced flies. Scientists such as the Italian physician and poet Francesco Redi, however, doubted these assumptions. Redi proposed that flies arose not by some mysterious transformation of decaying meat, but from other flies that laid their eggs on the meat. In 1668, Redi performed a scientific experiment—a relatively new concept at the time—to test his hypothesis. He set out several jars containing chunks of meat.

- One jar contained meat exposed to both air and flies.
- A second jar was covered with a fine cloth so that the meat was exposed to air, but not to flies.
- The third jar was sealed so the meat was exposed to neither air nor flies.

As he had hypothesized, Redi found maggots, which then hatched into flies, only in the first jar. This finding demonstrated that maggots could occur only where flies were present. The idea that a complex organism like a fly could appear *de novo* from a nonliving substance in the meat, or from "something in the air," was laid to rest. Well, perhaps not quite to rest.

In the 1660s, newly developed microscopes revealed a vast new biological world. Under microscopic observation, virtually every environment on Earth was found to be teeming with tiny organisms. Some scientists believed these organisms arose spontaneously from their rich chemical environment, by the action of a "life force." But experiments by the great French scientist Louis Pasteur showed that microorganisms can arise only from other microorganisms, and that an environment without life remains lifeless (**Figure 4.7**).

---
yourBioPortal.com

GO TO Animated Tutorial 4.1 • Pasteur's Experiment

Pasteur's and Redi's experiments showed that living organisms cannot arise from nonliving materials *under the conditions that existed on Earth during their lifetimes*. But their experiments did not prove that spontaneous generation never occured. Eons ago, conditions on Earth and in the atmosphere above it were vastly different. Indeed, conditions similar to those found on primitive Earth may have existed, or may exist now, on other bodies in our solar system and elsewhere. This has led scientists to ask whether life has originated on other bodies in space, as it did on Earth.

Life began in water

As we emphasize in Chapter 2 and in the opening story of this chapter, the presence of water on a planet or moon is a necessary prerequisite for life as we know it. Astronomers believe our solar system began forming about 4.6 billion years ago, when a

INVESTIGATING LIFE

4.7 Disproving the Spontaneous Generation of Life

Previous experiments disproving spontaneous generation were called into question in regard to microorganisms, whose abundance and diversity were appreciated but whose living processes were not understood. Louis Pasteur's classic experiments disproved the spontaneous generation of microorganisms.

HYPOTHESIS Microorganisms come only from other microorganisms and cannot arise by spontaneous generation.

METHOD

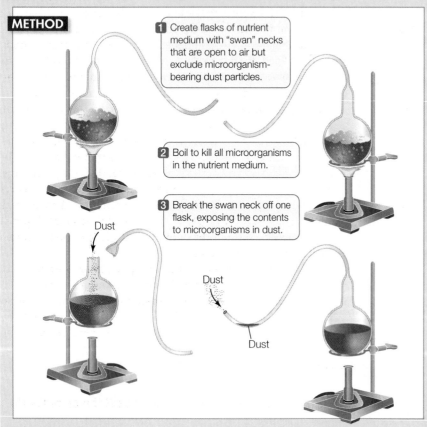

1. Create flasks of nutrient medium with "swan" necks that are open to air but exclude microorganism-bearing dust particles.

2. Boil to kill all microorganisms in the nutrient medium.

3. Break the swan neck off one flask, exposing the contents to microorganisms in dust.

Dust

Dust

Dust

RESULTS Microbial life grows only in the flasks exposed to microorganisms. There is no "spontaneous generation" of life in the sterile flask.

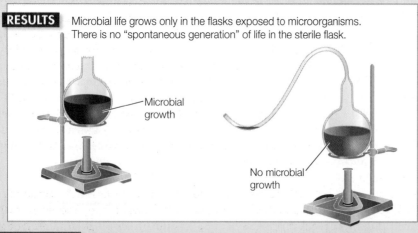

Microbial growth

No microbial growth

CONCLUSION All life comes from pre-existing life. An environment without life remains lifeless.

Go to **yourBioPortal.com** for original citations, discussions, and relevant links for all INVESTIGATING LIFE figures.

star exploded and collapsed to form the sun and 500 or so bodies, called planetesimals. These planetesimals collided with one another to form the inner planets, including Earth and Mars. The first chemical signatures indicating the presence of life on Earth appear to be about 4 billion years old. So it took 600 million years, during a geological time frame called the Hadean, for the chemical conditions on Earth to become just right for life. Key among those conditions was the presence of water.

Ancient Earth probably had a lot of water high in its atmosphere. But the new planet was hot, and the water remained in vapor form and dissipated into space. As Earth cooled, it became possible for water to condense on the planet's surface—but where did that water come from? One current view is that comets (loose agglomerations of dust and ice that have orbited the sun since the planets formed) struck Earth and Mars repeatedly, bringing to those planets not only water but other chemical components of life, such as nitrogen.

As the planets cooled and chemicals from their crusts dissolved in the water, simple chemical reactions would have taken place. Some of these reactions might have led to life, but impacts by large comets and rocky meteorites released enough energy to heat the developing oceans almost to boiling, thus destroying any early life. On Earth, these large impacts eventually subsided, and some time around 3.8 to 4 billion years ago life gained a foothold. There has been life on Earth ever since. Because Mars and some other celestial bodies have a similar geological history, the possibility exists that life exists or has existed on them. This possibility was an impetus for sending the Viking and Phoenix landers to Mars.

Several models have been proposed to explain the origin of life on Earth. The next sections discuss two alternative theories: that life came from outside of Earth, or that life arose on Earth through chemical evolution.

Life may have come from outside Earth

In 1969, a remarkable event led to the discovery that a meteorite from space carried molecules that were characteristic of life on Earth. On September 28 of that year, fragments of a meteorite fell around the town of Murchison, Australia. Using gloves to avoid Earth-derived contamination, scientists immediately shaved off tiny pieces of the rock, put them in

4.8 The Murchison Meteorite Pieces from a fragment of the meteorite that landed in Australia in 1969 were put into test tubes with water. Soluble molecules present in the rock, including amino acids, nucleotide bases, and sugars, dissolved in the water. Plastic gloves and sterile instruments were used to reduce the possibility of contamination with substances from Earth.

test tubes and extracted them in water (**Figure 4.8**). They found a number of the molecules that are unique to life, including purines, pyrimidines, sugars, and ten amino acids.

Were these molecules truly brought from space as part of the meteorite, or did they get there after the rock landed on Earth? There were a number of reasons to believe the molecules were not Earthly contaminants:

- The scientists took great care to avoid contamination. They used gloves and sterile instruments, took pieces from below the rock's surface, and did their work very soon after it landed (hopefully before Earth organisms could contaminate the samples).

- Amino acids found in living organisms on Earth are left-handed (see Figure 3.2). The amino acids in the meteorite were a mixture of right- and left-handed forms, with a slight preponderance of the left-handed. Thus the amino acids in the meteorite were not likely to have come from a living organism on Earth.

- In the story that opens Chapter 2, we describe how the ratio of isotopes in a living organism reflects that isotope ratio in the environment where the organism lives. The isotope ratios for carbon and hydrogen in the sugars from the meteorite were different from the ratios of those elements found on Earth.

In 1984, another informative meteorite, this one the size of a softball, was found in Antarctica. We know that the meteorite, ALH 84001, came from Mars because the composition of the gases trapped within the rock was identical to the composition found in the Martian atmosphere, which is quite different from Earth's atmosphere. Radioactive dating and mineral analyses determined that ALH 84001 was 4.5 billion years old and was blasted off the Martian surface 16 million years ago. It landed on Earth fairly recently, about 13,000 years ago.

Scientists found water trapped below the Martian meteorite's surface. This discovery was not surprising, given that surface observations had already shown that water was once abundant on Mars (see the chapter-opening story). Because water is essential for life, scientists wondered whether the meteorite might contain other signs of life as well. Their analysis revealed two substances related to living systems. First, simple carbon-containing molecules called polycyclic aromatic hydrocarbons were present in small but unmistakable amounts; these substances can be formed by living organisms. Second, crystals of magnetite, an iron oxide mineral made by many living organisms on Earth, were found in the interior of the rock.

ALH 84001 and the Murchison meteorite are not the only visitors from outer space that have been shown to contain chemical signatures of life. While the presence of such molecules in rocks may suggest that those rocks once harbored life, it does not prove that there were living organisms in the rocks when they landed on Earth. Most scientists find it hard to believe that an organism could survive thousands of years of traveling through space in a meteorite, followed by intense heat as the meteorite passed through Earth's atmosphere. But there is some evidence that the heat inside some meteorites may not have been severe. When weakly magnetized rock is heated, it reorients its magnetic field to align with the magnetic field around it. In the case of ALH 84001, this would have been Earth's powerful magnetic field, which would have affected the meteorite as it approached our planet.

Careful measurements indicate that, while reorientation did occur at the surface of the rock, it did not occur on the inside. The scientists who took these measurements, Benjamin Weiss and Joseph Kirschvink at the California Institute of Technology, concluded that the inside of ALH 84001 was never heated over 40°C as it entered Earth's atmosphere. This suggests that a long interplanetary trip by living organisms could be possible.

Prebiotic synthesis experiments model the early Earth

It is clear that other bodies in the solar system have, or once had, water and other simple molecules. Possibly, a meteorite was the source of the simple molecules that were the original building blocks for life on Earth. But a second theory for the origin of life on Earth, **chemical evolution**, holds that conditions on primitive Earth led to the formation of these simple molecules (prebiotic synthesis), and these molecules led to the formation of life forms. Scientists have sought to reconstruct those primitive conditions, both physically (hot or cold) and chemically (by re-creating the combinations and proportions of elements that may have been present).

HOT CHEMISTRY The amounts of trace metals such as molybdenum and rhenium in sediments under oceans and lakes is directly proportional to the amount of oxygen gas (O_2) present in and above the water. Measurements of dated sedimentary cores indicate that none of these rare metals was present prior to 2.5 billion years ago. This and other lines of evidence suggest that there was little oxygen gas in Earth's early atmosphere. Oxygen gas is thought to have accumulated about 2.5 billion yea as the by-product of photosynthesis by single-celled life forms; today 21 percent of our atmosphere is O_2.

In the 1950s, Stanley Miller and Harold Urey at the University of Chicago set up an experimental "atmosphere" containing the gases thought to have been present in Earth's early atmosphere: hydrogen gas, ammonia, methane gas, and water vapor. They passed an electric spark through these gases, to simulate lightning as a source of energy to drive chemical reactions. Then, they cooled the system so the gases would condense and collect in a watery solution, or "ocean" (**Figure 4.9**). After a few days of continuous operation, the system contained numerous complex molecules, including amino acids, purines, and pyrimidines—some of the building blocks of life.

---
yourBioPortal.com

GO TO Animated Tutorial 4.2 • Synthesis of Prebiotic Molecules

---

The results of this experiment were profoundly important in giving weight to speculations about the chemical origin of life on Earth and elsewhere in the universe. Decades of experimental work and critical evaluation followed. The experiments showed that, under the conditions used by Miller and Urey, many small molecular building blocks of life could be formed:

• All five bases that are present in DNA and RNA (i.e., A, T, C, G and U)

• 17 of the 20 amino acids used in protein synthesis

• 3- to 6-carbon sugars

However, the 5-carbon sugar ribose was not produced in these experiments.

In science, an experiment and its results must be repeated, reinterpreted, and refined as more knowledge accumulates. The results of the Miller–Urey experiments have undergone several such refinements.

The amino acids in living things are always L-isomers (see Figure 3.2 and p. 43). But a mixture of D- and L-isomers appeared in the amino acids formed in the Miller–Urey experiments. Recent experiments show that natural processes could have selected the L-amino acids from the mixture. Some minerals, especially calcite-based rocks, have unique crystal structures that selectively bind to D- or L-amino acids, separating the two. Such rocks were abundant on early Earth. This suggests that while both kinds of amino acid structures were made, binding to certain rocks may have eliminated the D- amino

acids. (Interestingly, some meteorites, such as the Murchison meteorite, also have this selectivity.)

Ideas about Earth's original atmosphere have changed since Miller and Urey did their experiments. There is abundant evidence indicating that major volcanic eruptions occurred 4 bil-

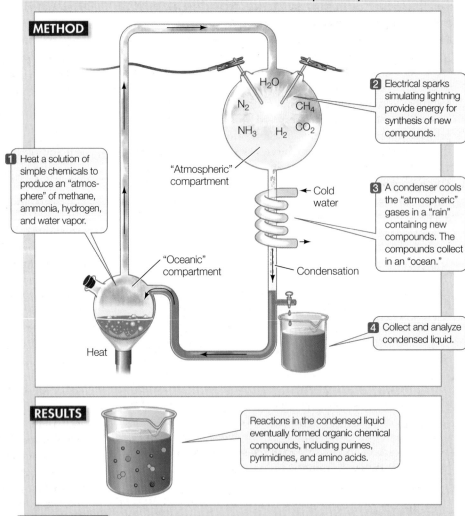

INVESTIGATING LIFE

4.9 Miller and Urey Synthesized Prebiotic Molecules in an Experimental Atmosphere

With an increased understanding of the atmospheric conditions that existed on primitive Earth, the researchers devised an experiment to see if these conditions could lead to the formation of organic molecules.

HYPOTHESIS Organic chemical compounds can be generated under conditions similar to those that existed in the atmosphere of primitive Earth.

METHOD

1 Heat a solution of simple chemicals to produce an "atmosphere" of methane, ammonia, hydrogen, and water vapor.

H_2O N_2 CH_4 NH_3 H_2 CO_2

"Atmospheric" compartment

2 Electrical sparks simulating lightning provide energy for synthesis of new compounds.

Cold water

3 A condenser cools the "atmospheric" gases in a "rain" containing new compounds. The compounds collect in an "ocean."

"Oceanic" compartment

Condensation

Heat

4 Collect and analyze condensed liquid.

RESULTS

Reactions in the condensed liquid eventually formed organic chemical compounds, including purines, pyrimidines, and amino acids.

CONCLUSION The chemical building blocks of life could have been generated in the probable atmosphere of early Earth.

FURTHER INVESTIGATION: What result would you predict if O_2 were present in the "atmosphere" in this experiment?

Go to **yourBioPortal.com** for original citations, discussions, and relevant links for all INVESTIGATING LIFE figures.

lion years ago, which would have released carbon dioxide (CO_2), nitrogen (N_2), hydrogen sulfide (H_2S), and sulfur dioxide (SO_2) into the atmosphere. Experiments using these gases in addition to the ones in the original experiment have produced more diverse molecules, including:

- Vitamin B_6, pantothenic acid (a component of coenzyme A), and nicotinamide (part of NAD, which is involved in energy metabolism).
- Carboxylic acids such as succinic and lactic acids (also involved in energy metabolism) and fatty acids.
- Ribose, a key component of RNA, which can be formed from formaldehyde gas (HCHO), evidence of which has been found in space.

COLD CHEMISTRY Stanley Miller also performed a long-term experiment in which the electric spark was not used. In 1972, he filled test tubes with ammonia gas, water vapor and cyanide (HCN), another molecule that is thought to have formed on primitive Earth. After checking that there were no contaminating substances or organisms that might confound the results, he sealed the tubes and cooled them to –100°C, the temperature of the ice that covers Europa, one of Jupiter's moons. Opening the tubes 25 years later, he found amino acids and nucleotide bases. Apparently, pockets of liquid water within the ice had allowed high concentrations of the starting materials to accumulate, thereby speeding up chemical reactions. The important conclusion is that the cold water within ice on ancient Earth, and other celestial bodies such as Mars, Europa, and Enceladus (one of Saturn's moons; satellite photos have revealed geysers of liquid water coming from its interior) may have provided environments for the prebiotic synthesis of molecules required for the subsequent formation of simple living systems.

4.2 RECAP

Life does not arise repeatedly through spontaneous generation, but comes from pre-existing life. Water is an essential ingredient for the emergence of life. Meteorites that have landed on Earth provide some evidence for an extraterrestrial origin of life. Prebiotic chemical synthesis experiments provide support for the idea that life's simple molecules formed in the primitive Earth environment.

- Explain how Redi's and Pasteur's experiments disproved spontaneous generation. **See p. 65 and Figure 4.7**

- What is the evidence that life on Earth came from other bodies in the solar system? **See pp. 66–67**

- What is the significance of the Miller–Urey experiment, what did it find, and what were its limitations? **See p. 68 and Figure 4.9**

Chemistry experiments using conditions modeling the ancient Earth's environment suggest an origin for the monomers (such as amino acids) that make up the polymers (such as proteins) that characterize life. How did these polymers develop on the ancient Earth?

4.3 How Did the Large Molecules of Life Originate?

The Miller–Urey experiment and other experiments that followed it provide a plausible scenario for the formation of the building blocks of life under conditions that prevailed on primitive Earth. The next step in forming and supporting a general theory on the origin of life on Earth would be an explanation of the formation of polymers from these monomers.

Chemical evolution may have led to polymerization

Scientists have used a number of model systems to try to simulate conditions under which polymers might have been made. Each of these systems is based on several observations and speculations:

- *Solid mineral surfaces*, such as powder-like clays, have large surface areas. Scientists speculate that the silicates within clay may have been catalytic (speeded up the reactions) in the formation of early carbon-based molecules.
- *Hydrothermal vents* deep in the ocean, where hot water emerges from beneath Earth's crust, lack oxygen gas and contain metals such as iron and nickel. In laboratory experiments, these metals have been shown to catalyze the polymerization of amino acids to polypeptides in the absence of oxygen.
- *Hot pools* at the edges of oceans may, through evaporation, have concentrated monomers to the point where polymerization was favored (the "primordial soup" hypothesis).

In whatever ways the earliest stages of chemical evolution occurred, they resulted in the emergence of monomers and polymers that have probably remained unchanged in their general structure and function for several billion years.

There are two theories for the emergence of nucleic acids, proteins, and complex chemistry

Earlier in this chapter, we described the key roles of nucleic acids as informational molecules that are passed on from one generation to the next. We also described how DNA is transcribed to RNA, which can then be translated into protein (see Figure 4.5). Chapter 3 describes the roles of proteins as catalysts, speeding up biochemical transformations (see Section 3.2). In existing life forms, nucleic acids and proteins require one another in order to perpetuate life. For the origin of life, this results in a chicken-or-egg problem. Which came first, the genetic material (nucleic acids) or proteins? Two ideas have emerged. One suggests that sequential catalytic changes (primitive metabolism) came first. The other suggests that replication by nucleic acids preceded metabolism (**Figure 4.10**).

CHEMICAL CHANGES (METABOLISM) FIRST In this model, life began in tiny droplets, or compartments, that concentrated and sepa-

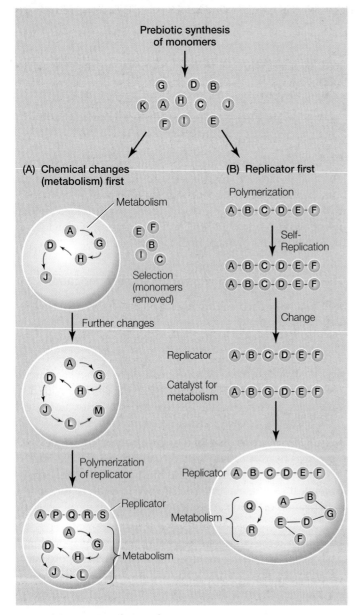

4.10 Two Pathways to Life Biologists have proposed two ways in which simple monomers could have become self-replicating systems capable of biological functions. (A) The chemical changes (metabolism) first pathway. (B) The replicator first pathway.

REPLICATOR FIRST In this model, the genetic material—nucleic acids—came first. The nucleotide building blocks made by prebiotic chemistry came together to form polymers. Some of these polymers might have had the right shape to be catalytic so that they could reproduce themselves and catalyze other chemical transformations. Such transformations might have included the synthesis of proteins, just as RNA is translated into proteins in living organisms today (see Figure 4.5). Along the way, those molecules that were best adapted to the environment would survive and reproduce. Eventually they would have become incorporated into living cells.

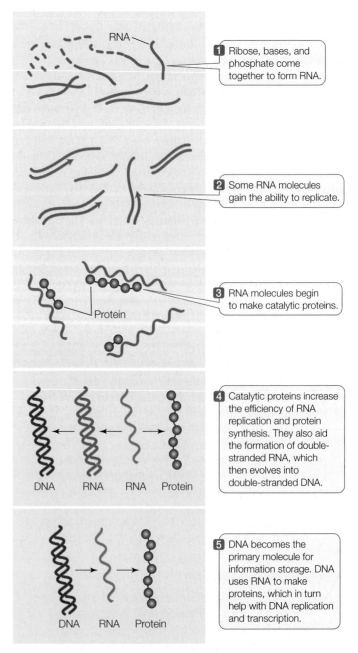

4.11 The "RNA World" Hypothesis In a world before DNA, this view postulates that RNA alone was both the blueprint for protein synthesis and a catalyst for its own replication. Eventually, the more compact information storage molecules of DNA could have evolved from RNA.

rated their contents from the external environment. Within such a chemically rich environment, some substances could occasionally and randomly undergo chemical changes. Proponents of this model speculate that those compartments where the changes were effective for survival in the environment might even have been selected for growth and some primitive form of reproduction. Could catalysis, the speeding up of reactions essential for life, occur in such an environment? The German scientist Günter Wächtershäuser proposed that catalysis and reproduction could have occurred without proteins on a mineral called pyrite (iron disulfide), which has been found at hydrothermal vents and which could serve as a source of energy for polymerization reactions. Over time, nucleic acids and eventually proteins might have formed in the concentrated droplets. Then, in some of these proteins, the ability to catalyze biochemical reactions—including the replication of nucleic acids—could have evolved.

4.12 An Early Catalyst for Life? In the laboratory, a ribozyme (a folded RNA molecule) can catalyze the polymerization of several short RNA strands into a longer molecule. Such a process could be a precursor for the copying of nucleic acids, which is essential for their replication and for gene expression.

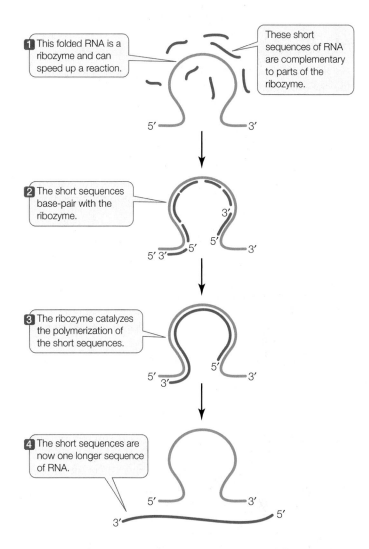

There are two major problems with the replicator first model:

- Nucleic acid polymers have not been observed in prebiotic chemistry simulations.

- DNA, the genetic material in almost all current organisms, is not self-catalytic.

The first problem remains, but the second has a plausible solution: RNA can be a catalyst and can catalyze its own synthesis.

RNA may have been the first biological catalyst

The three-dimensional structure of a folded RNA molecule presents a unique surface to the external environment (see Figure 4.3). The surfaces of RNA molecules can be every bit as specific as those of proteins. Just as the shapes of proteins allow them to function as catalysts, speeding up reactions that would ordinarily take place too slowly to be biologically useful, the three-dimensional shapes and other chemical properties of certain RNA molecules allow them to function as catalysts. Catalytic RNAs, called **ribozymes**, can catalyze reactions on their own nucleotides as well as in other cellular substances. Although in retrospect it is not too surprising, the discovery of catalytic RNAs was a major shock to a community of biologists who were convinced that all biological catalysts were proteins (enzymes). It took almost a decade for the work of the scientists involved, Thomas Cech and Sidney Altman, to be fully accepted by other scientists. Later, they were awarded the Nobel Prize.

Given that RNA can be both informational (in its nucleotide sequence) and catalytic (due to its ability to form unique three-dimensional shapes), it has been hypothesized that early life consisted of an "RNA world"—a world before DNA. It is thought that when RNA was first made, it could have acted as a catalyst for its own replication as well as for the synthesis of proteins. DNA could eventually have evolved from RNA (**Figure 4.11**). Some laboratory evidence supports this scenario:

- When certain short RNA sequences are added to a mixture of nucleotides, RNA polymers can be formed at a rate 7 million times greater than the formation of polymers without the added RNA. This added RNA is not a template, but a catalyst.

- In the test tube, a ribozyme can catalyze the assembly of short RNAs into a longer molecule (**Figure 4.12**). This may be how nucleic acid replication evolved.

- In living organisms today, the formation of peptide linkages (see Figure 3.6) is catalyzed by ribozymes.

- In certain viruses called retroviruses, there is an enzyme called reverse transcriptase that catalyzes the synthesis of DNA from RNA.

4.3 RECAP

The emergence of the chemical reactions characteristic of life (metabolism), and the polymerization of monomers to polymers, may have occurred on the surfaces of hydrothermal vents. One theory proposes that metabolism came before polymerization; another suggests that the reverse occurred. RNA may have been the first genetic material and catalyst.

- What are the two theories for the emergence of metabolism and polymers? See pp. 69–71 and **Figure 4.10**

- How does RNA self-replicate? See p. 71 and **Figure 4.12**

The discovery of mechanisms for the formation of small and large molecules is essential to answering questions about the origin of life on Earth. But we also need to understand how organized systems formed that include these molecules and display the characteristic properties of life, such as reproduction, energy processing, and responsiveness to the environment. These properties are present in cells, and we now turn to ideas on their origin.

4.4 How Did the First Cells Originate?

As you have seen from many of the theories for the origin of life, the evolution of biochemistry occurred under localized conditions. That is, the chemical reactions of metabolism, polymerization, and replication could not occur in a dilute aqueous environment. There had to be a compartment of some sort that brought together and concentrated the compounds involved in these events. Biologists have proposed that initially this compartment may have simply been a tiny droplet of water on the surface of a rock. But another major event in the origin of life was necessary.

Life as we know it is separated from the environment within structurally defined units called **cells**. The internal contents of a cell are separated from the nonbiological environment by a special barrier—a **membrane**. The membrane is not just a barrier; it regulates what goes into and out of the cell, as we describe in Chapter 6. This role of the surface membrane is very important because it permits the interior of the cell to maintain a chemical composition that is different from its external environment. How did the first cells with membranes come into existence?

Experiments describe the origin of cells

Jack Szostak and his colleagues at Harvard University built a laboratory model that gives insights into the origin of cells. To do this, they first put fatty acids (which can be made in prebiotic experiments) into water. Recall from Chapter 3 that fatty acids are *amphipathic*: they have a hydrophilic polar end and a long, nonpolar tail that is hydrophobic (see Figure 3.20). When placed in water, fatty acids will arrange themselves in a round "huddle" much like a football team: the hydrophilic ends point outward to interact with the aqueous environment and the fatty acid tails point inward, away from the water molecules.

What if some water becomes trapped in the interior of this "huddle"? Now the layer of hydrophobic fatty acid tails is in water, which is an unstable situation. To stabilize this, a second layer of fatty acids forms. This *lipid bilayer* has the polar ends of the fatty acids facing both outward and inward, because they are attracted to the polar water molecules present on each side of the double layer. The nonpolar tails form the interior of the bilayer (**Figure 4.13**). These prebiotic, water-filled structures, defined by a lipid bilayer membrane, very much resemble living cells. Scientists refer to these compartments as **protocells**. Examining their properties revealed that

- Large molecules such as DNA or RNA could not pass through the bilayer to enter the protocells, but small molecules such as sugars and individual nucleotides could.

- Nucleic acids inside the protocells could replicate using the nucleotides from outside. When the investigators placed a short nucleic acid strand capable of self-replication inside protocells and added nucleotides to the watery environment outside, the nucleotides crossed the barrier, entered the protocells, and became incorporated into new polynucleotide

chains. This may have been the first step toward cell reproduction, and it took place without protein catalysis.

Were these protocells truly cells, and was the lipid bilayer produced in these experiments a true cell membrane? Certainly not. The protocells could not fully reproduce, nor could they carry out all the metabolic reactions that take place in modern cells. The simple lipid bilayer had few of the sophisticated functions

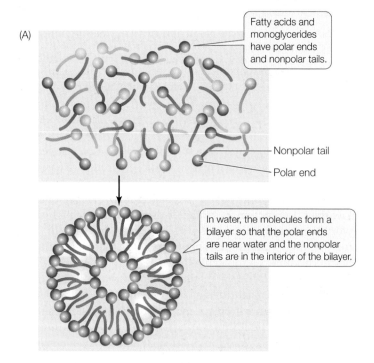

(A)

Fatty acids and monoglycerides have polar ends and nonpolar tails.

Nonpolar tail

Polar end

In water, the molecules form a bilayer so that the polar ends are near water and the nonpolar tails are in the interior of the bilayer.

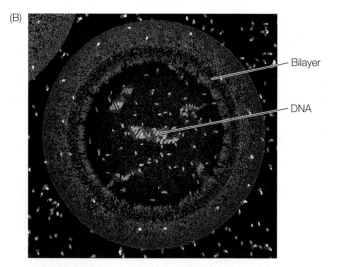

(B)

Bilayer

DNA

4.13 Protocells (A) In a series of experiments in the Szostak lab, researchers mixed fatty acid molecules in water. The molecules formed bilayers that have some of the properties of a cell membrane. The bilayers and the water "trapped" inside them are essential to form a protocell. (B) A model of the protocell. Nutrients and nucleotides (blue and white particles) pass through the "membrane" and enter the protocell, where they copy an already present DNA template. The new copies of DNA remain in the protocell.

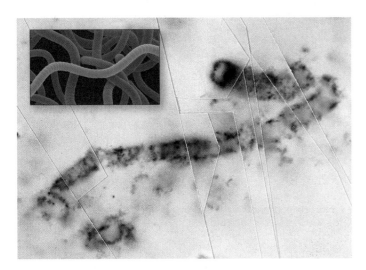

4.14 The Earliest Cells? This fossil from Western Australia is 3.5 billion years old. Its form is similar to that of modern filamentous cyanobacteria (inset).

of modern cell membranes. Nevertheless, the protocell may be a reasonable facsimile of a cell as it evolved billions of years ago:

- It can act as a system of interacting parts
- It is capable of organization and self-catalysis
- It includes an interior that is distinct from the exterior environment.

These are all fundamental characteristics of living cells.

Some ancient cells left a fossil imprint

In the 1990s, scientists made an extremely rare find: a formation of ancient rocks in Australia that had remained relatively unchanged since they first formed 3.5 billion years ago. In one of these rock samples, geologist J. William Schopf of the University of California, Los Angeles, saw chains and clumps of what looked tantalizingly like contemporary cyanobacteria, or "blue-green" bacteria (**Figure 4.14**). Cyanobacteria are believed to

have been among the first organisms, because they can perform photosynthesis, converting CO_2 from the atmosphere and water into carbohydrates. Schopf needed to prove that the chains were once alive, not just the results of simple chemical reactions. He and his colleagues looked for chemical evidence of photosynthesis in the rock samples.

The use of carbon dioxide in photosynthesis is a hallmark of life and leaves a unique chemical signature—a specific ratio of isotopes of carbon (^{13}C:^{12}C) in the resulting carbohydrates. Schopf showed that the Australian material had this isotope signature. Furthermore, microscopic examination of the chains revealed *internal* substructures that are characteristic of living systems and were not likely to be the result of simple chemical reactions. Schopf's evidence suggests that the Australian sample is indeed the remains of a truly ancient living organism.

Taking geological, chemical, and biological evidence into account, it is plausible that it took about 500 million to a billion years from the formation of the Earth until the appearance of the first cells (**Figure 4.15**). Life has been cellular ever since. In the next chapter, we begin our study of cell structure and function.

4.4 RECAP

The chemical reactions that preceded living organisms probably occurred in specialized compartments, such as water droplets on the surfaces of minerals. Life as we know it did not begin until the emergence of cells. Protocells made in the laboratory have some of the properties of modern cells. Cell-like structures fossilized in ancient rocks date the first cells to about 3.5 billion years ago.

- Explain the importance of the cell membrane to the evolution of living organisms. See p. 72

- What is the evidence that ancient rocks contain the fossils of cells? See p. 73

4.15 The Origin of Life This highly simplified timeline gives a sense of the major events that culminated in the origin of life more than 3.5 billion years ago.

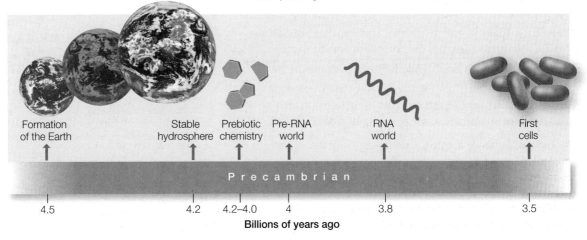

Formation of the Earth — Stable hydrosphere — Prebiotic chemistry — Pre-RNA world — RNA world — First cells

Precambrian

4.5   4.2   4.2–4.0   4   3.8   3.5

Billions of years ago

CHAPTER SUMMARY

4.1 What Are the Chemical Structures and Functions of Nucleic Acids?

- The unique function of the nucleic acids—DNA and RNA—is information storage. They form the hereditary material that passes genetic information from one generation to the next.

- Nucleic acids are polymers of nucleotides. A **nucleotide** consists of a phosphate group, a pentose sugar (**ribose** in RNA and **deoxyribose** in DNA), and a nitrogen-containing **base**. Review Figure 4.1

- In DNA, the nucleotide bases are **adenine**, **guanine**, **cytosine**, and **thymine**. **Uracil** replaces thymine in RNA. The nucleotides are joined by **phosphodiester linkages** between the sugar of one and the phosphate of the next, forming a nucleic acid polymer. **WEB ACTIVITY 4.1**

- DNA is a **double helix** with two separate strands in which there is **complementary base pairing** based on hydrogen bonds between adenine and thymine (A-T) and between guanine and cytosine (G-C). The two strands of the DNA double helix run in opposite directions. RNA consists of one chain of nucleotides. Hydrogen bonding can occur within the single strand of RNA, forming double-stranded regions and giving the molecule a three-dimensional surface shape. Review Figures 4.2 and 4.3; **WEB ACTIVITY 4.2**

- The information content of DNA and RNA resides in their **base sequences**.

- DNA is expressed as RNA in **transcription**. RNA can then specify the amino acid sequence of a protein in **translation**. Review Figures 4.5 and 4.6

4.2 How and Where did the Small Molecules of Life Originate?

- Historically, many cultures believed that life originates repeatedly by **spontaneous generation**. This was disproven experimentally. Review Figure 4.7; **ANIMATED TUTORIAL 4.1**

- Life probably originated from chemical reactions. A prerequisite for life is the presence of water.

- The presence of chemical traces of life on meteorites that have landed on Earth suggests that life might have originated extraterrestrially.

- Chemical experiments modeling the prebiotic conditions on Earth have shown that the small molecules that characterize life could have been formed from atmospheric chemicals. Review Figure 4.9; **ANIMATED TUTORIAL 4.2**

4.3 How Did the Large Molecules of Life Originate?

- Polymerization of small molecules to polymers could occur in small compartments such as droplets or on surfaces. Both of these conditions concentrate molecules such that reactions are favored.

- The "metabolism first" theory of poylmerization proposes that chemical reactions involving small molecules evolved first, and some of them formed polymers that acted as genetic information and catalysts.

- The "replicator first" theory proposes that RNA formed early, and acted as both genetic material and catalyst. Then reactions involving small molecules could occur. Review Figure 4.10

- In contemporary organisms, RNA can act as both an information molecule and as a catalyst. This favors the replicator first model. The **RNA world** may have been an important step on the way to life. Review Figure 4.11

4.4 How Did Cells Originate?

- A key to the emergence of living cells was the prebiotic chemical generation of compartments enclosed by **membranes**. Such enclosed compartments permitted the generation and maintenance of internal chemical conditions that were different from those in the exterior environment.

- In the laboratory, fatty acids and related lipids assemble into **protocells** that have some of the characteristics of cells. Review Figure 4.13

- Ancient rocks (3.5 billion years old) have been found with imprints that are probably fossils of early cells.

SELF-QUIZ

1. A nucleotide in DNA is made up of
   a. four bases.
   b. a base plus a ribose sugar.
   c. a base plus a deoxyribose sugar plus phosphate.
   d. a sugar plus a phosphate.
   e. a sugar and a base.

2. Nucleotides in RNA are connected to one another in the polynucleotide chain by
   a. covalent bonds between bases.
   b. covalent bonds between sugars.
   c. covalent bonds between sugar and phosphate.
   d. hydrogen bonds between purines.
   e. hydrogen bonds between any bases.

3. Which is a difference between DNA and RNA?
   a. DNA is single-stranded and RNA is double-stranded.
   b. DNA is only informational and RNA is only catalytic.
   c. DNA contains deoxyribose and RNA contains ribose.
   d. DNA is transcribed and RNA is replicated.
   e. DNA contains uracil (U) and RNA contains thymine (T).

4. The nucleotide sequence of DNA
   a. is the same in all organisms of a species.
   b. contains only information for translation.
   c. evolved before RNA.
   d. contains the four bases, A, T, G, and C.
   e. is produced by prebiotic chemistry experiments.

5. Spontaneous generation of life from nonliving materials
   a. can occur in dark places.
   b. has not been a belief of humans.
   c. has never occurred.
   d. requires only nucleotides and fatty acids.
   e. was disproven for microorganisms by Pasteur's experiment.

6. The components in the atmosphere for the Miller–Urey experiment on prebiotic synthesis did not include
   a. H_2.
   b. H_2O.
   c. O_2.
   d. NH_3.
   e. CH_4.

7. All of the major building blocks of macromolecules were made in Miller–Urey prebiotic synthesis experiments *except*
   a. amino acids.
   b. hexose sugars.
   c. bases for nucleotides.
   d. fatty acids.
   e. ribose.

8. The "RNA world" hypothesis proposes that
   a. RNA formed from DNA.
   b. RNA was both a catalyst and genetic material.
   c. RNA was a catalyst only.
   d. RNA formed after proteins.
   e. DNA formed after RNA was broken down.

9. Ribozymes are
   a. enzymes that are made up of ribose sugar.
   b. ancient catalysts that no longer exist.
   c. RNA catalysts.
   d. present in bacterial cells only.
   e. less active than protein enzymes.

10. Findings in ancient rocks indicate cells first appeared
    a. about 4.5 billion years ago.
    b. about 3.5 billion years ago.
    c. about 2 billion years ago.
    d. before rocks were formed.
    e. before water arrived on Earth.

FOR DISCUSSION

1. Are the statements "all life comes from pre-existing life" and "life on Earth could have arisen from prebiotic molecules" truly paradoxical? What conditions existing on Earth today might preclude the origin of life from such molecules?

2. Why might RNA have preceded proteins in the evolution of biological macromolecules?

3. Do you consider the two alternative theories presented in this chapter as possible explanations of the origin of life on Earth (that life came from outside of Earth, or that life arose on Earth through chemical evolution) to be equally plausible? Which do you favor, and why?

4. Why was the evolution of a self-contained cell essential for life as we know it?

ADDITIONAL INVESTIGATION

1. The interpretation of Pasteur's experiment (see Figure 4.7) depended on the inactivation of microorganisms by heat. We now know of microorganisms that can survive extremely high temperatures (see Chapter 26). Does this change the interpretation of Pasteur's experiment? What experiments would you do to inactivate such microbes?

2. The Miller–Urey experiment (see Figure 4.9) showed that it was possible for amino acids to be formed from gases that were hypothesized to have been in Earth's early atmosphere. These amino acids were dissolved in water. Knowing what you do about the polymerization of amino acids into proteins (see Figure 3.6), how would you set up experiments to show that proteins can form under the conditions of early Earth? What properties would you expect of those proteins?

WORKING WITH DATA (GO TO yourBioPortal.com)

Synthesis of Prebiotic Molecules in an Experimental Atmosphere   In this hands-on exercise, you will examine the original research paper of Miller and Urey to see the experimental approach they used to show that amino acids could be made in a simulation of Earth's early atmosphere (Figure 4.9). You will also analyze more recent data using the same apparatus.

Disproving the Spontaneous Generation of Life   In this hands-on exercise, you will examine data from an experiment similar to Pasteur's famous experiments (Figure 4.7). By calculating growth rates in the different flasks, you will be able to see how Pasteur came to the conclusion he did.

5

Cells: The Working Units of Life

How to mend a broken heart

It is a day in the not-too-distant future. Decades of eating fatty foods, combined with an inherited tendency to deposit cholesterol in his arteries, have finally caught up with 70-year-old Don. A blood clot has closed off blood flow to part of his heart, leading to a heart attack and severe damage to that vital organ.

If this had happened today, Don would have been faced with a long period of rehabilitation, taking medications to manage his weakened heart. Instead, his physicians take a pinch of skin tissue from his arm and bring it to a laboratory. After certain DNA sequences are added, Don's skin cells no longer look and act like skin cells: They are undifferentiated (unspecialized) and reproduce continuously in the laboratory dish. These cells are also multipotent stem

cells, able to differentiate into almost any type of cell in the body if given the right environment. When they are injected directly into Don's heart, his stem cells soon become heart muscle cells, repairing the damage caused by the heart attack. Don leaves the hospital with full cardiac function and recommendations for a healthy diet.

You are probably familiar with another type of multi–potent cell, the fertilized human egg. This single cell ultimately produces the tens of trillions of cells that make up the human body. The fertilized egg is programmed to generate an entire organism—not just the heart and skin, but blood, nerves, liver, brain, and even bones—and for this reason is called totipotent ("toti" means all; "multi" means most). In contrast, the stem cells derived from Don's skin need specific external signals to differentiate into other kinds of cells, and could not develop into an entire person.

The potential uses of stem cells in medicine have generated a lot of excitement in recent years. Such widely read periodicals as *Time* have hailed advances in stem cell research as "breakthroughs of the year." Patients with the neurological disorder Parkinson's disease dream of the day when their skin cells can be turned into brain cells to fix their damaged nervous systems. People with diabetes hope for stem cells to repair their pancreases. The list is long.

Behind all of this hope and the research it inspires is a cornerstone of biological science: the cell theory. As you saw in the last

A New Heart Cell This cardiac stem cell is developing into a fully differentiated heart cell. The hope is to be able to coax stem cells to follow this path or to produce other cell types to repair damaged tissues.

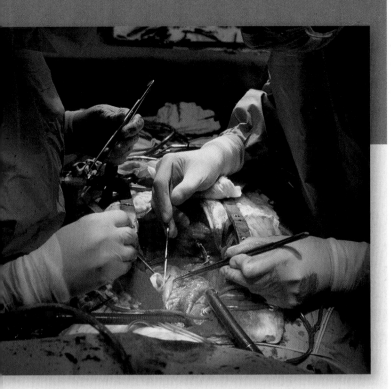

Open Heart Surgery Stem cell therapies may provide alternative approaches to treating heart disease in the future.

chapter, a key event in the emergence of life was the enclosure of biochemical reactions inside a cell, thus concentrating them and separating them from the external environment. These are the first two tenets of the cell theory, that the cell is the unit of life and that the activities of life either happen inside cells or are caused by them. Don's stem cells contain not just the activities of a living entity, but also the potential to change those activities in new directions. The third tenet of the cell theory—equally important—is that the cell is the unit of reproduction: all cells come from pre-existing cells. Stem cell therapy does not create new cells out of thin air; it coaxes existing ones to differentiate and reproduce along the desired path.

IN THIS CHAPTER we examine the structure and some of the functions of cells. We will begin with a fuller explanation of cell theory. Then, we will examine the relatively simple cells of prokaryotes. This is followed by a tour of the more complex eukaryotic cell and its various internal compartments, each of which performs specific functions. Finally, we discuss ideas on how complex cells evolved.

5.1 What Features Make Cells the Fundamental Units of Life?

In Chapter 1 we introduced some of the characteristics of life: chemical complexity, growth and reproduction, the ability to refashion substances from the environment, and the ability to determine what substances can move into and out of the organism. These characteristics are all demonstrated by cells. Just as atoms are the building blocks of chemistry, cells are the building blocks of life.

The **cell theory** is described in Section 1.1 as the first unifying principle of biology. There are three critical components of the cell theory:

- Cells are the fundamental units of life.
- All living organisms are composed of cells.
- All cells come from preexisting cells.

Cells contain water and the other small and large molecules, which we examined in Chapters 2–4. Each cell contains at least 10,000 different types of molecules, most of them present in many copies. Cells use these molecules to transform matter and energy, to respond to their environments, and to reproduce themselves.

The cell theory has three important implications:

- Studying cell biology is in some sense the same as studying life. The principles that underlie the functions of the single cell of a bacterium are similar to those governing the approximately 60 trillion cells of your body.
- Life is continuous. All those cells in your body came from a single cell, a fertilized egg. That egg came from the fusion of two cells, a sperm and an egg, from your parents. The cells of your parents' bodies were all derived from their parents, and so on back through generations and evolution to the first living cell.
- The origin of life on Earth was marked by the origin of the first cells (see Chapter 4).

Even the largest creatures on Earth are composed of cells, but the cells themselves are usually too small for the naked eye to see. Why are cells so small?

Cell size is limited by the surface area-to-volume ratio

Most cells are tiny. In 1665, the early microscopist Robert Hooke estimated that in one square inch of cork, which he examined under his magnifying lens, there were 1,259,712,000 cells! The volumes of cells range from 1 to 1,000 cubic micrometers (μm^3). There are some exceptions: the eggs of birds are single cells that

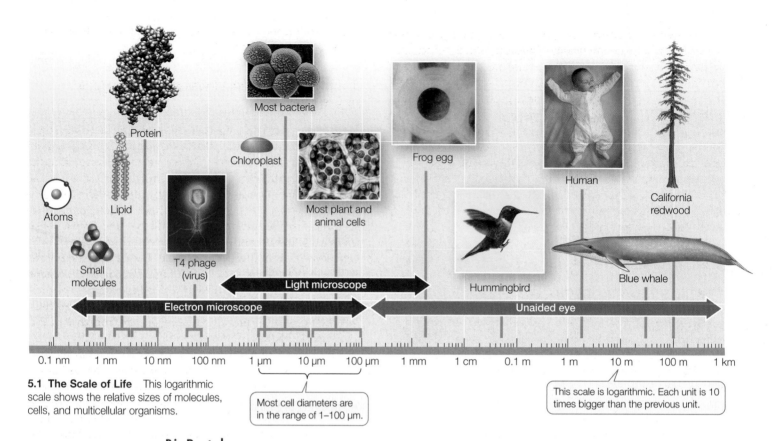

5.1 The Scale of Life This logarithmic scale shows the relative sizes of molecules, cells, and multicellular organisms.

Most cell diameters are in the range of 1–100 μm.

This scale is logarithmic. Each unit is 10 times bigger than the previous unit.

—— **yourBioPortal**.com ——
GO TO Web Activity 5.1 • The Scale of Life

are, relatively speaking, enormous, and individual cells of several types of algae and bacteria are large enough to be viewed with the unaided eye (**Figure 5.1**). And although neurons (nerve cells) have volumes that are within the "usual" range, they often have fine projections that may extend for meters, carrying signals from one part of a large animal to another. So there is enormous diversity among cells in their dimensions and volumes, but cells are usually very small.

Small cell size is a practical necessity arising from the change in the **surface area-to-volume ratio** of any object as it increases in size. As an object increases in volume, its surface area also increases, but not at the same rate (**Figure 5.2**). This phenomenon has great biological significance for two reasons:

- The volume of a cell determines the amount of chemical activity it carries out per unit of time.

- The surface area of a cell determines the amount of substances that can enter it from the outside environment, and the amount of waste products that can exit to the environment.

5.2 Why Cells Are Small Whether it is cuboid (A) or spheroid (B), as an object grows larger its volume increases more rapidly than its surface area. Cells must maintain a large surface area-to-volume ratio in order to function. This fact explains why large organisms must be composed of many small cells rather than a few huge ones.

Larger surface area compared to volume.

Smaller surface area compared to volume.

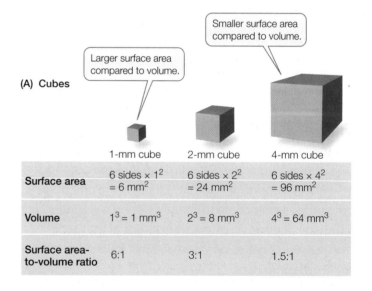

(A) Cubes

	1-mm cube	2-mm cube	4-mm cube
Surface area	6 sides × 1^2 = 6 mm^2	6 sides × 2^2 = 24 mm^2	6 sides × 4^2 = 96 mm^2
Volume	1^3 = 1 mm^3	2^3 = 8 mm^3	4^3 = 64 mm^3
Surface area-to-volume ratio	6:1	3:1	1.5:1

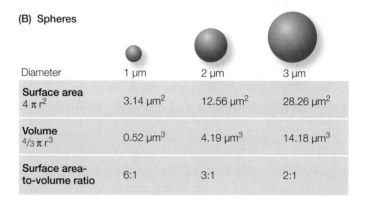

(B) Spheres

Diameter	1 μm	2 μm	3 μm
Surface area $4 \pi r^2$	3.14 μm^2	12.56 μm^2	28.26 μm^2
Volume $^4/_3 \pi r^3$	0.52 μm^3	4.19 μm^3	14.18 μm^3
Surface area-to-volume ratio	6:1	3:1	2:1

As a living cell grows larger, its chemical activity, and thus its need for resources and its rate of waste production, increases faster than its surface area. (The surface area increases in proportion to the square of the radius, while the volume increases much more—in proportion to the cube of the radius.) In addition, substances must move from one site to another within the cell; the smaller the cell, the more easily this is accomplished. This explains why large organisms must consist of many small cells: cells must be small in volume in order to maintain a large enough surface area-to-volume ratio and an ideal internal volume. The large surface area represented by the many small cells of a multicellular organism enables it to carry out the many different functions required for survival.

Microscopes reveal the features of cells

Microscopes do two different things to allow cells and details within them to be seen by the human eye. First, they increase the apparent size of the object: this is called *magnification*. But just increasing the magnification does not necessarily mean that the object will be seen clearly. In addition to being larger, a magnified object must be sharp, or clear. This is a property called *resolution*. Formally defined, resolution is the minimum distance two objects can be apart and still be seen as two objects. Resolution for the human eye is about 0.2 mm (200 μm). Most cells are much smaller than 200 μm, and thus are invisible to the human eye. Microscopes magnify and increase resolution so that cells and their internal structures can be seen clearly (**Figure 5.3**).

There are two basic types of microscopes—*light microscopes* and *electron microscopes*—that use different forms of radiation (see Figure 5.3). While the resolution is better in electron microscopy, we should emphasize that because cells are prepared in a vacuum, only dead, dehydrated cells are visualized. Therefore, the preparation of cells for electron microscopy may alter them, and this must be taken into consideration when interpreting the images produced. On the other hand, light microscopes can be used to visualize living cells (for example, by phase-contrast microscopy; see Figure 5.3).

Before we delve into the details of cell structure, it is useful to consider the many uses of microscopy. An entire branch of medicine, *pathology*, makes use of many different methods of microscopy to aid in the analysis of cells and the diagnosis of diseases. For instance, a surgeon might remove from a body some tissue suspected of being cancerous. The pathologist might:

- examine the tissue quickly by phase-contrast microscopy or interference-contrast microscopy to determine the size, shape, and spread of the cells

- stain the tissue with a general dye and examine it by bright-field microscopy to bring out features such as the shape of the nucleus, or cell division characteristics

- stain the tissue with a fluorescent dye and examine it by fluorescence microscopy or confocal microscopy for the presence of specific proteins that are diagnostic of a particular cancer

- examine the tissue under the electron microscope to observe its most minute internal structures, such as the shapes

of the mitochondria and the chromatin. (These structures are described in Section 5.3.)

The plasma membrane forms the outer surface of every cell

While the structural diversity of cells can often be observed using light microscopy, the **plasma membrane** is best observed with an electron microscope. This very thin structure forms the outer surface of every cell, and it has more or less the same thickness and molecular structure in all cells. Biochemical methods have shown that membranes have great functional diversity. These methods have revealed that the thin, almost invisible plasma membrane is actively involved in many cellular functions—it is not a static structure. The plasma membrane separates the interior of the cell from its outside environment, creating a segregated (but not isolated) compartment. The presence of this outer limiting membrane is a feature of all cells. What is the composition and molecular architecture of this amazing structure?

The plasma membrane is composed of a *phospholipid bilayer* (or simply *lipid bilayer*), with the hydrophilic "heads" of the lipids facing the cell's aqueous interior on one side of the membrane and the extracellular environment on the other (see Figure 3.20). Proteins and other molecules are embedded in the lipids. The membrane is not a rigid, static structure. Rather, it is an oily fluid, in which the proteins and lipids are in constant motion. This allows the membrane to move and change the shape of the cell. A detailed description of the structure and functions of the plasma membrane is given in Chapter 6. Here is a brief summary:

- The plasma membrane acts as a *selectively permeable barrier*, preventing some substances from crossing it while permitting other substances to enter and leave the cell. For example, macromolecules such as DNA and proteins cannot normally cross the plasma membrane, but some smaller molecules such as oxygen can. In addition to size, other factors such as polarity determine a molecule's ability to cross the plasma membrane: because the membrane is composed mostly of hydrophobic fatty acids, nonpolar molecules cross it more easily than polar or charged molecules.

- The plasma membrane allows the cell to maintain a more or less *constant internal environment*. A self-maintaining, constant internal environment (known as *homeostasis*) is a key characteristic of life that will be discussed in detail in Chapter 40. One way that the membrane does this is by actively regulating the transport of substances across it. This dynamic process is distinct from the more passive process of diffusion, which is dependent on the size of a molecule.

- As the cell's boundary with the outside environment, the plasma membrane is important in *communicating* with adjacent cells and receiving signals from the environment. We will describe this function in Chapter 7.

- The plasma membrane often has proteins protruding from it that are responsible for *binding and adhering* to adjacent

TOOLS FOR INVESTIGATING LIFE

The six images on this page show some techniques used in light microscopy. The three images on the following page were created using electron microscopes. All of these images are of a particular type of cultured cell known as HeLa cells. Note that the images in most cases are flat, two-dimensional views. As you look at images of cells, keep in mind that they are three-dimensional structures.

In a *light microscope*, glass lenses and visible light are used to form an image. The resolution is about 0.2 µm, which is 1,000 times greater than that of the human eye. Light microscopy allows visualization of cell sizes and shapes and some internal cell structures. Internal structures are hard to see under visible light, so cells are often chemically treated and stained with various dyes to make certain structures stand out by increasing contrast.

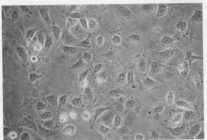

140 µm

In **bright-field microscopy**, light passes directly through these human cells. Unless natural pigments are present, there is little contrast and details are not distinguished.

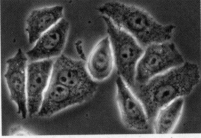

30 µm

In **phase-contrast microscopy**, contrast in the image is increased by emphasizing differences in refractive index (the capacity to bend light), thereby enhancing light and dark regions in the cell.

30 µm

Differential interference-contrast microscopy uses two beams of polarized light. The combined images look as if the cell is casting a shadow on one side.

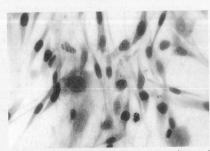

30 µm

In **stained bright-field microscopy**, a stain enhances contrast and reveals details not otherwise visible. Stains differ greatly in their chemistry and their capacity to bind to cell materials, so many choices are available.

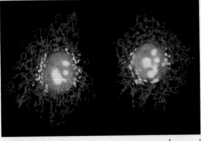

20 µm

In **fluorescence microscopy**, a natural substance in the cell or a fluorescent dye that binds to a specific cell material is stimulated by a beam of light, and the longer-wavelength fluorescent light is observed coming directly from the dye.

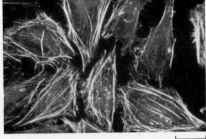

20 µm

Confocal microscopy uses fluorescent materials but adds a system of focusing both the stimulating and emitted light so that a single plane through the cell is seen. The result is a sharper two-dimensional image than with standard fluorescence microscopy.

cells. Thus the plasma membrane plays an important structural role and contributes to cell shape.

All cells are classified as either prokaryotic or eukaryotic

As we learned in Section 1.2, biologists classify all living things into three domains: Archaea, Bacteria, and Eukarya. The organisms in Archaea and Bacteria are collectively called **prokaryotes** because they have in common a prokaryotic cell organization. A prokaryotic cell does not typically have membrane-enclosed internal compartments; in particular, it does not have a nucleus. The first cells were probably similar in organization to those of modern prokaryotes.

TOOLS FOR INVESTIGATING LIFE

5.3 Looking at Cells (*continued*)

In an *electron microscope*, electromagnets are used to focus an electron beam, much as a light microscope uses glass lenses to focus a beam of light. Since we cannot see electrons, the electron microscope directs them through a vacuum at a fluorescent screen or photographic film to create a visible image. The resolution of electron microscopes is about 2 nm, which is about 100,000 times greater than that of the human eye. This resolution permits the details of many subcellular structures to be distinguished.

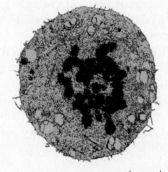

10 µm

In **transmission electron microscopy** (TEM), a beam of electrons is focused on the object by magnets. Objects appear darker if they absorb the electrons. If the electrons pass through they are detected on a fluorescent screen.

20 µm

Scanning electron microscopy (SEM) directs electrons to the surface of the sample, where they cause other electrons to be emitted. These electrons are viewed on a screen. The three-dimensional surface of the object can be visualized.

0.1 µm

In **freeze-fracture microscopy**, cells are frozen and then a knife is used to crack them open. The crack often passes through the interior of plasma and internal membranes. The "bumps" that appear are usually large proteins or aggregates embedded in the interior of the membrane.

yourBioPortal.com

GO TO Web Activity 5.2 • Know Your Techniques

Eukaryotic cell organization, on the other hand, is found in members of the domain Eukarya (**eukaryotes**), which includes the protists, plants, fungi, and animals. As we will discuss later in this chapter, eukaryotic cells probably evolved from prokaryotes. In contrast to the prokaryotes, the genetic material (DNA) of eukaryotic cells is contained in a special membrane-enclosed compartment called the **nucleus**. Eukaryotic cells also contain other membrane-enclosed compartments in which specific chemical reactions occur. For example, some of the key reactions that generate usable chemical energy for cells take place in mitochondria. The internal membranes that enclose these compartments have the same basic composition, structure and properties as the plasma membrane. The efficiency afforded by these compartments has led to the impressive functions that can occur in eukaryotic cells, and their specialization into tissues as diverse as the parts of a flower, muscles, and nerves.

5.1 RECAP

The cell theory is a unifying principle of biology. Surface area-to-volume ratios limit the sizes of cells. Both prokaryotic and eukaryotic cells are enclosed within a plasma membrane, but prokaryotic cells lack the membrane-enclosed internal compartments found in eukaryotes.

- How does cell biology embody all the principles of life? See p. 77

- Why are cells small? See pp. 77–79 and Figure 5.2

- Explain the importance of the plasma membrane to cells. See pp. 79–80

As we mentioned in this section, there are two structural themes in cell architecture: prokaryotic and eukaryotic. We now turn to the organization of prokaryotic cells.

5.2 What Features Characterize Prokaryotic Cells?

Prokaryotes can derive energy from more diverse sources than any other living organisms. They can tolerate environmental extremes—such as very hot springs with temperatures up to 100°C (*Thermus aquaticus*) or very salty water (*Halobacterium*)—that would kill other organisms. As we examine prokaryotic cells in this section, bear in mind that there are vast numbers of prokaryotic species, and that the Bacteria and Archaea are distinguished in numerous ways. These differences, and the vast diversity of organisms in these two domains, will be the subject of Chapter 26.

The volume of a prokaryotic cell is generally about one fiftieth of the volume of a eukaryotic cell. Prokaryotic cells range from about 1 to 10 μm in length or diameter. Each individual prokaryote is a single cell, but many types of prokaryotes are usually seen in chains or small clusters, and some occur in large clusters containing hundreds of cells. In this section we will first consider the features shared by cells in the domains Bacteria and Archaea. Then we will examine structural features that are found in some, but not all, prokaryotes.

Prokaryotic cells share certain features

All prokaryotic cells have the same basic structure (**Figure 5.4**):

- The plasma membrane encloses the cell, regulating the traffic of materials into and out of the cell, and separating its interior from the external environment.

- The **nucleoid** is a region in the cell where the DNA is located. As we described in Section 4.1, DNA is the hereditary material that controls cell growth, maintenance, and reproduction.

The rest of the material enclosed in the plasma membrane is called the **cytoplasm**. The cytoplasm has two components: the cytosol and insoluble suspended particles, including ribosomes:

- The **cytosol** consists mostly of water that contains dissolved ions, small molecules, and soluble macromolecules such as proteins.

- **Ribosomes** are complexes of RNA and proteins that are about 25 nm in diameter. They can only be visualized with the electron microscope. They are the sites of protein synthesis, where information coded for in nucleic acids directs the sequential linking of amino acids to form proteins.

The cytoplasm is not a static region. Rather, the substances in this environment are in constant motion. For example, a typical protein moves around the entire cell within a minute, and it collides with many other molecules along the way.

Although they are structurally less complex than eukaryotic cells, prokaryotic cells are functionally complex, carrying out thousands of biochemical reactions. Based on our current knowledge about the origins of the first cells (see Section 4.4), some prokaryotic cell lineages must stretch back in time for more than 3 billion years. Thus, prokaryotes are very successful organisms from an evolutionary perspective.

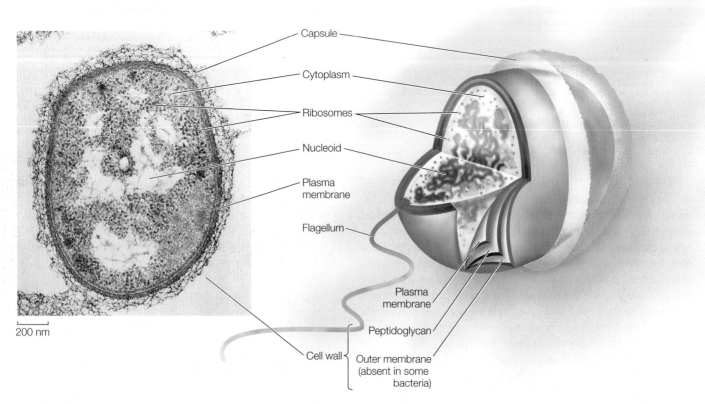

Capsule
Cytoplasm
Ribosomes
Nucleoid
Plasma membrane
Flagellum
Plasma membrane
Peptidoglycan
Cell wall
Outer membrane (absent in some bacteria)

200 nm

5.4 A Prokaryotic Cell   The bacterium *Pseudomonas aeruginosa* illustrates the typical structures shared by all prokaryotic cells. This bacterium also has a protective outer membrane that not all prokaryotes have. The flagellum and capsule are also structures found in some, but not all, prokaryotic cells.

Specialized features are found in some prokaryotes

As they evolved, some prokaryotes developed specialized structures that gave a selective advantage to those that had them: cells with these structures were better able to survive and reproduce in particular environments than cells lacking them. These structures include a protective cell wall, an internal membrane for compartmentalization of some chemical reactions, flagella for cell movement through the watery environment, and a rudimentary internal skeleton.

CELL WALLS Most prokaryotes have a cell wall located outside the plasma membrane. The rigidity of the cell wall supports the cell and determines its shape. The cell walls of most bacteria, but not archaea, contain peptidoglycan, a polymer of amino sugars that are cross-linked by covalent bonds to peptides, to form a single giant molecule around the entire cell. In some bacteria, another layer, the outer membrane (a polysaccharide-rich phospholipid membrane), encloses the peptidoglycan layer (see Figure 5.4). Unlike the plasma membrane, this outer membrane is not a major barrier to the movement of molecules across it.

Enclosing the cell wall in some bacteria is a slimy layer composed mostly of polysaccharides, and referred to as a capsule. In some cases these capsules protect the bacteria from attack by white blood cells in the animals they infect. Capsules also help to keep the cells from drying out, and sometimes they help bacteria attach to other cells. Many prokaryotes produce no capsule, and those that do have capsules can survive even if they lose them, so the capsule is not essential to prokaryotic life.

As you will see later in this chapter, eukaryotic plant cells also have a cell wall, but it differs in composition and structure from the cell walls of prokaryotes.

INTERNAL MEMBRANES Some groups of bacteria—including the cyanobacteria—carry out photosynthesis: they use energy from the sun to convert carbon dioxide and water into carbohydrates. These bacteria have an internal membrane system that contains molecules needed for photosynthesis. The development of photosynthesis, which requires membranes, was an important event in the early evolution of life on Earth. Other prokaryotes have internal membrane folds that are attached to the plasma membrane. These folds may function in cell division or in various energy-releasing reactions.

FLAGELLA AND PILI Some prokaryotes swim by using appendages called **flagella**, which sometimes look like tiny corkscrews (**Figure 5.5A**). In bacteria a single flagellum is made of a protein called flagellin. A complex motor protein spins the flagellum on its axis like a propeller, driving the cell along. The motor protein is anchored to the plasma membrane and, in some bacteria, to the outer membrane of the cell wall (**Figure 5.5B**). We know that the flagella cause the motion of cells because if they are removed, the cells do not move.

Pili are structures made of protein that project from the surfaces of some types of bacterial cells. These hairlike structures are shorter than flagella, and are used for adherence. The sex-pili help bacteria join to one another to exchange genetic material. The *fimbriae* are similar to pili but shorter, and help cells to adhere to surfaces such as animal cells, for food and protection.

(A)

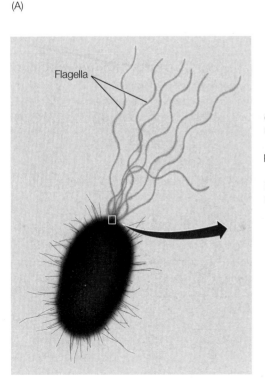

(B)

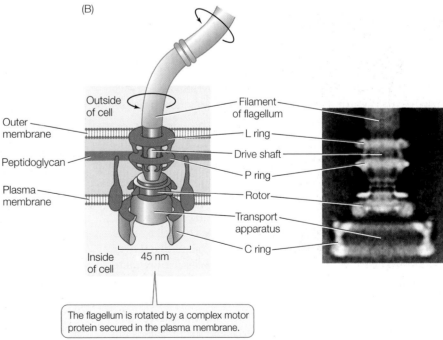

The flagellum is rotated by a complex motor protein secured in the plasma membrane.

5.5 Prokaryotic Flagella (A) Flagella contribute to the movement and adhesion of prokaryotic cells. (B) Complex protein ring structures anchored in the plasma membrane form a motor unit that rotates the flagellum and propels the cell.

CYTOSKELETON Some prokaryotes, especially rod-shaped bacteria, have a helical network of filamentous structures that extend down the length of the cell just inside the plasma membrane. The proteins that make up this structure are similar in amino acid sequence to actin in eukaryotic cells. Since actin is part of the cytoskeleton in eukaryotes (see Section 5.3), it has been suggested that the helical filaments in prokaryotes play a role in maintaining the rod-like cell shape.

5.2 RECAP

Prokaryotic organisms can live on diverse energy sources and in extreme environments. Unlike eukaryotic cells, prokaryotic cells do not have extensive internal compartments.

- What structures are present in all prokaryotic cells? See p. 82 and Figure 5.4

- Describe the structure and function of a specialized prokaryotic cell feature, such as the cell wall, capsule, flagellum, or pilus. See pp. 83–84 and Figure 5.5

As we mentioned earlier, the prokaryotic cell is one of two types of cell structure recognized in cell biology. The other is the eukaryotic cell. Eukaryotic cells, and multicellular eukaryotic organisms, are more structurally and functionally complex than prokaryotic cells.

5.3 What Features Characterize Eukaryotic Cells?

Eukaryotic cells generally have dimensions up to 10 times greater than those of prokaryotes; for example, the spherical yeast cell has a diameter of about 8 μm, in contrast to a typical bacterium with a diameter of 1 μm. Like prokaryotic cells, eukaryotic cells have a plasma membrane, cytoplasm, and ribosomes. But as you learned earlier in this chapter, eukaryotic cells also have compartments within the cytoplasm whose interiors are separated from the cytosol by membranes.

Compartmentalization is the key to eukaryotic cell function

The membranous compartments of eukaryotic cells are called **organelles**. Each type of organelle has a specific role in its particular cell. Some of the organelles have been characterized as factories that make specific products. Others are like power plants that take in energy in one form and convert it into a more useful form. These functional roles are defined by the chemical reactions each organelle can carry out:

- The *nucleus* contains most of the cell's genetic material (DNA). The replication of the genetic material and the first steps in expressing genetic information take place in the nucleus.

- The *mitochondrion* is a power plant and industrial park, where energy stored in the bonds of carbohydrates and

fatty acids is converted into a form that is more useful to the cell (ATP; see Section 9.1).

- The *endoplasmic reticulum* and *Golgi apparatus* are compartments in which some proteins synthesized by the ribosomes are packaged and sent to appropriate locations in the cell.

- *Lysosomes* and vacuoles are cellular digestive systems in which large molecules are hydrolyzed into usable monomers.

- *Chloroplasts* (found in only some cells) perform photosynthesis.

The membrane surrounding each organelle has two essential roles. First, it keeps the organelle's molecules away from other molecules in the cell, to prevent inappropriate reactions. Second, it acts as a traffic regulator, letting important raw materials into the organelle and releasing its products to the cytoplasm. In some organelles, the membrane also has proteins that have functional roles in chemical reactions that occur at the organelle surface.

There are a number of other structures in eukaryotic cells that have specialized functions, but are not generally called organelles because they lack membranes:

- Ribosomes, where protein synthesis takes place

- The cytoskeleton, composed of several types of protein-based filaments, which has both structural and functional roles

- The extracellular matrix, which also has structural and functional roles

The evolution of compartments was an important development that enabled eukaryotic cells to specialize, forming the organs and tissues of complex multicellular organisms.

Organelles can be studied by microscopy or isolated for chemical analysis

Cell organelles and structures were first detected by light and then by electron microscopy. The functions of the organelles could sometimes be inferred by observations and experiments, leading, for example, to the hypothesis (later confirmed) that the nucleus contained the genetic material. Later, the use of stains targeted to specific macromolecules allowed cell biologists to determine the chemical compositions of organelles (see Figure 5.17, which shows a single cell stained for three different proteins).

Another way to analyze cells is to take them apart in a process called cell fractionation. This process permits cell organelles and other cytoplasmic structures to be separated from each other and examined using chemical methods. Cell fractionation begins with the destruction of the plasma membrane, which allows the cytoplasmic components to flow out into a test tube. The various organelles can then be separated from one another on the basis of size or density (**Figure 5.6**). Biochemical analyses can then be done on the isolated organelles.

Microscopy and cell fractionation have complemented each other, giving us a more complete picture of the composition and function of each organelle and structure.

TOOLS FOR INVESTIGATING LIFE

5.6 Cell Fractionation

Organelles can be separated from one another after cells are broken open and their contents suspended in an aqueous medium. The medium is placed in a tube and spun in a centrifuge, which rotates about an axis at high speed. Centrifugal forces cause particles to sediment at the bottom of the tube where they may be collected for biochemical study. Heavier particles sediment at lower speeds than do lighter particles. By adjusting the speed of centrifugation, cellular organelles and even large particles like ribosomes can be separated and partially purified.

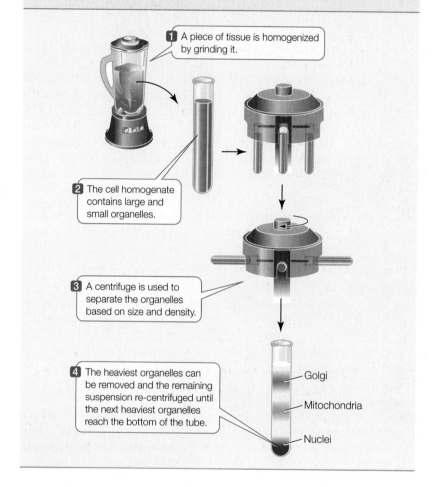

1 A piece of tissue is homogenized by grinding it.

2 The cell homogenate contains large and small organelles.

3 A centrifuge is used to separate the organelles based on size and density.

4 The heaviest organelles can be removed and the remaining suspension re-centrifuged until the next heaviest organelles reach the bottom of the tube.

Golgi

Mitochondria

Nuclei

Microscopy of plant and animal cells has revealed that many of the organelles are similar in appearance in each cell type (**Figure 5.7**). By comparing the illustrations in Figure 5.7 and Figure 5.4 you can see some of the prominent differences between eukaryotic cells and prokaryotic cells.

Ribosomes are factories for protein synthesis

The ribosomes of prokaryotes and eukaryotes are similar in that both types consist of two different-sized subunits. Eukaryotic ribosomes are somewhat larger than those of prokaryotes, but the structure of prokaryotic ribosomes is better understood. Chemically, ribosomes consist of a special type of RNA called ribosomal RNA (rRNA). Ribosomes also contain more than 50

different protein molecules, which are noncovalently bound to the rRNA.

In prokaryotic cells, ribosomes float freely in the cytoplasm. In eukaryotic cells they are found in two places: in the cytoplasm, where they may be free or attached to the surface of the endoplasmic reticulum (a membrane-bound organelle, see below), and inside mitochondria and chloroplasts. In each of these locations, the ribosomes are molecular factories where proteins are synthesized with their amino acid sequences specified by nucleic acids. Although they seem small in comparison to the cells that contain them, by molecular standards ribosomes are huge complexes (about 25 nm in diameter), made up of several dozen different molecules.

The nucleus contains most of the genetic information

Organisms depend on accurate information—internal signals, environmental cues, and stored instructions—in order to respond appropriately to changing conditions, to maintain a constant internal environment, and to reproduce. In the cell, hereditary information is stored in the sequence of nucleotides in DNA molecules. Most of the DNA in eukaryotic cells resides in the nucleus (see Figure 5.7). Information encoded in the DNA is *translated* into proteins at the ribosomes. This process is described in detail in Chapter 14.

Most cells have a single nucleus, which is usually the largest organelle (**Figure 5.8**). The nucleus of a typical animal cell is approximately 5 μm in diameter—substantially larger than most prokaryotic cells. The nucleus has several functions in the cell:

- It is the location of the DNA and the site of DNA replication.

- It is the site where gene transcription is turned on or off.

- A region within the nucleus, the **nucleolus**, is where ribosomes begin to be assembled from RNA and proteins.

The nucleus is surrounded by two membranes, which together form the *nuclear envelope* (see Figure 5.8). This structure separates the genetic material from the cytoplasm. Functionally, it separates DNA transcription (which occurs in the nucleus) from translation (in the cytoplasm) (see Figure 4.5). The two membranes of the nuclear envelope are perforated by thousands of nuclear pores, each measuring approximately 9 nm in diameter, which connect the interior of the nucleus with the cytoplasm (see Figure 5.8). The pores regulate the traffic between these two cellular compartments by allowing some molecules to enter the nucleus and blocking others. This allows the nucleus to regulate the information-processing functions.

At the nuclear pore, small substances, including ions and other molecules with molecular weights of less than 10,000 dal-

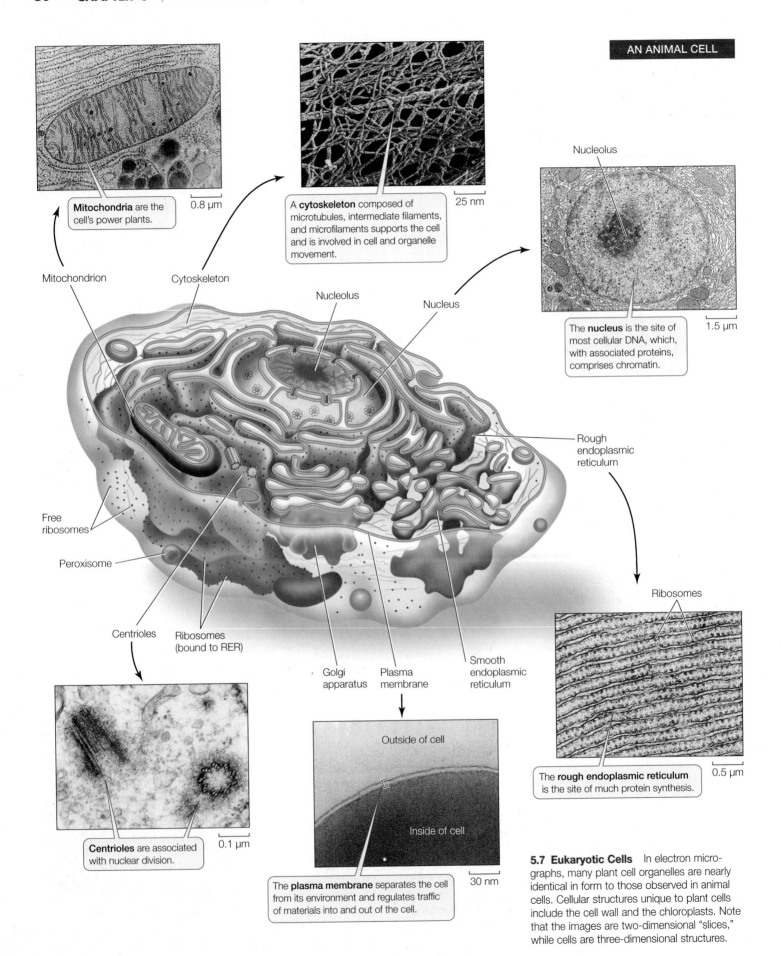

AN ANIMAL CELL

Mitochondria are the cell's power plants.

0.8 μm

A **cytoskeleton** composed of microtubules, intermediate filaments, and microfilaments supports the cell and is involved in cell and organelle movement.

25 nm

Nucleolus

The **nucleus** is the site of most cellular DNA, which, with associated proteins, comprises chromatin.

1.5 μm

Mitochondrion

Cytoskeleton

Nucleolus

Nucleus

Rough endoplasmic reticulum

Free ribosomes

Peroxisome

Centrioles

Ribosomes (bound to RER)

Golgi apparatus

Plasma membrane

Smooth endoplasmic reticulum

Ribosomes

Centrioles are associated with nuclear division.

0.1 μm

Outside of cell

Inside of cell

The **plasma membrane** separates the cell from its environment and regulates traffic of materials into and out of the cell.

30 nm

The **rough endoplasmic reticulum** is the site of much protein synthesis.

0.5 μm

5.7 Eukaryotic Cells In electron micrographs, many plant cell organelles are nearly identical in form to those observed in animal cells. Cellular structures unique to plant cells include the cell wall and the chloroplasts. Note that the images are two-dimensional "slices," while cells are three-dimensional structures.

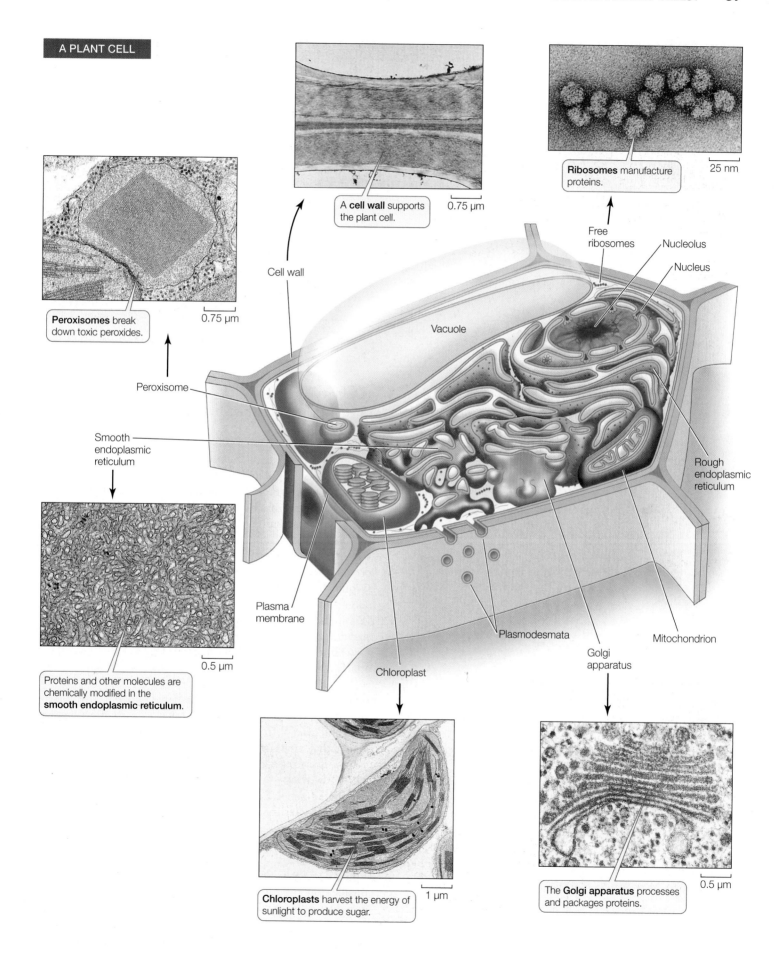

A PLANT CELL

A **cell wall** supports the plant cell.

0.75 μm

Ribosomes manufacture proteins.

25 nm

Peroxisomes break down toxic peroxides.

0.75 μm

Proteins and other molecules are chemically modified in the **smooth endoplasmic reticulum**.

0.5 μm

Chloroplasts harvest the energy of sunlight to produce sugar.

1 μm

The **Golgi apparatus** processes and packages proteins.

0.5 μm

Cell wall

Vacuole

Free ribosomes

Nucleolus

Nucleus

Peroxisome

Smooth endoplasmic reticulum

Rough endoplasmic reticulum

Plasma membrane

Plasmodesmata

Mitochondrion

Chloroplast

Golgi apparatus

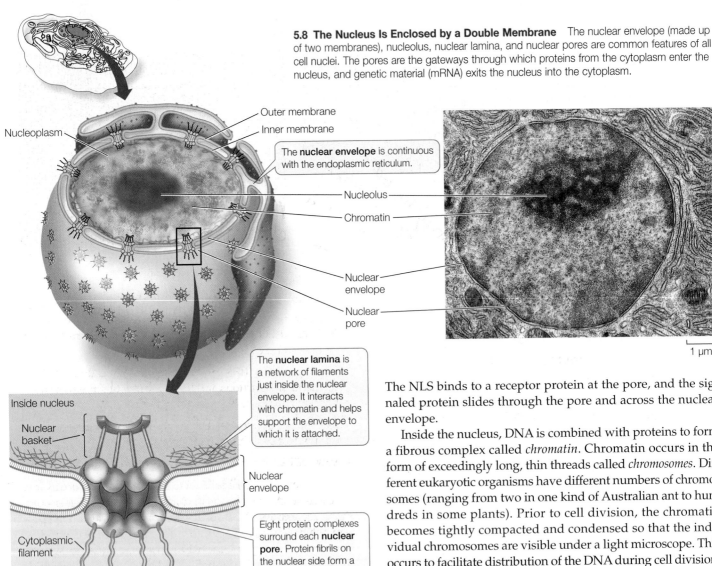

5.8 The Nucleus Is Enclosed by a Double Membrane The nuclear envelope (made up of two membranes), nucleolus, nuclear lamina, and nuclear pores are common features of all cell nuclei. The pores are the gateways through which proteins from the cytoplasm enter the nucleus, and genetic material (mRNA) exits the nucleus into the cytoplasm.

Nucleoplasm

Outer membrane

Inner membrane

The **nuclear envelope** is continuous with the endoplasmic reticulum.

Nucleolus

Chromatin

Nuclear envelope

Nuclear pore

The **nuclear lamina** is a network of filaments just inside the nuclear envelope. It interacts with chromatin and helps support the envelope to which it is attached.

Inside nucleus

Nuclear basket

Nuclear envelope

Cytoplasmic filament

Eight protein complexes surround each **nuclear pore**. Protein fibrils on the nuclear side form a basketlike structure.

Outside nucleus (cytoplasm)

1 µm

The NLS binds to a receptor protein at the pore, and the signaled protein slides through the pore and across the nuclear envelope.

Inside the nucleus, DNA is combined with proteins to form a fibrous complex called *chromatin*. Chromatin occurs in the form of exceedingly long, thin threads called *chromosomes*. Different eukaryotic organisms have different numbers of chromosomes (ranging from two in one kind of Australian ant to hundreds in some plants). Prior to cell division, the chromatin becomes tightly compacted and condensed so that the individual chromosomes are visible under a light microscope. This occurs to facilitate distribution of the DNA during cell division. (**Figure 5.9**). Surrounding the chromatin are water and dissolved substances collectively referred to as the *nucleoplasm*. Within the nucleoplasm, a network of structural proteins called the *nuclear matrix* helps organize the chromatin.

At the interior periphery of the nucleus, the chromatin is attached to a protein meshwork, called the *nuclear lamina*, which is formed by the polymerization of proteins called lamins into long thin structures called intermediate filaments. The nuclear lamina maintains the shape of the nucleus by its attachment to both the chromatin and the nuclear envelope. There is some evidence that the nuclear lamina may be involved with human aging. As people age, the nuclear lamina begins to disintegrate and in the process the structural integrity of the nucleus declines. In people with the rare disease called progeria, this decline begins very early in life and their aging is accelerated.

During most of a cell's life cycle, the nuclear envelope is a stable structure. When the cell reproduces, however, the nuclear envelope breaks down into small, membrane-bound droplets, called *vesicles*, containing pore complexes. The envelope reforms after the replicated DNA has been distributed to the daughter cells (see Section 11.3).

At certain sites, the outer membrane of the nuclear envelope folds outward into the cytoplasm and is continuous with the

tons, freely diffuse through the pore. Larger molecules, such as many proteins that are made in the cytoplasm and imported into the nucleus, cannot get through without a certain short sequence of amino acids that is part of the protein. We know that this sequence is the *nuclear localization signal* (NLS) from several lines of evidence (see also Figure 14.20):

- The NLS occurs in most proteins targeted to the nucleus, but not in proteins that remain in the cytoplasm.

- If the NLS is removed from a protein, the protein stays in the cytoplasm.

- If the NLS is added to a protein that normally stays in the cytoplasm, that protein moves into the nucleus.

- Some viruses have an NLS that allows them to enter the nucleus; viruses without the signal sequence do not enter the nucleus as virus particles.

5.9 Chromatin and Chromosomes
(A) When a cell is not dividing, the nuclear DNA is aggregated with proteins to form chromatin, which is dispersed throughout the nucleus. This two-dimensional image was made using a transmission electron microscope. (B) The chromosomes in dividing cells become highly condensed. This three-dimensional image of isolated metaphase chromosomes was produced by a scanning electron microscope.

(A)

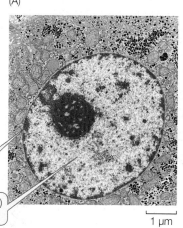

Dense chromatin (dark) near the nuclear envelope is attached to the nuclear lamina.

Diffuse chromatin (light) is in the nucleoplasm.

1 µm

(B)

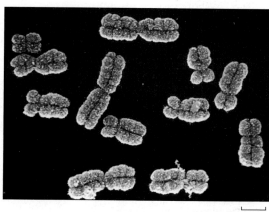

1.4 µm

membrane of another organelle, the endoplasmic reticulum, which we will discuss next.

The endomembrane system is a group of interrelated organelles

Much of the volume of some eukaryotic cells is taken up by an extensive **endomembrane system**. This is an interconnected system of membrane-enclosed compartments that are sometimes flattened into sheets and sometimes have other characteristic shapes (see Figure 5.7). The endomembrane system includes the plasma membrane, nuclear envelope, endoplasmic reticulum, Golgi apparatus, and lysosomes, which are derived from the Golgi. Tiny, membrane-surrounded droplets called vesicles shuttle substances between the various components of the endomembrane system (**Figure 5.10**). In drawings and electron microscope pictures this system appears static, fixed in space and time. But these depictions are just snapshots; in the living cell, membranes and the materials they contain are in constant motion. Membrane components have been observed to shift from one organelle to another within the endomembrane system. Thus, all these membranes must be functionally related.

ENDOPLASMIC RETICULUM Electron micrographs of eukaryotic cells reveal networks of interconnected membranes branching throughout the cytoplasm, forming tubes and flattened sacs. These membranes are collectively called the **endoplasmic reticulum**, or **ER**. The interior compartment of the ER, referred to as the lumen, is separate and distinct from the surrounding cytoplasm (see Figure 5.10). The ER can enclose up to 10 percent of the interior volume of the cell, and its foldings result in a surface area many times greater than that of the plasma membrane. There are two types of endoplasmic reticulum, the so-called rough and smooth.

Rough endoplasmic reticulum (RER) is called "rough" because of the many ribosomes attached to the outer surface of the membrane, giving it a "rough" appearance in electron microscopy (see Figure 5.7). The attached ribosomes are actively involved in protein synthesis, but that is not the entire story:

- The RER receives into its lumen certain newly synthesized proteins, segregating them away from the cytoplasm. The RER also participates in transporting these proteins to other locations in the cell.

- While inside the RER, proteins can be chemically modified to alter their functions and to chemically 'tag' them for delivery to specific cellular destinations.

- Proteins are shipped to cellular destinations enclosed within vesicles that pinch off from the ER.

- Most membrane-bound proteins are made in the RER.

A protein enters the lumen of the RER through a pore as it is synthesized. As with a protein passing through a nuclear pore, this is accomplished via a sequence of amino acids on the protein, which acts as a RER localization signal (see Section 14.6). Once in the lumen of the RER, proteins undergo several changes, including the formation of disulfide bridges and folding into their tertiary structures (see Figure 3.7).

Some proteins are covalently linked to carbohydrate groups in the RER, thus becoming glycoproteins. In the case of proteins directed to the lysosomes, the carbohydrate groups are part of an "addressing" system that ensures that the right proteins are directed to those organelles. This addressing system is very important because the enzymes within the lysosomes are some of the most destructive the cell makes. Were they not properly addressed and contained, they could destroy the cell.

The **smooth endoplasmic reticulum (SER)** lacks ribosomes and is more tubular (and less like flattened sacs) than the RER, but it shows continuity with portions of the RER (see Figure 5.10). Within the lumen of the SER, some proteins that have been synthesized on the RER are chemically modified. In addition, the SER has three other important roles:

- It is responsible for the chemical modification of small molecules taken in by the cell, including drugs and pesticides. These modifications make the targeted molecules more polar, so they are more water-soluble and more easily removed.

- It is the site for glycogen degradation in animal cells. We discuss this important process in Chapter 9.

- It is the site for the synthesis of lipids and steroids.

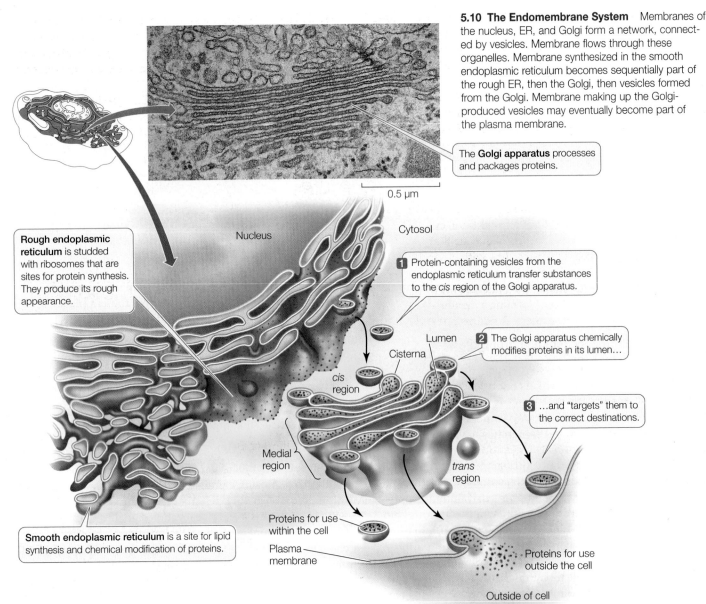

5.10 The Endomembrane System Membranes of the nucleus, ER, and Golgi form a network, connected by vesicles. Membrane flows through these organelles. Membrane synthesized in the smooth endoplasmic reticulum becomes sequentially part of the rough ER, then the Golgi, then vesicles formed from the Golgi. Membrane making up the Golgi-produced vesicles may eventually become part of the plasma membrane.

The **Golgi apparatus** processes and packages proteins.

0.5 μm

Nucleus

Cytosol

Rough endoplasmic reticulum is studded with ribosomes that are sites for protein synthesis. They produce its rough appearance.

1 Protein-containing vesicles from the endoplasmic reticulum transfer substances to the *cis* region of the Golgi apparatus.

Lumen

Cisterna

2 The Golgi apparatus chemically modifies proteins in its lumen…

cis region

3 …and "targets" them to the correct destinations.

Medial region

trans region

Smooth endoplasmic reticulum is a site for lipid synthesis and chemical modification of proteins.

Proteins for use within the cell

Plasma membrane

Proteins for use outside the cell

Outside of cell

Cells that synthesize a lot of protein for export are usually packed with RER. Examples include glandular cells that secrete digestive enzymes and white blood cells that secrete antibodies. In contrast, cells that carry out less protein synthesis (such as storage cells) contain less RER. Liver cells, which modify molecules (including toxins) that enter the body from the digestive system, have abundant SER.

GOLGI APPARATUS The **Golgi apparatus** (or Golgi complex), more often referred to merely as the Golgi, is another part of the diverse, dynamic, and extensive endomembrane system (see Figure 5.10). The exact appearance of the Golgi apparatus (named for its discoverer, Camillo Golgi) varies from species to species, but it almost always consists of two components: flattened membranous sacs called *cisternae* (singular *cisterna*) that are piled up like saucers, and small membrane-enclosed vesicles. The entire apparatus is about 1 μm long.

──── **yourBioPortal.com** ────

GO TO Animated Tutorial 5.1 • The Golgi Apparatus

The Golgi has several roles:

• When protein-containing vesicles from the RER fuse with the Golgi membranes, the Golgi receives the proteins and may further modify them.

• It concentrates, packages, and sorts proteins before they are sent to their cellular or extracellular destinations.

• It adds some carbohydrates to proteins and modifies others that were attached to proteins in the ER.

• It is where some polysaccharides for the plant cell wall are synthesized.

While there is a characteristic form for all Golgi, there are also variations in its size and appearance in different cell types. In the cells of plants, protists, fungi, and many invertebrate animals, the stacks of cisternae are individual units scattered throughout the cytoplasm. In vertebrate cells, a few such stacks usually form a single, larger, more complex Golgi apparatus.

The cisternae of the Golgi apparatus appear to have three functionally distinct regions: the *cis* region lies nearest to the nucleus or a patch of RER, the *trans* region lies closest to the plasma membrane, and the *medial* region lies in between (see Figure 5.10). (The terms *cis*, *trans*, and *medial* derive from Latin words meaning, respectively, "on the same side," "on the opposite side," and "in the middle.") These three parts of the Golgi apparatus contain different enzymes and perform different functions.

The Golgi apparatus receives proteins from the ER, packages them, and sends them on their way. Since there is often no direct membrane continuity between the ER and Golgi apparatus, how does a protein get from one organelle to the other? The protein could simply leave the ER, travel across the cytoplasm, and enter the Golgi apparatus. But that would expose the protein to interactions with other molecules in the cytoplasm. On the other hand, segregation from the cytoplasm could be maintained if a piece of the ER could "bud off," forming a membranous vesicle that contains the protein—and that is exactly what happens.

Proteins make the passage from the ER to the Golgi apparatus safely enclosed in vesicles. Once it arrives, a vesicle fuses with the *cis* membrane of the Golgi apparatus, releasing its cargo into the lumen of the Golgi cisterna. Other vesicles may move between the cisternae, transporting proteins, and it appears that some proteins move from one cisterna to the next through tiny channels. Vesicles budding off from the trans region carry their contents away from the Golgi apparatus. These vesicles go to the plasma membrane, or to another organelle in the endomembrane system called the lysosome.

LYSOSOMES The **primary lysosomes** originate from the Golgi apparatus. They contain digestive enzymes, and they are the sites where macromolecules—proteins, polysaccharides, nucleic acids, and lipids—are hydrolyzed into their monomers (see Figure 3.4). Lysosomes are about 1 μm in diameter; they are surrounded by a single membrane and have a densely staining, featureless interior (**Figure 5.11**). There may be dozens of lysosomes in a cell, depending on its needs.

Lysosomes are sites for the breakdown of food, other cells, or foreign objects that are taken up by the cell. These materials get into the cell by a process called *phagocytosis* (*phago*, "eat"; *cytosis*, "cellular"). In this process, a pocket forms in the plasma membrane and then deepens and encloses material from outside

5.11 Lysosomes Isolate Digestive Enzymes from the Cytoplasm Lysosomes are sites for the hydrolysis of material taken into the cell by phagocytosis.

── **yourBioPortal**.com ──
GO TO Web Activity 5.3 • Lysosomal Digestion

the cell. The pocket becomes a small vesicle called a phagosome, containing food or other material, which breaks free of the plasma membrane to move into the cytoplasm. The phagosome fuses with a primary lysosome to form a **secondary lysosome**, in which digestion occurs.

The effect of this fusion is rather like releasing hungry foxes into a chicken coop: the enzymes in the secondary lysosome quickly hydrolyze the food particles. These reactions are en-

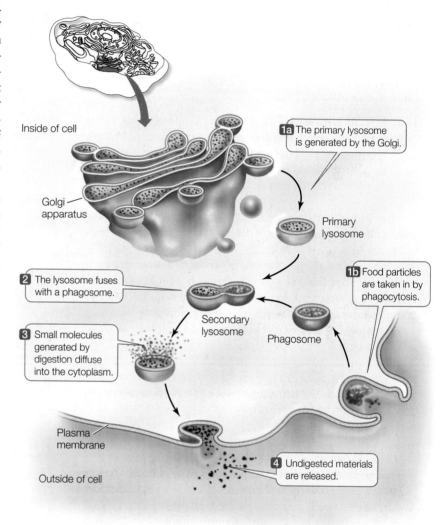

Inside of cell

Golgi apparatus

1a The primary lysosome is generated by the Golgi.

Primary lysosome

2 The lysosome fuses with a phagosome.

1b Food particles are taken in by phagocytosis.

Secondary lysosome

Phagosome

3 Small molecules generated by digestion diffuse into the cytoplasm.

Plasma membrane

Outside of cell

4 Undigested materials are released.

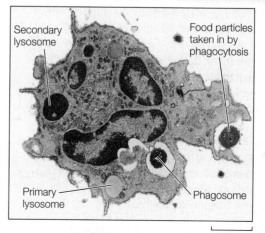

Secondary lysosome

Food particles taken in by phagocytosis

Primary lysosome

Phagosome

1 μm

hanced by the mild acidity of the lysosome's interior, where the pH is lower than in the surrounding cytoplasm. The products of digestion pass through the membrane of the lysosome, providing energy and raw materials for other cellular processes. The "used" secondary lysosome, now containing undigested particles, then moves to the plasma membrane, fuses with it, and releases the undigested contents to the environment.

Phagocytes are specialized cells that have an essential role in taking up and breaking down materials; they are found in nearly all animals and many protists. You will encounter them and their activities again at many places in this book, but at this point one example suffices: in the human liver and spleen, phagocytes digest approximately 10 billion aged or damaged blood cells each day! The digestion products are then used to make new cells to replace those that are digested.

Lysosomes are active even in cells that do not perform phagocytosis. Because cells are such dynamic systems, some cell components are frequently destroyed and replaced by new ones. The programmed destruction of cell components is called *autophagy*, and lysosomes are where the cell breaks down its own materials. With the proper signal, lysosomes can engulf entire organelles, hydrolyzing their constituents.

How important is autophagy? An entire class of human diseases called lysosomal storage diseases occur when lysosomes fail to digest internal components; these diseases are invariably very harmful or fatal. An example is Tay-Sachs disease, in which a particular lipid called a ganglioside is not broken down in lysosomes and instead accumulates in brain cells. In the most common form of this disease, a baby starts exhibiting neurological symptoms and becomes blind, deaf, and unable to swallow after six months of age. Death occurs before age 4.

Plant cells do not appear to contain lysosomes, but the central vacuole of a plant cell (which we will describe below) may function in an equivalent capacity because it, like lysosomes, contains many digestive enzymes.

Some organelles transform energy

All living things require external sources of energy. The energy from such sources must be transformed so that it can be used by cells. A cell requires energy to make the molecules it needs for activities such as growth, reproduction, responsiveness, and movement. Energy is transformed from one form to another in mitochondria (found in all eukaryotic cells) and in chloroplasts (found in eukaryotic cells that harvest energy from sunlight). In contrast, energy transformations in prokaryotic cells are associated with enzymes attached to the inner surface of the plasma membrane or to extensions of the plasma membrane that protrude into the cytoplasm.

MITOCHONDRIA In eukaryotic cells, the breakdown of fuel molecules such as glucose begins in the cytosol. The molecules that result from this partial degradation enter the **mitochondria** (singular *mitochondrion*), whose primary function is to convert the chemical energy of those fuel molecules into a form that the cell can use, namely the energy-rich molecule ATP (adenosine triphosphate) (see Section 8.2). The production of ATP in the mi-

tochondria, using fuel molecules and molecular oxygen (O_2), is called *cellular respiration.*

Typical mitochondria are somewhat less than 1.5 μm in diameter and 2–8 μm in length—about the size of many bacteria. They can divide independently of the central nucleus. The number of mitochondria per cell ranges from one gigantic organelle in some unicellular protists to a few hundred thousand in large egg cells. An average human liver cell contains more than a thousand mitochondria. Cells that are active in movement and growth require the most chemical energy, and these tend to have the most mitochondria per unit of volume.

Mitochondria have two membranes. The outer membrane is smooth and protective, and it offers little resistance to the movement of substances into and out of the organelle. Immediately inside the outer membrane is an inner membrane, which folds inward in many places, and thus has a surface area much greater than that of the outer membrane (**Figure 5.12**). The folds tend to be quite regular, giving rise to shelf-like structures called *cristae*.

5.12 A Mitochondrion Converts Energy from Fuel Molecules into ATP The electron micrograph is a two-dimensional slice through a three-dimensional organelle. As the drawing emphasizes, the cristae are extensions of the inner mitochondrial membrane.

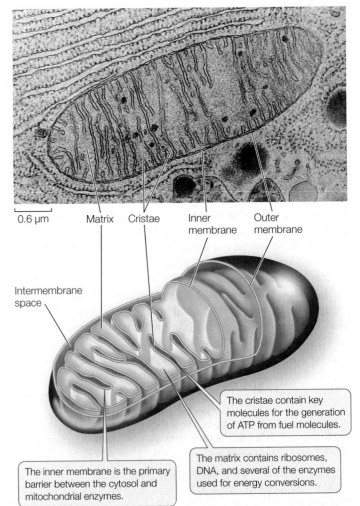

0.6 μm   Matrix   Cristae   Inner membrane   Outer membrane

Intermembrane space

The cristae contain key molecules for the generation of ATP from fuel molecules.

The inner membrane is the primary barrier between the cytosol and mitochondrial enzymes.

The matrix contains ribosomes, DNA, and several of the enzymes used for energy conversions.

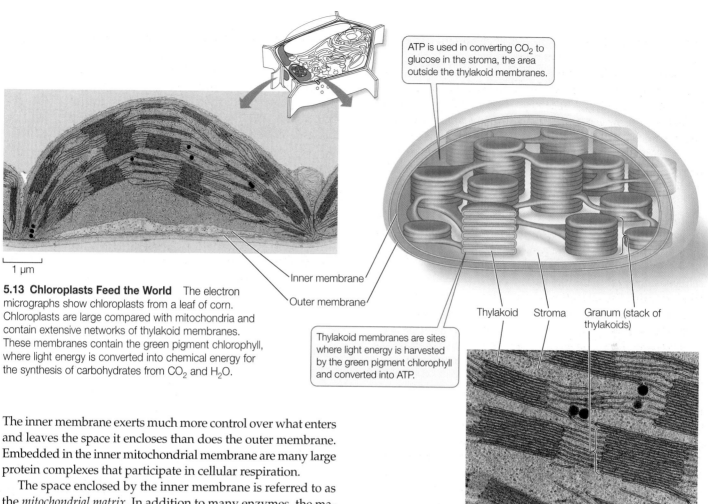

ATP is used in converting CO_2 to glucose in the stroma, the area outside the thylakoid membranes.

Inner membrane

Outer membrane

Thylakoid membranes are sites where light energy is harvested by the green pigment chlorophyll and converted into ATP.

Thylakoid   Stroma   Granum (stack of thylakoids)

1 μm

5.13 Chloroplasts Feed the World The electron micrographs show chloroplasts from a leaf of corn. Chloroplasts are large compared with mitochondria and contain extensive networks of thylakoid membranes. These membranes contain the green pigment chlorophyll, where light energy is converted into chemical energy for the synthesis of carbohydrates from CO_2 and H_2O.

0.5 μm

The inner membrane exerts much more control over what enters and leaves the space it encloses than does the outer membrane. Embedded in the inner mitochondrial membrane are many large protein complexes that participate in cellular respiration.

The space enclosed by the inner membrane is referred to as the *mitochondrial matrix*. In addition to many enzymes, the matrix contains ribosomes and DNA that are used to make some of the proteins needed for cellular respiration. As you will see later in this chapter, this DNA is the remnant of a much larger, complete chromosome of a prokaryote that may have been the mitochondrion's progenitor (see Figure 5.26). In Chapter 9 we discuss how the different parts of the mitochondrion work together in cellular respiration.

PLASTIDS One class of organelles—the plastids—is present only in the cells of plants and certain protists. Like mitochondria, plastids can divide autonomously. There are several types of plastids, with different functions.

Chloroplasts contain the green pigment chlorophyll and are the sites of photosynthesis (**Figure 5.13**). In photosynthesis, light energy is converted into the chemical energy of bonds between atoms. The molecules formed by photosynthesis provide food for the photosynthetic organism and for other organisms that eat it. Directly or indirectly, photosynthesis is the energy source for most of the living world.

Chloroplasts are variable in size and shape (**Figure 5.14**). Like a mitochondrion, a chloroplast is surrounded by two membranes. In addition, there is a series of internal membranes whose structure and arrangement vary from one group of photosynthetic organisms to another. Here we concentrate on the chloroplasts of the flowering plants.

The internal membranes of chloroplasts look like stacks of flat, hollow pita bread. Each stack, called a *granum* (plural

grana), consists of a series of flat, closely packed, circular compartments called **thylakoids** (see Figure 5.13). Thylakoid lipids are distinctive: only 10 percent are phospholipids, while the rest are galactose-substituted diglycerides and sulfolipids. Because of the abundance of chloroplasts, these are the most abundant lipids in the biosphere.

In addition to lipids and proteins, the membranes of the thylakoids contain chlorophyll and other pigments that harvest light energy for photosynthesis (we see how they do this in Section 10.2). The thylakoids of one granum may be connected to those of other grana, making the interior of the chloroplast a highly developed network of membranes, much like the ER.

The fluid in which the grana are suspended is called the *stroma*. Like the mitochondrial matrix, the chloroplast stroma contains ribosomes and DNA, which are used to synthesize some, but not all, of the proteins that make up the chloroplast.

Animal cells typically do not contain chloroplasts, but some do contain functional photosynthetic organisms. The green color of some corals and sea anemones comes from chloroplasts in algae that live within those animals (see Figure 5.14C). The animals derive some of their nutrition from the photosynthesis that their chloroplast-containing "guests" carry out. Such an intimate relationship between two different organisms is called *symbiosis*.

(A) Chloroplasts   Leaf cell

(B)

Algal cell

The chloroplasts in these filamentous green algae have assembled into spirals.

(C)

Chloroplast-filled green algae live in the tissues of this sea anemone.

50 µm

150 µm

5.14 Chloroplasts Are Everywhere (A) In green plants, chloroplasts are concentrated in the leaf cells. (B) Green algae are photosynthetic and filled with chloroplasts. (C) No animal species produces its own chloroplasts, but this sea anemone (an animal) is nourished by the chloroplasts of unicellular green algae living within its tissues, in what is termed a symbiotic relationship.

Other types of plastids such as *chromoplasts* and *leucoplasts* have functions different from those of chloroplasts (**Figure 5.15**). Chromoplasts make and store red, yellow, and orange pigments, especially in flowers and fruits. Leucoplasts are storage organelles that do not contain pigments. An amyloplast is a leucoplast that stores starch.

There are several other membrane-enclosed organelles

There are several other organelles whose boundary membranes separate their specialized chemical reactions and contents from the cytoplasm: peroxisomes, glyoxysomes, and vacuoles, including contractile vacuoles.

Peroxisomes are organelles that accumulate toxic peroxides, such as hydrogen peroxide (H_2O_2), that occur as byproducts of some biochemical reactions. These peroxides can be safely broken down inside the peroxisomes without mixing with other parts of the cell. Peroxisomes are small organelles, about 0.2 to 1.7 µm in diameter. They have a single membrane and a granular interior containing specialized enzymes. Peroxisomes are found in at least some of the cells of almost every eukaryotic species.

(A)

Chromoplast

5 µm

(B)

Leucoplast

Starch grains

5.15 Chromoplasts and Leucoplasts (A) Colorful pigments stored in the chromoplasts of flowers like this poppy may help attract pollinating insects. (B) Leucoplasts in the cells of a potato are filled with white starch grains.

1 µm

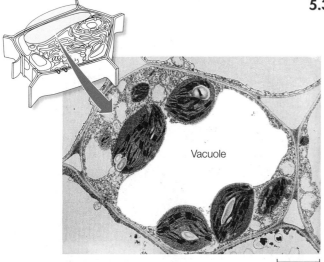

5.16 Vacuoles in Plant Cells Are Usually Large The large central vacuole in this cell is typical of mature plant cells. Smaller vacuoles are visible toward each end of the cell.

Glyoxysomes are similar to peroxisomes and are found only in plants. They are most abundant in young plants, and are the locations where stored lipids are converted into carbohydrates for transport to growing cells.

Vacuoles occur in many eukaryotic cells, but particularly those of plants and protists. Plant vacuoles (**Figure 5.16**) have several functions:

- *Storage*: Plant cells produce a number of toxic by-products and waste products, many of which are simply stored within vacuoles. Because they are poisonous or distasteful, these stored materials deter some animals from eating the plants, and may thus contribute to plant defenses and survival.

- *Structure*: In many plant cells, enormous vacuoles take up more than 90 percent of the cell volume and grow as the cell grows. The presence of dissolved substances in the vacuole causes water to enter it from the cytoplasm, making the vacuole swell like a balloon. The plant cell does not swell when the vacuole fills with water, since it has a rigid cell wall. Instead, it stiffens from the increase in water pressure (called turgor), which supports the plant (see Figure 6.10).

- *Reproduction*: Some pigments (especially blue and pink ones) in the petals and fruits of flowering plants are contained in vacuoles. These pigments—the anthocyanins—are visual cues that help attract the animals that assist in pollination or seed dispersal.

- *Digestion*: In some plants, vacuoles in seeds contain enzymes that hydrolyze stored seed proteins into monomers that the developing plant embryo can use as food.

Contractile vacuoles are found in many freshwater protists. Their function is to get rid of the excess water that rushes into the cell because of the imbalance in solute concentration between the interior of the cell and its freshwater environment. The contractile vacuole enlarges as water enters, then abruptly contracts, forcing the water out of the cell through a special pore structure.

So far, we have discussed numerous membrane-enclosed organelles. Now we turn to a group of cytoplasmic structures without membranes.

The cytoskeleton is important in cell structure and movement

From the earliest observations, light microscopy revealed distinctive shapes of cells that would sometimes change, and within cells rapid movements were observed. With the advent of electron microscopy, a new world of cellular substructure was revealed, including a meshwork of filaments inside cells. Experimentation showed that this **cytoskeleton** fills several important roles:

- It supports the cell and maintains its shape.

- It holds cell organelles in position within the cell.

- It moves organelles within the cell.

- It is involved with movements of the cytoplasm, called cytoplasmic streaming.

- It interacts with extracellular structures, helping to anchor the cell in place.

There are three components of the cytoskeleton: microfilaments (smallest diameter), intermediate filaments, and microtubules (largest diameter). These filaments have very different functions.

MICROFILAMENTS **Microfilaments** can exist as single filaments, in bundles, or in networks. They are about 7 nm in diameter and up to several micrometers long. Microfilaments have two major roles:

- They help the entire cell or parts of the cell to move.

- They determine and stabilize cell shape.

Microfilaments are assembled from *actin* monomers, a protein that exists in several forms and has many functions, especially in animals. The actin found in microfilaments (which are also known as actin filaments) has distinct ends designated "plus" and "minus." These ends permit actin monomers to interact with one another to form long, double helical chains (**Figure 5.17A**). Within cells, the polymerization of actin into microfilaments is reversible, and the microfilaments can disappear from cells by breaking down into monomers of free actin. Special actin-binding proteins mediate these events.

In the muscle cells of animals, actin filaments are associated with another protein, the "motor protein" *myosin*, and the interactions of these two proteins account for the contraction of muscles (described in Section 48.1). In non-muscle cells, actin filaments are associated with localized changes in cell shape. For example, microfilaments are involved in the flowing movement of the cytoplasm called cytoplasmic streaming, in amoeboid movement, and in the "pinching" contractions that divide an animal cell into two daughter cells. Microfilaments are also involved in the formation of cellular extensions called pseudopodia (*pseudo*, "false"; *podia*, "feet") that enable some cells to move (**Figure 5.18**). As you will see in Chapter 42, cells of the immune system must move toward other cells during the immune response.

In some cell types, microfilaments form a meshwork just inside the plasma membrane. Actin-binding proteins then cross-link the microfilaments to form a rigid net-like structure that supports the cell. For example, microfilaments support the tiny

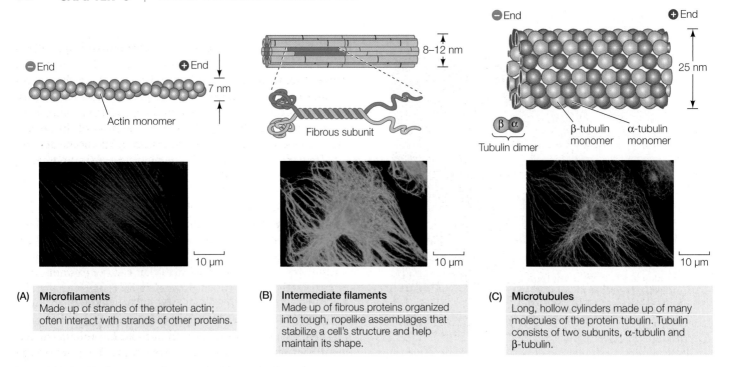

(A) **Microfilaments**
Made up of strands of the protein actin; often interact with strands of other proteins.

(B) **Intermediate filaments**
Made up of fibrous proteins organized into tough, ropelike assemblages that stabilize a cell's structure and help maintain its shape.

(C) **Microtubules**
Long, hollow cylinders made up of many molecules of the protein tubulin. Tubulin consists of two subunits, α-tubulin and β-tubulin.

5.17 The Cytoskeleton Three highly visible and important structural components of the cytoskeleton are shown here in detail. These structures maintain and reinforce cell shape and contribute to cell movement.

microvilli that line the human intestine, giving it a larger surface area through which to absorb nutrients (**Figure 5.19**).

INTERMEDIATE FILAMENTS There are at least 50 different kinds of **intermediate filaments**, many of them specific to a few cell types. They generally fall into six molecular classes (based on amino acid sequence) that share the same general structure. One of these classes consists of fibrous proteins of the keratin family, which also includes the proteins that make up hair and fingernails. The intermediate filaments are tough, ropelike protein assemblages 8 to 12 nm in diameter (**Figure 5.17B**). Intermediate filaments are more permanent than the other two types; in cells they do not form and re-form, as the microtubules and microfilaments do.

Intermediate filaments have two major structural functions:

• They anchor cell structures in place. In some cells, intermediate filaments radiate from the nuclear envelope and help maintain the positions of the nucleus and other organelles in the cell. The lamins of the nuclear lamina are intermediate filaments (see Figure 5.8). Other kinds of intermediate filaments help hold in place the complex apparatus of microfilaments in the microvilli of intestinal cells (see Figure 5.19).

• They resist tension. For example, they maintain rigidity in body surface tissues by stretching through the cytoplasm and connecting specialized membrane structures called desmosomes (see Figure 6.7).

5.18 Microfilaments and Cell Movements Microfilaments mediate the movement of whole cells (as illustrated here for amoebic movement), as well as the movement of cytoplasm within a cell.

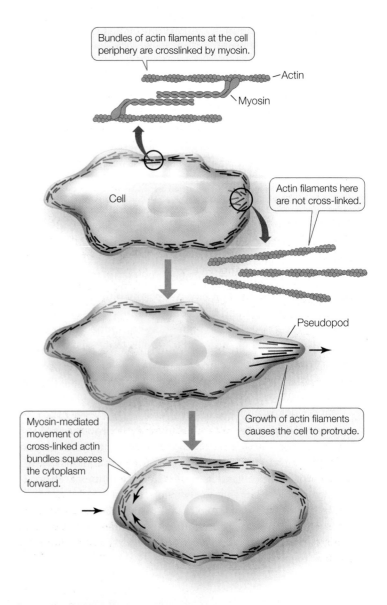

Bundles of actin filaments at the cell periphery are crosslinked by myosin.

Actin

Myosin

Cell

Actin filaments here are not cross-linked.

Pseudopod

Growth of actin filaments causes the cell to protrude.

Myosin-mediated movement of cross-linked actin bundles squeezes the cytoplasm forward.

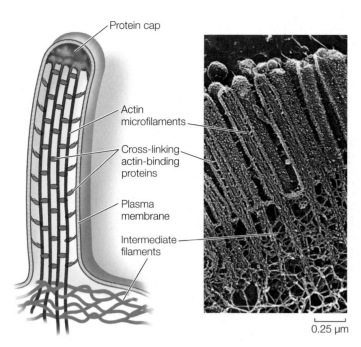

Protein cap

Actin microfilaments

Cross-linking actin-binding proteins

Plasma membrane

Intermediate filaments

0.25 μm

5.19 Microfilaments for Support Cells that line the intestine are folded into tiny projections called microvilli, which are supported by microfilaments. The microfilaments interact with intermediate filaments at the base of each microvillus. The microvilli increase the surface area of the cells, facilitating their absorption of small molecules.

MICROTUBULES The largest diameter components of the cytoskeletal system, **microtubules**, are long, hollow, unbranched cylinders about 25 nm in diameter and up to several micrometers long. Microtubules have two roles in the cell:

• They form a rigid internal skeleton for some cells.

• They act as a framework along which motor proteins can move structures within the cell.

Microtubules are assembled from dimers of the protein *tubulin*. A dimer is a molecule made up of two monomers. The polypeptide monomers that make up a tubulin dimer are known as α-tubulin and β-tubulin. Thirteen chains of tubulin dimers surround the central cavity of the microtubule (**Figure 5.17C**; see also Figure 5.20). The two ends of a microtubule are different: one is designated the plus (+) end, and the other the minus (–) end. Tubulin dimers can be rapidly added or subtracted, mainly at the plus end, lengthening or shortening the microtubule. This capacity to change length rapidly makes microtubules dynamic structures, permitting some animal cells to rapidly change shape.

Many microtubules radiate from a region of the cell called the microtubule organizing center. Tubulin polymerization results in a rigid structure, and tubulin depolymerization leads to its collapse.

In plants, microtubules help control the arrangement of the cellulose fibers of the cell wall. Electron micrographs of plants frequently show microtubules lying just inside the plasma membranes of cells that are forming or extending their cell walls. Experimental alteration of the orientation of these microtubules leads to a similar change in the cell wall and a new shape for the cell.

Microtubules serve as tracks for **motor proteins**, specialized molecules that use cellular energy to change their shape and move. Motor proteins bond to and move along the microtubules, carrying materials from one part of the cell to another. Microtubules are also essential in distributing chromosomes to daughter cells during cell division. Because of this, drugs such as vincristine and taxol that disrupt microtubule dynamics also disrupt cell division. These drugs are useful for treating cancer, where cell division is excessive.

CILIA AND FLAGELLA Microtubules are also intimately associated with movable cell appendages: the **cilia** and **flagella**. Many eukaryotic cells have one or both of these appendages. Cilia are smaller than flagella—only 0.25 μm in length. They may move surrounding fluid over the surface of the cell (for example, protists or cells lining tubes through which eggs move, the oviducts). Eukaryotic flagella are 0.25 μm in diameter and 100–200 μm in length. (The structure and operation of eukaryotic flagella are very different from those of prokaryotic flagella; see Figure 5.5.) They may push or pull the cell through its aqueous environment (for example, protists or sperm). Cilia and eukaryotic flagella are both assembled from specialized microtubules and have identical internal structures, but differ in their length and pattern of beating:

• Cilia (singular *cilium*) are usually present in great numbers (**Figure 5.20A**). They beat stiffly in one direction and recover flexibly in the other direction (like a swimmer's arm), so that the recovery stroke does not undo the work of the power stroke.

• Eukaryotic flagella are usually found singly or in pairs. Waves of bending propagate from one end of a flagellum to the other in a snakelike undulation. Forces exerted by these waves on the surrounding fluid medium move the cell.

In cross section, a typical *cilium* or eukaryotic flagellum is surrounded by the plasma membrane and contains a "9 + 2" array of microtubules. As **Figure 5.20B** shows, nine fused pairs of microtubules—called doublets—form an outer cylinder, and one pair of unfused microtubules runs up the center. A spoke radiates from one microtubule of each doublet and connects the doublet to the center of the structure. These structures are essential to the bending motions of both cilia and flagella.

In the cytoplasm at the base of every eukaryotic flagellum and cilium is an organelle called a **basal body**. The nine microtubule doublets extend into the basal body. In the basal body, each doublet is accompanied by another microtubule, making nine sets of three microtubules. The central, unfused microtubules in the cilium do not extend into the basal body.

Centrioles are almost identical to the basal bodies of cilia and flagella. Centrioles are found in the microtubule organizing centers (sites of tubulin storage where microtubules polymerize) of all eukaryotes except the seed plants and some protists. Under the light microscope, a centriole looks like a small, featureless particle, but the electron microscope reveals that it contains a precise bundle of microtubules arranged in nine sets of three. Centrioles are involved in the formation of the mitotic spindle, to which the chromosomes attach during cell division (see Figure 11.10).

(A)

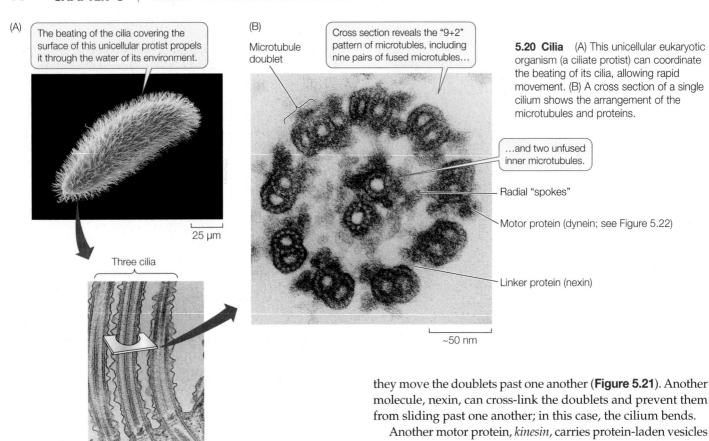

The beating of the cilia covering the surface of this unicellular protist propels it through the water of its environment.

25 μm

Three cilia

250 nm

(B)

Cross section reveals the "9+2" pattern of microtubules, including nine pairs of fused microtubles...

Microtubule doublet

...and two unfused inner microtubules.

Radial "spokes"

Motor protein (dynein; see Figure 5.22)

Linker protein (nexin)

~50 nm

5.20 Cilia (A) This unicellular eukaryotic organism (a ciliate protist) can coordinate the beating of its cilia, allowing rapid movement. (B) A cross section of a single cilium shows the arrangement of the microtubules and proteins.

MOTOR PROTEINS AND MOVEMENT The nine microtubule doublets of cilia and flagella are linked by proteins. The motion of cilia and flagella results from the sliding of the microtubule doublets past each other. This sliding is driven by a motor protein called *dynein*, which can change its three-dimensional shape. All motor proteins work by undergoing reversible shape changes powered by energy from ATP hydrolysis. Dynein molecules that are attached to one microtubule doublet bind to a neighboring doublet. As the dynein molecules change shape,

they move the doublets past one another (**Figure 5.21**). Another molecule, nexin, can cross-link the doublets and prevent them from sliding past one another; in this case, the cilium bends.

Another motor protein, *kinesin*, carries protein-laden vesicles from one part of the cell to another (**Figure 5.22**). Kinesin and similar motor proteins bind to a vesicle or other organelle, then "walk" it along a microtubule by a repeated series of shape changes. Recall that microtubules are directional, with a plus end and a minus end. Dynein moves attached organelles toward the minus end, while kinesin moves them toward the plus end (see Figure 5.17).

DEMONSTRATING CYTOSKELETON FUNCTIONS How do we know that the structural fibers of the cytoskeleton can achieve all these dynamic functions? We can observe an individual structure under the microscope and a function in a living cell that contains that structure. These observations may suggest that the structure carries out that function, but in science mere correlation does not show cause and effect. For example, light microscopy of living cells reveals that the cytoplasm is actively streaming around the cell, and that cytoplasm flows into an extended portion of an amoeboid cell during movement. The observed presence of cytoskeletal components *suggests, but does not prove*, their role in this process. Science seeks to show the specific links that relate one process, "A," to a function, "B." In cell biology, there are two ways to show that a structure or process "A" causes function "B":

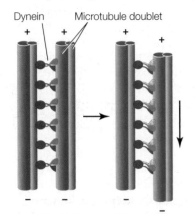

Dynein   Microtubule doublet

In isolated cilia without nexin cross-links, movement of dynein motor proteins causes microtubule doublets to slide past one another.

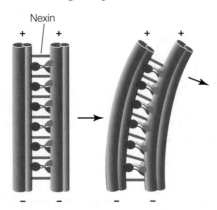

Nexin

When nexin is present to cross-link the doublets, they cannot slide and the force generated by dynein movement causes the cilium to bend.

5.21 A Motor Protein Moves Microtubules in Cilia and Flagella A motor protein, dynein, causes microtubule doublets to slide past one another. In a flagellum or cilium, anchorage of the microtubule doublets to one another results in bending.

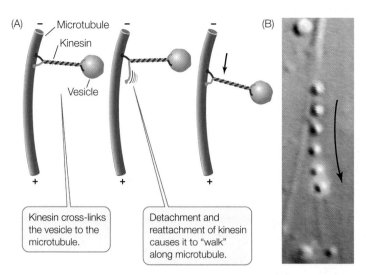

5.22 A Motor Protein Drives Vesicles along Microtubules
(A) Kinesin delivers vesicles or organelles to various parts of the cell by moving along microtubule "railroad tracks." Kinesin moves things from the minus toward the plus end of a microtubule; dynein works similarly, but moves from the plus toward the minus end. (B) Powered by kinesin, a vesicle moves along a microtubule track in the protist *Dictyostelium*. The time sequence (time-lapse micrography at half-second intervals) is shown by the color gradient of purple to blue.

- *Inhibition*: use a drug that inhibits A and see if B still occurs. If it does not, then A is probably a causative factor for B. **Figure 5.23** shows an experiment with such a drug (an inhibitor) that demonstrates cause and effect in the case of the cytoskeleton and cell movement.

- *Mutation*: examine a cell that lacks the gene (or genes) for A and see if B still occurs. If it does not, then A is probably a causative factor for B. Part Four of this book describes many experiments using this genetic approach.

---

yourBioPortal.com
GO TO Animated Tutorial 5.2 • Eukaryotic Cell Tour

5.3 RECAP

The hallmark of eukaryotic cells is compartmentalization. Membrane-enclosed organelles process information, transform energy, form internal compartments for transporting proteins, and carry out intracellular digestion. An internal cytoskeleton plays several structural roles.

- What are some advantages of organelle compartmentalization? **See p. 84**

- Describe the structural and functional differences between rough and smooth endoplasmic reticulum. **See pp. 89–90 and Figure 5.10**

- Explain how motor proteins and microtubules move materials within the cell. **See pp. 95–98 and Figures 5.21 and 5.22**

INVESTIGATING LIFE

5.23 The Role of Microfilaments in Cell Movement— Showing Cause and Effect in Biology
After a test tube demonstration that the drug cytochalasin B prevented microfilament formation from monomeric precursors, the question was asked: Will the drug work like this in living cells and inhibit cell movement in *Amoeba*? Complementary experiments showed that the drug did not poison other cellular processes.

HYPOTHESIS Amoeboid cell movements are caused by the cytoskeleton.

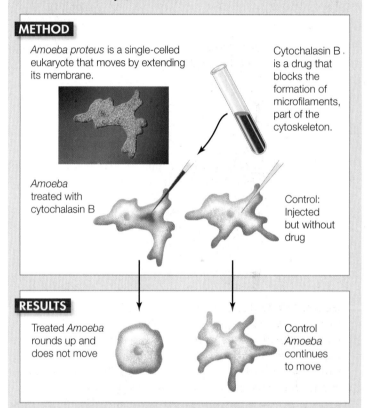

CONCLUSION Microfilaments of the cytoskeleton are essential for amoeboid cell movement.

FURTHER INVESTIGATION: The drug colchicine breaks apart microtubules. How would you show that these components of the cytoskeleton are not involved in cell movement in *Amoeba*?

Go to **yourBioPortal.com** for original citations, discussions, and relevant links for all INVESTIGATING LIFE figures.

All cells interact with their environments, and many eukaryotic cells are parts of multicellular organisms and must interact, and closely coordinate activities, with other cells. The plasma membrane plays a crucial role in these interactions, but other structures outside that membrane are involved as well.

5.4 What Are the Roles of Extracellular Structures?

Although the plasma membrane is the functional barrier between the inside and the outside of a cell, many structures are produced by cells and secreted to the outside of the plasma membrane, where they play essential roles in protecting, supporting, or attaching cells to each other. Because they are outside the plasma membrane, these structures are said to be *extracellular*. The peptidoglycan cell wall of bacteria is an example of an extracellular structure (see Figure 5.4). In eukaryotes, other extracellular structures—the cell walls of plants and the extracellular matrices found between the cells of animals—play similar roles. Both of these structures are made up of two components:

- a prominent fibrous macromolecule
- a gel-like medium in which the fibers are embedded

The plant cell wall is an extracellular structure

The plant **cell wall** is a semirigid structure outside the plasma membrane (**Figure 5.24**). We consider the structure and role of the cell wall in more detail in Chapter 34. For now, we note that it is typical of a two-component extracellular matrix, with cellulose fibers (see Figure 3.16) embedded in other complex polysaccharides and proteins. The plant cell wall has three major roles:

- It provides support for the cell and limits its volume by remaining rigid.
- It acts as a barrier to infection by fungi and other organisms that can cause plant diseases.
- It contributes to plant form by growing as plant cells expand.

Because of their thick cell walls, plant cells viewed under a light microscope appear to be entirely isolated from one another. But electron microscopy reveals that this is not the case. The cytoplasms of adjacent plant cells are connected by numerous plasma membrane–lined channels, called **plasmodesmata**, that are about 20–40 nm in diameter and extend through the cell walls (see Figures 5.7 and 6.7). Plasmodesmata permit the diffusion of water, ions, small molecules, RNA, and proteins between connected cells, allowing for utilization of these substances far from their site of synthesis.

The extracellular matrix supports tissue functions in animals

Animal cells lack the semirigid wall that is characteristic of plant cells, but many animal cells are surrounded by, or in contact with, an **extracellular matrix**. This matrix is composed of three types of molecules: fibrous proteins such as **collagen** (the most abundant protein in mammals, constituting over 25 percent of the protein in the human body); a matrix of glycoproteins termed **proteoglycans**, consisting primarily of sugars; and a third group of proteins that link the fibrous proteins and the gel-like proteoglycan matrix together (**Figure 5.25**). These proteins and proteoglycans are secreted, along with other substances that are specific to certain body tissues, by cells that are present in or near the matrix. The functions of the extracellular matrix are many:

- It holds cells together in tissues. In Chapter 6 we see how there is an intercellular "glue" that is involved in both cell recognition and adhesion.
- It contributes to the physical properties of cartilage, skin, and other tissues. For example, the mineral component of bone is laid down on an organized extracellular matrix.
- It helps filter materials passing between different tissues. This is especially important in the kidney.
- It helps orient cell movements during embryonic development and during tissue repair.
- It plays a role in chemical signaling from one cell to another. Proteins connect the cell's plasma membrane to the extracellular matrix. These proteins (for example, *integrin*) span the plasma membrane and are involved with transmitting signals to the interior of the cell. This allows communication between the extracellular matrix and the cytoplasm of the cell.

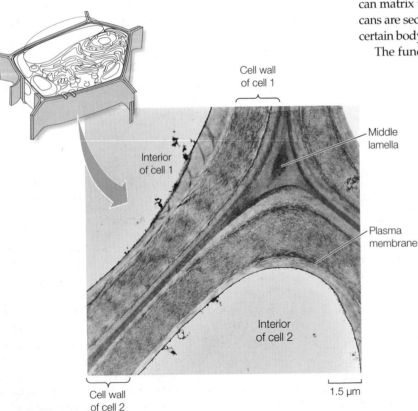

Cell wall of cell 1

Interior of cell 1

Middle lamella

Plasma membrane

Interior of cell 2

Cell wall of cell 2

1.5 μm

5.24 The Plant Cell Wall The semirigid cell wall provides support for plant cells. It is composed of cellulose fibrils embedded in a matrix of polysaccharides and proteins.

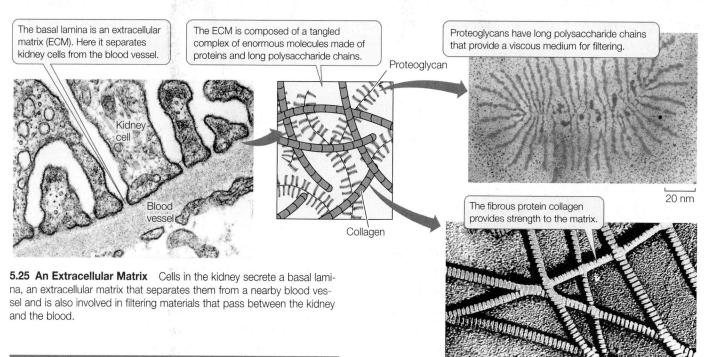

The basal lamina is an extracellular matrix (ECM). Here it separates kidney cells from the blood vessel.

The ECM is composed of a tangled complex of enormous molecules made of proteins and long polysaccharide chains.

Proteoglycans have long polysaccharide chains that provide a viscous medium for filtering.

Proteoglycan

Kidney cell

Blood vessel

Collagen

20 nm

The fibrous protein collagen provides strength to the matrix.

100 nm

5.25 An Extracellular Matrix Cells in the kidney secrete a basal lamina, an extracellular matrix that separates them from a nearby blood vessel and is also involved in filtering materials that pass between the kidney and the blood.

5.4 RECAP

Extracellular structures are produced by cells and secreted outside the plasma membrane. Most consist of a fibrous component in a gel-like medium.

- What are the functions of the cell wall in plants and the extracellular matrix in animals? See p. 100

We have now discussed the structures and some functions of prokaryotic and eukaryotic cells. Both exemplify the cell theory, showing that cells are the basic units of life and of biological continuity. Much of the rest of this part of the book will deal with these two aspects of cells. There is abundant evidence that the simpler prokaryotic cells are more ancient than eukaryotic cells, and that the first cells were probably prokaryotic. We now turn to the next step in cellular evolution, the origin of eukaryotic cells.

5.5 How Did Eukaryotic Cells Originate?

For about 2 billion years, life on Earth was entirely prokaryotic—from the time when prokaryotic cells first appeared until about 1.5 billion years ago, when eukaryotic cells arrived on the scene. The advent of compartmentalization—the hallmark of eukaryotes—was a major event in the history of life, as it permitted many more biochemical functions to coexist in the same cell than had previously been possible. Compared to the typical eukaryote, a single prokaryotic cell is often biochemically specialized, limited in the resources it can use and the functions it can perform.

What is the origin of compartmentalization? We will describe the evolution of eukaryotic organelles in more detail in Section 27.1. Here, we outline two major themes in this process.

Internal membranes and the nuclear envelope probably came from the plasma membrane

We noted earlier that some bacteria contain internal membranes. How could these arise? In electron micrographs, the internal membranes of prokaryotes often appear to be inward folds of the plasma membrane. This has led to a theory that the endomembrane system and cell nucleus originated by a related process (**Figure 5.26A**). The close relationship between the ER and the nuclear envelope in today's eukaryotes is consistent with this theory.

A bacterium with enclosed compartments would have several evolutionary advantages. Chemicals could be concentrated within particular regions of the cell, allowing chemical reactions to proceed more efficiently. Biochemical activities could be segregated within organelles with, for example, a different pH from the rest of the cell, creating more favorable conditions for certain metabolic processes. Finally, gene transcription could be separated from translation, providing more opportunities for separate control of these steps in gene expression.

Some organelles arose by endosymbiosis

Symbiosis means "living together," and often refers to two organisms that coexist, each one supplying something that the other needs. Biologists have proposed that some organelles—the mitochondria and the plastids—arose not by an infolding of the plasma membrane but by one cell ingesting another cell, giving rise to a symbiotic relationship. Eventually, the ingested cell lost its autonomy and some of its functions. In addition, many of the ingested cell's genes were transferred to the host's DNA. Mitochondria and plastids in today's eukaryotic cells are the remnants of these *symbionts*, retaining some specialized functions that benefit their host cells. This is the essence of the **endosymbiosis theory** for the origin of organelles.

Consider the case of the plastid. About 2.5 billion years ago some prokaryotes (the cyanobacteria) developed photosynthesis (see Figure 1.9). The emergence of these prokaryotes was a key event in the evolution of complex organisms, because they increased the O_2 concentration in Earth's atmosphere (see Section 1.2).

(A)

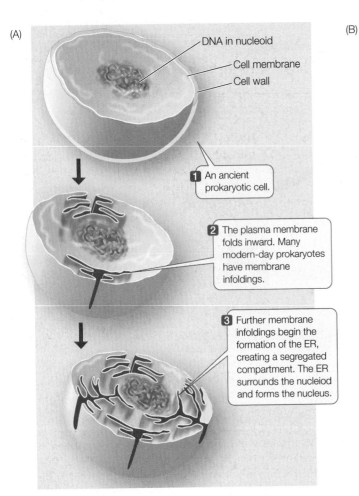

1 An ancient prokaryotic cell.

2 The plasma membrane folds inward. Many modern-day prokaryotes have membrane infoldings.

3 Further membrane infoldings begin the formation of the ER, creating a segregated compartment. The ER surrounds the nucleiod and forms the nucleus.

(B)

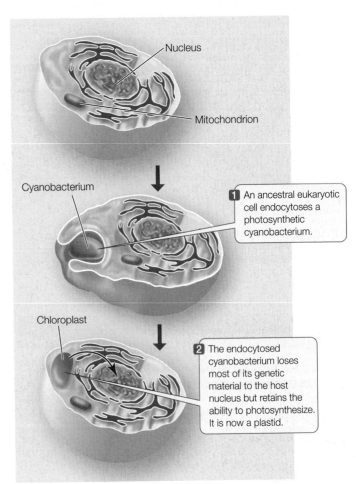

1 An ancestral eukaryotic cell endocytoses a photosynthetic cyanobacterium.

2 The endocytosed cyanobacterium loses most of its genetic material to the host nucleus but retains the ability to photosynthesize. It is now a plastid.

5.26 The Origin of Organelles   (A) The endomembrane system and cell nucleus may have been formed by infolding and then fusion of the plasma membrane. (B) The endosymbiosis theory proposes that some organelles may be descended from prokaryotes that were engulfed by other, larger cells.

According to endosymbiosis theory, photosynthetic prokaryotes also provided the precursor of the modern-day plastid. Cells without cell walls can engulf relatively large particles by phagocytosis (see Figure 5.11). In some cases, such as that of phagocytes in the human immune system, the engulfed particle can be an entire cell, such as a bacterium. Plastids may have arisen by a similar event involving an ancestral eukaryote and a cyanobacterium (**Figure 5.26B**).

Among the abundant evidence supporting the endosymbiotic origin of plastids (see Section 27.1), perhaps the most remarkable comes from a sandy beach in Japan. Noriko Okamoto and Isao Inouye recently discovered a single-celled eukaryote that contains a large "chloroplast," and named it *Hatena* (**Figure 5.27**). It turns out that the "chloroplast" is the remains of a green alga, *Nephroselmis*, which lives among the *Hatena* cells. When living autonomously, this algal cell has flagella, a cytoskeleton, ER, Golgi, and mitochondria in addition to a plastid. Once ingested by *Hatena*, all of these structures, and presumably their associated functions, are lost. What remains is essentially a plastid.

When *Hatena* divides, only one of the two daughter cells ends up with the "chloroplast." The other cell finds and ingests its own *Nephroselmis* alga—almost like a "replay" of what may have occurred in the evolution of eukaryotic cells. No wonder the Japanese scientists call the host cell *Hatena*: in Japanese, it means "how odd"!

5.5 RECAP

Eukaryotic cells arose long after prokaryotic cells. Some organelles may have evolved by infolding of the plasma membrane, while others evolved by endosymbiosis.

- How could membrane infolding in a prokaryotic cell lead to the endomembrane system? See p. 101 and Figure 5.26A

- Explain the endosymbiosis theory for the origin of chloroplasts. See Figure 5.26B

In this chapter, we presented an overview of the structures of cells, with some ideas about their relationships and origins. As you now embark on the study of major cell functions, keep in

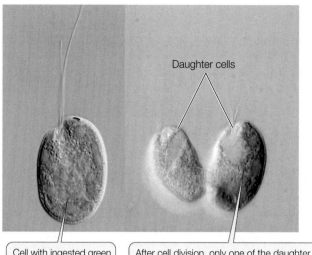

Daughter cells

Cell with ingested green photosynthetic plastid.

After cell division, only one of the daughter cells inherits the plastid; the other cell must ingest a new one from the environment.

5.27 Endosymbiosis in Action A *Hatena* cell engulfs an algal cell, which then loses most of its cellular functions other than photosynthesis. This re-enacts a possible event in the origin of plastids in eukaryotic cells.

mind that the structures in a cell do not exist in isolation. They are part of a dynamic, interacting cellular system. In Chapter 6 we show that the plasma membrane is far from a passive barrier, but instead is a multi-functional system that connects the inside of the cell with its extracellular environment.

CHAPTER SUMMARY

5.1 What Features Make Cells the Fundamental Units of Life?

SEE WEB ACTIVITIES 5.1 AND 5.2

- All cells come from preexisting cells.

- Cells are small because a cell's surface area must be large compared with its volume to accommodate exchanges with its environment. **Review Figure 5.2**

- All cells are enclosed by a selectively permeable **plasma membrane** that separates their contents from the external environment.

- While certain biochemical processes, molecules, and structures are shared by all kinds of cells, two categories of cells—**prokaryotes** and **eukaryotes**—are easily distinguished.

5.2 What Features Characterize Prokaryotic Cells?

- Prokaryotic cells have no internal compartments, but have a **nucleoid** region containing DNA, and a **cytoplasm** containing **cytosol**, **ribosomes**, proteins, and small molecules. Some prokaryotes have additional protective structures, including a **cell wall**, an **outer membrane**, and a **capsule**. **Review Figure 5.4**

- Some prokaryotes have folded membranes that may be photosynthetic membranes, and some have **flagella** or **pili** for motility or attachment. **Review Figure 5.5**

5.3 What Features Characterize Eukaryotic Cells?

- Eukaryotic cells are larger than prokaryotic cells and contain many membrane-enclosed **organelles**. The membranes that envelop organelles ensure compartmentalization of their functions. **Review Figure 5.7**

- **Ribosomes** are sites of protein synthesis.

- The **nucleus** contains most of the cell's DNA and participates in the control of protein synthesis. **Review Figure 5.8**

- The **endomembrane system**—consisting of the **endoplasmic reticulum** and **Golgi apparatus**—is a series of interrelated compartments enclosed by membranes. It segregates proteins and

modifies them. **Lysosomes** contain many digestive enzymes. **Review Figures 5.10 and 5.11, WEB ACTIVITY 5.3, ANIMATED TUTORIAL 5.1**

- **Mitochondria** and **chloroplasts** are semi-autonomous organelles that process energy. Mitochondria are present in most eukaryotic organisms and contain the enzymes needed for cellular respiration. The cells of photosynthetic eukaryotes contain chloroplasts that harvest light energy for photosynthesis. **Review Figures 5.12 and 5.13**

- **Vacuoles** are prominent in many plant cells and consist of a membrane-enclosed compartment full of water and dissolved substances.

- The **microfilaments**, **intermediate filaments**, and **microtubules** of the **cytoskeleton** provide the cell with shape, strength, and movement. **Review Figure 5.18**

SEE ANIMATED TUTORIAL 5.2

5.4 What Are the Roles of Extracellular Structures?

- The plant **cell wall** consists principally of **cellulose**. Cell walls are pierced by **plasmodesmata** that join the cytoplasms of adjacent cells.

- In animals, the **extracellular matrix** consists of different kinds of proteins, including collagen and proteoglycans. **Review Figure 5.25**

5.5 How Did Eukaryotic Cells Originate?

- Infoldings of the plasma membrane could have led to the formation of some membrane-enclosed organelles, such as the endomembrane system and the nucleus. **Review Figure 5.26A**

- The **endosymbiosis theory** states that mitochondria and chloroplasts originated when larger prokaryotes engulfed, but did not digest, smaller prokaryotes. Mutual benefits permitted this symbiotic relationship to be maintained, allowing the smaller cells to evolve into the eukaryotic organelles observed today. **Review Figure 5.26B**

SELF-QUIZ

1. Which structure is generally present in both prokaryotic cells and eukaryotic plant cells?
   a. Chloroplasts
   b. Cell wall
   c. Nucleus
   d. Mitochondria
   e. Microtubules

2. The major factor limiting cell size is the
   a. concentration of water in the cytoplasm.
   b. need for energy.
   c. presence of membrane-enclosed organelles.
   d. ratio of surface area to volume.
   e. composition of the plasma membrane.

3. Which statement about mitochondria is *not* true?
   a. The inner mitochondrial membrane folds to form cristae.
   b. The outer membrane is relatively permeable to macromolecules.
   c. Mitochondria are green because they contain chlorophyll.
   d. Fuel molecules from the cytosol are used for respiration in mitochondria.
   e. ATP is synthesized in mitochondria.

4. Which statement about plastids is true?
   a. They are found in prokaryotes.
   b. They are surrounded by a single membrane.
   c. They are the sites of cellular respiration.
   d. They are found only in fungi.
   e. They may contain several types of pigments or polysaccharides.

5. If all the lysosomes within a cell suddenly ruptured, what would be the most likely result?
   a. The macromolecules in the cytosol would break down.
   b. More proteins would be made.
   c. The DNA within mitochondria would break down.
   d. The mitochondria and chloroplasts would divide.
   e. There would be no change in cell function.

6. The Golgi apparatus
   a. is found only in animals.
   b. is found in prokaryotes.
   c. is the appendage that moves a cell around in its environment.
   d. is a site of rapid ATP production.
   e. modifies and packages proteins.

7. Which structure is *not* surrounded by one or more membranes?
   a. Ribosome
   b. Chloroplast
   c. Mitochondrion
   d. Peroxisome
   e. Vacuole

8. The cytoskeleton consists of
   a. cilia, flagella, and microfilaments.
   b. cilia, microtubules, and microfilaments.
   c. internal cell walls.
   d. microtubules, intermediate filaments, and microfilaments.
   e. calcified microtubules.

9. Microfilaments
   a. are composed of polysaccharides.
   b. are composed of actin.
   c. allow cilia and flagella to move.
   d. make up the spindle that aids the movement of chromosomes.
   e. maintain the position of the chloroplast in the cell.

10. Which statement about the plant cell wall is *not* true?
    a. Its principal chemical components are polysaccharides.
    b. It lies outside the plasma membrane.
    c. It provides support for the cell.
    d. It completely isolates adjacent cells from one another.
    e. It is semirigid.

FOR DISCUSSION

1. The drug vincristine is used to treat many cancers. It apparently works by causing microtubules to depolymerize. Vincristine use has many side effects, including loss of dividing cells and nerve problems. Explain why this might be so.

2. Through how many membranes would a molecule have to pass in moving from the interior (stroma) of a chloroplast to the interior (matrix) of a mitochondrion? From the interior of a lysosome to the outside of a cell? From one ribosome to another?

3. How does the possession of double membranes by chloroplasts and mitochondria relate to the endosymbiosis theory of the origins of these organelles? What other evidence supports the theory?

4. Compare the extracellular matrix of the animal cell with the plant cell wall, with respect to composition of the fibrous and nonfibrous components, rigidity, and connectivity of cells.

ADDITIONAL INVESTIGATION

The pathway of newly synthesized proteins can be followed through the cell using a "pulse-chase" experiment. During synthesis, proteins are tagged with a radioactive isotope (the "pulse"), and then the cell is allowed to process the proteins for varying periods of time. The locations of the radioactive proteins are then determined by isolating cell organelles and quantifying their radioactivity. How would you use this method, and what results would you expect for (*a*) a lysosomal enzyme and (*b*) a protein that is released from the cell?

6 Cell Membranes

Membranes and memory

James noticed the changes in his grandfather when he was home from college for the winter holiday. He and grandpa John had always joked about grandpa John's missing keys and glasses; the old man, who had lived with James' family since his wife died, was forever searching for them. Now the memory lapses had become more pronounced. When James introduced his new girlfriend to the family, he was relieved (as was she) when she was welcomed with open arms. But an hour later, grandpa John just stared at her, unable to remember who she was. By the time James came home for the summer, his grandfather had become withdrawn; he could no longer talk about current events, and often he became confused and lashed out in anger.

James' grandfather had Alzheimer's disease. This condition is most common in (but not limited to) the elderly, and as more people today are living to advanced ages, more and more Alzheimer's cases are diagnosed. But the symptoms are not new to human experience or to medicine. The condition was first recognized as a disease in 1901. That year, the family of 51-year-old Frau Auguste D. brought her to Dr. Alois Alzheimer at the Frankfurt hospital in Germany. She had severe memory lapses, accused her husband of infidelity, and had difficulty communicating. These symptoms got worse before she died several years later. When Alzheimer autopsied her brain, he saw that the parts of the brain that are important in thought and speech were shrunken. Moreover, when he examined these areas through the microscope he saw abnormal protein deposits in and around the brain cells.

In the century since Alzheimer's original case, cell biologists have investigated the nature of these abnormal deposits, now known as *plaques*. It turns out that the key events that produce plaques take place in the plasma membrane of nerve cells in the brain. Plaques are clumps of the protein amyloid beta, which at high levels is toxic to brain cells. Amyloid beta is a small piece of a larger amyloid precursor protein (APP), which is embedded in the nerve cell plasma membrane; APP is cut twice by two other membrane proteins, β-secretase and γ-secretase, to produce amyloid beta, which is released from the membrane to fall outside of the cell. All these proteins are present in a variety of animal cells and have multiple important

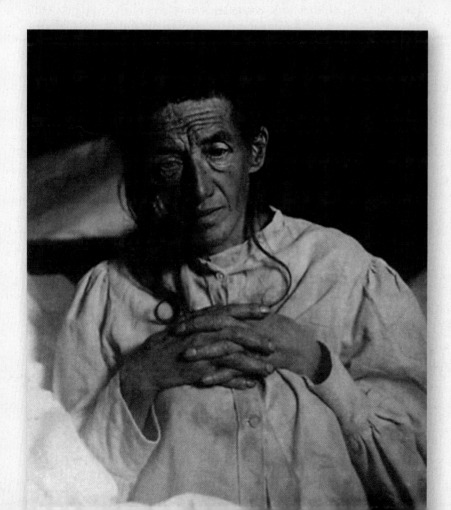

Dr. Alzheimer's Patient Frau Auguste D., who died in 1906, was the first patient described with progressive dementia by Dr. Alois Alzheimer.

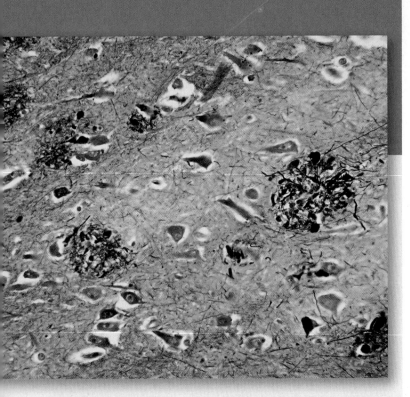

Plaques in the Brain As seen at autopsy, the brain of an Alzheimer's disease patient accumulates plaques (dark fibers in this micrograph) composed of protein fragments produced by an enzyme in the nerve cell membrane.

roles in the dynamic cell membrane; they may even be essential for normal nervous system development and function.

So what goes wrong in Alzheimer's disease? Cells in the diseased brain might be producing too much amyloid beta (e.g., because γ-secretase is too active) or producing it at the wrong time (e.g., in old age instead of infancy). One form of the disease is caused by a mutant form of γ-secretase, which has a tendency to cut APP in the "wrong" place, thereby producing a particularly toxic form of amyloid beta. Because of their role in producing plaques, APP and γ-secretase are potential targets for Alzheimer's disease therapies.

Learning how membranes are made and how they work has been a key to understanding, and perhaps treating, this increasingly prevalent disease.

IN THIS CHAPTER we focus on the structure and functions of biological membranes. First we describe the composition and structure of biological membranes. We go on to discuss their functions—how membranes are involved in intercellular interactions, and how membranes regulate which substances enter and leave the cell.

6.1 What Is the Structure of a Biological Membrane?

The physical organization and functioning of all biological membranes depend on their constituents: lipids, proteins, and carbohydrates. You are already familiar with these molecules from Chapter 3; it may be useful to review that chapter now. The lipids establish the physical integrity of the membrane and create an effective barrier to the rapid passage of hydrophilic materials such as water and ions. In addition, the phospholipid bilayer serves as a lipid "lake" in which a variety of proteins "float" (**Figure 6.1**). This general design is known as the **fluid mosaic model**.

In the fluid mosaic model for biological membranes, the proteins are noncovalently embedded in the phospholipid bilayer by their hydrophobic regions (or *domains*), but their hydrophilic domains are exposed to the watery conditions on either side of the bilayer. These membrane proteins have a number of functions, including moving materials through the membrane and receiving chemical signals from the cell's external environment. Each membrane has a set of proteins suitable for the specialized functions of the cell or organelle it surrounds.

The carbohydrates associated with membranes are attached either to the lipids or to protein molecules. In plasma membranes, carbohydrates are located on the outside of the cell, where they may interact with substances in the external environment. Like some of the membrane proteins, carbohydrates are crucial in recognizing specific molecules, such as those on the surfaces of adjacent cells.

Although the fluid mosaic model is largely valid for membrane structure, it does not say much about membrane composition. As you read about the different molecules in membranes in the next sections, keep in mind that some membranes have more protein than lipids, others are lipid-rich, others have significant amounts of cholesterol or other sterols, and still others are rich in carbohydrates.

Lipids form the hydrophobic core of the membrane

The lipids in biological membranes are usually *phospholipids*. Recall from Section 2.2 that some compounds are hydrophilic ("water-loving") and others are hydrophobic ("water-hating"), and from Section 3.4 that a phospholipid molecule has regions of both kinds:

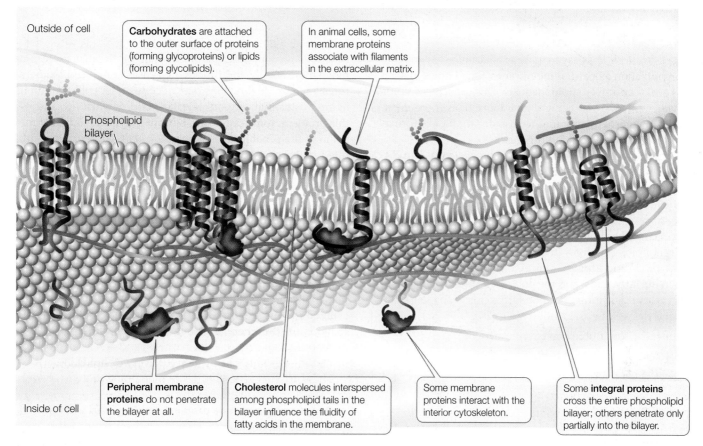

Outside of cell

Carbohydrates are attached to the outer surface of proteins (forming glycoproteins) or lipids (forming glycolipids).

In animal cells, some membrane proteins associate with filaments in the extracellular matrix.

Phospholipid bilayer

Inside of cell

Peripheral membrane proteins do not penetrate the bilayer at all.

Cholesterol molecules interspersed among phospholipid tails in the bilayer influence the fluidity of fatty acids in the membrane.

Some membrane proteins interact with the interior cytoskeleton.

Some **integral proteins** cross the entire phospholipid bilayer; others penetrate only partially into the bilayer.

6.1 The Fluid Mosaic Model The general molecular structure of biological membranes is a continuous phospholipid bilayer which has proteins embedded in or associated with it.

yourBioPortal.com
GO TO Web Activity 6.1 • The Fluid-Mosaic Model

- *Hydrophilic regions*: The phosphorus-containing "head" of the phospholipid is electrically charged and therefore associates with polar water molecules.

- *Hydrophobic regions*: The long, nonpolar fatty acid "tails" of the phospholipid associate with other nonpolar materials, but they do not dissolve in water or associate with hydrophilic substances.

Because of these properties, one way in which phospholipids can coexist with water is to form a *bilayer*, with the fatty acid "tails" of the two layers interacting with each other and the polar "heads" facing the outside aqueous environment (**Figure 6.2**). The thickness of a biological membrane is about 8 nm (0.008 μm), which is twice the length of a typical phospholipid—another indication that the membrane consists of a lipid bilayer. This thickness is about 8,000 times thinner than a piece of paper.

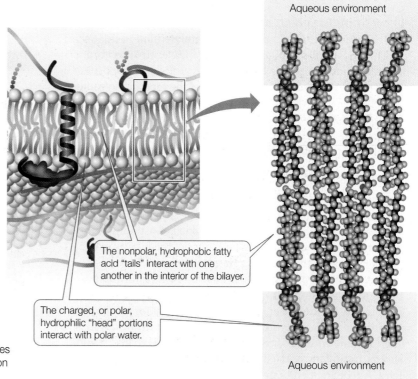

Aqueous environment

The nonpolar, hydrophobic fatty acid "tails" interact with one another in the interior of the bilayer.

The charged, or polar, hydrophilic "head" portions interact with polar water.

Aqueous environment

6.2 A Phospholipid Bilayer The phospholipid bilayer separates two aqueous regions. The eight phospholipid molecules shown on the right represent a small cross section of a membrane bilayer.

In the laboratory, it is easy to make artificial bilayers with the same organization as natural membranes. Small holes in such bilayers seal themselves spontaneously. This capacity of lipids to associate with one another and maintain a bilayer organization helps biological membranes to fuse during vesicle formation, phagocytosis, and related processes.

All biological membranes have a similar structure, but differ in the kinds of proteins and lipids they contain. Membranes from different cells or organelles may differ greatly in their *lipid composition*. Not only are phospholipids highly variable, but a significant proportion of the lipid content in an animal cell membrane may be cholesterol.

Phospholipids can differ in terms of fatty acid chain length (number of carbon atoms), degree of unsaturation (double bonds) in the fatty acids, and the polar (phosphate-containing) groups present. The most common fatty acids with their chain length and degree of unsaturation are:

- Palmitic: C_{14}, no double bonds, saturated
- Palmitoleic: C_{16}, one double bond
- Stearic: C_{18}, no double bonds, saturated
- Oleic: C_{18}, one double bond
- Linoleic: C_{18}, two double bonds
- Linolenic: C_{18}, three double bonds

The saturated fatty acid chains allow close packing of fatty acids in the bilayer, while the "kinks" in unsaturated fatty acids (see Figure 3.19) make for a less dense, more fluid packing. These less-dense membranes in animal cells can accommodate cholesterol molecules.

Up to 25 percent of the lipid content of an animal cell plasma membrane may be cholesterol. When present, cholesterol is important for membrane integrity; the cholesterol in your membranes is not hazardous to your health. A molecule of cholesterol is usually situated next to an unsaturated fatty acid.

The phospholipid bilayer stabilizes the entire membrane structure, but leaves it flexible. The fatty acids of the phospholipids make the hydrophobic interior of the membrane somewhat fluid—about as fluid as lightweight machine oil. This fluidity permits some molecules to move laterally within the plane of the membrane. A given phospholipid molecule in the plasma membrane can travel from one end of the cell to the other in a little more than a second! On the other hand, seldom does a phospholipid molecule in one half of the bilayer spontaneously flip over to the other side. For that to happen, the polar part of the molecule would have to move through the hydrophobic interior of the membrane. Since spontaneous phospholipid flip-flops are rare, the inner and outer halves of the bilayer may be quite different in the kinds of phospholipids they contain.

The fluidity of a membrane is affected by its lipid composition and by its temperature. Long-chain, saturated fatty acids pack tightly beside one another, with little room for movement. Cholesterol interacts hydrophobically with the fatty acid chains. A membrane with these components is less fluid than one with shorter-chain fatty acids, unsaturated fatty acids, or less cholesterol. Adequate membrane fluidity is essential for many of the functions we will describe in this chapter. Because molecules move more slowly and fluidity decreases at reduced temperatures, membrane functions may decline under cold conditions in organisms that cannot keep their bodies warm. To address this problem, some organisms simply change the lipid composition of their membranes when they get cold, replacing saturated with unsaturated fatty acids and using fatty acids with shorter tails. These changes play a role in the survival of plants, bacteria, and hibernating animals during the winter.

Membrane proteins are asymmetrically distributed

All biological membranes contain proteins. Typically, plasma membranes have one protein molecule for every 25 phospholipid molecules. This ratio varies depending on membrane function. In the inner membrane of the mitochondrion, which is specialized for energy processing, there is one protein for every 15 lipids. On the other hand, myelin—a membrane that encloses portions of some neurons (nerve cells) and acts as an electrical insulator—has only one protein for every 70 lipids.

There are two general types of membrane proteins: peripheral proteins and integral proteins.

Peripheral membrane proteins lack exposed hydrophobic groups and are not embedded in the bilayer. Instead, they have polar or charged regions that interact with exposed parts of integral membrane proteins, or with the polar heads of phospholipid molecules (see Figure 6.1).

Integral membrane proteins are at least partly embedded in the phospholipid bilayer (see Figure 6.1). Like phospholipids, these proteins have both hydrophilic and hydrophobic regions (**Figure 6.3**).

- *Hydrophilic domains*: Stretches of amino acids with hydrophilic side chains (see Table 3.1) give certain regions of the

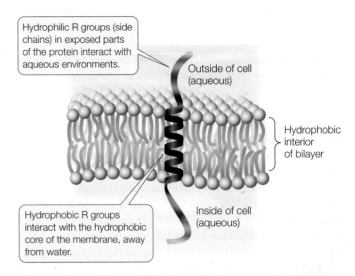

Hydrophilic R groups (side chains) in exposed parts of the protein interact with aqueous environments.

Outside of cell (aqueous)

Hydrophobic interior of bilayer

Hydrophobic R groups interact with the hydrophobic core of the membrane, away from water.

Inside of cell (aqueous)

6.3 Interactions of Integral Membrane Proteins An integral membrane protein is held in the membrane by the distribution of the hydrophilic and hydrophobic side chains on its amino acids. The hydrophilic parts of the protein extend into the aqueous cell exterior and the internal cytoplasm. The hydrophobic side chains interact with the hydrophobic lipid core of the membrane.

protein a polar character. These hydrophilic domains interact with water and stick out into the aqueous environment inside or outside the cell.

- *Hydrophobic domains*: Stretches of amino acids with hydrophobic side chains give other regions of the protein a nonpolar character. These domains interact with the fatty acids in the interior of the phospholipid bilayer, away from water.

A special preparation method for electron microscopy, called **freeze-fracturing**, reveals proteins that are embedded in the phospholipid bilayers of cellular membranes (**Figure 6.4**). When the two lipid *leaflets* (or layers) that make up the bilayer are separated, the proteins can be seen as bumps that protrude from the interior of each membrane. The bumps are not observed when artificial bilayers of pure lipid are freeze-fractured.

According to the fluid mosaic model, the proteins and lipids in a membrane are somewhat independent of each other and interact only noncovalently. The polar ends of proteins can interact with the polar ends of lipids, and the nonpolar regions of both molecules can interact hydrophobically.

However, some membrane proteins have fatty acids or other lipid groups covalently attached to them. Proteins in this subgroup of integral membrane proteins are referred to as *anchored membrane proteins*, because their hydrophobic lipid components allow them to insert themselves into the phospholipid bilayer.

Proteins are asymmetrically distributed on the inner and outer surfaces of membranes. An integral protein that extends all the way through the phospholipid bilayer and protrudes on both sides is known as a **transmembrane protein**. In addition to one or more *transmembrane domains* that extend through the bilayer, such a protein may have domains with other specific functions on the inner and outer sides of the membrane. Peripheral membrane proteins are localized on one side of the membrane or the other. This asymmetrical arrangement of membrane proteins gives the two surfaces of the membrane different properties. As we will soon see, these differences have great functional significance.

Like lipids, some membrane proteins move around relatively freely within the phospholipid bilayer. Experiments that involve the technique of cell fusion illustrate this migration dramatically. When two cells are fused, a single continuous membrane forms and surrounds both cells, and some proteins from each cell distribute themselves uniformly around this membrane (**Figure 6.5**).

Although some proteins are free to migrate in the membrane, others are not, but rather appear to be "anchored" to a specific region of the membrane. These membrane regions are like a corral of horses on a farm: the horses are free to move around within the fenced area, but not outside it. An example is the protein in the plasma membrane of a muscle cell that recognizes a chemical signal from a neuron. This protein is normally found only at the specific region where the neuron meets the muscle cell.

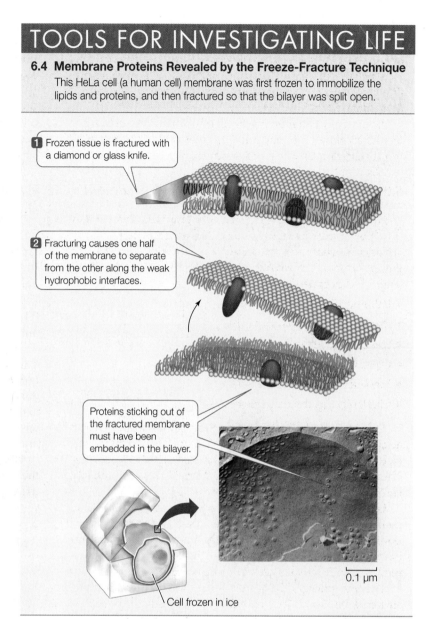

TOOLS FOR INVESTIGATING LIFE

6.4 Membrane Proteins Revealed by the Freeze-Fracture Technique

This HeLa cell (a human cell) membrane was first frozen to immobilize the lipids and proteins, and then fractured so that the bilayer was split open.

1 Frozen tissue is fractured with a diamond or glass knife.

2 Fracturing causes one half of the membrane to separate from the other along the weak hydrophobic interfaces.

Proteins sticking out of the fractured membrane must have been embedded in the bilayer.

0.1 μm

Cell frozen in ice

Proteins inside the cell can restrict the movement of proteins within a membrane. The cytoskeleton may have components just below the inner face of the membrane that are attached to membrane proteins protruding into the cytoplasm. The stability of the cytoskeletal components may thus restrict movement of attached membrane proteins.

Membranes are constantly changing

Membranes in eukaryotic cells are constantly forming, transforming from one type to another, fusing with one another, and breaking down. As we discuss in Chapter 5, fragments of membrane move, in the form of vesicles, from the endoplasmic reticulum (ER) to the Golgi, and from the Golgi to the plasma membrane (see Figure 5.10). Secondary lysosomes form when primary lysosomes from the Golgi fuse with phagosomes from the plasma membrane (see Figure 5.11).

6.5 Rapid Diffusion of Membrane Proteins

Two animal cells can be fused together in the laboratory, forming a single large cell (heterokaryon). This phenomenon was used to test whether membrane proteins can diffuse independently in the plane of the plasma membrane.

HYPOTHESIS Proteins embedded in a membrane can diffuse freely within the membrane.

METHOD

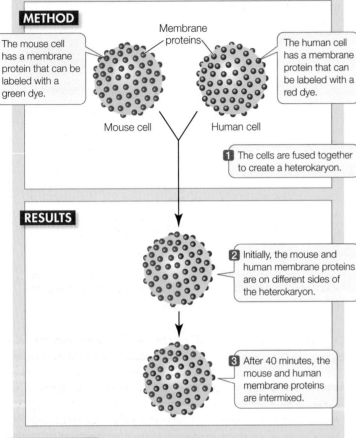

Membrane proteins

The mouse cell has a membrane protein that can be labeled with a green dye.

The human cell has a membrane protein that can be labeled with a red dye.

Mouse cell   Human cell

1 The cells are fused together to create a heterokaryon.

RESULTS

2 Initially, the mouse and human membrane proteins are on different sides of the heterokaryon.

3 After 40 minutes, the mouse and human membrane proteins are intermixed.

CONCLUSION Membrane proteins can diffuse rapidly in the plane of the membrane.

Go to **yourBioPortal.com** for original citations, discussions, and relevant links for all INVESTIGATING LIFE figures.

Because all membranes appear similar under the electron microscope, and because they interconvert readily, we might expect all subcellular membranes to be chemically identical. However, that is not the case, for there are major chemical differences among the membranes of even a single cell. Membranes are changed chemically when they form parts of certain organelles. In the Golgi apparatus, for example, the membranes of the *cis* face closely resemble those of the endoplasmic reticulum in chemical composition, but those of the *trans* face are more similar to the plasma membrane.

Plasma membrane carbohydrates are recognition sites

In addition to lipids and proteins, the plasma membrane contains carbohydrates. The carbohydrates are located on the outer surface of the plasma membrane and serve as recognition sites for other cells and molecules, as you will see in Section 6.2.

Membrane-associated carbohydrates may be covalently bonded to lipids or to proteins:

- A **glycolipid** consists of a carbohydrate covalently bonded to a lipid. Extending outside the cell surface, the carbohydrate may serve as a recognition signal for interactions between cells. For example, the carbohydrates on some glycolipids change when cells become cancerous. This change may allow white blood cells to target cancer cells for destruction.

- A **glycoprotein** consists of a carbohydrate covalently bonded to a protein. The bound carbohydrate is an oligosaccharide, usually not exceeding 15 monosaccharide units in length (see Section 3.3). The oligosaccharides of glycoproteins often function as signaling sites, as do the carbohydrates attached to glycolipids.

The "alphabet" of monosaccharides on the outer surfaces of membranes can generate a large diversity of messages. Recall from Section 3.3 that sugar molecules consist of three to seven carbons that are attached at different sites to one another. They may form linear or branched oligosaccharides with many different three-dimensional shapes. An oligosaccharide of a specific shape on one cell can bind to a complementary shape on an adjacent cell. This binding is the basis of cell–cell adhesion.

6.1 RECAP

The fluid mosaic model applies to both the plasma membrane and the membranes of organelles. An integral membrane protein has both hydrophilic and hydrophobic domains, which affect its position and function in the membrane. Carbohydrates that attach to lipids and proteins on the outside of the membrane serve as recognition sites.

- What are some of the features of the fluid mosaic model of biological membranes? See p. 106

- Explain how the hydrophobic and hydrophilic regions of phospholipids cause a membrane bilayer to form. **See Figures 6.1 and 6.2**

- What differentiates an integral protein from a peripheral protein? **See p. 108 and Figure 6.1**

- What is the experimental evidence that membrane proteins can diffuse in the plane of the membrane? **See pp. 109–110 and Figure 6.5**

Now that you understand the structure of biological membranes, let's see how their components function. In the next section we'll focus on the membrane that surrounds individual cells: the plasma membrane. We'll look at how the plasma membrane allows individual cells to be grouped together into multicellular systems of tissues.

6.2 How Is the Plasma Membrane Involved in Cell Adhesion and Recognition?

Some organisms, such as bacteria, are unicellular; that is, the entire organism is a single cell. Others, such as plants and animals, are multicellular—composed of many cells. Often these cells exist in specialized groups with similar functions, called tissues. Your body has about 60 trillion cells, arranged in different kinds of tissues (such as muscle, nerve, and epithelium).

Two processes allow cells to arrange themselves in groups:

- **Cell recognition**, in which one cell specifically binds to another cell of a certain type

- **Cell adhesion**, in which the connection between the two cells is strengthened

Both processes involve the plasma membrane. They are most easily studied if a tissue is separated into its individual cells, which are then allowed to adhere to one another again. Simple organisms provide a good model for studying processes that also occur in the complex tissues of larger species. Studies of sponges, for example, have revealed how cells associate with one another.

A sponge is a multicellular marine animal with a simple body plan that consists of only a few distinct tissues (see Section 31.5). The cells of a sponge adhere to one another, but can be separated mechanically by passing the animal several times through a fine wire screen (**Figure 6.6**). Through this process, what was a single animal becomes hundreds of individual cells suspended in seawater. Remarkably, if the cell suspension is shaken for a few hours, the cells bump into one another and stick together in the same shape and organization as the original sponge! The cells recognize and adhere to one another, and re-form the original tissues.

There are many different species of sponges. If disaggregated sponge cells from two different species are placed in the same container and shaken, individual cells will stick only to other cells of the same species. Two different sponges form, just like the ones at the start of the experiment. This demonstrates not just adhesion, but species-specific cell recognition.

Such tissue-specific and species-specific cell recognition and cell adhesion are essential to the formation and maintenance of tissues and multicellular organisms. Think of your own body. What keeps muscle cells bound to muscle cells and skin to skin? Specific cell adhesion is so obvious a characteristic of complex organisms that it is easy to overlook. You will see many examples of specific cell adhesion throughout this book; here, we describe its general principles. As you will see, cell recognition and cell adhesion depend on plasma membrane proteins.

Cell recognition and cell adhesion involve proteins at the cell surface

The molecule responsible for cell recognition and adhesion in sponges is a huge integral membrane glycoprotein (which is 80 percent carbohydrate by molecular weight) that is partly embedded in the plasma membrane, with the carbohydrate part sticking out and exposed to the environment (and to other

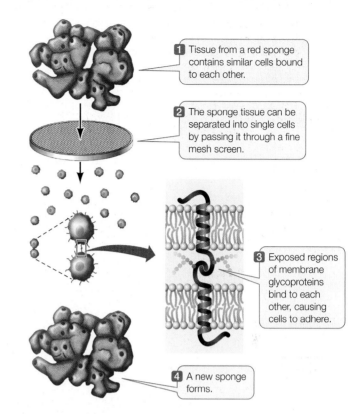

1 Tissue from a red sponge contains similar cells bound to each other.

2 The sponge tissue can be separated into single cells by passing it through a fine mesh screen.

3 Exposed regions of membrane glycoproteins bind to each other, causing cells to adhere.

4 A new sponge forms.

6.6 Cell Recognition and Adhesion In most cases (including the aggregation of animal cells into tissues), protein binding is homotypic.

sponge cells). As we describe in Section 3.2, a protein not only has a specific shape, but also has specific chemical groups exposed on its surface where they can interact with other substances, including other proteins. Both of these features allow binding to other specific molecules. The cells of the disaggregated sponge in Figure 6.6 find one another again through the recognition of exposed chemical groups on their membrane glycoproteins. Adhesion proteins are not restricted to animal cells. In most plant cells, the plasma membrane is covered with a thick cell wall, but this structure also has adhesion proteins that allow cells to bind to one another.

In most cases, the binding of cells in a tissue is **homotypic**; that is, the same molecule sticks out of both cells, and the exposed surfaces bind to each other. But **heterotypic** binding (of cells with different proteins) can also occur. In this case, different chemical groups on different surface molecules have an affinity for one another. For example, when the mammalian sperm meets the egg, different proteins on the two types of cells have complementary binding surfaces. Similarly, some algae form male and female reproductive cells (analogous to sperm and eggs) that have flagella to propel them toward each other. Male and female cells can recognize each other by heterotypic proteins on their flagella.

Three types of cell junctions connect adjacent cells

In a complex multicellular organism, cell recognition proteins allow specific types of cells to bind to one another. Often, after

(A)

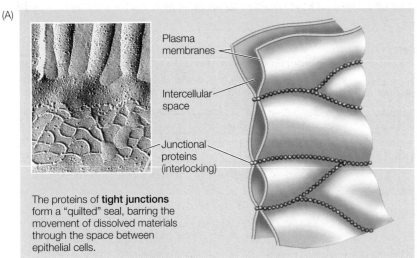

Plasma membranes

Intercellular space

Junctional proteins (interlocking)

The proteins of **tight junctions** form a "quilted" seal, barring the movement of dissolved materials through the space between epithelial cells.

(B)

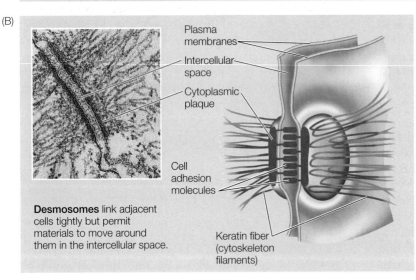

Plasma membranes

Intercellular space

Cytoplasmic plaque

Cell adhesion molecules

Keratin fiber (cytoskeleton filaments)

Desmosomes link adjacent cells tightly but permit materials to move around them in the intercellular space.

(C)

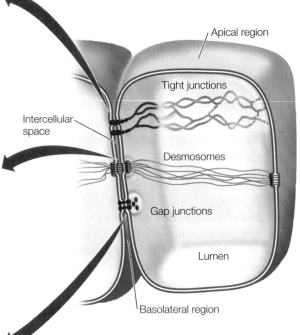

Plasma membranes

Intercellular space

Hydrophilic channel

Molecules pass between cells

Connexins (channel proteins)

Gap junctions let adjacent cells communicate.

6.7 Junctions Link Animal Cells Together   Tight junctions (A) and desmosomes (B) are abundant in epithelial tissues. Gap junctions (C) are also found in some muscle and nerve tissues, in which rapid communication between cells is important. Although all three junction types are shown in the cell at the right, all three are not necessarily seen at the same time in actual cells.

yourBioPortal.com

GO TO Web Activity 6.2 • Animal Cell Junctions

Apical region

Tight junctions

Intercellular space

Desmosomes

Gap junctions

Lumen

Basolateral region

pressure, or both, so it is particularly important that their cells adhere tightly. We will examine three types of cell junctions that enable animal cells to seal intercellular spaces, reinforce attachments to one another, and communicate with each other. Tight junctions, desmosomes, and gap junctions, respectively, perform these three functions.

TIGHT JUNCTIONS SEAL TISSUES   **Tight junctions** are specialized structures that link adjacent epithelial cells, and they result from the mutual binding of specific proteins in the plasma membranes of the cells. These proteins are arrayed in bands so that they form a series of joints encircling each cell (**Figure 6.7A**). Tight junctions are found in the lining of lumens (cavities) in organs such as the stomach and intestine. They have two major functions:

• They prevent substances from moving from the lumen through the spaces between cells. For example, the presence of tight junctions means that substances must pass through, rather than between, the epithelial cells that form

the initial binding, both cells contribute material to form additional membrane structures that connect them to one another. These specialized structures, called **cell junctions**, are most evident in electron micrographs of *epithelial* tissues, which are layers of cells that line body cavities or cover body surfaces. These surfaces often receive stresses, or must retain contents under

the lining of the digestive tract. In another example, the cells lining the bladder have tight junctions so urine cannot leak out into the body cavity. Thus, tight junctions help to establish cellular control over what enters and leaves the body.

- They define specific functional regions of membranes by restricting the migration of membrane proteins and phospholipids from one region of the cell to another. Thus the membrane proteins and phospholipids in the apical ("tip") region of an intestinal epithelial cell (facing the lumen) are different from those in the basolateral (*basal*, "bottom"; *lateral*, "side") regions of the cell (facing the body cavity or blood capillary outside the lumen).

By forcing materials to enter certain cells, and by allowing different areas of the same cell to have different membrane proteins with different functions, tight junctions in the digestive tract help ensure the directional movement of materials into the body.

DESMOSOMES HOLD CELLS TOGETHER   **Desmosomes** connect adjacent plasma membranes. Desmosomes hold neighboring cells firmly together, acting like spot welds or rivets (**Figure 6.7B**). Each desmosome has a dense structure called a plaque on the cytoplasmic side of the plasma membrane. To this plaque are attached special cell adhesion molecules that stretch from the plaque through the plasma membrane of one cell, across the intercellular space, and through the plasma membrane of the adjacent cell, where they bind to the plaque proteins in that adjacent cell.

The plaque is also attached to fibers in the cytoplasm. These fibers, which are intermediate filaments of the cytoskeleton (see Figure 5.18), are made of a protein called keratin. They stretch from one cytoplasmic plaque across the cell to another plaque on the other side of the cell. Anchored thus on both sides of the cell, these extremely strong fibers provide great mechanical stability to epithelial tissues. This stability is needed for these tissues, which often receive rough wear while protecting the integrity of the organism's body surface, or the surface of an organ.

GAP JUNCTIONS ARE A MEANS OF COMMUNICATION   Whereas tight junctions and desmosomes have mechanical roles, **gap junctions** facilitate communication between cells. Each gap junction is made up of specialized channel proteins, called *connexins*, which interact to form a structure (called a *connexon*) that spans the plasma membranes of adjacent cells and the intercellular space between them (**Figure 6.7C**). Water, dissolved small molecules, and ions can pass from cell to cell through these junctions. This allows groups of cells to coordinate their activities. In Chapter 7 we discuss cell communication and signaling, and in that chapter we describe in more detail the roles of gap junctions and plasmodesmata, which perform a similar role in plants.

Cell membranes adhere to the extracellular matrix

In Section 5.4 we describe the extracellular matrix of animal cells, which is composed of collagen protein arranged in fibers

in a gelatinous matrix of proteoglycans. The attachment of a cell to the extracellular matrix is important in maintaining the integrity of a tissue. In addition, some cells can detach from their neighbors, move, and attach to other cells; this is often mediated by interactions with the extracellular matrix.

A transmembrane protein called **integrin** often mediates the attachment of epithelial cells to the extracellular matrix (**Figure 6.8**). More than 24 different integrins have been described in human cells. All of them bind to a protein in the extracellular matrix on the outside of the cell, and to actin filaments, which are part of the cytoskeleton, on the inside of the cell. So, in addition to adhesion, integrin has a role in maintaining cell structure via its interaction with the cytoskeleton.

The binding of integrin to the extracellular matrix is noncovalent and reversible. When a cell moves its location within a tissue or organism, the first step is detachment of the cell's integrin from the matrix. The integrin protein changes its three-dimensional structure and no longer maintains its link to the matrix. These events are important for cell movement within the developing embryo, and for the spread of cancer cells.

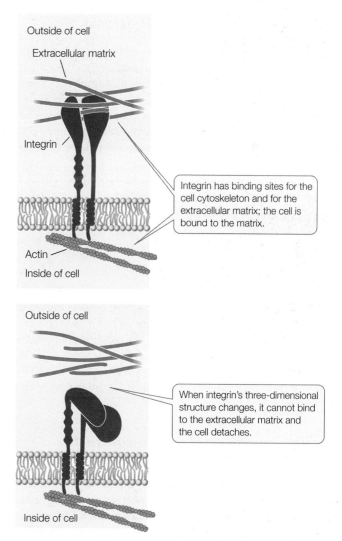

Outside of cell

Extracellular matrix

Integrin

Integrin has binding sites for the cell cytoskeleton and for the extracellular matrix; the cell is bound to the matrix.

Actin

Inside of cell

Outside of cell

When integrin's three-dimensional structure changes, it cannot bind to the extracellular matrix and the cell detaches.

Inside of cell

6.8 Integrins Mediate the Attachment of Animal Cells to the Extracellular Matrix

We have just examined how the plasma membrane structure accommodates the binding and maintenance of cell adhesion. We turn now to another major function of membranes: regulating the substances that enter or leave a cell or organelle.

6.3 What Are the Passive Processes of Membrane Transport?

As you have already learned, biological membranes have many functions, and control of the cell's internal composition is one of the most significant. Biological membranes allow some substances, but not others, to pass through them. This characteristic of membranes is called **selective permeability**. Selective permeability allows the membrane to determine what substances enter or leave a cell or organelle.

There are two fundamentally different processes by which substances cross biological membranes:

- The processes of **passive transport** do not require any input of outside energy to drive them (no metabolic energy).
- The processes of **active transport** require the input of chemical energy from an outside source (metabolic energy).

This section focuses on the passive processes by which substances cross membranes. The energy for the passive transport of a substance is found in the difference between its concentration on one side of the membrane and its concentration on the other. Passive transport processes include two types of diffusion: simple diffusion through the phospholipid bilayer, and facilitated diffusion through *channel proteins* or by means of *carrier proteins*.

Diffusion is the process of random movement toward a state of equilibrium

Nothing in this world is ever absolutely at rest. Everything is in motion, although the motions may be very small. An important consequence of all this random vibration, rotation and translocation (moving from one location to another) of molecules is that all the components of a solution tend eventually to become evenly distributed. For example, if a drop of ink is allowed to fall into a container of water, the pigment molecules of the ink are initially very concentrated. Without human intervention, such as stirring, the pigment molecules move about at random, spreading slowly through the water until eventually the concentration of pigment—and thus the intensity of color—is exactly the same in every drop of liquid in the container.

A solution in which the solute particles are uniformly distributed is said to be at *equilibrium* because there will be no future net change in their concentration. Equilibrium does not mean that the particles have stopped moving; it just means that they are moving in such a way that their overall distribution does not change.

Diffusion is the process of random movement toward a state of equilibrium. Although the motion of each individual particle is absolutely random, the net movement of particles is directional until equilibrium is reached. Diffusion is thus a net movement from regions of greater concentration to regions of lesser concentration (**Figure 6.9**).

In a complex solution (one with many different solutes), the diffusion of each solute is independent of those of the others. How fast a substance diffuses depends on three factors:

- The *diameter* of the molecules or ions: smaller molecules diffuse faster.
- The *temperature* of the solution: higher temperatures lead to faster diffusion because ions or molecules have more energy, and thus move more rapidly, at higher temperatures.
- The *concentration gradient* in the system—that is, the change in solute concentration with distance in a given direction: the greater the concentration gradient, the more rapidly a substance diffuses.

We'll see how these factors influence membrane transport in the detailed discussions that follow.

DIFFUSION WITHIN CELLS AND TISSUES Within cells, or wherever distances are very short, solutes distribute themselves rapidly by diffusion. Small molecules and ions may move from one end of an organelle to another in a millisecond (10^{-3} s, or one-thousandth of a second). However, the usefulness of diffusion as a transport mechanism declines drastically as distances become greater. In the absence of mechanical stirring, diffusion across more than a centimeter may take an hour or more, and diffusion across meters may take years! Diffusion would not be adequate to distribute materials over the length of a human body, much less that of a larger organism. But within our cells or across layers of one or two cells, diffusion is rapid enough to distribute small molecules and ions almost instantaneously.

DIFFUSION ACROSS MEMBRANES In a solution without barriers, all the solutes diffuse at rates determined by temperature, their physical properties, and their concentration gradients. If a biological membrane divides the solution into separate compartments, then the movement of the different solutes can be affected by the properties of the membrane. The membrane is said to be *permeable* to solutes that can cross it more or less easily, but *impermeable* to substances that cannot move across it.

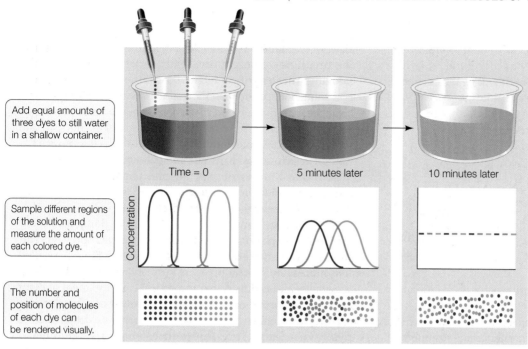

Add equal amounts of three dyes to still water in a shallow container.

Sample different regions of the solution and measure the amount of each colored dye.

The number and position of molecules of each dye can be rendered visually.

Time = 0   5 minutes later   10 minutes later

6.9 Diffusion Leads to Uniform Distribution of Solutes A simple experiment demonstrates that solutes move from regions of greater concentration to regions of lesser concentration until equilibrium is reached.

Molecules to which the membrane is impermeable remain in separate compartments, and their concentrations may be different on the two sides of the membrane. Molecules to which the membrane is permeable diffuse from one compartment to the other until their concentrations are equal on both sides of the membrane. When the concentrations of a diffusing substance on the two sides of the permeable membrane are identical, equilibrium is reached. Individual molecules continue to pass through the membrane after equilibrium is established, but equal numbers of molecules move in each direction, so *at equilibrium there is no net change in concentration.*

Simple diffusion takes place through the phospholipid bilayer

In **simple diffusion**, small molecules pass through the phospholipid bilayer of the membrane. A molecule that is itself hydrophobic, and is therefore soluble in lipids, enters the membrane readily and is able to pass through it. The more lipid-soluble the molecule is, the more rapidly it diffuses through the membrane bilayer. This statement holds true over a wide range of molecular weights.

On the other hand, electrically charged or polar molecules, such as amino acids, sugars, and ions, do not pass readily through a membrane for two reasons. First, such charged or polar molecules are not very soluble in the hydrophobic interior of the bilayer. Second, such charged and polar substances form many hydrogen bonds with water and ions in the aqueous environment, be it the cytoplasm or the cell exterior. The multiplicity of these hydrogen bonds prevent the substances from moving into the hydrophobic interior of the membrane.

Consider two molecules: a small protein made up of a few polar amino acids, and a cholesterol-based steroid of equivalent size. If a membrane separates high and low concentrations of these substances, the protein, being polar, will diffuse only very slowly through the membrane, while the nonpolar steroid will diffuse through it readily.

Osmosis is the diffusion of water across membranes

Water molecules pass through specialized channels in membranes (see below) by a diffusion process called **osmosis**. This completely passive process uses no metabolic energy and can be understood in terms of solute concentrations. Recall that a solute dissolves in a solvent and the solute's constituents are dispersed throughout the solution. Osmosis depends on the *number* of solute particles present, not on the *kinds* of particles. We will describe osmosis using red blood cells and plant cells as examples. In these examples, the plasma membranes are considered to be permeable to water and impermeable to most solutes.

Red blood cells are normally suspended in a fluid called plasma, which contains salts, proteins, and other solutes. Examining a drop of blood under the light microscope reveals that these red cells have a characteristic flattened disk shape with a depressed center, sometimes called "biconcave." If pure water is added to the drop of blood, drastically reducing the solute concentration of the plasma, the red cells quickly swell and burst. Similarly, if slightly wilted lettuce is placed in pure water, it soon becomes crisp; by weighing it before and after, we can show that it has taken up water. If, on the other hand, red blood cells or crisp lettuce leaves are placed in a relatively concentrated solution of salt or sugar, the leaves become limp (they wilt), and the red blood cells pucker and shrink.

From such observations we know that the difference in solute concentration between a cell and its surrounding environment determines whether water will move from the environment into the cell or out of the cell into the environment. Other things being equal, if two different solutions are separated by a membrane that allows water, *but not solutes*, to pass through, water molecules will move across the membrane toward the solution with a higher solute concentration. In other words, water will diffuse from a region of its higher concentration (with a lower concentration of solutes) to a region of its lower concentration (with a higher concentration of solutes).

Three terms are used to compare the solute concentrations of two solutions separated by a membrane:

• A **hypertonic** solution has a higher solute concentration than the other solution with which it is being compared (**Figure 6.10A**).

• **Isotonic** solutions have equal solute concentrations (**Figure 6.10B**).

• A **hypotonic** solution has a lower solute concentration than the other solution with which it is being compared (**Figure 6.10C**).

Water moves from a hypotonic solution across a membrane to a hypertonic solution.

When we say that "water moves," bear in mind that we are referring to the net movement of water. Since it is so abundant, water is constantly moving through protein channels across the plasma membrane into and out of cells. What concerns us here is whether the overall movement is greater in one direction or the other.

The concentration of solutes in the environment determines the direction of osmosis in all animal cells. A red blood cell takes up water from a solution that is hypotonic to the cell's contents.

The cell bursts because its plasma membrane cannot withstand the pressure created by the water entry and the resultant swelling. The integrity of red blood cells (and other blood cells) is absolutely dependent on the maintenance of a constant solute concentration in the blood plasma: the plasma must be isotonic to the blood cells if the cells are not to burst or shrink. Regulation of the solute concentration of body fluids is thus an important process for organisms without cell walls.

In contrast to animal cells, the cells of plants, archaea, bacteria, fungi, and some protists have cell walls that limit their volumes and keep them from bursting. Cells with sturdy walls take up a limited amount of water, and in so doing they build up internal pressure against the cell wall, which prevents further water from entering. This pressure within the cell is called **turgor pressure**. Turgor pressure keeps plants upright (and lettuce crisp) and is the driving force for the enlargement of plant cells. It is a normal and essential component of plant growth. If enough water leaves the cells, turgor pressure drops and the plant wilts. Turgor pressure reaches about 100 pounds per square inch (0.7 kg/cm^2)—several times greater than the pres-

6.10 Osmosis Can Modify the Shapes of Cells In a solution that is isotonic with the cytoplasm (center column), a plant or animal cell maintains a consistent, characteristic shape because there is no net movement of water into or out of the cell. In a solution that is hypotonic to the cytoplasm (right), water enters the cell. An environment that is hypertonic to the cytoplasm (left) draws water out of the cell.

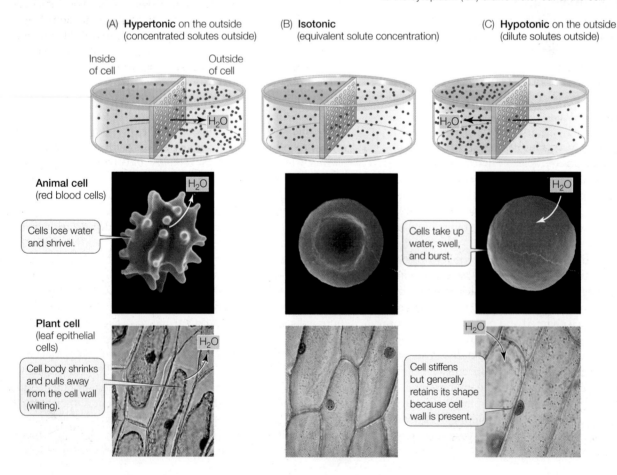

(A) **Hypertonic** on the outside (concentrated solutes outside)

(B) **Isotonic** (equivalent solute concentration)

(C) **Hypotonic** on the outside (dilute solutes outside)

Inside of cell / Outside of cell

H_2O

Animal cell (red blood cells)

Cells lose water and shrivel. / H_2O

Cells take up water, swell, and burst. / H_2O

Plant cell (leaf epithelial cells)

Cell body shrinks and pulls away from the cell wall (wilting). / H_2O

Cell stiffens but generally retains its shape because cell wall is present. / H_2O

sure in automobile tires. This pressure is so great that the cells would change shape and detach from one another, were it not for adhesive molecules in the plant cell wall.

Diffusion may be aided by channel proteins

As we saw earlier, polar or charged substances such as water, amino acids, sugars and ions do not readily diffuse across membranes. But they can cross the hydrophobic phospholipid bilayer passively (that is, without the input of energy) in one of two ways, depending on the substance:

- **Channel proteins** are integral membrane proteins that form channels across the membrane through which certain substances can pass.
- Some substances can bind to membrane proteins called **carrier proteins** that speed up their diffusion through the phospholipid bilayer.

Both of these processes are forms of **facilitated diffusion**. That is, the substances diffuse according to their concentration gradients, but their diffusion is facilitated by protein channels or carriers.

ION CHANNELS The best-studied channel proteins are the **ion channels**. As you will see in later chapters, the movement of ions across membranes is important in many biological processes, ranging from respiration within the mitochondria, to the electrical activity of the nervous system and the opening of the pores in leaves that allow gas exchange with the environment. Several types of ion channels have been identified, each of them specific for a particular ion. All of them show the same basic structure of a hydrophilic pore that allows a particular ion to move through it (**Figure 6.11**).

Just as a fence may have a gate that can be opened or closed, most ion channels are gated: they can be opened or closed to ion passage. A **gated channel** opens when a stimulus causes a change in the three-dimensional shape of the channel. In some cases, this stimulus is the binding of a chemical signal, or **ligand** (see Figure 6.11). Channels controlled in this way are called *ligand-gated channels*. In contrast, a *voltage-gated channel* is stimulated to open or close by a change in the voltage (electrical charge difference) across the membrane.

THE MEMBRANE POTENTIAL All living cells maintain an imbalance of ion concentrations across the plasma membrane, and consequently a small voltage or **membrane potential** exists across that membrane. When a gated ion

channel opens, millions of ions can rush through it per second. How fast the ions move, and in which direction (into or out of the cell), depends on two factors, the concentration gradient and the magnitude of the voltage. Let's consider how these factors affect the concentration of potassium ions (K^+) inside an animal cell:

- *The concentration gradient*: Because of active transport (discussed below), the concentration of K^+ is usually much higher inside the cell than outside, so K^+ will tend to diffuse out of the cell through an open potassium channel.
- *The distribution of electrical charge*: As K^+ diffuses out of the cell it leaves behind an excess of chloride (Cl^-) and other negatively charged ions. These negatively charged substances cannot readily diffuse through the plasma membrane to follow K^+ out of the cell, and this results in a charge difference (negative inside) across the membrane. K^+ is attracted to the negative charge inside the cell, creating a tendency for K^+ to stay inside the cell, even though it is more concentrated there than outside.

Now, consider what happens when the K^+ channel is opened. Two forces are at work: diffusion draws K^+ out of the cell through the channel, and electrical attraction keeps K^+ inside the cell. The system exists in a state of equilibrium, in which the ion's rate of diffusion out through the channel is balanced by the rate of movement in through the channel due to electrical attraction. Obviously, the concentrations of K^+ on each side of the membrane will not be equal, as we would expect if diffusion were the only force involved. Instead, the attraction of electrical charges keeps some extra K^+ inside the cell. This imbalance in K^+ is a major factor in generating a voltage across the plasma membrane called the *membrane potential*.

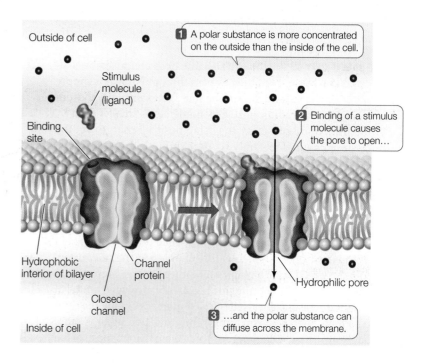

6.11 A Gated Channel Protein Opens in Response to a Stimulus The channel protein has a pore of polar amino acids and water. It is anchored in the hydrophobic bilayer interior by its outer coating of nonpolar R groups of its amino acids. The protein changes its three-dimensional shape when a stimulus molecule (ligand) binds to it, opening the pore so that hydrophilic polar substances can pass through. Other gated channels open in response to an electrical potential (voltage).

The membrane potential is related to the concentration imbalance of K^+ by the Nernst equation:

$$E_K = 2.3 \frac{RT}{zF} \log \frac{[K]_o}{[K]_i}$$

where R is the gas constant, F is the Faraday constant (both familiar to chemistry students), T is the temperature, and z is the charge on the ion (+1). Solving for $2.3\, RT/zF$ at 20°C ("room temperature"), the equation becomes much simpler:

$$E_K = 58 \log \frac{[K]_o}{[K]_i}$$

where E_K is the membrane potential (in millivolts, mV) that results from the ratio of K^+ concentrations outside the cell $[K]_o$ and inside the cell $[K]_i$.

What does this equation tell us about cells? It shows that a small change in K^+ concentration, due to the opening of a ligand-gated K^+ channel, for example, can have a large effect on the electrical potential (E) across the membrane. This change in potential might be enough to cause other proteins in the membrane, such as voltage-gated channels, to change configuration. As we discuss in Chapter 45, this is exactly what happens in the nervous system. Many drugs that act on electrically sensitive tissues work as ligands that open ion channels and thereby affect membrane potential. And as you will see shortly, membrane potential drives secondary active transport.

Actual measurements from animal cells give a total membrane potential between –60 and –70 mV across the membrane, where the inside is negative with respect to the outside (see Figure 45.5). Cells have a tremendous amount of potential energy stored in their membrane potentials. In fact, the brain cells you are using to read this book have more potential energy—about 200,000 volts per centimeter—than the high-voltage electric lines powering your reading light, which carry about 2 volts per centimeter.

THE SPECIFICITY OF ION CHANNELS How does an ion channel allow one ion, but not another, to pass through? It is not simply a matter of charge and size of the ion. For example, a sodium ion (Na^+), with a radius of 0.095 nanometers, is smaller than K^+ (0.130 nm), and both carry the same positive charge. Yet the potassium channel lets only K^+ pass through the membrane, and not the smaller Na^+. Nobel Laureate Roderick MacKinnon at The Rockefeller University found an elegant explanation for this when he deciphered the structure of a potassium channel from a bacterium (**Figure 6.12**).

Being charged, both Na^+ and K^+ are attracted to water molecules. They are surrounded by water "shells" in solution, held by the attraction of their positive charges to the negatively charged oxygen atoms on the water molecules (see Figure 2.10). The potassium channel contains highly polar oxygen atoms at its opening. The gap enclosed by these atoms is exactly the right size so that when a K^+ ion approaches the opening, it is more strongly attracted to the oxygen atoms there than to those of the

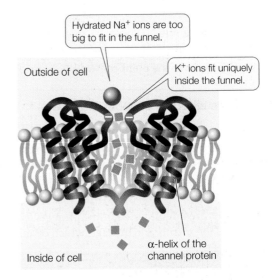

6.12 The Potassium Channel The positively charged potassium ions are attracted by the polar (negatively charged) oxygen atoms in the R groups (side chains) of the channel protein, and the ions funnel through the channel. This channel is a "custom fit" for K^+; other ions do not pass through.

water molecules in its shell. It sheds its water shell and passes through the channel. The smaller Na^+ ion, on the other hand, is kept a bit more distant from the oxygen atoms at the opening of the channel because extra water molecules can fit between the ion (with its shell) and the oxygen atoms at the opening. So Na^+ does not enter the potassium channel. The gate that opens or closes the channel appears to be an interaction between positively charged arginine residues on the protein and negative charges on membrane phospholipids. This is an example of the functional interactions between membrane proteins and lipids.

AQUAPORINS FOR WATER Water crosses membranes at a much faster rate than would be expected for simple diffusion through the hydrophobic phospholipid bilayer. One way that water can do this is by "hitchhiking" with some ions, such as Na^+, as they pass through ion channels. Up to 12 water molecules may coat an ion as it traverses a channel. But there is an even faster way to get water across membranes. Plant cells and some animal cells, such as red blood cells and kidney cells, have membrane channels called **aquaporins**. These channels function as a cellular plumbing system for moving water. Like the K^+ channel, the aquaporin channel is highly specific. Water molecules move in single file through the channel, which excludes ions so that the electrical properties of the cell are maintained.

Aquaporins were first identified by Peter Agre at Duke University, who shared the Nobel Prize with Rod McKinnon (see above). Agre noticed a membrane protein that was present in red blood cells, kidney cells, and plant cells but did not know its function. A colleague suggested that it might be a water channel, because these cell types show rapid diffusion of water across their membranes. Agre inserted the protein into the membrane of an oocyte, which normally does not permit much diffusion of water. He injected the oocyte with mRNA for aquaporin, from which the protein was produced and inserted into the membrane. Remarkably, the oocyte began swelling immediately after being transferred to a hypotonic solution, indicating rapid diffusion of water into the cell (**Figure 6.13**).

Carrier proteins aid diffusion by binding substances

As we described earlier, another kind of facilitated diffusion involves not just the opening of a channel, but also the actual binding of the transported substance to a membrane protein called a carrier protein. Like channel proteins, carrier proteins allow diffusion both into and out of the cell or organelle. In other words, carrier proteins operate in both directions. Carrier proteins transport polar molecules such as sugars and amino acids.

INVESTIGATING LIFE

6.13 Aquaporin Increases Membrane Permeability to Water

A protein was isolated from the membranes of cells in which water diffuses rapidly across the membranes. When the protein was inserted into oocytes, which do not normally have it, the water permeability of the oocytes was greatly increased.

HYPOTHESIS Aquaporin increases membrane permeability to water.

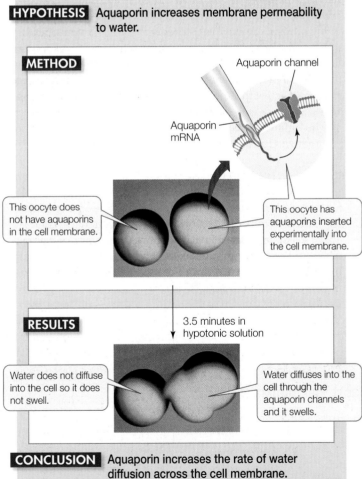

METHOD

Aquaporin channel

Aquaporin mRNA

This oocyte does not have aquaporins in the cell membrane.

This oocyte has aquaporins inserted experimentally into the cell membrane.

RESULTS

3.5 minutes in hypotonic solution

Water does not diffuse into the cell so it does not swell.

Water diffuses into the cell through the aquaporin channels and it swells.

CONCLUSION Aquaporin increases the rate of water diffusion across the cell membrane.

Go to **yourBioPortal.com** for original citations, discussions, and relevant links for all INVESTIGATING LIFE figures.

Glucose is the major energy source for most mammalian cells, and they require a great deal of it. Their membranes contain a carrier protein—the glucose transporter—that facilitates glucose uptake into the cell. Binding of glucose to a specific three-dimensional site on one side of the transporter protein causes the protein to change its shape and release glucose on the other side of the membrane (**Figure 6.14A**). Since glucose is broken down almost as soon as it enters a cell, there is almost always a strong concentration gradient favoring glucose entry (that is, a higher concentration outside the cell than inside). The transporter allows glucose molecules to cross the membrane and enter the cell much faster than they would by simple diffusion through the bilayer. This rapid entry is necessary to ensure that the cell receives enough glucose for its energy needs.

Transport by carrier proteins is different from simple diffusion. In both processes, the rate of movement depends on the concentration gradient across the membrane. However, in carrier-mediated transport, a point is reached at which increases in the concentration gradient are not accompanied by an increased rate of diffusion. At this point, the facilitated diffusion system is said to be *saturated* (**Figure 6.14B**). Because there are only a limited number of carrier protein molecules per unit of membrane area, the rate of diffusion reaches a maximum when all the carrier molecules are fully loaded with solute molecules. Think of waiting for the elevator on the ground floor of a hotel with 50 other people. They can't all get in the elevator (carrier) at once, so the rate of transport (say 10 people at a time) is saturated.

yourBioPortal.com

GO TO Animated Tutorial 6.1 • Passive Transport

6.3 RECAP

Diffusion is the movement of ions or molecules from a region of greater concentration to a region of lesser concentration. Water can diffuse through cell membranes by a process called osmosis. Channel proteins, which can be open or closed, and carrier proteins facilitate diffusion of charged and polar substances, including water. The diffusion of ions across cell membranes sets up an electrochemical potential gradient across the membranes.

- What properties of a substance determine whether, and how fast, it will diffuse across a membrane? See p. 114
- Describe osmosis and explain the terms hypertonic, hypotonic, and isotonic. See p. 116 and Figure 6.10
- How does a channel protein facilitate diffusion? See p. 118 and Figures 6.11 and 6.12

The process of diffusion tends to equalize the concentrations of substances outside and inside cells. However, one hallmark of

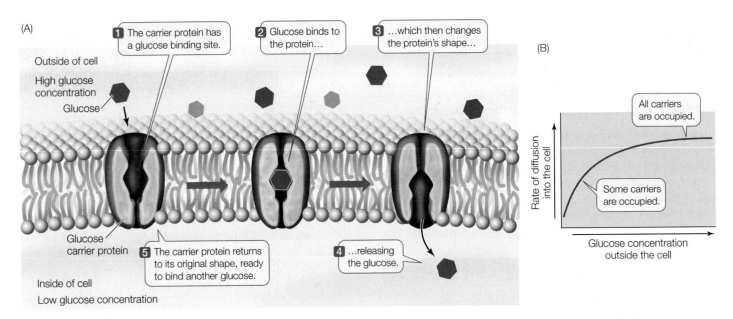

(A)

1 The carrier protein has a glucose binding site.

2 Glucose binds to the protein...

3 ...which then changes the protein's shape...

(B)

Outside of cell

High glucose concentration

Glucose

Glucose carrier protein

5 The carrier protein returns to its original shape, ready to bind another glucose.

4 ...releasing the glucose.

Inside of cell

Low glucose concentration

All carriers are occupied.

Some carriers are occupied.

Rate of diffusion into the cell

Glucose concentration outside the cell

6.14 A Carrier Protein Facilitates Diffusion The glucose transporter is a carrier protein that allows glucose to enter the cell at a faster rate than would be possible by simple diffusion. (A) The transporter binds to glucose, brings it into the membrane interior, then changes shape, releasing glucose into the cell cytoplasm. (B) The graph shows the rate of glucose entry via a carrier versus the concentration of glucose outside the cell. As the glucose concentration increases, the rate of diffusion increases until the point at which all the available transporters are being used (the system is saturated).

a living thing is that it can have an internal composition quite different from that of its environment. To achieve this it must sometimes move substances in opposite directions from the ones in which they would naturally tend to diffuse. That is, substances must sometimes be moved against concentration gradients and/or against the cell's membrane potential (electrical gradient). This process requires work—the input of energy—and is known as *active transport*.

6.4 What are the Active Processes of Membrane Transport?

In many biological situations, there is a different concentration of a particular ion or small molecule inside compared with outside a cell. In these cases, the imbalance is maintained by a pro-

tein in the plasma membrane that moves the substance against its concentration and/or electrical gradient. This is called *active transport*, and because it is acting "against the normal flow," it requires the expenditure of energy. Often the energy source is adenosine triphosphate (ATP). In eukaryotes, ATP is produced in the mitochondria and has chemical energy stored in its terminal phosphate bond. This energy is released when ATP is converted to adenosine diphosphate (ADP) in a hydrolysis reaction that breaks the terminal phosphate bond. This is one source of energy for active transport. (We give the details of how ATP provides energy to cells in Section 8.2.)

The differences between diffusion and active transport are summarized in **Table 6.1**.

Active transport is directional

Simple and facilitated diffusion follow concentration gradients and can occur in both directions across a membrane. In contrast, active transport is directional, and moves a substance either into or out of the cell or organelle, depending on need. There are three types of active transport, each involving its own type of membrane protein (**Figure 6.15**):

- A **uniporter** moves a single substance in one direction. For example, a calcium-binding protein found in the plasma

TABLE 6.1
Membrane Transport Mechanisms

	SIMPLE DIFFUSION	DIFFUSION THROUGH CHANNEL	FACILITATED DIFFUSION	ACTIVE TRANSPORT
Cellular energy required?	No	No	No	Yes
Driving force	Concentration gradient	Concentration gradient	Concentration gradient	ATP hydrolysis (against concentration gradient)
Membrane protein required?	No	Yes	Yes	Yes
Specificity	No	Yes	Yes	Yes

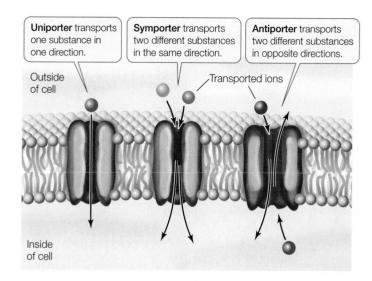

6.15 Three Types of Proteins for Active Transport   Note that in each of the three cases, transport is directional. Symporters and antiporters are examples of coupled transporters. All three types of transporters are coupled to energy sources in order to move substances against their concentration gradients.

membrane and endoplasmic reticulum of many cells actively transports Ca^{2+} to locations where it is more highly concentrated, either outside the cell or inside the ER.

- A **symporter** moves two substances in the same direction. For example, a symporter in the cells that line the intestine must bind Na^+ in addition to an amino acid in order to absorb amino acids from the intestine.

- An **antiporter** moves two substances in opposite directions, one into the cell (or organelle) and the other out of the cell (or organelle). For example, many cells have a sodium–potassium pump that moves Na^+ out of the cell and K^+ into it.

Symporters and antiporters are also known as *coupled transporters* because they move two substances at once.

Different energy sources distinguish different active transport systems

There are two basic types of active transport:

- **Primary active transport** involves the direct hydrolysis of ATP, which provides the energy required for transport.

- **Secondary active transport** does not use ATP directly. Instead, its energy is supplied by an ion concentration and electrical gradient established by primary active transport. This transport system uses the energy of ATP indirectly to set up the gradient.

In primary active transport, energy released by the hydrolysis of ATP drives the movement of specific ions against their concentration gradients. For example, we mentioned earlier that concentrations of potassium ions (K^+) inside a cell are often much higher than in the fluid bathing the cell. On the other hand, the concentration of sodium ions (Na^+) is often much higher outside the cell. A protein in the plasma membrane pumps Na^+ out of the cell and K^+ into the cell against these concentration and electrochemical gradients, ensuring that the gradients are maintained (**Figure 6.16**). This **sodium–potassium (Na^+–K^+) pump** is

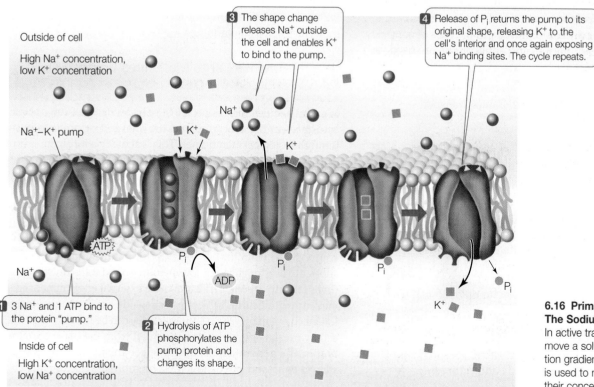

6.16 Primary Active Transport: The Sodium–Potassium Pump
In active transport, energy is used to move a solute against its concentration gradient. Here, energy from ATP is used to move Na^+ and K^+ against their concentration gradients.

6.17 Secondary Active Transport The Na$^+$ concentration gradient established by primary active transport (left) powers the secondary active transport of glucose (right). A symporter protein couples the movement of glucose across the membrane against its concentration gradient to the passive movement of Na$^+$ into the cell.

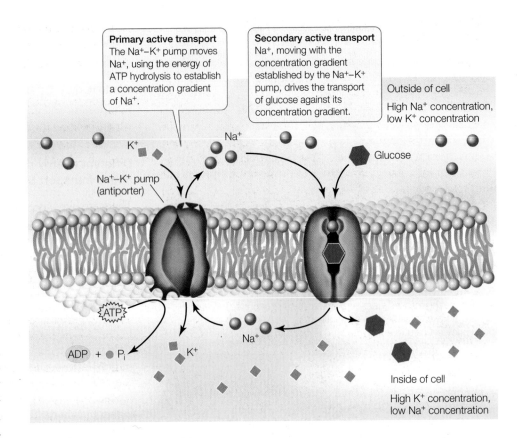

Primary active transport The Na$^+$–K$^+$ pump moves Na$^+$, using the energy of ATP hydrolysis to establish a concentration gradient of Na$^+$.

Secondary active transport Na$^+$, moving with the concentration gradient established by the Na$^+$–K$^+$ pump, drives the transport of glucose against its concentration gradient.

Outside of cell
High Na$^+$ concentration, low K$^+$ concentration

K$^+$

Na$^+$

Glucose

Na$^+$–K$^+$ pump (antiporter)

ATP

ADP + P$_i$

K$^+$

Na$^+$

Inside of cell
High K$^+$ concentration, low Na$^+$ concentration

found in all animal cells. The pump is an integral membrane glycoprotein. It breaks down a molecule of ATP to ADP and a free phosphate ion (P$_i$) and uses the energy released to bring two K$^+$ ions into the cell and export three Na$^+$ ions. The Na$^+$–K$^+$ pump is thus an antiporter because it moves two substances in different directions.

In secondary active transport, the movement of a substance against its concentration gradient is accomplished using energy "regained" by letting ions move across the membrane with their electrochemical and concentration gradients. For example, once the sodium–potassium pump establishes a concentration gradient of sodium ions, the passive diffusion of some Na$^+$ back into the cell can provide energy for the secondary active transport of glucose into the cell (**Figure 6.17**). This occurs when glucose is absorbed into the bloodstream from the digestive tract. Secondary active transport aids in the uptake of amino acids and sugars, which are essential raw materials for cell maintenance and growth. Both types of coupled transport proteins—symporters and antiporters—are used for secondary active transport.

yourBioPortal.com
GO TO Animated Tutorial 6.2 • Active Transport

6.4 RECAP

Active transport across a membrane is directional and requires an input of energy to move substances against their concentration gradients. Active transport allows a cell to maintain small molecules and ions at concentrations very different from those in the surrounding environment.

- Why is energy required for active transport? **See p. 120**

- Explain the difference between primary active transport and secondary active transport. **See p. 121**

- Why is the sodium–potassium (Na$^+$–K$^+$) pump classified as an antiporter? **See p. 122 and Figure 6.16**

We have examined a number of passive and active ways in which ions and small molecules can enter and leave cells. But what about large molecules such as proteins? Many proteins are so large that they diffuse very slowly, and their bulk makes it difficult for them to pass through the phospholipid bilayer. It takes a completely different mechanism to move intact large molecules across membranes.

6.5 How Do Large Molecules Enter and Leave a Cell?

Macromolecules such as proteins, polysaccharides, and nucleic acids are simply too large and too charged or polar to pass through biological membranes. This is actually a fortunate property—think of the consequences if such molecules diffused out of cells. A red blood cell would not retain its hemoglobin! Indeed, as we discuss in Chapter 5, the development of a selectively permeable membrane was essential for the functioning of the first cells when life on Earth began. The interior of a cell can be maintained as a separate compartment with a different composition from that of the exterior environment, which is subject to abrupt changes. On the other hand, cells must sometimes take up or *secrete* (release to the external environment) intact large molecules. In Section 5.3 we describe phagocytosis, the mechanism by which solid particles can be brought into the cell by means of vesicles that pinch off from the plasma membrane. The general terms for the mechanisms by which substances enter and leave the cell via membrane vesicles are *endocytosis* and *exocytosis*.

Macromolecules and particles enter the cell by endocytosis

Endocytosis is a general term for a group of processes that bring small molecules, macromolecules, large particles, and even small cells into the eukaryotic cell (**Figure 6.18A**). There are three types of endocytosis: phagocytosis, pinocytosis, and receptor-mediated endocytosis. In all three, the plasma membrane invaginates (folds inward), forming a small pocket around materials from the environment. The pocket deepens, forming a vesicle. This vesicle separates from the plasma membrane and migrates with its contents to the cell's interior.

- In **phagocytosis** ("cellular eating"), part of the plasma membrane engulfs large particles or even entire cells. Unicellular protists use phagocytosis for feeding, and some white blood cells use phagocytosis to defend the body by engulfing foreign cells and substances. The food vacuole or phagosome that forms usually fuses with a lysosome, where its contents are digested (see Figure 5.11).

- In **pinocytosis** ("cellular drinking"), vesicles also form. However, these vesicles are smaller, and the process operates to bring dissolved substances, including proteins or fluids, into the cell. Like phagocytosis, pinocytosis can be relatively nonspecific regarding what it brings into the cell. For example, pinocytosis goes on constantly in the endothelium, the single layer of cells that separates a tiny blood capillary from the surrounding tissue. Pinocytosis allows cells of the endothelium to rapidly acquire fluids and dissolved solutes from the blood.

- In **receptor-mediated endocytosis**, molecules at the cell surface recognize and trigger the uptake of specific materials.

Let's take a closer look at this last process.

Receptor-mediated endocytosis is highly specific

Receptor-mediated endocytosis is used by animal cells to capture specific macromolecules from the cell's environment. This process depends on **receptor proteins**, which are proteins that can bind to specific molecules within the cell or in the cell's external environment. In receptor-mediated endocytosis, the receptors are integral membrane proteins located at particular regions on the extracellular surface of the plasma membrane. These membrane regions are called *coated pits* because they form slight depressions in the plasma membrane and their cytoplasmic surfaces are coated by other proteins, such as clathrin. The uptake process is similar to that in phagocytosis.

When a receptor protein binds to its specific ligand (in this case, the macromolecule to be taken into the cell), its coated pit invaginates and forms a coated vesicle around the bound macromolecule. The clathrin molecules strengthen and stabilize the vesicle, which carries the macromolecule away from the plasma membrane and into the cytoplasm (**Figure 6.19**). Once inside, the vesicle loses its clathrin coat and may fuse with a lysosome, where the engulfed material is digested (by the hydrolysis of polymers to monomers) and the products released into the cytoplasm. Because of its specificity for particular macromolecules, receptor-mediated endocytosis is an efficient method of taking up substances that may exist at low concentrations in the cell's environment.

Receptor-mediated endocytosis is the method by which cholesterol is taken up by most mammalian cells. Water-insoluble cholesterol and triglycerides are packaged by liver cells into lipoprotein particles. Most of the cholesterol is packaged into a type of lipoprotein particle called *low-density lipoprotein*, or LDL, which is circulated via the bloodstream. When a particular cell requires cholesterol, it produces specific LDL receptors, which are inserted into the plasma membrane in clathrin-coated pits. Binding of LDLs to the receptor proteins triggers the uptake of the LDLs via receptor-mediated endocytosis. Within the resulting vesicle, the LDL particles are freed from the receptors. The receptors segregate to a region that buds off and forms a new vesicle, which is recycled to the plasma membrane. The freed LDL particles remain in the original vesicle, which fuses with a lysosome. There, the LDLs are digested and the cholesterol made available for cell use.

In healthy individuals, the liver takes up unused LDLs for recycling. People with the inherited disease *familial hypercholesterolemia* have a deficient LDL receptor in their livers. This prevents receptor-mediated endocytosis of LDLs, resulting in

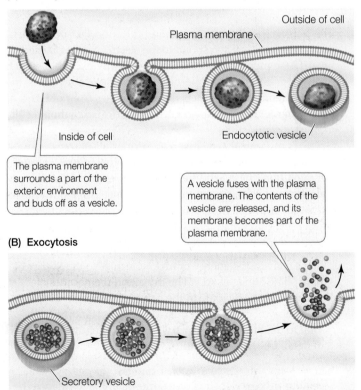

(A) Endocytosis

Outside of cell

Plasma membrane

Inside of cell

Endocytotic vesicle

The plasma membrane surrounds a part of the exterior environment and buds off as a vesicle.

A vesicle fuses with the plasma membrane. The contents of the vesicle are released, and its membrane becomes part of the plasma membrane.

(B) Exocytosis

Secretory vesicle

6.18 Endocytosis and Exocytosis Endocytosis (A) and exocytosis (B) are used by eukaryotic cells to take up and release large molecules and particles, and small cells.

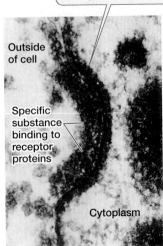

The protein clathrin coats the cytoplasmic side of the plasma membrane at a coated pit.

Outside of cell

Specific substance binding to receptor proteins

Cytoplasm

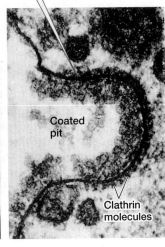

Coated pit

Clathrin molecules

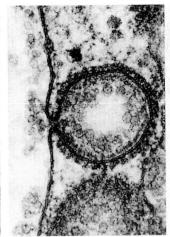

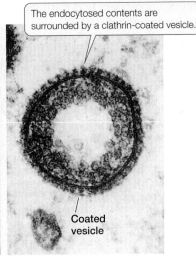

The endocytosed contents are surrounded by a clathrin-coated vesicle.

Coated vesicle

6.19 Receptor-Mediated Endocytosis The receptor proteins in a coated pit bind specific macromolecules, which are then carried into the cell by a coated vesicle.

dangerously high levels of cholesterol in the blood. The cholesterol builds up in the arteries that nourish the heart and causes heart attacks. In extreme cases where only the deficient receptor is present, children and teenagers can have severe cardiovascular disease.

Exocytosis moves materials out of the cell

Exocytosis is the process by which materials packaged in vesicles are secreted from a cell when the vesicle membrane fuses with the plasma membrane (see Figure 6.18B). This fusing makes an opening to the outside of the cell. The contents of the vesicle are released into the environment, and the vesicle membrane is smoothly incorporated into the plasma membrane.

In Chapter 5 we encounter exocytosis as the last step in the processing of material engulfed by phagocytosis—the release of undigested materials back to the extracellular environment. Exocytosis is also important in the secretion of many different substances, including digestive enzymes from the pancreas, neurotransmitters from neurons, and materials for the construction of the plant cell wall. You will encounter these processes in later chapters.

— **yourBioPortal.com** —
GO TO **Animated Tutorial 6.3 • Endocytosis and Exocytosis**

6.5 RECAP

Endocytosis and exocytosis are the processes by which large particles and molecules are transported into and out of the cell. Endocytosis may be mediated by a receptor protein in the plasma membrane.

- Explain the difference between phagocytosis and pinocytosis. See p. 123

- Describe an example of receptor-mediated endocytosis. See p. 123 and Figure 6.19

We have now examined the structures and some of the functions of biological membranes. We have seen how macromolecules on the plasma membrane surface allow cells to recognize and adhere to each other, so that tissues and organs can form. We have also seen how membranes selectively regulate the traffic of small and large molecules, and how large particles such as LDLs can be taken up by cells. These are crucial functions, but they are not the only functions of biological membranes.

6.6 What Are Some Other Functions of Membranes?

The plasma membranes of certain types of cells, such as neurons and muscle cells, respond to the electric charges carried by ions. These membranes are thus electrically excitable, which gives them important properties. For example, in neurons, the plasma membrane conducts nerve impulses from one end of the cell to the other. In muscle cells, electrical excitation results in muscle contraction.

Other biological activities and properties associated with membranes are discussed in the chapters that follow. Throughout evolution, these activities have been essential for the specialization of cells, tissues, and organisms. Three of these activities are especially important:

- *Some organelle membranes help transform energy* (**Figure 6.20A**). For example, the inner mitochondrial membrane helps convert the energy of fuel molecules to the energy of phosphate bonds in ATP. The thylakoid membranes of chloroplasts participate in the conversion of light energy to the energy of chemical bonds. These important processes, vital to the life of most eukaryotic organisms, are discussed in detail in Chapters 9 and 10.

- *Some membrane proteins organize chemical reactions.* Often a cellular process depends on a series of enzyme-catalyzed

(A) Energy transformation

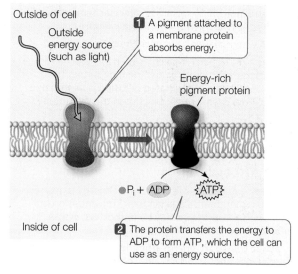

Outside of cell

Outside energy source (such as light)

1 A pigment attached to a membrane protein absorbs energy.

Energy-rich pigment protein

P_i + ADP → ATP

Inside of cell

2 The protein transfers the energy to ADP to form ATP, which the cell can use as an energy source.

(C) Information processing

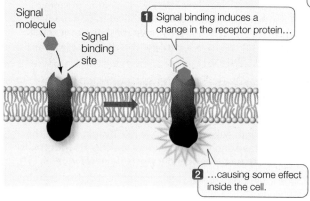

Signal molecule

Signal binding site

1 Signal binding induces a change in the receptor protein…

2 …causing some effect inside the cell.

6.20 Other Membrane Functions The compartmentation afforded by a lipid bilayer or protein membrane was a key event in the emergence of cells. Functions such as energy transformation (A), organization of chemical reactions (B), and signaling (C) probably evolved later and conferred a selective advantage on cells and organisms that had them.

(B) Organizing chemical reactions

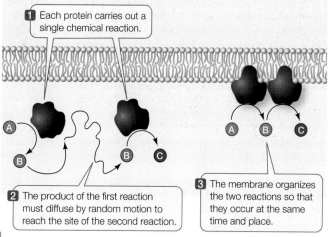

1 Each protein carries out a single chemical reaction.

A
B
B
C

2 The product of the first reaction must diffuse by random motion to reach the site of the second reaction.

A
B
C

3 The membrane organizes the two reactions so that they occur at the same time and place.

reactions, in which the products of one reaction serve as reactants in the next. For such a series of reactions to occur, all the necessary molecules must come together. In a solution, reactant and enzyme molecules are randomly distributed and collisions among them are random. Because these collisions are necessary for chemical reactions to occur, a complete series of chemical reactions may occur only very slowly in a solution. However, if the different enzymes are bound to a membrane in sequential order, the product of one reaction can be released close to the enzyme for the next reaction. Such an "assembly line" allows reactions to proceed rapidly and efficiently (**Figure 6.20B**).

- *Some membrane proteins process information.* As we have seen, biological membranes may have integral membrane proteins or attached carbohydrates that can bind to specific substances in the environment. Without entering a cell, a specific ligand can bind to a receptor and serve as a signal to initiate, modify, or turn off a cell function (**Figure 6.20C**). In this type of information processing, specificity in binding is essential.

We have seen the informational role of the LDL receptor protein in the recognition and endocytosis of LDL, with its cargo of cholesterol. Another example is the binding of a hormone such as insulin to specific receptors on a target cell. When insulin binds to receptors on a liver cell, it elicits the uptake of glucose. In Chapter 7 there are many other examples of the role of membrane proteins in information processing.

CHAPTER SUMMARY

6.1 What Is the Structure of a Biological Membrane?

- Biological membranes consist of lipids, proteins, and carbohydrates. The **fluid mosaic model** of membrane structure describes a phospholipid bilayer in which proteins can move about within the plane of the membrane. **SEE WEB ACTIVITY 6.1**
- The two leaflets of a membrane may have different properties because of their different phospholipid compositions, exposed

domains of **integral membrane proteins**, and **peripheral membrane proteins**. Some proteins, called **transmembrane proteins**, span the membrane. Review Figure 6.1

- Carbohydrates, attached to proteins in **glycoproteins** or to phospholipids in **glycolipids**, project from the external surface of the plasma membrane and function as recognition signals.
- Membranes are not static structures, but are constantly forming, exchanging, and breaking down.

6.2 How Is the Plasma Membrane Involved in Cell Adhesion and Recognition?

- In order for cells to assemble into tissues they must recognize and adhere to one another. **Cell recognition** and **cell adhesion** depend on integral membrane proteins that protrude from the cell surface. Binding can be between the same proteins from two cells (**homotypic**) or different proteins (**heterotypic**). Review Figure 6.6

- Cell junctions connect adjacent cells. **Tight junctions** prevent the passage of molecules through the intercellular spaces between cells, and they restrict the migration of membrane proteins over the cell surface. **Desmosomes** cause cells to adhere firmly to one another. **Gap junctions** provide channels for communication between adjacent cells. Review Figure 6.7, **WEB ACTIVITY 6.2**

- **Integrins** mediate the attachment of animal cells to the extracellular matrix. Review Figure 6.8

6.3 What Are the Passive Processes of Membrane Transport?

SEE ANIMATED TUTORIAL 6.1

- Membranes exhibit **selective permeability**, regulating which substances pass through them.

- A substance can diffuse passively across a membrane by one of two processes: **simple diffusion** through the phospholipid bilayer or **facilitated diffusion** either through a **channel** or by means of a **carrier protein**.

- A solute diffuses across a membrane from a region with a greater concentration of that solute to a region with a lesser concentration of that solute. Equilibrium is reached when the solute concentrations on both sides of the membrane show no net change over time. Review Figure 6.9

- In **osmosis**, water diffuses from a region of higher water concentration to a region of lower water concentration.

- Most cells are in an **isotonic** environment, where total solute concentrations on both sides of the plasma membrane are equal. If the solution surrounding a cell is **hypotonic** to the cell interior, more water enters the cell than leaves it. In plant cells, this leads to **turgor pressure**. In a **hypertonic** solution, more water leaves the cell than enters it. Review Figure 6.10

- **Ion channels** are membrane proteins that allow the rapid facilitated diffusion of ions through membranes. **Gated channels** can be opened or closed by certain conditions or chemicals. The opening or closing of channels, as well as an asymmetric distribution of charged molecules, sets up an **electrochemical gradient** on different sides of a membrane. Review Figure 6.11

- **Aquaporins** are water channels. Review Figure 6.13

- **Carrier proteins** bind to polar molecules such as sugars and amino acids and transport them across the membrane. The maximum rate of this type of facilitated diffusion is limited by the number of carrier (transporter) proteins in the membrane. Review Figure 6.14

6.4 What Are the Active Processes of Membrane Transport?

SEE ANIMATED TUTORIAL 6.2

- **Active transport** requires the use of chemical energy to move substances across membranes against their concentration gradients. Active transport proteins may be **uniporters**, **symporters**, or **antiporters**. Review Figure 6.15

- In **primary active transport**, energy from the hydrolysis of ATP is used to move ions into or out of cells. The **sodium-potassium pump** is an important example. Review Figure 6.16

- **Secondary active transport** couples the passive movement of one substance down its concentration gradient to the movement of another substance against its concentration gradient. Energy from ATP is used indirectly to establish the concentration gradient that results in the movement of the first substance. Review Figure 6.17

6.5 How Do Large Molecules Enter and Leave a Cell?

SEE ANIMATED TUTORIAL 6.3

- **Endocytosis** is the transport of macromolecules, large particles, and small cells into eukaryotic cells via the invagination of the plasma membrane and the formation of vesicles. **Phagocytosis** and **pinocytosis** are types of endocytosis. Review Figure 6.18A

- In **receptor-mediated endocytosis**, a specific **receptor protein** on the plasma membrane binds to a particular macromolecule.

- In **exocytosis**, materials in vesicles are secreted from the cell when the vesicles fuse with the plasma membrane. Review Figure 6.18B

6.6 What Are Some Other Functions of Membranes?

- Membranes function as sites for energy transformations, for organizing chemical reactions, and for recognition and initial processing of extracellular signals. Review Figure 6.20

SELF-QUIZ

1. Which statement about membrane phospholipids is *not* true?
   a. They associate to form bilayers.
   b. They have hydrophobic "tails."
   c. They have hydrophilic "heads."
   d. They give the membrane fluidity.
   e. They flip-flop readily from one side of the membrane to the other.

2. When a hormone molecule binds to a specific protein on the plasma membrane, the protein it binds to is called a
   a. ligand.
   b. clathrin.
   c. receptor protein.
   d. hydrophobic protein.
   e. cell adhesion molecule.

3. Which statement about membrane proteins is *not* true?
   a. They all extend from one side of the membrane to the other.
   b. Some serve as channels for ions to cross the membrane.
   c. Many are free to migrate laterally within the membrane.
   d. Their position in the membrane is determined by their tertiary structure.
   e. Some play roles in photosynthesis.

4. Which statement about membrane carbohydrates is *not* true?
   a. Some are bound to proteins.
   b. Some are bound to lipids.
   c. They are added to proteins in the Golgi apparatus.
   d. They show little diversity.
   e. They are important in recognition reactions at the cell surface.

5. Which statement about animal cell junctions is *not* true?
   a. Tight junctions are barriers to the passage of molecules between cells.
   b. Desmosomes allow cells to adhere firmly to one another.
   c. Gap junctions block communication between adjacent cells.
   d. Connexons are made of protein.
   e. The fibers associated with desmosomes are made of protein.

6. You are studying how the protein transferrin enters cells. When you examine cells that have taken up transferrin, you find it inside clathrin-coated vesicles. Therefore, the most likely mechanism for uptake of transferrin is
   a. facilitated diffusion.
   b. an antiporter.
   c. receptor-mediated endocytosis.
   d. gap junctions.
   e. ion channels.

7. Which statement about ion channels is *not* true?
   a. They form pores in the membrane.
   b. They are proteins.
   c. All ions pass through the same type of channel.
   d. Movement through them is from regions of high concentration to regions of low concentration.
   e. Movement through them is by simple diffusion.

8. Facilitated diffusion and active transport both
   a. require ATP.
   b. require the use of proteins as carriers or channels.
   c. carry solutes in only one direction.
   d. increase without limit as the concentration gradient increases.
   e. depend on the solubility of the solute in lipids.

9. Primary and secondary active transport both
   a. generate ATP.
   b. are based on passive movement of Na^+ ions.
   c. include the passive movement of glucose molecules.
   d. use ATP directly.
   e. can move solutes against their concentration gradients.

10. Which statement about osmosis is *not* true?
   a. It obeys the laws of diffusion.
   b. In animal tissues, water moves into cells if they are hypertonic to their environment.
   c. Red blood cells must be kept in a plasma that is hypotonic to the cells.
   d. Two cells with identical solute concentrations are isotonic to each other.
   e. Solute concentration is the principal factor in osmosis.

FOR DISCUSSION

1. Muscle function requires calcium ions (Ca^{2+}) to be pumped into a subcellular compartment against a concentration gradient. What types of molecules are required for this to happen?

2. Section 27.5 describes the diatoms, which are protists that have complex glassy structures in their cell walls (see Figure 27.7B). These structures form within the Golgi apparatus. How do these structures reach the cell wall without having to pass through a membrane?

3. Organisms that live in fresh water are almost always hypertonic to their environment. In what way is this a serious problem? How do some organisms cope with this problem?

4. Contrast nonspecific endocytosis and receptor-mediated endocytosis.

5. The emergence of the phospholipid membrane was important to the origin of cells. Describe the properties of membranes that might have allowed cells to thrive in comparison with molecular aggregates without membranes.

ADDITIONAL INVESTIGATION

When a normal lung cell becomes a lung cancer cell, there are several important changes in plasma membrane properties. How would you investigate the following phenomena? *(a)* The cancer cell membrane is more fluid, with more rapid diffusion in the plane of the membrane of both lipids and proteins. *(b)* The cancer cell has altered cell adhesion properties, binding to other tissues in addition to lung cells.

WORKING WITH DATA (GO TO yourBioPortal.com)

Aquaporin Increases Membrane Permeability to Water In this hands-on exercise based on Figure 6.13, you will investigate how Agre and colleagues used an egg cell to show that expression of aquaporin results in rapid water uptake when the cell is placed in a hypotonic medium. Analyzing their experimental design and data, you will see how this model cell system and control experiments confirmed the important role of aquaporin as a water channel.

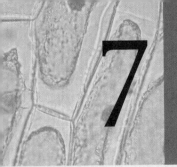

7 Cell Signaling and Communication

Love signals

Prairie voles (*Microtus ochrogaster*) are small rodents that live in temperate climates, where they dig tunnels in fields. When a male prairie vole encounters a female, mating often ensues. After mating (which can take as long as a day), the couple stays together, building a nest and raising their pups together. The two voles bond so tightly that they stay together for life. Contrast this behavior with that of the montane vole (*M. montanus*), which is closely related to the prairie vole and lives in the hills not far away. In this species, mating is quick, and afterwards the couple separates. The male looks for new mates and the female abandons her young soon after they are born.

The explanation for these dramatic behavioral differences lies in the brains of these two species. Neuroscientist Thomas Insel and his colleagues found that when prairie voles mate for all those hours, their brains release a 9-amino-acid peptide. In females, this peptide is oxytocin; in males, it is vasopressin. The peptide is circulated in the bloodstream and reaches all tissues in the body, but it binds to only a few cell types. These cells have surface proteins, called receptors, that specifically bind the peptide, like a key inserting into a lock.

The interaction of peptide and receptor causes the receptor, which extends across the plasma membrane, to change shape. Within the cytoplasm, this change sets off a series of events called a signal transduction pathway. Such a pathway can cause many different cellular responses, but in this case, the notable changes are in behavior. The receptors for oxytocin and vasopressin in prairie voles are most concentrated in the regions of the brain that are responsible for behaviors such as bonding and caring for the young. In montane voles, there are far fewer receptors and as a result, fewer postmating behaviors.

These cause-and-effect relationships between peptides, receptors, and behavior have been established through experiments. For example, a female prairie vole that is injected before mating with a molecule that blocks oxytocin does not bond with the male. Also, a female injected with oxytocin will bond with a male even without mating. Experiments with vasopressin in males give similar results. Furthermore, promiscuous vole males that were genetically manipulated to express prairie vole amounts of the vasopressin receptor grew up to behave more like prairie vole males. These experiments show that oxytocin and

Voles Prairie voles display extensive bonding behaviors after mating. These behaviors are mediated by peptides acting as intercellular signals.

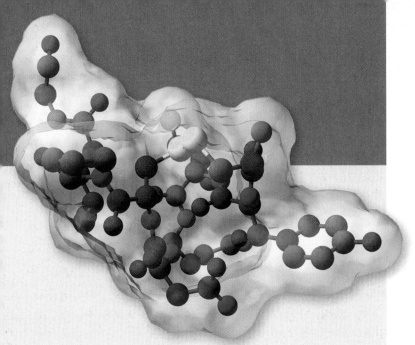

Oxytocin This peptide with 9 amino acids acts as a signal for postmating behaviors.

7.1 What Are Signals, and How Do Cells Respond to Them?

Both prokaryotic and eukaryotic cells process information from their environments. This information can be in the form of a physical stimulus, such as the light reaching your eyes as you read this book, or chemicals that bathe a cell, such as lactose in a bacterial growth medium. It may come from outside the organism, such as the scent of a female moth seeking a mate in the dark, or from a neighboring cell within the organism, such as in the heart, where thousands of muscle cells contract in unison by transmitting signals to one another.

Of course, the mere presence of a signal does not mean that a cell will respond to it, just as you do not pay close attention to every image in your environment as you study. To respond to a signal, the cell must have a specific receptor that can detect it. This section provides examples of some types of cellular signals and one model of *signal transduction*. A **signal transduction pathway** is a sequence of molecular events and chemical reactions that lead to a cell's response to a signal. After discussing signals in this section, we will consider their receptors in Section 7.2.

Cells receive signals from the physical environment and from other cells

The physical environment is full of signals. Our sense organs allow us to respond to light, odors and tastes (chemical signals), temperature, touch, and sound. Bacteria and protists can respond to minute chemical changes in their environments. Plants respond to light as a signal as well as an energy source. The amount and wavelengths of light reaching a plant's surface differ from day to night and in direct sunlight versus shade. These variations act as signals that affect plant growth and reproduction. Some plants also respond to temperature: when the weather gets cold, they may respond either by becoming tolerant to cold or by accelerating flowering.

A cell deep inside a large multicellular organism is far away from the exterior environment. Such a cell's environment consists of other cells and extracellular fluids. Cells receive their nutrients from, and pass their wastes into, extracellular fluids. Cells also receive signals—mostly chemical signals—from their extracellular fluid environment. Most of these chemical signals come from other cells, but they can also come from the environment via the digestive and respiratory systems. And cells can respond to changes in the extracellular concentrations of cer-

vasopressin are signals that induce bonding and caring behaviors in voles. Could this also be true of humans?

Neuroeconomist Paul Zaks thinks so. He has done experiments with human volunteers, who were asked to "invest" funds with a stranger. A group of investors that was given a nasal spray containing oxytocin was more trusting of the stranger (and invested more funds) than a group that got an inert spray. So the oxytocin signaling pathway is important in human behavior too.

A cell's response to any signal molecule takes place in three sequential steps. First, the signal binds to a receptor in the cell, often on the outside surface of the plasma membrane. Second, signal binding conveys a message to the cell. Third, the cell changes its activity in response to the signal. And in a multicellular organism, this leads to changes in that organism's functioning.

IN THIS CHAPTER we first describe the types of signals that affect cells. These include chemicals produced by other cells and substances from outside the body, as well as physical and environmental factors such as light. Then we show how a signal affects only those cells that have the specific receptor to recognize that signal. Next, we describe the steps of signal transduction in which the receptor communicates to the cell that a signal has been received, thus causing a change in cell function.

tain chemicals, such as CO_2 and H^+, which are affected by the metabolic activities of other cells.

Inside a large multicellular organism, chemical signals made by the body itself reach a target cell by local diffusion or by circulation within the blood. These signals are usually in tiny concentrations (as low as $10^{-10} M$) (see Chapter 2 for an explanation of *molar* concentrations). **Autocrine** signals diffuse to and affect the cells that make them; for example, part of the reason many tumor cells reproduce uncontrollably is because they self-stimulate cell division by making their own division signals. **Paracrine** signals diffuse to and affect nearby cells; an example is a neurotransmitter made by one nerve cell that diffuses to an adjacent cell and stimulates it. (**Figure 7.1A**). Signals to distant cells called hormones travel through the circulatory system (**Figure 7.1B**).

A signal transduction pathway involves a signal, a receptor, and responses

For the information from a signal to be transmitted to a cell, the target cell must be able to receive or sense the signal and respond to it, and the response must have some effect on the func-

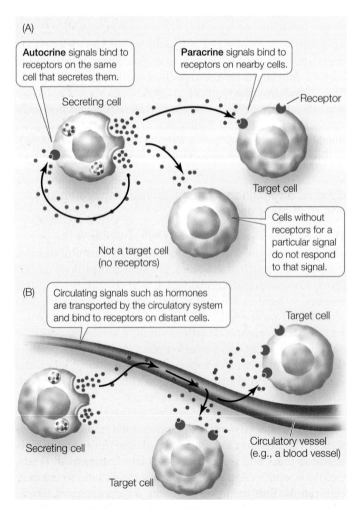

7.1 Chemical Signaling Systems (A) A signal molecule can diffuse to act on the cell that produces it, or on a nearby cell. (B) Many signals act on distant cells and must be transported by the organism's circulatory system.

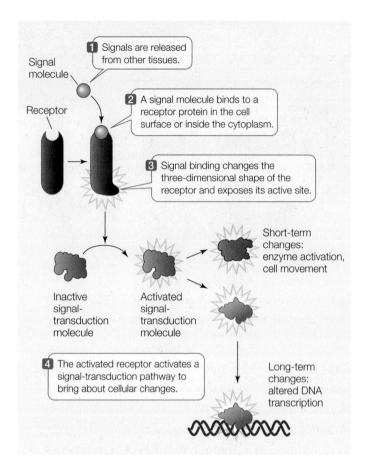

7.2 A Signal Transduction Pathway This general pathway is common to many cells and situations. The ultimate effects on the cell are either short-term or long-term molecular changes, or both.

tion of the cell. In a multicellular organism, all cells may receive chemical signals that are circulated in the blood, such as the peptides oxytocin and vasopressin that are released following mating in voles (see the opening of this chapter), but most body cells are not capable of responding to the signals. Only the cells with the necessary receptors can respond.

The kinds of responses vary greatly depending on the signal and the target cell. Just a few examples are: a skin cell initiating cell division to heal a wound; a cell moving to a new location in the embryo to form a tissue; a cell releasing enzymes to digest food; a plant cell loosening bonds that hold its cell wall polymers together so that it can expand; and a cell in the eye sending messages to the brain about the book you are reading. A signal transduction pathway involves a signal, a receptor, and a response (**Figure 7.2**).

Let's look at an example of such a pathway in the bacterium *Escherichia coli* (*E. coli*). Follow the features of this pathway in general (see Figure 7.2) and in particular (**Figure 7.3**).

SIGNAL As a prokaryotic cell, a bacterium is very sensitive to changes in its environment. One thing that can change is the total solute concentration (osmotic concentration—see Section 6.3) in the environment surrounding the cell. In the mammalian intestine where *E. coli* lives, the solute concentration around

7.3 A Model Signal Transduction Pathway *E. coli* responds to the signal of an increase in solute concentration in its environment. The basic steps of such a signal transduction pathway occur in all living organisms.

1. Signal
Solutes enter the space between the two membranes through large pores in the outer membrane of *E. coli.*

2. Receptor
The EnvZ receptor protein changes shape in response to the high solute concentration, catalyzing the addition of a phosphate from ATP.

3. Responder
The phosphate from EnvZ is transferred to the responding OmpR protein…

…and the phosphorylated OmpR changes shape, enabling it to bind to DNA and stimulate transcription of the *ompC* gene.

4. Effects
OmpC protein inserts into the outer membrane, preventing solute entry and keeping the cell's exterior osmotically balanced.

the bacterium often rises far above the solute concentration inside the cell. A fundamental characteristic of all living cells is that they maintain a constant internal environment, or homeostasis. To do this, the bacterium must perceive and quickly respond to this environmental signal (**Figure 7.3, step 1**). The cell does this by a signal transduction pathway involving two major components: a receptor and a responder.

RECEPTOR The *E. coli* receptor protein for changes in solute concentration is called EnvZ. EnvZ is a transmembrane protein that extends across the bacterium's plasma membrane into the space between the plasma membrane and the highly porous outer membrane, which forms a complex with the cell wall. When the solute concentration of the extracellular environment rises, so does the solute concentration in the space between the two membranes. This change in the aqueous solution causes the part of the receptor protein that sticks out into the intermembrane space to undergo a change in conformation (its three-dimensional shape).

The conformational change in the intermembrane domain (a *domain* is a sequence of amino acids folded into a particular shape) causes a conformational change in the domain that lies in the cytoplasm and initiates the events of signal transduction. The cytoplasmic domain of EnvZ can act as an *enzyme*. As you will see in more detail in Chapter 8, an enzyme is a biological catalyst that greatly speeds up a chemical reaction, and the active site is the region where the reaction actually takes place. The conformational change in EnvZ exposes an active site that was previously buried within the protein, so that EnvZ becomes a **protein kinase**—an enzyme that catalyzes the transfer of a phosphate group from ATP to another molecule. EnvZ transfers the phosphate group to one of its own histidine amino acids. In other words, EnvZ *phosphorylates* itself (**Figure 7.3, step 2**).

$$\text{EnvZ} + \text{ATP} \xrightarrow{\text{EnvZ}} \text{EnvZ–P} + \text{ADP}$$

What does phosphorylation do to a protein? As discussed in Section 3.2, proteins can have both hydrophilic regions (which tend to interact with water on the outside of the protein macromolecule) and hydrophobic regions (which tend to interact with one another on the inside of the macromolecule). These regions are important in giving a protein its three-dimensional shape. Phosphate groups are charged, so an amino acid with such a group tends to be on the outside of the protein. Thus phosphorylation leads to a change in the shape and function of a protein by changing its charge.

RESPONDER A **responder** is the second component of a signal transduction pathway. The charged phosphate group added to the histidine of the EnvZ protein causes its cytoplasmic domain to change its shape again. It now binds to a second protein, OmpR, and transfers the phosphate to it. In turn, this phosphorylation changes the shape of OmpR (**Figure 7.3, step 3**). The change in the responder is a key event in signaling, for three reasons:

- The signal on the outside of the cell has now been *transduced* to a protein that lies totally within the cell's cytoplasm.

- The altered responder can *do something*. In the case of the phosphorylated OmpR, that "something" is to bind to DNA to alter the expression of many genes; in particular, it increases the expression of the protein OmpC. This binding begins the final phase of the signaling pathway: the effect of the signal, which is an alteration in cell function.

- The signal has been *amplified*. Because a single enzyme can catalyze the conversion of many substrate molecules, one EnvZ molecule alters the structure of many OmpR molecules.

Phosphorylated OmpR has the correct three-dimensional structure to bind to the *ompC* DNA, resulting in an increase in the transcription of that gene. This results in the production of OmpC protein, which enables the cell to respond to the increase in osmotic concentration in its environment (**Figure 7.3, step 4**). The OmpC protein is inserted into the outer membrane of the cell, where it blocks pores and prevents solutes from entering the intermembrane space. As a result, the solute concentration in the intermembrane space is lowered, and homeostasis is restored. Thus the EnvZ-OmpR signal transduction pathway allows the *E. coli* cell to function just as if the external environment had a normal solute concentration.

Many of the elements that we have highlighted in this prokaryotic signal transduction pathway also exist in the signal transduction pathways of eukaryotic organisms. A typical eukaryotic signal transduction pathway has the following general steps:

- A receptor protein changes its conformation upon interaction with a signal. This receptor protein may or may not be in a membrane.

- A conformational change in the receptor protein activates its protein kinase activity, resulting in the transfer of a phosphate group from ATP to a target protein.

- This phosphorylation alters the function of a responder protein.

- The signal is amplified.

- A protein that binds to DNA is activated.

- The expression of one or more specific genes is turned on or off.

- Cell activity is altered.

7.1 RECAP

Cells are constantly exposed to molecular signals that can come from the external environment or from within the body of a multicellular organism. To respond to a signal, the cell must have a specific receptor that detects the signal and activates some cellular response.

- What are the differences between an autocrine signal, a paracrine signal, and a hormone? **See p. 130 and Figure 7.1**

- Describe the three components in a cell's response to a signal. **See pp. 130–132 and Figure 7.2**

- What are the elements of signal transduction that are described at the close of this section?

The general features of signal transduction pathways described in this section will recur in more detail throughout the chapter. First let's consider more closely the nature of the receptors that bind signal molecules.

7.2 How Do Signal Receptors Initiate a Cellular Response?

Any given cell in a multicellular organism is bombarded with many signals. However, it responds to only some of them, because no cell makes receptors for all signals. A receptor protein that binds to a chemical signal does so very specifically, in much the same way that a membrane transport protein binds to the substance it transports. This *specificity* of binding ensures that only those cells that make a specific receptor will respond to a given signal.

Receptors have specific binding sites for their signals

A specific chemical signal molecule fits into a three-dimensional site on its protein receptor (**Figure 7.4A**). A molecule that binds to a receptor site on another molecule in this way is called a **ligand**. Binding of the signaling ligand causes the receptor protein to change its three-dimensional shape, and that conformational change initiates a cellular response. The ligand does not contribute further to this response. In fact, the ligand is usually not metabolized into a useful product; its role is purely to "knock on the door." (This is in sharp contrast to the enzyme–substrate interaction, which is described in Chapter 8. The whole purpose of that interaction is to change the substrate into a useful product.)

Receptors bind to their ligands according to chemistry's *law of mass action*:

$$R + L \rightleftharpoons RL$$

This means that the binding is reversible, although for most ligand–receptor complexes, the equilibrium point is far to the right—that is, binding is favored. Reversibility is important, however, because if the ligand were never released, the receptor would be continuously stimulated.

7.4 A Signal and Its Receptor (A) The adenosine 2A receptor occurs in the human brain, where it is involved in inhibiting arousal. (B) Adenosine is the normal ligand for the receptor. Caffeine has a similar structure to that of adenosine and can act as an antagonist that binds the receptor and prevents its normal functioning.

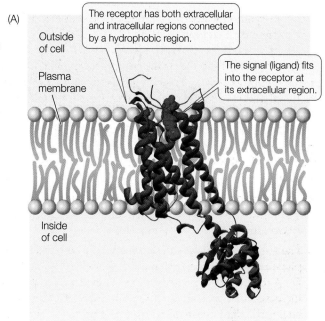

(A)

Outside of cell

Plasma membrane

Inside of cell

The receptor has both extracellular and intracellular regions connected by a hydrophobic region.

The signal (ligand) fits into the receptor at its extracellular region.

An inhibitor (or *antagonist*) can also bind to a receptor protein, instead of the normal ligand. There are both natural and artificial antagonists of receptor binding. For example, many substances that alter human behavior bind to specific receptors in the brain, and prevent the binding of the receptors' specific ligands. An example is caffeine, which is probably the world's most widely consumed stimulant. In the brain, the nucleoside adenosine acts as a ligand that binds to a receptor on nerve cells, initiating a signal transduction pathway that reduces brain activity, especially arousal. Because caffeine has a similar molecular structure to that of adenosine, it also binds to the adenosine receptor (**Figure 7.4B**). But in this case binding does not initiate a signal transduction pathway. Rather, it "ties up" the receptor, preventing adenosine binding and thereby allowing nerve cell activity and arousal.

Receptors can be classified by location and function

The chemistry of ligand signals is quite variable, but they can be divided into two groups, based on whether or not they can diffuse through membranes. Correspondingly, a receptor can be classified by its location in the cell, which largely depends on the nature of its ligand (**Figure 7.5**):

- *Cytoplasmic receptors*: Small or nonpolar ligands can diffuse across the nonpolar phospholipid bilayer of the plasma membrane and enter the cell. Estrogen, for example, is a lipid-soluble steroid hormone that can easily diffuse across the plasma membrane; it binds to a receptor in the cytoplasm.

- *Membrane receptors*: Large or polar ligands cannot cross the lipid bilayer. Insulin, for example, is a protein hormone that cannot diffuse through the plasma membrane; instead, it binds to a transmembrane receptor with an extracellular binding domain.

In complex eukaryotes such as mammals and higher plants, there are three well-studied categories of plasma membrane receptors that are grouped according to their functions: ion channels, protein kinase receptors, and G protein-linked receptors.

ION CHANNEL RECEPTORS As described in Section 6.3, the plasma membranes of many types of cells contain gated **ion channels** for ions such as Na^+, K^+, Ca^{2+}, or Cl^- to enter or leave the cell (see Figure 6.11). The gate-opening mechanism is an alteration in the three-dimensional shape of the channel protein upon ligand binding; thus these proteins function as receptors. Each type of ion channel has its own signal, and these include sensory stimuli such as light, sound, and electric charge

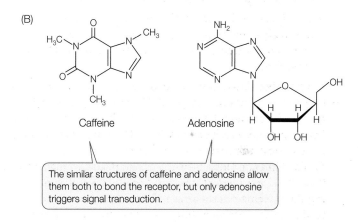

(B)

Caffeine

Adenosine

The similar structures of caffeine and adenosine allow them both to bond the receptor, but only adenosine triggers signal transduction.

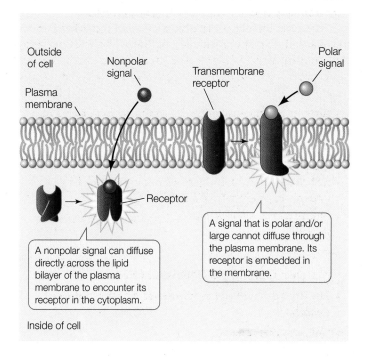

Outside of cell

Plasma membrane

Nonpolar signal

Transmembrane receptor

Polar signal

Receptor

A nonpolar signal can diffuse directly across the lipid bilayer of the plasma membrane to encounter its receptor in the cytoplasm.

A signal that is polar and/or large cannot diffuse through the plasma membrane. Its receptor is embedded in the membrane.

Inside of cell

7.5 Two Locations for Receptors Receptors can be located in the cytoplasm or in the plasma membrane of the cell.

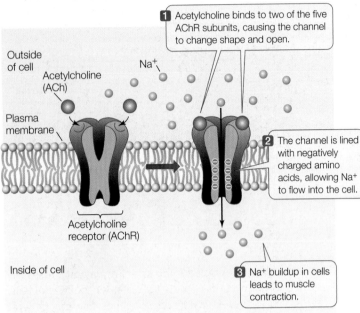

1 Acetylcholine binds to two of the five AChR subunits, causing the channel to change shape and open.

Outside of cell

Acetylcholine (ACh)

Na^+

Plasma membrane

2 The channel is lined with negatively charged amino acids, allowing Na^+ to flow into the cell.

Acetylcholine receptor (AChR)

Inside of cell

3 Na^+ buildup in cells leads to muscle contraction.

7.6 A Gated Ion Channel The acetylcholine receptor (AChR) is a ligand-gated ion channel for sodium ions. It is made up of five polypeptide subunits. When acetylcholine molecules (ACh) bind to two of the subunits, the gate opens and Na^+ flows into the cell. This channel helps regulate membrane polarity (see Chapter 6).

differences across the plasma membrane, as well as chemical ligands such as hormones and neurotransmitters.

The *acetylcholine receptor*, which is located in the plasma membrane of skeletal muscle cells, is an example of a gated ion channel. This receptor protein is a sodium channel that binds the ligand acetylcholine, which is a neurotransmitter—a chemical signal released from neurons (nerve cells) (**Figure 7.6**). When two molecules of acetylcholine bind to the receptor, it opens for about a thousandth of a second. That is enough time for Na^+, which is more concentrated outside the cell than inside, to rush into the cell, moving in response to both concentration and electrical potential gradients. The change in Na^+ concentration in the cell initiates a series of events that result in muscle contraction.

PROTEIN KINASE RECEPTORS Like the EnvZ receptor of *E. coli*, some eukaryotic receptor proteins become protein kinases when they are activated. They catalyze the phosphorylation of themselves and/or other proteins, thus changing their shapes and therefore their functions.

The receptor for insulin is an example of a protein kinase receptor. Insulin is a protein hormone made by the mammalian pancreas. Its receptor has two copies each of two different polypeptide subunits (**Figure 7.7**). When insulin binds to the receptor, the receptor becomes activated and able to phosphorylate itself and certain cytoplasmic proteins that are appropriately called *insulin response substrates*. These proteins then initiate many cellular responses, including the insertion of glucose transporters (see Figure 6.14) into the plasma membrane.

G PROTEIN-LINKED RECEPTORS A third category of eukaryotic plasma membrane receptors is the G protein-linked receptors, also referred to as the seven transmembrane domain receptors. This descriptive name identifies a fascinating group of receptors, each of which is composed of a single protein with seven transmembrane domains. These seven domains pass through the phospholipid bilayer and are separated by short loops that extend either outside or inside the cell. Ligand binding on the extracellular side of the receptor changes the shape of its cytoplasmic region, exposing a site that binds to a mobile membrane protein called a **G protein**. The G protein is partially inserted into the lipid bilayer and partially exposed on the cytoplasmic surface of the membrane.

Many G proteins have three polypeptide subunits and can bind three different molecules (**Figure 7.8A**):

- The receptor
- GDP and GTP (guanosine diphosphate and triphosphate, respectively; these are nucleoside phosphates like ADP and ATP)
- An effector protein

When the G protein binds to an activated receptor protein, GDP is exchanged for GTP (**Figure 7.8B**). At the same time, the ligand is usually released from the extracellular side of the receptor. GTP binding causes a conformational change in the G protein. The GTP-bound subunit then separates from the rest of the protein, diffusing in the plane of the phospholipid bilayer until it encounters an **effector protein** to which it can bind. An effector protein is just what its name implies: it causes an effect in the cell. The binding of the GTP-bearing G protein

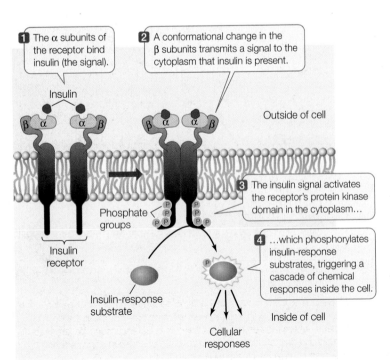

1 The α subunits of the receptor bind insulin (the signal).

Insulin

2 A conformational change in the β subunits transmits a signal to the cytoplasm that insulin is present.

Outside of cell

β α   α β

Phosphate groups

3 The insulin signal activates the receptor's protein kinase domain in the cytoplasm...

Insulin receptor

Insulin-response substrate

4 ...which phosphorylates insulin-response substrates, triggering a cascade of chemical responses inside the cell.

Cellular responses

Inside of cell

7.7 A Protein Kinase Receptor The mammalian hormone insulin binds to a receptor on the outside surface of the cell and initiates a response.

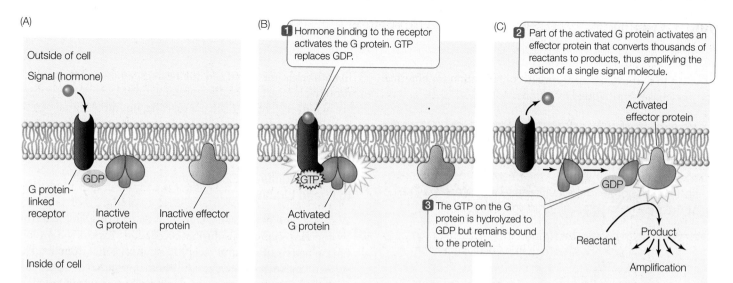

7.8 A G Protein-Linked Receptor The G protein is an intermediary between the receptor and an effector.

Panel (A):
Outside of cell
Signal (hormone)
G protein-linked receptor
GDP
Inactive G protein
Inactive effector protein
Inside of cell

Panel (B):
1 Hormone binding to the receptor activates the G protein. GTP replaces GDP.
GTP
Activated G protein

Panel (C):
2 Part of the activated G protein activates an effector protein that converts thousands of reactants to products, thus amplifying the action of a single signal molecule.
Activated effector protein
3 The GTP on the G protein is hydrolyzed to GDP but remains bound to the protein.
GDP
Reactant
Product
Amplification

─── **yourBioPortal**.com ───
GO TO Animated Tutorial 7.1 • Signal Transduction Pathway

subunit activates the effector—which may be an enzyme or an ion channel—thereby causing changes in cell function (**Figure 7.8C**).

After activation of the effector protein, the GTP on the G protein is hydrolyzed to GDP. The now inactive G protein subunit separates from the effector protein and diffuses in the membrane to collide with and bind to the other two G protein subunits. When the three components of the G protein are reassembled, the protein is capable of binding again to an activated receptor. After binding, the activated receptor exchanges the GDP on the G protein for a GTP, and the cycle begins again.

There are variations in all three G protein subunits, giving different G protein complexes different functions. A G protein can either activate or inhibit an effector protein. An example in humans of an *activating* response involves the receptor for epinephrine (adrenaline), which is a hormone made by the adrenal gland in response to stress or heavy exercise. In heart muscle, this hormone binds to its G protein-linked receptor, activating a G protein. The GTP-bound subunit then activates a membrane-bound enzyme to produce a small molecule, cyclic adenosine monophosphate (cAMP). This molecule, in turn, has many effects on the cell (as we will see below), including the mobilization of glucose for energy and muscle contraction.

G protein-mediated *inhibition* occurs when the same hormone, epinephrine, binds to its receptor in the smooth muscle cells surrounding blood vessels lining the digestive tract. Again, the epinephrine-bound receptor changes its shape and activates a G protein, and the GTP-bound subunit binds to a target enzyme. But in this case, the enzyme is inhibited instead of being activated. As a result, the muscles relax and the blood vessel diameter increases, allowing more nutrients to be carried away from the digestive system to the rest of the body. Thus the same signal and signaling mechanism can have different consequences in different cells, depending on the presence of specific receptor and effector molecules.

CYTOPLASMIC RECEPTORS **Cytoplasmic receptors** are located inside the cell and bind to signals that can diffuse across the plasma membrane. Binding to the signaling ligand causes the receptor to change its shape so that it can enter the cell nucleus, where it affects expression of specific genes. But this general view is somewhat simplified. The receptor for the steroid hormone cortisol, for example, is normally bound to a chaperone protein, which blocks it from entering the nucleus. Binding of the hormone causes the receptor to change its shape so that the chaperone is released (**Figure 7.9**). This release allows the

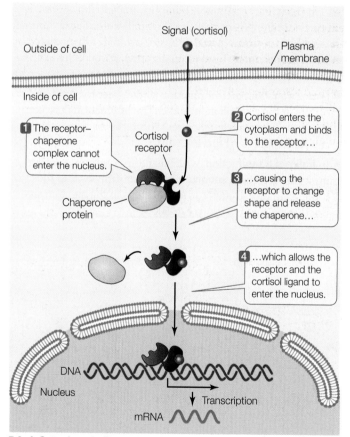

Signal (cortisol)
Outside of cell
Plasma membrane
Inside of cell
1 The receptor–chaperone complex cannot enter the nucleus.
Cortisol receptor
2 Cortisol enters the cytoplasm and binds to the receptor…
Chaperone protein
3 …causing the receptor to change shape and release the chaperone…
4 …which allows the receptor and the cortisol ligand to enter the nucleus.
DNA
Nucleus
Transcription
mRNA

7.9 A Cytoplasmic Receptor The receptor for cortisol is bound to a chaperone protein. Binding of the signal to the receptor releases the chaperone and allows the ligand–receptor complex to enter the cell's nucleus, where it binds to DNA. Changes in DNA transcription are long-term in comparison to the more immediate changes in enzyme activity observed in other pathways (see Figure 7.20).

receptor to fold into an appropriate conformation for entering the nucleus and initiating DNA transcription.

7.2 RECAP

Receptors are proteins that bind, or are changed by, specific signals or ligands; the changed receptor initiates a response in the cell. These receptors may be at the plasma membrane or inside the cell.

- What are the nature and importance of specificity in the binding of receptors to their particular ligands? **See pp. 132–133**

- What are three important categories of plasma membrane receptors seen in complex eukaryotes? See pp. **133–134 and Figures 7.6, 7.7, and 7.8**

Now that we have discussed signals and receptors, let's examine the characteristics of the molecules (*transducers*) that mediate between the receptor and the cellular response.

7.3 How Is the Response to a Signal Transduced through the Cell?

As we have just seen with epinephrine, the same signal may produce different responses in different tissues. These different responses to the same signal–receptor complex are mediated by the components of different signal transduction pathways. Signal transduction may be either direct or indirect:

- **Direct transduction** is a function of the receptor itself and occurs at the plasma membrane. The interaction between the signal (primary messenger) and receptor results in the cellular response. (**Figure 7.10A**).

- In **indirect transduction**, which is more common, another molecule termed a **second messenger** diffuses into the cytoplasm and mediates additional steps in the signal transduction pathway (**Figure 7.10B**).

In both cases, the signal can initiate a *cascade* of events, in which proteins interact with other proteins until the final responses are achieved. Through such a cascade, an initial signal can be both amplified and distributed to cause several different responses in the target cell.

A protein kinase cascade amplifies a response to ligand binding

We have seen that when a signal binds to a protein kinase receptor, the receptor's conformation changes, exposing a protein kinase active site on the receptor's cytoplasmic domain. The protein kinase then catalyzes the phosphorylation of target proteins. This process is an example of direct signal transduction, because the amplifying enzyme is the receptor itself. Protein kinase receptors are important in binding signals called growth factors that stimulate cell division in both plants and animals.

A complete signal transduction pathway that occurs after a protein kinase receptor binds a growth factor was discovered in studies on a cell that went wrong. Many human bladder cancers contain an abnormal form of a protein called Ras (so named because a similar protein was previously isolated from a *rat* sarcoma tumor). Investigations of these bladder cancers showed that Ras was a G protein, and the abnormal form was always active because it was permanently bound to GTP, and thus caused continuous cell division (**Figure 7.11**). If this abnormal form of Ras was inhibited, the cells stopped dividing. This discovery has led to a major effort to develop specific Ras inhibitors for cancer treatment.

Other cancers have abnormalities in different aspects of signal transduction. Biologists have compared the defects in these cells with the normal signaling process in non-cancerous cells, and thus worked out the entire signaling pathway. It is an ex-

7.10 Direct and Indirect Signal Transduction   (A) All the events of direct transduction occur at or near the receptor (in this case, at the plasma membrane). (B) In indirect transduction, a second messenger mediates the events inside the cell. The signal is considered to be the first messenger.

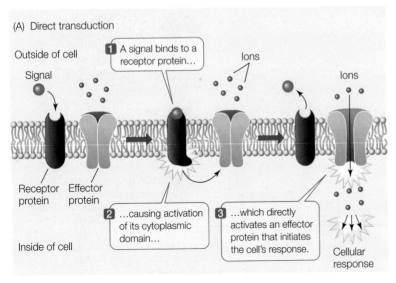

(A) Direct transduction

Outside of cell

Signal

1 A signal binds to a receptor protein...

Ions

Ions

Receptor protein   Effector protein

2 ...causing activation of its cytoplasmic domain...

3 ...which directly activates an effector protein that initiates the cell's response.

Inside of cell

Cellular response

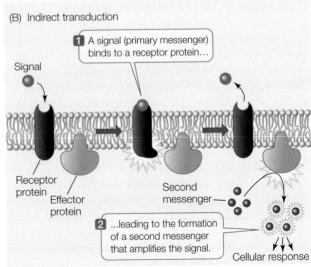

(B) Indirect transduction

1 A signal (primary messenger) binds to a receptor protein...

Signal

Receptor protein

Effector protein

Second messenger

2 ...leading to the formation of a second messenger that amplifies the signal.

Cellular response

(A) Normal cell

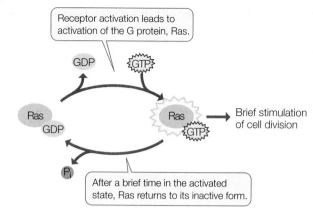

(B) Cancer cell

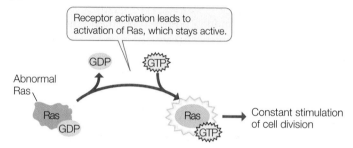

7.11 Signal Transduction and Cancer (A) Ras is a G protein that regulates cell division. (B) In some tumors, the Ras protein is permanently active, resulting in uncontrolled cell division.

ample of a more general phenomenon, called a **protein kinase cascade**, where one protein kinase activates the next, and so on (**Figure 7.12**). Such cascades are key to the external regulation of many cellular activities. Indeed, the eukaryotic genome codes for hundreds, even thousands, of such kinases.

Protein kinase cascades are useful signal transducers for four reasons:

- At each step in the cascade of events, the signal is *amplified*, because each newly activated protein kinase is an enzyme that can catalyze the phosphorylation of many target proteins.
- The information from a signal that originally arrived at the plasma membrane is *communicated* to the nucleus.
- The multitude of steps provides some *specificity* to the process.
- Different target proteins at each step in the cascade can provide *variation* in the response.

─── **yourBioPortal.com** ───
GO TO Animated Tutorial 7.2 • Signal Transduction and Cancer

Second messengers can stimulate protein kinase cascades

As we have just seen, protein kinase receptors initiate protein kinase cascades right at the plasma membrane. However, the stimulation of events in the cell is more often indirect. In a series of clever experiments, Earl Sutherland and his colleagues at Case Western Reserve University discovered that a small water-solu-

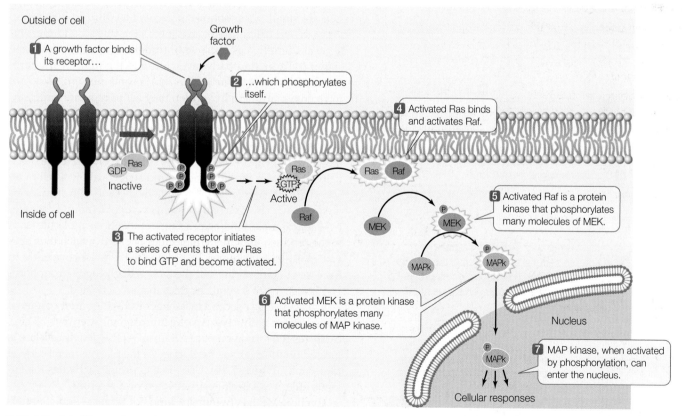

7.12 A Protein Kinase Cascade In a protein kinase cascade, a series of proteins are sequentially activated.

INVESTIGATING LIFE

7.13 The Discovery of a Second Messenger

Glycogen phosphorylase is activated in liver cells after epinephrine binds to a membrane receptor. Sutherland and his colleagues observed that this activation could occur in vivo only if fragments of the plasma membrane were present. They designed experiments to show that a second messenger caused the activation of glycogen phosphorylase.

HYPOTHESIS A second messenger mediates between receptor activation at the plasma membrane and enzyme activation in the cytoplasm.

METHOD

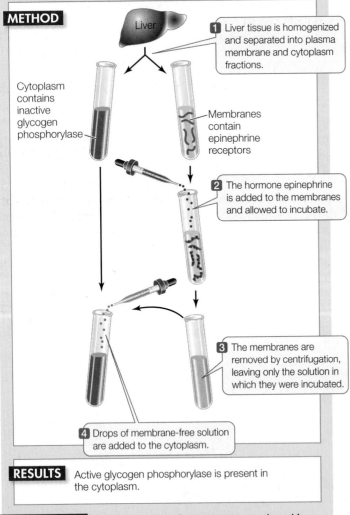

Liver

Cytoplasm contains inactive glycogen phosphorylase

1 Liver tissue is homogenized and separated into plasma membrane and cytoplasm fractions.

Membranes contain epinephrine receptors

2 The hormone epinephrine is added to the membranes and allowed to incubate.

3 The membranes are removed by centrifugation, leaving only the solution in which they were incubated.

4 Drops of membrane-free solution are added to the cytoplasm.

RESULTS Active glycogen phosphorylase is present in the cytoplasm.

CONCLUSION A soluble second messenger, produced by hormone-activated membranes, is present in the solution and activates enzymes in the cytoplasm.

FURTHER INVESTIGATION: The soluble molecule produced in this experiment was later identified as cAMP. How would you show that cAMP, and not ATP, is the second messenger in this system?

ble chemical messenger mediates the cytoplasmic events initiated by a plasma membrane receptor. These researchers were investigating the activation of the liver enzyme glycogen phosphorylase by the hormone epinephrine. The enzyme is released when an animal faces life-threatening conditions and needs energy fast for the fight-or-flight response. Glycogen phosphorylase catalyzes the breakdown of glycogen stored in the liver so that the resulting glucose molecules can be released to the blood. The enzyme is present in the liver cell cytoplasm, but is inactive except in the presence of epinephrine.

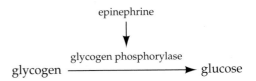

epinephrine

glycogen phosphorylase

glycogen ⟶ glucose

The researchers found that epinephrine could activate glycogen phosphorylase in liver cells that had been broken open, but only if the entire cell contents, including plasma membrane fragments, were present. Under these circumstances epinephrine bound to the plasma membranes, but the active phosphorylase was present in the solution. The researchers hypothesized that there must be a second "messenger" that transmits the signal of epinephrine (the "first messenger," which binds to a receptor at the plasma membrane) to the phosphorylase (in the cytoplasm). To investigate the production of this messenger, they separated plasma membrane fragments from the cytoplasms of broken liver cells and followed the sequence of steps described in **Figure 7.13**. This experiment confirmed their hypothesis that hormone binding to the membrane receptor causes the production of a small, water-soluble molecule that diffuses into the cytoplasm and activates the enzyme. Later, this second messenger was identified as **cyclic AMP** (**cAMP**). (We will describe the signal transduction pathway leading to the fight-or-flight response in more detail in Section 7.4.) Second messengers do not have enzymatic activity; rather, they act to regulate target enzymes (see Chapter 8).

A second messenger is a small molecule that mediates later steps in a signal transduction pathway after the first messenger—the signal or ligand—binds to its receptor. In contrast to the specificity of receptor binding, second messengers allow a cell to respond to a single event at the plasma membrane with *many events inside the cell*. Thus, second messengers serve to amplify the signal—for example, binding of a single epinephrine molecule leads to the production of many molecules of cAMP, which then activate many enzyme targets by binding to them noncovalently. In the case of epinephrine and the liver cell, glycogen phosphorylase is just one of several enzymes that are activated.

Cyclic AMP is a second messenger in a wide variety of signal transduction pathways. An effector protein, adenylyl cyclase, catalyzes the formation of cAMP from ATP. Adenylyl cyclase is located on the cytoplasmic surface of the plasma membrane of target cells (**Figure 7.14**). Usually a G protein activates the enzyme after it has itself been activated by a receptor.

Cyclic AMP has two major kinds of targets—ion channels and protein kinases. In many sensory cells, cAMP binds to ion channels and thus opens them. Cyclic AMP may also bind to a

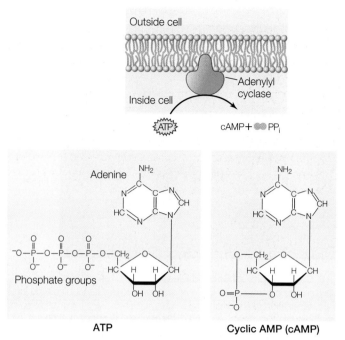

7.14 The Formation of Cyclic AMP The formation of cAMP from ATP is catalyzed by adenylyl cyclase, an enzyme that is activated by G proteins.

protein kinase in the cytoplasm, activating its catalytic function. A protein kinase cascade (see Figure 7.12) ensues, leading to the final effects in the cell.

Second messengers can be derived from lipids

In addition to their role as structural components of the plasma membrane, phospholipids are also involved in signal transduction. When certain phospholipids are hydrolyzed into their component parts by enzymes called **phospholipases**, second messengers are formed.

The best-studied examples of lipid-derived second messengers come from the hydrolysis of the phospholipid **phosphatidyl inositol-bisphosphate** (**PIP2**). Like all phospholipids, PIP2 has a hydrophobic portion embedded in the plasma membrane: two fatty acid tails attached to a molecule of glycerol, which together form **diacylglycerol**, or **DAG**. The hydrophilic portion of PIP2 is **inositol trisphosphate**, or **IP$_3$**, which projects into the cytoplasm.

As with cAMP, the receptors involved in this second-messenger system are often G protein-linked receptors. A G protein subunit is activated by the receptor, then diffuses within the plasma membrane and activates phospholipase C, an enzyme that is also located in the membrane. This enzyme cleaves off the IP$_3$ from PIP2, leaving the diacylglycerol (DAG) in the phospholipid bilayer:

$$\text{PIP2} \xrightarrow{\text{phospholipase C}} \text{IP}_3 \quad + \quad \text{DAG}$$

PIP2	IP$_3$	DAG
in membrane	released to cytoplasm	in membrane

IP$_3$ and DAG, both second messengers, have different modes of action that build on each other, activating protein kinase C (PKC) (**Figure 7.15**). PKC refers to a family of protein kinases that can phosphorylate a wide variety of target proteins, leading to a multiplicity of cellular responses that vary depending on the tissue or cell type.

The IP$_3$/DAG pathway is apparently a target for the ion lithium (Li$^+$), which was used for many years as a psychoactive drug to treat bipolar (manic-depressive) disorder. This serious illness occurs in about 1 in every 100 people. In these patients, an overactive IP$_3$/DAG signal transduction pathway in the

7.15 The IP$_3$/DAG Second-Messenger System Phospholipase C hydrolyzes the phospholipid PIP2 into its components, IP$_3$ and DAG, both of which are second messengers. Lithium ions (Li$^+$) block this pathway and are used to treat bipolar disorder (red type).

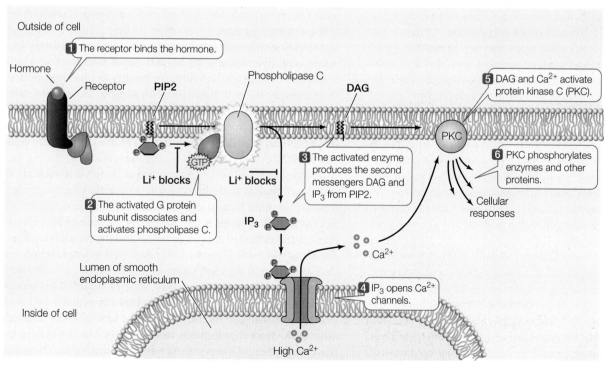

brain leads to excessive brain activity in certain regions. Lithium "tones down" this pathway in two ways, as indicated by the red notations in Figure 7.15. It inhibits G protein activation of phospholipase C, and also inhibits the synthesis of IP_3. The overall result is that brain activity returns to normal.

Calcium ions are involved in many signal transduction pathways

Calcium ions (Ca^{2+}) are scarce inside most cells, which have cytosolic Ca^{2+} concentrations of only about 0.1 mM. Ca^{2+} concentrations outside cells and within the endoplasmic reticulum are usually much higher. Active transport proteins in the plasma and ER membranes maintain this concentration difference by pumping Ca^{2+} out of the cytosol. In contrast to cAMP and the lipid-derived second messengers, Ca^{2+} cannot be made in order to increase the intracellular Ca^{2+} concentration. Instead, Ca^{2+} ion levels are regulated via the opening and closing of ion channels, and the action of membrane pumps.

There are many signals that can cause calcium channels to open, including IP_3 (see Figure 7.15). The entry of a sperm into an egg is a very important signal that causes a massive opening of calcium channels, resulting in numerous and dramatic changes that prepare the now fertilized egg for cell divisions and development (**Figure 7.16**). Whatever the initial signal that causes the calcium channels to open, their opening results in a dramatic increase in cytosolic Ca^{2+} concentration, which can increase up to one hundredfold within a fraction of a second. As we saw earlier, this increase activates protein kinase C. In addition, Ca^{2+} controls other ion channels and stimulates secretion by exocytosis in many cell types.

Nitric oxide can act in signal transduction

Most signaling molecules and second messengers are solutes that remain dissolved in either the aqueous or hydrophobic components of cells. It was a great surprise to find that a gas could also be active in signal transduction. Pharmacologist Robert Furchgott, at the State University of New York in Brooklyn, was investigating the mechanisms that cause the smooth muscles lining blood vessels in mammals to relax, thus allowing more blood to flow to certain organs. The neurotransmitter acetylcholine (see Section 7.2) appeared to stimulate the IP_3/DAG signal transduction pathway to produce an influx of Ca^{2+}, leading to an increase in the level of another second messenger, cyclic guanosine monophosphate (cGMP). Cyclic GMP then binds to a protein kinase, stimulating a protein kinase cascade that leads to muscle relaxation. So far, the pathway seemed to conform to what was generally understood about signal transduction in general.

While this signal transduction pathway seemed to work in intact animals, it did not work on isolated strips of artery tissue. However, when Furchgott switched to tubular sections of artery, signal transduction did occur. What accounted for the different results between tissue strips and tubular sections? Furchgott realized that the endothelium, the delicate inner layer of cells lining the blood vessels, was lost during preparation of

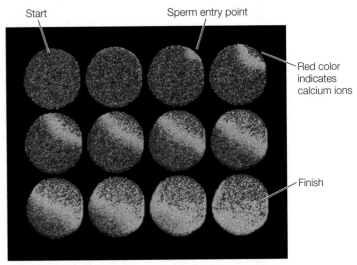

7.16 Calcium Ions as Second Messengers The concentration of Ca^{2+} can be measured using a dye that fluoresces when it binds the ion. Here, fertilization in a starfish egg causes a rush of Ca^{2+} from the environment into the cytoplasm. Areas of high calcium ion concentration are indicated by the red color and the events are photographed at 5-second intervals. Calcium signaling occurs in virtually all animal groups and triggers cell division in fertilized eggs, initiating the development of new individuals.

the tissue strips. He hypothesized that the endothelium was producing some chemical that diffused into the smooth muscle cells and was needed for their response to acetylcholine. However, the substance was not easy to isolate. It seemed to break down quickly, with a half-life (the time in which half of it disappeared) of 5 seconds in living tissue.

Furchgott's elusive substance turned out to be a gas, **nitric oxide** (**NO**), which formerly had been recognized only as a toxic air pollutant! In the body, NO is made from the amino acid arginine by the enzyme NO synthase. When the acetylcholine receptor on the surface of an endothelial cell is activated, IP_3 is released, causing a calcium channel on the ER membrane to open and a subsequent increase in cytosolic Ca^{2+}. The Ca^{2+} then activates NO synthase to produce NO. NO is chemically very unstable, readily reacting with oxygen gas as well as other small molecules. Although NO diffuses readily, it does not get far. Conveniently, the endothelial cells are close to the smooth muscle cells, where NO acts as a paracrine signal. In smooth muscle, NO activates an enzyme called guanylyl cyclase, catalyzing the formation of cGMP, which in turn relaxes the muscle cells (**Figure 7.17**).

The discovery of NO as a participant in signal transduction explained the action of nitroglycerin, a drug that has been used for over a century to treat angina, the chest pain caused by insufficient blood flow to the heart. Nitroglycerin releases NO, which results in relaxation of the blood vessels and increased blood flow. The drug sildenafil (Viagra) was developed to treat angina via the NO signal transduction pathway, but was only modestly useful for that purpose. However, men taking it reported more pronounced penile erections. During sexual stimulation, NO acts as a signal causing an increase in cGMP and a subsequent relaxation of the smooth muscles surrounding the arteries in the corpus cavernosum of the penis. As a result of this signal, the penis

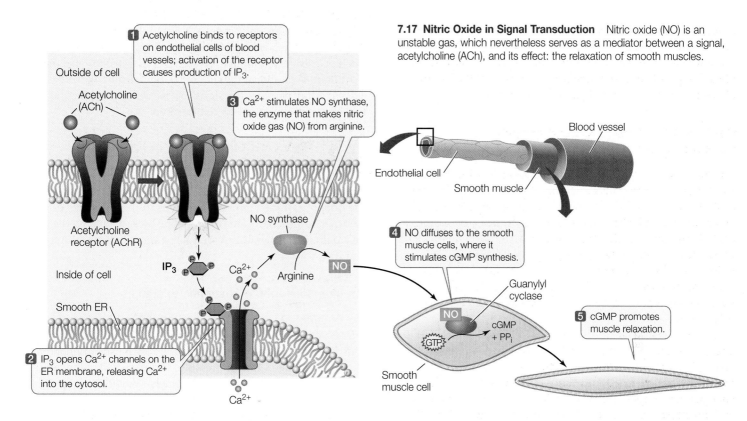

7.17 Nitric Oxide in Signal Transduction Nitric oxide (NO) is an unstable gas, which nevertheless serves as a mediator between a signal, acetylcholine (ACh), and its effect: the relaxation of smooth muscles.

1 Acetylcholine binds to receptors on endothelial cells of blood vessels; activation of the receptor causes production of IP_3.

3 Ca^{2+} stimulates NO synthase, the enzyme that makes nitric oxide gas (NO) from arginine.

2 IP_3 opens Ca^{2+} channels on the ER membrane, releasing Ca^{2+} into the cytosol.

4 NO diffuses to the smooth muscle cells, where it stimulates cGMP synthesis.

5 cGMP promotes muscle relaxation.

fills with blood, producing an erection. Sildenafil acts by inhibiting an enzyme (a phosphodiesterase) that breaks down cGMP—resulting in more cGMP and better erections.

Signal transduction is highly regulated

There are several ways in which cells can regulate the activity of a transducer. The concentration of NO, which breaks down quickly, can be regulated only by how much of it is made. On the other hand, membrane pumps and ion channels regulate the concentration of Ca^{2+}, as we have seen. To regulate protein kinase cascades, G proteins, and cAMP, there are enzymes that convert the activated transducer back to its inactive precursor (**Figure 7.18**).

The balance between the activities of enzymes that activate transducers (for example, protein kinase) and enzymes that inactivate them (for example, protein phosphatase) is what determines the ultimate cellular response to a signal. Cells can alter this balance in several ways:

- *Synthesis or breakdown of the enzymes involved.* For example, synthesis of adenylyl cyclase and breakdown of phosphodiesterase (which breaks down cAMP) would tilt the balance in favor of more cAMP in the cell.

- *Activation or inhibition of the enzymes by other molecules.* Examples include the activation of a G protein-linked receptor by ligand binding, and inhibition of phosphodiesterase (which also breaks down cGMP) by sildenafil.

Because cell signaling is so important in diseases such as cancer, a search is under way for new drugs that can modulate the activities of enzymes that participate signal transduction pathways.

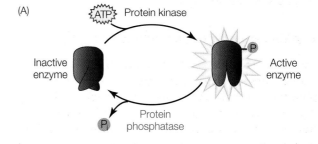

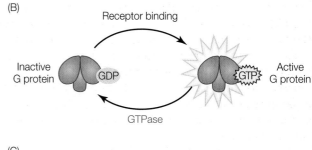

7.18 Regulation of Signal Transduction Some signals lead to the production of active transducers such as (A) protein kinases, (B) G proteins, and (C) cAMP. Other enzymes (red type) inactivate or remove these transducers.

We have seen how the binding of a signal to its receptor initiates the response of a cell to the signal, and how signal transduction pathways amplify the signal and distribute its effects to numerous targets in the cell. In the next section we will consider the third step in the signal transduction process, the actual effects of the signal on cell function.

7.4 How Do Cells Change in Response to Signals?

The effects of a signal on cell function take three primary forms: the opening of ion channels, changes in the activities of enzymes, or differential gene expression. These events set the cell on a path for further and sometimes dramatic changes in form and function.

Ion channels open in response to signals

The opening of ion channels is a key step in the response of the nervous system to signals. In the sense organs, specialized cells have receptors that respond to external stimuli such as light, sound, taste, odor, or pressure. The alteration of the receptor results in the opening of ion channels. We will focus here on one such signal transduction pathway, that for the sense of smell, which responds to gaseous molecules in the environment.

The sense of smell is well developed in mammals. Each of the thousands of neurons in the nose expresses one of many different odorant receptors. The identification of which chemical signal, or odorant, activates which receptor is just getting under way. Humans have the genetic capacity to make about 950 different odorant receptor proteins, but very few people express more than 400 of them. Some express far fewer, which may explain why you are able to smell certain things that your roommate cannot, or vice versa.

Odorant receptors are G-protein linked, and signal transduction leads to the opening of ion channels for sodium and calcium ions, which have higher concentrations outside the cell than in the cytosol (**Figure 7.19**). The resulting influx of Na⁺ and

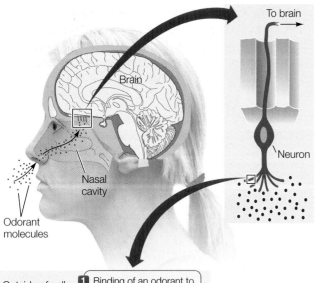

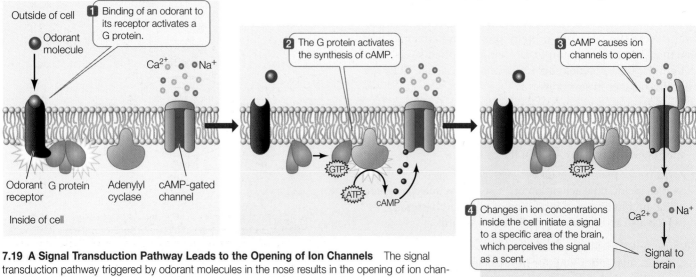

7.19 A Signal Transduction Pathway Leads to the Opening of Ion Channels The signal transduction pathway triggered by odorant molecules in the nose results in the opening of ion channels. The resulting influx of Na⁺ and Ca²⁺ into the neuron cells of the nose stimulates the transmission of a scent message to a specific region of the brain.

Ca^{2+} causes the neuron to become stimulated so that it sends a signal to the brain that a particular odor is present.

Enzyme activities change in response to signals

Proteins will change their shapes if they are modified either covalently or noncovalently. We have seen examples of both types of modification in our description of signal transduction. A protein kinase adds a phosphate group to a target protein, and this covalent change alters the protein's conformation and activates or inhibits a function. Cyclic AMP binds noncovalently to a target protein, and this changes the protein's shape, activating or inhibiting its function. In the case of activation, a previously inaccessible active site is exposed, and the target protein goes on to perform a new cellular role.

The G protein-mediated protein kinase cascade that is stimulated by epinephrine in liver cells results in the activation by cAMP of a key signaling molecule, protein kinase A. In turn, protein kinase A phosphorylates two other enzymes, with opposite effects:

- *Inhibition*. Glycogen synthase, which catalyzes the joining of glucose molecules to synthesize the energy-storing molecule glycogen, is inactivated when a phosphate group is added to it by protein kinase A. Thus the epinephrine signal *prevents glucose from being stored* in glycogen (**Figure 7.20, step 1**).

- *Activation*. Phosphorylase kinase is activated when a phosphate group is added to it. It is part of a protein kinase cascade that ultimately leads to the activation of glycogen phosphorylase, another key enzyme in glucose metabolism. This enzyme results in the *liberation of glucose molecules* from glycogen (**Figure 7.20, steps 2 and 3**).

The amplification of the signal in this pathway is impressive; as detailed in Figure 7.20, each molecule of epinephrine that arrives at the plasma membrane ultimately results in 10,000 molecules of blood glucose:

    1 molecule of epinephrine bound to the membrane activates
    20 molecules of cAMP, which activate
    20 molecules of protein kinase A, which activate
    100 molecules of phosphorylase kinase, which activate
  1,000 molecules of glycogen phosphorylase, which produce
10,000 molecules of glucose 1-phosphate, which produce
10,000 molecules of blood glucose

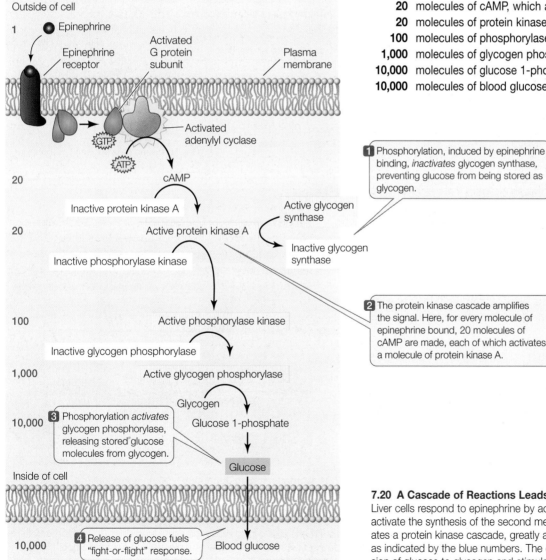

1 Phosphorylation, induced by epinephrine binding, *inactivates* glycogen synthase, preventing glucose from being stored as glycogen.

2 The protein kinase cascade amplifies the signal. Here, for every molecule of epinephrine bound, 20 molecules of cAMP are made, each of which activates a molecule of protein kinase A.

3 Phosphorylation *activates* glycogen phosphorylase, releasing stored glucose molecules from glycogen.

4 Release of glucose fuels "fight-or-flight" response.

7.20 A Cascade of Reactions Leads to Altered Enzyme Activity
Liver cells respond to epinephrine by activating G proteins, which in turn activate the synthesis of the second messenger cAMP. Cyclic AMP initiates a protein kinase cascade, greatly amplifying the epinephrine signal, as indicated by the blue numbers. The cascade both inhibits the conversion of glucose to glycogen and stimulates the release of previously stored glucose.

Signals can initiate DNA transcription

As we introduce in Section 4.1, the genetic material, DNA, is expressed by transcription as RNA, which is then translated into a protein whose amino acid sequence is specified by the original DNA sequence. Proteins are important in all cellular functions, so a key way to regulate specific functions in a cell is to regulate which proteins are made, and therefore, which DNA sequences are transcribed.

Signal transduction plays an important role in determining which DNA sequences are transcribed. Common targets of signal transduction are proteins called transcription factors, which bind to specific DNA sequences in the cell nucleus and activate or inactivate transcription of the adjacent DNA regions. For example, the Ras signaling pathway ends in the nucleus (see Figure 7.12). The final protein kinase in the Ras signaling cascade, MAPk, enters the nucleus and phosphorylates a protein which stimulates the expression of a number of genes involved in cell proliferation.

In this chapter we have concentrated on signaling pathways that occur in animal cells. However, as you will see in Part Eight of this book, plants also have signal transduction pathways, with equally important roles.

7.4 RECAP

Cells respond to signal transduction by activating enzymes, opening membrane channels, or initiating gene transcription.

- What role does cAMP play in the sense of smell? **See pp. 142–143 and Figure 7.19**

- How does amplification of a signal occur and why is it important in a cell's response to changes in its environment? **See p. 143 and Figure 7.20**

We have described how signals from a cell's environment can influence the cell. But the environment of a cell in a multicellular organism is more than the extracellular medium—it includes neighboring cells as well. In the next section we'll look at specialized junctions between cells that allow them to signal one another directly.

7.5 How Do Cells Communicate Directly?

Most cells are in contact with their neighbors. Section 6.2 describes various ways in which cells adhere to one another, such as via recognition proteins that protrude from the cell surface, or via tight junctions and desmosomes. But as we know from our own experience with our neighbors (and roommates), just being in proximity does not necessarily mean that there is functional communication. Neither tight junctions nor desmosomes are specialized for intercellular communication. However, many multicellular organisms have specialized cell junctions that allow their cells to communicate directly. In animals, these structures are gap junctions; in plants, they are plasmodesmata.

Animal cells communicate by gap junctions

Gap junctions are channels between adjacent cells that occur in many animals, occupying up to 25 percent of the area of the plasma membrane (**Figure 7.21A**). Gap junctions traverse the narrow space between the plasma membranes of two cells (the "gap") by means of channel structures called **connexons**. The walls of a connexon are composed of six subunits of the integral membrane protein connexin. In adjacent cells, two connexons come together to form a gap junction that links the cytoplasms of the two cells. There may be hundreds of these channels between a cell and its neighbors. The channel pores are about 1.5 nm in diameter—far too narrow for the passage of large molecules such as proteins. But they are wide enough to allow small mol-

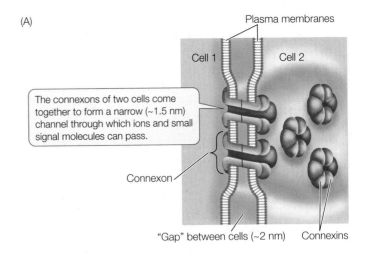

(A)

Plasma membranes

Cell 1   Cell 2

The connexons of two cells come together to form a narrow (~1.5 nm) channel through which ions and small signal molecules can pass.

Connexon

"Gap" between cells (~2 nm)   Connexins

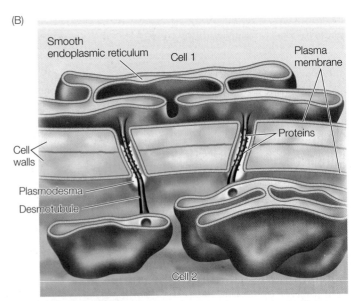

(B)

Smooth endoplasmic reticulum   Cell 1   Plasma membrane

Cell walls

Proteins

Plasmodesma

Desmotubule

Cell 2

7.21 Communicating Junctions (A) An animal cell may contain hundreds of gap junctions connecting it to neighboring cells. The pores of gap junctions allow small molecules to pass from cell to cell, assuring similar concentrations of important signaling molecules in adjacent cells so that the cells can carry out the same functions. (B) Plasmodesmata connect plant cells. The desmotubule, derived from the smooth endoplasmic reticulum, fills up most of the space inside a plasmodesma, leaving a tiny gap through which small metabolites and ions can pass.

ecules to pass between the cells. Experiments in which labeled signal molecules or ions are injected into one cell show that they can readily pass into adjacent cells if the cells are connected by gap junctions. Why is it necessary to have these linkages between the cytoplasms of adjacent cells?

Gap junctions permit *metabolic cooperation* between the linked cells. Such cooperation ensures the sharing between cells of important small molecules such as ATP, metabolic intermediates, amino acids, and coenzymes (see Section 8.4). In some tissues, metabolic cooperation is needed so that signals and metabolic products can be passed from cells at the edges of tissues to cells in the interior and vice versa. It is not clear how important this function is in many tissues, but it is known to be vital in some. For example, in the lens of the mammalian eye only the cells at the periphery are close enough to the blood supply to allow diffusion of nutrients and wastes. But because lens cells are connected by large numbers of gap junctions, material can diffuse between them rapidly and efficiently.

As mentioned above, there is evidence that signal molecules such as hormones and second messengers such as cAMP can move through gap junctions. If this is true, then only a few cells would need receptors for a signal in order for the signal to be transduced throughout the tissue. In this way, a tissue can have a coordinated response to the signal.

Plant cells communicate by plasmodesmata

Instead of gap junctions, plants have **plasmodesmata** (singular *plasmodesma*), which are membrane-lined tunnels that traverse the thick cell walls separating plant cells from one another. A typical plant cell has several thousand plasmodesmata.

Plasmodesmata differ from gap junctions in one fundamental way: unlike gap junctions, in which the wall of the channel is made of integral proteins from the adjacent plasma membranes,

plasmodesmata are lined by the fused plasma membranes themselves. Plant biologists are so familiar with the notion of a tissue as cells interconnected in this way that they refer to these continuous cytoplasms as a *symplast* (see Figure 35.6).

The diameter of a plasmodesma is about 6 nm, far larger than a gap junction channel. But the actual space available for diffusion is about the same—1.5 nm. Examination of the interior of the plasmodesma by transmission electron microscopy reveals that a tubule called the **desmotubule**, apparently derived from the endoplasmic reticulum, fills up most of the opening of the plasmodesma (**Figure 7.21B**). Typically, only small metabolites and ions can move between plant cells. This fact is important in plant physiology because the bulk transport system in plants, the vascular system, lacks the tiny circulatory vessels (capillaries) that many animals have for bringing gases and nutrients to every cell. Diffusion from cell to cell across plasma membranes is probably inadequate to account for the movement of a plant hormone from the site of production to the site of action. Instead, plants rely on more rapid diffusion through plasmodesmata to ensure that all cells of a tissue respond to a signal at the same time. There are cases in which larger molecules or particles can pass between cells via plasmodesmata. For example, some viruses can move through plasmodesmata by using "movement proteins" to assist their passage.

7.5 RECAP

Cells can communicate with their neighbors through specialized cell junctions. In animals, these structures are gap junctions; in plants, they are plasmodesmata.

- What are the roles that gap junctions and plasmodesmata play in cell signaling?

CHAPTER SUMMARY

7.1 What Are Signals, and How Do Cells Respond to Them?

- Cells receive many signals from the physical environment and from other cells. Chemical signals are often at very low concentrations. **Autocrine** signals affect the cells that make them; **paracrine** signals diffuse to and affect nearby cells. **Review Figure 7.1, WEB ACTIVITY 7.1**

- A **signal transduction pathway** involves the interaction of a signal molecule with a **receptor**; the transduction and amplification of the signal via a series of steps within the cell; and effects on the function of the cell. **Review Figure 7.2**

7.2 How Do Signal Receptors Initiate a Cellular Response?

- Cells respond to signals only if they have specific receptor proteins that can bind those signals. Depending on the nature of its signal or **ligand**, a receptor may be located in the plasma membrane or in the cytoplasm of the target cell. **Review Figure 7.5**

- Receptors located in the plasma membrane include **ion channels**, **protein kinases**, and **G protein-linked receptors**.

- Ion channel receptors are "gated": the gate "opens" when the three-dimensional structure of the channel protein is altered by ligand binding. **Review Figure 7.6**

- A **G protein** has three important binding sites, which bind a G protein-linked receptor, GDP or GTP, and an **effector protein**. A G protein can either activate or inhibit an effector protein. **Review Figure 7.8, ANIMATED TUTORIAL 7.1**

- Lipid-soluble signals, such as steroid hormones, can diffuse through the plasma membrane and meet their receptors in the cytoplasm; the ligand–receptor complex may then enter the nucleus to affect gene expression. **Review Figure 7.9**

7.3 How Is the Response to a Signal Transduced through the Cell?

- **Direct signal transduction** is a function of the receptor itself and occurs at the plasma membrane. **Indirect transduction** involves a soluble **second messenger**. **Review Figure 7.10**

- A **protein kinase cascade** amplifies the response to receptor binding. **Review Figure 7.12, ANIMATED TUTORIAL 7.2**

- Second messengers include **cyclic AMP (cAMP)**, **inositol trisphosphate (IP₃)**, **diacylglycerol (DAG)**, and **calcium** ions. IP_3 and DAG are derived from the phospholipid **phosphatidyl inositol-bisphosphate (PIP2)**.

- The gas **nitric oxide (NO)** is involved in signal transduction in human smooth muscle cells. **Review Figure 7.17**

- Signal transduction can be regulated in several ways. The balance between activating and inactivating the molecules involved determines the ultimate cellular response to a signal. **Review Figure 7.18**

7.4 How Do Cells Change in Response to Signals?

- The cellular responses to signals may be the opening of ion channels, the alteration of enzyme activities, or changes in gene expression. **Review Figure 7.19**

- Protein kinases covalently add phosphate groups to target proteins; cAMP binds target proteins noncovalently. Both kinds of binding change the target protein's conformation to expose or hide its active site.

- Activated enzymes may activate other enzymes in a signal transduction pathway, leading to impressive amplification of a signal. **Review Figure 7.20**

7.5 How Do Cells Communicate Directly?

- Many adjacent animal cells can communicate with one another directly through small pores in their plasma membranes called **gap junctions**. Protein structures called **connexons** form thin channels between two adjacent cells through which small signal molecules and ions can pass. **Review Figure 7.21A**

- Plant cells are connected by somewhat larger pores called **plasmodesmata**, which traverse both plasma membranes and cell walls. The **desmotubule** narrows the opening of the plasmodesma. **Review Figure 7.21B**

SEE WEB ACTIVITY 7.2 for a concept review of this chapter.

SELF-QUIZ

1. What is the correct order for the following events in the interaction of a cell with a signal? (1) Alteration of cell function; (2) signal binds to receptor; (3) signal released from source; (4) signal transduction.
   a. 1234
   b. 2314
   c. 3214
   d. 3241
   e. 3421

2. Why do some signals ("first messengers") trigger "second messengers" to activate target cells?
   a. The first messenger requires activation by ATP.
   b. The first messenger is not water soluble.
   c. The first messenger binds to many types of cells.
   d. The first messenger cannot cross the plasma membrane.
   e. There are no receptors for the first messenger.

3. Steroid hormones such as estrogen act on target cells by
   a. initiating second messenger activity.
   b. binding to membrane proteins.
   c. initiating gene expression.
   d. activating enzymes.
   e. binding to membrane lipids.

4. The major difference between a cell that responds to a signal and one that does not is the presence of a
   a. DNA sequence that binds to the signal.
   b. nearby blood vessel.
   c. receptor.
   d. second messenger.
   e. transduction pathway.

5. Which of the following is *not* a consequence of a signal binding to a receptor?
   a. Activation of receptor enzyme activity
   b. Diffusion of the receptor in the plasma membrane
   c. Change in conformation of the receptor protein
   d. Breakdown of the receptor to amino acids
   e. Release of the signal from the receptor

6. A nonpolar molecule such as a steroid hormone usually binds to a
   a. cytoplasmic receptor.
   b. protein kinase.
   c. ion channel.
   d. phospholipid.
   e. second messenger.

7. Which of the following is *not* a common type of receptor?
   a. Ion channel
   b. Protein kinase
   c. G protein–linked receptor
   d. Cytoplasmic receptor
   e. Adenylyl cyclase

8. Which of the following is *not* true of a protein kinase cascade?
   a. The signal is amplified.
   b. A second messenger is formed.
   c. Target proteins are phosphorylated.
   d. The cascade ends up at the mitochondrion.
   e. The cascade begins at the plasma membrane.

9. Which of the following is *not* a second messenger?
   a. Calcium ion
   b. Inositol trisphosphate
   c. ATP
   d. Cyclic AMP
   e. Diacylglycerol

10. Plasmodesmata and gap junctions
   a. allow small molecules and ions to pass rapidly between cells.
   b. are both membrane-lined channels.
   c. are channels about 1 mm in diameter.
   d. are present only once per cell.
   e. are involved in cell recognition.

FOR DISCUSSION

1. Like the Ras protein itself, the various components of the Ras signaling pathway were discovered when cancer cells showed changes (mutations) in the genes encoding one or another of the components. What might be the biochemical consequences of mutations in the genes coding for (*a*) Raf and (*b*) MAP kinase that resulted in rapid cell division?

2. Cyclic AMP is a second messenger in many different responses. How can the same messenger act in different ways in different cells?

3. Compare direct communication via plasmodesmata or gap junctions with receptor-mediated communication between cells. What are the advantages of one method over the other?

4. The tiny invertebrate Hydra has an apical region with tentacles and a long, slender body. Hydra can reproduce asexually when cells on the body wall differentiate and form a bud, which then breaks off as a new organism. Buds form only at certain distances from the apex, leading to the idea that the apex releases a signal molecule that diffuses down the body and, at high concentrations (i.e., near the apex), inhibits bud formation. Hydra lacks a circulatory system, so this inhibitor must diffuse from cell to cell. If you had an antibody that binds to connexons and plugs up the gap junctions, how would you test the hypothesis that Hydra's inhibitory factor passes through these junctions?

ADDITIONAL INVESTIGATION

Endosymbiotic bacteria in the marine invertebrate *Begula neritina* synthesize bryostatins, a name derived from the invertebrate's animal group Ectoprocta, once known as *bryo*zoans ("moss animals"), and *stat* (stop). When used as drugs, bryostatins curtail cell division in many cell types, including several cancers. It has been proposed that bryostatins inhibit protein kinase C (see Figure 7.15). How would you investigate this hypothesis, and how would you relate this inhibition to cell division?

WORKING WITH DATA (GO TO yourBioPortal.com)

The Discovery of a Second Messenger In this hands-on exercise, you will examine the experiments that Sutherland and his colleagues performed (Figure 7.13) using liver tissue to demonstrate that there can be a second, soluble chemical messenger between a hormone binding to a receptor and its eventual effects in the cell. By analyzing their data, you will see how controls were important in their reasoning.

8 Energy, Enzymes, and Metabolism

Lactase deficiency

United Nations officials first noticed the problem during the 1950s when massive food relief efforts were made to alleviate famines in Asia and Africa. The conventional wisdom was that donated food should provide a balanced diet, and that an important component of the diet (one of the "four major food groups") was dairy products. Reports started coming in of people developing bloating, nausea, and diarrhea after consuming donated dairy products.

At first, this problem was attributed to contamination by bacteria during shipping, or to errors in the preparation of powdered milk products by the recipients. It never occurred to the donors that the scientific principles of nutrition that they had so carefully developed did not apply to people everywhere. But it soon became apparent that the donors in Europe and other wealthy countries, who were usually of European descent, were atypical of humanity in their ability to hydrolyze the disaccharide lactose, the major milk sugar, to its constituent monosaccharides, glucose and galactose. Their small intestines make a protein called lactase (β-galactosidase) that acts to speed up the hydrolysis reaction millions-fold. Such catalytic proteins are called enzymes, and their names often end with the suffix "ase." Most people around the world are born with the ability to make the enzyme lactase, but soon after infancy they lose it. People of European descent are unusual in that they do not lose their lactase production after infancy.

When many non-European adults consume lactose it does not get hydrolyzed in their small intestine, because they do not produce lactase. Disaccharides such as lactose are not absorbed into the blood stream by cells lining the small intestine. So the lactose remains intact and travels onward to the colon (large intestine). Among the billions of bacteria in the colon, there are species that make lactase. But as a side product, these bacteria produce the gases that cause all the discomfort. The condition of discomfort after eating lactose is called lactose intolerance.

Why does lactase production go down after infancy in most humans? The explanation lies with diet: an infant first consumes mother's milk, which contains abundant lactose. This stimulates the intestinal cells to make lactase. But many humans—and other mammals—consume little or no milk after weaning, and the ability to make lactase in the small intestine is not needed. So most mammals have evolved to produce lactase only during infancy. Lactose intolerance is not a problem in many human societies because the people simply don't consume dairy products—

A Precursor to Trouble   Many adults do not produce the enzyme lactase in their small intestines. When they consume dairy products, these people have ill effects.

Maasai Herders The Maasai are unusual among Africans in that they consume milk throughout their lives. They can do this because they produce lactase after weaning.

unless they are given them by well-meaning donors! They get their carbohydrates from other sources.

Then why are many people of European descent still able to make lactase as adults? It turns out that they carry a mutation (a change in their DNA sequence) that eliminates the shutdown in lactase production after weaning. This mutation became predominant in European (and some east African) populations after those people began to keep grazing animals and to use their milk.

Lactase activity is an example of an enzyme-catalyzed biochemical transformation. The hydrolysis of lactose is the beginning of its transformation to simpler molecules—ultimately CO_2—and this transformation releases energy.

IN THIS CHAPTER we begin our study of biochemical transformations, focusing on the role of energy. We first describe the physical principles that underlie energy transformations and how these principles apply to biology. Then, we go on to show how the energy carrier ATP plays an important role in the cell. Finally, we follow up on the lactase story by describing the nature, activities, and regulation of enzymes, which speed up biochemical transformations and are essential for life.

8.1 What Physical Principles Underlie Biological Energy Transformations?

Metabolic reactions and catalysts are essential to the biochemical transformation of energy by living things. Whether it is a plant using light energy to produce carbohydrates or a cat transforming food energy so it can leap to a countertop (where it hopes to find food so it can obtain more energy), the transformation of energy is a hallmark of life.

Physicists define energy as the capacity to do work. Work occurs when a force operates on an object over a distance. In biochemistry, it is more useful to consider energy as *the capacity for change*. In biochemical reactions these energy changes are usually associated with changes in the chemical composition and properties of molecules. No cell creates energy; all living things must obtain energy from the environment. Indeed one of the fundamental laws of physics is that energy can neither be created nor destroyed. However, energy can be transformed from one form into another, and living cells carry out many such transformations. For example, green plant cells convert light energy into chemical energy; the jumping cat transforms chemical energy into movement. Energy transformations are linked to the chemical transformations that occur in cells—the breaking and creating of chemical bonds, the movement of substances across membranes, cell reproduction, and so forth.

There are two basic types of energy and of metabolism

Energy comes in many forms: chemical, electrical, heat, light, and mechanical. But all forms of energy can be considered as one of two basic types:

- *Potential energy* is the energy of state or position—that is, stored energy. It can be stored in many forms: in chemical bonds, as a concentration gradient, or even as an electric charge imbalance (as in the membrane potential; see Section 6.3). Think of a crouching cat, holding still as it prepares to pounce.

- *Kinetic energy* is the energy of movement—that is, the type of energy that does work, that makes things change. Think of the cat leaping as some of the potential energy stored in its muscles is converted into the kinetic energy of muscle contractions.

Potential energy can be converted into kinetic energy and vice versa, and the form that the energy takes can also be converted. The potential energy in the cat's muscles is in covalent bonds (chemical energy), while the kinetic energy of the pouncing cat is mechanical (**Figure 8.1**). You can think of many other such

Potential chemical energy is converted into kinetic mechanical energy when the cat leaps.

Potential chemical energy is stored in the muscles of the cat.

8.1 Energy Conversions and Work A leaping cat illustrates both the conversion between potential and kinetic energy and the conversion of energy from one form (chemical) to another (mechanical).

conversions: while reading this book, for example, light energy is converted into chemical energy in your eyes, and then is converted into electric energy in the nerve cells that carry messages to your brain. When you decide to turn a page, the electrical and chemical energy of nerve and muscle are converted into kinetic energy.

In any living organism, chemical reactions are occurring continuously. **Metabolism** is defined as the totality of these reactions. While particular cells carry out many reactions at any given instant, scientists usually focus on a few reactions at a time. Two broad categories of metabolic reactions occur in all cells of all organisms:

- **Anabolic reactions** (anabolism) link simple molecules to form more complex molecules (for example, the synthesis of a protein from amino acids). Anabolic reactions require an input of energy and capture it in the chemical bonds that are formed.

- **Catabolic reactions** (catabolism) break down complex molecules into simpler ones and release the energy stored in chemical bonds. For example, when the polysaccharide starch is hydrolyzed to simpler molecules, energy is released.

Catabolic and anabolic reactions are often linked. The energy released in catabolic reactions is often used to drive anabolic re-

actions—that is, to do biological work. For example, the energy released by the breakdown of glucose (catabolism) is used to drive anabolic reactions such as the synthesis of nucleic acids and proteins.

Catabolic reactions also provide energy for movement: muscle contraction is driven by the catabolism (hydrolysis) of ATP (see Section 8.2). In this case, the potential energy released by catabolism is converted to kinetic energy.

The **laws of thermodynamics** (thermo, "energy"; dynamics, "change") were derived from studies of the fundamental physical properties of energy, and the ways it interacts with matter. The laws apply to all matter and all energy transformations in the universe. Their application to living systems helps us to understand how organisms and cells harvest and transform energy to sustain life.

The first law of thermodynamics: Energy is neither created nor destroyed

The first law of thermodynamics states that in any conversion of energy, it is neither created nor destroyed. Another way of saying this is: in any conversion of energy, the total energy before and after the conversion is the same (**Figure 8.2A**). As you will see in the next two chapters, the potential energy present in the chemical bonds of carbohydrates and lipids can be converted to potential energy in the form of ATP. This can then be converted into kinetic energy to do mechanical work (such as in muscle contractions), or used to do biochemical work (such as protein synthesis).

The second law of thermodynamics: Disorder tends to increase

Although energy cannot be created or destroyed, the second law of thermodynamics states that when energy is converted from one form to another, some of that energy becomes unavailable for doing work (**Figure 8.2B**). In other words, no physical process or chemical reaction is 100 percent efficient; some of the released energy is lost to a form associated with disorder. Think of disorder as a kind of randomness due to the thermal motion of particles; this energy is of such a low value and so dispersed that it is unusable. *Entropy* is a measure of the disorder in a system.

It takes energy to impose order on a system. Unless energy is applied to a system, it will be randomly arranged or disordered. The second law applies to all energy transformations, but we will focus here on chemical reactions in living systems.

NOT ALL ENERGY CAN BE USED In any system, the total energy includes the usable energy that can do work and the unusable energy that is lost to disorder:

total energy = usable energy + unusable energy

In biological systems, the total energy is called **enthalpy** (**H**). The usable energy that can do work is called **free energy** (**G**). Free energy is what cells require for all the chemical reactions needed for growth, cell division, and maintenance. The unusable en-

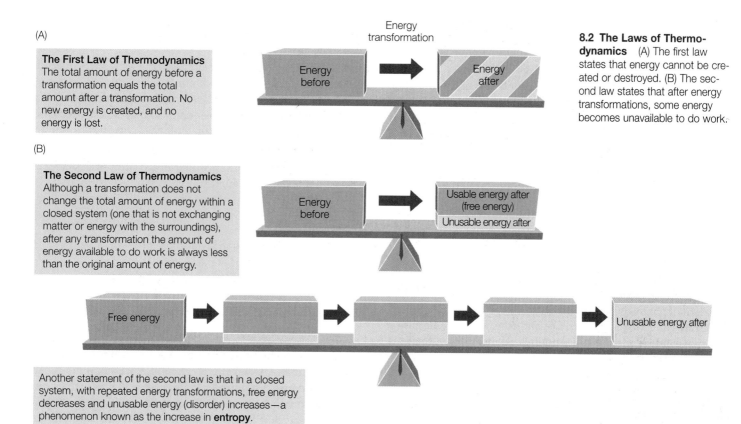

(A)

The First Law of Thermodynamics
The total amount of energy before a transformation equals the total amount after a transformation. No new energy is created, and no energy is lost.

Energy transformation

Energy before

Energy after

(B)

The Second Law of Thermodynamics
Although a transformation does not change the total amount of energy within a closed system (one that is not exchanging matter or energy with the surroundings), after any transformation the amount of energy available to do work is always less than the original amount of energy.

Energy before

Usable energy after (free energy)

Unusable energy after

Free energy

Unusable energy after

Another statement of the second law is that in a closed system, with repeated energy transformations, free energy decreases and unusable energy (disorder) increases—a phenomenon known as the increase in **entropy**.

8.2 The Laws of Thermodynamics (A) The first law states that energy cannot be created or destroyed. (B) The second law states that after energy transformations, some energy becomes unavailable to do work.

ergy is represented by **entropy (S)** multiplied by the absolute temperature (T). Thus we can rewrite the word equation above more precisely as:

$$H = G + TS$$

Because we are interested in usable energy, we rearrange this expression:

$$G = H - TS$$

Although we cannot measure G, H, or S absolutely, we can determine the change in each at a constant temperature. Such energy changes are measured in calories (cal) or joules (J).* A change in energy is represented by the Greek letter delta (Δ). The change in free energy (ΔG) of any chemical reaction is equal to the difference in free energy between the products and the reactants:

$$\Delta G_{reaction} = G_{products} - G_{reactants}$$

Such a change can be either positive or negative; that is, the free energy of the products can be more or less than the free energy of the reactants. If the products have more free energy than the reactants, then there must have been some input of energy

into the reaction. (Remember that energy cannot be created, so some energy must have been added from an external source.) At a constant temperature, ΔG is defined in terms of the change in total energy (ΔH) and the change in entropy (ΔS):

$$\Delta G = \Delta H - T\Delta S$$

This equation tells us whether free energy is released or consumed by a chemical reaction:

- If ΔG is negative (ΔG < 0), free energy is released.

- If ΔG is positive (ΔG > 0), free energy is required (consumed).

If the necessary free energy is not available, the reaction does not occur. The sign and magnitude of ΔG depend on the two factors on the right of the equation:

- ΔH: In a chemical reaction, ΔH is the total amount of energy added to the system (ΔH > 0) or released (ΔH < 0).

- ΔS: Depending on the sign and magnitude of ΔS, the entire term, TΔS, may be negative or positive, large or small. In other words, in living systems at a constant temperature (no change in T), the magnitude and sign of ΔG can depend a lot on changes in entropy.

If a chemical reaction increases entropy, its products are more disordered or random than its reactants. If there are more products than reactants, as in the hydrolysis of a protein to its amino acids, the products have considerable freedom to move around. The disorder in a solution of amino acids will be large compared with that in the protein, in which peptide bonds and other forces prevent free movement. So in hydrolysis, the change in entropy (ΔS) will be positive. Conversely, if there are fewer products and they are more restrained in their movements than the reac-

---

*A calorie is the amount of heat energy needed to raise the temperature of 1 gram of pure water from 14.5°C to 15.5°C. In the SI system, energy is measured in joules. 1 J = 0.239 cal; conversely, 1 cal = 4.184 J. Thus, for example, 486 cal = 2,033 J, or 2.033 kJ. Although defined here in terms of heat, the calorie and the joule are measures of any form of energy—mechanical, electrical, or chemical. When you compare data on energy, always compare joules with joules and calories with calories.

tants (as for amino acids being joined in a protein), ΔS will be negative.

DISORDER TENDS TO INCREASE The second law of thermodynamics also predicts that, as a result of energy transformations, disorder tends to increase; some energy is always lost to random thermal motion (entropy). Chemical changes, physical changes, and biological processes all tend to increase entropy (see Figure 8.2B), and this tendency gives direction to these processes. It explains why some reactions proceed in one direction rather than another.

How does the second law apply to organisms? Consider the human body, with its highly organized tissues and organs composed of large, complex molecules. This level of complexity appears to be in conflict with the second law but is not for two reasons. First, the construction of complexity also generates disorder. Constructing 1 kg of a human body requires the metabolism of about 10 kg of highly ordered biological materials, which are converted into CO_2, H_2O, and other simple molecules that move independently and randomly. So metabolism creates far more disorder (more energy is lost to entropy) than the amount of order (total energy; enthalpy) stored in 1 kg of flesh. Second, life requires a constant input of energy to maintain order. Without this energy, the complex structures of living systems would break down. Because energy is used to generate and maintain order, there is no conflict with the second law of thermodynamics.

Having seen that the laws of thermodynamics apply to living things, we will now turn to a consideration of how these laws apply to biochemical reactions.

Chemical reactions release or consume energy

Since anabolic reactions link simple molecules to form more complex molecules, they tend to increase complexity (order) in the cell. On the other hand, catabolic reactions break down complex molecules into simpler ones, so they tend to decrease complexity (generate disorder).

- *Catabolic* reactions may break down an ordered reactant into smaller, more randomly distributed products. Reactions that release free energy ($-\Delta G$) are called **exergonic** reactions (**Figure 8.3A**). For example:

    complex molecules → free energy + small molecules

- *Anabolic* reactions may make a single product (a highly ordered substance) out of many smaller reactants (less ordered). Reactions that require or consume free energy ($+\Delta G$) are called **endergonic** reactions (**Figure 8.3B**). For example:

    free energy + small molecules → complex molecules

In principle, chemical reactions are reversible and can run both forward and backward. For example, if compound A can be converted into compound B (A → B), then B, in principle, can be converted into A (B → A), although *the concentrations of A and B determine which of these directions will be favored*. Think of the overall reaction as resulting from competition between forward and reverse reactions (A ⇌ B). Increasing the concentration of A speeds up the forward reaction, and increasing the concentration of B favors the reverse reaction.

At some concentration of A and B, the forward and reverse reactions take place at the same rate. At this concentration, no further net change in the system is observable, although individual molecules are still forming and breaking apart. This balance between forward and reverse reactions is known as **chemical equilibrium**. Chemical equilibrium is a state of no net change, and a state in which $\Delta G = 0$.

(A) Exergonic reaction

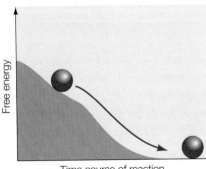

In an exergonic reaction, *energy is released* as the reactants form lower-energy products. ΔG is negative.

(B) Endergonic reaction

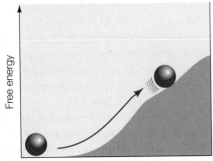

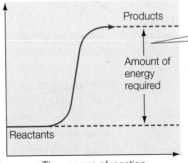

Energy must be added for an endergonic reaction, in which reactants are converted to products with a higher energy level. ΔG is positive.

8.3 Exergonic and Endergonic Reactions (A) In an exergonic reaction, the reactants behave like a ball rolling down a hill, and energy is released. (B) A ball will not roll uphill by itself. Driving an endergonic reaction, like moving a ball uphill, requires the addition of free energy.

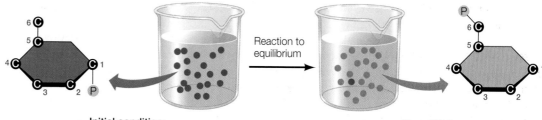

8.4 Chemical Reactions Run to Equilibrium No matter what quantities of glucose 1-phosphate and glucose 6-phosphate are dissolved in water, when equilibrium is attained, there will always be 95 percent glucose 6-phosphate and 5 percent glucose 1-phosphate.

Reaction to equilibrium

Initial condition:
100% Glucose 1-phosphate
(0.02 *M* concentration)

At equilibrium:
95% Glucose 6-phosphate (0.019 *M* concentration)
5% Glucose 1-phosphate (0.001 *M* concentration)

Chemical equilibrium and free energy are related

Every chemical reaction proceeds to a certain extent, but not necessarily to completion (all reactants converted into products). Each reaction has a specific equilibrium point, which is related to the free energy released by the reaction under specified conditions. To understand the principle of equilibrium, consider the following example.

Most cells contain glucose 1-phosphate, which is converted into glucose 6-phosphate.

glucose 1-phosphate $\rightleftharpoons$ glucose 6-phosphate

Imagine that we start out with an aqueous solution of glucose 1-phosphate that has a concentration of 0.02 *M*. (*M* stands for molar concentration; see Section 2.4). The solution is maintained under constant environmental conditions (25°C and pH 7). As the reaction proceeds to equilibrium, the concentration of the product, glucose 6-phosphate, rises from 0 to 0.019 *M*, while the concentration of the reactant, glucose 1-phosphate, falls to 0.001 *M*. At this point, equilibrium is reached (**Figure 8.4**). At equilibrium, the reverse reaction, from glucose 6-phosphate to glucose 1-phosphate, progresses at the same rate as the forward reaction.

At equilibrium, then, this reaction has a product-to-reactant ratio of 19:1 (0.019/0.001), so the forward reaction has gone 95 percent of the way to completion ("to the right," as written above). This result is obtained every time the experiment is run under the same conditions.

The change in free energy (ΔG) for any reaction is related directly to its point of equilibrium. The further toward completion the point of equilibrium lies, the more free energy is released. In an exergonic reaction, such as the conversion of glucose 1-phosphate to glucose 6-phosphate, ΔG is a negative number (in this example, $\Delta G = -1.7$ kcal/mol, or -7.1 kJ/mol).

A large, positive ΔG for a reaction means that it proceeds hardly at all to the right (A $\rightarrow$ B). If the concentration of B is initially high relative to that of A, such a reaction runs "to the left" (A $\leftarrow$ B), and nearly all B is converted into A. A ΔG value near zero is characteristic of a readily reversible reaction: reactants and products have almost the same free energies.

In Chapters 9 and 10 we examine the metabolic reactions that harvest energy from food and light. In turn, this energy is used to synthesize carbohydrates, lipids and proteins. All of the chemical reactions carried out by living organisms are governed by the principles of thermodynamics and equilibrium.

8.1 RECAP

Two laws of thermodynamics govern energy transformations in biological systems. A biochemical reaction can release or consume energy, and it may not run to completion, but instead end up at a point of equilibrium.

- What is the difference between potential energy and kinetic energy? Between anabolism and catabolism? **See pp. 149–150**

- What are the laws of thermodynamics? How do they relate to biology? **See pp. 150–152 and Figure 8.2**

- What is the difference between endergonic and exergonic reactions and what is the importance of ΔG? **See p. 152 and Figure 8.3**

The principles of thermodynamics that we have been discussing apply to all energy transformations in the universe, so they are very powerful and useful. Next, we'll apply them to reactions in cells that involve the currency of biological energy, ATP.

8.2 What Is the Role of ATP in Biochemical Energetics?

Cells rely on adenosine triphosphate (ATP) for the capture and transfer of the free energy they need to do chemical work. ATP operates as a kind of "energy currency." Just as it is more effective, efficient, and convenient for you to trade money for a lunch than to trade your actual labor, it is useful for cells to have a single currency for transferring energy between different reactions and cell processes. So some of the free energy that is released by exergonic reactions is captured in the formation of ATP from adenosine diphosphate (ADP) and inorganic phosphate (HPO_4^{2-}, which is commonly abbreviated to P_i). The ATP can then be hydrolyzed at other sites in the cell to release free energy to drive endergonic reactions. (In some reactions, guanosine triphosphate [GTP] is used as the energy transfer molecule instead of ATP, but we will focus on ATP here.)

ATP has another important role in the cell beyond its use as an energy currency: it is a nucleotide that can be converted into a building block for nucleic acids (see Chapter 4). The structure of ATP is similar to those of other nucleotides, but two things about ATP make it especially useful to cells. First, ATP releases a relatively large amount of energy when hydrolyzed to ADP and P_i. Second, ATP can phosphorylate (donate a phosphate group to) many different molecules, which gain some of the energy that was stored in the ATP. We will examine these two properties in the discussion that follows.

ATP hydrolysis releases energy

An ATP molecule consists of the nitrogenous base adenine bonded to ribose (a sugar), which is attached to a sequence of three phosphate groups (**Figure 8.5A**). The hydrolysis of a molecule of ATP yields free energy, as well as ADP and the inorganic phosphate ion (P_i). Thus:

$$ATP + H_2O \rightarrow ADP + P_i + \text{free energy}$$

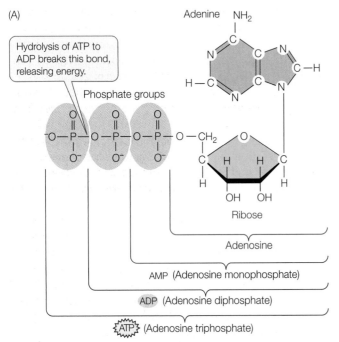

(A)

Hydrolysis of ATP to ADP breaks this bond, releasing energy.

Phosphate groups

Ribose

Adenosine

AMP (Adenosine monophosphate)

ADP (Adenosine diphosphate)

ATP (Adenosine triphosphate)

(B) *Luciola cruciata*

8.5 ATP (A) ATP is richer in energy than its relatives ADP and AMP. (B) Fireflies use ATP to initiate the oxidation of luciferin. This process converts chemical energy into light energy, emitting rhythmic flashes that signal the insect's readiness to mate.

The important property of this reaction is that it is exergonic, releasing free energy. Under standard laboratory conditions, the change in free energy for this reaction (ΔG) is about −7.3 kcal/mol (−30 kJ/mol). However, under cellular conditions, the value can be as much as −14 kcal/mol. We give both values here because you will encounter both values, and you should be aware of their origins. Both are correct, but in different conditions.

Two characteristics of ATP account for the free energy released by the loss of one or two of its phosphate groups:

- The free energy of the P—O bond between phosphate groups (called a phosphoric acid anhydride bond) is much higher than the energy of the O—H bond that forms after hydrolysis. So some usable energy is released by hydrolysis.

- Because phosphate groups are negatively charged and so repel each other, it takes energy to get phosphates near enough to each other to make the covalent bond that links them together (e.g., to add a phosphate to ADP to make ATP). Some of this energy is conserved when the third phosphate is attached.

A molecule of ATP can be hydrolyzed either to ADP and P_i, or to adenosine monophosphate (AMP) and a pyrophosphate ion ($P_2O_7^{4-}$; commonly abbreviated to PP_i). Cells use the energy released by ATP hydrolysis to fuel endergonic reactions (such as the biosynthesis of complex molecules), for active transport, and for movement. Another interesting example of the use of ATP involves converting its chemical energy into light energy.

BIOLUMINESCENCE The production of light by living organisms is referred to as **bioluminesence** (**Figure 8.5B**). It is an example of an endergonic reaction driven by ATP hydrolysis that involves an interconversion of energy forms (chemical to light). The chemical that becomes luminescent is called luciferin (after the light-bearing fallen angel, Lucifer):

$$\text{luciferin} + O_2 + ATP \xrightarrow{\text{luciferase}} \text{oxyluciferin} + AMP + PP_i + \text{light}$$

This reaction and the enzyme that catalyzes it (luciferase) occur in a wide variety of organisms in addition to the familiar firefly. These include a variety of marine organisms, microorganisms, worms, and mushrooms. The light is generally used to avoid predators or for signaling to mates.

Soft-drink companies use the firefly proteins luciferin and luciferase to detect bacterial contamination. Where there are

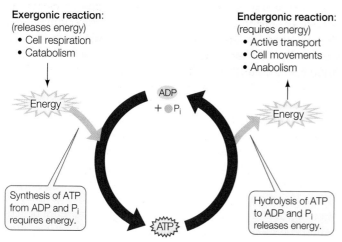

8.6 Coupling of Reactions Exergonic cellular reactions release the energy needed to make ATP from ADP. The energy released from the conversion of ATP back to ADP can be used to fuel endergonic reactions.

— yourBioPortal.com —
GO TO Web Activity 8.1 • ATP and Coupled Reactions

living cells there is ATP, and when the firefly proteins encounter ATP and oxygen, they give off light. Thus, a sample of soda that lights up in the test is contaminated with bacteria and is discarded.

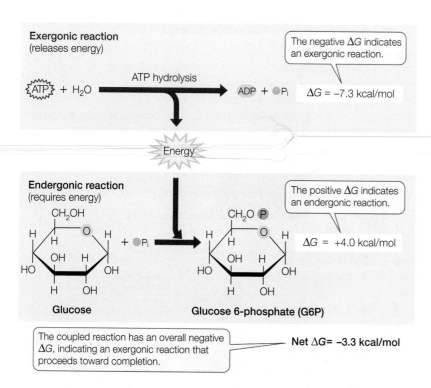

8.7 Coupling of ATP Hydrolysis to an Endergonic Reaction The addition of phosphate derived from the hydrolysis of ATP to glucose forms the molecule glucose 6-phosphate (in a reaction catalyzed by hexokinase). ATP hydrolysis is exergonic and the energy released drives the second reaction, which is endergonic.

ATP couples exergonic and endergonic reactions

As we have just seen, the hydrolysis of ATP is exergonic and yields ADP, P_i, and free energy. The reverse reaction, the formation of ATP from ADP and P_i, is endergonic and consumes as much free energy as is released by the hydrolysis of ATP:

$$ADP + P_i + \text{free energy} \rightarrow ATP + H_2O$$

Many different exergonic reactions in the cell can provide the energy to convert ADP into ATP. For eukaryotes and many prokaryotes, the most important of these reactions is cellular respiration, in which some of the energy released from fuel molecules is captured in ATP. The formation and hydrolysis of ATP constitute what might be called an "energy-coupling cycle," in which ADP picks up energy from exergonic reactions to become ATP, which then donates energy to endergonic reactions. ATP is the common component of these reactions and is the agent of coupling, as illustrated in **Figure 8.6**.

Coupling of exergonic and endergonic reactions is very common in metabolism. Free energy is captured and retained in the P—O bonds of ATP. ATP then diffuses to another site in the cell, where its hydrolysis releases the free energy to drive an endergonic reaction. For example, the formation of glucose 6-phosphate from glucose (**Figure 8.7**), which has a positive ΔG (is endergonic), will not proceed without the input of free energy from ATP hydrolysis, which has a negative ΔG (is exergonic). The overall ΔG for the coupled reactions (when the two ΔGs are added together) is negative. Hence the reactions proceed exergonically when they are coupled, and glucose 6-phosphate is synthesized. As you will see in Chapter 9, this is the initial reaction in the catabolism of glucose.

An active cell requires the production of millions of molecules of ATP per second to drive its biochemical machinery. An ATP molecule is typically consumed within a second of its formation. At rest, an average person produces and hydrolyzes about 40 kg of ATP per day—as much as some people weigh. This means that each ATP molecule undergoes about 10,000 cycles of synthesis and hydrolysis every day!

8.2 RECAP

ATP is the "energy currency" of cells. Some of the free energy released by exergonic reactions can be captured in the form of ATP. This energy can then be released by ATP hydrolysis and used to drive endergonic reactions.

● How does ATP store energy? See p. 153

● What are coupled reactions? See p. 155 and Figure 8.7

ATP is synthesized and used up very rapidly. But these biochemical reactions could not proceed so rapidly without the help of *enzymes*.

8.3 What Are Enzymes?

When we know the change in free energy (ΔG) of a reaction, we know where the equilibrium point of the reaction lies: the more negative ΔG is, the further the reaction proceeds toward completion. However, ΔG tells us nothing about the *rate* of a reaction—the speed at which it moves toward equilibrium. The reactions that cells depend on have spontaneous rates that are so slow that the cells would not survive without a way to speed up the reactions. That is the role of catalysts: substances that speed up reactions without themselves being permanently altered. A catalyst does not cause a reaction to occur that would not proceed without it, *but merely increases the rate of the reaction,* allowing equilibrium to be approached more rapidly. This is an important point: *no catalyst makes a reaction occur that cannot otherwise occur.*

Most biological catalysts are proteins called *enzymes.* Although we will focus here on proteins, some catalysts—perhaps the earliest ones in the origin of life—are RNA molecules called ribozymes (see Section 4.3). A biological catalyst, whether protein or RNA, is a framework or scaffold within which chemical catalysis takes place. This molecular framework binds the reactants and can participate in the reaction itself; however, such participation does not permanently change the enzyme. The catalyst ends up in exactly the same chemical condition after a reaction as before it. Over time, cells have evolved to utilize proteins rather than RNA as catalysts in most biochemical reactions, probably because of the great diversity in the three-dimensional structures of proteins, and because of the variety of chemical functions provided by their functional groups (see Figure 3.1).

In this section we will discuss the energy barrier that controls the rate of a chemical reaction. Then we will focus on the roles of enzymes: how they interact with specific reactants, how they lower the energy barrier, and how they permit reactions to proceed more quickly.

To speed up a reaction, an energy barrier must be overcome

An exergonic reaction may release a great deal of free energy, but take place very slowly. Such reactions are slow because there is an energy barrier between reactants and products. Think about the propane stove we describe in Section 2.3.

The burning of propane ($C_3H_8 + 5\,O_2 \rightarrow 3\,CO_2 + 4\,H_2O$ + energy) is an exergonic reaction—energy is released in the form of heat and light. Once started, the reaction goes to completion: all of the propane reacts with oxygen to form carbon dioxide and water vapor.

Because burning propane liberates so much energy, you might expect this reaction to proceed rapidly whenever propane is exposed to oxygen. But this does not happen; propane will start burning only if a spark, an input of energy such as a burning match, is provided. A spark is needed because there is an energy barrier between the reactants and the products.

In general, exergonic reactions proceed only after the reactants are pushed over the energy barrier by some added energy.

The energy barrier thus represents the amount of energy needed to start the reaction, known as the **activation energy** (E_a) (**Figure 8.8A**). Recall the ball rolling down the hill in Figure 8.3. The ball has a lot of potential energy at the top of the hill. However, if it is stuck in a small depression, it will not roll down the hill, even though that action is exergonic. To start the ball rolling, a small amount of energy (activation energy) is needed to push it out of the depression (**Figure 8.8B**). In a chemical reaction, the activation energy is the energy needed to change the reactants into unstable molecular forms called transition-state intermediates.

Transition-state intermediates have higher free energies than either the reactants or the products. Their bonds may be stretched and therefore unstable. Although the amount of activation energy needed for different reactions varies, it is often small compared with the change in free energy of the reaction. The activation energy put in to start a reaction is recovered during the ensuing "downhill" phase of the reaction, so it is not a part of the net free energy change, ΔG (see Figure 8.8A).

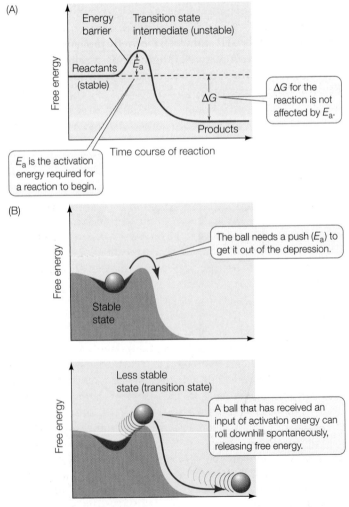

8.8 Activation Energy Initiates Reactions (A) In any chemical reaction, an initial stable state must become less stable before change is possible. (B) A ball on a hillside provides a physical analogy to the biochemical principle graphed in (A).

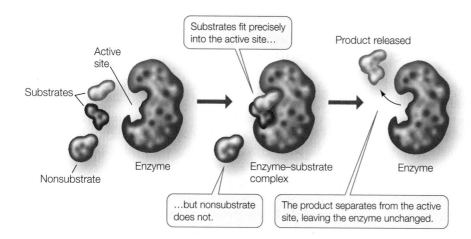

Substrates fit precisely into the active site…

…but nonsubstrate does not.

Active site

Substrates

Nonsubstrate

Enzyme

Enzyme–substrate complex

Product released

The product separates from the active site, leaving the enzyme unchanged.

Enzyme

8.9 Enzyme and Substrate An enzyme is a protein catalyst with an active site capable of binding one or more substrate molecules.

Where does the activation energy come from? In any collection of reactants at room or body temperatures, the molecules are moving around. A few are moving fast enough that their kinetic energy can overcome the energy barrier, enter the transition state, and react. However, the reaction takes place very slowly at room or body temperatures. If the system were heated, all the reactant molecules would move faster and have more kinetic energy, and the reaction would speed up. You have probably used this technique in the chemistry laboratory.

However, adding enough heat to increase the average kinetic energy of the molecules would not work in living systems. Such a nonspecific approach would accelerate all reactions, including destructive ones such as the denaturation of proteins (see Figure 3.9). A more effective way to speed up a reaction in a living system is to lower the energy barrier by bringing the reactants close together. In living cells, enzymes and ribozymes accomplish this task.

Enzymes bind specific reactants at their active sites

Catalysts increase the rates of chemical reactions. Most nonbiological catalysts are nonspecific. For example, powdered platinum catalyzes virtually any reaction in which molecular hydrogen (H_2) is a reactant. In contrast, most biological catalysts are highly specific. An enzyme or ribozyme usually recognizes and binds to only one or a few closely related reactants, and it catalyzes only a single chemical reaction. In the discussion that follows, we focus on enzymes, but remember that similar rules of chemical behavior apply to ribozymes as well.

In an enzyme-catalyzed reaction, the reactants are called **substrates**. Substrate molecules bind to a particular site on the enzyme, called the active site, where catalysis takes place (**Figure 8.9**). The specificity of an enzyme results from the exact three-dimensional shape and structure of its active site, into which only a narrow range of substrates can fit. Other molecules—with different shapes, different functional groups, and different properties—cannot fit properly and bind to the active site. This specificity is comparable to the specific binding of a membrane transport protein or receptor protein to its specific ligand, as described in Chapters 6 and 7.

The names of enzymes reflect their functions and often end with the suffix "ase." For example the enzyme lactase, which you encountered in the opening story for this chapter, catalyzes the hy-

drolysis of lactose but not another disaccharide, sucrose. The enzyme hexokinase accelerates the phosphorylation of glucose, but not ribose, to make glucose 6-phosphate (see Figure 8.7).

The binding of a substrate to the active site of an enzyme produces an **enzyme–substrate complex (ES)** that is held together by one or more means, such as hydrogen bonding, electrical attraction, or temporary covalent bonding. The enzyme–substrate complex gives rise to product and free enzyme:

$$E + S \rightarrow ES \rightarrow E + P$$

where E is the enzyme, S is the substrate, P is the product, and ES is the enzyme–substrate complex. The free enzyme (E) is in the same chemical form at the end of the reaction as at the beginning. While bound to the substrate, it may change chemically, but by the end of the reaction it has been restored to its initial form and is ready to bind more substrate.

Enzymes lower the energy barrier but do not affect equilibrium

When reactants are bound to the enzyme, forming an enzyme–substrate complex, they require less activation energy than the transition-state species of the corresponding uncatalyzed reaction (**Figure 8.10**). Thus the enzyme lowers the energy barrier for the reaction—it offers the reaction an easier path, speeding it up. When an enzyme lowers the energy bar-

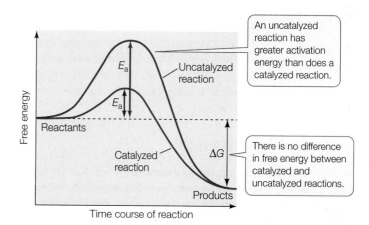

An uncatalyzed reaction has greater activation energy than does a catalyzed reaction.

There is no difference in free energy between catalyzed and uncatalyzed reactions.

E_a

E_a

Free energy

Reactants

Uncatalyzed reaction

Catalyzed reaction

ΔG

Products

Time course of reaction

8.10 Enzymes Lower the Energy Barrier Although the activation energy is lower in an enzyme-catalyzed reaction than in an uncatalyzed reaction, the energy released is the same with or without catalysis. In other words, E_a is lower, but ΔG is unchanged. A lower activation energy means the reaction will take place at a faster rate.

yourBioPortal.com

GO TO Web Activity 8.2 • Free Energy Changes

rier, both the forward and the reverse reactions speed up, so the enzyme-catalyzed overall reaction proceeds toward equilibrium more rapidly than the uncatalyzed reaction. *The final equilibrium is the same with or without the enzyme.* Similarly, adding an enzyme to a reaction does not change the difference in free energy (ΔG) between the reactants and the products (see Figure 8.10).

Enzymes can change the rate of a reaction substantially. For example, if 600 molecules of a protein with arginine as its terminal amino acid just sit in solution, the protein molecules tend toward disorder and the terminal peptide bonds break, releasing the arginines (ΔS increases). Without an enzyme this is a very slow reaction—it takes about 7 years for half (300) of the proteins to undergo this reaction. However, with the enzyme carboxypeptidase A catalyzing the reaction, the 300 arginines are released in half a second! The important consequence of this for living cells is not difficult to imagine. Such speeds make new realities possible.

8.3 RECAP

A chemical reaction requires a "push" over the energy barrier to get started. Enzymes provide this activation energy by binding specific reactants (substrates).

- Explain how the structure of an enzyme makes that enzyme specific. **See p. 157 and Figure 8.9**

- What is the relationship between an enzyme and the equilibrium point of a reaction? **See pp. 157–158**

Now that you have a general understanding of the structures, functions, and specificities of enzymes, let's see how they work to speed up chemical reactions between the substrate molecules.

8.4 How Do Enzymes Work?

During and after the formation of the enzyme–substrate complex, chemical interactions occur. These interactions contribute directly to the breaking of old bonds and the formation of new ones. In catalyzing a reaction, an enzyme may use one or more mechanisms.

Enzymes can orient substrates

When free in solution, substrates are moving from place to place randomly while at the same time vibrating, rotating, and tumbling around. They may not have the proper orientation to interact when they collide. Part of the activation energy needed to start a reaction is used to bring together specific atoms so that bonds can form (**Figure 8.11A**). For example, if acetyl coenzyme A (acetyl CoA) and oxaloacetate are to form citrate (a step in the metabolism of glucose; see Section 9.2), the two substrates must be oriented so that the carbon atom of the methyl group of acetyl CoA can form a covalent bond with the carbon atom of the carbonyl group of oxaloacetate. The active site of the enzyme

citrate synthase has just the right shape to bind these two molecules so that these atoms are adjacent.

Enzymes can induce strain in the substrate

Once a substrate has bound to its active site, an enzyme can cause bonds in the substrate to stretch, putting it in an unstable transition state (**Figure 8.11B**). For example, lysozyme is a protective enzyme abundant in tears and saliva that destroys invading bacteria by cleaving polysaccharide chains in their cell walls. Lysozyme's active site "stretches" the bonds of the bacterial polysaccharide, rendering the bonds unstable and more reactive to lysozyme's other substrate, water.

Enzymes can temporarily add chemical groups to substrates

The side chains (R groups) of an enzyme's amino acids may be direct participants in making its substrates more chemically reactive (**Figure 8.11C**).

- In *acid–base catalysis*, the acidic or basic side chains of the amino acids in the active site transfer H^+ to or from the substrate, destabilizing a covalent bond in the substrate, and permitting it to break.

- In *covalent catalysis*, a functional group in a side chain forms a temporary covalent bond with a portion of the substrate.

- In *metal ion catalysis,* metal ions such as copper, iron, and manganese, which are often firmly bound to side chains of enzymes, can lose or gain electrons without detaching from the enzymes. This ability makes them important participants in oxidation–reduction reactions, which involve the loss or gain of electrons.

Molecular structure determines enzyme function

Most enzymes are much larger than their substrates. An enzyme is typically a protein containing hundreds of amino acids and may consist of a single folded polypeptide chain or of several subunits (see Section 3.2). Its substrate is generally a small molecule or a small part of a large molecule. The active site of the enzyme is usually quite small, not more than 6–12 amino acids. Two questions arise from these observations:

- What features of the active site allow it to recognize and bind the substrate?

- What is the role of the rest of the huge protein?

THE ACTIVE SITE IS SPECIFIC TO THE SUBSTRATE   The remarkable ability of an enzyme to select exactly the right substrate depends on a precise interlocking of molecular shapes and interactions of chemical groups at the active site. The binding of the substrate to the active site depends on the same kinds of forces that maintain the tertiary structure of the enzyme: hydrogen bonds, the attraction and repulsion of electrically charged groups, and hydrophobic interactions.

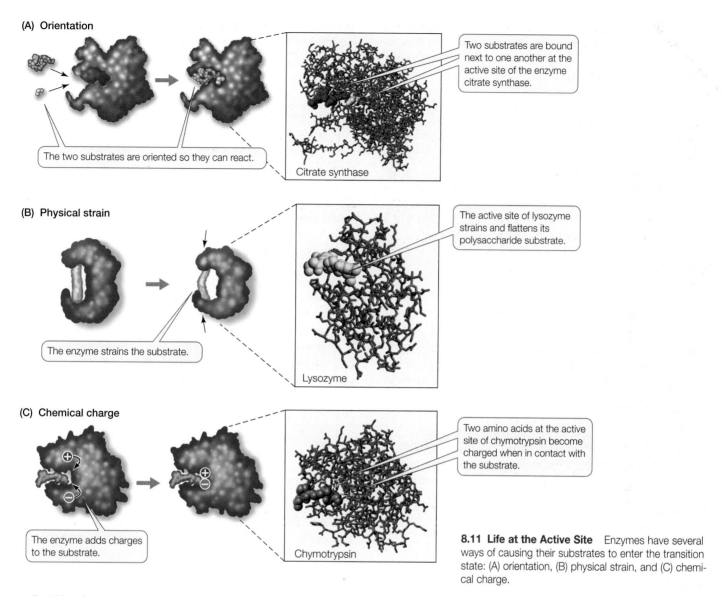

(A) Orientation

The two substrates are oriented so they can react.

Two substrates are bound next to one another at the active site of the enzyme citrate synthase.

Citrate synthase

(B) Physical strain

The enzyme strains the substrate.

The active site of lysozyme strains and flattens its polysaccharide substrate.

Lysozyme

(C) Chemical charge

The enzyme adds charges to the substrate.

Two amino acids at the active site of chymotrypsin become charged when in contact with the substrate.

Chymotrypsin

8.11 Life at the Active Site Enzymes have several ways of causing their substrates to enter the transition state: (A) orientation, (B) physical strain, and (C) chemical charge.

In 1894, the German chemist Emil Fischer compared the fit between an enzyme and its substrate to that of a lock and key. Fischer's model persisted for more than half a century with only indirect evidence to support it. The first direct evidence came in 1965, when David Phillips and his colleagues at the Royal Institution in London crystallized the enzyme lysozyme and determined its tertiary structure using the technique of X-ray crystallography (described in Section 13.2). They observed a pocket in lysozyme that neatly fits its substrate (see Figure 8.11B).

AN ENZYME CHANGES SHAPE WHEN IT BINDS A SUBSTRATE Just as a membrane receptor protein may undergo precise changes in conformation upon binding to its ligand (see Chapter 7), some enzymes change their shapes when they bind their substrate(s). These shape changes, which are called **induced fit**, expose the active site (or sites) of the enzyme.

An example of induced fit can be seen in the enzyme hexokinase (see Figure 8.7), which catalyzes the reaction

$$\text{glucose} + \text{ATP} \rightarrow \text{glucose 6-phosphate} + \text{ADP}$$

Induced fit brings reactive side chains from the hexokinase active site into alignment with the substrates (**Figure 8.12**), facilitating its catalytic mechanisms. Equally important, the folding of hexokinase to fit around the substrates (glucose and ATP) excludes water from the active site. This is essential, because if water were present, the ATP could be hydrolyzed to ADP and P_i. But since water is absent, the transfer of a phosphate from ATP to glucose is favored.

Induced fit at least partly explains why enzymes are so large. The rest of the macromolecule may have three roles:

- It provides a framework so that the amino acids of the active site are properly positioned in relation to the substrate(s).

- It participates in significant changes in protein shape and structure that result in induced fit.

- It provides binding sites for regulatory molecules (see Section 8.5).

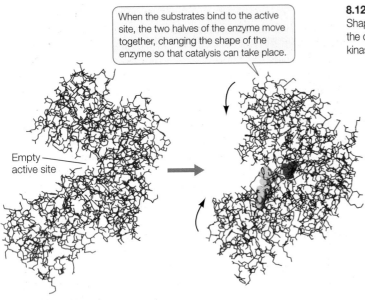

When the substrates bind to the active site, the two halves of the enzyme move together, changing the shape of the enzyme so that catalysis can take place.

Empty active site

8.12 Some Enzymes Change Shape When Substrate Binds to Them
Shape changes result in an induced fit between enzyme and substrate, improving the catalytic ability of the enzyme. Induced fit can be observed in the enzyme hexokinase, seen here with and without its substrates, glucose (red) and ATP (yellow).

Some enzymes require other molecules in order to function

As large and complex as enzymes are, many of them require the presence of nonprotein chemical "partners" in order to function (**Table 8.1**):

- *Prosthetic groups* are distinctive, non-amino acid atoms or molecular groupings that are permanently bound to their enzymes. An example is a flavin nucleotide, which binds to succinate dehydrogenase, an important enzyme in cellular respiration (see Section 9.2).

- *Cofactors* are inorganic ions such as copper, zinc, and iron that bind to certain enzymes. For example, the cofactor zinc binds to the enzyme alcohol dehydrogenase.

TABLE 8.1
Some Examples of Nonprotein "Partners" of Enzymes

TYPE OF MOLECULE	ROLE IN CATALYZED REACTIONS
COFACTORS	
Iron (Fe^{2+} or Fe^{3+})	Oxidation/reduction
Copper (Cu^+ or Cu^{2+})	Oxidation/reduction
Zinc (Zn^{2+})	Helps bind NAD
COENZYMES	
Biotin	Carries $-COO^-$
Coenzyme A	Carries $-CO-CH_3$
NAD	Carries electrons
FAD	Carries electrons
ATP	Provides/extracts energy
PROSTHETIC GROUPS	
Heme	Binds ions, O_2, and electrons; contains iron cofactor
Flavin	Binds electrons
Retinal	Converts light energy

- A *coenzyme* is a carbon-containing molecule that is required for the action of one or more enzymes. It is usually relatively small compared with the enzyme to which it temporarily binds.

A coenzyme moves from enzyme to enzyme, adding or removing chemical groups from the substrate. A coenzyme is like a substrate in that it does not permanently bind to the enzyme; it binds to the active site, changes chemically during the reaction, and then separates from the enzyme to participate in other reactions. ATP and ADP, as energy carriers, can be considered coenzymes, even though they are really substrates. The term coenzyme was coined before the functions of these molecules were fully understood. Biochemists continue to use the term, and to be consistent with the field, we will use the term in this book.

In the next chapter we will encounter other coenzymes that function in energy-harvesting reactions by accepting or donating electrons or hydrogen atoms. In animals, some coenzymes are produced from vitamins—substances that must be obtained from food because they cannot be synthesized by the body. For example, the B vitamin niacin is used to make the coenzyme nicotinamide adenine dinucleotide (NAD).

The substrate concentration affects the reaction rate

For a reaction of the type $A \rightarrow B$, the rate of the uncatalyzed reaction is directly proportional to the concentration of A. The higher the concentration of substrate, the more reactions per unit of time. Addition of the appropriate enzyme speeds up the reaction, of course, but it also changes the shape of a plot of rate versus substrate concentration (**Figure 8.13**). For a given concentration of enzyme, the rate of the enzyme-catalyzed reaction initially increases as the substrate concentration increases from zero, but then it levels off. At some point, further increases in the substrate concentration do not significantly increase the reaction rate—the maximum rate has been reached.

Since the concentration of an enzyme is usually much lower than that of its substrate and does not change as substrate concentration changes, what we see is a saturation phenomenon like the one that occurs in facilitated diffusion (see Figure 6.14). When all the enzyme molecules are bound to substrate molecules, the enzyme is working as fast as it can—at its maximum rate. Nothing is gained by adding more substrate, because no free enzyme molecules are left to act as catalysts. Under these conditions the active sites are said to be saturated.

The maximum rate of a catalyzed reaction can be used to measure how efficient the enzyme is—that is, how many molecules of substrate are converted into product per unit of time when there is an excess of substrate present. This *turnover number* ranges from one molecule every two seconds for lysozyme to an amazing 40 million molecules per second for the liver enzyme catalase.

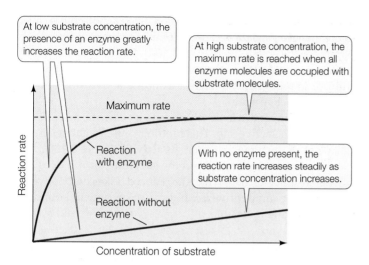

At low substrate concentration, the presence of an enzyme greatly increases the reaction rate.

At high substrate concentration, the maximum rate is reached when all enzyme molecules are occupied with substrate molecules.

Maximum rate

Reaction with enzyme

With no enzyme present, the reaction rate increases steadily as substrate concentration increases.

Reaction without enzyme

Reaction rate

Concentration of substrate

8.13 Catalyzed Reactions Reach a Maximum Rate Because there is usually less enzyme than substrate present, the reaction rate levels off when the enzyme becomes saturated.

8.4 RECAP

Enzymes orient their substrates to bring together specific atoms so that bonds can form. An enzyme can participate in the reaction it catalyzes by temporarily changing shape or destabilizing the enzyme-substrate complex. Some enzymes require cofactors, coenzymes, or prosthetic groups in order to function.

- What are three mechanisms of enzyme catalysis? **See p. 158 and Figure 8.11**

- What are the chemical roles of coenzymes in enzymatic reactions? **See p. 160**

Now that you understand more about how enzymes function, let's see how different enzymes work together in a complex organism.

8.5 How Are Enzyme Activities Regulated?

A major characteristic of life is homeostasis—the maintenance of stable internal conditions (see Chapter 40). How does a cell maintain a relatively constant internal environment while thousands of chemical reactions are going on? These chemical reactions operate within *metabolic pathways* in which the product of one reaction is a reactant for the next. The pathway for the metabolism of lactose begins with lactase (as we described in the chapter's opening story), and is just one of many pathways that regulate the internal environment of the cell. These pathways have such diverse functions as the catabolism of glucose to yield energy, CO_2, and H_2O, and the anabolism of amino acids to yield proteins. Metabolic pathways do not exist in isolation, but interact extensively, and each reaction in each pathway is catalyzed by a specific enzyme.

Within a cell or organism, the presence and activity of enzymes determine the "flow" of chemicals through different metabolic pathways. The amount of enzyme activity, in turn, is controlled in part via the regulation of gene expression. Many signal transduction pathways (described in Chapter 7) end with changes in gene expression, and often the genes that are switched on or off encode enzymes. But the simple presence of an enzyme does not ensure that it is functioning. Another means by which cells can control which pathways are active at a particular time is by the activation or inactivation of enzymes. If one enzyme in the pathway is inactive, that step and all subsequent steps shut down. Thus, enzymes are target points for the regulation of entire sequences of chemical reactions.

Regulation of the rates at which thousands of different enzymes operate contributes to homeostasis within an organism. Such control permits cells to make orderly changes in their functions in response to changes in the external environment. In Chapter 7 we describe a number of enzymes that become activated in signal transduction pathways, illustrating how enzyme activation can dramatically alter cell functions. (For example, see the activation of glycogen phosphorylase in Figure 7.20.)

The flow of chemicals such as carbon atoms through interacting metabolic pathways can be studied, but this process becomes complicated quickly, because each pathway influences the others. Computer algorithms are used to model these pathways and show how they mesh in an interdependent system (**Figure 8.14**). Such models can help predict what will happen if the concentration of one molecule or another is altered. This new field of biology is called **systems biology**, and it has numerous applications.

In this section we will investigate the roles of enzymes in organizing and regulating metabolic pathways. In living cells, enzymes can be activated or inhibited in various ways, and there are also mechanisms for controlling the rates at which some enzymes catalyze reactions. We will also examine how the environment—particularly temperature and pH—affects enzyme activity.

Enzymes can be regulated by inhibitors

Various chemical inhibitors can bind to enzymes, slowing down the rates of the reactions they catalyze. Some inhibitors occur naturally in cells; others are artificial. Naturally occurring inhibitors regulate metabolism; artificial ones can be used to treat disease, to kill pests, or to study how enzymes work. In some cases the inhibitor binds the enzyme irreversibly, and the enzyme becomes permanently inactivated. In other cases the inhibitor has reversible effects; it can separate from the enzyme, allowing the enzyme to function fully as before. The removal of a natural reversible inhibitor increases an enzyme's rate of catalysis.

IRREVERSIBLE INHIBITION If an inhibitor covalently binds to certain side chains at the active site of an enzyme, it will permanently inactivate the enzyme by destroying its capacity to interact with its normal substrate. An example of an irreversible inhibitor is DIPF (diisopropyl phosphorofluoridate), which

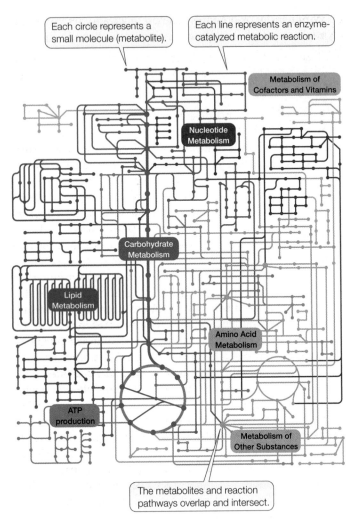

Each circle represents a small molecule (metabolite).

Each line represents an enzyme-catalyzed metabolic reaction.

Metabolism of Cofactors and Vitamins

Nucleotide Metabolism

Carbohydrate Metabolism

Lipid Metabolism

Amino Acid Metabolism

ATP production

Metabolism of Other Substances

The metabolites and reaction pathways overlap and intersect.

8.14 Metabolic Pathways The complex interactions of metabolic pathways can be modeled by the tools of systems biology. In cells, the main elements controlling these pathways are enzymes.

reacts with serine (**Figure 8.15**). DIPF is an irreversible inhibitor of acetylcholinesterase, whose operation is essential for the normal functioning of the nervous system. Because of their effect on acetylcholinesterase, DIPF and other similar compounds are classified as nerve gases, and were developed for biological warfare. One of these compounds, Sarin, was used in an attack on the Tokyo subway in 1995, resulting in a dozen deaths and the hospitalization of hundreds more. The widely used insecticide malathion is a derivative of DIPF that inhibits only insect acetylcholinesterase, not the mammalian enzyme. The irreversible inhibition of enzymes is of practical use to humans, but this form of regulation is not common in the cell, because the enzyme is permanently inactivated and cannot be recycled. Instead, cells use reversible inhibition.

REVERSIBLE INHIBITION In some cases an inhibitor is similar enough to a particular enzyme's natural substrate to bind non-covalently to its active site, yet different enough that the enzyme catalyzes no chemical reaction. While such a molecule is bound to the enzyme, the natural substrate cannot enter the active site

and the enzyme is unable to function. Such a molecule is called a **competitive inhibitor** because it competes with the natural substrate for the active site (**Figure 8.16A**). In this case, the inhibition is reversible. When the concentration of the competitive inhibitor is reduced, it detaches from the active site, and the enzyme is active again.

A **noncompetitive inhibitor** binds to an enzyme at a site distinct from the active site. This binding causes a change in the shape of the enzyme that alters its activity (**Figure 8.16B**). The active site may no longer bind the substrate, or if it does, the rate of product formation may be reduced. Like competitive inhibitors, noncompetitive inhibitors can become unbound, so their effects are reversible.

Allosteric enzymes control their activity by changing shape

The change in enzyme shape due to noncompetitive inhibitor binding is an example of allostery (*allo*, "different"; *stereos*, "shape"). **Allosteric regulation** occurs when an effector molecule binds to a site other than the active site of an enzyme, *inducing the enzyme to change its shape*. The change in shape alters the affinity of the active site for the substrate, and so the rate of the reaction is changed.

Often, an enzyme will exist in the cell in more than one possible shape (**Figure 8.17**):

- The *active form* of the enzyme has the proper shape for substrate binding.

- The *inactive form* of the enzyme has a shape that cannot bind the substrate.

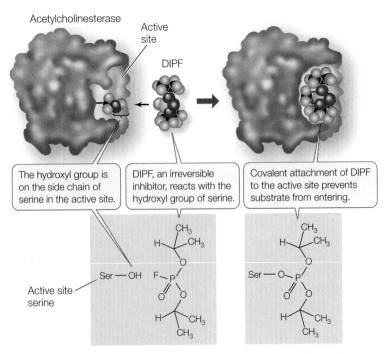

Acetylcholinesterase

Active site

DIPF

The hydroxyl group is on the side chain of serine in the active site.

DIPF, an irreversible inhibitor, reacts with the hydroxyl group of serine.

Covalent attachment of DIPF to the active site prevents substrate from entering.

Active site serine

Ser — OH

Ser — O

8.15 Irreversible Inhibition DIPF forms a stable covalent bond with the side chain of the amino acid serine at the active site of the enzyme acetylcholinesterase, thus irreversibly disabling the enzyme.

(A) Competitive inhibition

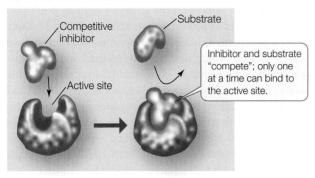

(B) Noncompetitive inhibition

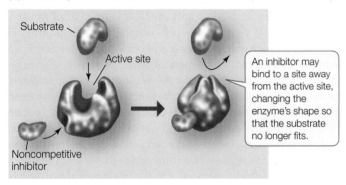

8.16 Reversible Inhibition (A) A competitive inhibitor binds temporarily to the active site of an enzyme. (B) A noncompetitive inhibitor binds temporarily to the enzyme at a site away from the active site. In both cases, the enzyme's function is disabled for only as long as the inhibitor remains bound.

yourBioPortal.com
GO TO Animated Tutorial 8.1 • Enzyme Catalysis

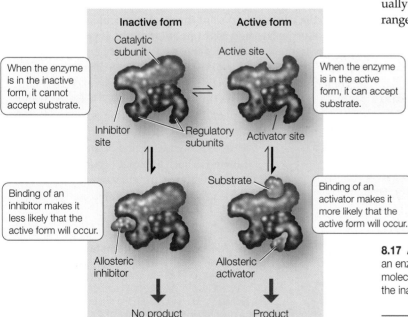

Other molecules, collectively referred to as effectors, can influence which form the enzyme takes:

- Binding of an inhibitor to a site separate from the active site can stabilize the inactive form of the enzyme, making it less likely to convert to the active form.
- The active form can be stabilized by the binding of an activator to another site on the enzyme.

Like substrate binding, the binding of inhibitors and activators to their regulatory sites (also called allosteric sites) is highly specific. Most (but not all) enzymes that are allosterically regulated are proteins with quaternary structure; that is, they are made up of multiple polypeptide subunits. The polypeptide that has the active site is called the catalytic subunit. The allosteric sites are often on different polypeptides, called the regulatory subunits.

Some enzymes have multiple subunits containing active sites, and the binding of substrate to one of the active sites causes allosteric effects. When substrate binds to one subunit, there is a slight change in protein structure that influences the adjacent subunit. The slight change to the second subunit makes its active site more likely to bind to the substrate. So the reaction speeds up as the sites become sequentially activated.

As a result, an allosteric enzyme with multiple active sites and a nonallosteric enzyme with a single active site differ greatly in their reaction rates when the substrate concentration is low. Graphs of reaction rates plotted against substrate concentrations show this relationship. For a nonallosteric enzyme, the plot looks like that in **Figure 8.18A**. The reaction rate first increases sharply with increasing substrate concentration, then tapers off to a constant maximum rate as the supply of enzyme becomes saturated.

The plot for a multisubunit allosteric enzyme is radically different, having a sigmoid (S-shaped) appearance (**Figure 8.18B**). At low substrate concentrations, the reaction rate increases gradually as substrate concentration increases. But within a certain range, the reaction rate is extremely sensitive to relatively small changes in substrate concentration. In addition, allosteric enzymes are very sensitive to low concentrations of inhibitors. Because of this sensitivity, allosteric enzymes are important in regulating entire metabolic pathways.

Allosteric effects regulate metabolism

Metabolic pathways typically involve a starting material, various intermediate products, and an end product that is used for some purpose by the cell. In each pathway there are a number of reactions, each

8.17 Allosteric Regulation of Enzymes Active and inactive forms of an enzyme can be interconverted, depending on the binding of effector molecules at sites other than the active site. Binding an inhibitor stabilizes the inactive form and binding an activator stabilizes the active form.

yourBioPortal.com
GO TO Animated Tutorial 8.2 • Allosteric Regulation of Enzymes

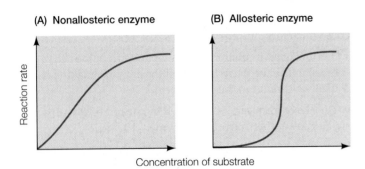

(A) Nonallosteric enzyme (B) Allosteric enzyme

Reaction rate

Concentration of substrate

8.18 Allostery and Reaction Rate The number of active sites on an enzyme determines how the rate of the enzyme-catalyzed reaction changes as substrate concentration increases. A sigmoid curve (B) is typical for an enzyme with multiple subunits, each with an active site. After one subunit binds the substrate, changes in structure make it more likely that the next subunit will also bind substrate. So the reaction speeds up more rapidly than in the case of an enzyme with a single active site (A).

forming an intermediate product and each catalyzed by a different enzyme. The first step in a pathway is called the commitment step, meaning that once this enzyme-catalyzed reaction occurs, the "ball is rolling," and the other reactions happen in sequence, leading to the end product. But what if the cell has no need for that product—for example, if that product is available from its environment in adequate amounts? It would be energetically wasteful for the cell to continue making something it does not need.

One way to avoid this problem is to shut down the metabolic pathway by having the final product inhibit the enzyme that catalyzes the commitment step (**Figure 8.19**). Often this inhibition occurs allosterically. When the end product is present at a high concentration, some of it binds to an allosteric site on the commitment step enzyme, thereby causing it to become inactive. Thus, the final product acts as a *noncompetitive inhibitor* (described earlier in this section) of the first enzyme in the pathway. This mechanism is known as feedback inhibition or end-product inhibition. We will describe many other examples of such inhibition in later chapters.

Enzymes are affected by their environment

Enzymes enable cells to perform chemical reactions and carry out complex processes rapidly without using the extremes of temperature and pH employed by chemists in the laboratory. However, because of their three-dimensional structures and the chemistry of the side chains in their active sites, enzymes (and their substrates) are highly sensitive to changes in temperature and pH. In Section 3.2 we describe the general effects of these environmental factors on proteins. Here we will examine their effects on enzyme function (which, of course, depends on enzyme structure and chemistry).

pH AFFECTS ENZYME ACTIVITY The rates of most enzyme-catalyzed reactions depend on the pH of the solution in which they occur. While the water inside cells is generally at a neutral pH of 7, the presence of acids, bases, and buffers can alter this. Each enzyme is most active at a particular pH; its activity decreases as the solution is made more acidic or more basic than the ideal (optimal) pH (**Figure 8.20**). As an example, consider the human digestive system (see Section 51.3). The pH inside the human stomach is highly acidic, around pH 1.5. Many enzymes that hydrolyze macromolecules, such as proteases, have pH optima in the neutral range. So when food enters the small intestine, a buffer (bicarbonate) is secreted into the intestine to raise the pH to 6.5. This allows the hydrolytic enzymes to be active and digest the food.

Several factors contribute to this effect. One factor is ionization of the carboxyl, amino, and other groups on either the substrate or the enzyme. In neutral or basic solutions, carboxyl groups (—COOH) release H^+ to become negatively charged carboxylate groups (—COO$^-$). On the other hand, in neutral or acidic solutions, amino groups (—NH$_2$) accept H^+ to become positively charged —NH$_3^+$ groups (see the discussion of acids and bases in Section 2.4). Thus, in a neutral solution, an amino group is electrically attracted to a carboxyl group on another molecule or another part of the same molecule, because both groups are ionized and have opposite charges. If the pH changes, however, the ionization of these groups may change. For example, at a low pH (high H^+ concentration, such as the stomach contents where the enzyme pepsin is active), the excess H^+ may react with —COO$^-$ to form —COOH. If this happens, the group is no longer charged and cannot interact with other charged groups in the protein, so the folding of the protein may be altered. If such a change occurs at the active site of an enzyme, the enzyme may no longer be able to bind to its substrate.

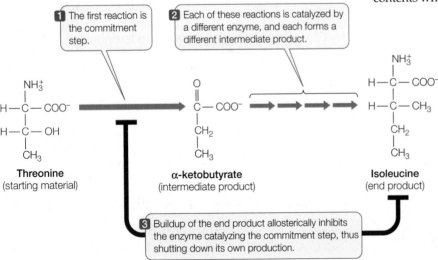

① The first reaction is the commitment step.

② Each of these reactions is catalyzed by a different enzyme, and each forms a different intermediate product.

NH$_3^+$
|
H—C—COO$^-$
|
H—C—OH
|
CH$_3$

Threonine
(starting material)

O
‖
C—COO$^-$
|
CH$_2$
|
CH$_3$

α-ketobutyrate
(intermediate product)

NH$_3^+$
|
H—C—COO$^-$
|
H—C—CH$_3$
|
CH$_2$
|
CH$_3$

Isoleucine
(end product)

③ Buildup of the end product allosterically inhibits the enzyme catalyzing the commitment step, thus shutting down its own production.

8.19 Feedback Inhibition of Metabolic Pathways The first reaction in a metabolic pathway is referred to as the commitment step. It is often catalyzed by an enzyme that can be allosterically inhibited by the end product of the pathway. The specific pathway shown here is the synthesis of isoleucine from threonine in bacteria. It is typical of many enzyme-catalyzed biosynthetic pathways.

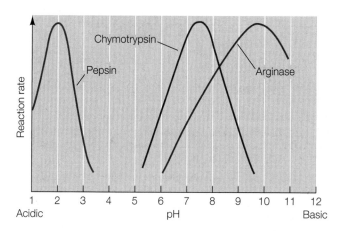

8.20 pH Affects Enzyme Activity An enzyme catalyzes its reaction at a maximum rate. The activity curve for each enzyme peaks at its optimal pH. For example, pepsin is active in the acidic environment of the stomach, while chymotrypsin is active in the small intestine.

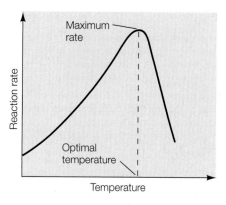

8.21 Temperature Affects Enzyme Activity Each enzyme is most active at a particular optimal temperature. At higher temperatures the enzyme becomes denatured and inactive; this explains why the activity curve falls off abruptly at temperatures above the optimal.

TEMPERATURE AFFECTS ENZYME ACTIVITY In general, warming increases the rate of a chemical reaction because a greater proportion of the reactant molecules have enough kinetic energy to provide the activation energy for the reaction. Enzyme-catalyzed reactions are no different (**Figure 8.21**). However, temperatures that are too high inactivate enzymes, because at high temperatures enzyme molecules vibrate and twist so rapidly that some of their noncovalent bonds break. When an enzyme's tertiary structure is changed by heat it loses its function. Some enzymes denature at temperatures only slightly above that of the human body, but a few are stable even at the boiling point (or freezing point) of water. All enzymes, however, have an optimal temperature for activity.

Individual organisms adapt to changes in the environment in many ways, one of which is based on groups of enzymes, called *isozymes*, that catalyze the same reaction but have different chemical compositions and physical properties. Different isozymes within a given group may have different optimal temperatures. The rainbow trout, for example, has several isozymes of the enzyme acetylcholinesterase. If a rainbow trout is transferred from warm water to near-freezing water (2°C), the fish produces an isozyme of acetylcholinesterase that is different from the one it produces at the higher temperature. The new isozyme has a lower optimal temperature, allowing the fish's nervous system to perform normally in the colder water.

In general, enzymes adapted to warm temperatures do not denature at those temperatures because their tertiary structures are held together largely by covalent bonds, such as charge interactions or disulfide bridges, instead of the more heat-sensitive weak chemical interactions. Most enzymes in humans are more stable at high temperatures than those of the bacteria that infect us, so that a moderate fever tends to denature bacterial enzymes, but not our own.

8.5 RECAP

The rates of most enzyme-catalyzed reactions are affected by interacting molecules (such as inhibitors and activators) and by environmental factors (such as temperature and pH).

- What is the difference between reversible and irreversible enzyme inhibition? **See pp. 161–162**

- How are allosteric enzymes regulated? **See pp. 162–163 and Figure 8.17**

- Explain the concept of feedback inhibition. How might the reactions shown in Figure 8.19 fit into a systems diagram such as the one shown in Figure 8.14?

CHAPTER SUMMARY

8.1 What Physical Principles Underlie Biological Energy Transformations?

- Energy is the capacity to do work. In a biological system, the usable energy is called **free energy** (*G*). The unusable energy is **entropy**, a measure of the disorder in the system.

- **Potential energy** is the energy of state or position; it includes the energy stored in chemical bonds. **Kinetic energy** is the energy of motion; it is the type of energy that can do work.

- The **laws of thermodynamics** apply to living organisms. The first law states that energy cannot be created or destroyed. The second law states that energy transformations decrease the amount of energy available to do work (free energy) and increase disorder. **Review Figure 8.2**

- The **change in free energy** (Δ*G*) of a reaction determines its point of **chemical equilibrium**, at which the forward and reverse reactions proceed at the same rate.

- An **exergonic reaction** releases free energy and has a negative ΔG. An **endergonic reaction** consumes or requires free energy and has a positive ΔG. Endergonic reactions proceed only if free energy is provided. **Review Figure 8.3**

- **Metabolism** is the sum of all the biochemical (metabolic) reactions in an organism. **Catabolic reactions** are associated with the breakdown of complex molecules and release energy (are exergonic). **Anabolic reactions** build complexity in the cell and are endergonic.

8.2 What Is the Role of ATP in Biochemical Energetics?

- **Adenosine triphosphate** (ATP) serves as an energy currency in cells. Hydrolysis of ATP releases a relatively large amount of free energy.

- The **ATP cycle** couples exergonic and endergonic reactions, harvesting free energy from exergonic reactions, and providing free energy for endergonic reactions. **Review Figure 8.6, WEB ACTIVITY 8.1**

8.3 What Are Enzymes?

- The rate of a chemical reaction is independent of ΔG, but is determined by the **energy barrier**. **Enzymes** are protein catalysts that affect the rates of biological reactions by lowering the energy barrier, supplying the **activation energy** (E_a) needed to initiate reactions. **Review Figure 8.10, WEB ACTIVITY 8.2**

- A **substrate** binds to the enzyme's **active site**—the site of catalysis—forming an **enzyme–substrate complex**. Enzymes are highly specific for their substrates.

8.4 How Do Enzymes Work?

- At the active site, a substrate can be oriented correctly, chemically modified, or strained. As a result, the substrate readily forms its **transition state**, and the reaction proceeds. **Review Figure 8.11**

- Binding substrate causes many enzymes to change shape, exposing their active site(s) and allowing catalysis. The change in enzyme shape caused by substrate binding is known as **induced fit**. **Review Figure 8.12**

- Some enzymes require other substances, known as **cofactors**, to carry out catalysis. **Prosthetic groups** are permanently bound to enzymes; **coenzymes** are not. A coenzyme can be considered a substrate, as it is changed by the reaction and then released from the enzyme.

- Substrate concentration affects the rate of an enzyme-catalyzed reaction.

8.5 How Are Enzyme Activities Regulated?

- Metabolism is organized into pathways in which the product of one reaction is a reactant for the next reaction. Each reaction in the pathway is catalyzed by an enzyme.

- Enzyme activity is subject to regulation. Some inhibitors bind irreversibly to enzymes. Others bind reversibly. **Review Figures 8.15 and 8.16, ANIMATED TUTORIAL 8.1**

- An **allosteric effector** binds to a site other than the active site and stabilizes the active or inactive form of an enzyme. **Review Figure 8.17, ANIMATED TUTORIAL 8.2**

- The end product of a metabolic pathway may inhibit an enzyme that catalyzes the **commitment step** of that pathway. **Review Figure 8.19**

- Enzymes are sensitive to their environments. Both pH and temperature affect enzyme activity. **Review Figures 8.20 and 8.21**

SELF-QUIZ

1. Coenzymes differ from enzymes in that coenzymes are
   a. only active outside the cell.
   b. polymers of amino acids.
   c. smaller molecules, such as vitamins.
   d. specific for one reaction.
   e. always carriers of high-energy phosphate.

2. Which statement about thermodynamics is true?
   a. Free energy is used up in an exergonic reaction.
   b. Free energy cannot be used to do work.
   c. The total amount of energy can change after a chemical transformation.
   d. Free energy can be kinetic but not potential energy.
   e. Entropy has a tendency to increase.

3. In a chemical reaction,
   a. the rate depends on the value of ΔG.
   b. the rate depends on the activation energy.
   c. the entropy change depends on the activation energy.
   d. the activation energy depends on the value of ΔG.
   e. the change in free energy depends on the activation energy.

4. Which statement about enzymes is *not* true?
   a. They usually consist of proteins.
   b. They change the rate of the catalyzed reaction.

   c. They change the ΔG of the reaction.
   d. They are sensitive to heat.
   e. They are sensitive to pH.

5. The active site of an enzyme
   a. never changes shape.
   b. forms no chemical bonds with substrates.
   c. determines, by its structure, the specificity of the enzyme.
   d. looks like a lump projecting from the surface of the enzyme.
   e. changes the ΔG of the reaction.

6. The molecule ATP is
   a. a component of most proteins.
   b. high in energy because of the presence of adenine.
   c. required for many energy-transforming biochemical reactions.
   d. a catalyst.
   e. used in some exergonic reactions to provide energy.

7. In an enzyme-catalyzed reaction,
   a. a substrate does not change.
   b. the rate decreases as substrate concentration increases.
   c. the enzyme can be permanently changed.
   d. strain may be added to a substrate.
   e. the rate is not affected by substrate concentration.

8. Which statement about enzyme inhibitors is *not* true?
   a. A competitive inhibitor binds the active site of the enzyme.
   b. An allosteric inhibitor binds a site on the active form of the enzyme.
   c. A noncompetitive inhibitor binds a site other than the active site.
   d. Noncompetitive inhibition cannot be completely overcome by the addition of more substrate.
   e. Competitive inhibition can be completely overcome by the addition of more substrate.

9. Which statement about the feedback inhibition of enzymes is *not* true?
   a. It is usually exerted through allosteric effects.

   b. It is directed at the enzyme that catalyzes the commitment step in a metabolic pathway.
   c. It affects the rate of reaction, not the concentration of enzyme.
   d. It acts by permanently modifying the active site.
   e. It is an example of reversible inhibition.

10. Which statement about temperature effects is *not* true?
   a. Raising the temperature may reduce the activity of an enzyme.
   b. Raising the temperature may increase the activity of an enzyme.
   c. Raising the temperature may denature an enzyme.
   d. Some enzymes are stable at the boiling point of water.
   e. All enzymes have the same optimal temperature.

FOR DISCUSSION

1. What makes it possible for endergonic reactions to proceed in organisms?

2. Consider two proteins: one is an enzyme dissolved in the cytosol of a cell, the other is an ion channel in its plasma membrane. Contrast the structures of the two proteins, indicating at least two important differences.

3. Plot free energy versus the time course of an endergonic reaction, and the same for an exergonic reaction. Include the activation energy on both plots. Label E_a and ΔG on both graphs.

4. Consider an enzyme that is subject to allosteric regulation. If a competitive inhibitor (not an allosteric inhibitor) is added to a solution containing such an enzyme, the ratio of enzyme molecules in the active form to those in the inactive form increases. Explain this observation.

ADDITIONAL INVESTIGATION

In humans, hydrogen peroxide (H_2O_2) is a dangerous toxin produced as a by-product of several metabolic pathways. The accumulation of H_2O_2 is prevented by its conversion to harmless H_2O, a reaction catalyzed by the appropriately named enzyme catalase. Air pollutants can inhibit this enzyme and leave individuals susceptible to tissue damage by H_2O_2. How would you investigate whether catalase has an allosteric or a nonallosteric mechanism, and whether the pollutants are acting as competitive or noncompetitive inhibitors?

Pathways that Harvest Chemical Energy

Of mice and marathons

Like success in your biology course, winning a prestigious marathon comes only after a lot of hard work. Distance runners have more mitochondria in the leg muscles than most of us. The chemical energy stored in the bonds of ATP in those mitochondria is converted into mechanical energy to move the muscles.

There are two types of muscle fibers. Most people have about equal proportions of each type. But in a top marathon racer, 90 percent of the body's muscle is made up of so-called *slow-twitch* fibers. Cells of these fibers have lots of mitochondria and use oxygen to break down fats and carbohydrates, forming ATP. In contrast, the muscles of sprinters are about 80 percent *fast-twitch* fibers, which have fewer mitochondria. Fast-twitch fibers generate short bursts of ATP in the absence of O_2, but the ATP is soon used up. Extensive research with athletes has shown that training can improve the efficiency of blood circulation to the muscle fibers, providing more oxygen, and can even change the ratio of fast-twitch to slow-twitch fibers.

Now enter Marathon Mouse. No, this is not a cartoon character or a computer game, but a very real mouse that was genetically programmed by Ron Evans at the Salk Institute to express high levels of the protein PPARδ in its muscles. This protein is a receptor located inside cell nuclei, where it regulates the transcription of genes involved with the breakdown of fat to yield ATP. Evans's mouse was supposed to break down fats better, and thus be leaner—but there was an unexpected bonus. With high levels of PPARδ came an increase in slow-twitch fibers and a decrease in fast-twitch ones. It was as if the mouse had been in marathon training for a long time!

Marathon mice are leaner and meaner than ordinary mice. Leaner, because they are good at burning fat; and meaner in terms of their ability to run long distances. On an exercise wheel, a normal mouse can run for 90 minutes and about a half-mile (900 meters) before it gets tired. PPARδ-enhanced mice can run almost twice as long and twice as far—marks of true distance runners. Could we also manipulate genes to enhance performance (and fat burning) in humans?

The genetic engineering of people, if it is feasible, is probably far in the future. But implanting genetically altered muscle tissue is actually not such a farfetched idea, and has already raised concerns over improper athletic enhancement. More likely in the near term is the use of an experimental drug called Aicar, which activates the PPARδ

Marathon Men It takes a lot of training to run a marathon. One of the results of all that training is that the leg muscles become packed with slow-twitch muscle fibers, containing cells rich in energy-metabolizing mitochondria.

Marathon Mouse This mouse can run for much longer than a normal mouse because its energy metabolism has been genetically altered.

protein. When Evans and colleagues gave the drug to normal mice, they achieved the same results as with the genetically modified mice. A test for Aicar in blood and urine has been developed to prevent its use by human athletes to gain a competitive advantage. Of more importance is the drug's potential in the treatment of obesity and diabetes, since the drug stimulates fat breakdown. Obesity is a key part of a disorder called metabolic syndrome, which also includes high blood pressure, heart disease, and diabetes.

The free energy trapped in ATP is the energy you use all the time to fuel both conscious actions, like running a marathon or turning the pages of a book, and your body's automatic actions, such as breathing or contracting your heart muscles.

IN THIS CHAPTER we will describe how cells extract usable energy from food, usually in the form of ATP. We describe the general principles of energy transformations in cells, and illustrate these principles by describing the pathways for the catabolism of glucose in the presence and absence of O_2. Finally, we describe the relationships between the metabolic pathways that use and produce the four biologically important classes of molecules—carbohydrates, fats, proteins, and nucleic acids.

9.1 How Does Glucose Oxidation Release Chemical Energy?

Energy is stored in the covalent bonds of fuels, and it can be released and transformed. Wood burning in a campfire releases energy as heat and light. In cells, fuel molecules release chemical energy that is used to make ATP, which in turn drives endergonic reactions. ATP is central to the energy transformations of all living organisms. Photosynthetic organisms use energy from sunlight to synthesize their own fuels, as we describe in Chapter 10. In nonphotosynthetic organisms, the most common chemical fuel is the sugar glucose ($C_6H_{12}O_6$). Other molecules, including other carbohydrates, fats, and proteins, can also supply energy. However, to release their energy they must be converted into glucose or intermediate compounds that can enter into the various pathways of glucose metabolism.

In this section we explore how cells obtain energy from glucose by the chemical process of oxidation, which is carried out through a series of metabolic pathways. Five principles govern metabolic pathways:

- A complex chemical transformation occurs in a series of separate reactions that form a metabolic pathway.
- Each reaction is catalyzed by a specific enzyme.
- Most metabolic pathways are similar in all organisms, from bacteria to humans.
- In eukaryotes, many metabolic pathways are compartmentalized, with certain reactions occurring inside specific organelles.
- Each metabolic pathway is regulated by key enzymes that can be inhibited or activated, thereby determining how fast the reactions will go.

Cells trap free energy while metabolizing glucose

As we saw in Section 2.3, the familiar process of combustion (burning) is very similar to the chemical processes that release energy in cells. If glucose is burned in a flame, it reacts with oxygen gas (O_2), forming carbon dioxide and water and releasing energy in the form of heat. The balanced equation for the complete combustion reaction is

$$C_6H_{12}O_6 + 6\ O_2 \rightarrow 6\ CO_2 + 6\ H_2O + \text{free energy}$$
$$(\Delta G = -686\ \text{Kcal/mol})$$

This is an oxidation-reduction reaction. Glucose ($C_6H_{12}O_6$) becomes completely oxidized and six molecules of O_2 are reduced to six molecules of water. The energy that is released can be used to do work. The same equation applies to the overall metabolism of glucose in cells. However, in contrast to combustion, the metabolism of glucose is a multistep pathway—each step is catalyzed by an enzyme, and the process is compartmentalized. Unlike combustion, glucose metabolism is tightly regulated and occurs at temperatures compatible with life.

The glucose metabolism pathway "traps" the energy stored in the covalent bonds of glucose and stores it instead in ATP molecules, via the phosphorylation reaction:

$$ADP + P_i + \text{free energy} \rightarrow ATP$$

As we introduce in Chapter 8, ATP is the energy currency of cells. The energy trapped in ATP can be used to do cellular work—such as movement of muscles or active transport across membranes—just as the energy captured from combustion can be used to do work.

The change in free energy (ΔG) resulting from the complete conversion of glucose and O_2 to CO_2 and water, whether by combustion or by metabolism, is –686 kcal/mol (–2,870 kJ/mol). Thus the overall reaction is highly exergonic and can drive the endergonic formation of a great deal of ATP from ADP and phosphate. Note that in the discussion that follows, "energy" means free energy.

Three metabolic processes harvest the energy in the chemical bonds of glucose: glycolysis, cellular respiration, and fermentation (**Figure 9.1**). All three processes involve pathways made up of many distinct chemical reactions.

- **Glycolysis** begins glucose metabolism in all cells. Through a series of chemical rearrangements, glucose is converted to two molecules of the three-carbon product **pyruvate**, and a small amount of energy is captured in usable forms. Glycolysis is an **anaerobic** process because it does not require O_2.

- **Cellular respiration** uses O_2 from the environment, and thus it is **aerobic**. Each pyruvate molecule is completely converted into three molecules of CO_2 through a set of metabolic pathways including pyruvate oxidation, the citric acid cycle, and an electron transport system (the respiratory chain). In the process, a great deal of the energy stored in the covalent bonds of pyruvate is captured to form ATP.

- **Fermentation** does not involve O_2 (it is anaerobic). Fermentation converts pyruvate into lactic acid or ethyl alcohol (ethanol), which are still relatively energy-rich molecules. Because the breakdown of glucose is incomplete, much less energy is released by fermentation than by cellular respiration.

Redox reactions transfer electrons and energy

As is illustrated in Figure 8.6, the addition of a phosphate group to ADP to make ATP is an endergonic reaction that can extract and transfer energy from exergonic to endergonic reactions. Another way of transferring energy is to transfer electrons. A reaction in which one substance transfers one or more electrons to another substance is called an oxidation–reduction reaction, or **redox** reaction.

- **Reduction** is the gain of one or more electrons by an atom, ion, or molecule.

- **Oxidation** is the loss of one or more electrons.

Oxidation and reduction *always occur together*: as one chemical is oxidized, the electrons it loses are transferred to another chemical, reducing it. In a redox reaction, we call the reactant that becomes reduced an oxidizing agent and the one that becomes oxidized a reducing agent:

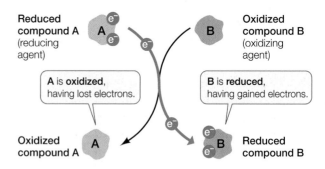

In both the combustion and the metabolism of glucose, glucose is the reducing agent (electron donor) and O_2 is the oxidizing agent (electron acceptor).

Although oxidation and reduction are always defined in terms of traffic in electrons, it is often helpful to think in terms of the gain or loss of hydrogen atoms. Transfers of hydrogen atoms involve transfers of electrons ($H = H^+ + e^-$). So when a molecule loses hydrogen atoms, it becomes oxidized.

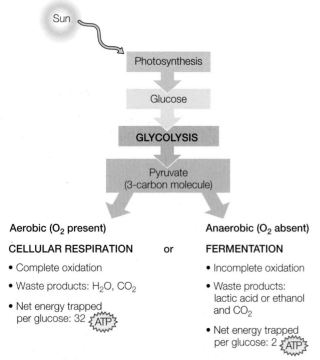

9.1 Energy for Life Living organisms obtain their energy from the food compounds produced by photosynthesis. They convert these compounds into glucose, which they metabolize to trap energy in ATP.

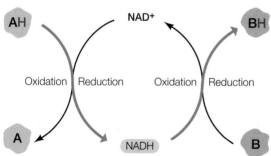

Methane (CH₄)   Methanol (CH₃OH)   Formaldehyde (CH₂O)   Formic acid (HCOOH)   Carbon dioxide (CO₂)

Most reduced state
Highest free energy

Most oxidized state
Lowest free energy

9.2 Oxidation, Reduction, and Energy The more oxidized a carbon atom in a molecule is, the less its free energy.

In general, the more reduced a molecule is, the more energy is stored in its covalent bonds (**Figure 9.2**). In a redox reaction, some energy is transferred from the reducing agent to the reduced product. The rest remains in the reducing agent or is lost to entropy. As we will see, some of the key reactions of glycolysis and cellular respiration are highly exergonic redox reactions.

The coenzyme NAD⁺ is a key electron carrier in redox reactions

Section 8.4 describes the role of coenzymes, small molecules that assist in enzyme-catalyzed reactions. ADP acts as a coenzyme when it picks up energy released in an exergonic reaction and packages it to form ATP. On the other hand, the coenzyme nicotinamide adenine dinucleotide (NAD⁺) acts as an electron carrier in redox reactions:

As you can see, NAD⁺ exists in two chemically distinct forms, one oxidized (NAD⁺) and the other reduced (NADH) (**Figure 9.3**). Both forms participate in redox reactions. The reduction reaction

$$NAD^+ + H^+ + 2\,e^- \rightarrow NADH$$

is actually the transfer of a proton (the hydrogen ion, H⁺) and two electrons, which are released by the accompanying oxidization reaction.

The electrons do not remain with the coenzyme. Oxygen is highly electronegative and readily accepts electrons from NADH. The oxidation of NADH by O₂ (which occurs in several steps)

$$NADH + H^+ + \tfrac{1}{2}\,O_2 \rightarrow NAD^+ + H_2O$$

is highly exergonic, with a ΔG of −52.4 kcal/mol (−219 kJ/mol). Note that the oxidizing agent appears here as "½ O₂" instead of "O." This notation emphasizes that it is molecular oxygen, O₂, that acts as the oxidizing agent.

Just as a molecule of ATP can be thought of as a package of about 12 kcal/mol (50 kJ/mol) of free energy, NADH can be thought of as a larger package of free energy (approximately 50 kcal/mol, or 200 kJ/mol). NAD⁺ is a common electron carrier in cells, but not the only one. Another carrier, flavin adenine dinucleotide (FAD), also transfers electrons during glucose metabolism.

An overview: Harvesting energy from glucose

The energy-harvesting processes in cells use different combinations of metabolic pathways depending on the presence or absence of O₂:

- Under aerobic conditions, when O₂ is available as the final electron acceptor, four pathways operate (**Figure 9.4A**). Glycolysis is followed by the three pathways of cellular respiration: pyruvate oxidation, the citric acid cycle (also called the Krebs cycle or the tricarboxylic acid cycle), and electron transport/ATP synthesis (also called the respiratory chain).

- Under anaerobic conditions when O₂ is unavailable, pyruvate oxidation, the citric acid cycle, and the respiratory chain do not function, and the pyruvate produced by glycolysis is further metabolized by fermentation (**Figure 9.4B**).

These five metabolic pathways occur in different locations in the cell (**Table 9.1**).

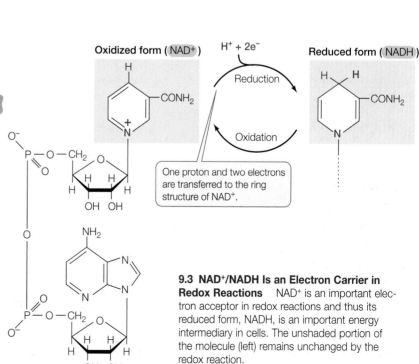

9.3 NAD⁺/NADH Is an Electron Carrier in Redox Reactions NAD⁺ is an important electron acceptor in redox reactions and thus its reduced form, NADH, is an important energy intermediary in cells. The unshaded portion of the molecule (left) remains unchanged by the redox reaction.

TABLE 9.1

Cellular Locations for Energy Pathways in Eukaryotes and Prokaryotes

EUKARYOTES	PROKARYOTES
External to mitochondrion	**In cytoplasm**
Glycolysis	Glycolysis
Fermentation	Fermentation
	Citric acid cycle
Inside mitochondrion	**On plasma membrane**
Inner membrane	Pyruvate oxidation
Respiratory chain	Respiratory chain
Matrix	
Citric acid cycle	
Pyruvate oxidation	

yourBioPortal.com

GO TO Web Activity 9.1 • Energy Pathways in Cells

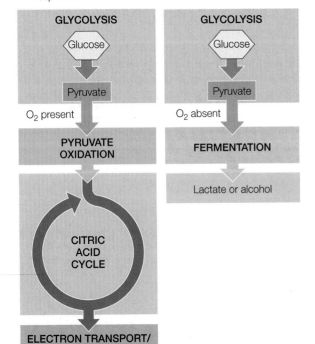

(A) Glycolysis and cellular respiration

(B) Glycolysis and fermentation

9.4 Energy-Producing Metabolic Pathways Energy-producing reactions can be grouped into five metabolic pathways: glycolysis, pyruvate oxidation, the citric acid cycle, the respiratory chain/ATP synthesis, and fermentation. (A) The three lower pathways occur only in the presence of O_2 and are collectively referred to as cellular respiration. (B) When O_2 is unavailable, glycolysis is followed by fermentation.

yourBioPortal.com

GO TO Web Activity 9.2 • Glycolysis and Fermentation

9.1 RECAP

The free energy released from the oxidation of glucose is trapped in the form of ATP. Five metabolic pathways combine in different ways to produce ATP, which supplies the energy for myriad other reactions in living cells.

- What principles govern metabolic pathways in cells? **See p. 169**

- Describe how the coupling of oxidation and reduction transfers energy from one molecule to another. **See pp. 170–171**

- Explain the roles of NAD^+ and O_2 with respect to electrons in a redox reaction. **See p. 171 and Figure 9.3**

Now that you have an overview of the metabolic pathways that harvest energy from glucose, let's take a closer look at the three pathways involved in aerobic catabolism: glycolysis, pyruvate oxidation, and the citric acid cycle.

9.2 What Are the Aerobic Pathways of Glucose Metabolism?

The aerobic pathways of glucose metabolism oxidize glucose completely to CO_2 and H_2O. Initially, the glycolysis reactions convert the six-carbon glucose molecule to two 3-carbon pyruvate molecules (**Figure 9.5**). Pyruvate is then converted to CO_2 in a second series of reactions beginning with pyruvate oxidation and followed by the citric acid cycle. In addition to generating CO_2, the oxidation events are coupled with the reduction of electron carriers, mostly NAD^+. So much of the chemical energy in the C—C and C—H bonds of glucose is transferred to NADH. Ultimately, this energy will be transferred to ATP, but this comes in a separate series of reactions involving electron transport, called the respiratory chain. In the respiratory chain, redox reactions result in the oxidative phosphorylation of ADP by ATP synthase. We will begin our consideration of the metabolism of glucose with a closer look at glycolysis.

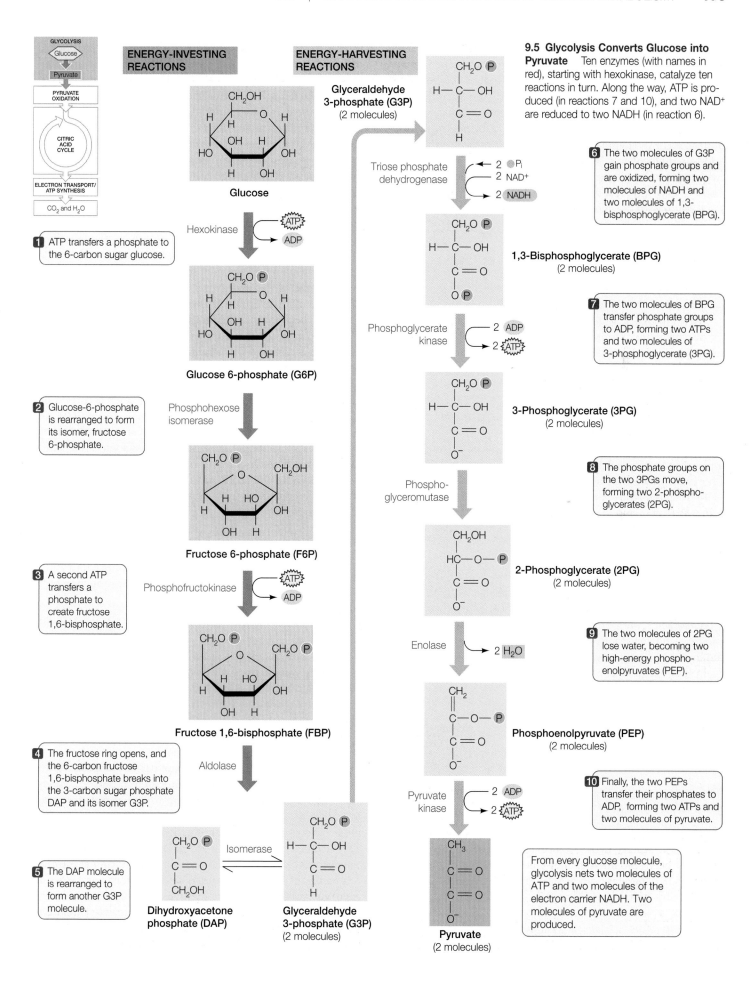

ENERGY-INVESTING REACTIONS

ENERGY-HARVESTING REACTIONS

9.5 Glycolysis Converts Glucose into Pyruvate Ten enzymes (with names in red), starting with hexokinase, catalyze ten reactions in turn. Along the way, ATP is produced (in reactions 7 and 10), and two NAD^+ are reduced to two NADH (in reaction 6).

Glucose

1 ATP transfers a phosphate to the 6-carbon sugar glucose.

Hexokinase

Glucose 6-phosphate (G6P)

2 Glucose-6-phosphate is rearranged to form its isomer, fructose 6-phosphate.

Phosphohexose isomerase

Fructose 6-phosphate (F6P)

3 A second ATP transfers a phosphate to create fructose 1,6-bisphosphate.

Phosphofructokinase

Fructose 1,6-bisphosphate (FBP)

4 The fructose ring opens, and the 6-carbon fructose 1,6-bisphosphate breaks into the 3-carbon sugar phosphate DAP and its isomer G3P.

Aldolase

Dihydroxyacetone phosphate (DAP)

Isomerase

Glyceraldehyde 3-phosphate (G3P) (2 molecules)

5 The DAP molecule is rearranged to form another G3P molecule.

Glyceraldehyde 3-phosphate (G3P) (2 molecules)

Triose phosphate dehydrogenase

2 P_i
2 NAD^+
2 NADH

6 The two molecules of G3P gain phosphate groups and are oxidized, forming two molecules of NADH and two molecules of 1,3-bisphosphoglycerate (BPG).

1,3-Bisphosphoglycerate (BPG) (2 molecules)

Phosphoglycerate kinase

2 ADP
2 ATP

7 The two molecules of BPG transfer phosphate groups to ADP, forming two ATPs and two molecules of 3-phosphoglycerate (3PG).

3-Phosphoglycerate (3PG) (2 molecules)

Phospho-glyceromutase

8 The phosphate groups on the two 3PGs move, forming two 2-phospho-glycerates (2PG).

2-Phosphoglycerate (2PG) (2 molecules)

Enolase

2 H_2O

9 The two molecules of 2PG lose water, becoming two high-energy phospho-enolpyruvates (PEP).

Phosphoenolpyruvate (PEP) (2 molecules)

Pyruvate kinase

2 ADP
2 ATP

10 Finally, the two PEPs transfer their phosphates to ADP, forming two ATPs and two molecules of pyruvate.

Pyruvate (2 molecules)

From every glucose molecule, glycolysis nets two molecules of ATP and two molecules of the electron carrier NADH. Two molecules of pyruvate are produced.

Glycolysis takes place in the cytosol. It converts glucose into pyruvate, produces a small amount of energy, and generates no CO_2. During glycolysis, some of the covalent bonds between carbon and hydrogen in the glucose molecule are oxidized, releasing some of the stored energy. The ten enzyme-catalyzed reactions of glycolysis result in the net production of two molecules of pyruvate (pyruvic acid), two molecules of ATP, and two molecules of NADH. Glycolysis can be divided into two stages: energy-investing reactions that consume ATP, and energy-harvesting reactions that produce ATP (see Figure 9.5). We'll begin with the energy-investing reactions.

The energy-investing reactions 1–5 of glycolysis require ATP

Using Figure 9.5 as a guide, let's work our way through the glycolytic pathway.

Two of the reactions (1 and 3 in Figure 9.5), involve the transfer of phosphate groups from ATP to form phosphorylated intermediates. The second of these intermediates, fructose 1,6-bisphosphate, has a free energy substantially higher than that of glucose. Later in the pathway, these phosphate groups are transferred to ADP to make new molecules of ATP. Although both of these steps use ATP as a substrate, each is catalyzed by a different, specific enzyme.

In reaction 1, the enzyme hexokinase catalyzes the transfer of a phosphate group from ATP to glucose, forming the sugar phosphate glucose 6-phosphate.

In reaction 2, the six-membered glucose ring is rearranged into a five-membered fructose ring.

In reaction 3, the enzyme phosphofructokinase adds a second phosphate to the fructose ring, forming fructose 1,6-bisphosphate.

Reaction 4 opens up the ring and cleaves it to produce two different three-carbon sugar (triose) phosphates: dihydroxyacetone phosphate and glyceraldehyde 3-phosphate.

In reaction 5, one of those products, dihydroxyacetone phosphate, is converted into a second molecule of the other, glyceraldehyde 3-phosphate (G3P).

In summary, by the halfway point of the glycolytic pathway, two things have happened:

- Two molecules of ATP have been invested.
- The six-carbon glucose molecule has been converted into two molecules of a three-carbon sugar phosphate, glyceraldehyde 3-phosphate (G3P).

The energy-harvesting reactions 6–10 of glycolysis yield NADH and ATP

In the discussion that follows, remember that each reaction occurs twice for each glucose molecule because each glucose molecule has been split into two molecules of G3P. The transformation of G3P generates both NADH and ATP. Again, follow the sequence by referring to Figure 9.5.

PRODUCING NADH   *Reaction 6* is catalyzed by the enzyme triose phosphate dehydrogenase, and its end product is a phosphate

ester, 1,3-bisphosphoglycerate (BPG). This is an exergonic oxidation reaction, and it is accompanied by a large drop in free energy—more than 100 kcal of energy is released per mole of glucose (**Figure 9.6, left**). The free energy released in this reaction is not lost to heat, but is captured by the accompanying reduction reaction. For each molecule of G3P that is oxidized, one molecule of NAD^+ is reduced to make a molecule of NADH.

NAD^+ is present in only small amounts in the cell, and it must be recycled to allow glycolysis to continue. As we will see, NADH is oxidized back to NAD^+ in the metabolic pathways that follow glycolysis.

PRODUCING ATP   *In reactions 7–10* of glycolysis, the two phosphate groups of BPG are transferred one at a time to molecules of ADP, with a rearrangement in between. More than 20 kcal (83.6 kJ/mol) of free energy is stored in ATP for every mole of BPG broken down. Finally, we are left with two moles of pyruvate for every mole of glucose that entered glycolysis.

The enzyme-catalyzed transfer of phosphate groups from donor molecules to ADP to form ATP is called **substrate-level phosphorylation**. (Phosphorylation is the addition of a phosphate group to a molecule.) Substrate-level phosphorylation is distinct from oxidative phosphorylation, which is carried out by the respiratory chain and ATP synthase, and will be discussed later in this chapter. Reaction 7 is an example of substrate-level phosphorylation, in which phosphoglycerate kinase catalyzes the transfer of a phosphate group from BPG to ADP, forming ATP. It is exergonic, even though a substantial amount of energy is consumed in the formation of ATP.

To summarize:

- The energy-investing steps of glycolysis use the energy of hydrolysis of two ATP molecules per glucose molecule.
- The energy-releasing steps of glycolysis produce four ATP molecules per glucose molecule, so the net production of ATP is two molecules.
- The energy-releasing steps of glycolysis produce two molecules of NADH.

If O_2 is present, glycolysis is followed by the three stages of cellular respiration: pyruvate oxidation, the citric acid cycle, and the respiratory chain/ATP synthesis.

Pyruvate oxidation links glycolysis and the citric acid cycle

In the process of **pyruvate oxidation**, pyruvate is oxidized to the two-carbon acetate molecule, which is then converted to acetyl CoA. This is the link between glycolysis and all the other reactions of cellular respiration. **Coenzyme A (CoA)** is a complex molecule responsible for binding the two-carbon acetate molecule. Acetyl CoA formation is a multi-step reaction catalyzed by the pyruvate dehydrogenase complex, an enormous complex containing 60 individual proteins and 5 different coenzymes. In eukaryotic cells, pyruvate dehydrogenase is located in the mitochondrial matrix (see Figure 5.12). Pyruvate enters the mitochondrion by active transport, and then a series of coupled reactions takes place:

9.6 Changes in Free Energy During Glycolysis and the Citric Acid Cycle The first five reactions of glycolysis (left) consume free energy, and the remaining five glycolysis reactions release energy. Pyruvate oxidation (middle) and the citric acid cycle (right) both release considerable energy. Refer to Figures 9.5 and 9.7 for the reaction numbers.

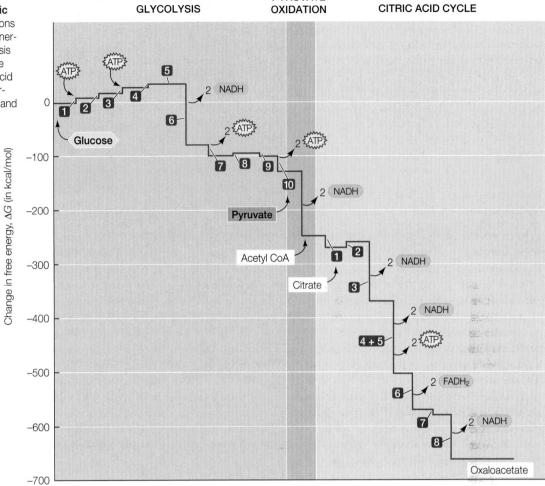

Each glucose yields:
6 CO_2
10 NADH
2 $FADH_2$
4 ATP

1. Pyruvate is oxidized to a two-carbon acetyl group (acetate), and CO_2 is released (decarboxylation).

2. Part of the energy from this oxidation is captured by the reduction of NAD^+ to NADH.

3. Some of the remaining energy is stored temporarily by combining the acetyl group with CoA, forming acetyl CoA:

pyruvate + NAD^+ + CoA + H^+ → acetyl CoA + NADH + CO_2

(In this reaction, the proton and electrons used to reduce NAD^+ are derived from the oxidation of both pyruvate and CoA.) Acetyl CoA has 7.5 kcal/mol (31.4 kJ/mol) more energy than simple acetate. Acetyl CoA can donate its acetyl group to various acceptor molecules, much as ATP can donate phosphate groups to various acceptors. But the main role of acetyl CoA is to donate its acetyl group to the four-carbon compound oxaloacetate, forming the six-carbon molecule citrate. This initiates the citric acid cycle, one of life's most important energy-harvesting pathways.

Arsenic, the classic poison of rodent exterminators and murder mysteries, acts by inhibiting pyruvate dehydrogenase, thus decreasing acetyl CoA production. The lack of acetyl CoA stops the citric acid cycle and all the subsequent reactions that de-pend on it. Consequently, cells eventually run out of ATP and cannot perform essential processes that are powered by ATP hydrolysis.

The citric acid cycle completes the oxidation of glucose to CO_2

Acetyl CoA is the starting point for the **citric acid cycle**. This pathway of eight reactions completely oxidizes the two-carbon acetyl group to two molecules of carbon dioxide. The free energy released from these reactions is captured by ADP and the electron carriers NAD^+ and FAD. **Figure 9.6 right** shows the free energy changes during each step of the pathway.

The citric acid cycle is maintained in a steady state—that is, although the intermediate compounds in the cycle enter and leave it, the concentrations of those intermediates do not change much. Refer to the numbered reactions in **Figure 9.7** as you read the description of each reaction.

9.7 Pyruvate Oxidation and the Citric Acid Cycle Pyruvate enters the mitochondrion and is oxidized to acetyl CoA, which enters the citric acid cycle. Reactions 3, 4, 5, 6, and 8 accomplish the major overall effects of the cycle—the trapping of energy. This is accomplished by reducing NAD^+ or FAD, or by producing GTP (reaction 5), whose energy is then transferred to ATP. Each reaction is catalyzed by a specific enzyme, although the enzymes are not shown in this figure.

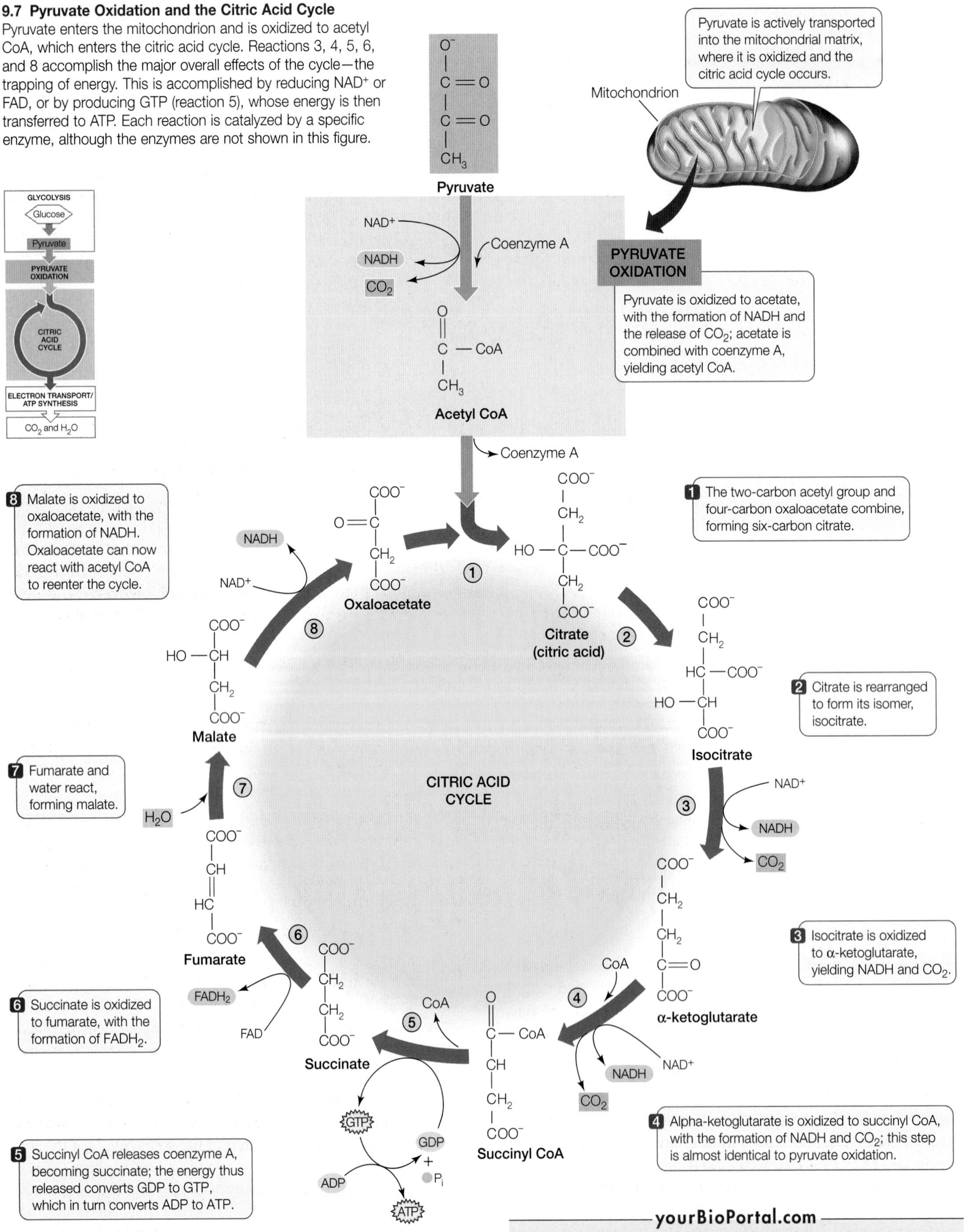

GLYCOLYSIS
Glucose
Pyruvate
PYRUVATE OXIDATION
CITRIC ACID CYCLE
ELECTRON TRANSPORT/ ATP SYNTHESIS
CO2 and H2O
Pyruvate is actively transported into the mitochondrial matrix, where it is oxidized and the citric acid cycle occurs.
Mitochondrion
Pyruvate
NAD+
NADH
CO2
Coenzyme A
Acetyl CoA
PYRUVATE OXIDATION
Pyruvate is oxidized to acetate, with the formation of NADH and the release of CO2; acetate is combined with coenzyme A, yielding acetyl CoA.
Coenzyme A
1 The two-carbon acetyl group and four-carbon oxaloacetate combine, forming six-carbon citrate.
Citrate (citric acid)
2 Citrate is rearranged to form its isomer, isocitrate.
Isocitrate
NAD+
NADH
CO2
3 Isocitrate is oxidized to α-ketoglutarate, yielding NADH and CO2.
α-ketoglutarate
CoA
NAD+
NADH
CO2
4 Alpha-ketoglutarate is oxidized to succinyl CoA, with the formation of NADH and CO2; this step is almost identical to pyruvate oxidation.
Succinyl CoA
CoA
GTP
GDP
+
Pi
ADP
ATP
5 Succinyl CoA releases coenzyme A, becoming succinate; the energy thus released converts GDP to GTP, which in turn converts ADP to ATP.
Succinate
FADH2
FAD
6 Succinate is oxidized to fumarate, with the formation of FADH2.
Fumarate
H2O
7 Fumarate and water react, forming malate.
Malate
NAD+
NADH
8 Malate is oxidized to oxaloacetate, with the formation of NADH. Oxaloacetate can now react with acetyl CoA to reenter the cycle.
Oxaloacetate
CITRIC ACID CYCLE

yourBioPortal.com
GO TO Web Activity 9.3 • The Citric Acid Cycle

In reaction 1, the energy temporarily stored in acetyl CoA drives the formation of citrate from oxaloacetate. During this reaction, the CoA molecule is removed and can be reused by pyruvate dehydrogenase.

In reaction 2, the citrate molecule is rearranged to form isocitrate.

In reaction 3, a CO_2 molecule, a proton, and two electrons are removed, converting isocitrate into α-ketoglutarate. This reaction releases a large amount of free energy, some of which is stored in NADH.

In reaction 4, α-ketoglutarate is oxidized to succinyl CoA. This reaction is similar to the oxidation of pyruvate to form acetyl CoA. Like that reaction, it is catalyzed by a multi-enzyme complex and produces CO_2 and NADH.

In reaction 5, some of the energy in succinyl CoA is harvested to make GTP (guanosine triphosphate) from GDP and P_i. This is another example of substrate-level phosphorylation. GTP is then used to make ATP from ADP and P_i.

In reaction 6, the succinate released from succinyl CoA in reaction 5 is oxidized to fumarate. In the process, free energy is released and two hydrogens are transferred to the electron carrier FAD, forming $FADH_2$.

Reaction 7 is a molecular rearrangement in which water is added to fumarate, forming malate.

In *reaction 8,* one more NAD^+ reduction occurs, producing oxaloacetate from malate. Reactions 7 and 8 illustrate a common biochemical mechanism: in reaction 7, water (H_2O) is added to form a hydroxyl (—OH) group, and then in reaction 8 the H from the hydroxyl group is removed, generating a carbonyl group and reducing NAD^+ to NADH.

The final product, oxaloacetate, is ready to combine with another acetyl group from acetyl CoA and go around the cycle again. The citric acid cycle operates twice for each glucose molecule that enters glycolysis (once for each pyruvate that enters the mitochondrion).

To summarize:

- The *inputs* to the citric acid cycle are acetate (in the form of acetyl CoA), water, and the oxidized electron carriers NAD^+, FAD, and GDP.

- The *outputs* are carbon dioxide, reduced electron carriers (NADH and $FADH_2$), and a small amount of GTP. Overall, the citric acid cycle releases two carbons as CO_2 and produces four reduced electron carrier molecules.

The citric acid cycle is regulated by the concentrations of starting materials

We have seen how pyruvate, a three-carbon molecule, is completely oxidized to CO_2 by pyruvate dehydrogenase and the citric acid cycle. For the cycle to continue, the starting molecules—acetyl CoA and oxidized electron carriers—must all be replenished. The electron carriers are reduced during the cycle and in reaction 6 of glycolysis (see Figure 9.5), and they must be reoxidized:

$$NADH \rightarrow NAD^+ + H^+ + 2\,e^-$$

$$FADH_2 \rightarrow FAD + 2\,H^+ + 2\,e^-$$

The oxidation of these electron carriers take place in coupled redox reactions, in which other molecules get reduced. When it is present, O_2 is the molecule that eventually accepts these electrons and gets reduced to form H_2O.

9.2 RECAP

The oxidation of glucose in the presence of O_2 involves glycolysis, pyruvate oxidation, and the citric acid cycle. In glycolysis, glucose is converted to pyruvate with some energy capture. Following the initial oxidation of pyruvate, the citric acid cycle completes its oxidation to CO_2 and more energy is captured in the form of reduced electron carriers.

- What is the net energy yield of glycolysis in terms of energy invested and energy harvested? **See p. 174 and Figure 9.6**

- What role does pyruvate oxidation play in the citric acid cycle? **See pp. 174–175 and Figure 9.7**

- Explain why reoxidation of NADH is crucial for the continuation of the citric acid cycle. **See p. 177**

Pyruvate oxidation and the citric acid cycle cannot continue operating unless O_2 is available to receive electrons during the reoxidation of reduced electron carriers. However, these electrons are not passed directly to O_2, as you will learn next.

9.3 How Does Oxidative Phosphorylation Form ATP?

The overall process of ATP synthesis resulting from the reoxidation of electron carriers in the presence of O_2 is called **oxidative phosphorylation**. Two components of the process can be distinguished:

1. *Electron transport.* The electrons from NADH and $FADH_2$ pass through the **respiratory chain**, a series of membrane-associated electron carriers. The flow of electrons along this pathway results in the active transport of protons out of the mitochondrial matrix and across the inner mitochondrial membrane, creating a proton concentration gradient.

2. *Chemiosmosis.* The protons diffuse back into the mitochondrial matrix through a channel protein, **ATP synthase**, which couples this diffusion to the synthesis of ATP. The inner mitochondrial membrane is otherwise impermeable to protons, so the only way for them to follow their concentration gradient is through the channel.

Before we proceed with the details of these pathways, let's consider an important question: Why should the respiratory chain be such a complex process? Why don't cells use the following single step?

$$NADH + H^+ + \tfrac{1}{2}\,O_2 \rightarrow NAD^+ + H_2O$$

The answer is that this reaction would be untamable. It is extremely exergonic—and would be rather like setting off a stick of dynamite in the cell. There is no biochemical way to harvest that burst of energy efficiently and put it to physiological use (that is, no single metabolic reaction is so endergonic as to consume a significant fraction of that energy in a single step). To control the release of energy during the oxidation of glucose, cells have evolved a lengthy respiratory chain: a series of reactions, each of which releases a small, manageable amount of energy, one step at a time.

The respiratory chain transfers electrons and releases energy

The respiratory chain is located in the inner mitochondrial membrane and contains several interactive components, including large integral proteins, smaller mobile proteins, and a small lipid molecule. **Figure 9.8** shows a plot of the free energy released as electrons are passed between the carriers.

- Four large protein complexes (I, II, III, and IV) contain electron carriers and associated enzymes. In eukaryotes they are integral proteins of the inner mitochondrial membrane (see Figure 5.12), and three are transmembrane proteins.

- Cytochrome *c* is a small peripheral protein that lies in the intermembrane space. It is loosely attached to the outer surface of the inner mitochondrial membrane.

- Ubiquinone (abbreviated Q) is a small, nonpolar, lipid molecule that moves freely within the hydrophobic interior

of the phospholipid bilayer of the inner mitochondrial membrane.

As illustrated in Figure 9.8, NADH passes electrons to protein complex I (called NADH-Q reductase), which in turn passes the electrons to Q. This electron transfer is accompanied by a large drop in free energy. Complex II (succinate dehydrogenase) passes electrons to Q from $FADH_2$, which was generated in reaction 6 of the citric acid cycle (see Figure 9.7). These electrons enter the chain later than those from NADH and will ultimately produce less ATP.

Complex III (cytochrome *c* reductase) receives electrons from Q and passes them to cytochrome *c*. Complex IV (cytochrome *c* oxidase) receives electrons from cytochrome *c* and passes them to oxygen. Finally the reduction of oxygen to H_2O occurs:

$$\tfrac{1}{2} O_2 + 2 H^+ + 2 e^- \rightarrow H_2O$$

Notice that two protons (H$^+$) are also consumed in this reaction. This contributes to the proton gradient across the inner mitochondrial membrane.

During electron transport, protons are also actively transported across the membrane—electron transport within each of the three transmembrane complexes (I, III, and IV) results in the transfer of protons from the matrix to the intermembrane space (**Figure 9.9**). So an imbalance of protons is set up, with the impermeable inner mitochondrial membrane as a barrier. The concentration of H$^+$ in the intermembrane space is higher than in the matrix, and this gradient represents a source of potential energy. The diffusion of those protons across the membrane is coupled with the formation of ATP. Thus the energy originally contained in glucose and other fuel molecules is finally captured in the cellular energy currency, ATP. For each pair of electrons passed along the chain from NADH to oxygen, about 2.5 molecules of ATP are formed. $FADH_2$ oxidation produces about 1.5 ATP molecules.

Proton diffusion is coupled to ATP synthesis

All the electron carriers and enzymes of the respiratory chain, except cytochrome *c*, are embedded in the inner mitochondrial membrane. As we have just seen, the operation of the respiratory chain results in the active transport of protons from the mitochon-

9.8 The Oxidation of NADH and FADH$_2$ in the Respiratory Chain Electrons from NADH and FADH$_2$ are passed along the respiratory chain, a series of protein complexes in the inner mitochondrial membrane containing electron carriers and enzymes. The carriers gain free energy when they become reduced and release free energy when they are oxidized.

yourBioPortal.com
GO TO Web Activity 9.4 • Respiratory Chain

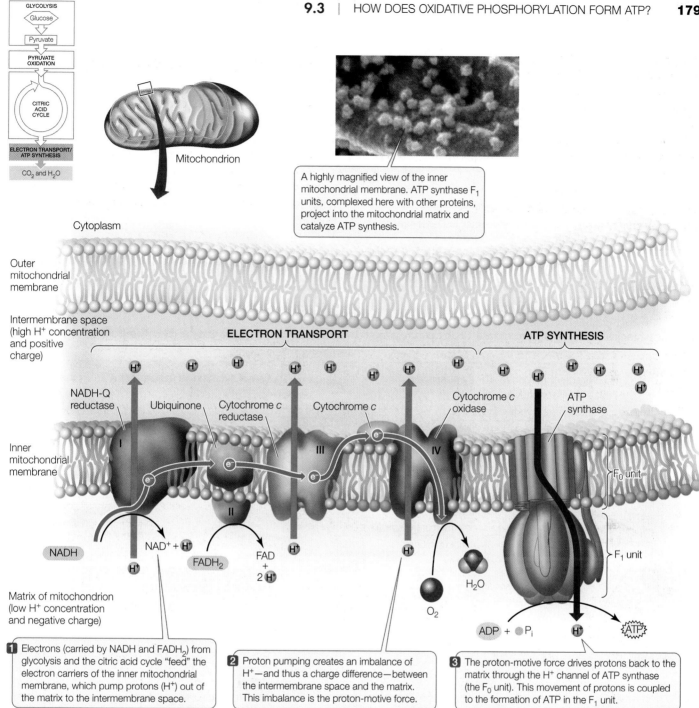

GLYCOLYSIS
Glucose
Pyruvate

PYRUVATE
OXIDATION

CITRIC
ACID
CYCLE

ELECTRON TRANSPORT/
ATP SYNTHESIS

CO_2 and H_2O

Mitochondrion

A highly magnified view of the inner mitochondrial membrane. ATP synthase F_1 units, complexed here with other proteins, project into the mitochondrial matrix and catalyze ATP synthesis.

Cytoplasm

Outer mitochondrial membrane

Intermembrane space (high H^+ concentration and positive charge)

ELECTRON TRANSPORT

ATP SYNTHESIS

NADH-Q reductase

Ubiquinone

Cytochrome *c* reductase

Cytochrome *c*

Cytochrome *c* oxidase

ATP synthase

Inner mitochondrial membrane

F_0 unit

F_1 unit

NADH

$NAD^+ + H^+$

FADH$_2$

FAD + 2 H^+

H_2O

O_2

ADP + P_i

Matrix of mitochondrion (low H^+ concentration and negative charge)

1 Electrons (carried by NADH and FADH$_2$) from glycolysis and the citric acid cycle "feed" the electron carriers of the inner mitochondrial membrane, which pump protons (H^+) out of the matrix to the intermembrane space.

2 Proton pumping creates an imbalance of H^+—and thus a charge difference—between the intermembrane space and the matrix. This imbalance is the proton-motive force.

3 The proton-motive force drives protons back to the matrix through the H^+ channel of ATP synthase (the F_0 unit). This movement of protons is coupled to the formation of ATP in the F_1 unit.

9.9 The Respiratory Chain and ATP Synthase Produce ATP by a Chemiosmotic Mechanism As electrons pass through the transmembrane protein complexes in the respiratory chain, protons are pumped from the mitochondrial matrix into the intermembrane space. As the protons return to the matrix through ATP synthase, ATP is formed.

─── **yourBioPortal.com** ───

GO TO Animated Tutorial 9.1 • Electron Transport and ATP Synthesis

drial matrix to the intermembrane space. The transmembrane protein complexes (I, III, and IV) act as proton pumps, and as a result, the intermembrane space is more acidic than the matrix.

Because of the positive charge carried by a proton (H^+), this pumping creates not only a concentration gradient but also a difference in electric charge across the inner mitochondrial membrane, making the mitochondrial matrix more negative than the intermembrane space. Together, the proton concentration gradient and the electrical charge difference constitute a source of potential energy called the **proton-motive force**. This force tends to drive the protons back across the membrane, just as the charge on a battery drives the flow of electrons to discharge the battery.

The hydrophobic lipid bilayer is essentially impermeable to protons, so the potential energy of the proton-motive force cannot be discharged by simple diffusion of protons across the membrane. However, protons can diffuse across the membrane by passing through a specific proton channel, called ATP synthase, which couples proton movement to the synthesis of ATP. This coupling of proton-motive force and ATP synthesis is called the chemiosmotic mechanism (or **chemiosmosis**) and is found in all respiring cells.

THE CHEMIOSMOTIC MECHANISM FOR ATP SYNTHESIS The chemiosmotic mechanism involves transmembrane proteins, including a proton channel and the enzyme ATP synthase, that couple proton diffusion to ATP synthesis. The potential energy of the H^+

---

yourBioPortal.com

GO TO **Animated Tutorial 9.2 • Two Experiments Demonstrate the Chemiosmotic Mechanism**

---

gradient, or the proton-motive force (described above), is harnessed by ATP synthase. This protein complex has two roles: it acts as a channel allowing H^+ to diffuse back into the matrix, and it uses the energy of that diffusion to make ATP from ADP and P_i.

ATP synthesis is a reversible reaction, and ATP synthase can also act as an ATPase, hydrolyzing ATP to ADP and P_i:

$$ATP \rightleftharpoons ADP + P_i + \text{free energy}$$

If the reaction goes to the right, free energy is released and is used to pump H^+ out of the mitochondrial matrix—not the usual mode

INVESTIGATING LIFE

9.10 Two Experiments Demonstrate the Chemiosmotic Mechanism

The chemiosmosis hypothesis was a bold departure for the conventional scientific thinking of the time. It required an intact compartment separated by a membrane. Could a proton gradient drive the synthesis of ATP? And was this capacity entirely due to the ATP synthase enzyme?

HYPOTHESIS A H^+ gradient can drive ATP synthesis by isolated mitochondria.

HYPOTHESIS ATP synthase is needed for ATP synthesis.

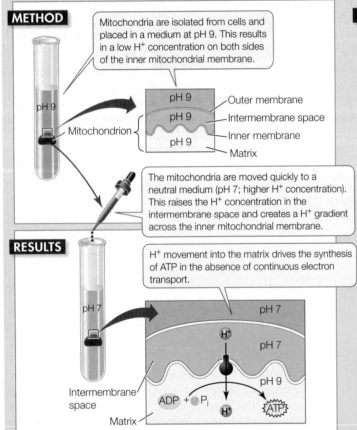

METHOD

Mitochondria are isolated from cells and placed in a medium at pH 9. This results in a low H^+ concentration on both sides of the inner mitochondrial membrane.

pH 9

Mitochondrion — Outer membrane / Intermembrane space / Inner membrane / Matrix (all pH 9)

The mitochondria are moved quickly to a neutral medium (pH 7; higher H^+ concentration). This raises the H^+ concentration in the intermembrane space and creates a H^+ gradient across the inner mitochondrial membrane.

RESULTS

H^+ movement into the matrix drives the synthesis of ATP in the absence of continuous electron transport.

pH 7

Intermembrane space — pH 7 / pH 7 / pH 9 — Matrix

$ADP + P_i \longrightarrow ATP$

CONCLUSION In the absence of electron transport, an artificial H^+ gradient is sufficient for ATP synthesis by mitochondria.

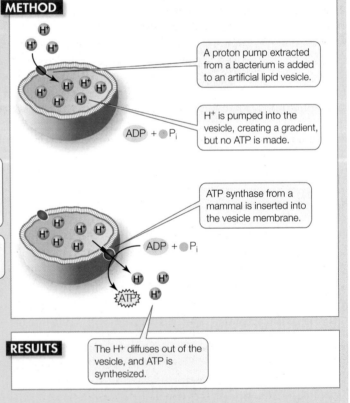

METHOD

A proton pump extracted from a bacterium is added to an artificial lipid vesicle.

H^+ is pumped into the vesicle, creating a gradient, but no ATP is made.

$ADP + P_i$

ATP synthase from a mammal is inserted into the vesicle membrane.

$ADP + P_i$

RESULTS The H^+ diffuses out of the vesicle, and ATP is synthesized.

CONCLUSION ATP synthase, acting as a H^+ channel, is necessary for ATP synthesis.

FURTHER INVESTIGATION: What would happen in the experiment on the right if a second ATP synthase, oriented in the opposite way to the one originally inserted in the membrane, were added?

Go to **yourBioPortal.com** for original citations, discussions, and relevant links for all INVESTIGATING LIFE figures.

of operation. If the reaction goes to the left, it uses the free energy from H^+ diffusion into the matrix to make ATP. What makes it prefer ATP synthesis? There are two answers to this question:

- ATP leaves the mitochondrial matrix for use elsewhere in the cell as soon as it is made, keeping the ATP concentration in the matrix low, and driving the reaction toward the left.

- The H^+ gradient is maintained by electron transport and proton pumping.

Every day a person hydrolyzes about 10^{25} ATP molecules to ADP. This amounts to 9 kg, a significant fraction of almost everyone's entire body weight! The vast majority of this ADP is "recycled"—converted back to ATP—using free energy from the oxidation of glucose.

EXPERIMENTS DEMONSTRATE CHEMIOSMOSIS When it was first proposed almost a half-century ago, the idea that a proton gradient was the energy intermediate linking electron transport to ATP synthesis was a departure from the current conventional thinking. Scientists had been searching for a mitochondrial intermediate that they believed would carry energy in much the same way as the ATP produced by substrate level phosphorylation. The search for this intermediate was not successful, and this led to the idea that chemiosmosis was the mechanism of oxidative phosphorylation. Experimental evidence was needed to support this hypothesis. Two key experiments demonstrated (1) that a proton (H^+) gradient across a membrane can drive ATP synthesis; and (2) that the enzyme ATP synthase is the catalyst for this reaction (**Figure 9.10**).

In the first experiment, mitochondria without a food source were "fooled" into making ATP by raising the H^+ concentration in their environment. In the second experiment, a light-driven proton pump isolated from bacteria was inserted into artificial lipid vesicles. This generated a proton gradient, but since ATP synthase was absent, ATP was not made. Then, ATP synthase was inserted into the vesicles and ATP was generated.

UNCOUPLING PROTON DIFFUSION FROM ATP PRODUCTION The tight coupling between H^+ diffusion and the formation of ATP provides further evidence for the chemiosmotic mechanism. If a second type of H^+ diffusion channel (that does not synthesize ATP) is present in the mitochondrial membrane, the energy of the H^+ gradient is released as heat rather than being coupled to ATP synthesis. Such uncoupling molecules actually exist in the mitochondria of some organisms to generate heat instead of ATP. For example, the natural uncoupling protein thermogenin plays an important role in regulating the temperatures of newborn human infants, who lack hair to keep warm, and in hibernating animals.

A popular weight loss drug in the 1930s was the uncoupler molecule, dinitrophenol. There were claims of dramatic weight loss when the drug was administered to obese patients. Unfortunately, the heat that was released caused fatally high fevers, and the effective dose and fatal dose were quite close. So the use of this drug was discontinued in 1938. However, the general strategy of using an uncoupler for weight loss remains a subject of research.

HOW ATP SYNTHASE WORKS: A MOLECULAR MOTOR Now that we have established that the H^+ gradient is needed for ATP synthesis, a question remains: how does the enzyme actually make ATP from ADP and P_i? This is certainly a fundamental question in biology, as it underlies energy harvesting in most cells. Look at the structure of ATP synthase in Figure 9.9. It is a molecular motor composed of two parts: the F_0 unit, a transmembrane region that is the H^+ channel, and the F_1 unit, which contains the active sites for ATP synthesis. F_1 consists of six subunits (three each of two polypeptide chains), arranged like the segments of an orange around a central polypeptide. ATP synthesis is coupled with conformational changes in the ATP synthase enzyme, which are induced by proton movement through the complex. The potential energy set up by the proton gradient across the inner membrane drives the passage of protons through the ring of polypeptides that make up the F_0 component. This ring rotates as the protons pass through the membrane, causing the F_1 unit to rotate as well. ADP and P_i bind to active sites that become exposed on the F_1 unit as it rotates, and ATP is made. The structure and function of ATP synthase are shared by living organisms as diverse as bacteria and humans. These molecular motors make ATP at rates up to 100 molecules per second.

9.3 RECAP

The oxidation of reduced electron carriers in the respiratory chain drives the active transport of protons across the inner mitochondrial membrane, generating a proton-motive force. Diffusion of protons down their electrochemical gradient through ATP synthase is coupled to the synthesis of ATP.

- What are the roles of oxidation and reduction in the respiratory chain? **See Figures 9.8 and 9.9**

- What is the proton motive force and how does it drive chemiosmosis? **See pp. 179–180**

- Explain how the two experiments described in Figure 9.10 demonstrate the chemiosmotic mechanism. **See p. 181**

Oxidative phosphorylation captures a great deal of energy in ATP. But it does not occur if O_2 is absent. We turn now to the metabolism of glucose in anaerobic conditions.

9.4 How Is Energy Harvested from Glucose in the Absence of Oxygen?

In the absence of O_2 (anaerobic conditions), a small amount of ATP can be produced by glycolysis and fermentation. Like glycolysis, fermentation pathways occur in the cytoplasm. There are many different types of fermentation, but they all operate to regenerate NAD^+ so that the NAD-requiring reaction of glycolysis can continue (see reaction 6 in Figure 9.5). Of course, if a necessary reactant such as NAD^+ is not present, the reaction will not take place. How do fermentation reactions regenerate NAD^+ and permit ATP formation to continue?

Prokaryotic organisms often live in O_2-deficient environments and are known to use many different fermentation pathways. But the two best understood fermentation pathways are found in a wide variety of organisms including eukaryotes. These two short pathways are lactic acid fermentation, whose end product is lactic acid (lactate); and alcoholic fermentation, whose end product is ethyl alcohol (ethanol).

In lactic acid fermentation, pyruvate serves as the electron acceptor and lactate is the product (**Figure 9.11**). This process takes place in many microorganisms and complex organisms, including higher plants and vertebrates. A notable example of lactic acid fermentation occurs in vertebrate muscle tissue. Usually, vertebrates get their energy for muscle contraction aerobically, with the circulatory system supplying O_2 to muscles. In small vertebrates, this is almost always adequate: for example, birds can fly long distances without resting. But in larger vertebrates such as humans, the circulatory system is not up to the task of delivering enough O_2 when the need is great, such as during high activity. At this point, the muscle cells break down glycogen (a stored polysaccharide) and undergo lactic acid fermentation.

Lactic acid buildup becomes a problem after prolonged periods because the acid ionizes, forming H^+ and lowering the pH of the cell. This affects cellular activities and causes muscle cramps, resulting in muscle pain, which abates upon resting. Lactate dehydrogenase, the enzyme that catalyzes the fermentation reaction, works in both directions. That is, it can catalyze the oxidation of lactate as well as the reduction of pyruvate. When lactate levels are decreased, muscle activity can resume.

Alcoholic fermentation takes place in certain yeasts (eukaryotic microbes) and some plant cells under anaerobic conditions. This process requires two enzymes, pyruvate decarboxylase and alcohol dehydrogenase, which metabolize pyruvate to ethanol (**Figure 9.12**). As with lactic acid fermentation, the reactions are essentially reversible. For thousands of years, humans have used anaerobic fermentation by yeast cells to produce alcoholic beverages. The cells use sugars from plant sources (glucose from grapes or maltose from barley) to produce the end product, ethanol, in wine and beer.

By recycling NAD^+, fermentation allows glycolysis to continue, thus producing small amounts of ATP through substrate-level phosphorylation. The net yield of two ATPs per glucose

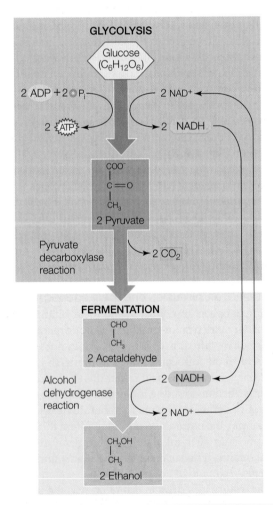

Summary of reactants and products:
$C_6H_{12}O_6 + 2\ ADP + 2\ P_i \rightarrow 2$ lactic acid $+ 2$ ATP

9.11 Lactic Acid Fermentation Glycolysis produces pyruvate, ATP, and NADH from glucose. Lactic acid fermentation uses NADH as a reducing agent to reduce pyruvate to lactic acid (lactate), thus regenerating NAD^+ to keep glycolysis operating.

Summary of reactants and products:
$C_6H_{12}O_6 + 2\ ADP + 2\ P_i \rightarrow 2$ ethanol $+ 2\ CO_2 + 2$ ATP

9.12 Alcoholic Fermentation In alcoholic fermentation, pyruvate from glycolysis is converted into acetaldehyde, and CO_2 is released. NADH from glycolysis is used to reduce acetaldehyde to ethanol, thus regenerating NAD^+ to keep glycolysis operating.

molecule is much lower than the energy yield from cellular respiration. For this reason, most organisms existing in anaerobic environments are small microbes that grow relatively slowly.

Cellular respiration yields much more energy than fermentation

The total net energy yield from glycolysis plus fermentation is two molecules of ATP per molecule of glucose oxidized. The maximum yield of ATP that can be harvested from a molecule of glucose through glycolysis followed by cellular respiration is much greater—about 32 molecules of ATP (**Figure 9.13**). (Review Figures 9.5, 9.7, and 9.9 to see where all the ATP molecules come from.)

Why do the metabolic pathways that operate in aerobic environments produce so much more ATP? Glycolysis and fermentation only partially oxidize glucose, as does fermentation. Much more energy remains in the end products of fermentation (lactic acid and ethanol) than in CO_2, the end product of cellular respiration. In cellular respiration, carriers (mostly NAD^+) are reduced in pyruvate oxidation and the citric acid cycle. Then the reduced carriers are oxidized by the respiratory chain, with the accompanying production of ATP by chemiosmosis (2.5 ATP for each NADH and 1.5 ATP for each $FADH_2$). In an aerobic environment, a cell or organism capable of aerobic metabolism will have the advantage over one that is limited to fermentation, in terms of its ability to harvest chemical energy. Two key events in the evolution of multicellular organisms were the rise in atmospheric O_2 levels (see Chapter 1) and the development of metabolic pathways to utilize that O_2.

The yield of ATP is reduced by the impermeability of some mitochondria to NADH

The total gross yield of ATP from the oxidation of one molecule of glucose to CO_2 is 32. However, in some animal cells the inner mitochondrial membrane is impermeable to NADH, and a "toll" of one ATP must be paid for each NADH molecule that is produced in glycolysis and must be "shuttled" into the mitochondrial matrix. So in these animals, the net yield of ATP is 30.

NADH *shuttle systems* transfer the electrons captured by glycolysis onto substrates that are capable of movement across the mitochondrial membranes. In muscle and liver tissues, an important shuttle involves glycerol 3-phosphate. In the cytosol,

NADH (from glycolysis) + dihydroxyacetone phosphate (DHAP) → NAD^+ + glycerol 3-phosphate

Glycerol 3-phosphate crosses both mitochondrial membranes. In the mitochondrial matrix,

FAD + glycerol 3-phosphate → $FADH_2$ + DHAP

DHAP is able to move back to the cytosol, where it is available to repeat the process. Note that the reducing electrons are transferred from NADH outside the mitochondrion to $FADH_2$ inside the mitochondrion. As you know from Figures 9.8 and 9.9, the energy yield in terms of ATP from $FADH_2$ is lower than that from NADH. This lowers the overall energy yield.

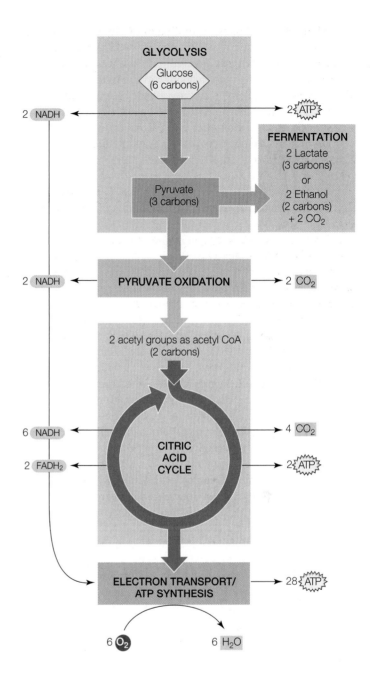

GLYCOLYSIS AND FERMENTATION

Summary of reactants and products:
$C_6H_{12}O_6$ ⟶ 2 lactate (or 2 ethanol + 2 CO_2) + 2 ATP

GLYCOLYSIS AND CELLULAR RESPIRATION

Summary of reactants and products:
$C_6H_{12}O_6$ + 6 O_2 ⟶ 6 CO_2 + 6 H_2O + 32 ATP

9.13 Cellular Respiration Yields More Energy Than Fermentation Electron carriers are reduced in pyruvate oxidation and the citric acid cycle, then oxidized by the respiratory chain. These reactions produce ATP via chemiosmosis.

━━━━━ **yourBioPortal.com** ━━━━━
GO TO Web Activity 9.5 • Energy Levels

9.4 RECAP

In the absence of O_2, fermentation pathways use NADH formed by glycolysis to reduce pyruvate and regenerate NAD^+. The energy yield of fermentation is low because glucose is only partially oxidized. When O_2 is present, the electron carriers of cellular respiration allow for the full oxidation of glucose, so the energy yield from glucose is much higher.

- Why is replenishing NAD^+ crucial to cellular metabolism? See pp. 182–183
- What is the total energy yield from glucose in human cells in the presence versus the absence of O_2? See p. 183 and **Figure 9.13**

Now that you've seen how cells harvest energy, let's see how that energy moves through other metabolic pathways in the cell.

9.5 How Are Metabolic Pathways Interrelated and Regulated?

Glycolysis and the pathways of cellular respiration do not operate in isolation. Rather, there is an interchange of molecules into and out of these pathways, to and from the metabolic pathways for the synthesis and breakdown of amino acids, nucleotides, fatty acids, and other building blocks of life. Carbon skeletons can enter the catabolic pathways and be broken down to release their energy, or they can enter anabolic pathways to be used in the formation of the macromolecules that are the major constituents of the cell. These relationships are summarized in **Figure 9.14**. In this section we will explore how pathways are interrelated by the sharing of intermediate substances, and we will see how pathways are regulated by the inhibitors of key enzymes.

Catabolism and anabolism are linked

A hamburger or veggie burger on a bun contains three major sources of carbon skeletons: carbohydrates, mostly in the form of starch (a polysaccharide); lipids, mostly as triglycerides (three fatty acids attached to glycerol); and proteins (polymers of amino acids). Look at Figure 9.14 to see how each of these three types of macromolecules can be hydrolyzed and used in catabolism or anabolism.

CATABOLIC INTERCONVERSIONS Polysaccharides, lipids, and proteins can all be broken down to provide energy:

- *Polysaccharides* are hydrolyzed to glucose. Glucose then passes through glycolysis and cellular respiration, where its energy is captured in ATP.
- *Lipids* are broken down into their constituents, glycerol and fatty acids. Glycerol is converted into dihydroxyacetone phosphate (DHAP), an intermediate in glycolysis. Fatty acids are highly reduced molecules that are converted to acetyl CoA inside the mitochondrion by a series of oxidation enzymes, in a process known as β-oxidation. For example, the β-oxidation of a C_{16} fatty acid occurs in several steps:

$$C_{16} \text{ fatty acid} + CoA \rightarrow C_{16} \text{ fatty acyl CoA}$$

$$C_{16} \text{ fatty acyl CoA} + CoA \rightarrow C_{14} \text{ fatty acyl CoA} + \text{acetyl CoA}$$

$$\text{repeat 6 times} \rightarrow 8 \text{ acetyl CoA}$$

The acetyl CoA can then enter the citric acid cycle and be catabolized to CO_2.

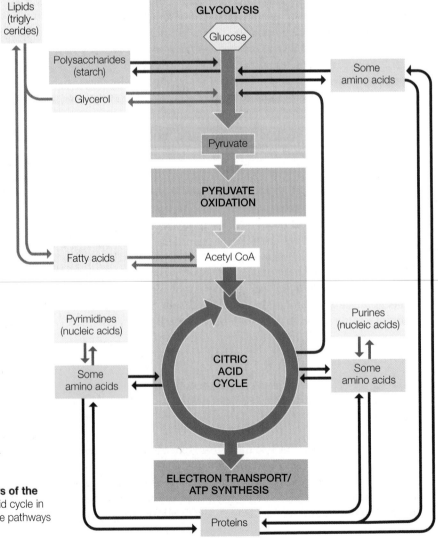

9.14 Relationships among the Major Metabolic Pathways of the Cell Note the central positions of glycolysis and the citric acid cycle in this network of metabolic pathways. Also note that many of the pathways can operate essentially in reverse.

• *Proteins* are hydrolyzed to their amino acid building blocks. The 20 different amino acids feed into glycolysis or the citric acid cycle at different points. For example, the amino acid glutamate is converted into α-ketoglutarate, an intermediate in the citric acid cycle.

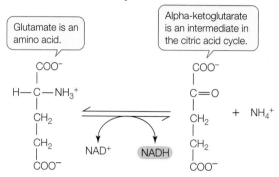

ANABOLIC INTERCONVERSIONS Many catabolic pathways can operate essentially in reverse, with some modifications. Glycolytic and citric acid cycle intermediates, instead of being oxidized to form CO_2, can be reduced and used to form glucose in a process called **gluconeogenesis** (which means "new formation of glucose"). Likewise, acetyl CoA can be used to form fatty acids. The most common fatty acids have even numbers of carbons: 14, 16, or 18. These are formed by the addition of two-carbon acetyl CoA "units" one at a time until the appropriate chain length is reached. Acetyl CoA is also a building block for various pigments, plant growth substances, rubber, steroid hormones, and other molecules.

Some intermediates in the citric acid cycle are reactants in pathways that synthesize important components of nucleic acids. For example, α-ketoglutarate is a starting point for purines, and oxaloacetate for pyrimidines. In addition, α-ketoglutarate is a starting point for the synthesis of chlorophyll (used in photosynthesis; see Chapter 10) and the amino acid glutamate (used in protein synthesis).

Catabolism and anabolism are integrated

A carbon atom from a protein in your burger can end up in DNA, fat, or CO_2, among other fates. How does the organism "decide" which metabolic pathways to follow, in which cells? With all of the possible interconversions, you might expect that cellular concentrations of various biochemical molecules would vary widely. Remarkably, the levels of these substances in what is called the metabolic pool—the sum total of all the biochemical molecules in a cell—are quite constant. Organisms regulate the enzymes of catabolism and anabolism in various cells in order to maintain a balance. This metabolic homeostasis gets upset only in unusual circumstances. Let's look one such unusual circumstance: undernutrition.

Glucose is an excellent source of energy, but lipids and proteins can also be broken down and their constituents used as energy sources. Any one or all three of these types of molecules could be used to provide the energy your body needs. But normally these substances are not equally available for energy me-

tabolism and ATP formation. Proteins, for example, have essential roles as enzymes and as structural elements, providing support and movement; they are not stored for energy, and using them for energy might deprive the body of other vital functions.

Fats (triglycerides) do not have catalytic roles. Because they are nonpolar, fats do not bind water, and they are therefore less dense than polysaccharides in aqueous environments. In addition, fats are more reduced than carbohydrates (have more C—H bonds and fewer C—OH bonds) and thus have more energy stored in their bonds (see Figure 9.2). So it is not surprising that fats are the preferred energy store in many organisms. The human body stores fats and carbohydrates; fats are stored in adipose tissue, and glucose is stored as the polysaccharide glycogen in muscles and the liver. A typical person has about one day's worth of food energy stored as glycogen (a polysaccharide) and over a month's food energy stored as fats.

What happens if a person does not eat enough to produce sufficient ATP and NADH for anabolism and biological activities? This situation can be deliberate (to lose weight), but for too many people, it is forced upon them because not enough food is available, resulting in undernutrition and starvation. Initially, homeostasis can be maintained. The first energy stores to be used are the glycogen stores in muscle and liver cells. These stores do not last long, and next come the fats.

In cells that have access to fatty acids, their breakdown produces acetyl CoA for cellular respiration. However, a problem remains: because fatty acids cannot cross from the blood to the brain, the brain can use only glucose as its energy source. With glycogen already depleted, the body must convert something else to make glucose for the brain. This is accomplished by the breakdown of proteins and the conversion of their amino acids to glucose by gluconeogenesis. Without sufficient food intake, proteins and fats are used up. After several weeks of starvation, fat stores become depleted, and the only energy source left is protein. At this point, essential structural proteins, enzymes, and antibodies get broken down. The loss of such proteins can lead to severe illness and eventual death.

Metabolic pathways are regulated systems

We have described the relationships between metabolic pathways and noted that these pathways work together to provide homeostasis in the cell and organism. But how does the cell regulate the interconversions between pathways to maintain constant metabolic pools? This is a problem of systems biology, which seeks to understand how biochemical pathways interact (see Figure 8.15). It is a bit like trying to predict traffic patterns in a city: if an accident blocks traffic on a major road, drivers take alternate routes, where the traffic volume consequently changes.

Consider what happens to the starch in your burger bun. In the digestive system, starch is hydrolyzed to glucose, which enters the blood for distribution to the rest of the body. But before the glucose is distributed, a regulatory check must be made: if there is already enough glucose in the blood to supply the body's needs, the excess glucose is converted into glycogen and

Compound G provides positive feedback to the enzyme catalyzing the step from D to E.

Compound G inhibits the enzyme catalyzing the conversion of C to F, blocking that reaction and ultimately its own synthesis.

Negative feedback

Positive feedback

9.15 Regulation by Negative and Positive Feedback
Allosteric feedback regulation plays an important role in metabolic pathways. The accumulation of some products can shut down their synthesis, or can stimulate other pathways that require the same raw materials.

stored in the liver. If not enough glucose is supplied by food, glycogen is broken down, or other molecules are used to make glucose by gluconeogenesis.

The end result is that the level of glucose in the blood is remarkably constant. How does the body accomplish this?

Glycolysis, the citric acid cycle, and the respiratory chain are subject to *allosteric regulation* (see Section 8.5) of the enzymes involved. An example of allosteric regulation is feedback inhibition, illustrated in Figure 8.19. In a metabolic pathway, a high concentration of the final product can inhibit the action of an enzyme that catalyzes an earlier reaction. On the other hand, an excess of the product of one pathway can speed up reactions in another pathway, diverting raw materials away from synthesis of the first product (**Figure 9.15**). These negative and positive feedback mechanisms are used at many points in the energy-harvesting pathways, and are summarized in **Figure 9.16**.

• The main control point in glycolysis is the enzyme *phosphofructokinase* (reaction 3 in Figure 9.5). This enzyme is allosterically inhibited by ATP or citrate, and activated by ADP or AMP. Under anaerobic conditions, fermentation yields a relatively small amount of ATP, and phosphofructokinase operates at a high rate. However when conditions are aerobic, respiration makes 16 times more ATP than fermentation does, and the abundant ATP allosterically inhibits phosphofructokinase. Consequently, the conversion of fructose 6-phosphate to fructose 1,6-bisphosphate declines, and so does the rate of glucose utilization.

• The main control point in the citric acid cycle is the enzyme *isocitrate dehydrogenase*, which converts isocitrate to α-ketoglutarate (reaction 3 in Figure 9.7). NADH and

ATP are feedback inhibitors of this reaction, while ADP and NAD⁺ are activators. If too much ATP or NADH accumulates, the conversion of isocitrate is slowed, and the citric acid cycle shuts down. A shutdown of the citric acid cycle would cause large amounts of isocitrate and citrate to accumulate if the production of citrate were not also slowed. But, as mentioned above, an excess of citrate acts as a feedback inhibitor of phosphofructokinase. Thus, if the citric acid cycle has been slowed or shut down because of abun-

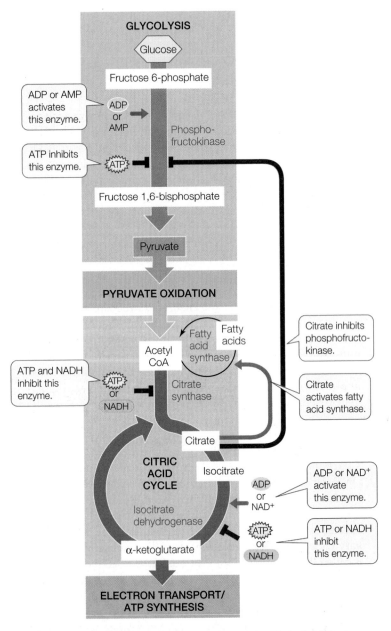

9.16 Allosteric Regulation of Glycolysis and the Citric Acid Cycle
Allosteric regulation controls glycolysis and the citric acid cycle at crucial early steps, increasing their efficiency and preventing the excessive buildup of intermediates.

yourBioPortal.com
GO TO Web Activity 9.6 • Regulation of Energy Pathways

dant ATP (and not because of a lack of oxygen), glycolysis is slowed as well. Both processes resume when the ATP level falls and they are needed again. Allosteric regulation keeps these processes in balance.

- Another control point involves *acetyl CoA*. If the level of ATP is high and the citric acid cycle shuts down, the accumulation of citrate activates fatty acid synthase, diverting acetyl CoA to the synthesis of fatty acids for storage. That is one reason why people who eat too much accumulate fat. These fatty acids may be metabolized later to produce more acetyl CoA.

9.5 RECAP

Glucose can be made from intermediates in glycolysis and the citric acid cycle by a process called gluconeogenesis. The metabolic pathways for the production and breakdown of lipids and amino acids are tied to those of glucose metabolism. Reaction products regulate key enzymes in the various pathways.

- Give examples of a catabolic interconversion of a lipid and of an anabolic interconversion of a protein. See pp. 184–185 and Figure 9.14

- How does phosphofructokinase serve as a control point for glycolysis? See p. 186 and Figure 9.16

- Describe what would happen if there was no allosteric mechanism for modulating the level of acetyl CoA.

CHAPTER SUMMARY

9.1 How Does Glucose Oxidation Release Chemical Energy?

- As a material is **oxidized**, the electrons it loses are transferred to another material, which is thereby **reduced**. Such **redox reactions** transfer large amounts of energy. Review Figure 9.2, **WEB ACTIVITIES 9.1 and 9.2**

- The coenzyme **NAD+** is a key electron carrier in biological redox reactions. It exists in two forms, one oxidized (NAD^+) and the other reduced (NADH).

- **Glycolysis** operates in the presence or absence of O_2. Under **aerobic** conditions, **cellular respiration** continues the process of breaking down glucose. Under **anaerobic** conditions, **fermentation** occurs. Review Figure 9.4

- The pathways of cellular respiration after glycolysis are **pyruvate oxidation**, the **citric acid cycle**, and the **electron transport/ATP synthesis**.

9.2 What Are the Aerobic Pathways of Glucose Metabolism?

- Glycolysis consists of 10 enzyme-catalyzed reactions that occur in the cell cytoplasm. Two **pyruvate** molecules are produced for each partially oxidized molecule of glucose, providing the starting material for both cellular respiration and fermentation. Review Figure 9.5

- The first five reactions of glycolysis require an investment of energy; the last five produce energy. The net gain is two molecules of ATP. Review Figure 9.6

- The enzyme-catalyzed transfer of phosphate groups to ADP by enzymes other than ATPase is called **substrate-level phosphorylation** and produces ATP.

- Pyruvate oxidation follows glycolysis and links glycolysis to the citric acid cycle. This pathway converts pyruvate into **acetyl CoA**.

- Acetyl CoA is the starting point of the citric acid cycle. It reacts with oxaloacetate to produce citrate. A series of eight enzyme-catalyzed reactions oxidize citrate and regenerate oxaloacetate, continuing the cycle. Review Figure 9.7, **WEB ACTIVITY 9.3**

9.3 How Does Oxidative Phosphorylation Form ATP?

- Oxidation of electron carriers in the presence of O_2 releases energy that can be used to form ATP in a process called **oxidative phosphorylation**.

- The NADH and $FADH_2$ produced in glycolysis, pyruvate oxidation, and the citric acid cycle are oxidized by the respiratory chain, regenerating NAD^+ and FAD. Oxygen (O_2) is the final acceptor of electrons and protons, forming water (H_2O). Review Figure 9.8, **WEB ACTIVITY 9.4**

- The respiratory chain not only transports electrons, but also pumps protons across the inner mitochondrial membrane, creating the **proton-motive force**.

- Protons driven by the proton-motive force can return to the mitochondrial matrix via **ATP synthase**, a molecular motor that couples this movement of protons to the synthesis of ATP. This process is called **chemiosmosis**. Review Figure 9.9, **ANIMATED TUTORIALS 9.1 and 9.2**

9.4 How Is Energy Harvested from Glucose in the Absence of Oxygen?

- In the absence of O_2, glycolysis is followed by fermentation. Together, these pathways partially oxidize pyruvate and generate end products such as **lactic acid** or **ethanol**. In the process, NAD^+ is regenerated from NADH so that glycolysis can continue, thus generating a small amount of ATP. Review Figures 9.11 and 9.12

- For each molecule of glucose used, fermentation yields 2 molecules of ATP. In contrast, glycolysis operating with pyruvate oxidation, the citric acid cycle, and the respiratory chain/ATP synthase yields up to 32 molecules of ATP per molecule of glucose. Review Figure 9.13, **WEB ACTIVITY 9.5**

9.5 How Are Metabolic Pathways Interrelated and Regulated?

- The **catabolic pathways** for the breakdown of carbohydrates, fats, and proteins feed into the energy-harvesting metabolic pathways. Review Figure 9.14

- **Anabolic pathways** use intermediate components of the energy-harvesting pathways to synthesize fats, amino acids, and other essential building blocks.

- The formation of glucose from intermediates of glycolysis and the citric acid cycle is called **gluconeogenesis**.

- The rates of glycolysis and the citric acid cycle are controlled by **allosteric regulation** and by the diversion of excess acetyl CoA into fatty acid synthesis. Key regulated enzymes include phosphofructokinase, citrate synthase, isocitrate dehydrogenase, and fatty acid synthase. See Figure 9.16, **WEB ACTIVITY 9.6**

SELF-QUIZ

1. The role of oxygen gas in our cells is to
   a. catalyze reactions in glycolysis.
   b. produce CO_2.
   c. form ATP.
   d. accept electrons from the respiratory chain.
   e. react with glucose to split water.

2. Oxidation and reduction
   a. entail the gain or loss of proteins.
   b. are defined as the loss of electrons.
   c. are both endergonic reactions.
   d. always occur together.
   e. proceed only under aerobic conditions.

3. NAD^+ is
   a. a type of organelle.
   b. a protein.
   c. present only in mitochondria.
   d. a part of ATP.
   e. formed in the reaction that produces ethanol.

4. Glycolysis
   a. takes place in the mitochondrion.
   b. produces no ATP.
   c. has no connection with the respiratory chain.
   d. is the same thing as fermentation.
   e. reduces two molecules of NAD^+ for every glucose molecule processed.

5. Fermentation
   a. takes place in the mitochondrion.
   b. takes place in all animal cells.
   c. does not require O_2.
   d. requires lactic acid.
   e. prevents glycolysis.

6. Which statement about pyruvate is *not* true?
   a. It is the end product of glycolysis.
   b. It becomes reduced during fermentation.
   c. It is a precursor of acetyl CoA.
   d. It is a protein.
   e. It contains three carbon atoms.

7. The citric acid cycle
   a. has no connection with the respiratory chain.
   b. is the same thing as fermentation.
   c. reduces two NAD^+ for every glucose processed.
   d. produces no ATP.
   e. takes place in the mitochondrion.

8. The respiratory chain
   a. is located in the mitochondrial matrix.
   b. includes only peripheral membrane proteins.
   c. always produces ATP.
   d. reoxidizes reduced coenzymes.
   e. operates simultaneously with fermentation.

9. Compared with fermentation, the aerobic pathways of glucose metabolism produce
   a. more ATP.
   b. pyruvate.
   c. fewer protons for pumping in the mitochondria.
   d. less CO_2.
   e. more oxidized coenzymes.

10. Which statement about oxidative phosphorylation is *not* true?
    a. It forms ATP by the respiratory chain/ATP synthesis.
    b. It is brought about by chemiosmosis.
    c. It requires aerobic conditions.
    d. It takes place in mitochondria.
    e. Its functions can be served equally well by fermentation.

FOR DISCUSSION

1. Trace the sequence of chemical changes that occurs in mammalian tissue when the oxygen supply is cut off. The first change is that the cytochrome *c* oxidase system becomes totally reduced, because electrons can still flow from cytochrome *c*, but there is no oxygen to accept electrons from cytochrome *c* oxidase. What are the remaining steps?

2. Some cells that use the aerobic pathways of glucose metabolism can also thrive by using fermentation under anaerobic conditions. Given the lower yield of ATP (per molecule of glucose) in fermentation, how can these cells function so efficiently under anaerobic conditions?

3. The drug antimycin A blocks electron transport in mitochondria. Explain what would happen if the experiment on the left in Figure 9.10 were repeated in the presence of this drug.

4. You eat a burger that contains polysaccharides, proteins, and lipids. Using what you know of the integration of biochemical pathways, explain how the amino acids in the proteins and the glucose in the polysaccharides can end up as fats.

ADDITIONAL INVESTIGATION

A protein in the fat of newborns uncouples the synthesis of ATP from electron transport and instead generates heat. How would you investigate the hypothesis that this uncoupling protein adds a second proton channel to the mitochondrial membrane?

WORKING WITH DATA (GO TO *yourBioPortal.com*)

Two Experiments Demonstrate the Chemiosmotic Mechanism
In this real-life exercise, you will examine the background and data from the original research paper by Jagendorf and Uribe in which they showed that an artificially induced H^+ gradient could drive ATP synthesis (Figure 9.10). You will see how they measured ATP by two different methods, and what control experiments they performed to confirm their interpretation.

Photosynthesis: Energy from Sunlight

Photosynthesis and global climate change

If all the carbohydrates produced by photosynthesis in a year were in the form of sugar cubes, there would be 300 quadrillion of them. Lined up, these cubes would extend from Earth to Pluto—a lot of photosynthesis! As you may have learned from previous courses, photosynthetic organisms use atmospheric carbon dioxide (CO_2) to produce carbohydrates. The simplified equation says it all:

$$CO_2 + H_2O \rightarrow O_2 + \text{carbohydrates}$$

Given the role of CO_2, how will photosynthesis change with increasing levels of atmospheric CO_2? Over the past 200 years, the concentration of atmospheric CO_2 has increased—from 280 parts per million (ppm) in 1800 to 386 ppm in 2008. This increase is correlated with industrialization and the accompanying use of fossil fuels such as coal and oil, which release CO_2 into the atmosphere when they are burned. The Intergovernmental Panel of Climate Change, sponsored by the United Nations, estimates that atmospheric CO_2 will continue to rise over the next century.

Carbon dioxide is a "greenhouse gas" that traps heat in the atmosphere, and the rising CO_2 level is predicted to result in global climate change. Policy makers concerned about climate change are asking plant biologists to answer two questions about the rise in CO_2: will it lead to increased photosynthesis, and if so, will it lead to increased plant growth? To answer these questions, scientists initially measured the rate of photosynthesis of plants grown in greenhouses with elevated concentrations of CO_2. The results were surprising: at first, the rate of photosynthesis went up, but then it returned to near normal as the plants adapted to the higher CO_2 levels.

To determine how plants might respond under more realistic conditions, scientists developed a way to expose plants to high levels of CO_2 in the field. *Free-air concentration enrichment* (FACE) involves the use of rings of pipes that release CO_2 to the air surrounding plants in fields or forests. Wind speed and direction are monitored by a computer, which constantly controls which pipes release CO_2. Data from these experiments confirm that photosynthetic rates increase as the concentration of CO_2 rises—although generally the increase is not as high as that seen initially in the greenhouse experiments. Nevertheless, these measurements indicate that as atmospheric CO_2 rises globally, there will be an increase in photosynthesis.

Will this increase in photosynthesis result in an increase in plant growth? Keep in mind that plants, like all organisms, use carbohydrates as an energy source. They perform cellular respiration with the general equation:

$$\text{carbohydrates} + O_2 \rightarrow CO_2 + H_2O$$

Primary Producers Covering less than 2 percent of Earth's surface, rainforests are photosynthetic dynamos. They may act as a "sink" for increasing atmospheric CO_2.

FACE Free-air carbon dioxide enrichment uses pipes to release CO_2 around plants in the field, to estimate the effects of rising atmospheric CO_2 on photosynthesis and plant growth.

The challenge facing plant biologists is to determine the balance between photosynthesis and respiration and how this affects the rate of plant growth. The FACE experiments indicate that crop yields increase under higher CO_2 concentrations, suggesting that the overall increase in photosynthesis is greater than the increase in respiration. But climate change alters rainfall patterns as well as temperatures. These changes affect where plants grow, and could shift the balance between plant growth and cellular respiration.

As with much in science, the initial questions at first appeared amenable to simple answers. Instead, they led to more questions, and more data are needed. An understanding of the processes of photosynthesis, described in this chapter, provides us with a foundation for asking and answering these urgent questions about climate change and its effects on our world.

IN THIS CHAPTER we begin with a consideration of light energy, and move on to describe how photosynthesis converts light energy into chemical energy, in the form of reduced electron carriers and ATP. Then, we show how these two sources of chemical energy are used to drive the synthesis of carbohydrates from CO_2. Finally, we describe how these processes relate to plant metabolism and growth.

10.1 What Is Photosynthesis?

Photosynthesis (literally, "synthesis from light") is a metabolic process by which the energy of sunlight is captured and used to convert carbon dioxide (CO_2) and water (H_2O) into carbohydrates (which we will represent as a six-carbon sugar, $C_6H_{12}O_6$) and oxygen gas (O_2). By early in the nineteenth century, scientists had grasped these broad outlines of photosynthesis and had established several facts about the way the process works:

- The water for photosynthesis in land plants comes primarily from the soil, and must travel from the roots to the leaves.
- Plants take in carbon dioxide, producing carbohydrates (sugars) for growth, and plants release O_2 (**Figure 10.1**).
- Light is absolutely necessary for the production of oxygen and sugars.

By 1804, scientists had summarized photosynthesis as follows:

$$\text{carbon dioxide} + \text{water} + \text{light energy} \rightarrow \text{sugar} + \text{oxygen}$$

In molecular terms, this equation seems to be the reverse of the overall equation for cellular respiration (see Section 9.1). More precisely, photosynthesis can be written as:

$$6\,CO_2 + 6\,H_2O \rightarrow C_6H_{12}O_6 + 6\,O_2$$

While this equation and the one for cellular respiration (given in the chapter opening story) are essentially correct, they are too general for a real understanding of the processes involved. A number of questions arise: What are the precise chemical reactions of photosynthesis? What role does light play in these reactions? How do carbons become linked to form carbohydrates? What carbohydrates are formed? And where does the oxygen gas come from: CO_2 or H_2O?

Experiments with isotopes show that in photosynthesis O_2 comes from H_2O

In 1941 Samuel Ruben and Martin Kamen, at the University of California, Berkeley, performed experiments using the isotopes ^{18}O and ^{16}O to identify the source of the O_2 produced during photosynthesis (**Figure 10.2**). Their results showed that all the oxygen gas produced during photosynthesis comes from water, as is reflected in the revised balanced equation:

$$6\,CO_2 + 12\,H_2O \rightarrow C_6H_{12}O_6 + 6\,O_2 + 6\,H_2O$$

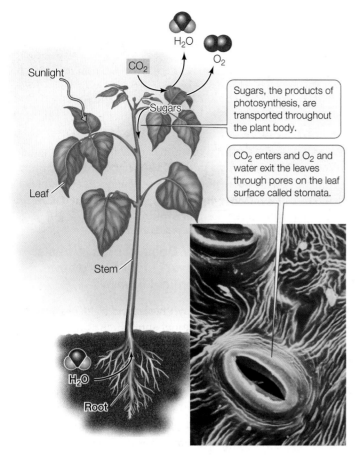

Sunlight

H_2O

CO_2

O_2

Sugars

Sugars, the products of photosynthesis, are transported throughout the plant body.

CO_2 enters and O_2 and water exit the leaves through pores on the leaf surface called stomata.

Leaf

Stem

H_2O

Root

10.1 The Ingredients for Photosynthesis A typical terrestrial plant uses light from the sun, water from the soil, and carbon dioxide from the atmosphere to form organic compounds by photosynthesis.

Water appears on both sides of the equation because it is both used as a reactant (the twelve molecules on the left) and released as a product (the six new ones on the right). This revised equation accounts for all the water molecules needed for all the oxygen gas produced.

The realization that water was the source of photosynthetic O_2 led to an understanding of photosynthesis in terms of *oxidation and reduction*. As we describe in Chapter 9, oxidation–reduction (redox) reactions are coupled: when one molecule becomes oxidized in a reaction, another gets reduced. In this case, oxygen atoms in the reduced state in H_2O get oxidized to O_2:

$$12\ H_2O \rightarrow 24\ H^+ + 24\ e^- + 6\ O_2$$

while carbon atoms in the oxidized state in CO_2 get reduced to carbohydrate, with the simultaneous production of water:

$$6\ CO_2 + 24\ H^+ + 24\ e^- \rightarrow C_6H_{12}O_6 + 6\ H_2O$$

Adding these two equations (chemistry students will recognize them as *half-cell reactions*) gives the overall equation shown above. As you will see, there is an intermediary carrier of the H^+ and electrons between these two processes—the redox coenzyme, nicotinamide adenine dinucleotide phosphate (NADP$^+$).

Photosynthesis involves two pathways

The equations above summarize the overall process of photosynthesis, but not the stages by which it is completed. Like gly-

colysis and the other metabolic pathways that harvest energy in cells, photosynthesis is a process consisting of many reactions. These reactions are commonly divided into two main pathways:

- The **light reactions** convert light energy into chemical energy in the form of ATP and the reduced electron carrier NADPH. This molecule is similar to NADH (see Section 9.1) but with an additional phosphate group attached to the sugar of its adenosine. In general, NADPH acts as a reducing agent in photosynthesis and other anabolic reactions.

- The **light-independent reactions** (carbon-fixation reactions) do not use light directly, but instead use ATP, NADPH (*made by the light reactions*), and CO_2 to produce carbohydrate.

INVESTIGATING LIFE

10.2 The Source of the Oxygen Produced by Photosynthesis

Although it was clear that O_2 was made during photosynthesis, its molecular source was not known. Two possibilities were the reactants, CO_2 and H_2O. In two separate experiments, Samuel Ruben and Martin Kamen labeled the oxygen in these molecules with the isotope ^{18}O, then tested the O_2 produced by a green plant to find out which molecule contributed the oxygen.

HYPOTHESIS The oxygen released by photosynthesis comes from water rather than CO_2.

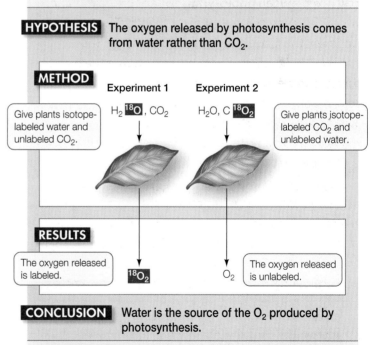

METHOD

Experiment 1

Experiment 2

$H_2\ ^{18}O,\ CO_2$

$H_2O,\ C\ ^{18}O_2$

Give plants isotope-labeled water and unlabeled CO_2.

Give plants isotope-labeled CO_2 and unlabeled water.

RESULTS

The oxygen released is labeled.

$^{18}O_2$

O_2

The oxygen released is unlabeled.

CONCLUSION Water is the source of the O_2 produced by photosynthesis.

FURTHER INVESTIGATION: How would you test for the source of oxygen atoms in the carbohydrates made by photosynthesis?

Go to **yourBioPortal.com** for original citations, discussions, and relevant links for all INVESTIGATING LIFE figures.

── **yourBioPortal.com** ──
GO TO Animated Tutorial 10.1 • The Source of the Oxygen Produced by Photosynthesis

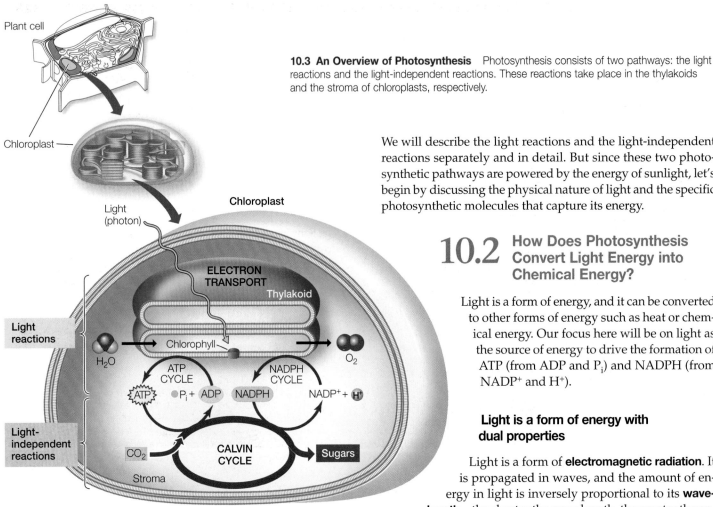

10.3 An Overview of Photosynthesis Photosynthesis consists of two pathways: the light reactions and the light-independent reactions. These reactions take place in the thylakoids and the stroma of chloroplasts, respectively.

We will describe the light reactions and the light-independent reactions separately and in detail. But since these two photosynthetic pathways are powered by the energy of sunlight, let's begin by discussing the physical nature of light and the specific photosynthetic molecules that capture its energy.

10.2 How Does Photosynthesis Convert Light Energy into Chemical Energy?

Light is a form of energy, and it can be converted to other forms of energy such as heat or chemical energy. Our focus here will be on light as the source of energy to drive the formation of ATP (from ADP and P_i) and NADPH (from $NADP^+$ and H^+).

Light is a form of energy with dual properties

Light is a form of **electromagnetic radiation**. It is propagated in waves, and the amount of energy in light is inversely proportional to its **wavelength**—the shorter the wavelength, the greater the energy. The visible portion of the electromagnetic spectrum (**Figure 10.4**) encompasses a wide range of wavelengths and energy levels. In addition to traveling in waves, light also behaves as particles, called **photons**, which have no mass. In plants and other photosynthetic organisms, receptive molecules absorb photons in order to harvest their energy for biological processes. Because these receptive molecules absorb only specific wavelengths of light, the photons must have the correct amount of energy—they must be of the appropriate wavelength.

Molecules become excited when they absorb photons

When a photon meets a molecule, one of three things can happen:

- The photon may bounce off the molecule—it may be scattered or reflected.

- The photon may pass through the molecule—it may be transmitted.

- The photon may be absorbed by the molecule, adding energy to the molecule.

Neither of the first two outcomes causes any change in the molecule. However, in the case of absorption, the photon disappears and its energy is absorbed by the molecule. The photon's energy cannot disappear, because according to the first law of thermodynamics, energy is neither created nor destroyed. When the molecule acquires the energy of the photon it is raised from a ground state (with lower energy) to an excited state (with higher energy) (**Figure 10.5A**).

The light-independent reactions are sometimes called the *dark reactions* because they do not directly require light energy. They are also called the *carbon-fixation reactions*. However, both the light reactions and the light-independent reactions stop in the dark because ATP synthesis and $NADP^+$ reduction require light. The reactions of both pathways proceed within the chloroplast, but they occur in different parts of that organelle (**Figure 10.3**).

As we describe these two series of reactions in more detail, you will see that they conform to the principles of biochemistry that we discuss in Chapters 8 and 9: energy transformations, oxidation-reduction, and the stepwise nature of biochemical pathways.

10.1 RECAP

The light reactions of photosynthesis convert light energy into chemical energy. The light-independent reactions use that chemical energy to reduce CO_2 to carbohydrates.

- What is the experimental evidence that water is the source of the O_2 produced during photosynthesis? **See pp. 190–191 and Figure 10.2**

- What is the relationship between the light reactions and the light-independent reactions of photosynthesis? **See pp. 191–192 and Figure 10.3**

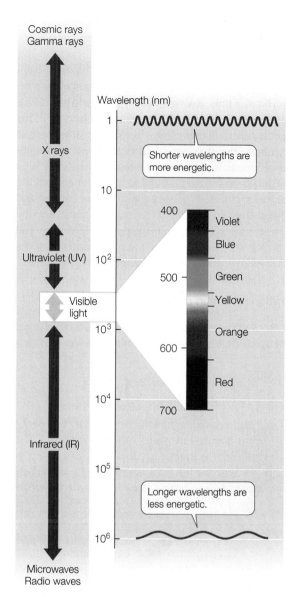

10.4 The Electromagnetic Spectrum The portion of the electromagnetic spectrum that is visible to humans as light is shown in detail at the right.

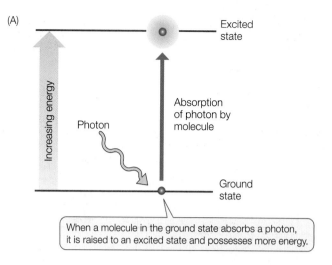

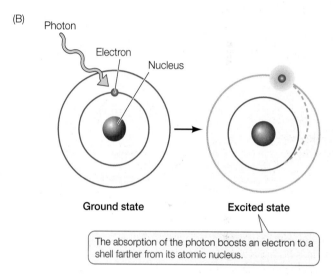

10.5 Exciting a Molecule (A) When a molecule absorbs the energy of a photon, it is raised from a ground state to an excited state. (B) In the excited state, an electron is boosted to a shell more distant from the atomic nucleus, where it is held less firmly.

The difference in free energy between the molecule's excited state and its ground state is approximately equal to the free energy of the absorbed photon (a small amount of energy is lost to entropy). The increase in energy boosts one of the electrons within the molecule into a shell farther from its nucleus; this electron is now held less firmly (**Figure 10.5B**), making the molecule unstable and more chemically reactive.

Absorbed wavelengths correlate with biological activity

The specific wavelengths absorbed by a particular molecule are characteristic of that type of molecule. Molecules that absorb wavelengths in the visible spectrum are called **pigments**.

When a beam of white light (containing all the wavelengths of visible light) falls on a pigment, certain wavelengths are absorbed. The remaining wavelengths, which are scattered or trans-mitted, make the pigment appear to us as colored. For example, if a pigment absorbs both blue and red light (as does chlorophyll) what we see is the remaining light, which is primarily green. If we plot light absorbed by a purified pigment against wavelength, the result is an **absorption spectrum** for that pigment.

In contrast to the absorption spectrum, an **action spectrum** is a plot of the *biological activity* of an organism as a function of the wavelengths of light to which it is exposed. The experimental determination of an action spectrum might be performed as follows:

1. Place a plant (a water plant with thin leaves is convenient) in a closed container.

2. Expose the plant to light of a certain wavelength for a period of time.

3. Measure photosynthesis by the amount of O_2 released.

4. Repeat with light of other wavelengths.

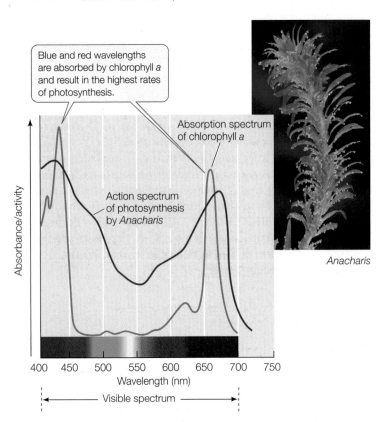

> Blue and red wavelengths are absorbed by chlorophyll *a* and result in the highest rates of photosynthesis.

Absorption spectrum of chlorophyll *a*

Action spectrum of photosynthesis by *Anacharis*

Anacharis

10.6 Absorption and Action Spectra The absorption spectrum of the purified pigment chlorophyll *a* from the aquatic plant *Anacharis* is similar to the action spectrum obtained when different wavelengths of light are shone on the intact plant and the rate of photosynthesis is measured. In the thicker leaves of land plants, the action spectra show less of a dip in the green region (500–650 nm).

Figure 10.6 shows the absorption spectrum of the pigment chlorophyll *a*, which was isolated from the leaves of *Anacharis*, a common aquarium plant. Also shown is the action spectrum for photosynthetic activity by the same plant. A comparison of the two spectra shows that the wavelengths at which photosynthesis is highest are the same wavelengths at which chlorophyll *a* absorbs light.

Several pigments absorb energy for photosynthesis

The light energy used for photosynthesis is not absorbed by just one type of pigment. Instead, several different pigments with different absorption spectra collect the energy that is eventually used for photosynthesis. In photosynthetic organisms as diverse as green algae, protists, and bacteria, these pigments include chlorophylls, carotenoids, and phycobilins.

CHLOROPHYLLS In plants, two chlorophylls are responsible for absorbing the light energy that is used to drive the light reactions: **chlorophyll *a*** and **chlorophyll *b***. These two molecules differ only slightly in their molecular structures. Both have a complex ring structure similar to that of the heme group of hemoglobin (**Figure 10.7**). In the center of the chlorophyll ring is a magnesium atom. Attached at a peripheral location on the ring is a long hydrocarbon "tail," which anchors the chlorophyll molecule to in-

tegral proteins in the thylakoid membrane of the chloroplast. (See Figure 5.13 to review the anatomy of a chloroplast.)

ACCESSORY PIGMENTS We saw in Figure 10.6 that chlorophyll absorbs blue and red light, which are near the two ends of the visible spectrum. Thus, if only chlorophyll were active in photosynthesis, much of the visible spectrum would go unused. This appears to be the case in higher plants. But lower plants (such as algae) and cyanobacteria possess accessory pigments, which absorb photons intermediate in energy between the red and the blue wavelengths and then transfer a portion of that energy to the chlorophylls. Among these accessory pigments are **carotenoids** such as β-carotene (see Figure 3.21), which absorb photons in the blue and blue-green wavelengths and appear deep yellow. The **phycobilins**, which are found in red algae and in cyanobacteria, absorb various yellow-green, yellow, and orange wavelengths.

Light absorption results in photochemical change

Any pigment molecule can become excited when its absorption spectrum matches the energies of incoming photons. After a

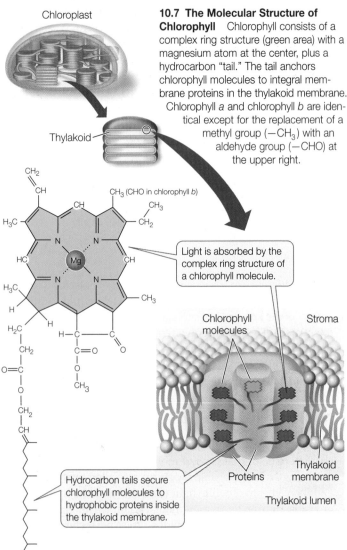

Chloroplast

Thylakoid

10.7 The Molecular Structure of Chlorophyll Chlorophyll consists of a complex ring structure (green area) with a magnesium atom at the center, plus a hydrocarbon "tail." The tail anchors chlorophyll molecules to integral membrane proteins in the thylakoid membrane. Chlorophyll *a* and chlorophyll *b* are identical except for the replacement of a methyl group ($-CH_3$) with an aldehyde group ($-CHO$) at the upper right.

> Light is absorbed by the complex ring structure of a chlorophyll molecule.

Chlorophyll molecules

Stroma

Proteins

Thylakoid membrane

Thylakoid lumen

> Hydrocarbon tails secure chlorophyll molecules to hydrophobic proteins inside the thylakoid membrane.

pigment molecule absorbs a photon and enters an excited state (see Figure 10.5), there are several alternative fates for the absorbed energy:

- It can be released as heat and/or light.
- It may be rapidly transferred to a neighboring pigment molecule.
- It can be used as free energy to drive a chemical reaction.

When the excited molecule gives up the absorbed energy it returns to the ground state.

Sometimes the absorbed energy is given off as heat and light, in a process called *fluorescence*. Because some of the energy of the original absorbed photon is lost as heat, the photon that is released as fluorescence has less energy and a longer wavelength than the absorbed light. When there is fluorescence, there are no permanent chemical changes made or biological functions performed—no chemical work is done.

On the other hand, the excited pigment molecule may pass the absorbed energy along to another molecule—provided that the target molecule is very near, has the right orientation, and has the appropriate structure to receive the energy. This is what happens in photosynthesis.

The pigments in photosynthetic organisms are arranged into energy-absorbing **antenna systems**, also called *light-harvesting complexes*. These form part of a large multi-protein complex called a **photosystem**. The photosystem spans the thylakoid membrane, and consists of multiple antenna systems, with their associated pigment molecules, all surrounding a **reaction center**. The pigment molecules in the antenna systems are packed together in such a way that the excitation energy from an absorbed photon can be passed along from one pigment molecule to another (**Figure 10.8**). Excitation energy moves from pigments that absorb shorter wavelengths (higher energy) to pigments that absorb

longer wavelengths (lower energy). Thus the excitation ends up in the pigment molecules that absorb the longest wavelengths. These pigment molecules are in the reaction center of the photosystem, and form special associations with the photosystem proteins (see Figure 10.8). The ratio of antenna pigments to reaction center pigments can be quite high (over 300:1).

The reaction center converts the absorbed light energy into chemical energy. A pigment molecule in the reaction center absorbs sufficient energy that it actually gives up its excited electron (is chemically oxidized) and becomes positively charged. In plants, the reaction center contains a pair of chlorophyll *a* molecules. There are many other chlorophyll *a* molecules in the antenna systems, but because of their interactions with antenna proteins, all of them absorb light at shorter wavelengths than the pair in the reaction center.

Excited chlorophylls in the reaction center act as electron donors

Chlorophyll has two vital roles in photosynthesis:

- It absorbs light energy and transforms it into excited electrons.
- It transfers those electrons to other molecules, initiating chemical changes.

We have dealt with the first role; now we turn to the second.

Photosynthesis harvests chemical energy by using the excited chlorophyll molecules in the reaction center as electron donors (reducing agents) to reduce a stable electron acceptor (see Figure 10.8). Ground-state chlorophyll (symbolized by Chl) is not much of a reducing agent, but excited chlorophyll (Chl*) is a good one. This is because in the excited molecule, one of the electrons has moved to a shell that is farther away from the nucleus than the shell it normally occupies. This electron is held less tightly than in the normal state, and it can be transferred in a redox reaction to an electron acceptor (an oxidizing agent):

$$\text{Chl* + acceptor} \rightarrow \text{Chl}^+ + \text{acceptor}^-$$

This, then, is the first consequence of light absorption by chlorophyll: *a reaction center chlorophyll (Chl*) loses its excited electron in a redox reaction and becomes Chl*+* (because it gives up a negative charge—it gets oxidized).

Reduction leads to electron transport

The electron acceptor that is reduced by Chl* is the first in a chain of electron carriers in the thylakoid membrane that participate in a process termed *electron transport*. This energetically "downhill" series of reductions and oxidations is similar to what occurs in the respiratory chain of

The energized electron from the chlorophyll molecules can be passed on to an electron acceptor to reduce it.

Excited state

Electron acceptor

Photon

Reaction center

Chlorophyll molecule

Hydrocarbon tail

Proteins

Light energy is absorbed by antenna chlorophylls and passed on to the reaction center.

Photosystem embedded in thylakoid membrane

10.8 Energy Transfer and Electron Transport Rather than being lost as fluorescence, energy from a photon may be transferred from one pigment molecule to another. In a photosystem, energy is transferred through a series of molecules to one or more pigment molecules in the reaction center. If a reaction center molecule becomes sufficiently excited, it will give up its excited electron to an electron acceptor.

mitochondria (see Section 9.3). The final electron acceptor is $NADP^+$ (nicotinamide adenine dinucleotide phosphate), which gets reduced:

$$NADP^+ + H^+ + 2\,e^- \rightarrow NADPH$$

The energy-rich NADPH is a stable, reduced coenzyme.

There are two different systems of electron transport in photosynthesis:

- **Noncyclic electron transport** produces NADPH and ATP. Essentially, the excited electron is "lost" from chlorophyll and the transport process ends up with a reduced coenzyme.
- **Cyclic electron transport** produces only ATP. Essentially, the transport process ends up with the excited electron returning to chlorophyll, after giving up energy to make ATP.

We'll consider these two systems before describing the production of ATP from ADP and P_i.

Noncyclic electron transport produces ATP and NADPH

In noncyclic electron transport, light energy is used to oxidize water, forming O_2, H^+, and electrons. In quantitative terms this would be

$$H_2O \rightarrow 2\,H^+ + 1/2\,O_2 + 2\,e^-$$

We saw above that a key reaction in photosynthesis occurs when chlorophyll that is excited by absorbing light (Chl*) gives up its excited electron, becoming oxidized:

$$Chl^* \rightarrow Chl^+ + e^-$$

Because it lacks an electron, Chl^+ is very unstable; it has a very strong tendency to "grab" an electron from another molecule to replenish the one it lost.

$$Chl^+ + e^- \rightarrow Chl$$

So in chemical terms, Chl^+ is a strong oxidizing agent. The replenishing electrons come from water, splitting the H–O–H bonds.

$$H_2O \rightarrow 1/2\,O_2 + 2\,H^+ + 2\,e^-$$
$$2\,e^- + 2\,Chl^+ \rightarrow 2\,Chl$$
$$\text{Overall: } 2\,Chl^* + H_2O \rightarrow 2\,Chl + 2\,H^+ + 1/2\,O_2$$

Notice that this is a more precise description of what Ruben and Kamen had found, namely that the source of O_2 in photosynthesis is H_2O (see Figure 10.2).

The electrons are passed from chlorophyll to $NADP^+$ through a chain of electron carriers in the thylakoid membrane. These redox reactions are exergonic, and some of the released free energy is ultimately used to form ATP by *chemiosmosis* (see p. 180).

TWO PHOTOSYSTEMS ARE REQUIRED Noncyclic electron transport requires the participation of two different photosystems in the thylakoid membrane. What is the evidence of the existence of these two photosystems? In 1957, Robert Emerson at the University of Illinois shone light of various wavelengths onto cells of *Chlorella*, a freshwater protist. Both red light (wavelength 680 nm) and far-red light (700 nm) resulted in modest rates of photosynthesis, as measured by O_2 production. But when the two lights

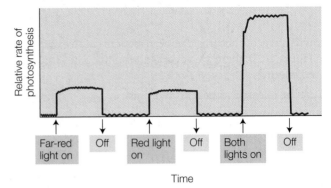

10.9 Two Photosystems The absorption and action spectra for chlorophyll and photosynthesis indicated that the rate of photosynthesis would increase in red light. Robert Emerson shone red (660 nm) and far-red (>700 nm) light both separately and together on algal cells to look for cooperative effects.

were combined, the rate of photosynthesis was much greater than the rates under either red light or far red light. In fact it was greater than the two rates added together. This phenomenon was termed photo enhancement (**Figure 10.9**). A few years later, photo enhancement was explained by the existence of *not one but two reaction centers*, which act together to enhance photosynthesis.

- **Photosystem I** uses light energy to pass an excited electron to $NADP^+$, reducing it to NADPH.
- **Photosystem II** uses light energy to oxidize water molecules, producing electrons, protons (H^+), and O_2.

The reaction center for photosystem I contains a pair of chlorophyll *a* molecules called P_{700} because it can best absorb light with a wavelength of 700 nm. Similarly, the pair of chlorophyll *a* molecules in the photosystem II reaction center is called P_{680} because it absorbs light maximally at 680 nm. Thus photosystem II requires photons that are somewhat more energetic (i.e., have shorter wavelengths) than those required by photosystem I. To keep noncyclic electron transport going, both photosystems must be constantly absorbing light, thereby boosting electrons to higher shells from which they may be captured by specific electron acceptors. A model for the way photosystems I and II interact and complement each other is called the "Z scheme," because when the path of the electrons is placed along an axis of rising energy level, it resembles a sideways letter Z (**Figure 10.10**).

ELECTRON TRANSPORT: THE Z SCHEME In the Z scheme model, which describes the reactions of noncyclic electron transport from water to $NADP^+$, photosystem II comes before photosystem I. When photosystem II absorbs photons, electrons pass from P_{680} to the primary electron acceptor and P_{680}^* is oxidized to P_{680}^+. Then an electron from the oxidation of water is passed to P_{680}^+, reducing it to P_{680} once again, so that it can receive more energy from neighboring chlorophyll molecules in the antenna systems. The electrons from photosystem II pass through a series of transfer reactions, one of which is directly responsible for the physical movement of protons from the stroma (the matrix outside the thylakoids) across the thylakoid membrane and into the lumen (see Figure 10.12). In addition to these protons, the protons derived from the splitting of water are deposited into the thylakoid lumen. Furthermore, protons in the stroma are

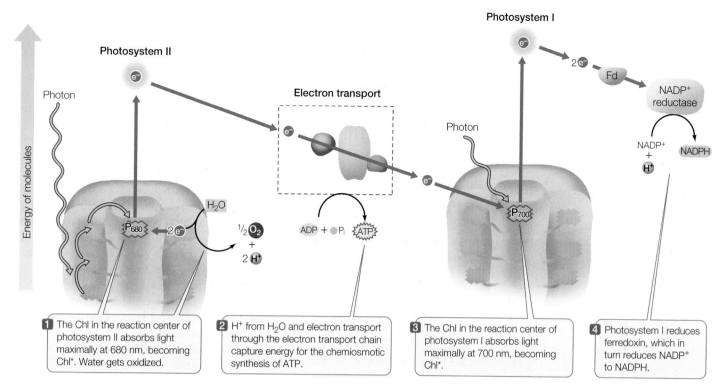

1 The Chl in the reaction center of photosystem II absorbs light maximally at 680 nm, becoming Chl*. Water gets oxidized.

2 H^+ from H_2O and electron transport through the electron transport chain capture energy for the chemiosmotic synthesis of ATP.

3 The Chl in the reaction center of photosystem I absorbs light maximally at 700 nm, becoming Chl*.

4 Photosystem I reduces ferredoxin, which in turn reduces $NADP^+$ to NADPH.

10.10 Noncyclic Electron Transport Uses Two Photosystems
Absorption of light energy by chlorophyll molecules in the reaction centers of photosystems I and II allows them to pass electrons into a series of redox reactions. The term "Z scheme" describes the path (blue arrows) of electrons as they travel through the two photosystems. On this scheme the vertical positions represent the energy levels of the molecules in the electron transport system.

consumed during the reduction of $NADP^+$, and together these reactions create a proton gradient across the thylakoid membrane, which provides the energy for ATP synthesis.

In photosystem I, the P_{700} molecules in the reaction center become excited to P_{700}^*, leading to the reduction of an electron carrier called ferredoxin (Fd) and the production of P_{700}^+. P_{700}^+ returns to the reduced state by accepting electrons passed through the electron transport system from photosystem II. Having identified the role of the electrons produced by photosystem II, we can now ask, "What is the role of the electrons transferred to Fd from photosystem I?" These electrons are used in the last step of noncyclic electron transport, in which two electrons and a proton are used to reduce a molecule of $NADP^+$ to NADPH.

In summary:

- Noncyclic electron transport extracts electrons from water and passes them ultimately to NADPH, utilizing energy absorbed by photosystems I and II, and resulting in ATP synthesis.

- Noncyclic electron transport yields NADPH, ATP, and O_2.

Cyclic electron transport produces ATP but no NADPH

Noncyclic electron transport results in the production of ATP and NADPH. However, as we will see, the light-independent reactions of photosynthesis require more ATP than NADPH + H^+. If only the noncyclic pathway is operating, there is the possibility that there will not be enough ATP formed. **Cyclic electron transport** makes up for the imbalance. This pathway, which produces only ATP, is called *cyclic* because an electron passed from an excited chlorophyll molecule at the outset cycles back to the same chlorophyll molecule at the end of the chain of reactions (**Figure 10.11**).

Cyclic electron transport begins and ends in photosystem I. A P_{700} chlorophyll molecule in the reaction center absorbs a photon and enters the excited state, P_{700}^*. The excited electron is passed from P_{700}^* to a primary acceptor, and then to oxidized ferredoxin (Fd_{ox}), producing reduced ferredoxin (Fd_{red}). Fd_{red} passes its added electron to a different oxidizing agent, plastoquinone (PQ, a small organic molecule), resulting in the transfer of two H^+ from the stroma to the thylakoid lumen. The electron passes from reduced PQ through the electron transport system until it completes its cycle by returning to P_{700}^+, restoring it to its uncharged form, P_{700}. This electron transport is carried out by plastocyanin (PC) and cytochromes that are similar to those of the mitochondrial respiratory chain.

By the time the electron from P_{700}^* travels through the electron transport system and comes back to reduce P_{700}^+, all the energy from the original photon has been released. The released energy is stored in the form of a proton gradient that can be used to produce ATP.

Chemiosmosis is the source of the ATP produced in photophosphorylation

In Chapter 9 we describe the chemiosmotic mechanism for ATP formation in the mitochondrion. A similar mechanism, called **photophosphorylation**, operates in the chloroplast, where electron transport is coupled to the transport of protons (H^+) across

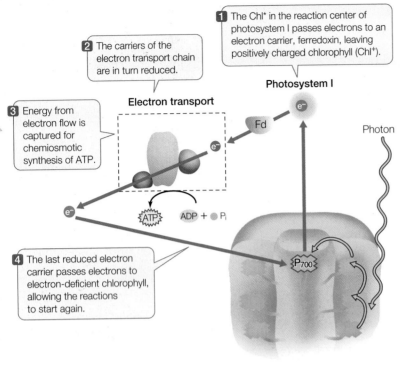

1 The Chl* in the reaction center of photosystem I passes electrons to an electron carrier, ferredoxin, leaving positively charged chlorophyll (Chl⁺).

2 The carriers of the electron transport chain are in turn reduced.

3 Energy from electron flow is captured for chemiosmotic synthesis of ATP.

4 The last reduced electron carrier passes electrons to electron-deficient chlorophyll, allowing the reactions to start again.

Photosystem I

Electron transport

Fd

Photon

Energy of molecules

P₇₀₀

ATP   ADP + Pᵢ

10.11 Cyclic Electron Transport Traps Light Energy as ATP Cyclic electron transport produces ATP, but no NADPH.

the thylakoid membrane, resulting in a proton gradient across the membrane (**Figure 10.12**).

The electron carriers in the thylakoid membrane are oriented so that protons are actively pumped from the stroma into the lumen of the thylakoid. Thus the lumen becomes acidic with respect to the stroma, resulting in an electrochemical gradient across the thylakoid membrane, whose bilayer is not permeable to H^+. Water oxidation and $NADP^+$ reduction also contribute to this gradient, which drives the movement of protons back out of the thylakoid lumen through specific protein channels in the thylakoid membrane. These channels are enzymes—ATP synthases—that couple the movement of protons to the formation of ATP, as they do in mitochondria (see Figure 9.9). Indeed, chloroplast ATP synthase is about 60 percent identical to human mitochondrial ATP synthase—a remarkable similarity, given that plants and animals had their most recent

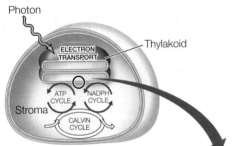

Photon

Thylakoid

ELECTRON TRANSPORT

ATP CYCLE   NADPH CYCLE

Stroma

CALVIN CYCLE

10.12 Chloroplasts Form ATP Chemiosmotically Compare this illustration with Figure 9.9, where a similar process is depicted in mitochondria.

yourBioPortal.com
GO TO Animated Tutorial 10.2 • Photophosphorylation

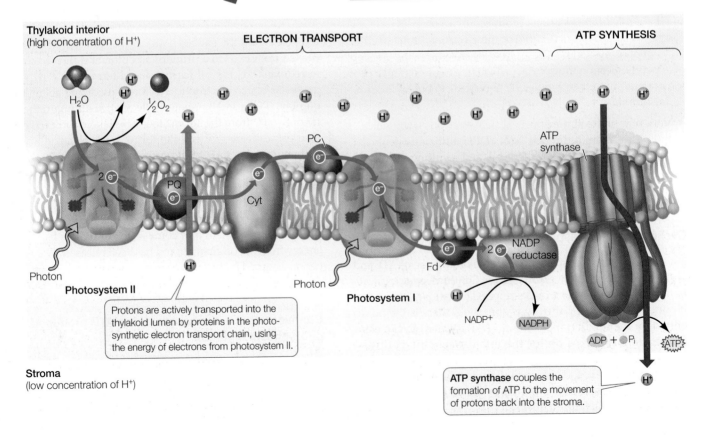

Thylakoid interior
(high concentration of H⁺)

ELECTRON TRANSPORT

ATP SYNTHESIS

H_2O H⁺   H⁺   ½ O_2 H⁺   H⁺   H⁺   H⁺   H⁺   H⁺   H⁺   H⁺   H⁺   H⁺   H⁺   H⁺   H⁺

PC

ATP synthase

2 e⁻   PQ   e⁻   Cyt   e⁻   e⁻

e⁻

e⁻   2 e⁻   NADP reductase

Fd

Photon   Photon

Photosystem II   **Photosystem I**

H⁺

H⁺

NADP⁺   NADPH

ADP + Pᵢ   ATP

Protons are actively transported into the thylakoid lumen by proteins in the photosynthetic electron transport chain, using the energy of electrons from photosystem II.

Stroma
(low concentration of H⁺)

ATP synthase couples the formation of ATP to the movement of protons back into the stroma.

H⁺

common ancestor more than a billion years ago. This is testimony to the evolutionary unity of life.

The mechanisms of the two enzymes are similar, but their orientations differ. In chloroplasts, protons flow through the ATP synthase out of the thylakoid lumen into the stroma (where the ATP is synthesized) but in mitochondria they flow out of the cytosol into the mitochondrial matrix.

10.2 RECAP

Conversion of light energy into chemical energy occurs when pigments absorb photons. Light energy is used to drive a series of protein-associated redox reactions in the thylakoid membranes of the chloroplast.

- How does chlorophyll absorb and transfer light energy? **See pp. 194–195 and Figure 10.8**
- How are electrons produced in photosystem II and how do they flow to photosystem I? **See pp. 196–197 and Figure 10.10**
- How does cyclic electron transport in photosystem I result in the production of ATP? **See p. 197 and Figure 10.11**

We have seen how light energy drives the synthesis of ATP and NADPH in the stroma of chloroplasts. We now turn to the light-independent reactions of photosynthesis, which use energy-rich ATP and NADPH to reduce CO_2 and form carbohydrates.

10.3 How Is Chemical Energy Used to Synthesize Carbohydrates?

Most of the enzymes that catalyze the reactions of CO_2 fixation are dissolved in the stroma of the chloroplast, where those reactions take place. These enzymes use the energy in ATP and NADPH to reduce CO_2 to carbohydrates. Therefore, with some exceptions, CO_2 fixation occurs only in the light, when ATP and NADPH are being generated.

Radioisotope labeling experiments revealed the steps of the Calvin cycle

To identify the reactions by which the carbon from CO_2 ends up in carbohydrates, scientists found a way to label CO_2 so that they could isolate and identify the compounds formed from it during photosynthesis. In the 1950s, Melvin Calvin, Andrew Benson, and their colleagues used radioactively labeled CO_2 in which some of the carbon atoms were the radioisotope ^{14}C rather than the normal ^{12}C. Although ^{14}C emits radiation, its chemical behavior is virtually identical to that of nonradioactive ^{12}C.

Calvin and his colleagues exposed cultures of the unicellular green alga *Chlorella* to $^{14}CO_2$ for various lengths of time. Then they rapidly killed the cells and extracted the organic compounds. They separated the different compounds from one another by paper chromatography and exposed the paper to X-ray film (**Figure 10.13**). When the film was developed, dark

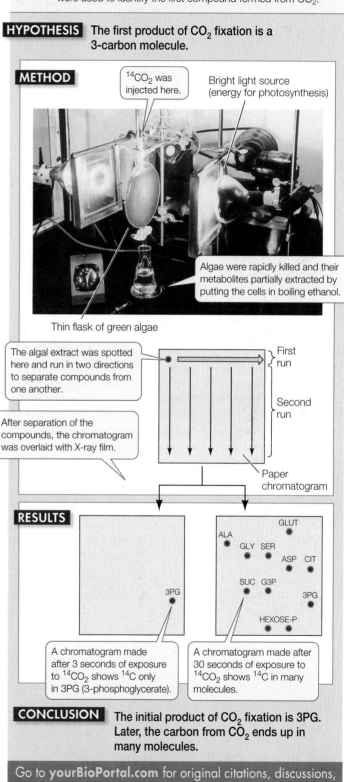

INVESTIGATING LIFE

10.13 Tracing the Pathway of CO_2

How is CO_2 incorporated into carbohydrate during photosynthesis? What is the first stable covalent linkage that forms with the carbon of CO_2? Short exposures to $^{14}CO_2$ were used to identify the first compound formed from CO_2.

HYPOTHESIS The first product of CO_2 fixation is a 3-carbon molecule.

METHOD

$^{14}CO_2$ was injected here.

Bright light source (energy for photosynthesis)

Algae were rapidly killed and their metabolites partially extracted by putting the cells in boiling ethanol.

Thin flask of green algae

The algal extract was spotted here and run in two directions to separate compounds from one another.

First run

Second run

After separation of the compounds, the chromatogram was overlaid with X-ray film.

Paper chromatogram

RESULTS

3PG

A chromatogram made after 3 seconds of exposure to $^{14}CO_2$ shows ^{14}C only in 3PG (3-phosphoglycerate).

GLUT
ALA
GLY SER
ASP CIT
SUC G3P
3PG
HEXOSE-P

A chromatogram made after 30 seconds of exposure to $^{14}CO_2$ shows ^{14}C in many molecules.

CONCLUSION The initial product of CO_2 fixation is 3PG. Later, the carbon from CO_2 ends up in many molecules.

Go to **yourBioPortal.com** for original citations, discussions, and relevant links for all INVESTIGATING LIFE figures.

yourBioPortal.com
GO TO Animated Tutorial 10.3 • Tracing the Pathway of CO_2

spots indicated the locations of compounds containing ^{14}C in the paper.

To discover the first compound in the pathway of CO_2 fixation, Calvin and his team exposed the algae to $^{14}CO_2$ for shorter and shorter periods of time. The 3-second exposure revealed that only one compound was labeled—a 3-carbon sugar phosphate called 3-phosphoglycerate (3PG) (the ^{14}C is shown in red):

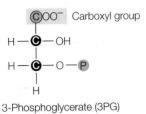

3-Phosphoglycerate (3PG)

With successive exposures longer than 3 seconds, Calvin and his colleagues were able to trace the route of ^{14}C as it moved through a series of compounds, including monosaccharides and amino acids. It turned out that the pathway the ^{14}C moved through was a cycle. In this cycle, the CO_2 initially bonds covalently to a larger five-carbon acceptor molecule, which then breaks into two three-carbon molecules. As the cycle repeats a carbohydrate is produced and the initial CO_2 acceptor is regenerated. This was appropriately named the **Calvin cycle**.

The initial reaction in the Calvin cycle adds the 1-carbon CO_2 to an acceptor molecule, the 5-carbon compound ribulose 1,5-bisphosphate (RuBP). The product is an intermediate 6-carbon compound, which quickly breaks down and forms two molecules of 3PG (**Figure 10.14**). The intermediate compound is broken down so rapidly that Calvin did not observe radioactive label appearing in it first. But the enzyme that catalyzes its formation, **ribulose bisphosphate carboxylase/oxygenase (rubisco)**, is the most abundant protein in the world! It constitutes up to 50 percent of all the protein in every plant leaf.

The Calvin cycle is made up of three processes

The Calvin cycle uses the ATP and NADPH made in the light to reduce CO_2 in the stroma to a carbohydrate. Like all biochem-ical pathways, each reaction is catalyzed by a specific enzyme. The cycle is composed of three distinct processes (**Figure 10.15**):

- *Fixation* of CO_2. As we have seen, this reaction is catalyzed by rubisco, and its stable product is 3PG.

- *Reduction* of 3PG to form glyceraldehyde 3-phosphate (G3P). This series of reactions involves a phosphorylation (using the ATP made in the light reactions) and a reduction (using the NADPH made in the light reactions).

- *Regeneration* of the CO_2 acceptor, RuBP. Most of the G3P ends up as ribulose monophosphate (RuMP), and ATP is used to convert this compound into RuBP. So for every "turn" of the cycle, with one CO_2 fixed, the CO_2 acceptor is regenerated.

The product of this cycle is **glyceraldehyde 3-phosphate (G3P)**, which is a 3-carbon sugar phosphate, also called triose phosphate:

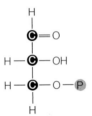

Glyceraldehyde 3-phosphate (G3P)

In a typical leaf, five-sixths of the G3P is recycled into RuBP. There are two fates for the remaining G3P, depending on the time of day and the needs of different parts of the plant:

- Some of it is exported out of the chloroplast to the cytosol, where it is converted to hexoses (glucose and fructose). These molecules may be used in glycolysis and mitochondrial respiration to power the activities of photosynthetic cells (see Chapter 9) or they may be converted into the disaccharide sucrose, which is transported out of the leaf to other organs in the plant. There it is hydrolyzed to its constituent monosaccharides, which can be used as sources of energy or as building blocks for other molecules.

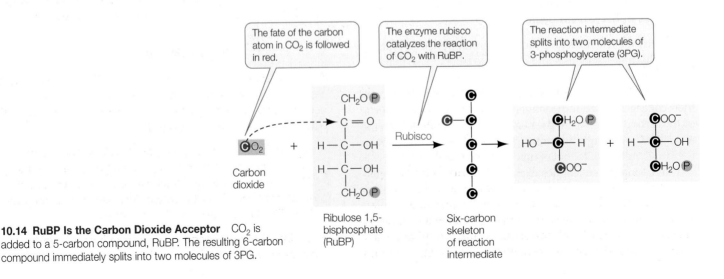

10.14 RuBP Is the Carbon Dioxide Acceptor CO_2 is added to a 5-carbon compound, RuBP. The resulting 6-carbon compound immediately splits into two molecules of 3PG.

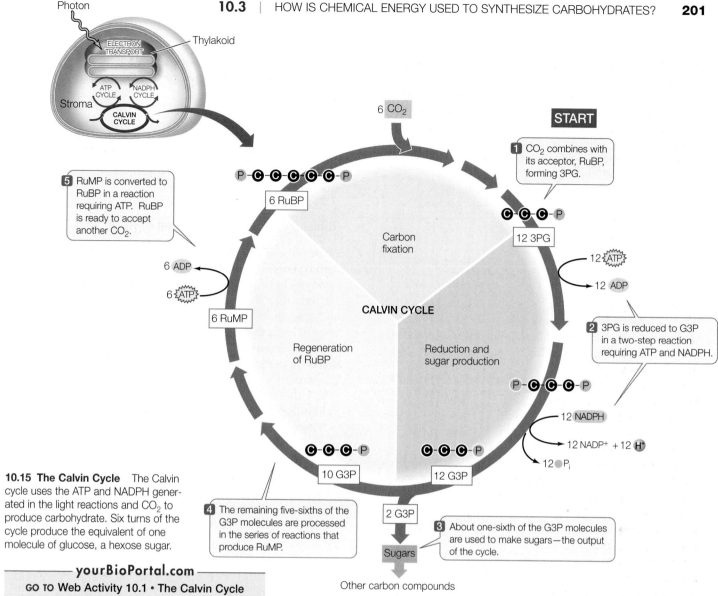

10.15 The Calvin Cycle The Calvin cycle uses the ATP and NADPH generated in the light reactions and CO_2 to produce carbohydrate. Six turns of the cycle produce the equivalent of one molecule of glucose, a hexose sugar.

yourBioPortal.com
GO TO Web Activity 10.1 • The Calvin Cycle

• As the day wears on, glucose accumulates inside of the chloroplast, and these glucose units are linked to form the polysaccharide starch. This stored carbohydrate can then be drawn upon during the night so that the photosynthetic tissues can continue to export sucrose to the rest of the plant, even when photosynthesis is not taking place. In addition, starch is abundant in nonphotosynthetic organs such as roots, underground stems and seeds, where it provides a ready supply of glucose to fuel cellular activities, including plant growth.

The carbohydrates produced in photosynthesis are used by the plant to make other compounds. The carbon molecules are incorporated into amino acids, lipids, and the building blocks of nucleic acids—in fact all the organic molecules in the plant.

The products of the Calvin cycle are of crucial importance to the Earth's entire biosphere. For the majority of living organisms on Earth, the C—H covalent bonds generated by the cycle provide almost all of the energy for life. Photosynthetic organisms, which are also called **autotrophs** ("self-feeders"), release most of this energy by glycolysis and cellular respiration, and use it to support their own growth, development, and repro-

duction. But plants are also the source of energy for other organisms. Much plant matter ends up being consumed by **heterotrophs** ("other-feeders"), such as animals, which cannot photosynthesize. Heterotrophs depend on autotrophs for both raw materials and energy. Free energy is released from food by glycolysis and cellular respiration in heterotroph cells.

Light stimulates the Calvin cycle

As we have seen, the Calvin cycle uses NADPH and ATP, which are generated using energy from light. Two other processes connect the light reactions with this CO_2 fixation pathway. Both connections are indirect but significant:

• Light-induced pH changes in the stroma activate some Calvin cycle enzymes. Proton pumping from the stroma into the thylakoid lumen causes an increase in the pH of the stroma from 7 to 8 (a tenfold decrease in H^+ concentration). This favors the activation of rubisco.

• The light-induced electron transport reduces disulfide bonds in four of the Calvin cycle enzymes, thereby activat-

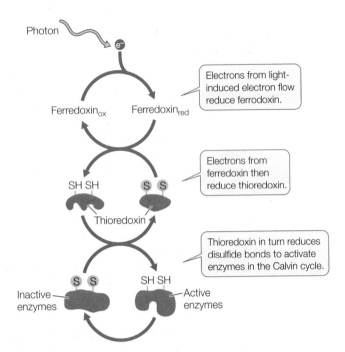

10.16 The Photochemical Reactions Stimulate the Calvin Cycle
By reducing (breaking) disulfide bridges, electrons from the light reactions activate enzymes in CO_2 fixation.

ing them (**Figure 10.16**). When ferredoxin is reduced in photosystem I (see Figure 10.10), it passes some electrons to a small, soluble protein called thioredoxin, and this protein passes electrons to four enzymes in the CO_2 fixation pathway. Reduction of the sulfurs in the disulfide bridges of these enzymes (see Figure 3.5) forms SH groups and breaks the bridges. The resulting changes in their three-dimensional shapes activate the enzymes and increase the rate at which the Calvin cycle operates.

10.3 RECAP

ATP and NADPH produced in the light reactions power the synthesis of carbohydrates by the Calvin cycle. This cycle fixes CO_2, reduces it, and regenerates the acceptor, RuBP, for further fixation.

- Describe the experiments that led to the identification of RuBP as the initial CO_2 acceptor in photosynthesis. See pp. 199–200 and Figure 10.13
- What are the three processes of the Calvin cycle? See pp. 200–201 and Figure 10.15
- In what ways does light stimulate the Calvin cycle? See pp. 201–202 and Figure 10.16

Although all green plants carry out the Calvin cycle, some plants have evolved variations on, or additional steps in, the light-independent reactions. These variations and additions have permitted plants to adapt to and thrive in certain environmental conditions. Let's look at these environmental limitations and the metabolic bypasses that have evolved to circumvent them.

10.4 How Do Plants Adapt to the Inefficiencies of Photosynthesis?

In addition to fixing CO_2 during photosynthesis, rubisco can react with O_2. This reaction leads to a process called photorespiration, which lowers the overall rate of CO_2 fixation in some plants. After examining this problem, we'll look at some biochemical pathways and features of plant anatomy that compensate for the limitations of rubisco.

Rubisco catalyzes the reaction of RuBP with O_2 or CO_2

As its full name indicates, rubisco is an **oxygenase** as well as a **carboxylase**—it can add O_2 to the acceptor molecule RuBP instead of CO_2. The affinity of rubisco for CO_2 is about ten times stronger than its affinity for O_2. This means that inside a leaf with a normal exchange of air with the outside, CO_2 fixation is favored even though the concentration of CO_2 in the air is far less than that of O_2. But if there is an even higher concentration of O_2 in the leaf, it acts as a competitive inhibitor, and RuBP reacts with O_2 rather than CO_2. This reduces the overall amount of CO_2 that is converted into carbohydrates, and may play a role in limiting plant growth.

When O_2 is added to RuBP, one of the products is a 2-carbon compound, phosphoglycolate:

$$RuBP + O_2 \rightarrow \text{phosphoglycolate} + \text{3-phosphoglycerate (3PG)}$$

The 3PG formed by oxygenase activity enters the Calvin cycle but the phosphoglycolate does not. Plants have evolved a metabolic pathway that can partially recover the carbon in phosphoglycolate. The phosphoglycolate is hydrolyzed to glycolate, which diffuses into membrane-enclosed organelles called peroxisomes (**Figure 10.17**). There, a series of reactions converts it into the amino acid glycine:

$$\text{glycolate} + O_2 \rightarrow \text{glycine}$$

The glycine then diffuses into a mitochondrion, where two glycine molecules are converted in a series of reactions into the amino acid serine, which goes back to the peroxisome and is converted into glycerate (a 3-carbon molecule) and CO_2:

$$2 \text{ glycine} \rightarrow \rightarrow \text{glycerate} + CO_2$$

The glycerate moves into the chloroplast, where it is phosphorylated to make 3PG, which enters the Calvin cycle. So overall:

$$\text{phosphoglycolate (4 carbons)} + O_2 \rightarrow \text{3PG (3 carbons)} + CO_2$$

This pathway thus reclaims 75 percent of the carbons from phosphoglycolate for the Calvin cycle. In other words, the reaction of RuBP with O_2 instead of CO_2 reduces the net carbon fixed by the Calvin cycle by 25 percent. The pathway is called **photorespiration** because it consumes O_2 and releases CO_2 and because it occurs only in the light (due to the same enzyme activation processes that were mentioned above with regard to the Calvin cycle).

Why does rubisco act as an oxygenase as well as a carboxylase? Several factors are involved: active site affinities, concentrations of CO_2 and O_2, and temperature.

(A)

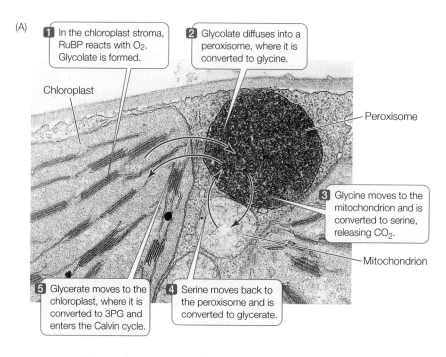

1 In the chloroplast stroma, RuBP reacts with O_2. Glycolate is formed.

Chloroplast

2 Glycolate diffuses into a peroxisome, where it is converted to glycine.

Peroxisome

3 Glycine moves to the mitochondrion and is converted to serine, releasing CO_2.

Mitochondrion

5 Glycerate moves to the chloroplast, where it is converted to 3PG and enters the Calvin cycle.

4 Serine moves back to the peroxisome and is converted to glycerate.

(B)

Carbon gained   Carbon lost

2-PG

O_2

CO_2

Carboxylase reaction

Photorespiration

Rubisco

Calvin cycle

Oxygenase reaction

O_2

3-PGA

CO_2

10.17 Organelles of Photorespiration (A) The reactions of photorespiration take place in the chloroplasts, peroxisomes, and mitochondria. (B) Overall, photorespiration consumes O_2 and releases CO_2.

- As noted above, rubisco has a ten times higher affinity for CO_2 than for O_2, and this favors CO_2 fixation.

- In the leaf, the relative concentrations of CO_2 and O_2 vary. If O_2 is relatively abundant, rubisco acts as an oxygenase and photorespiration ensues. If CO_2 predominates, rubisco fixes it for the Calvin cycle.

- Photorespiration is more likely at high temperatures. On a hot, dry day, small pores in the leaf surface called **stomata** close to prevent water from evaporating from the leaf (see Figure 10.1). But this also prevents gases from entering and leaving the leaf. The CO_2 concentration in the leaf falls as CO_2 is used up in photosynthetic reactions, and the O_2 concentration rises because of these same reactions. As the ratio of CO_2 to O_2 falls, the oxygenase activity of rubisco is favored, and photorespiration proceeds.

C_3 plants undergo photorespiration but C_4 plants do not

Plants differ in how they fix CO_2, and can be distinguished as C_3 or C_4 plants, based on whether the first product of CO_2 fixation is a 3- or 4-carbon molecule. In **C_3 plants** such as roses, wheat, and rice, the first product is the 3-carbon molecule 3PG—as we have just described for the Calvin cycle. In these plants the cells of the mesophyll, which makes up the main body of the leaf, are full of chloroplasts containing rubisco (**Figure 10.18A**). On a hot day, these leaves close their stomata to conserve water, and as a result, rubisco acts as an oxygenase as well as a carboxylase, and photorespiration occurs.

C_4 plants, which include corn, sugarcane, and tropical grasses, make the 4-carbon molecule **oxaloacetate** as the first product of CO_2 fixation (**Figure 10.18B**). On a hot day, they partially close their stomata to conserve water, but their rate of photosynthesis does not fall, nor does photorespiration occur. What do they do differently?

(A) Arrangement of cells in a C_3 leaf

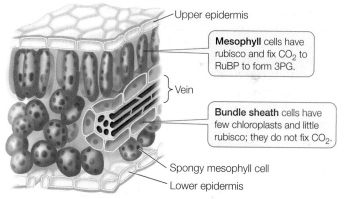

Upper epidermis

Mesophyll cells have rubisco and fix CO_2 to RuBP to form 3PG.

Vein

Bundle sheath cells have few chloroplasts and little rubisco; they do not fix CO_2.

Spongy mesophyll cell

Lower epidermis

(B) Arrangement of cells in a C_4 leaf

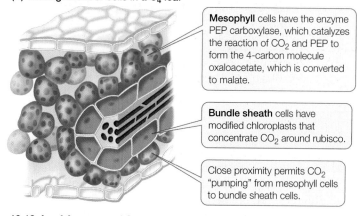

Mesophyll cells have the enzyme PEP carboxylase, which catalyzes the reaction of CO_2 and PEP to form the 4-carbon molecule oxaloacetate, which is converted to malate.

Bundle sheath cells have modified chloroplasts that concentrate CO_2 around rubisco.

Close proximity permits CO_2 "pumping" from mesophyll cells to bundle sheath cells.

10.18 Leaf Anatomy of C_3 and C_4 Plants Carbon dioxide fixation occurs in different organelles and cells of the leaves in (A) C_3 plants and (B) C_4 plants. Cells that are tinted blue have rubisco.

yourBioPortal.com

GO TO Web Activity 10.2 • C_3 and C_4 Leaf Anatomy

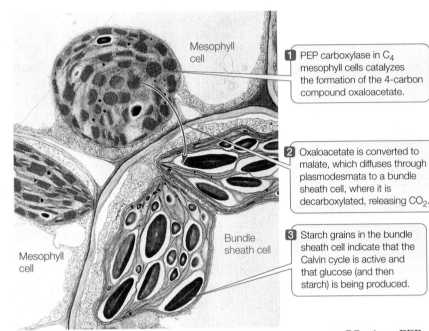

(A)

Mesophyll cell

1 PEP carboxylase in C$_4$ mesophyll cells catalyzes the formation of the 4-carbon compound oxaloacetate.

2 Oxaloacetate is converted to malate, which diffuses through plasmodesmata to a bundle sheath cell, where it is decarboxylated, releasing CO$_2$.

3 Starch grains in the bundle sheath cell indicate that the Calvin cycle is active and that glucose (and then starch) is being produced.

Bundle sheath cell

Mesophyll cell

10.19 The Anatomy and Biochemistry of C$_4$ Carbon Fixation (A) Carbon dioxide is fixed initially in the mesophyll cells, but enters the Calvin cycle in the bundle sheath cells. (B) The two cell types share an interconnected biochemical pathway for CO$_2$ assimilation.

(B)

Plasma membrane

Cell wall

Mesophyll cell

CO$_2$

PEP

Carboxylation

C$_4$ cycle

Regeneration

4C compound

3C compound

4C compound

3C compound

Decarboxylation

Bundle sheath cell

CO$_2$

5C sugar

Carboxylation

Regeneration

Calvin cycle

Triose-P

3C sugar

Reduction

C$_4$ plants perform the normal Calvin cycle, but they have an additional early reaction that fixes CO$_2$ without losing carbon to photorespiration. Because this initial CO$_2$ fixation step can function even at low levels of CO$_2$ and high temperatures, C$_4$ plants very effectively optimize photosynthesis under conditions that inhibit it in C$_3$ plants. C$_4$ plants have *two separate enzymes for CO$_2$ fixation, located in different parts of the leaf* (**Figure 10.19**; see also Figure 10.18B). The first enzyme, called **PEP carboxylase**, is present in the cytosols of mesophyll cells near the

surface of the leaf. It fixes CO$_2$ to a 3-carbon acceptor compound, **phosphoenolpyruvate** (**PEP**), to produce the 4-carbon fixation product, oxaloacetate. PEP carboxylase has two advantages over rubisco:

- It does not have oxygenase activity.
- It fixes CO$_2$ even at very low CO$_2$ levels.

So even on a hot day when the stomata are partially closed and the ratio of O$_2$ to CO$_2$ rises, PEP carboxylase just keeps on fixing CO$_2$.

Oxaloacetate is converted to malate, which diffuses out of the mesophyll cells and into the **bundle sheath cells** (see Figure 10.18B), located in the interior of the leaf. (Some C$_4$ plants convert the oxaloacetate to aspartate instead of malate, but we will only discuss the malate pathway here.) The bundle sheath cells contain modified chloroplasts that are designed to concentrate CO$_2$ around the rubisco. There, the 4-carbon malate loses one carbon (is decarboxylated), forming CO$_2$ and pyruvate. The latter moves back to the mesophyll cells where the 3-carbon acceptor compound, PEP, is regenerated at the expense of ATP. Thus the role of PEP is to bind CO$_2$ from the air in the leaf so that it can be transferred to the bundle sheath cells, where it is delivered to rubisco. This process essentially "pumps up" the CO$_2$ concentration around rubisco, so that it acts as a carboxylase and begins the Calvin cycle.

C$_3$ plants have an advantage over C$_4$ plants in that they don't expend extra ATP to "pump up" the concentration of CO$_2$ near rubisco. But this advantage begins to be outweighed under conditions that favor photorespiration, such as warmer seasons and climates. Under these conditions C$_4$ plants have the advantage. For example, Kentucky bluegrass is a C$_3$ plant that thrives on lawns in April and May. But in the heat of summer it does not do as well and Bermuda grass, a C$_4$ plant, takes over the lawn. The same is true on a global scale for crops: C$_3$ plants such as soybean, wheat, and barley have been adapted for human food production in temperate climates, while C$_4$ plants such as corn and sugarcane originated and are grown in the tropics.

THE EVOLUTION OF CO$_2$ FIXATION PATHWAYS C$_3$ plants are certainly more ancient than C$_4$ plants. While C$_3$ photosynthesis appears to have begun about 3.5 billion years ago, C$_4$ plants appeared about 12 million years ago. A possible factor in the emergence of the C$_4$ pathway is the decline in atmospheric CO$_2$. When dinosaurs dominated Earth 100 million years ago, the concentration of CO$_2$ in the atmosphere was four times what it is now. As CO$_2$ levels declined thereafter, the more efficient C$_4$

TABLE 10.1
Comparison of Photosynthesis in C_3, C_4, and CAM Plants

	C_3 PLANTS	C_4 PLANTS	CAM PLANTS
Calvin cycle used?	Yes	Yes	Yes
Primary CO_2 acceptor	RuBP	PEP	PEP
CO_2-fixing enzyme	Rubisco	PEP carboxylase	PEP carboxylase
First product of CO_2 fixation	3PG (3-carbon)	Oxaloacetate (4-carbon)	Oxaloacetate (4-carbon)
Affinity of carboxylase for CO_2	Moderate	High	High
Photosynthetic cells of leaf	Mesophyll	Mesophyll and bundle sheath	Mesophyll with large vacuoles
Photorespiration	Extensive	Minimal	Minimal

plants would have gained an advantage over their C_3 counterparts.

As we described in the opening essay of this chapter, CO_2 levels have been increasing over the past 200 years. Currently, the level of CO_2 is not enough for maximal CO_2 fixation by rubisco, so photorespiration occurs, reducing the growth rates of C_3 plants. Under hot conditions, C_4 plants are favored. But if CO_2 levels in the atmosphere continue to rise, the reverse will occur and C_3 plants will have a comparative advantage. The overall growth rates of crops such as rice and wheat should increase. This may or may not translate into more food, given that other effects of the human-spurred CO_2 increase (such as global warming) will also alter Earth's ecosystems.

CAM plants also use PEP carboxylase

Other plants besides the C_4 plants use PEP carboxylase to fix and accumulate CO_2. They include some water-storing plants (called succulents) of the family Crassulaceae, many cacti, pineapples, and several other kinds of flowering plants. The CO_2 metabolism of these plants is called **crassulacean acid metabolism**, or **CAM**, after the family of succulents in which it was discovered. CAM is much like the metabolism of C_4 plants in that CO_2 is initially fixed into a 4-carbon compound. But in CAM plants the initial CO_2 fixation and the Calvin cycle are separated in time rather than space.

- At night, when it is cooler and water loss is minimized, the stomata open. CO_2 is fixed in mesophyll cells to form the 4-carbon compound oxaloacetate, which is converted into malate and stored in the vacuole.

- During the day, when the stomata close to reduce water loss, the accumulated malate is shipped to the chloroplasts, where its decarboxylation supplies the CO_2 for the Calvin cycle and the light reactions supply the necessary ATP and NADPH.

CAM benefits the plant by allowing it to close its stomata during the day. As you will learn in Chapter 35, plants lose most of the water that they take up in their roots by evaporation through the leaves (transpiration). In dry climates, closing stomata is a key to water conservation and survival.

Table 10.1 compares photosynthesis in C_3, C_4, and CAM plants.

10.4 RECAP

Rubisco catalyzes the carboxylation of RuBP to form two 3PG, and the oxygenation of RuBP to form one 3PG and one phosphoglycolate. The diversion of rubisco to its oxygenase function decreases net CO_2 fixation. C_4 photosynthesis and CAM allow plants to adapt to environmental conditions that result in a limited availability of CO_2 inside the leaf.

- Explain how photorespiration recovers some of the carbon that is channeled away from the Calvin cycle. **See pp. 202–203 and Figure 10.17**

- What do C_4 plants do to keep the concentration of CO_2 around rubisco high, and why? **See pp. 203–204 and Figure 10.19**

- What is the pathway for CO_2 fixation in CAM plants? **See p. 205**

Now that we understand how photosynthesis produces carbohydrates, let's see how the pathways of photosynthesis are connected to other metabolic pathways.

10.5 How Does Photosynthesis Interact with Other Pathways?

Green plants are autotrophs and can synthesize all the molecules they need from simple starting materials: CO_2, H_2O, phosphate, sulfate, ammonium ions (NH_4^+), and small quantities of other mineral nutrients. The NH_4^+ is needed to synthesize amino acids and nucleotides, and it comes from either the conversion of nitrogen-containing molecules in soil water or the conversion of N_2 gas from the atmosphere by bacteria, as we will see in Chapter 36.

Plants use the carbohydrates generated in photosynthesis to provide energy for processes such as active transport and anabolism. Both cellular respiration and fermentation can occur in plants, although the former is far more common. Unlike photosynthesis, plant cellular respiration takes place both in the light and in the dark.

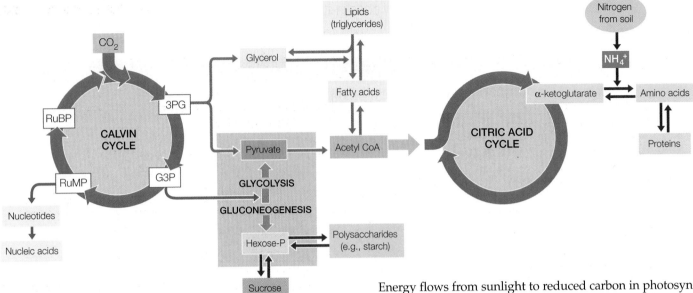

10.20 Metabolic Interactions in a Plant Cell The products of the Calvin cycle are used in the reactions of cellular respiration (glycolysis and the citric acid cycle).

Photosynthesis and respiration are closely linked through the Calvin cycle (**Figure 10.20**). The partitioning of G3P is particularly important:

- Some G3P from the Calvin cycle takes part in the glycolysis pathway and is converted into pyruvate in the cytosol. This pyruvate can be used in cellular respiration for energy, or its carbon skeletons can be used in anabolic reactions to make lipids, proteins, and other carbohydrates (see Figure 9.14).

- Some G3P can enter a pathway that is the reverse of glycolysis (gluconeogenesis; see Section 9.5). In this case, hexose-phosphates and then sucrose are formed and transported to the nonphotosynthetic tissues of the plant (such as the root).

Energy flows from sunlight to reduced carbon in photosynthesis, then to ATP in respiration. Energy can also be stored in the bonds of macromolecules such as polysaccharides, lipids, and proteins. For a plant to grow, energy storage (as body structures) must exceed energy release; that is, overall carbon fixation by photosynthesis must exceed respiration. This principle is the basis of the ecological food chain, as we will see in later chapters.

Photosynthesis provides most of the energy that we need for life. Given the uncertainties about the future of photosynthesis (due to changes in CO_2 levels and climate change), it would be wise to seek ways to improve photosynthetic efficiency. **Figure 10.21** shows the various ways in which solar energy is utilized by plants or lost. In essence, only about 5 percent of the sunlight that reaches Earth is converted into plant growth. The inefficiencies of photosynthesis involve basic chemistry and physics (some light energy is not absorbed by photosynthetic pigments) as well as biology (plant anatomy and leaf exposure, photorespiration, and inefficiencies in metabolic pathways). While it is hard to change chemistry and physics, biologists might be able to use their knowledge of plants to improve on the basic biology of photosynthesis. This could result in a more efficient use of resources and better food production.

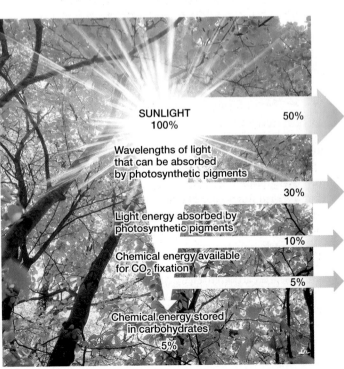

SUNLIGHT 100%

Wavelengths of light that can be absorbed by photosynthetic pigments

50%

Light energy absorbed by photosynthetic pigments

30%

Chemical energy available for CO_2 fixation

10%

Chemical energy stored in carbohydrates

5%

ENERGY LOSS

Wavelengths of light not part of absorption spectrum of photosynthetic pigments (e.g., green light)

Light energy not absorbed due to plant structure (e.g., leaves not properly oriented to sun)

Inefficiency of light reactions converting light to chemical energy

Inefficiency of CO_2 fixation pathways

5%

10.21 Energy Losses During Photosynthesis As we face an increasingly uncertain future for photosynthesis on Earth, understanding its inefficiencies becomes increasingly important. Photosynthetic pathways preserve at most about 5 percent of the sun's energy input as chemical energy in carbohydrates.

10.5 RECAP

The products of photosynthesis are utilized in glycolysis and the citric acid cycle, as well as in the synthesis of lipids, proteins, and other large molecules.

- How do common intermediates link the pathways of glycolysis, the citric acid cycle, and photosynthesis? **See p. 206 and Figure 10.20**

CHAPTER SUMMARY

10.1 What Is Photosynthesis?

- In the process of **photosynthesis**, plants and other organisms take in CO_2, water, and light energy, producing O_2 and carbohydrates. **SEE ANIMATED TUTORIAL 10.1**

- The **light reactions** of photosynthesis convert light energy into chemical energy. They produce ATP and reduce $NADP^+$ to NADPH. Review Figure 10.3

- The **light-independent reactions** do not use light directly but instead use ATP and NADPH to reduce CO_2, forming carbohydrates.

10.2 How Does Photosynthesis Convert Light Energy into Chemical Energy?

- Light is a form of electromagnetic radiation. It is emitted in particle-like packets called **photons** but has wavelike properties.

- Molecules that absorb light in the visible spectrum are called **pigments**. Photosynthetic organisms have several pigments, most notably **chlorophylls**, but also accessory pigments such as carotenoids and phycobilins.

- Absorption of a photon puts a pigment molecule in an excited state that has more energy than its ground state. Review Figure 10.5

- Each compound has a characteristic **absorption spectrum**. An **action spectrum** reflects the biological activity of a photosynthetic organism for a given wavelength of light. Review Figure 10.6

- The pigments in photosynthetic organisms are arranged into **antenna systems** that absorb energy from light and funnel this energy to a pair of chlorophyll *a* molecules in the reaction center of the **photosystem**. Chlorophyll can act as a reducing agent, transferring excited electrons to other molecules. Review Figure 10.8

- **Noncyclic electron transport** uses photosystems I and II to produce ATP, NADPH and O_2. **Cyclic electron transport** uses only photosystem I and produces only ATP. Review Figures 10.10 and 10.11

- Chemiosmosis is the mechanism of ATP production in **photophosphorylation**. Review Figure 10.12, **ANIMATED TUTORIAL 10.2**

10.3 How Is Chemical Energy Used to Synthesize Carbohydrates?

- The **Calvin cycle** makes carbohydrates from CO_2. The cycle consists of three processes: fixation of CO_2, reduction and carbohydrate production, and regeneration of RuBP. **SEE ANIMATED TUTORIAL 10.3**

- **RuBP** is the initial CO_2 acceptor, and **3PG** is the first stable product of CO_2 fixation. The enzyme **rubisco** catalyzes the reaction of CO_2 and RuBP to form 3PG. Review Figure 10.14, **WEB ACTIVITY 10.1**

- ATP and NADPH formed by the light reactions are used in the reduction of 3PG to form **G3P**. Review Figure 10.15

- Light stimulates enzymes in the Calvin cycle, further integrating the light-dependent and light-independent pathways.

10.4 How Do Plants Adapt to the Inefficiencies of Photosynthesis?

- Rubisco can catalyze a reaction between O_2 and RuBP in addition to the reaction between CO_2 and RuBP. At high temperatures and low CO_2 concentrations, the **oxygenase** function of rubisco is favored over its **carboxylase** function.

- When rubisco functions as an oxygenase, the result is **photorespiration**, which significantly reduces the efficiency of photosynthesis.

- In C_4 **plants**, CO_2 reacts with **PEP** to form a 4-carbon intermediate in mesophyll cells. The 4-carbon product releases its CO_2 to rubisco in the **bundle sheath** cells in the interior of the leaf. Review Figure 10.18, **WEB ACTIVITY 10.2**

- **CAM** plants operate much like C_4 plants, but their initial CO_2 fixation by PEP carboxylase is temporally separated from the Calvin cycle, rather than spatially separated as in C_4 plants.

10.5 How Does Photosynthesis Interact with Other Pathways?

- Photosynthesis and cellular respiration are linked through the **Calvin cycle**, the **citric acid cycle**, and **glycolysis**. Review Figure 10.20

- To survive, a plant must photosynthesize more than it respires.

- Photosynthesis utilizes only a small portion of the energy of sunlight. Review Figure 10.21

SELF-QUIZ

1. In noncyclic photosynthetic electron transport, water is used to
   - *a.* excite chlorophyll.
   - *b.* hydrolyze ATP.
   - *c.* reduce P_i.
   - *d.* oxidize NADPH.
   - *e.* reduce chlorophyll.

2. Which statement about light is true?
   - *a.* An absorption spectrum is a plot of biological effectiveness versus wavelength.
   - *b.* An absorption spectrum may be a good means of identifying a pigment.
   - *c.* Light need not be absorbed to produce a biological effect.
   - *d.* A given kind of molecule can occupy any energy level.
   - *e.* A pigment loses energy as it absorbs a photon.

3. Which statement about chlorophylls is *not* true?
   - *a.* Chlorophylls absorb light near both ends of the visible spectrum.
   - *b.* Chlorophylls can accept energy from other pigments, such as carotenoids.
   - *c.* Excited chlorophyll can either reduce another substance or release light energy.
   - *d.* Excited chlorophyll cannot be an oxidizing agent.
   - *e.* Chlorophylls contain magnesium.

4. In cyclic electron transport,
   a. oxygen gas is released.
   b. ATP is formed.
   c. water donates electrons and protons.
   d. NADPH forms.
   e. CO_2 reacts with RuBP.

5. Which of the following does *not* happen in noncyclic electron transport?
   a. Oxygen gas is released.
   b. ATP forms.
   c. Water donates electrons and protons.
   d. NADPH forms.
   e. CO_2 reacts with RuBP.

6. In chloroplasts,
   a. light leads to the flow of protons out of the thylakoids.
   b. ATP is formed when protons flow into the thylakoid lumen.
   c. light causes the thylakoid lumen to become less acidic than the stroma.
   d. protons return passively to the stroma through protein channels.
   e. proton pumping requires ATP.

7. Which statement about the Calvin cycle is *not* true?
   a. CO_2 reacts with RuBP to form 3PG.
   b. RuBP forms by the metabolism of 3PG.

c. ATP and NADPH form when 3PG is oxidized.
   d. The concentration of 3PG rises if the light is switched off.
   e. Rubisco catalyzes the reaction of CO_2 and RuBP.

8. In C_4 photosynthesis,
   a. 3PG is the first product of CO_2 fixation.
   b. rubisco catalyzes the first step in the pathway.
   c. 4-carbon acids are formed by PEP carboxylase in bundle sheath cells.
   d. photosynthesis continues at lower CO_2 levels than in C_3 plants.
   e. CO_2 released from RuBP is transferred to PEP.

9. Photosynthesis in green plants occurs only during the day. Respiration in plants occurs
   a. only at night.
   b. only when there is enough ATP.
   c. only during the day.
   d. all the time.
   e. in the chloroplast after photosynthesis.

10. Photorespiration
    a. takes place only in C_4 plants.
    b. includes reactions carried out in peroxisomes.
    c. increases the yield of photosynthesis.
    d. is catalyzed by PEP carboxylase.
    e. is independent of light intensity.

FOR DISCUSSION

1. Both photosynthetic electron transport and the Calvin cycle stop in the dark. Which specific reaction stops first? Which stops next? Continue answering the question "Which stops next?" until you have explained why both pathways have stopped.

2. In what principal ways are the reactions of electron transport in photosynthesis similar to the reactions of oxidative phosphorylation discussed in Section 9.3?

3. Differentiate between cyclic and noncyclic electron transport in terms of (1) the products and (2) the source of electrons for the reduction of oxidized chlorophyll.

4. If water labeled with ^{18}O is added to a suspension of photosynthesizing chloroplasts, which of the following compounds will first become labeled with ^{18}O: ATP, NADPH, O_2, or 3PG? If water labeled with ^3H is added, which of the

same compounds will first become radioactive? Which will be first if CO_2 labeled with ^{14}C is added?

5. The Viking lander was sent to Mars in 1976 to detect signs of life. Explain the rationale behind the following experiments this unmanned probe performed:
   a. A scoop of dirt was inserted into a container and $^{14}CO_2$ was added. After a while during the Martian day, the $^{14}CO_2$ was removed and the dirt was heated to a high temperature. Scientists monitoring the experiment back on Earth looked for the release of $^{14}CO_2$ as a sign of life.
   b. The same experiment was performed, except that the dirt was heated to a high temperature for 30 minutes and then allowed to cool to Martian temperature right after scooping, and before the $^{14}CO_2$ was added. If living things were present, then $^{14}CO_2$ would be released in experiment (*a*), but not this one.

ADDITIONAL INVESTIGATION

Calvin's experiment (see Figure 10.13) laid the foundations for a full description of the pathway of CO_2 fixation. Given the interrelationships between metabolic pathways in plants, how

would you do an experiment to follow the pathway of fixed carbon from photosynthesis to proteins?

WORKING WITH DATA (GO TO yourBioPortal.com)

Water is the Source of the Oxygen Produced by Photosynthesis   The proposal that the source of O_2 in photosynthesis was H_2O rather than CO_2 was first made in 1932. But it took the invention of isotope tracing a decade later to prove this. In this exercise, you will examine the methods that Ruben and Kamen used (Figure 10.2) to identify the isotopes of oxygen and the data they obtained.

Tracing the Pathway of CO_2   Studies of radioactive isotopes were intensified during World War II as an offshoot of the development of nuclear weapons. This led Calvin and his colleagues to perform the experiments designed to trace the path of carbon in photosynthesis (Figure 10.13). In this hands-on exercise, you will examine their data and see the reasoning that led to the CO_2 fixation pathway.

11

The Cell Cycle and Cell Division

An enemy of the cell reproduction cycle

Ruth felt healthy and was surprised when she was called back to her physician's office a week after her annual checkup. "Your lab report indicates you have early cervical cancer," said the doctor. "I ordered a follow-up test, and it came back positive—at some point, you were infected with HPV."

Ruth felt numb as soon as she heard the word "cancer." Her mother had died of breast cancer in the previous year. The doctor's statement about HPV (human papillomavirus) did not register in her consciousness. Sensing Ruth's discomfort, the doctor quickly reassured her that the cancer was caught at an early stage, and that a simple surgical procedure would remove it. Two weeks later, the cancer was removed and Ruth remains cancer-free. She was fortunate that her annual medical exam included a Papanicolau (Pap) test, in which the cells lining the cervix are examined for abnormalities. Since they were begun almost 50 years ago in Europe, Pap tests have resulted in the early detection and removal of millions of early cervical cancers, and the death rate from this potentially lethal disease has plummeted.

Only recently was HPV found to be the cause of most cervical cancers. The German physician Harald Zur-Hansen was awarded the Nobel Prize in 2008 for this discovery and it has led to a vaccine to prevent future infections. There are many different types of HPV and many of the ones that infect humans cause warts, which are small, rough growths on the skin. The types of HPV that infect tissues at the cervix get there by sexual transmission, and this is a common infection in Western societies.

When HPV arrives at the tissues lining the cervix, it has one of two fates. Most of the time, it gets into the cells, turning them into HPV factories, producing a lot of HPV in the mucus outside the uterus. These viruses can infect another person during a sexual encounter. In some cases, however, the virus follows a different and—for the host cells—more sinister path. It infects the cervical cells and causes them to make a viral protein called E7, a protein that can deregulate human cell reproduction.

Cell reproduction in healthy humans is tightly controlled, and one of the strongest regulators that prevent a cell from dividing is the

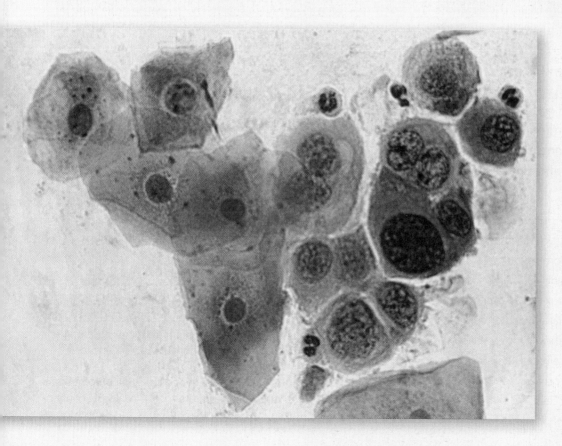

Abnormal Cells In this Pap test, cervical cancer cells at right differ from the normal cells at left. The cancer cells have larger nuclei.

E7, RB, and Cell Reproduction The E7 protein (blue) from human papillomavirus binds to the RB protein (red) to inhibit RB's ability to block cell division. This results in cancer.

retinoblastoma protein (RB), which you will encounter later in the chapter. One of the viral gene products is the protein E7, which has a three-dimensional shape that just fits into the protein-binding site of RB, thereby inactivating it. With no active RB to put the brakes on, cell division proceeds. As you know, uncontrolled cell reproduction is a hallmark of cancer—and so cervical cancer begins.

Understanding the cell division cycle and its control is clearly an important subject for understanding cancer. But cell division is not just important in medicine. It underlies the growth, development and reproduction of all organisms.

IN THIS CHAPTER we will see how cells give rise to more cells. We first describe how prokaryotic cells divide to produce new, single-celled organisms. Then we turn to the two types of nuclear division in eukaryotes—mitosis and meiosis—and relate them to asexual and sexual reproduction. Cell reproduction is linked to cell death, so we then consider the process of programmed cell death, also known as apoptosis. Finally, we relate these processes to the loss of cell reproduction control in cancer cells.

11.1 How Do Prokaryotic and Eukaryotic Cells Divide?

The life cycle of an organism, from birth to death, is intimately linked to cell division. Cell division plays important roles in the growth and repair of tissues in multicellular organisms, as well as in the reproduction of all organisms (**Figure 11.1**).

In order for any cell to divide, four events must occur:

- There must be a reproductive signal. This signal initiates cell division and may originate from either inside or outside the cell.

- **Replication** of DNA (the genetic material) must occur so that each of the two new cells will have identical genes and complete cell functions.

- The cell must distribute the replicated DNA to each of the two new cells. This process is called **segregation**.

- In addition to synthesizing needed enzymes and organelles, new material must be added to the plasma membrane (and the cell wall, in organisms that have one), in order to separate the two new cells by a process called **cytokinesis**.

These four events proceed somewhat differently in prokaryotes and eukaryotes.

Prokaryotes divide by binary fission

In prokaryotes, cell division results in the reproduction of the entire single-celled organism. The cell grows in size, replicates its DNA, and then separates the cytoplasm and DNA into two new cells by a process called **binary fission**.

REPRODUCTIVE SIGNALS The reproductive rates of many prokaryotes respond to conditions in the environment. The bacterium *Escherichia coli*, a species commonly used in genetic studies, is a "cell division machine"; if abundant sources of carbohydrates and mineral nutrients are available, it can divide as often as every 20 minutes. Another bacterium, *Bacillus subtilis*, does not just slow its growth when nutrients are low but stops dividing and then resumes dividing when conditions improve. Clearly, external factors such as environmental conditions and nutrient concentrations are signals for the initiation of cell division in prokaryotes.

REPLICATION OF DNA As we saw in Section 5.3, a **chromosome** can be defined in molecular terms as a DNA molecule containing genetic information. When a cell divides, all of its chromo-

11.1 Important Consequences of Cell Division Cell division is the basis for (A) reproduction, (B) growth, and (C) repair and regeneration of tissues.

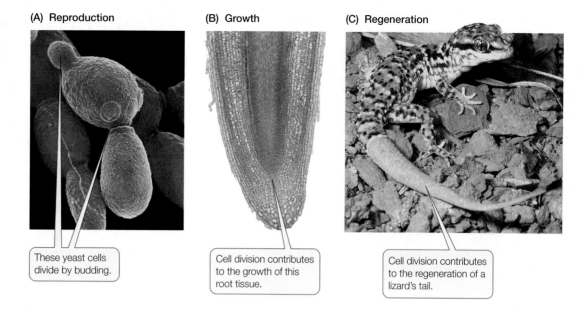

(A) Reproduction

These yeast cells divide by budding.

(B) Growth

Cell division contributes to the growth of this root tissue.

(C) Regeneration

Cell division contributes to the regeneration of a lizard's tail.

somes must be replicated, and one copy of each chromosome must find its way into one of the two new cells.

Most prokaryotes have only one chromosome—a single long DNA molecule with proteins bound to it. In *E. coli*, the ends of the DNA molecule are joined to create a circular chromosome. Circular chromosomes are characteristic of most prokaryotes as well as some viruses, and are also found in the chloroplasts and mitochondria of eukaryotic cells.

If the *E. coli* DNA were spread out into an actual circle, it would be about 500 μm in diameter. The bacterium itself is only about 2 μm long and 1μm in diameter. Thus if the bacterial DNA were fully extended, it would form a circle over 200 times larger than the cell! To fit into the cell, bacterial DNA must be compacted. The DNA folds in on itself, and positively charged (basic) proteins bound to the negatively charged (acidic) DNA contribute to this folding.

Two regions of the prokaryotic chromosome play functional roles in cell reproduction:

- *ori*: the site where replication of the circular chromosome starts (the *ori*gin of replication)

- *ter*: the site where replication ends (the *ter*minus of replication)

Chromosome replication takes place as the DNA is threaded through a "replication complex" of proteins near the center of the cell. (These proteins include the enzyme DNA polymerase, whose important role in replication is discussed further in Section 13.3.) Replication begins at the *ori* site and moves toward the *ter* site. While the DNA replicates, anabolic metabolism is active and the cell grows. When replication is complete, the two daughter DNA molecules separate and segregate from one another at opposite ends of the cell. In rapidly dividing prokaryotes, DNA replication occupies the entire time between cell divisions.

SEGREGATION OF DNAS Replication begins near the center of the cell, and as it proceeds, the *ori* regions move toward opposite ends of the cell (**Figure 11.2A**). DNA sequences adjacent to the

ori region bind proteins that are essential for this segregation. This is an active process, since the binding proteins hydrolyze ATP. The prokaryotic cytoskeleton (see Section 5.2) may be involved in DNA segregation, either actively moving the DNA along, or passively acting as a "railroad track" along which DNA moves.

CYTOKINESIS The actual division of a single cell and its contents into two cells is called cytokinesis and begins immediately after chromosome replication is finished in rapidly growing cells. The first event of cytokinesis is a pinching in of the plasma membrane to form a ring of fibers similar to a purse string. The major component of these fibers is a protein that is related to eukaryotic tubulin (which makes up microtubules). As the membrane pinches in, new cell wall materials are deposited, which finally separate the two cells (**Figure 11.2B**).

Eukaryotic cells divide by mitosis or meiosis followed by cytokinesis

As in prokaryotes, cell reproduction in eukaryotes entails reproductive signals, DNA replication, segregation, and cytokinesis. The details, however, are quite different:

- *Signal.* Unlike prokaryotes, eukaryotic cells do not constantly divide whenever environmental conditions are adequate. In fact, most eukaryotic cells that are part of a multicellular organism and have become specialized seldom divide. In a eukaryotic organism, the signals for cell division are related not to the environment of a single cell, but to the needs of the entire organism.

- *Replication.* While most prokaryotes have a single main chromosome, eukaryotes usually have many (humans have 46). Consequently the processes of replication and segregation are more intricate in eukaryotes than in prokaryotes. In eukaryotes, DNA replication is usually limited to a portion of the period between cell divisions.

(A)

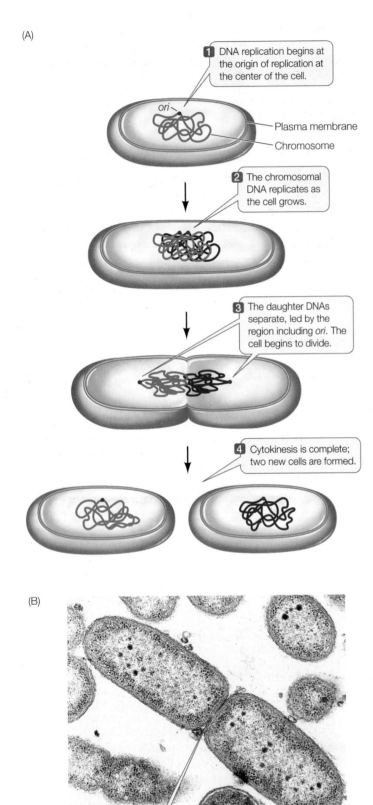

1 DNA replication begins at the origin of replication at the center of the cell.

ori

Plasma membrane

Chromosome

2 The chromosomal DNA replicates as the cell grows.

3 The daughter DNAs separate, led by the region including *ori*. The cell begins to divide.

4 Cytokinesis is complete; two new cells are formed.

(B)

Plasma membranes have completely formed, separating the cytoplasm of one cell from that of the other. Only a small gap of cell wall remains to be filled in.

Chromosome

11.2 Prokaryotic Cell Division (A) The process of cell division in a bacterium. (B) These two cells of the bacterium *Pseudomonas aeruginosa* have almost completed cytokinesis.

- *Segregation.* In eukaryotes, the newly replicated chromosomes are closely associated with each other (thus they are known as **sister chromatids**), and a mechanism called **mitosis** segregates them into two new nuclei.
- *Cytokinesis.* Cytokinesis proceeds differently in plant cells (which have a cell wall) than in animal cells (which do not).

The cells resulting from mitosis are identical to the parent cell in the amount and kind of DNA that they contain. This contrasts with the second mechanism of nuclear division, meiosis.

Meiosis is the process of nuclear division that occurs in cells involved with sexual reproduction. While the two products of mitosis are genetically identical to the cell that produced them—they both have the same DNA—the products of meiosis are not. As we will see in Section 11.5, meiosis generates diversity by shuffling the genetic material, resulting in new gene combinations. Meiosis plays a key role in the sexual life cycle.

11.1 RECAP

Four events are required for cell division: a reproductive signal, replication of the genetic material (DNA), segregation of replicated DNA, and separation of the two daughter cells (cytokinesis). In prokaryotes, cell division can be rapid; in eukaryotes, the process is more intricate, and the chromosomes must be duplicated before cell division can occur.

- What is the reproductive signal that leads the bacterium *Bacillus subtilis* to divide? **See p. 210**

- Explain why DNA must be replicated and segregated *before* a cell can divide. **See p. 210**

- What are the differences between cell division in prokaryotes (binary fission) and mitosis in eukaryotes? **See pp. 211–212**

What determines whether a cell will divide? How does mitosis lead to identical cells, and meiosis to diversity? Why do most eukaryotic organisms reproduce sexually? In the sections that follow, we will describe the details of mitosis and meiosis, and discuss their roles in development and evolution.

11.2 How Is Eukaryotic Cell Division Controlled?

As you will see throughout the book, different cells have different rates of cell division. Some cells, such as those in an early embryo, stem cells in bone marrow, or cells in the growing tip of a plant root, divide rapidly and continuously. Others, such as neurons in the brain or phloem cells in a plant stem, don't divide at all. Clearly, the signaling pathways for cells to divide are highly controlled.

The period between cell divisions is referred to as the **cell cycle**. The cell cycle can be divided into mitosis/cytokinesis and interphase. During **interphase**, the cell nucleus is visible and typical cell functions occur, including DNA replication. This phase of

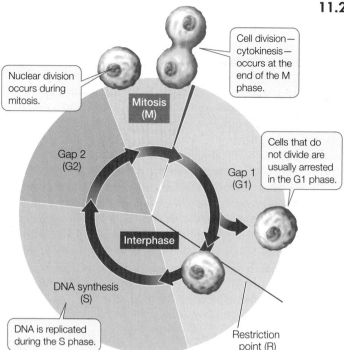

11.3 The Eukaryotic Cell Cycle The cell cycle consists of a mitotic (M) phase, during which mitosis and cytokinesis take place, and a long period of growth known as interphase. Interphase has three subphases (G1, S, and G2) in cells that divide.

the cell cycle begins when cytokinesis is completed and ends when mitosis begins (**Figure 11.3**). In this section we will describe the events of interphase, especially those that trigger mitosis.

Cells, even when rapidly dividing, spend most of their time in interphase. So if we take a snapshot through the microscope of a cell population, most of the cells will be in interphase; only a small percentage will be in mitosis or cytokinesis at any given moment.

Interphase has three subphases, called G1, S, and G2. The cell's DNA replicates during **S phase** (the *S* stands for synthesis) (see Figure 11.3). The period between the end of cytokinesis and the onset of S phase is called **G1**, or Gap 1. Another gap phase—**G2**—separates the end of S phase and the beginning of mitosis. Mitosis and cytokinesis are referred to as the **M phase** of the cell cycle.

Let's look at the events of interphase in more detail:

- *G1 phase.* During G1, a cell is preparing for S phase, so at this stage each chromosome is a single, unreplicated structure. G1 is quite variable in length in different cell types. Some rapidly dividing embryonic cells dispense with it entirely, while other cells may remain in G1 for weeks or even years. In many cases these cells enter a resting phase called G0. Special internal and external signals are needed to prompt a cell to leave G0 and reenter the cell cycle at G1.

- *The G1-to-S transition.* At the G1-to-S transition, called the **restriction point** (**R**), the commitment is made to DNA replication and subsequent cell division (and thus another cell cycle).

- *S phase.* DNA replication occurs during S phase (see Section 13.3 for a detailed description). Each chromosome is duplicated and thereafter consists of two sister chromatids joined together and awaiting segregation into two new cells.

- *G2 phase.* During G2, the cell makes preparations for mitosis—for example, by synthesizing components of the microtubules that will move the chromatids to opposite ends of the dividing cell.

The initiation, termination, and operations of these phases are regulated by specific signals.

Specific signals trigger events in the cell cycle

What events cause a cell to enter the S or M phases? A first indication that there were substances that control these transitions came from experiments involving *cell fusion.* Polyethylene glycol can be used to make different cells fuse together. Membrane lipids tend to partially dissolve in this nonpolar solvent, so that when it is present, cells will fuse their plasma membranes. Experiments involving the fusion of mammalian cells at different phases of the cell cycle showed that a cell in S phase produces a substance that activates DNA replication (**Figure 11.4**).

INVESTIGATING LIFE

11.4 Regulation of the Cell Cycle
Nuclei in G1 do not undergo DNA replication, but nuclei in S phase do. To determine if there is some signal in the S cells that stimulates G1 cells to replicate their DNA, cells in G1 and S phases were fused together, creating cells with both G1 and S properties.

HYPOTHESIS A cell in S phase contains an activator of DNA replication.

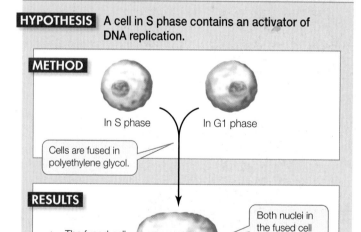

CONCLUSION The S phase cell produces a substance that diffuses to the G1 nucleus and activates DNA replication.

FURTHER INVESTIGATION: How would you use this method to show that a cell in M phase produces an activator of mitosis?

Go to **yourBioPortal.com** for original citations, discussions, and relevant links for all INVESTIGATING LIFE figures.

Similar experiments point to a molecular activator for entry into M phase. As you will see, the signals that control progress through the cell cycle act through protein kinases.

Progress through the cell cycle depends on the activities of **cyclin-dependent kinases**, or **Cdk's**. Recall from Section 7.1 that a *protein kinase* is an enzyme that catalyzes the transfer of a phosphate group from ATP to a target protein; this phosphate transfer is called *phosphorylation*.

$$\text{protein} + \text{ATP} \xrightarrow{\text{protein kinase}} \text{protein -P} + \text{ADP}$$

By catalyzing the phosphorylation of certain target proteins, Cdk's play important roles at various points in the cell cycle. The discovery that Cdk's induce cell division is a beautiful example of how research on different organisms and different cell types can converge on a single mechanism. One group of scientists, led by James Maller at the University of Colorado, was studying immature sea urchin eggs, trying to find out how they are stimulated to divide and eventually form a mature egg. A protein called *maturation promoting factor* was purified from maturing eggs, which by itself prodded immature egg cells to divide.

Meanwhile, Leland Hartwell at the University of Washington was studying the cell cycle in yeast (a single-celled eukaryote, see Figure 11.1A), and found a strain that was stalled at the G1–S boundary because it lacked a Cdk. It turned out that this yeast Cdk and the sea urchin maturation promoting factor had similar properties, and further work confirmed that the sea urchin protein was indeed a Cdk. Similar Cdk's were soon found to control the G1-to-S transition in many other organisms, including humans. Then others were found to control other parts of the cell cycle.

Cdk's are not active by themselves. As their name implies, cyclin-dependent kinases need to be activated by binding to a second type of protein, called **cyclin**. This binding—an example of *allosteric regulation* (see Section 8.5)—activates the Cdk by altering its shape and exposing its active site (**Figure 11.5**).

The cyclin–Cdk that controls passage from G1 to S phase is not the only such complex involved in regulating the eukaryotic cell cycle. There are different cyclin–Cdk's that act at different stages of the cycle (**Figure 11.6**). Let's take a closer look at G1–S cyclin–Cdk, which was the first to be discovered.

G1–S cyclin–Cdk catalyzes the phosphorylation of a protein called *retinoblastoma protein* (*RB*). In many cells, RB or a protein like it acts as an inhibitor of the cell cycle at the R (for "restriction") point in late G1. To begin S phase, a cell must get by the RB block. Here is where G1–S cyclin–Cdk comes in: it catalyzes the addition of a phosphate to RB. This causes a change in the three-dimensional structure of RB, thereby inactivating it. With RB out of the way, the cell cycle can proceed. To summarize:

$$\text{RB} \xrightarrow{\text{G1–S cyclin–Cdk}} \text{RB-P}$$

(active—blocks cell cycle)   (inactive—allows cell cycle)

Progress through the cell cycle is regulated by the activities of Cdk's, and so regulating *them* is a key to regulating cell division. An effective way to regulate Cdk's is to regulate the presence or absence of cyclins (**Figure 11.7**). Simply put, if a cyclin is not present, its partner Cdk is not active. As their name suggests, the presence of cyclins is cyclic: they are made only at certain times of in the cell cycle.

The different cyclin–Cdk's act at **cell cycle checkpoints**, points at which a cell cycle's progress is regulated. For example, if a cell's DNA is substantially damaged by radiation or

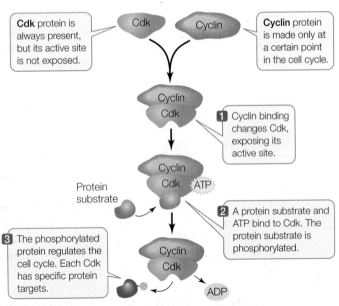

11.5 Cyclin Binding Activates Cdk Binding of a cyclin changes the three-dimensional structure of an inactive Cdk, making it an active protein kinase. Each cyclin–Cdk complex phosphorylates a specific target protein in the cell cycle.

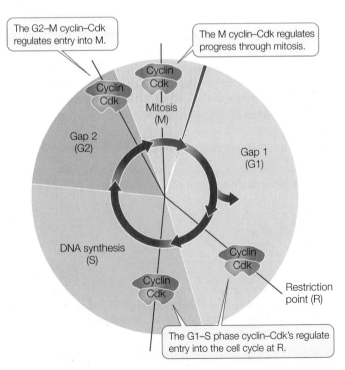

11.6 Cyclin-Dependent Kinases Regulate Progress Through the Cell Cycle By acting at checkpoints (red lines), different cyclin–Cdk complexes regulate the orderly sequence of events in the cell cycle.

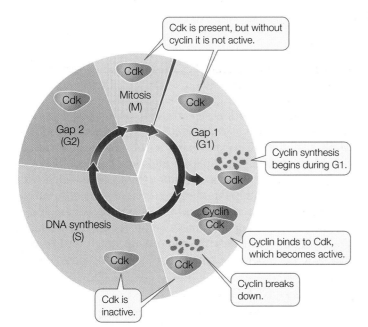

11.7 Cyclins Are Transient in the Cell Cycle Cyclins are made at a particular time and then break down. In this case, the cyclin is present during G1 and activates a Cdk at that time.

toxic chemicals, it may be prevented from successfully completing a cell cycle. For DNA damage, there are three checkpoints:

- During G1, before the cell enters S phase (restriction point)
- During S phase
- After S phase, during G2

Let's consider the G1 checkpoint. If DNA is damaged by radiation during G1, a protein called p21 is made. (The *p* stands for "protein" and the 21 stands for its molecular weight—about 21,000 daltons.) The p21 protein can bind to the G1–S Cdk, preventing its activation by cyclin. So the cell cycle stops while repairs are made to the DNA (you will learn more about DNA repair in Section 13.4). The p21 protein breaks down after the DNA is repaired, allowing cyclin to bind to the Cdk so that the cell cycle can proceed. If DNA damage is severe and it cannot be repaired, the cell will undergo programmed cell death (apoptosis, which we will discuss in Section 11.6).

In addition to these internal signals, the cell cycle is influenced by signals from the extracellular environment.

Growth factors can stimulate cells to divide

Cyclin–Cdk's provide cells with internal controls of their progress through the cell cycle. Not all cells in an organism go through the cell cycle on a regular basis. Some cells either no longer go through the cell cycle and enter G0, or go through it slowly and divide infrequently. If such cells are to divide, they must be stimulated by external chemical signals called **growth factors**. These proteins activate a signal transduction pathway that often ends up with the activation of Cdk's (signal transduction is discussed in Chapter 7):

- If you cut yourself and bleed, specialized cell fragments called *platelets* gather at the wound to initiate blood clotting. The platelets produce and release a protein called *platelet-*

derived growth factor that diffuses to the adjacent cells in the skin and stimulates them to divide and heal the wound.

- Red and white blood cells have limited lifetimes and must be replaced through the division of immature, unspecialized blood cell precursors in the bone marrow. Two types of growth factors, *interleukins* and *erythropoietin*, stimulate the division and specialization, respectively, of precursor cells.

In these and other examples, growth factors bind to specific receptors on target cells, and activate signal transduction pathways the end with cyclin synthesis, thereby activating Cdk's and the cell cycle. As you can see from the examples, growth factors are important in maintaining homeostasis.

11.2 RECAP

The eukaryotic cell cycle is under both external and internal control. Cdk's control the eukaryotic cell cycle and are themselves controlled by cyclins. External signals such as growth factors can initiate the cell cycle.

- Draw a cell cycle diagram showing the various stages of interphase. **See pp. 212–213 and Figure 11.3**
- How do cyclin–Cdk's control the progress of the cell cycle? **See pp. 214–215 and Figure 11.6**
- What are the differences between external and internal controls of the cell cycle? **See p. 215**

11.3 What Happens during Mitosis?

The third essential step in the process of cell division—segregation of the replicated DNA—occurs during mitosis. Prior to segregation, the huge DNA molecules and their associated proteins in each chromosome become condensed into more compact structures. After segregation by mitosis, cytokinesis separates the two cells. Let's now look at these steps more closely.

Prior to mitosis, eukaryotic DNA is packed into very compact chromosomes

A eukaryotic chromosome consists of one or two gigantic, linear, double-stranded DNA molecules complexed with many proteins (the complex of DNA and proteins is referred to as **chromatin**). Before S phase, each chromosome contains only one double-stranded DNA molecule. After it replicates during S phase, however, there are two double-stranded DNA molecules, known as sister chromatids. The sister chromatids are held together along most of their length by a protein complex called *cohesin*. They stay this way throughout interphase G2 until mitosis, when most of the cohesin is removed, except in a region called the **centromere** at which the chromatids remain held together. At the end of G2, a second group of proteins called *condensins* coat the DNA molecules and makes them more compact (**Figure 11.8**).

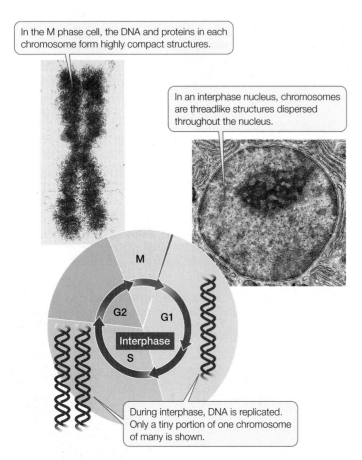

In the M phase cell, the DNA and proteins in each chromosome form highly compact structures.

In an interphase nucleus, chromosomes are threadlike structures dispersed throughout the nucleus.

During interphase, DNA is replicated. Only a tiny portion of one chromosome of many is shown.

11.8 Chromosomes, Chromatids, and Chromatin DNA in the interphase nucleus is diffuse and becomes compacted as mitosis begins.

If all the DNA in a typical human cell were put end to end, it would be nearly 2 meters long. Yet the nucleus is only 5 μm (0.000005 meters) in diameter. So eukaryotic DNA, like that in prokaryotes, is extensively packaged in a highly organized way (**Figure 11.9**). This packing is achieved largely by proteins that are closely associated with the DNA; these proteins are called **histones** (*histos*, "web" or "loom"). They are positively charged at cellular pH levels because of their high content of the basic amino acids lysine and arginine. These positive charges attract the negative phosphate groups on DNA. These DNA–histone interactions, as well as histone–histone interactions, result in the formation of beadlike units called **nucleosomes**.

During interphase, the chromatin that makes up each chromosome is much less densely packaged, and consists of single DNA molecules running around vast numbers of nucleosomes like beads on a string. During this phase of the cell cycle, the DNA is accessible to proteins involved in replication and transcription. Once a mitotic chromosome is formed its compact nature makes it inaccessible to replication and transcription factors, and so these processes cannot occur.

During the early stages of both mitosis and meiosis, the chromatin becomes ever more tightly coiled and condensed as the nucleosomes pack together. Further coiling of the chromatin continues up to the time at which the chromatids begin to move apart.

Overview: Mitosis segregates copies of genetic information

In mitosis, a single nucleus gives rise to two nuclei that are genetically identical to each other and to the parent nucleus. Mitosis (the M phase of the cell cycle) ensures the accurate segregation of the eukaryotic cell's multiple chromosomes into the daughter nuclei. While mitosis is a continuous process in which each event flows smoothly into the next, it is convenient to subdivide it into a series of stages: prophase, prometaphase, metaphase, anaphase, and telophase. Before we consider each of these stages, we will describe two cellular structures that contribute to the orderly segregation of the chromosomes during mitosis—the centrosome and the spindle.

The centrosomes determine the plane of cell division

Before the spindle apparatus for chromosome segregation forms, the orientation of this spindle is determined. This is accomplished by the **centrosome** ("central body"), an organelle in the cytoplasm near the nucleus. In many organisms, each centrosome consists of a pair of **centrioles**, each one a hollow tube formed by nine triplets of microtubules. The two tubes are at right angles to each other.

During S phase the centrosome doubles to form a pair of centrosomes. At the G2-to-M transition, the two centrosomes separate from one another, moving to opposite ends of the nuclear envelope. Eventually these will identify "poles" toward which chromosomes will move during segregation. The positions of the centrosomes determine the plane at which the cell will divide; therefore they determine the spatial relationship between the two new cells. This relationship may be of little consequence to single free-living cells such as yeasts, but it is important for cells in a multicellular organism. For example, during development from a fertilized egg to an embryo, the daughter cells from some divisions must be positioned correctly to receive signals to form new tissues.

The centrioles are surrounded by high concentrations of tubulin dimers, and these proteins aggregate to form the microtubules that orchestrate chromosomal movement. (Plant cells lack centrosomes, but distinct microtubule organizing centers at each end of the cell play the same role.) These microtubules are the major part of the spindle structure, which is required for the orderly segregation of the chromosomes.

The spindle begins to form during prophase

During interphase, only the nuclear envelope, the nucleoli (see Section 5.3), and a barely discernible tangle of chromatin are visible under the light microscope. The appearance of the nucleus changes as the cell enters **prophase**—the beginning of mitosis. Most of the cohesin that has held the two products of DNA replication together since S phase is removed, so the individual chromatids become visible. They are still held together by a small amount of cohesin at the centromere. Late in prophase, specialized three-layered structures called **kinetochores** develop in the centromere region, one on each chro-

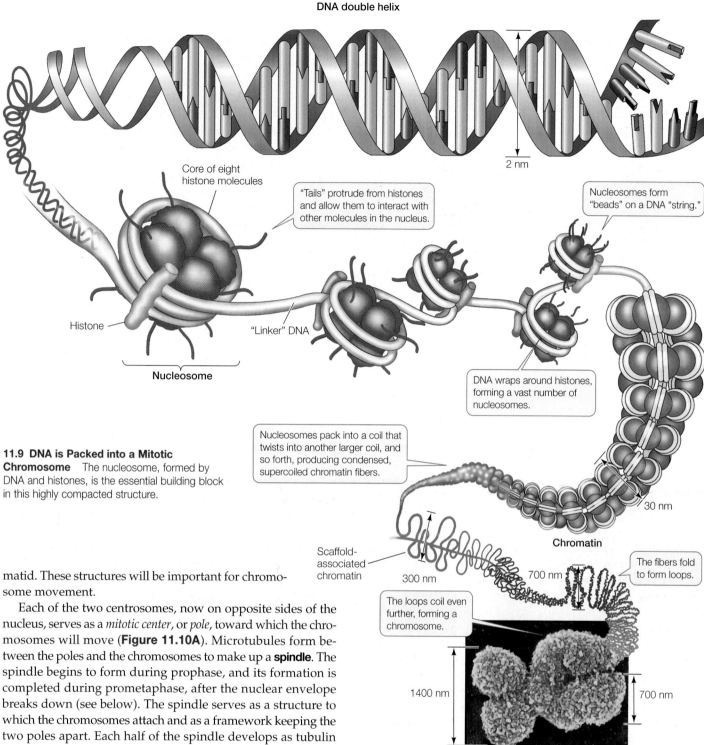

DNA double helix

2 nm

Core of eight histone molecules

"Tails" protrude from histones and allow them to interact with other molecules in the nucleus.

Nucleosomes form "beads" on a DNA "string."

Histone

"Linker" DNA

Nucleosome

DNA wraps around histones, forming a vast number of nucleosomes.

Nucleosomes pack into a coil that twists into another larger coil, and so forth, producing condensed, supercoiled chromatin fibers.

30 nm

Chromatin

11.9 DNA is Packed into a Mitotic Chromosome The nucleosome, formed by DNA and histones, is the essential building block in this highly compacted structure.

Scaffold-associated chromatin

300 nm

700 nm

The fibers fold to form loops.

The loops coil even further, forming a chromosome.

1400 nm

700 nm

Metaphase chromosome

matid. These structures will be important for chromosome movement.

Each of the two centrosomes, now on opposite sides of the nucleus, serves as a *mitotic center*, or *pole*, toward which the chromosomes will move (**Figure 11.10A**). Microtubules form between the poles and the chromosomes to make up a **spindle**. The spindle begins to form during prophase, and its formation is completed during prometaphase, after the nuclear envelope breaks down (see below). The spindle serves as a structure to which the chromosomes attach and as a framework keeping the two poles apart. Each half of the spindle develops as tubulin dimers aggregate from around the centrioles and form long fibers that extend into the middle region of the cell. The microtubules are initially unstable, constantly forming and falling apart, until they contact kinetochores or microtubules from the other half-spindle and become more stable.

There are two types of microtubule in the spindle:

- *Polar microtubules* form the framework of the spindle, and run from one pole to the other.

- *Kinetochore microtubules*, which form later, attach to the kinetochores on the chromosomes. The two sister chromatids in each chromosome pair become attached to kine-

tochore microtubules in opposite halves of the spindle (**Figure 11.10B**). This ensures that the two chromatids will eventually move to opposite poles.

Movement of the chromatids is the central feature of mitosis. It accomplishes the segregation that is needed for cell division and completion of the cell cycle. Prophase prepares for this movement, and the actual segregation takes place in the next three phases of mitosis.

(A)

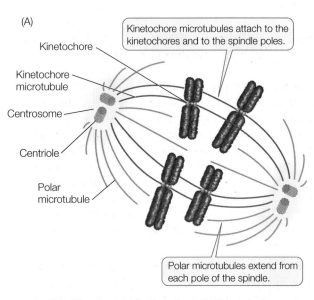

Kinetochore

Kinetochore microtubule

Centrosome

Centriole

Polar microtubule

Kinetochore microtubules attach to the kinetochores and to the spindle poles.

Polar microtubules extend from each pole of the spindle.

(B)

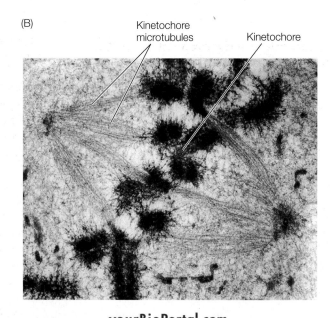

Kinetochore microtubules

Kinetochore

11.10 The Mitotic Spindle Consists of Microtubules
(A) The spindle apparatus in an animal cell at metaphase. In plant cells, centrioles are not present. (B) An electron micrograph of metaphase emphasizing the kinetochore microtubules.

yourBioPortal.com
GO TO Web Activity 11.1 • The Mitotic Spindle

11.11 The Phases of Mitosis Mitosis results in two new nuclei that are genetically identical to each other and to the nucleus from which they were formed. In the micrographs, the green dye stains microtubules (and thus the spindle); the red dye stains the chromosomes. The chromosomes in the diagrams are stylized to emphasize the fates of the individual chromatids.

yourBioPortal.com
GO TO Web Activity 11.2 • Images of Mitosis

Interphase

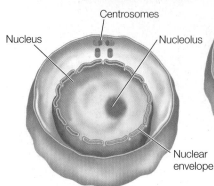

Nucleus

Centrosomes

Nucleolus

Nuclear envelope

1 During the S phase of interphase, the nucleus replicates its DNA and centrosomes.

Prophase

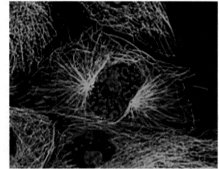

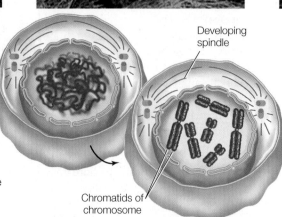

Developing spindle

Chromatids of chromosome

2 The chromatin coils and supercoils, becoming more and more compact and condensing into visible chromosomes. The chromosomes consist of identical, paired sister chromatids. Centrosomes move to opposite poles.

Prometaphase

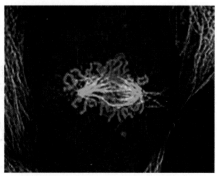

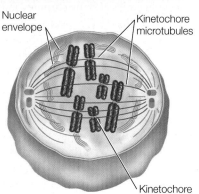

Nuclear envelope

Kinetochore microtubules

Kinetochore

3 The nuclear envelope breaks down. Kinetochore microtubules appear and connect the kinetochores to the poles.

Chromosome separation and movement are highly organized

During the next three phases, prometaphase, metaphase, and anaphase, dramatic changes take place in the cell and the chromosomes.

- **Prometaphase**. The nuclear envelope breaks down and the compacted chromosomes consisting of two chromatids attach to the kinetochore microtubules.
- **Metaphase**. The chromosomes line up at the midline of the cell (equatorial position).
- **Anaphase**. The chromatids separate and move away from each other toward the poles.

You will find these events depicted and described in **Figure 11.11**. Here, we will consider two key processes: separation of the chromatids, and the mechanism of their actual movement toward the poles.

CHROMATID SEPARATION The separation of chromatids occurs at the beginning of anaphase. It is controlled by M phase cyclin–Cdk (see Figure 11.6), which activates another protein complex called the *anaphase-promoting complex* (APC). Separation occurs because one subunit of the cohesin protein holding the sister chromatids together is hydrolyzed by a specific protease,

appropriately called *separase* (**Figure 11.12**). After they separate, the chromatids are called **daughter chromosomes**.

CHROMOSOME MOVEMENT The migration of the two sets of daughter chromosomes to the poles of the cell is a highly organized, active process. Two mechanisms operate to move the chromosomes along. First, the kinetochores contain a protein called *cytoplasmic dynein* that acts as a "molecular motor." It hydrolyzes ATP to ADP and phosphate, thus releasing energy that may move the chromosomes along the microtubules toward the poles. This accounts for about 75 percent of the force of motion. Second, the kinetochore microtubules shorten from the poles, drawing the chromosomes toward them, accounting for about 25 percent of the force of motion.

- **Telophase** occurs after the chromosomes have separated and is the last phase of mitosis. During this period, a nuclear envelope forms around each set of chromosomes, nucleoli appear, and the chromosomes become less compact. The spindle also disappears at this time. As a result, there are two new nuclei in a single cell.

Cytokinesis is the division of the cytoplasm

Mitosis refers only to the division of the nucleus. The division of the cell's cytoplasm, which follows mitosis, is called cytoki-

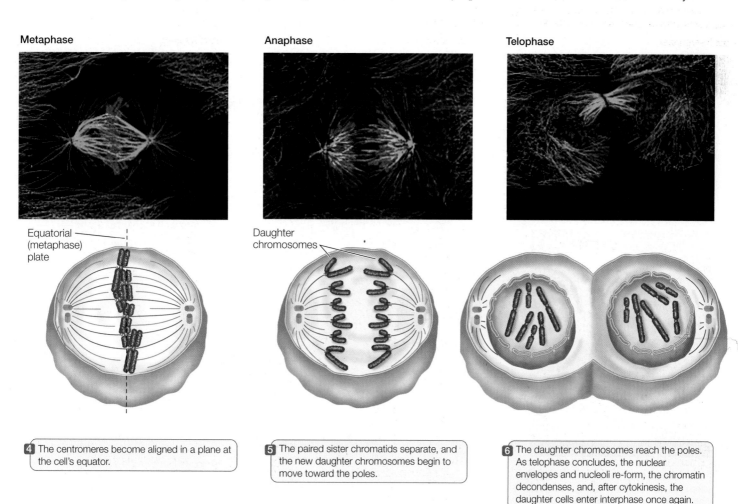

Metaphase

Anaphase

Telophase

Equatorial (metaphase) plate

Daughter chromosomes

4 The centromeres become aligned in a plane at the cell's equator.

5 The paired sister chromatids separate, and the new daughter chromosomes begin to move toward the poles.

6 The daughter chromosomes reach the poles. As telophase concludes, the nuclear envelopes and nucleoli re-form, the chromatin decondenses, and, after cytokinesis, the daughter cells enter interphase once again.

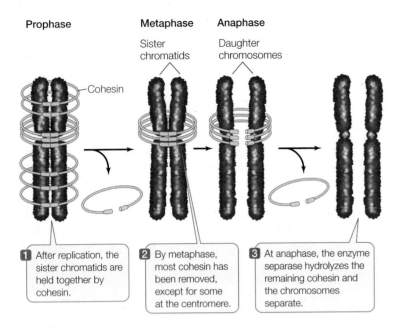

Prophase   Metaphase   Anaphase

Sister chromatids   Daughter chromosomes

Cohesin

1 After replication, the sister chromatids are held together by cohesin.

2 By metaphase, most cohesin has been removed, except for some at the centromere.

3 At anaphase, the enzyme separase hydrolyzes the remaining cohesin and the chromosomes separate.

11.12 Chromatid Attachment and Separation   The cohesin protein complex holds sister chromatids together at the centromere. The enzyme separase hydrolyzes cohesin at the onset of anaphase, allowing the chromatids to separate into daughter chromosomes.

mic surface of the plasma membrane. These two proteins interact to produce a contraction, just as they do in muscles, thus pinching the cell in two. The microfilaments assemble rapidly from actin monomers that are present in the interphase cytoskeleton. Their assembly is under the control of calcium ions that are released from storage sites in the center of the cell.

The plant cell cytoplasm divides differently because plants have cell walls that are rigid. In plant cells, as the spindle breaks down after mitosis, membranous vesicles derived from the Golgi apparatus appear along the plane of cell division, roughly midway between the two daughter nuclei. The vesicles are propelled along microtubules by the motor protein kinesin, and fuse to form a new plasma membrane. At the same time they contribute their contents to a *cell plate*, which is the beginning of a new cell wall (**Figure 11.13B**).

Following cytokinesis, each daughter cell contains all the components of a complete cell. A precise distribution of chromosomes is ensured by mitosis. In contrast, organelles such as ribosomes, mitochondria, and chloroplasts need not be distributed equally between daughter cells as long as some of each are present in each cell. Accordingly, there is no mechanism with a precision comparable to that of mitosis to provide for their equal allocation to daughter cells. As we will see in Chapter 19, the unequal distribution of cytoplasmic components during development can have functional significance for the two new cells.

nesis. Cytokinesis occurs in different ways, depending on the type of organism. In particular there are substantial differences between the process in plants and in animals.

In animal cells, cytokinesis usually begins with a furrowing of the plasma membrane, as if an invisible thread were cinching the cytoplasm between the two nuclei (**Figure 11.13A**). This *contractile ring* is composed of microfilaments of actin and myosin (see Figure 5.18), which form a ring on the cytoplas-

yourBioPortal.com
GO TO Animated Tutorial 11.1 • Mitosis

(A)

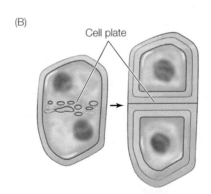

Contractile ring

The contractile ring has completely separated the cytoplasms of these two daughter cells, although their surfaces remain in contact.

(B)

Cell plate

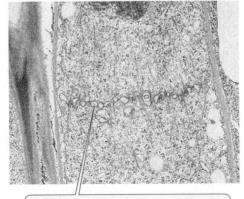

This row of vesicles will fuse to form a cell plate between the cell above and the cell below.

11.13 Cytokinesis Differs in Animal and Plant Cells   (A) A sea urchin zygote (fertilized egg) that has just completed cytokinesis at the end of the first cell division of its development into an embryo. (B) A dividing plant cell in late telophase. Plant cells divide differently from animal cells because plant cells have cell walls.

11.3 RECAP

Mitosis is the division of the nucleus of a eukaryotic cell into two nuclei identical to each other and to the parent nucleus. The process of mitosis, while continuous, can be viewed as a series of events (prophase, prometaphase, metaphase, anaphase, and telophase). Once two identical nuclei have formed, the cell divides into two cells by cytokinesis.

- What is the difference between a chromosome, a chromatid, and a daughter chromosome? **See Figures 11.8 and 11.11**

- What are the various levels of "packing" by which the genetic information contained in linear DNA is condensed during prophase? **See p. 216 and Figure 11.9**

- Describe how chromosomes move during mitosis. **See p. 219 and Figure 11.11**

- What are the differences in cytokinesis between plant and animal cells? **See p. 220 and Figure 11.13**

The intricate process of mitosis results in two cells that are genetically identical. But, as mentioned earlier, there is another eukaryotic cell division process, called meiosis, that results in genetic diversity. What is the role of that process?

11.4 What Role Does Cell Division Play in a Sexual Life Cycle?

The mitotic cell cycle repeats itself and by this process, a single cell can give rise to a vast number of cells with identical nuclear DNA. Meiosis, on the other hand, produces just four daughter cells. Mitosis and meiosis are both involved in reproduction, but they have different roles: asexual reproduction involves only mitosis, while sexual reproduction involves both mitosis and meiosis.

Asexual reproduction by mitosis results in genetic constancy

Asexual reproduction, sometimes called *vegetative reproduction*, is based on the mitotic division of the nucleus. An organism that reproduces asexually may be single-celled like yeast, reproducing itself with each cell cycle, or it may be multicellular like the cholla cactus, that breaks off a piece to produce a new multicellular organism (**Figure 11.14**). Asexual reproduction is a rapid and effective means of making new individuals, and it is common in nature. In asexual reproduction, the offspring are **clones** of the parent organism; that is, the offspring are *genetically identical* to the parent. Any genetic variation among the offspring is most likely due to small environmentally caused changes in the DNA, called *mutations*. As you will see, this small amount of variation contrasts with the extensive variation possible in sexually reproducing organisms.

Sexual reproduction by meiosis results in genetic diversity

Unlike asexual reproduction, **sexual reproduction** results in an organism that is not identical to its parents. Sexual reproduction requires **gametes** created by meiosis; two parents each contribute one gamete to each of their offspring. Meiosis can produce gametes—and thus offspring—that differ genetically from each other and from the parents. Because of this genetic variation, some offspring may be better adapted than others to sur-

(A)

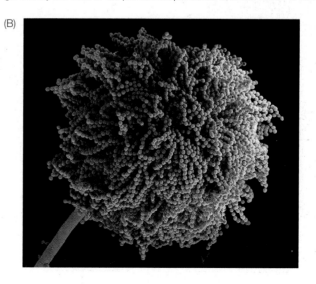

(B)

11.14 Asexual Reproduction in the Large and the Small (A) Some cacti like this cholla have brittle stems that break off easily. Fragments on the ground set down roots and develop by mitotic cell divisions into new plants that are genetically identical to the plant they came from. (B) These strings of cells are asexual spores formed by a fungus. Each spore contains a nucleus produced by a mitotic division and is genetically identical to the parent that produced it. It can divide to form a new fungus.

vive and reproduce in a particular environment. Meiosis thus generates the genetic diversity that is the raw material for natural selection and evolution.

In most multicellular organisms, the body cells that are *not* specialized for reproduction, called **somatic cells**, each contain two sets of chromosomes, which are found in pairs. One chromosome of each pair comes from each of the organism's two parents; for example, in humans with 46 chromosomes, 23 come from the mother and 23 from the father. The members of such a **homologous pair** are similar in size and appearance, except for the sex chromosomes found in some species (see Section 12.4). The two chromosomes in a homologous pair (called **homologs**) bear corresponding, though often not identical, genetic information. For example, a homologous pair of chromosomes in a plant may carry different versions of a gene that controls seed shape. One homolog may carry the version for wrinkled seeds while the other may carry the version for smooth seeds.

Gametes, on the other hand, contain only a single set of chromosomes—that is, one homolog from each pair. The number of chromosomes in a gamete is denoted by n, and the cell is said to be **haploid**. Two haploid gametes fuse to form a **zygote**, in a process called **fertilization**. The zygote thus has two sets of chromosomes, just as somatic cells do. Its chromosome number is denoted by $2n$, and the zygote is said to be **diploid**. Depending on the organism, the zygote may divide by either meiosis or mi-

tosis. Either way, a new mature organism develops that is capable of sexual reproduction.

All sexual life cycles involve meiosis to produce gametes or cells that are haploid. Eventually, the haploid cells or gametes fuse to produce a zygote, beginning the diploid stage of the life-cycle. Since the origin of sexual reproduction, evolution has generated many different versions of the sexual life cycle. **Figure 11.15** presents three examples.

- In *haplontic* organisms, including most protists, fungi, and some green algae, the tiny zygote is the only diploid cell in the life cycle. After it is formed it immediately undergoes meiosis to produce more haploid cells. These are usually *spores*, which are the dispersal units for the organism, like the seeds of a plant. A spore germinates to form a new haploid organism, which may be single-celled or multicellular. Cells of the mature haploid organism fuse to form the diploid zygote.

- Most plants and some fungi display **alternation of generations**. As for many haplontic organisms, meiosis gives rise

11.15 Fertilization and Meiosis Alternate in Sexual Reproduction In sexual reproduction, haploid (*n*) cells or organisms alternate with diploid (*2n*) cells or organisms.

yourBioPortal.com
GO TO **Web Activity 11.3 • Sexual Life Cycle**

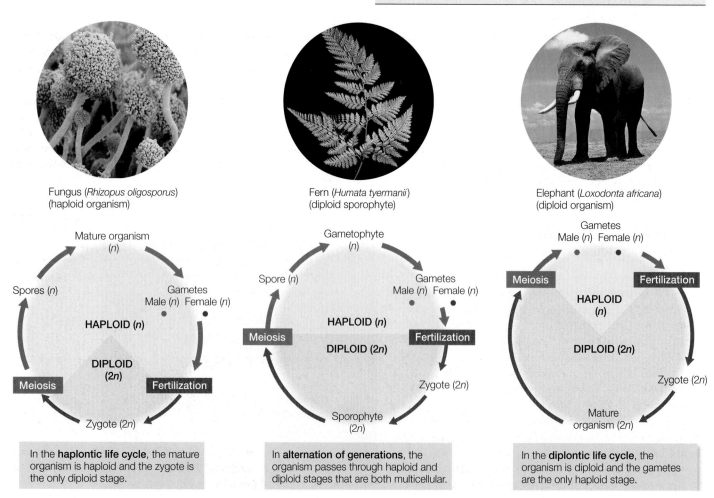

Fungus (*Rhizopus oligosporus*)
(haploid organism)

Fern (*Humata tyermanii*)
(diploid sporophyte)

Elephant (*Loxodonta africana*)
(diploid organism)

In the **haplontic life cycle**, the mature organism is haploid and the zygote is the only diploid stage.

In **alternation of generations**, the organism passes through haploid and diploid stages that are both multicellular.

In the **diplontic life cycle**, the organism is diploid and the gametes are the only haploid stage.

(A)

> Centromeres (arrows) occupy characteristic positions on homologous chromosomes.

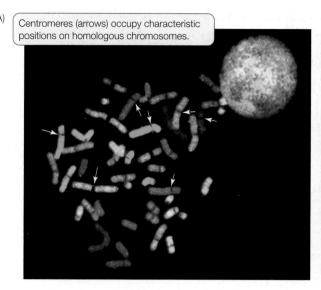

(B)

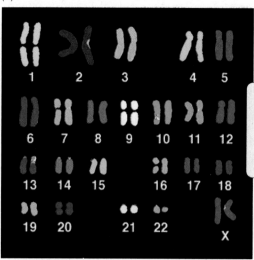

> Humans have 23 pairs of chromosomes, including the sex chromosomes. This female's sex chromosomes are X and X; a male would have X and Y chromosomes.

11.16 The Human Karyotype (A) Chromosomes from a human cell in metaphase. The DNA of each chromosome pair has a specific nucleotide sequence that is stained by a particular colored dye, so that the chromosomes in a homologous pair share a distinctive color. Each chromosome at this stage is composed of two chromatids, but they cannot be distinguished. At the upper right is an interphase nucleus. (B) This karyogram, produced by computerized analysis of the image on the left, shows homologous pairs lined up together and numbered, clearly revealing the individual's karyotype.

to haploid spores, which divide by mitosis to form a haploid life stage called the *gametophyte*. The gametophyte forms gametes by mitosis, which fuse to form a diploid zygote. The zygote divides by mitosis to become the diploid *sporophyte*, which in turn produces the gametes by meiosis.

- In *diplontic* organisms, which include animals, brown algae and some fungi, the gametes are the only haploid cells in the life cycle, and the mature organism is diploid.

These life cycles are described in greater detail in Part Seven. For now we will focus on the role of sexual reproduction in generating diversity among individual organisms.

The essence of sexual reproduction is the *random selection of half of the diploid chromosome set* to make a haploid gamete, followed by fusion of two haploid gametes to produce a diploid cell. Both of these steps contribute to a shuffling of genetic information in the population, so that no two individuals have exactly the same genetic constitution. The diversity provided by sexual reproduction opens up enormous opportunities for evolution.

The number, shapes, and sizes of the metaphase chromosomes constitute the karyotype

When cells are in metaphase of mitosis, it is often possible to count and characterize their individual chromosomes. If a photomicrograph of the entire set of chromosomes is made, the images of the individual chromosomes can be manipulated, pairing and placing them in an orderly arrangement. Such a re-arranged photomicrograph reveals the number, shapes, and sizes of the chromosomes in a cell, which together constitute its **karyotype** (**Figure 11.16**). In humans, karyotypes can aid in the diagnosis of certain diseases, and this has led to an entire branch of medicine called *cytogenetics*. However, as you will see in Chapter 15, chromosome analysis with the microscope is being replaced by direct analysis of DNA.

Individual chromosomes can be recognized by their lengths, the positions of their centromeres, and characteristic banding

patterns that are visible when the chromosomes are stained and observed at high magnification. In diploid cells, the karyotype consists of homologous pairs of chromosomes—for example, there are 23 pairs and a total of 46 chromosomes in humans. There is no simple relationship between the size of an organism and its chromosome number. A housefly has 5 chromosome pairs and a horse has 32, but the smaller carp (a fish) has 52 pairs. Probably the highest number of chromosomes in any organism is in the fern *Ophioglossum reticulatum*, which has 1,260 (630 pairs)!

Meiosis, unlike mitosis, results in daughter cells that have half as many chromosomes as the parent cell. Next we will look at the processes of meiosis.

Early prophase I

Centrosomes

1 The chromatin begins to condense following interphase.

Mid-prophase I

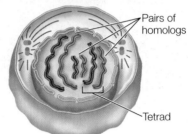

Pairs of homologs

Tetrad

2 Synapsis aligns homologs, and chromosomes condense further.

Late prophase I–Prometaphase

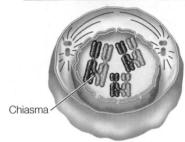

Chiasma

3 The chromosomes continue to coil and shorten. The chiasmata reflect crossing over, the exchange of genetic material between nonsister chromatids in a homologous pair. In prometaphase the nuclear envelope breaks down.

Prophase II

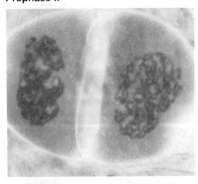

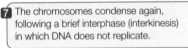

7 The chromosomes condense again, following a brief interphase (interkinesis) in which DNA does not replicate.

Metaphase II

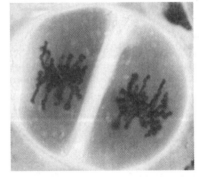

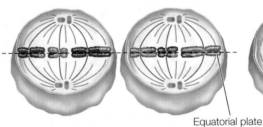

Equatorial plate

8 The centromeres of the paired chromatids line up across the equatorial plates of each cell.

Anaphase II

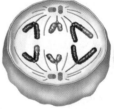

9 The chromatids finally separate, becoming chromosomes in their own right, and are pulled to opposite poles. Because of crossing over and independent assortment, each new cell will have a different genetic makeup.

11.5 What Happens during Meiosis?

In the last section we described the role and importance of meiosis in sexual reproduction. Now we will see how meiosis accomplishes the orderly and precise generation of haploid cells.

Meiosis consists of *two* nuclear divisions that reduce the number of chromosomes to the haploid number, in preparation for sexual reproduction. Although the *nucleus divides twice* during meiosis, the *DNA is replicated only once*. Unlike the products of mitosis, the products of meiosis are genetically different from one another and from the parent cell. To understand the process of meiosis and its specific details, it is useful to keep in mind the overall functions of meiosis:

Metaphase I

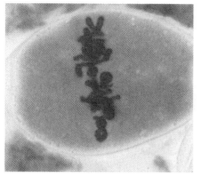

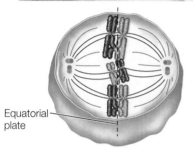

Equatorial plate

4 The homologous pairs line up on the equatorial (metaphase) plate.

Anaphase I

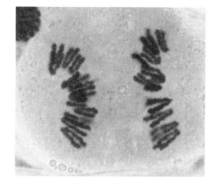

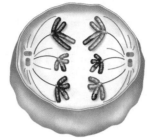

5 The homologous chromosomes (each with two chromatids) move to opposite poles of the cell.

Telophase I

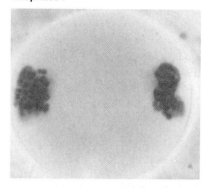

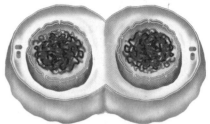

6 The chromosomes gather into nuclei, and the original cell divides.

Telophase II

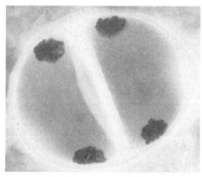

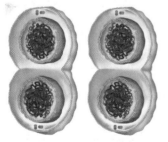

10 The chromosomes gather into nuclei, and the cells divide.

Products

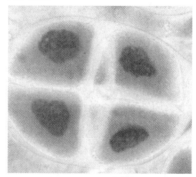

11 Each of the four cells has a nucleus with a haploid number of chromosomes.

11.17 Meiosis: Generating Haploid Cells
In meiosis, two sets of chromosomes are divided among four daughter nuclei, each of which has half as many chromosomes as the original cell. Four haploid cells are the result of two successive nuclear divisions. The micrographs show meiosis in the male reproductive organ of a lily; the diagrams show the corresponding phases in an animal cell. (For instructional purposes, the chromosomes from one parent are colored blue and those from the other parent are red.)

yourBioPortal.com
GO TO Web Activity 11.4 • Images of Meiosis

Meiotic division reduces the chromosome number

As noted above, meiosis consists of two nuclear divisions, *meiosis I* and *meiosis II*. Two unique features characterize **meiosis I**.

- *Homologous chromosomes come together to pair* along their entire lengths. No such pairing occurs in mitosis.

- *The homologous chromosome pairs separate,* but the individual chromosomes, each consisting of two sister chromatids, remain intact. (The chromatids will separate during meiosis II.)

Like mitosis, meiosis I is preceded by an interphase with an S phase, during which each chromosome is replicated. As a result, each chromosome consists of two sister chromatids, held together by cohesin proteins. At the end of meiosis I, two nuclei form, each with half of the original chromosomes (one member of each homologous pair). Since the centromeres did not

- To reduce the chromosome number from diploid to haploid
- To ensure that each of the haploid products has a complete set of chromosomes
- To generate genetic diversity among the products

The events of meiosis are illustrated in **Figure 11.17**. In this section, we discuss some of the key features that distinguish meiosis from mitosis.

separate, these chromosomes are still double—composed of two sister chromatids. The sister chromatids are separated during **meiosis II**, which is *not* preceded by DNA replication. As a result, the products of meiosis I and II are four cells, each containing the haploid number of chromosomes. But these four cells are not genetically identical.

Chromatid exchanges during meiosis I generate genetic diversity

Meiosis I begins with a long prophase I (the first three panels of Figure 11.17), during which the chromosomes change markedly. The homologous chromosomes pair by adhering along their lengths in a process called **synapsis**. (This does not happen in mitosis.) This pairing process lasts from prophase I to the end of metaphase I. The four chromatids of each pair of homologous chromosomes form a **tetrad**, or *bivalent*. For example, in a human cell at the end of prophase I there are 23 tetrads, each consisting of four chromatids. The four chromatids come from the two partners in each homologous pair of chromosomes.

Throughout prophase I and metaphase I, the chromatin continues to coil and compact, so that the chromosomes appear ever thicker. At a certain point, the homologous chromosomes appear to repel each other, especially near the centromeres, but they remain held together by physical attachments mediated by cohesins. Later in prophase, regions having these attachments take on an X-shaped appearance (**Figure 11.18**) and are called **chiasmata** (singular *chiasma*, "cross").

A chiasma reflects an *exchange of genetic material* between nonsister chromatids on homologous chromosomes—what geneticists call **crossing over** (**Figure 11.19**). The chromosomes usually

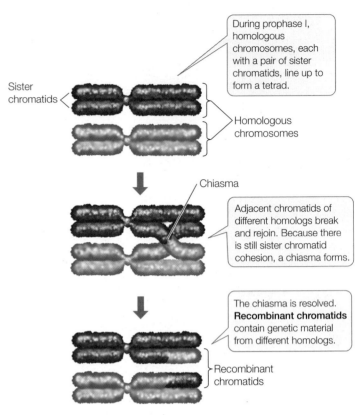

11.19 Crossing Over Forms Genetically Diverse Chromosomes
The exchange of genetic material by crossing over results in new combinations of genetic information on the recombinant chromosomes. The two different colors distinguish the chromosomes contributed by the male and female parents.

Callout boxes:
During prophase I, homologous chromosomes, each with a pair of sister chromatids, line up to form a tetrad.

Adjacent chromatids of different homologs break and rejoin. Because there is still sister chromatid cohesion, a chiasma forms.

The chiasma is resolved. **Recombinant chromatids** contain genetic material from different homologs.

Labels: Sister chromatids; Homologous chromosomes; Chiasma; Recombinant chromatids

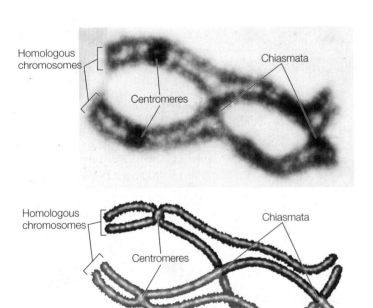

11.18 Chiasmata: Evidence of Genetic Exchange between Chromatids This micrograph shows a pair of homologous chromosomes, each with two chromatids, during prophase I of meiosis in a salamander. Two chiasmata are visible.

Labels: Homologous chromosomes; Chiasmata; Centromeres

begin exchanging material shortly after synapsis begins, but chiasmata do not become visible until later, when the homologs are repelling each other. Crossing over results in **recombinant** chromatids, and it increases genetic variation among the products of meiosis by reshuffling genetic information among the homologous pairs. In Chapter 12 we explore further the genetic consequences of crossing over. Mitosis seldom takes more than an hour or two, but meiosis can take *much* longer. In human males, the cells in the testis that undergo meiosis take about a week for prophase I and about a month for the entire meiotic cycle. In females, prophase I begins long before a woman's birth, during her early fetal development, and ends as much as decades later, during the monthly ovarian cycle.

During meiosis homologous chromosomes separate by independent assortment

A diploid organism has two sets of chromosomes (*2n*); one set derived from its male parent, and the other from its female parent. As the organism grows and develops, its cells undergo mitotic divisions. In mitosis, each chromosome behaves independently of its homolog, and its two chromatids are sent to opposite poles during anaphase. Each daughter nucleus ends up with *2n* chromosomes. In meiosis, things are very different. **Figure 11.20** compares the two processes.

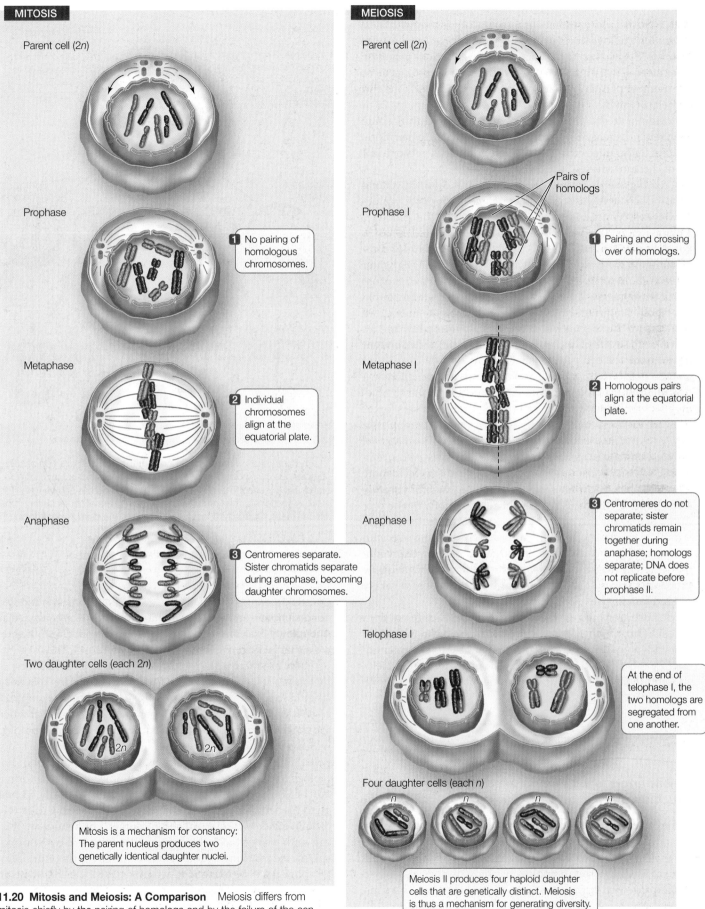

11.20 Mitosis and Meiosis: A Comparison Meiosis differs from mitosis chiefly by the pairing of homologs and by the failure of the centromeres to separate at the end of metaphase I.

MITOSIS

Parent cell (2n)

Prophase

1 No pairing of homologous chromosomes.

Metaphase

2 Individual chromosomes align at the equatorial plate.

Anaphase

3 Centromeres separate. Sister chromatids separate during anaphase, becoming daughter chromosomes.

Two daughter cells (each 2n)

2n   2n

Mitosis is a mechanism for constancy: The parent nucleus produces two genetically identical daughter nuclei.

MEIOSIS

Parent cell (2n)

Prophase I

Pairs of homologs

1 Pairing and crossing over of homologs.

Metaphase I

2 Homologous pairs align at the equatorial plate.

Anaphase I

3 Centromeres do not separate; sister chromatids remain together during anaphase; homologs separate; DNA does not replicate before prophase II.

Telophase I

At the end of telophase I, the two homologs are segregated from one another.

Four daughter cells (each n)

n   n   n   n

Meiosis II produces four haploid daughter cells that are genetically distinct. Meiosis is thus a mechanism for generating diversity.

In meiosis I, chromosomes of maternal origin pair with their paternal homologs during synapsis. *This pairing does not occur in mitosis.* Segregation of the homologs during meiotic anaphase I (see steps 4–6 of Figure 11.17) ensures that each pole receives one member of each homologous pair. For example, at the end of meiosis I in humans, each daughter nucleus contains 23 of the original 46 chromosomes. In this way, the chromosome number is decreased from diploid to haploid. Furthermore, meiosis I guarantees that each daughter nucleus gets one full set of chromosomes.

Crossing over is one reason for the genetic diversity among the products of meiosis. The other source of diversity is independent assortment. It is a matter of chance which member of a homologous pair goes to which daughter cell at anaphase I. For example, imagine there are two homologous pairs of chromosomes in the diploid parent nucleus. A particular daughter nucleus could receive the paternal chromosome 1 and the maternal chromosome 2. Or it could get paternal 2 and maternal 1, or both maternal, or both paternal chromosomes. It all depends on the way in which the homologous pairs line up at metaphase I. This phenomenon is termed **independent assortment**.

Note that of the four possible chromosome combinations just described, only two produce daughter nuclei with full complements of either maternal or paternal chromosome sets (apart from the material exchanged by crossing over). *The greater the number of chromosomes, the less probable that the original parental combinations will be reestablished, and the greater the potential for genetic diversity.* Most species of diploid organisms have more than two pairs of chromosomes. In humans, with 23 chromosome pairs, 2^{23} (8,388,608) different combinations can be produced just by the mechanism of independent assortment. Taking the extra genetic shuffling afforded by crossing over into account, the number of possible combinations is virtually infinite. Crossing over and independent assortment, along with the processes that result in mutations, provide the genetic diversity needed for evolution by natural selection.

We have seen how meiosis I is fundamentally different from mitosis. On the other hand, meiosis II is similar to mitosis, in that it involves the separation of chromatids into daughter nuclei (see steps 7–11 in Figure 11.17). The final products of meiosis I and meiosis II are four haploid daughter cells, each with one set (*n*) of chromosomes.

— **yourBioPortal.com** —
GO TO Animated Tutorial 11.2 • Meiosis

Meiotic errors lead to abnormal chromosome structures and numbers

In the complex process of meiosis, things occasionally go wrong. A pair of homologous chromosomes may fail to separate during meiosis I, or sister chromatids may fail to separate during meiosis II. Conversely, homologous chromosomes may fail to remain together during metaphase I, and then both may mi-

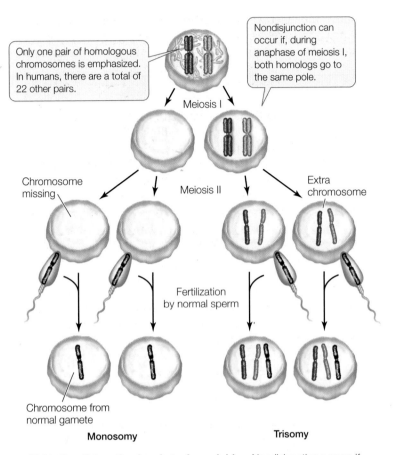

11.21 Nondisjunction Leads to Aneuploidy   Nondisjunction occurs if homologous chromosomes fail to separate during meiosis I, or if chromatids fail to separate during meiosis II. The first case is shown here. The result is aneuploidy: one or more chromosomes are either lacking or present in excess. Generally, aneuploidy is lethal to the developing embryo.

grate to the same pole in anaphase I. This phenomenon is called **nondisjunction** and it results in the production of *aneuploid* cells. **Aneuploidy** is a condition in which one or more chromosomes are either lacking or present in excess (**Figure 11.21**).

There are many different causes of aneuploidy, but one of them may result from a breakdown in the cohesins that keep sister chromatids and tetrads joined together during prophase I (see Figure 11.17). These and other proteins ensure that when the chromosomes line up at the equatorial plate, one homolog will face one pole and the other homolog will face the other pole. If the cohesins break down at the wrong time, both homologs may go to one pole. If, for example, during the formation of a human egg, both members of the chromosome 21 pair go to the same pole during anaphase I, the resulting eggs will contain either two copies of chromosome 21 or none at all. If an egg with two of these chromosomes is fertilized by a normal sperm, the resulting zygote will have three copies of the chromosome: it will be **trisomic** for chromosome 21. A child with an extra chromosome 21 has the symptoms of *Down syndrome*: impaired intelligence; characteristic abnormalities of the hands, tongue, and eyelids; and an increased susceptibility to cardiac abnormalities and diseases such as leukemia. If an egg that did not receive chromosome 21 is fertilized by a normal sperm, the zygote will have only one copy: it will be **monosomic** for chromosome 21, and this is lethal.

Trisomies and the corresponding monosomies are surprisingly common in human zygotes, with 10–30 percent of all conceptions showing aneuploidy. But most of the embryos that develop from such zygotes do not survive to birth, and those that do often die before the age of 1 year (trisomies for chromosome 21 are the viable exception). At least one-fifth of all recognized pregnancies are spontaneously terminated (miscarried) during the first 2 months, largely because of trisomies and monosomies. The actual proportion of spontaneously terminated pregnancies is certainly higher, because the earliest ones often go unrecognized.

Other abnormal chromosomal events can also occur. In a process called **translocation**, a piece of a chromosome may break away and become attached to another chromosome. For example, a particular large part of one chromosome 21 may be translocated to another chromosome. Individuals who inherit this translocated piece along with two normal chromosomes 21 will have Down syndrome.

Polyploids have more than two complete sets of chromosomes

As mentioned in Section 11.4, mature organisms are often either diploid (for example, most animals) or haploid (for example, most fungi). Under some circumstances, triploid (3n), tetraploid (4n), or higher-order **polyploid** nuclei may form. Each of these *ploidy levels* represents an increase in the number of complete chromosome sets present. Organisms with complete extra sets of chromosomes may sometimes be produced by artificial breeding or by natural accidents. Polyploidy occurs naturally in some animals and many plants, and it has probably led to speciation (the evolution of a new species) in some cases.

A diploid nucleus can undergo normal meiosis because there are two sets of chromosomes to make up homologous pairs, which separate during anaphase I. Similarly, a tetraploid nucleus has an even number of each kind of chromosome, so each chromosome can pair with its homolog. However, a triploid nucleus cannot undergo normal meiosis because one-third of the chromosomes would lack partners. Polyploidy has implications for agriculture, particularly in the production of hybrid plants. For example, ploidy must be taken into account in wheat breeding because there are diploid, tetraploid, and hexaploid wheat varieties. Polyploidy can be a desirable trait in crops and ornamental plants because it often leads to more robust plants with larger flowers, fruits, and seeds. Triploidy can be useful in some circumstances. For example, rivers and lakes can be stocked with triploid trout, which are sterile and will not escape to reproduce in waters where they might upset the natural ecology.

11.5 RECAP

Meiosis produces four daughter cells in which the chromosome number is reduced from diploid to haploid. Because of the independent assortment of chromosomes and the crossing over of homologous chromatids, the four products of meiosis are not genetically identical. Meiotic errors, such as the failure of a homologous chromosome pair to separate, can lead to abnormal numbers of chromosomes.

- How do crossing over and independent assortment result in unique daughter nuclei? **See p. 226 and Figure 11.19**

- What are the differences between meiosis and mitosis? **See pp. 224–228 and Figure 11.20**

- What is aneuploidy, and how can it arise from nondisjunction during meiosis? **See p. 228 and Figure 11.21**

An essential role of cell division in complex eukaryotes is to replace cells that die. What happens to those cells?

11.6 In a Living Organism, How Do Cells Die?

Cells die in one of two ways. The first type of cell death, **necrosis**, occurs when cells are damaged by mechanical means or toxins, or are starved of oxygen or nutrients. These cells usually swell up and burst, releasing their contents into the extracellular environment. This process often results in inflammation (see Section 42.2).

More typically, cell death is due to **apoptosis** (Greek, "falling apart"). Apoptosis is a programmed series of events that result in cell death. Why would a cell initiate apoptosis, which is essentially cell suicide? In animals, there are two possible reasons:

- *The cell is no longer needed by the organism.* For example, before birth, a human fetus has weblike hands, with connective tissue between the fingers. As development proceeds, this unneeded tissue disappears as its cells undergo apoptosis in response to specific signals.

- *The longer cells live, the more prone they are to genetic damage that could lead to cancer.* This is especially true of epithelial cells of the surface of an organism, which may be exposed to radiation or toxic substances. Such cells normally die after only days or weeks and are replaced by new cells.

The outward events of apoptosis are similar in many organisms. The cell becomes detached from its neighbors, cuts up its chromatin into nucleosome-sized pieces, and forms membranous lobes, or "blebs," that break up into cell fragments (**Figure 11.22A**). In a remarkable example of the economy of nature, the surrounding living cells usually ingest the remains of the dead cell by phagocytosis. Neighboring cells digest the apop-

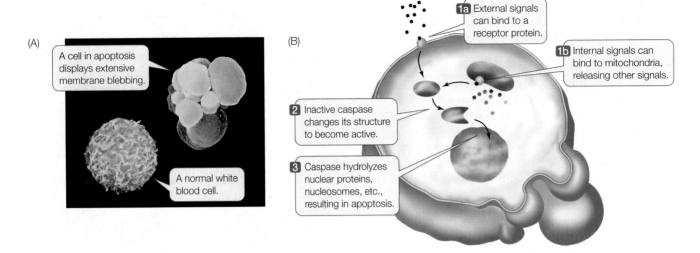

(A)

A cell in apoptosis displays extensive membrane blebbing.

A normal white blood cell.

(B)

1a External signals can bind to a receptor protein.

1b Internal signals can bind to mitochondria, releasing other signals.

2 Inactive caspase changes its structure to become active.

3 Caspase hydrolyzes nuclear proteins, nucleosomes, etc., resulting in apoptosis.

11.22 Apoptosis: Programmed Cell Death   (A) Many cells are programmed to "self-destruct" when they are no longer needed, or when they have lived long enough to accumulate a burden of DNA damage that might harm the organism. (B) Both external and internal signals stimulate caspases, the enzymes that break down specific cell constituents, resulting in apoptosis.

totic cell contents in their lysosomes and the digested components are recycled.

Apoptosis is also used by plant cells, in an important defense mechanism called the *hypersensitive response*. Plants can protect themselves from disease by undergoing apoptosis at the site of infection by a fungus or bacterium. With no living tissue to grow in, the invading organism is not able to spread to other parts of the plant. Because of their rigid cell wall, plant cells do not form blebs the way that animal cells do. Instead, they digest their own cell contents in the vacuole and then release the digested components into the vascular system.

Despite these differences between plant and animal cells, they share many of the signal transduction pathways that lead to apoptosis. Like the cell division cycle, programmed cell death is controlled by signals, which may come from inside or outside the cell (**Figure 11.22B**). Internal signals may be linked to the absence of mitosis or the recognition of damaged DNA. External signals (or a lack of them) can cause a receptor protein in the plasma membrane to change its shape, and in turn activate a signal transduction pathway. Both internal and external signals can lead to the activation of a class of enzymes called **caspases**. These enzymes are proteases that hydrolyze target molecules in a cascade of events. As a result, the cell dies as the caspases hydrolyze proteins of the nuclear envelope, nucleosomes, and plasma membrane.

11.6 RECAP

Cell death can occur either by necrosis or by apoptosis. Apoptosis is governed by precise molecular controls.

- What are some differences between apoptosis and necrosis? See p. 229
- In what situation is apoptosis necessary? See p. 229
- How is apoptosis regulated? See Figure 11.22

11.7 How Does Unregulated Cell Division Lead to Cancer?

Perhaps no malady affecting people in the industrialized world instills more fear than cancer, and most people realize that it involves an inappropriate increase in cell numbers. One in three Americans will have some form of cancer in their lifetimes, and at present, one in four will die of it. With 1.5 million new cases and half a million deaths in the United States annually, cancer ranks second only to heart disease as a killer.

Cancer cells differ from normal cells

Cancer cells differ from the normal cells from which they originate in two ways:

- Cancer cells lose control over cell division.
- Cancer cells can migrate to other locations in the body.

Most cells in the body divide only if they are exposed to extracellular signals such as growth factors. Cancer cells do not respond to these controls, and instead divide more or less continuously, ultimately forming **tumors** (large masses of cells). By the time a physician can feel a tumor or see one on an X-ray film or CAT scan, it already contains millions of cells. Tumors can be benign or malignant.

Benign tumors resemble the tissue they came from, grow slowly, and remain localized where they develop. For example, a lipoma is a benign tumor of fat cells that may arise in the armpit and remain there. Benign tumors are not cancers, but they must be removed if they impinge on an organ, obstructing its function.

Malignant tumors do not look like their parent tissue at all. A flat, specialized epithelial cell in the lung wall may turn into a relatively featureless, round, malignant lung cancer cell (**Figure 11.23**). Malignant cells often have irregular structures, such as variable nucleus sizes and shapes. Recall the opening story of this chapter, in which cervical cancer was diagnosed by cell structure.

The second and most fearsome characteristic of cancer cells is their ability to invade surrounding tissues and spread to other parts of the body by traveling through the bloodstream or lymphatic ducts. When malignant cells become lodged in some distant part of the body they go on dividing and growing, establishing a tumor at that new site. This spreading, called **metastasis**, results in organ failures and makes the cancer very hard to treat.

Cancer cells lose control over the cell cycle and apoptosis

Earlier in this chapter you learned about proteins that regulate the progress of a eukaryotic cell through the cell cycle:

- Positive regulators such as growth factors stimulate the cell cycle: they are like "gas pedals."

- Negative regulators such as RB inhibit the cell cycle: they are like "brakes."

Just as driving a car requires stepping on the gas pedal *and* releasing the brakes, a cell will go through a division cycle only if the positive regulators are active and the negative regulators are inactive.

In most cells, the two regulatory systems ensure that cells divide only when needed. In cancer cells, these two processes are abnormal.

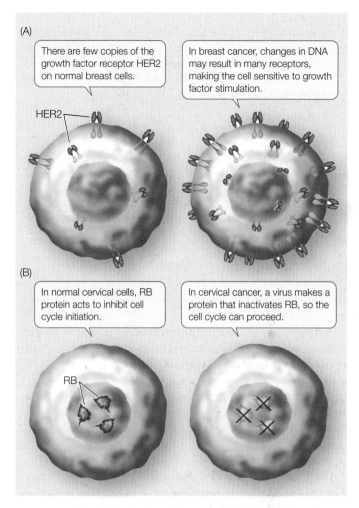

(A)

There are few copies of the growth factor receptor HER2 on normal breast cells.

In breast cancer, changes in DNA may result in many receptors, making the cell sensitive to growth factor stimulation.

HER2

(B)

In normal cervical cells, RB protein acts to inhibit cell cycle initiation.

In cervical cancer, a virus makes a protein that inactivates RB, so the cell cycle can proceed.

RB

11.24 Molecular Changes in Cancer Cells In cancer, oncogene proteins become active (A) and tumor suppressor proteins become inactive (B).

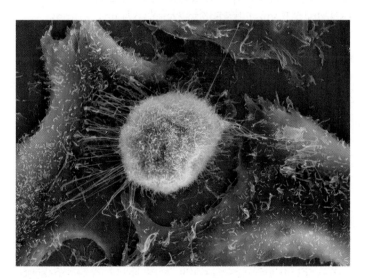

11.23 A Cancer Cell with its Normal Neighbors This lung cancer cell (yellow-green) is quite different from the normal lung cells surrounding it. The cancer cell can divide more rapidly than its normal counterparts, and it can spread to other organs. This form of small-cell cancer is lethal, with a 5-year survival rate of 10 percent. Most cases are caused by tobacco smoking.

- **Oncogene** proteins are positive regulators in cancer cells. They are derived from normal positive regulators that have become mutated to be overly active or that are present in excess, and they stimulate the cancer cells to divide more often. Oncogene products could be growth factors, their receptors, or other components in the signal transduction pathway. An example of an oncogene protein is the growth factor receptor in a breast cancer cell (**Figure 11.24A**). Normal breast cells have relatively low numbers of the growth factor receptor HER2. So when this growth factor is made, it doesn't find many breast cell receptors with which to bind and initiate cell division. In about 25 percent of breast cancers, a DNA change results in the increased production of the HER2 receptor. This results in positive stimulation of the cell cycle, and a rapid proliferation of cells with the altered DNA.

- **Tumor suppressors** are negative regulators in both cancer and normal cells, but in cancer cells they are inactive. An example is the RB protein that acts at R (the restriction point) in G1 (see Figure 11.6). When RB is active the cell cycle does not proceed, but it is inactive in cancer cells, allowing the cell cycle to occur. Some viral proteins can inactivate tumor suppressors. For example, in the opening story of this chapter we saw how HPV infects cells of the cervix and produces a protein called E7. E7 binds to the RB protein and prevents it from inhibiting the cell cycle (**Figure 11.24B**).

The discovery of apoptosis and its importance (see Section 11.6) has changed the way biologists think about cancer. In a population of organisms, the net increase in the number of individuals over time (the growth rate) is a function of the individuals added (the birth rate) and lost (the death rate). Cell populations behave the same way:

$$\text{growth rate of cell population} = \text{rate of cell division ("births")} - \text{rate of apoptosis ("deaths")}$$

Cancer cells may lose the ability to respond to positive regulators of apoptosis (see Figure 11.22). This lowers the cellular "death rate" so that the overall cell population grows rapidly.

Cancer treatments target the cell cycle

The most successful and widely used treatment for cancer is surgery. While physically removing a tumor is optimal, it is often difficult for a surgeon to get all of the tumor cells. (A tumor about 1 cm in size already has a billion cells!) Tumors are generally embedded in normal tissues. Added to this is the probability that cells of the tumor may have broken off and spread to other organs. This makes it unlikely that localized surgery will be curative. So other approaches are taken to treat or cure cancer, and these generally target the cell cycle (**Figure 11.25**).

An example of a cancer drug that targets the cell cycle is *5-fluorouracil*, which blocks the synthesis of thymine, one of the four bases in DNA. The drug *taxol* prevents the functioning of microtubules in the mitotic spindle. Both drugs inhibit the cell cycle, and apoptosis causes tumor shrinkage. More dramatic is radiation treatment, in which a beam of high-energy radiation is focused on the tumor. DNA damage is extensive, and the cell cycle checkpoint for DNA repair is overwhelmed. As a result, the cell undergoes apoptosis. A major problem with these treatments is that they target normal cells as well as the tumor cells. These treatments are toxic to tissues with large populations of normal dividing cells such as those in the intestine, skin and bone marrow (producing blood cells).

A major effort in cancer research is to find treatments that target only cancer cells. A promising recent example is *Herceptin*, which targets the HER2 growth factor receptor that is expressed at high levels on the surfaces of some breast cancer cells (see Figure 11.24A). Herceptin binds specifically to the HER2 receptor but does not stimulate it. This prevents the natural growth factor from binding, and so the cells are not stimulated to divide. As a result, the tumor shrinks because the apoptosis rate remains the same. More such treatments are on the way.

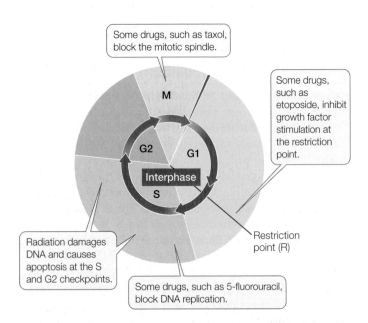

Some drugs, such as taxol, block the mitotic spindle.

Some drugs, such as etoposide, inhibit growth factor stimulation at the restriction point.

Radiation damages DNA and causes apoptosis at the S and G2 checkpoints.

Restriction point (R)

Some drugs, such as 5-fluorouracil, block DNA replication.

11.25 Cancer Treatment and the Cell Cycle To prevent cancer cells from dividing, physicians use combinations of therapies that attack the cell cycle at different points.

11.7 RECAP

Cancer cells differ from normal cells in terms of their rapid cell division and their ability to spread (metastasis). Many proteins regulate the cell cycle, either positively or negatively. In cancer, one or another of these proteins is altered in some way, making its activity abnormal. Radiation and many cancer drugs target proteins involved in the cell cycle.

- How are oncogene proteins and tumor suppressor proteins involved in cell cycle control in normal and cancer cells? **Review p. 231 and Figure 11.24**

- How does cancer treatment target the cell cycle? **Review p. 232 and Figure 11.25**

We have now looked at the cell cycle and at cell division by binary fission, mitosis, and meiosis. We have described the normal cell cycle and how it is upset in cancer. We have seen how meiosis produces haploid cells in sexual life cycles. In the coming chapters we examine heredity, genes, and DNA. In Chapter 12 we see how Gregor Mendel studied heredity in the nineteenth century and how the enormous power of his discoveries founded the science of genetics, and changed forever the science of biology.

CHAPTER SUMMARY

11.1 How Do Prokaryotic and Eukaryotic Cells Divide?

- Cell division is necessary for the reproduction, growth, and repair of organisms.

- Cell division must be initiated by a reproductive signal. Before a cell can divide, the genetic material (DNA) must be **replicated** and **segregated** to separate portions of the cell. **Cytokinesis** then divides the cytoplasm into two cells.

- In prokaryotes, most cellular DNA is a single molecule, usually in the form of a circular **chromosome**. Prokaryotes reproduce by **binary fission**. Review Figure 11.2

- In eukaryotes, cells divide by either **mitosis** or **meiosis**. Eukaryotic cell division follows the same general pattern as binary fission, but with significant differences. For example, a eukaryotic cell has a distinct nucleus that must be replicated prior to separating the two daughter cells.

- Cells that produce gametes undergo a special kind of nuclear division called meiosis; the four daughter cells produced by meiosis are not genetically identical.

11.2 How Is Eukaryotic Cell Division Controlled?

- The eukaryotic cell cycle has two main phases: **interphase**, during which cells are not dividing and the DNA is replicating, and mitosis or **M phase**, when the cells are dividing.

- During most of the eukaryotic cell cycle, the cell is in interphase, which is divided into three subphases: **S, G1**, and **G2**. DNA is replicated during the S phase. Mitosis (M phase) and cytokinesis follow. **Review Figure 11.3**

- **Cyclin–Cdk complexes** regulate the passage of cells through checkpoints in the cell cycle. The suppressor protein **RB** inhibits the cell cycle. The G1–S cyclin–Cdk functions by inactivating RB and allows the cell cycle to progress beyond the **restriction point**. Review Figures 11.5 and 11.6

- External controls such as **growth factors** can also stimulate the cell to begin a division cycle.

11.3 What Happens during Mitosis?

SEE ANIMATED TUTORIAL 11.1

- In mitosis, a single nucleus gives rise to two nuclei that are genetically identical to each other and to the parent nucleus.

- DNA is wrapped around proteins called **histones**, forming beadlike units called **nucleosomes**. A eukaryotic chromosome contains strings of nucleosomes bound to proteins in a complex called **chromatin**. Review Figure 11.9

- At mitosis, the replicated chromosomes, called **sister chromatids**, are held together at the **centromere**. Each chromatid consists of one double-stranded DNA molecule. **Review Figure 11.10, WEB ACTIVITY 11.1**

- Mitosis can be divided into several phases called **prophase**, **prometaphase**, **metaphase**, **anaphase**, and **telophase**.

- During mitosis sister chromatids, attached by **cohesin**, line up at the equatorial plate and attach to the **spindle**. The chromatids separate (becoming **daughter chromosomes**) and migrate to opposite ends of the cell. **Review Figure 11.11, WEB ACTIVITY 11.2**

- Nuclear division is usually followed by cytokinesis. Animal cell cytoplasms divide via a contractile ring made up of actin microfilaments. In plant cells, cytokinesis is accomplished by vesicles that fuse to form a cell plate. **Review Figure 11.13**

11.4 What Role Does Cell Division Play in a Sexual Life Cycle?

- **Asexual reproduction** produces clones, new organisms that are genetically identical to the parent. Any genetic variation is the result of mutations.

- In **sexual reproduction**, two **haploid** gametes—one from each parent—unite in **fertilization** to form a genetically unique, **diploid zygote**. There are many different sexual life cycles that can be **haplontic**, **diplontic**, or involve **alternation of generations**. Review Figure 11.15, **WEB ACTIVITY 11.3**

- In sexually reproducing organisms, certain cells in the adult undergo meiosis, a process by which a diploid cell produces haploid gametes.

- Each gamete contains of one of each pair of **homologous chromosomes** from the parent.

- The numbers, shapes, and sizes of the chromosomes constitute the **karyotype** of an organism.

11.5 What Happens during Meiosis?

SEE ANIMATED TUTORIAL 11.2

- Meiosis consists of two nuclear divisions, **meiosis I** and **meiosis II**, that collectively reduce the chromosome number from diploid to haploid. It ensures that each haploid cell contains one member of each chromosome pair, and results in four genetically diverse haploid cells, usually gametes. **Review Figure 11.17, WEB ACTIVITY 11.4**

- In meiosis I, entire chromosomes, each with two chromatids, migrate to the poles. In meiosis II, the sister chromatids separate.

- During prophase I, homologous chromosomes undergo **synapsis** to form pairs in a **tetrad**. Chromatids can form junctions called **chiasmata** and genetic material may be exchanged between the two homologs by **crossing over**. Review Figure 11.18

- Both crossing over during prophase I and **independent assortment** of the homologs as they separate during anaphase I ensure that the gametes are genetically diverse.

- In **nondisjunction**, two members of a homologous pair of chromosomes go to the same pole during meiosis I, or two chromatids go to the same pole during meiosis II. This leads to one gamete having an extra chromosome and another lacking that chromosome. **Review Figure 11.21**

- The union between a gamete with an abnormal chromosome number and a normal haploid gamete results in **aneuploidy**. Such genetic abnormalities are harmful or lethal to the organism.

11.6 In a Living Organism, How Do Cells Die?

- A cell may die by **necrosis**, or it may self-destruct by **apoptosis**, a genetically programmed series of events that includes the fragmentation of its nuclear DNA.

- Apoptosis is regulated by external and internal signals. These signals result in activation of a class of enzymes called **caspases** that hydrolyze proteins in the cell. **Review Figure 11.22**

11.7 How Does Unregulated Cell Division Lead to Cancer?

- Cancer cells divide more rapidly than normal cells and can be **metastatic**, spreading to distant organs in the body.

- Cancer can result from changes in either of two types of proteins that regulate the cell cycle. **Oncogene** proteins stimulate cell division and are activated in cancer. **Tumor suppressor** proteins normally inhibit the cell cycle but in cancer they are inactive. **Review Figure 11.24**

- Cancer treatment often targets the cell cycle in tumor cells. **Review Figure 11.25**

SELF-QUIZ

1. Which statement about eukaryotic chromosomes is *not* true?
   a. They sometimes consist of two chromatids.
   b. They sometimes consist only of a single chromatid.
   c. They normally possess a single centromere.
   d. They consist only of proteins.
   e. During metaphase they are visible under the light microscope.

2. Nucleosomes
   a. are made of chromosomes.
   b. consist entirely of DNA.
   c. consist of DNA wound around a histone core.
   d. are present only during mitosis.
   e. are present only during prophase.

3. Which statement about the cell cycle is *not* true?
   a. It consists of interphase, mitosis, and cytokinesis.
   b. The cell's DNA replicates during G1.
   c. A cell can remain in G1 for weeks or much longer.
   d. DNA is not replicated during G2.
   e. Cells enter the cell cycle as a result of internal or external signals.

4. Which statement about mitosis is *not* true?
   a. A single nucleus gives rise to two identical daughter nuclei.
   b. The daughter nuclei are genetically identical to the parent nucleus.
   c. The centromeres separate at the onset of anaphase.
   d. Homologous chromosomes synapse in prophase.
   e. The centrosomes organize the microtubules of the spindle fibers.

5. Which statement about cytokinesis is true?
   a. In animals, a cell plate forms.
   b. In plants, it is initiated by furrowing of the membrane.
   c. It follows mitosis.
   d. In plant cells, actin and myosin play an important part.
   e. It is the division of the nucleus.

6. Apoptosis
   a. occurs in all cells.
   b. involves the formation of the plasma membrane.
   c. does not occur in an embryo.
   d. is a series of programmed events resulting in cell death.
   e. is the same as necrosis.

7. In meiosis,
   a. meiosis II reduces the chromosome number from diploid to haploid.
   b. DNA replicates between meiosis I and meiosis II.
   c. the chromatids that make up a chromosome in meiosis II are identical.
   d. each chromosome in prophase I consists of four chromatids.
   e. homologous chromosomes separate from one another in anaphase I.

8. In meiosis,
   a. a single nucleus gives rise to two daughter nuclei.
   b. the daughter nuclei are genetically identical to the parent nucleus.
   c. the centromeres separate at the onset of anaphase I.
   d. homologous chromosomes synapse in prophase I.
   e. no spindle forms.

9. An animal has a diploid chromosome number of 12. An egg cell of that animal has 5 chromosomes. The most probable explanation is
   a. normal mitosis.
   b. normal meiosis.
   c. nondisjunction in meiosis I.
   d. nondisjunction in meiosis I or II.
   e. nondisjunction in mitosis.

10. The number of daughter chromosomes in a human cell (diploid number 46) in anaphase II of meiosis is
    a. 2.
    b. 23.
    c. 46.
    d. 69.
    e. 92.

FOR DISCUSSION

1. Compare the roles of cohesins in mitosis, meiosis I, and meiosis II.

2. Compare and contrast cell division in animals and plants.

3. Contrast mitotic prophase and prophase I of meiosis. Contrast mitotic anaphase and anaphase I of meiosis.

4. Compare the sequence of events in the mitotic cell cycle with the sequence of events in apoptosis.

5. Cancer-fighting drugs are rarely used alone. Usually, there are several drugs given in combination that target different stages of the cell cycle. Why might this be a better approach than single drugs?

ADDITIONAL INVESTIGATION

1. Suggest two ways in which one might use a microscope to determine the relative durations of the various phases of mitosis.

2. Studying the events and controls of the cell cycle is much easier if the cells under investigation are synchronous; that is, if they are all in the same stage of the cell cycle. This can be accomplished with various chemicals. But some populations of cells are naturally synchronous. The anther (male sex organ) of a lily plant contains cells that become pollen grains (male gametes). As anthers develop in the flower, their lengths correlate precisely with the stage of the meiotic cycle in those cells. These stages each take many days, so an anther that is 1.5 millimeters long, for example, contains cells in early prophase I. How would you use lily anthers to investigate the roles of cyclins and Cdk's in the meiotic cell cycle?

12 Inheritance, Genes, and Chromosomes

Genetic piracy

The Nazis were infamous for plundering the art collections of Europe during World War II, but why did they also spirit away a collection of seeds? The answer lies in the power and promise offered by the scientific understanding of genetics, the science of *heredity*.

A key step in the rise of human civilizations was the development of agriculture—the cultivation of plants and animals for food and other human needs. Some 10,000 years ago, early farmers began preferentially cultivating plants with certain traits (e.g., that survived drought better). Over time, the cultivated varieties (*cultivars*) became quite different from their wild relatives, an example of evolution by selection. In this case, it was not the result of natural selection (see Chapter 21), but of artificial selection by the practices of ancient farmers.

Early in the twentieth century, Russian scientist Nicolai Vavilov began systematically collecting seeds from thousands of cultivars and their wild relatives. He convinced Lenin, the leader of the new Communist regime, that his seed collection would be useful in breeding crops that would be more productive in the difficult Russian climate. Lenin put Vavilov in charge of a large research institute. But when Lenin died in 1929, his successor, Josef Stalin, had little interest in science.

A politically ambitious student of Vavilov's, Trofim Lysenko, proposed to Stalin that favorable characteristics in plants could be rendered heritable by manipulating the parent plant's *phenotype* (physical state). This idea was at odds with what scientists knew about heredity and evolution, but it appealed to Stalin's political ideology. Stalin put Lysenko in charge of Vavilov's institute and sent Vavilov to a prison camp, where he died in 1943. Vavilov's unique seed collection—a *gene bank*—was ignored.

Meanwhile, in Germany, the Nazi leader Heinrich Himmler learned of the collection and was convinced that Vavilov's seeds could be a valuable key to providing better crops for the expanding German empire. Himmler put Heinz Brücher, a young SS officer with a doctorate in botany, in charge of obtaining the seeds. When the German army invaded Russia, Brücher's team removed thousands of seeds to a castle in Austria that already housed a seed collection Brücher had brought from Tibet.

Brücher's aim was to cross-breed plants from Tibet with plants from Russia to develop new crops that would grow well at high elevations and in cold climates; these plans came to a halt with the end of World War II. However,

Genetics Pioneer Collecting thousands of crop plant varieties from all over the world, Nikolai Vavilov laid the foundations for theories about the genetic origins of modern crops.

Hardy Grain Early geneticists hoped to increase food production by breeding crop varieties adapted to harsh climates (such as those in Tibet) with varieties with other desirable traits.

Brücher ignored a superior's order to blow up the castle, thus preserving most of Vavilov's seed bank. The collection was returned to Russia, where it continued to be used in breeding programs.

The ideas of Vavilov and the breeding plans of Brücher depended on the principles of *genetics*, a science born in an Austrian monastery in the 1860s, where Gregor Mendel performed—and, importantly, correctly interpreted—experiments on pea plants. It was almost 50 years before the scientific community recognized the significance of Mendel's work, but once that recognition was achieved, science and medicine sprang forward at a rapid pace.

IN THIS CHAPTER we will discuss how the units of inheritance—genes—are transmitted from generation to generation. We will show that many of the rules that govern inheritance can be explained by the behavior of chromosomes during meiosis. We will describe the interactions of genes with one another and with the environment, and we will see how the specific positions of genes on chromosomes affect diversity.

12.1 What Are the Mendelian Laws of Inheritance?

Much of the early study of biological inheritance was done with plants and animals of economic importance. Records show that people were deliberately cross-breeding date palm trees and horses as early as 5,000 years ago. By the early nineteenth century plant breeding was widespread, especially for ornamental flowers such as tulips. Plant breeders of that time were operating under two key assumptions about how inheritance worked. Only one of those assumptions turned out to be supported by experimental evidence.

- *Each parent contributes equally to offspring* (supported by experiments). In the 1770s, the German botanist Josef Gottlieb Kölreuter studied the offspring of **reciprocal crosses**, in which plants were crossed (mated with each other) in both directions. For example, in one cross, plants with white flowers were used as males to pollinate related plants with red flowers. In the complementary crosses, the red-flowered plants were used as males in crosses with the white flowered plants. In Kölreuter's studies, such reciprocal crosses always gave identical results, showing that both parents contributed equally to the offspring.

- *Hereditary determinants blend in offspring* (not supported by experiments). Kölreuter and others proposed that there were hereditary determinants in the egg and sperm cells. When these determinants came together in a single cell after mating, they were believed to blend together. If a plant with one form of a character (say, red flowers) was crossed with a plant with a different form of that character (blue flowers), the offspring would have a blended combination of the two parents' characteristics (purple flowers). According to the blending theory, once heritable elements were combined, they could not be separated again (like inks of different colors mixed together). The red and blue hereditary determinants were thought to be forever blended into the new purple one.

In his experiments in the 1860s, Gregor Mendel confirmed the first of these two assumptions but refuted the second.

Mendel brought new methods to experiments on inheritance

Gregor Mendel was an Austrian monk, not an academic scientist (**Figure 12.1**). He was well qualified, however, to under-

12.1 Gregor Mendel and His Garden The Austrian monk Gregor Mendel (left) did his groundbreaking genetics experiments in a garden at the monastery at Brno, in what is now the Czech Republic.

take scientific investigations. In 1850 he failed an examination for a teaching certificate in natural science, so he undertook intensive studies in physics, chemistry, mathematics, and various aspects of biology at the University of Vienna. His studies in physics and mathematics under the famous physicist Christian Doppler strongly influenced his use of experimental and quantitative methods in his studies of heredity, and it was those quantitative experiments that were key to his successful deductions.

Over the seven years he spent working out some principles of inheritance in plants, Mendel made crosses with hundreds of plants and noted the resulting characteristics of 24,034 progeny. Analysis of his meticulously gathered data suggested to him a new theory of how inheritance might work. He presented this theory in a public lecture in 1865 and a detailed written publication in 1866. Mendel's paper appeared in a journal that was received by 120 libraries, and he sent reprinted copies (of which he had obtained 40) to several distinguished scholars, including Charles Darwin. However, his theory was not readily accepted. In fact, it was mostly ignored.

One reason Mendel's paper received so little attention was that most prominent biologists of his time were not in the habit of thinking in mathematical terms, even the simple terms Mendel used. Even Charles Darwin, whose theory of evolution by natural selection was predicated on heritable variations among individuals, failed to understand the significance of

Mendel's findings. In fact, Darwin performed breeding experiments on snapdragons that were similar to Mendel's work with peas. Although Darwin's data were similar to Mendel's, he failed to question the assumption that parental contributions blend in the offspring.

By 1900, the events of meiosis had been observed and described, and Mendel's discoveries burst into sudden prominence as a result of independent experiments by three plant geneticists: Hugo DeVries, Carl Correns, and Erich von Tschermak. Each carried out crossing experiments, each published his principal findings in 1900, and each cited Mendel's 1866 paper. These three men realized that chromosomes and meiosis provided a physical explanation for the theory that Mendel had proposed to explain the data from his crosses.

That Mendel was able to achieve his remarkable insights before the discovery of genes and meiosis was largely due to his experimental methods. His work is a definitive example of extensive preparation, meticulous execution, and imaginative yet logical interpretation. He was also fortunate in his choice of experimental subjects. Let's take a closer look at these experiments and the conclusions and hypotheses that emerged.

Mendel devised a careful research plan

Mendel chose to study the common garden pea because of its ease of cultivation, the feasibility of controlled pollination, and

TOOLS FOR INVESTIGATING LIFE

12.2 A Controlled Cross between Two Plants

Plants were widely used in early genetic studies because it is easy to control which individuals mate with which. Mendel used the garden pea (*Pisum sativum*) in many of his experiments.

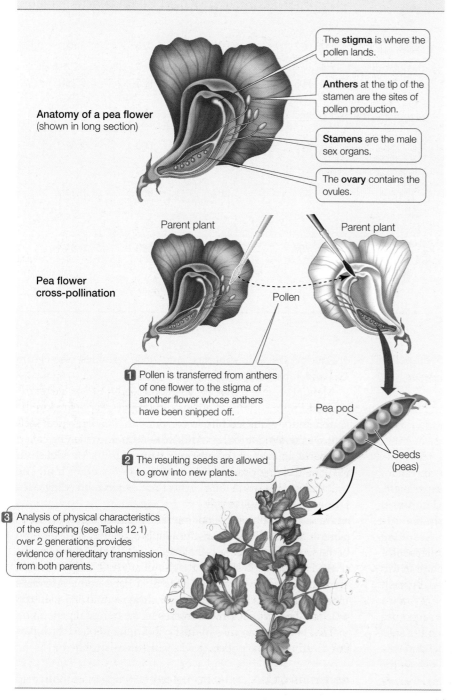

Anatomy of a pea flower
(shown in long section)

The **stigma** is where the pollen lands.

Anthers at the tip of the stamen are the sites of pollen production.

Stamens are the male sex organs.

The **ovary** contains the ovules.

Pea flower cross-pollination

Parent plant

Parent plant

Pollen

1 Pollen is transferred from anthers of one flower to the stigma of another flower whose anthers have been snipped off.

Pea pod

2 The resulting seeds are allowed to grow into new plants.

Seeds (peas)

3 Analysis of physical characteristics of the offspring (see Table 12.1) over 2 generations provides evidence of hereditary transmission from both parents.

the availability of varieties with contrasting traits. He controlled pollination, and thus fertilization, of his parent plants by manually moving pollen from one plant to another (**Figure 12.2**). Thus he knew the parentage of the offspring in his experiments. The pea plants Mendel studied produce male and female sex organs and gametes—sex cells such as eggs and sperm—in the

same flower. If untouched, they naturally self-pollinate—that is, the female organ of each flower receives pollen from the male organs of the same flower. Mendel made use of this natural phenomenon in some of his experiments.

Mendel began by examining different varieties of peas in a search for heritable characters and traits suitable for study:

- A **character** is an observable physical feature, such as flower color.

- A **trait** is a particular form of a character, such as purple flowers or white flowers.

- A **heritable trait** is one that is passed from parent to offspring.

Mendel looked for characters with well-defined, contrasting alternative traits, such as purple flowers versus white flowers. Furthermore, these traits had to be **true-breeding**, meaning that the observed trait was the only form present for many generations. In other words, if they were true-breeding, peas with white flowers would give rise only to progeny with white flowers when self-pollinated or crossed with one another for repeated generations. Similarly, tall plants bred with other tall plants would produce only tall progeny.

Mendel isolated each of his true-breeding strains by repeated inbreeding (done by crossing sibling plants that were seemingly identical or by allowing individuals to self-pollinate) and selection. In most of his work, Mendel concentrated on the seven pairs of contrasting traits shown in **Table 12.1** (left side). His use of true-breeding strains for experimental crosses was an essential feature of his work.

Mendel then performed his crosses in the following manner:

- He removed the anthers from the flowers of one parental strain so that it couldn't self-pollinate. Then he collected pollen from the other parental strain and placed it on the stigmas of flowers of the strain whose anthers had been removed. The plants providing and receiving the pollen were the **parental generation**, designated **P**.

- In due course, seeds formed and were planted. The seeds and the resulting new plants constituted the **first filial generation**, or **F₁**. (The word "filial" refers to the relationship between offspring and parents, from the Latin, *filius*, "son.") Mendel and his assistants examined each F₁ plant to see which traits it bore and then recorded the number of F₁ plants expressing each trait.

TABLE 12.1
Mendel's Results from Monohybrid Crosses

| PARENTAL GENERATION PHENOTYPES | | | F_2 GENERATION PHENOTYPES | | | |
DOMINANT	RECESSIVE		DOMINANT	RECESSIVE	TOTAL	RATIO
Spherical seeds ×	Wrinkled seeds		5,474	1,850	7,324	2.96:1
Yellow seeds ×	Green seeds		6,022	2,001	8,023	3.01:1
Purple flowers ×	White flowers		705	224	929	3.15:1
Inflated pods ×	Constricted pods		882	299	1,181	2.95:1
Green pods ×	Yellow pods		428	152	580	2.82:1
Axial flowers ×	Terminal flowers		651	207	858	3.14:1
Tall stems × (1 m)	Dwarf stems (0.3 m)		787	277	1,064	2.84:1

- In some experiments the F_1 plants were allowed to self-pollinate and produce a **second filial generation**, the **F_2**. Again, each F_2 plant was characterized and counted.

Mendel's first experiments involved monohybrid crosses

The term *hybrid* refers to the offspring of crosses between organisms differing in one or more traits. In Mendel's first experiment, he crossed two true-breeding parental (P) lineages differing in just one trait, producing monohybrids in the F_1 generation. He subsequently planted the F_1 seeds and allowed the resulting plants to self-pollinate to produce the F_2 generation. This technique is referred to as a **monohybrid cross**, even though in this case, the monohybrid plants were not literally crossed, but self-pollinated.

Mendel performed the same experiment for all seven pea traits. His method is illustrated in **Figure 12.3**, using the seed shape trait as an example. He took pollen from pea plants of a true-breeding strain with wrinkled seeds and placed it on the stigmas of flowers of a true-breeding strain with spherical seeds. He also performed the reciprocal cross, in which the parental source of each trait is reversed: he placed pollen from the spherical-seeded strain on the stigmas of flowers of the strain with wrinkled seeds. In all cases, the F_1 seeds were spherical—it was as if the wrinkled seed trait had disappeared completely.

The following spring, Mendel grew 253 F_1 plants from these spherical seeds. Each of the F_1 plants was allowed to self-pollinate to produce F_2 seeds. In all, 7,324 F_2 seeds were produced,

of which 5,474 were spherical, and 1,850 wrinkled (see Figure 12.3 and Table 12.1).

Mendel observed that the wrinkled seed trait was never expressed in the F_1 generation, even though it reappeared in the F_2 generation. This led him to conclude that the spherical seed trait was **dominant** to the wrinkled seed trait, which he called **recessive**. In each of the other six pairs of traits Mendel studied, one trait proved to be dominant over the other trait. The trait that disappears in the F_1 generation of a true-breeding cross is always the recessive trait.

Mendel also observed that the ratio of the two traits in the F_2 generation was always the same—approximately 3:1—for each of the seven pea-plant traits that he studied. That is, three-fourths of the F_2 generation showed the dominant trait and one-fourth showed the recessive trait (see Table 12.1). For example, Mendel's monohybrid cross for seed shape produced a ratio of 5,474:1,850 = 2.96:1. The two reciprocal crosses in the parental generation yielded similar outcomes in the F_2; it did not matter which parent contributed the pollen, just as Kölreuter had shown.

REJECTION OF THE BLENDING THEORY Mendel's monohybrid cross experiments showed that inheritance cannot be the result of a blending phenomenon. According to the blending theory, Mendel's F_1 seeds should have had an appearance that was intermediate between those of the two parents—for example, the F_1 seeds from the cross between strains with wrinkled and spherical seeds should have been slightly wrinkled. Furthermore, the blending theory offered no explanation for the reappearance of the recessive trait in the F_2 seeds after its absence in the F_1 seeds.

INVESTIGATING LIFE

12.3 Mendel's Monohybrid Experiments

Mendel performed crosses with pea plants and carefully analyzed the outcomes to show that genetic determinants are particulate.

HYPOTHESIS When two strains of peas with contrasting traits are bred, their characteristics are irreversibly blended in succeeding generations.

METHOD

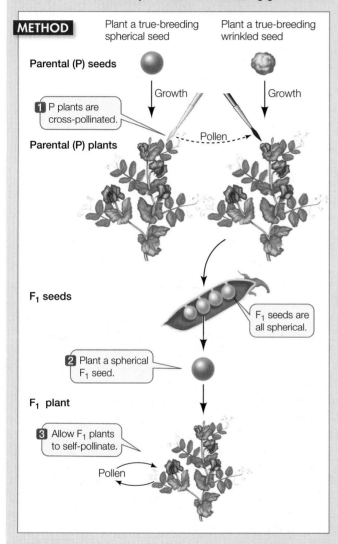

Plant a true-breeding spherical seed   Plant a true-breeding wrinkled seed

Parental (P) seeds

1 P plants are cross-pollinated.

↓ Growth   ↓ Growth

Pollen

Parental (P) plants

F₁ seeds

F₁ seeds are all spherical.

2 Plant a spherical F₁ seed.

F₁ plant

3 Allow F₁ plants to self-pollinate.

Pollen

RESULTS F₂ seeds from F₁ plant

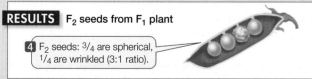

4 F₂ seeds: ³/₄ are spherical, ¹/₄ are wrinkled (3:1 ratio).

CONCLUSION The hypothesis is rejected. There is no irreversible blending of characteristics, and a recessive trait can reappear in succeeding generations.

Go to **yourBioPortal.com** for original citations, discussions, and relevant links for all INVESTIGATING LIFE figures.

SUPPORT FOR THE PARTICULATE THEORY Given the absence of blending and the reappearance of the recessive seed traits in the F₂ generations of his monohybrid cross experiments, Mendel proposed that the units responsible for the inheritance of specific traits are present as discrete particles that occur in pairs and segregate (separate) from one another during the formation of gametes. According to his **particulate theory**, the units of inheritance retain their integrity in the presence of other units. Mendel concluded that each pea plant has two units (particles) of inheritance for each character, one from each parent. We now use the term **diploid** to refer to the two copies of each heritable unit in an organism. Mendel proposed that during the production of gametes, only one of these paired units is given to a gamete. We now use the term **haploid** to refer to the single set of heritable units. Mendel concluded that while each gamete contains one unit, the resulting zygote contains two, because it is produced by the fusion of two gametes. This conclusion is the core of Mendel's model of inheritance. Mendel's unit of inheritance is now called a **gene**. The totality of all the genes of an organism is that organism's **genome**.

Mendel reasoned that in his experiments, the two true-breeding parent plants had different forms of the gene affecting a particular character, such as seed shape (although he did not use the term "gene"). The true-breeding spherical-seeded parent had two genes of the same form, which we will call *S*, and the parent with wrinkled seeds had two copies of an alternative form of the gene, which we will call *s*. The *SS* parent would produce gametes having a single *S* gene, and the *ss* parent would produce gametes having a single *s* gene. The cross producing the F₁ generation would donate an *S* from one parent and an *s* from the other to each seed; the F₁ offspring would thus be *Ss*. We say that *S* is dominant over *s* because the trait specified by *s* is not evident—is not expressed—when both forms of the gene are present.

Alleles are different forms of a gene

The different forms of a gene (*S* and *s* in this case) are called **alleles**. Individuals that are true-breeding for a trait contain two copies of the same allele. For example, all the individuals in a population of true-breeding peas with wrinkled seeds must have the allele pair *ss*; if the dominant *S* allele were present, some of the plants would produce spherical seeds.

We say that the individuals that produce wrinkled seeds are **homozygous** for the allele *s*, meaning that they have two copies of the same allele (*ss*). Some peas with spherical seeds—the ones with the genotype *SS*—are also homozygous. However, not all plants with spherical seeds have the *SS* genotype. Some spherical-seeded plants, like Mendel's F₁, are **heterozygous**: they have two different alleles of the gene in question (in this case, *Ss*). An individual that is homozygous for a character is sometimes called a **homozygote**; an individual that is heterozygous for a character is termed a **heterozygote**.

As a somewhat more complex example of inheritance, let's consider three gene pairs. An individual with the three genes and alleles *AABbcc* is homozygous for the *A* and *C* genes, because it has two *A* alleles and two *c* alleles, but heterozygous for the *B* gene, because it contains the *B* and *b* alleles.

The physical appearance of an organism is its **phenotype**. Mendel correctly supposed the phenotype to be the result of the **genotype**, or genetic constitution, of the organism showing the phenotype. Spherical seeds and wrinkled seeds are two phenotypes, which are the result of three genotypes: the wrinkled seed phenotype is produced by the genotype ss, whereas the spherical seed phenotype is produced by either of the genotypes SS or Ss.

Mendel's first law says that the two copies of a gene segregate

How does Mendel's model of inheritance explain the ratios of traits seen in the F_1 and F_2 generations? Consider first the F_1, in which all progeny have the spherical seed phenotype. According to Mendel's model, *when any individual produces gametes, the two copies of a gene separate, so that each gamete receives only one copy*. This is Mendel's first law, the **law of segregation**. Thus, every individual in the offspring from a cross between the P generation parents inherits one gene copy from each parent, and has the genotype Ss (**Figure 12.4**).

Now let's consider the composition of the F_2 generation. Half of the gametes produced by the F_1 generation have the S allele and the other half the s allele. Since both SS and Ss plants produce spherical seeds while ss plants produce wrinkled seeds, in the F_2 generation there are three ways to get a spherical-seeded plant (SS, Ss, or sS), but only one way to get a plant with wrinkled seeds (ss). Therefore, we predict a 3:1 ratio, remarkably close to the values Mendel found experimentally for all seven of the traits he compared (see Table 12.1).

The allele combinations that will result from a cross can be predicted using a **Punnett square**, a method devised in 1905 by the British geneticist Reginald Crundall Punnett. This device ensures that we consider all possible combinations of gametes when calculating expected genotype frequencies. A Punnett square looks like this:

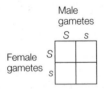

It is a simple grid with all possible male gamete (haploid sperm) genotypes shown along the top and all possible female gamete (haploid egg) genotypes along the left side. The grid is completed by filling in each square with the diploid genotype that can be generated from each combination of gametes (see Figure 12.4). In this example, to fill in the top right square, we put in the S from the egg (female gamete) and the s from the pollen (male gamete), yielding Ss.

Mendel did not live to see his theory placed on a sound physical footing with the discoveries of chromosomes and DNA. Genes are now known to be regions of the DNA molecules in chromosomes. More specifically, a gene is a sequence of DNA that resides at a particular site on a chromosome, called a **locus** (plural **loci**). Genes are expressed in the phenotype mostly as proteins with particular functions, such as enzymes.

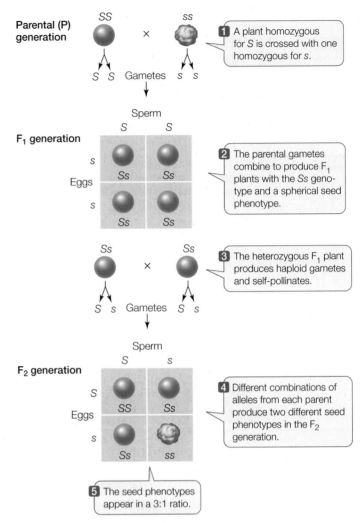

1 A plant homozygous for S is crossed with one homozygous for s.

2 The parental gametes combine to produce F_1 plants with the Ss genotype and a spherical seed phenotype.

3 The heterozygous F_1 plant produces haploid gametes and self-pollinates.

4 Different combinations of alleles from each parent produce two different seed phenotypes in the F_2 generation.

5 The seed phenotypes appear in a 3:1 ratio.

12.4 Mendel's Explanation of Inheritance Mendel concluded that inheritance depends on discrete factors from each parent that do not blend in the offspring.

So, in many cases, a dominant gene can be thought of as a region of DNA that is expressed as a functional protein, while a recessive gene typically expresses a nonfunctional protein, or a protein whose function is overshadowed by the dominant form. Mendel arrived at his law of segregation with no knowledge of chromosomes or meiosis, but today we can picture the different alleles of a gene segregating as the chromosomes separate during meiosis I (**Figure 12.5**).

Mendel verified his hypothesis by performing a test cross

Mendel set out to test his hypothesis that there were two possible allele combinations (SS and Ss) in the spherical-seeded F_1 generation. He did so by performing a **test cross**, which is a way of finding out whether an individual showing a dominant trait is homozygous or heterozygous. In a test cross, the individual in question is crossed with an individual that is known to be homozygous for the recessive trait—an easy individual to identify, because all individuals with the recessive phenotype are homozygous for the recessive trait.

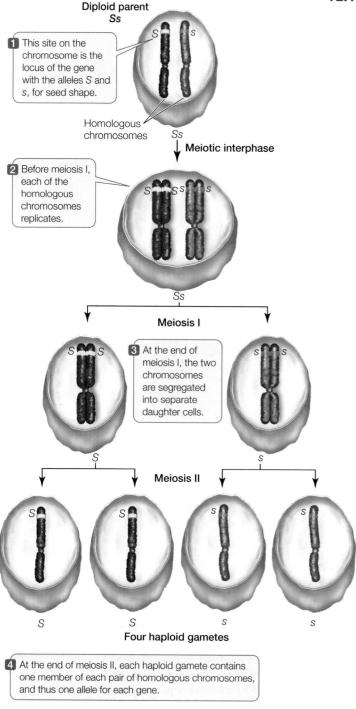

1 This site on the chromosome is the locus of the gene with the alleles *S* and *s*, for seed shape.

Diploid parent *Ss*

Homologous chromosomes

Meiotic interphase

2 Before meiosis I, each of the homologous chromosomes replicates.

Meiosis I

3 At the end of meiosis I, the two chromosomes are segregated into separate daughter cells.

Meiosis II

Four haploid gametes

4 At the end of meiosis II, each haploid gamete contains one member of each pair of homologous chromosomes, and thus one allele for each gene.

12.5 Meiosis Accounts for the Segregation of Alleles Although Mendel had no knowledge of chromosomes or meiosis, we now know that a pair of alleles resides on homologous chromosomes, and that those alleles segregate during meiosis.

For the seed shape gene that we have been considering, the recessive homozygote used for the test cross is *ss*. The individual being tested may be described initially as *S_* because we do not yet know the identity of the second allele. We can predict two possible results:

- If the individual being tested is homozygous dominant (*SS*), all offspring of the test cross will be *Ss* and show the dominant trait (spherical seeds) (**Figure 12.6, left**).

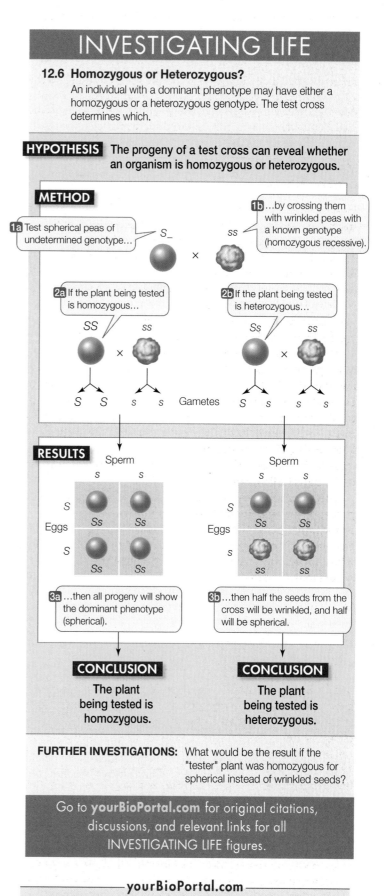

INVESTIGATING LIFE

12.6 Homozygous or Heterozygous?
An individual with a dominant phenotype may have either a homozygous or a heterozygous genotype. The test cross determines which.

HYPOTHESIS The progeny of a test cross can reveal whether an organism is homozygous or heterozygous.

METHOD

1a Test spherical peas of undetermined genotype... *S_*

1b ...by crossing them with wrinkled peas with a known genotype (homozygous recessive). *ss*

2a If the plant being tested is homozygous... *SS × ss*

2b If the plant being tested is heterozygous... *Ss × ss*

Gametes *S S s s* / *S s s s*

RESULTS

Sperm *s s*

Eggs *S S*

	s	*s*
S	*Ss*	*Ss*
S	*Ss*	*Ss*

Sperm *s s*

Eggs *S s*

	s	*s*
S	*Ss*	*Ss*
s	*ss*	*ss*

3a ...then all progeny will show the dominant phenotype (spherical).

3b ...then half the seeds from the cross will be wrinkled, and half will be spherical.

CONCLUSION
The plant being tested is homozygous.

CONCLUSION
The plant being tested is heterozygous.

FURTHER INVESTIGATIONS: What would be the result if the "tester" plant was homozygous for spherical instead of wrinkled seeds?

Go to **yourBioPortal.com** for original citations, discussions, and relevant links for all INVESTIGATING LIFE figures.

- If the individual being tested is heterozygous (*Ss*), then approximately half of the offspring of the test cross will be heterozygous and show the dominant trait (*Ss*), but the other half will be homozygous for, and will show, the recessive trait (*ss*) (**Figure 12.6, right**).

Mendel obtained results consistent with both of these predictions; thus Mendel's hypothesis accurately predicted the results of his test crosses.

With his first hypothesis confirmed, Mendel went on to ask another question: How do different pairs of genes behave in crosses when considered together?

Mendel's second law says that copies of different genes assort independently

Consider an organism that is heterozygous for two genes (*SsYy*), in which the *S* and *Y* alleles came from its mother, and the *s* and *y* alleles came from its father. When this organism produces gametes, do the alleles of maternal origin (*S* and *Y*) go together in one gamete and those of paternal origin (*s* and *y*) in another gamete? Or can a single gamete receive one maternal and one paternal allele, *S* and *y* (or *s* and *Y*)?

To answer these questions, Mendel performed another series of experiments. He began with peas that differed in two seed characters: seed shape and seed color. One true-breeding parental strain produced only spherical, yellow seeds (*SSYY*), and the other produced only wrinkled, green ones (*ssyy*). A cross between these two strains produced an F₁ generation in which all the plants were *SsYy*. Because the *S* and *Y* alleles are dominant, the F₁ seeds were all spherical and yellow.

Mendel continued this experiment into the F₂ generation by performing a **dihybrid cross** (a cross between individuals that are identical double heterozygotes) with the F₁ plants (although again, in this case, this was done by allowing the F₁ plants to self-pollinate). There are two possible ways in which such doubly heterozygous plants might produce gametes, as Mendel saw it (remember that he had never heard of chromosomes or meiosis):

1. *The alleles could maintain the associations they had in the parental generation (that is, they could be linked).*

In this case, the F₁ plants should produce two types of gametes (*SY* and *sy*), and the F₂ progeny resulting from self-pollination of the F₁ plants should consist of three times as many plants bearing spherical, yellow seeds as plants with wrinkled, green seeds. If such results were obtained, there might be no reason to suppose that two different genes regulated seed shape and seed color, because spherical seeds would always be yellow and wrinkled ones always green.

2. *The segregation of S from s could be independent of the segregation of Y from y (that is, the two genes could be unlinked).*

In this case, four kinds of gametes should be produced by the F₁ in equal numbers: *SY*, *Sy*, *sY*, and *sy*. When these gametes combine at random, they should produce an F₂ having nine different genotypes. The F₂ progeny could have any of three possible genotypes for shape (*SS*, *Ss*, or *ss*) and any of three possi-

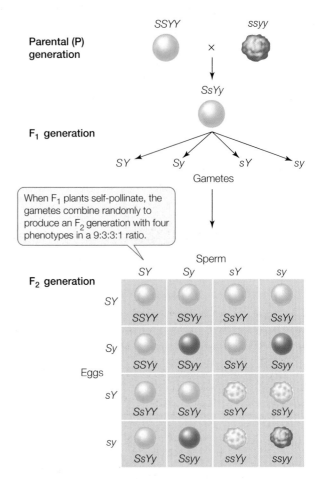

12.7 Independent Assortment The 16 possible combinations of gametes in this dihybrid cross result in nine different genotypes. Because *S* and *Y* are dominant over *s* and *y*, respectively, the nine genotypes result in four phenotypes in a ratio of 9:3:3:1. These results show that the two genes segregate independently.

ble genotypes for color (*YY*, *Yy*, or *yy*). The combined nine genotypes should produce four phenotypes (spherical yellow, spherical green, wrinkled yellow, wrinkled green). Putting these possibilities into a Punnett square, we can predict that these four phenotypes will occur in a ratio of 9:3:3:1 (**Figure 12.7**).

Mendel's dihybrid crosses supported the second prediction: four different phenotypes appeared in the F₂ generation in a ratio of about 9:3:3:1. The parental traits appeared in new combinations (spherical green and wrinkled yellow) in some progeny. Such new combinations are called **recombinant** phenotypes.

These results led Mendel to the formulation of what is now known as Mendel's second law: *alleles of different genes assort independently of one another during gamete formation.* That is, the segregation of gene A alleles is independent of the segregation of gene B alleles. We now know that this **law of independent assortment** is not as universal as the law of segregation, because it applies to genes located on separate chromosomes, but not to those located near one another on the same chromosome, as we will see in Section 12.4. However, it is correct to say that chromosomes segregate independently during the formation of gametes, and so do any two genes on separate homologous chromosome pairs (**Figure 12.8**).

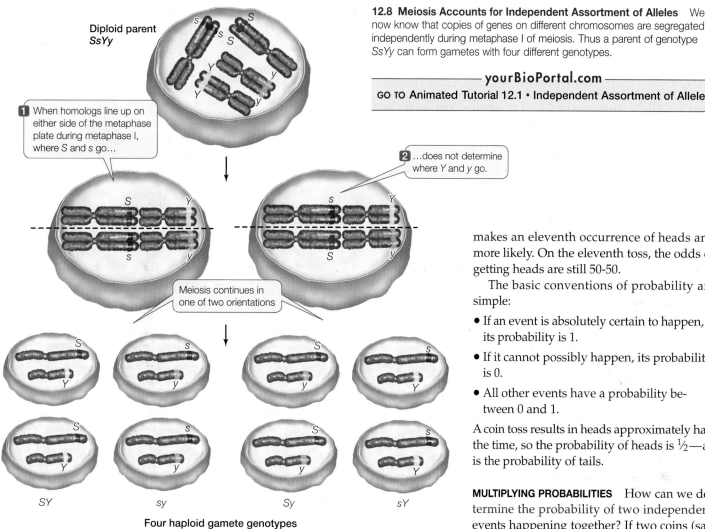

Diploid parent
SsYy

1 When homologs line up on either side of the metaphase plate during metaphase I, where *S* and *s* go...

2 ...does not determine where *Y* and *y* go.

Meiosis continues in one of two orientations

Four haploid gamete genotypes
SY, sy, Sy, sY

12.8 Meiosis Accounts for Independent Assortment of Alleles We now know that copies of genes on different chromosomes are segregated independently during metaphase I of meiosis. Thus a parent of genotype *SsYy* can form gametes with four different genotypes.

—— **yourBioPortal.com** ——
GO TO Animated Tutorial 12.1 • Independent Assortment of Alleles

makes an eleventh occurrence of heads any more likely. On the eleventh toss, the odds of getting heads are still 50-50.

The basic conventions of probability are simple:

- If an event is absolutely certain to happen, its probability is 1.

- If it cannot possibly happen, its probability is 0.

- All other events have a probability between 0 and 1.

A coin toss results in heads approximately half the time, so the probability of heads is $\frac{1}{2}$—as is the probability of tails.

MULTIPLYING PROBABILITIES How can we determine the probability of two independent events happening together? If two coins (say a penny and a dime) are tossed, each acts independently of the other. What is the probability of both coins coming up heads? In half of the tosses, the penny comes up heads; of that fraction, the dime also comes up heads half of the time. Therefore, the joint probability of both coins coming up heads is half of one-half, or $\frac{1}{2} \times \frac{1}{2} = \frac{1}{4}$. So, to find the joint probability of independent events, we multiply the probabilities of the individual events (**Figure 12.9**). How does this method apply to genetics?

To see how joint probability is calculated in genetics problems, let's consider the monohybrid cross. The probabilities of two events are involved: gamete formation and random fertilization.

Calculating the probabilities involved in gamete formation is straightforward. A homozygote can produce only one type of gamete, so, for example, the probability of an *SS* individual producing gametes with the genotype *S* is 1. The heterozygote *Ss* produces *S* gametes with a probability of $\frac{1}{2}$ and *s* gametes with a probability of $\frac{1}{2}$.

Now let's see how the rules of probability might predict the ratio of the F_2 progeny of the cross shown in Figure 12.4. These plants are obtained by the self-pollination of F_1 plants of genotype *Ss*. The probability that an F_2 plant will have the genotype *SS* must be $\frac{1}{2} \times \frac{1}{2} = \frac{1}{4}$, because there is a 50-50 chance that the

One of Mendel's major contributions to the science of genetics was his use of the rules of statistics and probability to analyze his masses of data from hundreds of crosses resulting in thousands of progeny plants. His mathematical analyses revealed clear patterns in the data that allowed him to formulate his hypotheses. Ever since his work became widely recognized, geneticists have used simple mathematics in the same ways that Mendel did.

Punnett squares or probability calculations: A choice of methods

Punnett squares provide one way of solving problems in genetics, and probability calculations provide another. Many people find it easier to use the principles of probability, some of which are intuitive and familiar. For example, when we flip a coin, the law of probability states that it has an equal probability of landing "heads" or "tails." For any given toss of a fair coin, the probability of heads is independent of what happened in all the previous tosses. A run of ten straight heads implies nothing about the next toss. No "law of averages" increases the likelihood that the next toss will come up tails, and no "momentum"

12.9 Using Probability Calculations in Genetics   Like the results of a coin toss, the probability of any given combination of alleles appearing in the offspring of a cross can be obtained by multiplying the probabilities of each event. Since a heterozygote can be formed in two ways, these two probabilities are added together.

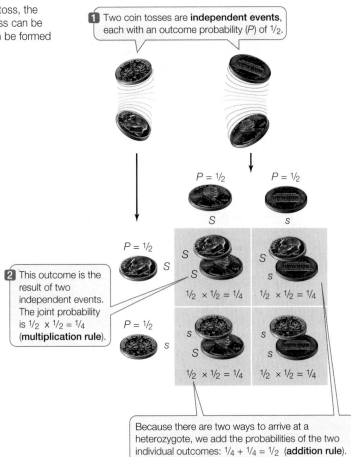

1 Two coin tosses are **independent events**, each with an outcome probability (P) of ½.

$P = ½$ $P = ½$

S s

2 This outcome is the result of two independent events. The joint probability is ½ × ½ = ¼ (**multiplication rule**).

$P = ½$ S S S S s
½ × ½ = ¼   ½ × ½ = ¼

$P = ½$ s S s s
½ × ½ = ¼   ½ × ½ = ¼

Because there are two ways to arrive at a heterozygote, we add the probabilities of the two individual outcomes: ¼ + ¼ = ½ (**addition rule**).

sperm will have the genotype S, and an independent chance of 50-50 that the egg will have the genotype S. Similarly, the probability of ss offspring is also $½ × ½ = ¼$.

ADDING PROBABILITIES   How are probabilities calculated when an event can happen in different ways? The probability of an F_2 plant getting an S allele from the sperm and an s allele from the egg is $¼$. In addition, there is a probability of $¼$ that the F_2 plant will get an s from the sperm and an S from the egg, resulting in the same genotype of Ss. The probability of an event that can occur in two or more different ways is the sum of the individual probabilities of those ways. Thus the probability that an F_2 plant will be a heterozygote is equal to the sum of the probabilities of the two ways of forming a heterozygote: $¼ + ¼ = ½$ (see Figure 12.9). The three genotypes are therefore expected to occur in the ratio $¼ SS : ½ Ss : ¼ ss$, resulting in the 1:2:1 ratio of genotypes and the 3:1 ratio of phenotypes seen in Figure 12.4.

PROBABILITY AND THE DIHYBRID CROSS   If F_1 plants heterozygous for two independent characters self-pollinate, the resulting F_2 plants express four different phenotypes. The proportions of these phenotypes are easily determined by probability calculations. Let's see how this works for the experiment shown in Figure 12.7.

Using the principles described above, we can calculate that the probability that an F_2 seed will be spherical is $¾$. This is found by adding the probability of an Ss heterozygote ($½$) and the probability of an SS homozygote ($¼$) = a total of $¾$. By the same reasoning, the probability that a seed will be yellow is also $¾$. The two characters are determined by separate genes and are independent of each other, so the joint probability that a seed will be both spherical and yellow is $¾ × ¾ = 9/16$. What is the probability of F_2 seeds being both wrinkled and yellow? The probability of being yellow is again $¾$; the probability of being wrinkled is $½ × ½ = ¼$. The joint probability that a seed will be *both* wrinkled and yellow is $¼ × ¾ = 3/16$.

The same probability applies, for similar reasons, to spherical, green F_2 seeds. Finally, the probability that F_2 seeds will be both wrinkled and green is $¼ × ¼ = 1/16$. Looking at all four phenotypes, we see that they are expected to occur in the ratio of 9:3:3:1.

Probability calculations and Punnett squares give the same results. Learn to do genetics problems both ways, and then decide which method you prefer.

Mendel's laws can be observed in human pedigrees

How are Mendel's laws of inheritance applied to humans? Mendel worked out his laws by performing many planned crosses and counting many offspring. Neither of these approaches is possible with humans, so human geneticists rely on **pedigrees**: family trees that show the occurrence of phenotypes (and alleles) in several generations of related individuals.

Because humans have such small numbers of offspring, human pedigrees do not show the clear proportions of offspring phenotypes that Mendel saw in his pea plants. For example, when a man and a woman who are both heterozygous for a recessive allele (say, Aa) have children together, each child has a 25 percent probability of being a recessive homozygote (aa). Thus if this couple were to have dozens of children, one-fourth of them would be recessive homozygotes. But the offspring of a single couple are likely to be too few to show the exact one-fourth proportion. In a family with only two children, for example, both could easily be aa (or Aa, or AA).

What if we want to know whether a recessive allele is carried by both the mother and the father? Human geneticists assume that any allele that causes an abnormal phenotype (such as a genetic disease) is rare in the human population. This means that if some members of a given family have a rare allele, it is highly unlikely that an outsider marrying into that family will have that same rare allele.

Human geneticists may wish to know whether a particular rare allele that causes an abnormal phenotype is dominant or recessive. **Figure 12.10A** is a pedigree showing the pattern of

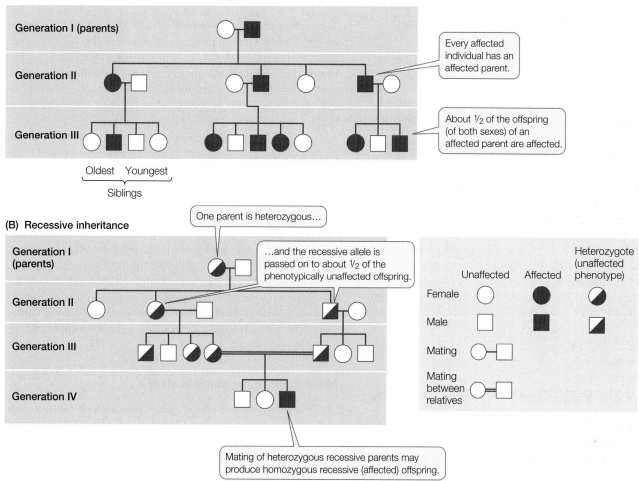

12.10 Pedigree Analysis and Inheritance (A) This pedigree represents a family affected by Huntington's disease, which results from a rare dominant allele. Everyone who inherits this allele is affected. (B) The family in this pedigree carries the allele for albinism, a recessive trait. Because the trait is recessive, heterozygotes do not have the albino phenotype, but they can pass the allele on to their offspring. Affected persons must inherit the allele from two heterozygous parents, or (rarely) from one homozygous recessive and one heterozygous parent, or (very rarely) two homozygous recessive parents. In this family, in generation III the heterozygous parents are cousins; however, the same result could occur if the parents were unrelated but heterozygous.

inheritance of a rare dominant allele. The following are the key features to look for in such a pedigree:

- Every affected person has an affected parent.
- About half of the offspring of an affected parent are also affected.
- The phenotype occurs equally in both sexes.

Compare this pattern with the one shown in **Figure 12.10B**, which is typical for the inheritance of a rare recessive allele:

- Affected people usually have two parents who are not affected.

- In affected families, about one-fourth of the children of unaffected parents are affected.
- The phenotype occurs equally in both sexes.

In pedigrees showing inheritance of a recessive phenotype, it is not uncommon to find a marriage of two relatives. This observation is a result of the rarity of recessive alleles that give rise to abnormal phenotypes. For two phenotypically normal parents to have an affected child (*aa*), the parents must both be heterozygous (*Aa*). If a particular recessive allele is rare in the general population, the chance of two people marrying who are both carrying that allele is quite low. On the other hand, if that allele is present in a family, two cousins might share it (see Figure 12.10B). For this reason, studies on populations that are isolated either culturally (by religion, as with the Amish in the United States) or geographically (as on islands) have been extremely valuable to human geneticists. People in these groups are more likely to marry relatives who may carry the same rare recessive alleles.

Because the major use of pedigree analysis is in the clinical evaluation and counseling of patients with inherited abnormalities in their families, a single pair of alleles is usually followed. However, pedigree analysis can also show independent assortment if two different allele pairs are considered.

12.1 RECAP

Mendel showed that genetic determinants are particulate and do not "blend" or disappear when the genes from two gametes combine. Mendel's first law of inheritance states that the two copies of a gene segregate during gamete formation. His second law states that genes assort independently during gamete formation. The frequencies with which different allele combinations will be expressed in offspring can be calculated with a Punnett square or using probability theory.

- What results seen in the F_1 and F_2 generations of Mendel's monohybrid cross experiments refuted the blending theory of inheritance? **See p. 240, Figures 12.3 and 12.4, and Table 12.1**

- How do events in meiosis explain Mendel's monohybrid cross results? **See pp. 242–244 and Figure 12.5**

- How do events in meiosis explain the independent assortment of alleles in Mendel's dihybrid cross experiments? **See p. 244 and Figures 12.7 and 12.8**

- Draw human pedigrees for dominant and recessive inheritance. **See pp. 246–247 and Figure 12.10**

The laws of inheritance as articulated by Mendel remain valid today; his discoveries laid the groundwork for all future studies of genetics. Inevitably, however, we have learned that things are more complicated. Let's take a look at some of these complications, beginning with the interactions between alleles at different loci.

12.2 How Do Alleles Interact?

Existing alleles are subject to change, and thus may give rise to new alleles, so there can be many alleles for a single character. In addition, alleles do not always show simple dominant-recessive relationships. Furthermore, a single allele may have multiple phenotypic effects.

New alleles arise by mutation

Genes are subject to **mutations**, which are rare, stable, and inherited changes in the genetic material. In other words, an allele can mutate to become a different allele. For example, you can envision that at one time all pea plants were tall and had the height allele T. A mutation occurred in that allele that resulted in a new allele, t (short). If this mutation was in a cell that underwent meiosis to form gametes, some of the resulting gametes would carry the t allele, and some offspring of this pea plant would carry the t allele. Mutation will be discussed in detail in Chapter 15. By creating variety, mutations are the raw material for evolution.

Geneticists usually define one particular allele of a gene as the **wild type**; this allele is the one that is present in most individuals in nature ("the wild") and gives rise to an expected trait or phenotype. Other alleles of that gene, often called mutant alleles, may produce a different phenotype. The wild-type and mutant alleles reside at the same locus and are inherited according to the rules set forth by Mendel. A genetic locus with a wild-type allele that is present less than 99 percent of the time (the rest of the alleles being mutant) is said to be **polymorphic** (Greek *poly*, "many"; *morph*, "form").

Many genes have multiple alleles

Because of random mutations, more than two alleles of a given gene may exist in a group of individuals. (Any one individual has only two alleles—one from its mother and one from its father. But different individuals may carry several different alleles.) In fact, there are many examples of such multiple alleles, and they often show a hierarchy of dominance.

Coat color in rabbits, for example, is determined by one gene with four alleles:

- C determines dark gray
- c determines albino
- c^{ch} determines chinchilla
- c^h determines light gray

12.11 Inheritance of Coat Color in Rabbits There are four alleles of the gene for coat color in these Netherlands dwarf rabbits. Different combinations of two alleles give different coat colors. The dominance hierarchy is $C > c^{ch} > c^h > c$.

Possible genotypes	CC, Cc^{ch}, Cc^h, Cc	$c^{ch}c^{ch}$	$c^{ch}c^h, c^{ch}c$	$c^h c^h, c^h c$	cc
Phenotype	Dark gray	Chinchilla	Light gray	Point restricted	Albino

Any rabbit with the *C* allele (paired with any of the four) is dark gray, and a rabbit with *cc* is albino. The intermediate colors result from the different allele combinations shown in **Figure 12.11**.

Multiple alleles increase the number of possible phenotypes. Each of Mendel's monohybrid crosses involved just one pair of alleles (for example, *S* and *s*) and two possible phenotypes (resulting from *SS* or *Ss* and *ss*). The four alleles of the rabbit coat color gene produce five different phenotypes.

Dominance is not always complete

In the pairs of alleles studied by Mendel, dominance is complete when an individual is heterozygous. That is, an *Ss* individual always expresses the *S* phenotype. However, many genes have alleles that are not dominant or recessive to one another. Instead, the heterozygotes show an intermediate phenotype—at first glance, like that predicted by the old blending theory of inheritance. For example, if a true-breeding red snapdragon is crossed with a true-breeding white one, all the F$_1$ flowers are pink. However, further crosses indicate that this apparent blending phenomenon can still be explained in terms of Mendelian genetics (**Figure 12.12**). The red and white alleles have not disappeared, as those colors reappear when the F$_1$ plants are interbred.

We can understand these results in terms of the Mendelian laws of inheritance. When heterozygotes show a phenotype that is intermediate between those of the two homozygotes, the gene is said to be governed by **incomplete dominance**. In other words, neither of the two alleles is dominant. Incomplete dominance is common in nature, and at the biochemical level, most examples of incomplete dominance are actually codominance (see below). In fact, Mendel's study of seven pea-plant traits is unusual in that all seven traits happened to be characterized by complete dominance.

In codominance, both alleles at a locus are expressed

Sometimes the two alleles at a locus produce two different phenotypes that *both* appear in heterozygotes, a phenomenon called **codominance**. Note that this is different from incomplete dominance, where the phenotype of a heterozygote is a blend of the phenotypes of the parents. A good example of codominance is seen in the ABO blood group system in humans.

There are numerous glycoproteins on the surfaces of red blood cells and they are all encoded by genes. One genetic locus is

called the ABO locus, with three alleles, I^A, I^B and I^O, that encode variants of a surface glycoprotein designated A, B, and O (the "ABO system"). Since people inherit one allele from each parent, they may have any combination of these alleles: $I^A I^B$, $I^A I^O$, $I^A I^A$, and so on. In terms of gene expression, it is important to note that in a codominant system, all alleles are expressed in a heterozygote. So people with $I^A I^B$ express both I^A and I^B alleles on their red blood cell surfaces.

Early attempts at blood transfusion frequently killed the patient. Around 1900, the Austrian scientist Karl Landsteiner mixed blood cells and serum (blood from which cells have been removed) from different individuals. He found that only certain combinations of blood and serum are compatible. In other combinations, the red blood cells from one individual form clumps in the presence of serum from the other individual. This discovery led to our ability to administer compatible blood transfusions that do not kill the recipient.

Incompatible transfusions result in the formation of clumps because of genetic systems like the ABO locus. People make specific proteins in the serum, called antibodies, that react with foreign, or "nonself," molecules called antigens. The A and B glycoproteins can act as antigens if present on the surfaces of red

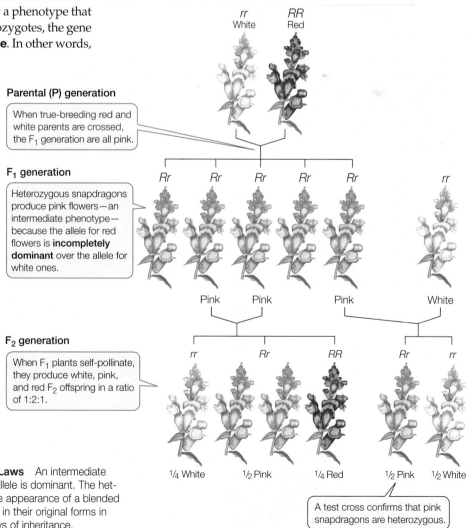

Parental (P) generation

When true-breeding red and white parents are crossed, the F$_1$ generation are all pink.

F$_1$ generation

Heterozygous snapdragons produce pink flowers—an intermediate phenotype—because the allele for red flowers is **incompletely dominant** over the allele for white ones.

F$_2$ generation

When F$_1$ plants self-pollinate, they produce white, pink, and red F$_2$ offspring in a ratio of 1:2:1.

A test cross confirms that pink snapdragons are heterozygous.

¼ White   ½ Pink   ¼ Red   ½ Pink   ½ White

12.12 Incomplete Dominance Follows Mendel's Laws An intermediate phenotype can occur in heterozygotes when neither allele is dominant. The heterozygous phenotype (here, pink flowers) may give the appearance of a blended trait, but the traits of the parental generation reappear in their original forms in succeeding generations, as predicted by Mendel's laws of inheritance.

blood cells in donated blood. If the person receiving the blood does not carry the I^A or I^B alleles, their antibodies will react with the nonself glycoproteins and the red blood cells will form clumps. The O glycoprotein does not act as an antigen. You can see these relationships in **Figure 12.13**. We will learn much more about the functions of antibodies and antigens in Chapter 42.

Interestingly, a recent development may make it possible to circumvent the ABO system of blood incompatibility. Enzymes have been isolated from bacteria that can convert the A and B glycoproteins into O glycoprotein. So blood from any genotype in the ABO system could be treated with these enzymes to make O-type blood, which is not antigenic. Since I^O is not a common allele in most human populations, this technology may be important in overcoming shortages of genetically suitable blood for transfusions.

Some alleles have multiple phenotypic effects

Mendel's principles were further extended when it was discovered that a single allele can influence more than one phenotype. When a single allele has more than one distinguishable phenotypic effect, we say that the allele is **pleiotropic**. A familiar example of pleiotropy involves the allele responsible for the coloration pattern (light body, darker extremities) of Siamese cats. The same allele is also responsible for the characteristic crossed eyes of Siamese cats. Although these effects appear to be unrelated, both are caused by the protein encoded by this allele.

12.13 ABO Blood Reactions Are Important in Transfusions This table shows the results of mixing red blood cells of types A, B, AB, and O with serum containing anti-A or anti-B antibodies. As you look down the columns, note that each of the types, when mixed separately with anti-A and with anti-B, gives a unique pair of results; this is the basic method by which blood is typed. People with type O blood are good blood donors because O cells do not react with either anti-A or anti-B antibodies. People with type AB blood are good recipients, since they make neither type of antibody. When blood transfusions are incompatible, the reaction (clumping of red blood cells) can have severely adverse consequences for the recipient.

Blood type of cells	Genotype	Antibodies made by body	Reaction to added antibodies	
			Anti-A	Anti-B
A	$I^A I^A$ or $I^A i^O$	Anti-B		
B	$I^B I^B$ or $I^B i^O$	Anti-A		
AB	$I^A I^B$	Neither anti-A nor anti-B		
O	$i^O i^O$	Both anti-A and anti-B		

Red blood cells that do not react with antibody remain evenly dispersed.

Red blood cells that react with antibody clump together (speckled appearance).

12.2 RECAP

Genes are subject to random mutations that give rise to new alleles; thus many genes have more than two alleles within a population. Dominance is not necessarily an all-or-nothing phenomenon.

- How does the experiment in Figure 12.12 demonstrate incomplete dominance? **See p. 249**

- Explain how blood type AB results from codominance. **See pp. 249–250 and Figure 12.13**

Thus far we have treated the phenotype of an organism, with respect to a given character, as a simple result of the alleles of a single gene. In many cases, however, several genes interact to determine a phenotype. To complicate things further, the physical environment may interact with the genetic constitution of an individual in determining the phenotype.

12.3 How Do Genes Interact?

We have just seen how two alleles of the same gene can interact to produce a phenotype. If you consider most complex phenotypes, such as human height, you will realize that they are influenced by the products of many genes. We now turn to the genetics of such gene interactions.

Epistasis occurs when the phenotypic expression of one gene is affected by another gene. For example, two genes (*B* and *E*) determine coat color in Labrador retrievers:

- Allele *B* (black pigment) is dominant to *b* (brown)

- Allele *E* (pigment deposition in hair) is dominant to *e* (no deposition, so hair is yellow)

So an *EE* or *Ee* dog with *BB* or *Bb* is black; one with *bb* is brown; and one with *ee* is yellow regardless of the *Bb* alleles present. Clearly, gene *E* determines the expression of *Bb* (**Figure 12.14**).

Hybrid vigor results from new gene combinations and interactions

In 1876, Charles Darwin reported that when he crossed two different true-breeding, homozygous genetic strains of corn, the offspring were 25 percent taller than either of the parent strains. Darwin's observation was largely ignored for the next 30 years. In 1908, George Shull "rediscovered" this idea, reporting that not just plant height but the weight of the corn grain produced was dramatically higher in the offspring. Agricultural scientists took note, and Shull's paper had a lasting impact on the field of applied genetics (**Figure 12.15**).

Farmers have known for centuries that matings among close relatives (known as **inbreeding**) can result in offspring of lower quality than matings between unrelated individuals. Agricultural scientists call this *in-*

A dog with alleles *B* and *E* is black.

A dog with alleles *bb* and *E* is brown.

A dog with *ee* is yellow, regardless of its *Bb* alleles.

(A) Black labrador (*B_E_*)

(B) Chocolate labrador (*bbE_*)

(C) Yellow labrador (*_ _ee*)

BbEe × *BbEe*

Sperm

Eggs

	BE	Be	bE	be
BE	Black *BBEE*	Black *BBEe*	Black *BbEE*	Black *BbEe*
Be	Black *BBEe*	Yellow *BBee*	Black *BbEe*	Yellow *Bbee*
bE	Black *BbEE*	Black *BbEe*	Brown *bbEE*	Brown *bbEe*
be	Black *BbEe*	Yellow *Bbee*	Brown *bbEe*	Yellow *bbee*

12.14 Genes May Interact Epistatically Epistasis occurs when one gene alters the phenotypic effect of another gene. In Labrador retrievers, the *Ee* gene determines the expression of the *Bb* gene.

breeding depression. The problems with inbreeding arise because close relatives tend to have the same recessive alleles, some of which may be harmful. The "hybrid vigor" after crossing inbred lines is called **heterosis** (short for heterozygosis). The cultivation of hybrid corn spread rapidly in the United States and all over the world, quadrupling grain production. Unfortunately, as we saw in the opening story, this scientific advance was not universally adopted, and regions such as the Russian empire fell far behind in corn production. The practice of hybridization has spread to many other crops and animals used in agriculture. For example, beef cattle that are crossbred are larger and live longer than cattle bred within their own genetic strain.

The mechanism by which heterosis works is not known. A widely accepted hypothesis is overdominance, in which the heterozygous condition in certain important genes whose products interact is superior to the homozygous condition in either or both genes. Another hypothesis is that the homozygotes have alleles that inhibit growth, and these are less active or absent in the heterozygote.

The environment affects gene action

The phenotype of an individual does not result from its genotype alone. *Genotype and environment interact to determine the phenotype of an organism.* This is especially important to remember in the era of genome sequencing (see Chapter 17). When the sequence of the human genome was completed in 2003, it was hailed as the "book of life," and public expectations of the benefits gained from this knowledge were (and are) high. But this kind of "genetic determinism" is wrong. Common knowledge tells us that environmental variables such as light, temperature, and nutrition can affect the phenotypic expression of a genotype.

B73 Hybrid Mo17

12.15 Hybrid Vigor in Corn Two homozygous parent lines of corn (cobs shown), B73 (left) and Mo17 (right), were crossed to produce the more vigorous hybrid line (center).

12.16 The Environment Influences Gene Expression This rabbit expresses a coat pattern known as "chocolate point." Its genotype specifies dark fur, but the enzyme for dark fur is inactive at normal body temperature, so only the rabbit's extremities—the coolest regions of the body—express this phenotype.

The temperature of the extremities is lower and allows expression of the black coat color gene.

The temperature of most of the body is too high for the expression of the black coat color gene.

A familiar example of this phenomenon involves "point restriction" coat patterns found in Siamese cats and certain rabbit breeds (**Figure 12.16**). These animals carry a mutant allele of a gene that controls the growth of black fur all over the body. As a result of this mutation, the enzyme encoded by the gene is inactive at temperatures above a certain point (usually around 35°C). The animals maintain a body temperature above this point, and so their fur is mostly light. However, the extremities—feet, ears, nose, and tail—are cooler, about 25°C, so the fur on these regions is dark. These animals are all white when they are born, because the extremities were kept warm in the mother's womb.

A simple experiment shows that the dark fur is temperature-dependent. If a patch of white fur on a point-restricted rabbit's back is removed and an ice pack is placed on the skin where the patch was, the fur that grows back will be dark. This indicates that although the gene for dark fur was expressed all along, the environment inhibited the activity of the mutant enzyme.

Two parameters describe the effects of genes and environment on the phenotype:

- **Penetrance** is the proportion of individuals in a group with a given genotype that actually show the expected phenotype.

- **Expressivity** is the degree to which a genotype is expressed in an individual.

Penetrance affects, for example, the incidence of Huntington's disease in humans. The disease results from the presence of a dominant allele, but 5 percent of people with the allele do not express the disease. So this allele is said to be 95 percent penetrant. For an example of environmental effects on expressivity, consider how Siamese cats kept indoors or outdoors in different climates might look.

Most complex phenotypes are determined by multiple genes and the environment

The differences between individual organisms in simple characters, such as those that Mendel studied in pea plants, are discrete and **qualitative**. For example, the individuals in a population of pea plants are either short or tall. For most complex characters, however, such as height in humans, the phenotype varies more or less continuously over a range. Some people are short, others are tall, and many are in between the two extremes. Such variation within a population is called **quantitative**, or continuous, variation (**Figure 12.17**).

12.17 Quantitative Variation Quantitative variation is produced by the interaction of genes at multiple loci and the environment. These students (women in white on the left are shorter; men in blue on the right are taller) show continuous variation in height that is the result of interactions between many genes and the environment.

Sometimes this variation is largely genetic. For instance, much of human eye color is the result of a number of genes controlling the synthesis and distribution of dark melanin pigment. Dark eyes have a lot of it, brown eyes less, and green, gray, and blue eyes even less. In the latter cases, the distribution of other pigments in the eye is what determines light reflection and color.

In most cases, however, quantitative variation is due to *both genes and environment*. Height in humans certainly falls into this category. If you look at families, you often see that parents and their offspring all tend to be tall or short. However, nutrition also plays a role in height: American 18-year-olds today are about 20 percent taller than their great-grandparents were at the same age. Three generations are not enough time for mutations that would exert such a dramatic effect to occur, so the height difference must not be due to genetics.

Geneticists call the genes that together determine such complex characters **quantitative trait loci**. Identifying these loci is a major challenge, and an important one. For example, the amount of grain that a variety of rice produces in a growing season is determined by many interacting genetic factors. Crop plant breeders have worked hard to decipher these factors in order to breed higher-yielding rice strains. In a similar way, human characteristics such as disease susceptibility and behavior are caused in part by quantitative trait loci. Recently, one of the many genes involved with human height was identified. The gene, *HMGA2*, has an allele that apparently has the potential to add 4 mm to human height.

12.3 RECAP

In epistasis, one gene affects the expression of another. Perhaps the most challenging problem for genetics is the explanation of complex phenotypes that are caused by many interacting genes and the environment.

- Explain the difference between penetrance and expressivity. See p. 252

- How is quantitative variation different from qualitative variation? See pp. 252–253

In the next section we'll see how the discovery that genes occupy specific positions on chromosomes enabled Mendel's successors to provide a physical explanation for his model of inheritance, and to provide an explanation for those cases where Mendel's second law does not apply.

12.4 What Is the Relationship between Genes and Chromosomes?

There are far more genes than chromosomes. Studies of different genes that are physically linked on the same chromosome reveal inheritance patterns that are not Mendelian. These patterns have been useful not only in detecting linkage of genes, but also in determining how far apart they are from one another on the chromosome.

The organism that revealed genetic linkage is the fruit fly *Drosophila melanogaster*. Its small size, the ease with which it can be bred, and its short generation time make this animal an attractive experimental subject. Beginning in 1909, Thomas Hunt Morgan and his students at Columbia University pioneered the study of *Drosophila*, and it remains a very important organism in studies of genetics.

---

yourBioPortal.com

GO TO Animated Tutorial 12.2 • Alleles That Do Not Sort Independently

---

Genes on the same chromosome are linked

Some of the crosses Morgan performed with fruit flies yielded phenotypic ratios that were not in accordance with those predicted by Mendel's law of independent assortment. Morgan crossed *Drosophila* with two known genotypes, $BbVgvg \times bbvgvg$,* for two different characters, body color and wing shape:

- *B* (wild-type gray body), is dominant over *b* (black body)
- *Vg* (wild-type wing) is dominant over *vg* (vestigial, a very small wing)

Morgan expected to see four phenotypes in a ratio of 1:1:1:1, but that is not what he observed. The body color gene and the wing size gene were not assorting independently; rather, they were, for the most part, inherited together (**Figure 12.18**).

These results became understandable to Morgan when he considered the possibility that the two loci are on the same chromosome—that is, that they might be linked. Suppose that the *B* and *Vg* loci are indeed located on the same chromosome. Why didn't all of Morgan's F_1 flies have the parental phenotypes—that is, why didn't his cross result in gray flies with normal wings (wild type) and black flies with vestigial wings, in a 1:1 ratio? If linkage were absolute—that is, if chromosomes always remained intact and unchanged—we would expect to see just those two types of progeny. However, this does not always happen.

Genes can be exchanged between chromatids

ABSOLUTE LINKAGE IS RARE If linkage were absolute, Mendel's law of independent assortment would apply only to loci on different chromosomes. What actually happens is more complex, and therefore more interesting. Genes at different loci on the same chromosome *do* sometimes separate from one another during meiosis. Genes may recombine when two homologous chromosomes physically exchange corresponding segments during prophase I of meiosis—that is, by crossing over (**Figure 12.19**; see also Figures 11.18 and 11.19). As described in Section 11.2, DNA is replicated during the S phase, so that by prophase I, when homologous chromosome pairs come together to form tetrads, each chromosome consists of two chromatids.

---

*Do you recognize this type of cross? It is a test cross for the two gene pairs; see Figure 12.6.

INVESTIGATING LIFE

12.18 Some Alleles Do Not Assort Independently

Morgan's studies showed that the genes for body color and wing size in *Drosophila* are linked, so that their alleles do not assort independently.

HYPOTHESIS Alleles for different characteristics always assort independently.

METHOD

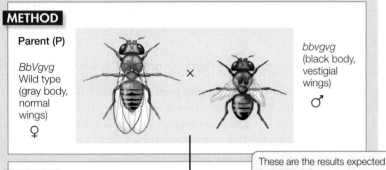

Parent (P)

BbVgvg
Wild type
(gray body,
normal
wings)
♀

×

bbvgvg
(black body,
vestigial
wings)
♂

RESULTS

F₁

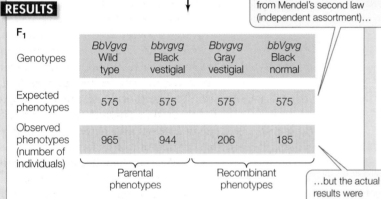

These are the results expected from Mendel's second law (independent assortment)…

Genotypes	*BbVgvg* Wild type	*bbvgvg* Black vestigial	*Bbvgvg* Gray vestigial	*bbVgvg* Black normal
Expected phenotypes	575	575	575	575
Observed phenotypes (number of individuals)	965	944	206	185

Parental phenotypes   Recombinant phenotypes

…but the actual results were inconsistent with the law.

CONCLUSION The hypothesis is rejected. These two genes do not assort independently, but are linked (on the same chromosome).

FURTHER INVESTIGATIONS: Look again at Mendel's dihybrid cross (see Figure 12.7). If the genes for seed shape and seed color were linked, what would the results be?

Go to **yourBioPortal.com** for original citations, discussions, and relevant links for all INVESTIGATING LIFE figures.

Note that the exchange event involves *only two of the four chromatids* in a tetrad, one from each member of the homologous pair, and can occur at any point along the length of the chromosome. The chromosome segments involved are exchanged reciprocally, so both chromatids involved in crossing over become recombinant (that is, each chromatid ends up with genes from both of

12.19 Crossing Over Results in Genetic Recombination

Recombination accounts for why linked alleles are not always inherited together. Alleles at different loci on the same chromosome can be recombined by crossing over, and separated from one another. Such recombination occurs during prophase I of meiosis.

the organism's parents). Usually several exchange events occur along the length of each homologous pair.

When crossing over takes place between two linked genes, not all the progeny of a cross have the parental phenotypes. Instead, recombinant offspring appear as well, as they did in Morgan's cross. They appear in proportions called **recombinant frequencies**, which are calculated by dividing the number of recombinant progeny by the total number of progeny (**Figure 12.20**). Recombinant frequencies will be *greater for loci that are farther apart* on the chromosome than for loci that are closer together because an exchange event is more likely to occur between genes that are far apart. Genetic recombination is another way to generate the diversity that is the raw material for natural selection and evolution.

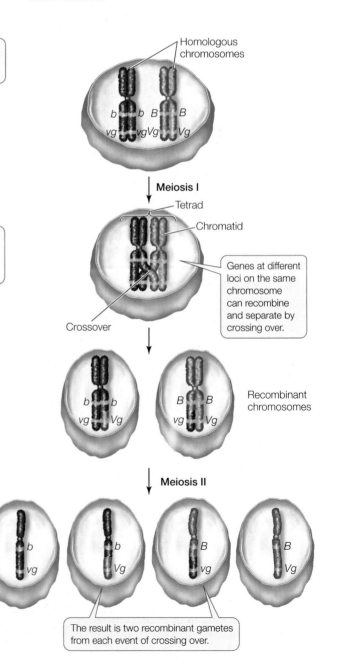

Homologous chromosomes

Meiosis I

Tetrad

Chromatid

Crossover

Genes at different loci on the same chromosome can recombine and separate by crossing over.

Recombinant chromosomes

Meiosis II

The result is two recombinant gametes from each event of crossing over.

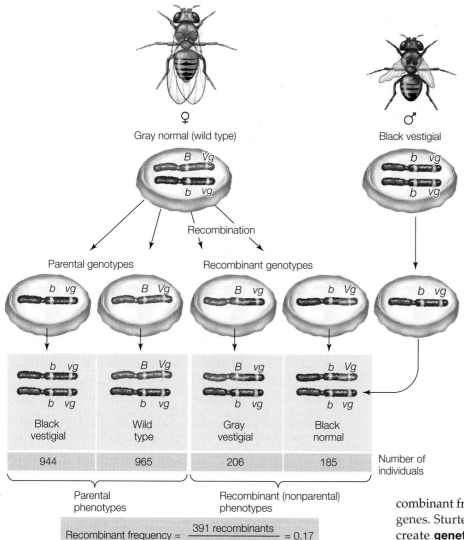

12.20 Recombinant Frequencies The frequency of recombinant offspring (those with a phenotype different from either parent) can be calculated.

Gray normal (wild type)

Black vestigial

Recombination

Parental genotypes   Recombinant genotypes

Black vestigial	Wild type	Gray vestigial	Black normal
944	965	206	185

Number of individuals

Parental phenotypes   Recombinant (nonparental) phenotypes

$$\text{Recombinant frequency} = \frac{391 \text{ recombinants}}{2{,}300 \text{ total offspring}} = 0.17$$

12.21 Steps toward a Genetic Map The chance of a crossing over between two loci on a chromosome increases with the distance between the loci. Thus, Sturtevant was able to derive this partial map of a *Drosophila* chromosome using the Morgan group's data on the recombinant frequencies of five recessive traits. He used an arbitrary unit of distance—the map unit, or centimorgan (cM)—equivalent to a recombinant frequency of 0.01.

Geneticists can make maps of chromosomes

If two loci are very close together on a chromosome, the odds of a crossover occurring between them are small. In contrast, if two loci are far apart, crossing over could occur between them at many points. This pattern is a consequence of the mechanism of crossing over: the farther apart two genes are, the more places there are in the chromosome for breakage and reunion of chromatids to occur. In a population of cells undergoing meiosis, a greater proportion of the cells will undergo recombination between two loci that are far apart than between two loci that are close together. In 1911, Alfred Sturtevant, then an undergraduate student in T. H. Morgan's fly room, realized how this simple insight could be used to show where different genes lie on a chromosome in relation to one another.

The Morgan group had determined recombinant frequencies for many pairs of linked *Drosophila* genes. Sturtevant used those recombinant frequencies to create **genetic maps** that showed the arrangements of genes along the chromosomes (**Figure 12.21**). Ever since Sturtevant demonstrated this method, geneticists have mapped the chromosomes of eukaryotes, prokaryotes, and viruses, assigning distances between genes in **map units**. A map unit corresponds to a recombinant frequency of 0.01; it is also referred to as a **centimorgan (cM)**, in honor of the founder of the fly room. You, too, can work out a genetic map (**Figure 12.22**).

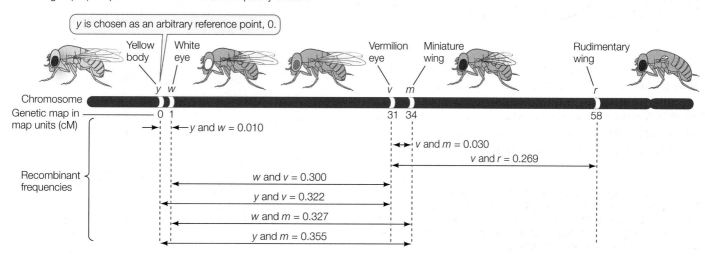

y is chosen as an arbitrary reference point, 0.

	Yellow body	White eye			Vermilion eye	Miniature wing		Rudimentary wing	

Chromosome

Genetic map in map units (cM): 0 1   31 34   58

y and *w* = 0.010

v and *m* = 0.030

v and *r* = 0.269

Recombinant frequencies

w and *v* = 0.300

y and *v* = 0.322

w and *m* = 0.327

y and *m* = 0.355

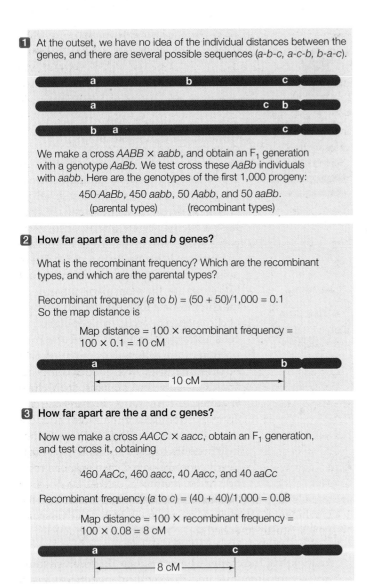

1 At the outset, we have no idea of the individual distances between the genes, and there are several possible sequences (a-b-c, a-c-b, b-a-c).

We make a cross *AABB* × *aabb*, and obtain an F₁ generation with a genotype *AaBb*. We test cross these *AaBb* individuals with *aabb*. Here are the genotypes of the first 1,000 progeny:

450 *AaBb*, 450 *aabb*, 50 *Aabb*, and 50 *aaBb*.
(parental types)   (recombinant types)

2 **How far apart are the *a* and *b* genes?**

What is the recombinant frequency? Which are the recombinant types, and which are the parental types?

Recombinant frequency (a to b) = (50 + 50)/1,000 = 0.1
So the map distance is

Map distance = 100 × recombinant frequency = 100 × 0.1 = 10 cM

3 **How far apart are the *a* and *c* genes?**

Now we make a cross *AACC* × *aacc*, obtain an F₁ generation, and test cross it, obtaining

460 *AaCc*, 460 *aacc*, 40 *Aacc*, and 40 *aaCc*

Recombinant frequency (a to c) = (40 + 40)/1,000 = 0.08

Map distance = 100 × recombinant frequency = 100 × 0.08 = 8 cM

12.22 Map These Genes The object of this exercise is to determine the order of three loci (*a*, *b*, and *c*) on a chromosome, as well as the map distances (in cM) between them.

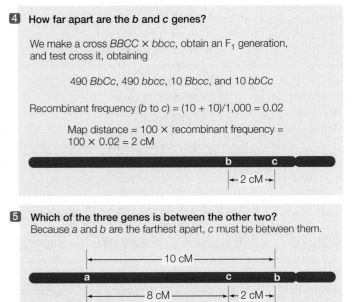

4 **How far apart are the *b* and *c* genes?**

We make a cross *BBCC* × *bbcc*, obtain an F₁ generation, and test cross it, obtaining

490 *BbCc*, 490 *bbcc*, 10 *Bbcc*, and 10 *bbCc*

Recombinant frequency (b to c) = (10 + 10)/1,000 = 0.02

Map distance = 100 × recombinant frequency = 100 × 0.02 = 2 cM

5 **Which of the three genes is between the other two?**
Because *a* and *b* are the farthest apart, *c* must be between them.

These numbers add up perfectly. In most real cases, they will not add up perfectly because of multiple crossovers.

Linkage is revealed by studies of the sex chromosomes

In Mendel's work, reciprocal crosses always gave identical results; it did not matter whether a dominant allele was contributed by the mother or by the father. But in some cases, the parental origin of a chromosome does matter. For example, human males inherit a bleeding disorder called hemophilia from their mothers, not from their fathers. To understand the types of inheritance in which the parental origin of an allele is important, we must consider the ways in which sex is determined in different species.

SEX DETERMINATION BY CHROMOSOMES In corn, every diploid adult has both male and female reproductive structures. The tissues in these two types of structure are genetically identical, just as roots and leaves are genetically identical. Plants such as corn, in which the same individual produces both male and female gametes, are said to be *monoecious* (Greek, "one house"). Other plants, such as date palms and oak trees, and most animals are *dioecious* ("two houses"), meaning that some individuals can produce only male gametes and the others can produce only fe-

male gametes. In other words, in dioecious organisms the different sexes are different individuals.

In most dioecious organisms, sex is determined by differences in the chromosomes, but such determination operates in different ways in different groups of organisms. For example, in many animals including mammals, sex is determined by a single **sex chromosome**, or by a pair of them. Both males and females have two copies of each of the rest of the chromosomes, which are called **autosomes**. In other animals, the chromosomal basis of sex determination is different from that of mammals (**Table 12.2**).

The sex chromosomes of female mammals consist of a pair of X chromosomes. Male mammals, on the other hand, have one X chromosome and a sex chromosome that is not found in fe-

TABLE 12.2
Sex Determination in Animals

ANIMAL GROUP	MECHANISM
Bees	Males are haploid, females are diploid
Fruit Flies	Fly is female if ratio of sex chromosomes to autosomes is ≥ 1
Birds	Males WW (homogametic), females WZ (heterogametic)
Mammals	Males XY (heterogametic), females XX (homogametic)

males, the Y chromosome. Females may be represented as XX and males as XY.

MALE MAMMALS PRODUCE TWO KINDS OF GAMETES Each gamete produced by a male mammal has a complete set of autosomes, but half the gametes carry an X chromosome, and the other half carry a Y. When an X-bearing sperm fertilizes an egg, the resulting XX zygote is female; when a Y-bearing sperm fertilizes an egg, the resulting XY zygote is male.

SEX CHROMOSOME ABNORMALITIES REVEALED THE GENE THAT DETERMINES SEX Can we determine which chromosome, X or Y, carries the sex-determining gene, and can the gene be identified? One way to determine cause (e.g., the presence of a gene on the Y chromosome) and effect (e.g., maleness) is to look at cases of biological error, in which the expected outcome does not happen.

Abnormal sex chromosome arrangements resulting from nondisjunction during meiosis (see Section 11.5) tell us something about the functions of the X and Y chromosomes. As you will recall, nondisjunction occurs when a pair of homologous chromosomes (in meiosis I) or sister chromatids (in meiosis II) fail to separate. As a result, a gamete may have one too few or one too many chromosomes. If this gamete fuses with another gamete that has the full haploid chromosome set, the resulting offspring will be aneuploid, with fewer or more chromosomes than normal.

In humans, XO individuals sometimes appear. (The O implies that a chromosome is missing—that is, individuals that are XO have only one sex chromosome.) Human XO individuals are females who are moderately abnormal physically but normal mentally; usually they are also sterile. The XO condition in humans is called Turner syndrome. It is the only known case in which a person can survive with only one member of a chromosome pair (here, the XY pair), although most XO conceptions are spontaneously terminated early in development. XXY individuals also occur; this condition, which affects males, is called Klinefelter syndrome, and results in overlong limbs and sterility.

These observations suggest that the gene controlling maleness is located on the Y chromosome. Observations of people with other types of chromosomal abnormalities helped researchers to pinpoint the location of that gene:

- Some women are genetically XY but lack a small portion of the Y chromosome.

- Some men are genetically XX but have a small piece of the Y chromosome attached to another chromosome.

The Y fragments that are respectively missing and present in these two cases are the same and contain the maleness-determining gene, which was named *SRY* (sex-determining *region* on the Y chromosome).

The *SRY* gene encodes a protein involved in **primary sex determination**—that is, the determination of the kinds of gametes that an individual will produce and the organs that will make them. In the presence of the functional SRY protein, an embryo develops sperm-producing testes. (Notice that *italic type* is used for the name of a gene, but roman type is used for the

name of a protein.) If the embryo has no Y chromosome, the *SRY* gene is absent, and thus the SRY protein is not made. In the absence of the SRY protein, the embryo develops egg-producing ovaries. In this case, a gene on the X chromosome called *DAX1* produces an anti-testis factor. So the role of SRY in a male is to inhibit the maleness inhibitor encoded by *DAX1*. The SRY protein does this in male cells, but since it is not present in females, DAX1 can act to inhibit maleness.

Primary sex determination is not the same as **secondary sex determination**, which results in the outward manifestations of maleness and femaleness (such as body type, breast development, body hair, and voice). These outward characteristics are not determined directly by the presence or absence of the Y chromosome. Instead, they are determined by genes that are scattered on the autosomes and the X chromosome. These genes control the actions of hormones, such as testosterone and estrogen.

Genes on sex chromosomes are inherited in special ways

Genes on sex chromosomes do not show the Mendelian patterns of inheritance. In *Drosophila* and in humans, the Y chromosome carries few known genes, but the X chromosome carries a substantial number of genes that affect a great variety of characters. These genes are present in two copies in females but only one copy in males. Therefore, males are always **hemizygous** for genes on the X chromosome—they have only one copy of each, and it is expressed. So reciprocal crosses do not give identical results for characters whose genes are carried on the sex chromosomes, and these characters do not show the usual Mendelian inheritance ratios.

Eye color in *Drosophila* is a good example of inheritance of a character that is governed by a locus on a sex chromosome (**sex-linked** inheritance). The wild-type eye color of these flies is red. In 1910, Morgan discovered a mutation that causes white eyes. He crossed flies of the wild-type and mutant phenotypes, and demonstrated that the eye color locus is on the X chromosome. If we abbreviate the eye color alleles as R (red eyes) and r (white eyes), the presence of the alleles on the X chromosome is designated by X^R and X^r.

When a homozygous red-eyed female ($X^R X^R$) was crossed with a (hemizygous) white-eyed male ($X^r Y$), all the sons and daughters had red eyes, because red (R) is dominant over white (r) and all the progeny had inherited a wild-type X chromosome (X^R) from their mothers (**Figure 12.23A**).

In the reciprocal cross, in which a white-eyed female ($X^r X^r$) was mated with a red-eyed male ($X^R Y$), all the sons were white-eyed and all the daughters were red-eyed (**Figure 12.23B**). The sons from the reciprocal cross inherited their only X chromosome from their white-eyed mother; the Y chromosome they inherited from their father did not carry the eye color locus. On the other hand, the daughters got an X chromosome bearing the white allele from their mother and an X chromosome bearing the red allele from their father; therefore they were red-eyed heterozygotes.

When heterozygous females were mated with red-eyed males, half their sons had white eyes, but all their daughters

(A)

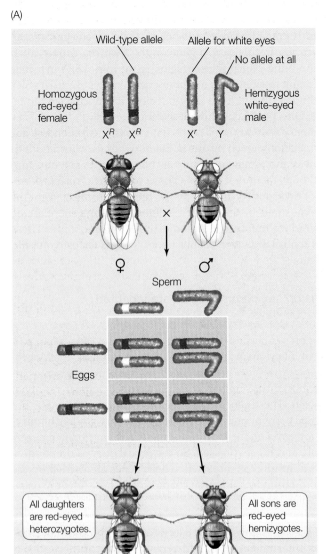

(B)

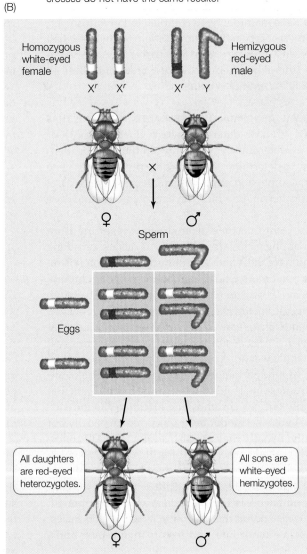

12.23 Eye Color Is a Sex-Linked Trait in *Drosophila* Morgan demonstrated that a mutant allele that causes white eyes in *Drosophila* is carried on the X chromosome. Note that in this case, the reciprocal crosses do not have the same results.

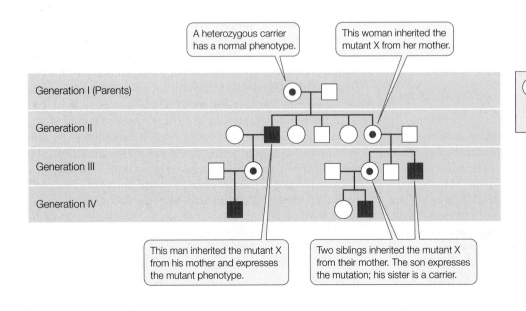

A heterozygous carrier has a normal phenotype.

This woman inherited the mutant X from her mother.

Female who carries allele for phenotype of interest on one X chromosome

This man inherited the mutant X from his mother and expresses the mutant phenotype.

Two siblings inherited the mutant X from their mother. The son expresses the mutation; his sister is a carrier.

12.24 Red-Green Color Blindness Is a Sex-Linked Trait in Humans The mutant allele for red-green color blindness is expressed as an X-linked recessive trait, and therefore is always expressed in males when they carry that allele.

had red eyes. Together, these three results showed that eye color was carried on the X chromosome and not on the Y.

Humans display many sex-linked characters

The human X chromosome carries about 2,000 known genes. The alleles at these loci follow the same pattern of inheritance as those for eye color in *Drosophila*. For example, one gene on the human X chromosome has a mutant recessive allele that leads to red–green color blindness, and it appears in individuals who are homozygous or hemizygous for the recessive mutant allele.

Pedigree analyses of X-linked recessive phenotypes (like the one in **Figure 12.24**) reveal the following patterns:

- The phenotype appears much more often in males than in females, because only one copy of the rare allele is needed for its expression in males, while two copies must be present in females.

- A male with the mutation can pass it on only to his daughters; all his sons get his Y chromosome.

- Daughters who receive one X-linked mutation are heterozygous **carriers**. They are phenotypically normal, but they can pass the mutant allele to either sons or daughters. (On average only half their children inherit the mutant allele, since half of their X chromosomes carry the normal allele.)

- The mutant phenotype can skip a generation if the mutation passes from a male to his daughter (who will be phenotypically normal) and thus to her son.

The small human Y chromosome carries several dozen genes. Among them is the maleness determinant, *SRY*. Interestingly, for some genes on the Y chromosome there are similar, but not identical, genes on the X chromosome. For example, one of the proteins that make up ribosomes is encoded by a gene on the Y chromosome that is expressed only in male cells, while the X-linked counterpart is expressed in both sexes. This means that there are "male" and "female" ribosomes; the significance of this phenomenon is unknown. Y-linked alleles are passed only from father to son. (Verify this with a Punnett square.)

12.4 RECAP

Simple Mendelian ratios are not observed when genes are linked on the same chromosome. Linkage is indicated by atypical frequencies of phenotypes in the offspring from a test cross. Sex linkage in humans refers to genes on the X chromosome that have no counterpart on the Y chromosome.

- What is the concept of linkage and what are its implications for the results of genetic crosses? **See pp. 253–254 and Figures 12.19 and 12.20**

- How does a sex-linked gene behave differently in genetic crosses than a gene on an autosome? **See pp. 257–259 and Figure 12.23**

The genes we've discussed so far in this chapter are all in the cell nucleus. But other organelles, including mitochondria and plastids, also carry genes. What are they, and how are they inherited?

12.5 What Are the Effects of Genes Outside the Nucleus?

The nucleus is not the only organelle in a eukaryotic cell that carries genetic material. As described in Section 5.5, mitochondria and plastids contain small numbers of genes, which are remnants of the entire genomes of colonizing prokaryotes that eventually gave rise to these organelles. For example, in humans, there are about 24,000 genes in the nuclear genome and 37 in the mitochondrial genome. Plastid genomes are about five times larger than those of mitochondria. In any case, several of the genes carried by cytoplasmic organelles are important for organelle assembly and function, so it is not surprising that mutations of these genes can have profound effects on the organism.

The inheritance of organelle genes differs from that of nuclear genes for several reasons:

- In most organisms, mitochondria and plastids are inherited only from the mother. As you will learn in Chapter 43, eggs contain abundant cytoplasm and organelles, but the only part of the sperm that survives to take part in the union of haploid gametes is the nucleus. So you have inherited your mother's mitochondria (with their genes), but not your father's.

- There may be hundreds of mitochondria or plastids in a cell. So a cell is not diploid for organelle genes.

- Organelle genes tend to mutate at much faster rates than nuclear genes, so there are multiple alleles of organelle genes.

The phenotypes resulting from mutations in organelle genes reflect the organelles' roles. For example, in plants and some photosynthetic protists, certain plastid gene mutations affect the proteins that assemble chlorophyll molecules into photosystems. These mutations result in a phenotype that is essentially white instead of green. The inheritance of this phenotype follows a non-Mendelian, maternal pattern (**Figure 12.25**).

Mitochondrial gene mutations that affect one of the complexes in the respiratory chain result in less ATP production. These mutations have particularly noticeable effects in tissues with high energy requirements, such as the nervous system, muscles, and kidneys. In 1995, Greg LeMond, a professional cyclist who had won the famous Tour de France three times, was forced to retire because of muscle weakness caused by a mitochondrial mutation.

12.5 RECAP

Genes in the genomes of organelles, specifically plastids and mitochondria, do not behave in a Mendelian fashion.

- Why are genes carried in the organelle genomes usually inherited only from the mother?

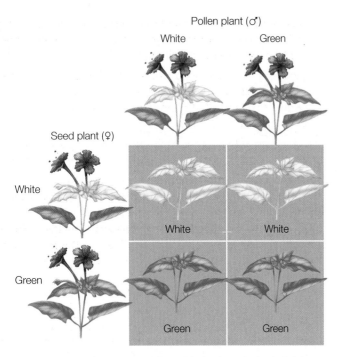

12.25 Cytoplasmic Inheritance In four o'clock plants, leaf color is inherited through the female plant only. The white leaf color is caused by a chloroplast mutation that occurs during the life of the parent plant; the leaves that form before the mutation occurs are green. The mutation is passed on to the germ cells, and the offspring that inherit the mutation are entirely white.

Mendel and those who followed him scientifically focused on eukaryotes, with diploid organisms and haploid gametes. A half-century after the rediscovery of Mendel's work, a sexual process that allows genetic recombination was discovered in prokaryotes as well. We now turn to that process.

12.6 How Do Prokaryotes Transmit Genes?

As you saw in Chapter 5, prokaryotic cells lack a nucleus but contain their genetic material as mostly a single chromosome in a central region of the cell. In Chapter 11, you saw that bacteria reproduce asexually by cell division, a process that gives rise to virtually genetically identical products. That is, the offspring of cell reproduction in bacteria constitute a clone. However, mutations occur in bacteria just as they do in eukaryotes; the resulting new alleles increase genetic diversity.

You might expect, therefore, that there is no way for individuals of these species to exchange genes, as in sexual reproduction. It turns out, though, that prokaryotes do have a sexual process.

Bacteria exchange genes by conjugation

The bacterial chromosome, like the bacterial cell, is considerably smaller than its eukaryotic counterpart. In humans, each of the 23 chromosomes in a haploid set may have thousands of linked genes and be a highly compacted linear strand several centimeters in length. In contrast, *E. coli* has a single, circular chromosome that carries a few thousand genes and is only about 1 μm in circumference. Genetic recombination in bacteria occurs after a chromosome is transferred from one cell to another, which brings the chromosomes of two cells into close proximity within a single cell.

Joshua Lederberg and Edward Tatum discovered this recombination process in 1946. They worked with two genetic strains of *E. coli* that had different alleles for each of six genes (each of the genes coded for the synthesis of certain small molecules). Simply put, the two strains had the following genotypes (remember that bacteria are haploid):

ABCdef   and   *abcDEF*

where capital letters stand for wild-type alleles and lower case letters stand for mutant alleles.

When the two strains were grown in the same environment in the laboratory, most of the cells produced clones. That is, almost all of the cells that grew had the original genotypes:

ABCdef   and   *abcDEF*

However, very rarely, Lederberg and Tatum detected bacteria that had the genotype

ABCDEF

How could these completely wild-type bacteria have arisen? One possibility was mutation: the *d* allele could have mutated to D, and so on for *e* and *f*. The problem with this explanation was that the probabilities of mutation from *d* to D, *e* to E, and *f* to F were each very low. The probability of all three events occurring in the same cell would be the product of the three individual probabilities—an extremely low number and millions of times lower than the actual rate of appearance of cells with the genotype *ABCDEF*.

Electron microscopy showed how sexual transmission in bacteria might happen, via physical contact between the cells (**Figure 12.26A**). Physical contact is initiated by a thin projection called a **sex pilus** (plural *pili*). Once sex pili bring the two cells together, the actual transfer of the chromosome occurs through a thin cytoplasmic bridge called a **conjugation tube** that forms between the cells.

The chromosome moves in a linear fashion from a donor cell to a recipient cell. Since the bacterial chromosome is circular, it must be made linear (cut) before it can pass through the tube. Contact between the cells is brief—only rarely long enough for the entire donor genome to enter the recipient cell. Therefore, the recipient cell usually receives only a portion of the donor chromosome. There is no reciprocal transfer of a chromosome from the recipient to the donor.

Once the donor chromosome fragment is inside the recipient cell, it can recombine with the recipient cell's chromosome. In much the same way that chromosomes pair up, gene for gene, in prophase I of meiosis, the donor chromosome can line up beside its homologous genes in the recipient, and crossing over can occur. Gene(s) from the donor can become integrated into the genome of the recipient, thus changing the recipient's genetic constitution (**Figure 12.26B**), although only about half the transferred genes become integrated in this way. When the recipient cells proliferate, the donor genes are passed on to all progeny cells.

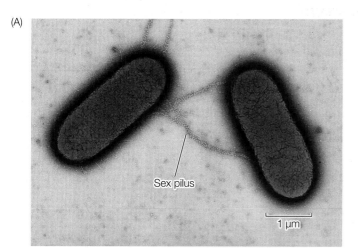

(A)

Sex pilus

1 μm

- Genes for unusual metabolic capacities, such as the ability to break down hydrocarbons; bacteria carrying these plasmids can be used to clean up oil spills.

- Genes for conjugation, including the ability to make a sex pilus; bacteria carrying this type of plasmid, called fertility factor, are designated F$^+$ and conjugate with bacteria that lack the plasmid (F$^-$).

- Genes for antibiotic resistance; bacteria carrying such gene(s)—the plasmids are called R factors—are a major threat to human health.

Plasmids can move between cells during conjugation, thereby transferring new genes to the recipient bacterium (**Figure 12.27**). Because plasmids can replicate independently of the main chromosome, they do not need to recombine with the main chromosome to add their genes to the recipient cell's genome.

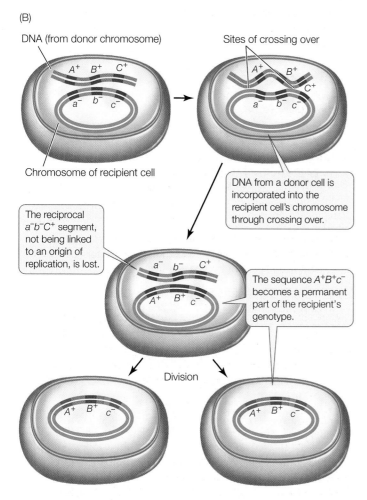

(B)

DNA (from donor chromosome)

Sites of crossing over

A^+ B^+ C^+

a^- b^- c^-

A^+ B^+ C^+

a^- b^- c^-

Chromosome of recipient cell

DNA from a donor cell is incorporated into the recipient cell's chromosome through crossing over.

The reciprocal $a^-b^-C^+$ segment, not being linked to an origin of replication, is lost.

a^- b^- C^+

A^+ B^+ c^-

The sequence $A^+B^+c^-$ becomes a permanent part of the recipient's genotype.

Division

A^+ B^+ c^-

A^+ B^+ c^-

12.26 Bacterial Conjugation and Recombination (A) Sex pili draw two bacteria into close contact, so that a cytoplasmic conjugation tube can form. DNA is transferred from one cell to the other via the conjugation tube. (B) DNA from a donor cell can become incorporated into a recipient cell's chromosome through crossing over.

Plasmids transfer genes between bacteria

In addition to their main chromosome, many bacteria harbor additional smaller, circular chromosomes called **plasmids**. They typically contain at most a few dozen genes, which, depending on the particular plasmid, may fall into one of several categories:

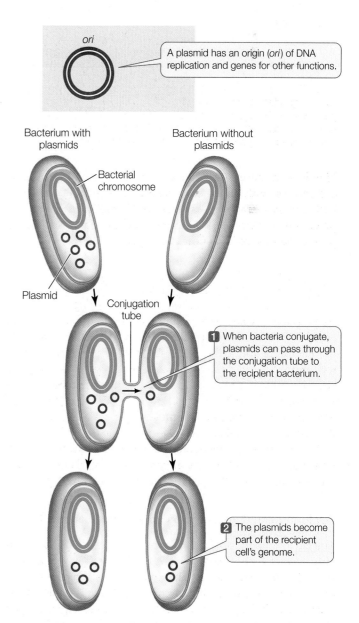

ori

A plasmid has an origin (*ori*) of DNA replication and genes for other functions.

Bacterium with plasmids

Bacterium without plasmids

Bacterial chromosome

Plasmid

Conjugation tube

1 When bacteria conjugate, plasmids can pass through the conjugation tube to the recipient bacterium.

2 The plasmids become part of the recipient cell's genome.

12.27 Gene Transfer by Plasmids When plasmids enter a cell via conjugation, their genes can be expressed in the recipient cell.

12.6 RECAP

Although they are haploid and reproduce primarily asexually, prokaryotes have the ability to transfer genes from one cell to another. These genes can be part of the main single chromosome or on a small chromosome called a plasmid.

- How were prokaryotic gene transfer and recombination discovered? **See p. 260**

- What are the differences between recombination after conjugation in prokaryotes and recombination during meiosis in eukaryotes?

CHAPTER SUMMARY

12.1 What Are the Mendelian Laws of Inheritance?

- Physical features of organisms, or **characters**, can exist in different forms, or **traits**. A **heritable trait** is one that can be passed from parent to offspring. A **phenotype** is the physical appearance of an organism; a **genotype** is the genetic constitution of the organism.

- The different forms of a **gene** are called **alleles**. Organisms that have two identical alleles for a trait are called **homozygous**; organisms that have two different alleles for a trait are called **heterozygous**. A gene resides at a particular site on a chromosome called a **locus**.

- Mendel's experiments included **reciprocal crosses** and **monohybrid crosses** between **true-breeding** pea plants. Analysis of his meticulously tabulated data led Mendel to propose a **particulate theory** of inheritance stating that discrete units (now called genes) are responsible for the inheritance of specific traits, to which both parents contribute equally.

- Mendel's first law, the **law of segregation**, states that when any individual produces gametes, the two copies of a gene separate, so that each gamete receives only one member of the pair. Thus every individual in the F_1 inherits one copy from each parent. **Review Figures 12.4 and 12.5**

- Mendel used a **test cross** to find out whether an individual showing a dominant phenotype was homozygous or heterozygous. **Review Figure 12.6, WEB ACTIVITY 12.1**

- Mendel's use of **dihybrid crosses** to study the inheritance of two characters led to his second law: the **law of independent assortment**. The independent assortment of genes in meiosis leads to **recombinant** phenotypes. **Review Figures 12.7 and 12.8, ANIMATED TUTORIAL 12.1**

- Probability calculations and **pedigrees** help geneticists trace Mendelian inheritance patterns. **Review Figures 12.9 and 12.10**

12.2 How Do Alleles Interact?

- New alleles arise by random **mutation**. Many genes have multiple alleles. A **wild-type** allele gives rise to the predominant form of a trait. When the wild-type allele is present at a locus less than 99 percent of the time, the locus is said to be **polymorphic**. **Review Figure 12.11**

- In **incomplete dominance**, neither of two alleles is dominant. The heterozygous phenotype is intermediate between the homozygous phenotypes. **Review Figure 12.12**

- **Codominance** exists when two alleles at a locus produce two different phenotypes that both appear in heterozygotes.

- An allele that affects more than one trait is said to be **pleiotropic**.

12.3 How Do Genes Interact?

- In **epistasis**, one gene affects the expression of another. **Review Figure 12.14**

- Environmental conditions can affect the expression of a genotype.

- **Penetrance** is the proportion of individuals in a group with a given genotype that show the expected phenotype. **Expressivity** is the degree to which a genotype is expressed in an individual.

- Variations in phenotypes can be **qualitative** (discrete) or **quantitative** (graduated, continuous). Most quantitative traits are the result of the effects of several genes and the environment. Genes that together determine quantitative characters are called **quantitative trait loci**.

12.4 What Is the Relationship between Genes and Chromosomes?

SEE ANIMATED TUTORIAL 12.2

- Each chromosome carries many genes. Genes on the same chromosome are referred to as a **linkage group**.

- Genes on the same chromosome can recombine by crossing over. The resulting recombinant chromosomes have new combinations of alleles. **Review Figures 12.19 and 12.20**

- **Sex chromosomes** carry genes that determine whether the organism will produce male or female gametes. All other chromosomes are called **autosomes**. The specific functions of X and Y chromosomes differ among different groups of organisms.

- **Primary sex determination** in mammals is usually a function of the presence or absence of the *SRY* gene. **Secondary sex determination** results in the outward manifestations of maleness or femaleness.

- In fruit flies and mammals, the X chromosome carries many genes, but the Y chromosome has only a few. Males have only one allele (are **hemizygous**) for X-linked genes, so recessive **sex-linked** mutations are expressed phenotypically more often in males than in females. Females may be unaffected **carriers** of such alleles.

12.5 What Are the Effects of Genes Outside the Nucleus?

- Cytoplasmic organelles such as plastids and mitochondria contain small numbers of genes. In many organisms, cytoplasmic genes are inherited only from the mother because the male gamete contributes only its nucleus (i.e., no cytoplasm) to the zygote at fertilization. **Review Figure 12.25**

12.6 How Do Prokaryotes Transmit genes?

- Prokaryotes reproduce primarily asexually but can exchange genes in a sexual process called conjugation. **Review Figure 12.26**
- Plasmids are small, extra chromosomes in bacteria that carry genes involved in important metabolic processes and that can be transmitted from one cell to another. **Review Figure 12.27**

SEE WEB ACTIVITIES 12.2 and 12.3 for a concept review of this chapter.

SELF-QUIZ

1. In a simple Mendelian monohybrid cross, true-breeding tall plants are crossed with short plants, and the F_1 plants, which are all tall, are allowed to self-pollinate. What fraction of the F_2 generation are both tall and heterozygous?
   a. 1/8
   b. 1/4
   c. 1/3
   d. 2/3
   e. 1/2

2. The phenotype of an individual
   a. depends at least in part on the genotype.
   b. is either homozygous or heterozygous.
   c. determines the genotype.
   d. is the genetic constitution of the organism.
   e. is either monohybrid or dihybrid.

3. The ABO blood groups in humans are determined by a multiple-allele system in which I^A and I^B are codominant and are both dominant to I^O. A newborn infant is type A. The mother is type O. Possible phenotypes of the father are
   a. A, B, or AB.
   b. A, B, or O.
   c. O only.
   d. A or AB.
   e. A or O.

4. Which statement about an individual that is homozygous for an allele is *not* true?
   a. Each of its cells possesses two copies of that allele.
   b. Each of its gametes contains one copy of that allele.
   c. It is true-breeding with respect to that allele.
   d. Its parents were necessarily homozygous for that allele.
   e. It can pass that allele to its offspring.

5. Which statement about a test cross is *not* true?
   a. It tests whether an unknown individual is homozygous or heterozygous.
   b. The test individual is crossed with a homozygous recessive individual.
   c. If the test individual is heterozygous, the progeny will have a 1:1 ratio.
   d. If the test individual is homozygous, the progeny will have a 3:1 ratio.
   e. Test cross results are consistent with Mendel's model of inheritance for unlinked genes.

6. Linked genes
   a. must be immediately adjacent to one another on a chromosome.
   b. have alleles that assort independently of one another.
   c. never show crossing over.
   d. are on the same chromosome.
   e. always have multiple alleles.

7. In the F_2 generation of a dihybrid cross
   a. four phenotypes appear in the ratio 9:3:3:1 if the loci are linked.
   b. four phenotypes appear in the ratio 9:3:3:1 if the loci are unlinked.
   c. two phenotypes appear in the ratio 3:1 if the loci are unlinked.
   d. three phenotypes appear in the ratio 1:2:1 if the loci are unlinked.
   e. two phenotypes appear in the ratio 1:1 whether or not the loci are linked.

8. The genetic sex of a human is determined by
   a. ploidy, with the male being haploid.
   b. the Y chromosome.
   c. X and Y chromosomes, the male being XX.
   d. the number of X chromosomes, the male being XO.
   e. Z and W chromosomes, the male being ZZ.

9. In epistasis
   a. nothing changes from generation to generation.
   b. one gene alters the effect of another.
   c. a portion of a chromosome is deleted.
   d. a portion of a chromosome is inverted.
   e. the behavior of two genes is entirely independent.

10. In humans, spotted teeth are caused by a dominant sex-linked gene. A man with spotted teeth whose father had normal teeth marries a woman with normal teeth. Therefore,
    a. all of their daughters will have normal teeth.
    b. all of their daughters will have spotted teeth.
    c. all of their children will have spotted teeth.
    d. half of their sons will have spotted teeth.
    e. all of their sons will have spotted teeth.

1. In guinea pigs, black body color (*B*) is completely dominant over albino (*b*). For the crosses below, give the genotypes of the parents:

Parental phenotypes	Black offspring	Albino offspring	Parental genotypes?
Black × albino	12	0	
Albino × albino	0	12	
Black × albino	5	7	
Black × black	9	3	

2. In the genetic cross, *AaBbCcDdEE × AaBBCcDdEe*, what fraction of the offspring will be heterozygous for all of these genes (*AaBbCcDdEe*)? Assume all genes are unlinked and the alleles show simple dominance.

3. The pedigree below shows the inheritance of a rare mutant phenotype in humans, congenital cataracts (black symbols).

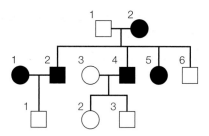

   a. Are cataracts inherited as an autosomal dominant trait? Autosomal recessive? Sex-linked dominant? Sex-linked recessive?
   b. Person #5 in the second generation marries a man who does not have cataracts. Two of their four children, a boy and a girl, develop cataracts. What is the chance that their next child will be a girl with cataracts?

4. In cats, black coat (*B*) is codominant with yellow (*b*). The coat color gene is on the X chromosome. Calico cats, which have coats with black and yellow patches, are heterozygous for the coat color alleles.
   a. Why are most calico cats females?
   b. A calico female, Pickle, had a litter with one yellow male, two black males, two yellow females and three calico females. What were the genotype and phenotype of the father?

5. In *Drosophila*, three autosomal genes have alleles as follows:
   Gray body color (*G*) is dominant over black (*g*)
   Full wings (*A*) is dominant over vestigial (*a*)
   Red eye (*R*) is dominant over sepia (*r*)

   Two crosses were performed, with the following results:

   Cross I: Parents: heterozygous red, full × sepia, vestigial
      Offspring: 131 red, full
         120 sepia, vestigial
         122 red, vestigial
         127 sepia, full

   Cross II: Parents: heterozygous gray, full × black, vestigial
      Offspring: 236 gray, full
         253 black, vestigial
         50 gray, vestigial
         61 black, full

Are any of the three genes linked on the same chromosome? If so, what is the map distance between the linked genes?

6. In a particular plant species, two alleles control flower color, which can be yellow, blue, or white. Crosses of these plants produce the following offspring:

Parental phenotypes	Offspring phenotypes (ratio)
Yellow × yellow	All yellow
Blue × yellow	Blue or yellow (1:1)
Blue × white	Blue or white (1:1)
White × white	All white

What will be the phenotype, and ratio, of the offspring of a cross of blue × blue?

7. In *Drosophila melanogaster*, the recessive allele *p*, when homozygous, determines pink eyes. *Pp* or *PP* results in wild-type eye color. Another gene on a different chromosome has a recessive allele, *sw*, that produces short wings when homozygous. Consider a cross between females of genotype *PPSwSw* and males of genotype *ppswsw*. Describe the phenotypes and genotypes of the F_1 generation and of the F_2 generation, produced by allowing the F_1 progeny to mate with one another.

8. On the same chromosome of *Drosophila melanogaster* that carries the *p* (pink eyes) locus, there is another locus that affects the wings. Homozygous recessives, *byby*, have blistery wings, while the dominant allele *By* produces wild-type wings. The *P* and *By* loci are very close together on the chromosome; that is, the two loci are tightly linked. In answering Questions 8*a* and 8*b*, assume that no crossing over occurs, and that the F_2 generation is produced by interbreeding the F_1 progeny.
   a. For the cross *PPByBy × ppbyby*, give the phenotypes and genotypes of the F_1 and F_2 generations.
   b. For the cross *PPbyby × ppByBy*, give the phenotypes and genotypes of the F_1 and F_2 generations.
   c. For the cross of Question 8*b*, what further phenotype(s) would appear in the F_2 generation if crossing over occurred?
   d. Draw a nucleus undergoing meiosis at the stage in which the crossing over (Question 8*c*) occurred. In which generation (P, F_1, or F_2) did this crossing over take place?

9. In chickens, when the dominant alleles of the genes for rose comb (*R*) and pea comb (*A*) are present together (*R_A_*), the result is a bird with a walnut comb. Chickens that are homozygous recessive for both genes produce a single comb. A rose-combed bird mated with a walnut-combed bird and produced offspring in the proportion:
   3/8 walnut:3/8 rose:1/8 pea:1/8 single
   What were the genotypes of the parents?

10. In *Drosophila melanogaster*, white (*w*), eosin (*w^e^*), and wild-type red (*w^+^*) are multiple alleles at a single locus for eye color. This locus is on the X chromosome. A female that has eosin (pale orange) eyes is crossed with a male that has wild-type eyes. All the female progeny are red-eyed; half the male progeny have eosin eyes, and half have white eyes.
   a. What is the order of dominance of these alleles?
   b. What are the genotypes of the parents and progeny?

11. In humans, red–green color blindness is determined by an X-linked recessive allele (a), while eye color is determined by an autosomal gene, where brown (B) is dominant over blue (b).
    a. What gametes can be formed with respect to these genes by a heterozygous, brown-eyed, color-blind male?
    b. If a blue-eyed mother with normal vision has a brown eyed, color-blind son and a blue-eyed, color-blind daughter, what are the genotypes of both parents and children?

12. If the dominant allele A is necessary for hearing in humans, and another allele, B, located on a different chromosome, results in deafness no matter what other genes are present, what percentage of the offspring of the marriage of $aaBb \times Aabb$ will be deaf?

13. The disease Leber's optic neuropathy is caused by a mutation in a gene carried on mitochondrial DNA. What would be the phenotype of their first child if a man with this disease married a woman who did not have the disease? What would be the result if the wife had the disease and the husband did not?

ADDITIONAL INVESTIGATION

Sometimes scientists get lucky. Consider Mendel's dihybrid cross shown in Figure 12.7. Peas have a haploid number of seven chromosomes, so many of their genes are linked. What would Mendel's results have been if the genes for seed color and seed shape were linked with a map distance of ten units? Now, consider Morgan's fruit flies (see Figure 12.21). Suppose that the genes for body color and wing shape were not linked? What results would Morgan have obtained?

WORKING WITH DATA (GO TO yourBioPortal.com)

Mendel's Monohybrid Experiments Mendel's experiments with pea plants (Figure 12.3) laid the foundations of genetics. In this real-world exercise, you will analyze Mendel's data from his published paper and see how he came to his conclusions about the nature of genes.

A structure for our times

Jurassic Park, in both its literary and film incarnations, features a fictional theme park populated with live dinosaurs. In the story, scientists isolate DNA from dinosaur blood extracted from the digestive tracts of fossil insects. The insects supposedly sucked the reptiles' blood right before being preserved in amber (fossilized tree resin). This DNA, according to the novel, could be manipulated to produce living individuals of long-extinct organisms such as velociraptors and the ever-memorable *Tyrannosaurus rex*.

The late Michael Crichton got the idea for his novel from an actual scientific paper in which the authors claimed to have detected reptilian DNA sequences in a fossil insect. Unfortunately, upon additional study, the "preserved" DNA turned out to be a contaminant from modern organisms.

Despite the facts that (1) the preservation of intact DNA over millions of years is highly improbable, and (2) DNA alone cannot generate a new organism, the huge success of Crichton's book brought DNA to the attention of millions. But even before *Jurassic Park*, the DNA double helix was a familiar secular icon.

The double helix first appeared in 1953, in a short paper by James Watson and Francis Crick in the journal *Nature*. An illustration of the molecule's structure drawn by Crick's wife, Odile, accompanied the article, and its simplicity and elegance caught the imagination of the general public as well as the intellect of scientists. As Watson later put it, "A structure this pretty just had to exist."

The double-helical structure of *deoxyribonucleic acid* is perhaps the most widely recognized symbol of modern science, and "DNA" has become part of everyday speech. One sees advertisements for a company whose customers get "into the DNA of business." A digital media software system is called the "DNA Server." A perfume called DNA bills itself as "the essence of life."

Salvador Dali was the first well-known artist to use the DNA double helix in his whimsical creations in 1958. Today, sculptures representing the DNA double helix abound, and it is not only DNA's appearance that stirs our imagination. The DNA nucleotide sequence itself, the "code for life," has inspired unique works of art that incorporate real DNA molecules. A portrait of Sir John Sulston, a Nobel prize-winning geneticist, is made of tiny bacterial colonies, each containing a piece of Sulston's DNA. The Brazilian artist Eduardo Kac translated a sentence from the Bible into Morse code, and from Morse

Reviving the Velociraptor   Scientists and artists have been creating inanimate reconstructions of dinosaurs for more than 100 years. Michael Crichton's novel *Jurassic Park* was based on the fictional premise that DNA retrieved from fossils could produce living dinosaurs, such as this velociraptor.

In the Nature of Things The double helix of DNA has become an iconic symbol of modern science and culture. Artists and designers make use of the widely recognized shape in many ways.

code into a DNA sequence. The sequence was synthesized and incorporated into bacteria. Viewers could turn on an ultraviolet lamp to create mutations in the DNA (and thus in the biblical verse it encoded).

For many people, DNA has come to symbolize the promise and perils of our rapidly expanding knowledge of genetics. Although DNA sequences alone cannot generate a new organism, *biotechnologies* using DNA can modify existing organisms into essentially new organisms. As we will see in Chapter 18, such use of this iconic molecule has generated both excitement and concern about potential risks.

IN THIS CHAPTER we will describe the key experiments that led to the identification of DNA as the genetic material. We will then describe the structure of the DNA molecule and how this structure determines its function. We will describe the processes by which DNA is replicated, repaired, and maintained. Finally, we present an important practical application arising from our knowledge of DNA replication: the polymerase chain reaction.

13.1 What Is the Evidence that the Gene is DNA?

By the early twentieth century, geneticists had associated the presence of genes with chromosomes. Research began to focus on exactly which chemical component of chromosomes comprised this genetic material.

By the 1920s, scientists knew that chromosomes were made up of DNA and proteins. At this time a new dye was developed by Robert Feulgen that could bind specifically to DNA and that stained cell nuclei red in direct proportion to the amount of DNA present in the cell. This technique provided circumstantial evidence that DNA was the genetic material:

- *It was in the right place.* DNA was confirmed to be an important component of the nucleus and the chromosomes, which were known to carry genes.

- *It varied among species.* When cells from different species were stained with the dye and their color intensity measured, each species appeared to have its own specific amount of nuclear DNA.

- *It was present in the right amounts.* The amount of DNA in somatic cells (body cells not specialized for reproduction) was twice that in reproductive cells (eggs or sperm)—as might be expected for diploid and haploid cells, respectively.

But circumstantial evidence is *not* a scientific demonstration of cause and effect. After all, proteins are also present in cell nuclei. Science relies on experiments to test hypotheses. The convincing demonstration that DNA is the genetic material came from two sets of experiments, one on bacteria and the other on viruses.

DNA from one type of bacterium genetically transforms another type

The history of biology is filled with incidents in which research on one specific topic has—with or without answering the question originally asked—contributed richly to another, apparently unrelated area. Such a case of serendipity is seen in the work of Frederick Griffith, an English physician.

In the 1920s, Griffith was studying the bacterium *Streptococcus pneumoniae*, or pneumococcus, one of the agents that cause pneumonia in humans. He was trying to develop a vaccine against this devastating illness (antibiotics had not yet been discovered). Griffith was working with two strains of pneumococcus:

- Cells of the S strain produced colonies that looked smooth (S). Covered by a polysaccharide capsule, these cells were

INVESTIGATING LIFE

13.1 Genetic Transformation

Griffith's experiments demonstrated that something in the virulent S strain of pneumococcus could transform nonvirulent R strain bacteria into a lethal form, even when the S strain bacteria had been killed by high temperatures.

HYPOTHESIS Material in dead bacterial cells can genetically transform living bacterial cells.

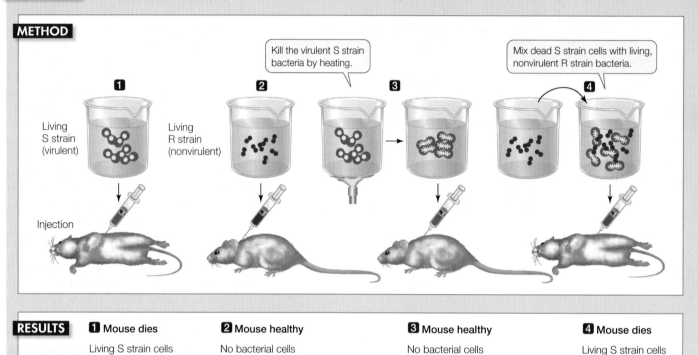

METHOD

Kill the virulent S strain bacteria by heating.

Mix dead S strain cells with living, nonvirulent R strain bacteria.

1 Living S strain (virulent)

2 Living R strain (nonvirulent)

3

4

Injection

RESULTS

1 Mouse dies	**2** Mouse healthy	**3** Mouse healthy	**4** Mouse dies
Living S strain cells found in heart	No bacterial cells found in heart	No bacterial cells found in heart	Living S strain cells found in heart

CONCLUSION A chemical substance from one cell is capable of genetically transforming another cell.

FURTHER INVESTIGATION: How would you show that heat-killed R strain bacteria can transform living S strain bacteria?

Go to **yourBioPortal.com** for original citations, discussions, and relevant links for all INVESTIGATING LIFE figures.

---

protected from attack by a host's immune system. When S cells were injected into mice, they reproduced and caused pneumonia (the strain was *virulent*).

- Cells of the R strain produced colonies that looked rough (R), lacked the protective capsule, and were not virulent.

Griffith inoculated some mice with heat-killed S-type pneumococcus cells. These heat-killed bacteria did not produce infection. However, when Griffith inoculated other mice with a mixture of living R bacteria and heat-killed S bacteria, to his astonishment, the mice died of pneumonia (**Figure 13.1**). When he examined blood from the hearts of these mice, he found it full of living bacteria—many of them with characteristics of the virulent S strain! Griffith concluded that in the presence of the dead S-type pneumococcus cells, some of the living R-type cells had been transformed into virulent S cells. The fact that these

S-type cells reproduced to make more S-type cells showed that the change from R-type to S-type was genetic.

Did this transformation of the bacteria depend on something that happened in the mouse's body? No. It was shown that simply incubating living R and heat-killed S bacteria together in a test tube yielded the same transformation. Years later, another group of scientists discovered that a cell-free extract of heat-killed S cells could also transform R cells. (A cell-free extract contains all the contents of ruptured cells, but no intact cells.) This result demonstrated that some substance—called at the time a chemical **transforming principle**—from the dead S pneumococcus cells could cause a heritable change in the affected R cells. This was an extraordinary discovery: treatment with a chemical substance could permanently change an inherited characteristic. Now it remained to identify the chemical structure of this substance.

INVESTIGATING LIFE

13.2 Genetic Transformation by DNA

Experiments by Avery, MacLeod, and McCarty showed that DNA from the virulent S strain of pneumococcus was responsible for the transformation in Griffith's experiments (see Figure 13.1).

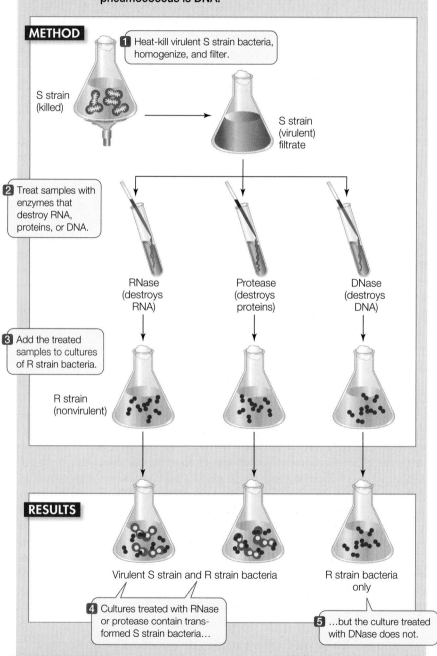

HYPOTHESIS The chemical nature of the transforming substance from pneumococcus is DNA.

METHOD

1 Heat-kill virulent S strain bacteria, homogenize, and filter.

S strain (killed)

S strain (virulent) filtrate

2 Treat samples with enzymes that destroy RNA, proteins, or DNA.

RNase (destroys RNA)

Protease (destroys proteins)

DNase (destroys DNA)

3 Add the treated samples to cultures of R strain bacteria.

R strain (nonvirulent)

RESULTS

Virulent S strain and R strain bacteria

R strain bacteria only

4 Cultures treated with RNase or protease contain transformed S strain bacteria...

5 ...but the culture treated with DNase does not.

CONCLUSION Because only DNase destroyed the transforming substance, the transforming substance is DNA.

Go to **yourBioPortal.com** for original citations, discussions, and relevant links for all INVESTIGATING LIFE figures.

The transforming principle is DNA

Identifying the transforming principle was a crucial step in the history of biology. Work on identifying the transforming principle was completed by Oswald Avery and his colleagues at what is now The Rockefeller University.

They treated samples known to contain the pneumococcal transforming principle in a variety of ways to destroy different types of molecules—proteins, nucleic acids, carbohydrates, and lipids—and tested the treated samples to see if they had retained the transforming activity. The answer was always the same: if the DNA in the sample was destroyed, transforming activity was lost, but there was no loss of activity when proteins, carbohydrates, or lipids were destroyed (**Figure 13.2**). As a final step, Avery and his colleagues Colin MacLeod and Maclyn McCarty isolated virtually pure DNA from a sample containing pneumococcal-transforming principle, and showed that it caused bacterial transformation. We now know that the gene for the enzyme that catalyzes the synthesis of the polysaccharide capsule, which makes the bacteria look "smooth," was transferred during transformation.

Genetic transformation occurs in nature, although only in certain species of bacteria such as *Pneumococcus*. It does not occur, for example, in *E. coli*. Cells can pick up DNA fragments released into the environment by dead and ruptured cells. Only a small part of the genome is taken up by the transformed cells. Once the new DNA enters the cell, a transforming event very similar to recombination occurs (see Figure 12.26), and new genes can be incorporated into the host chromosome.

The work of Avery's group was a milestone in establishing that DNA is the genetic material in bacterial cells. However, when it was first published in 1944, it had little impact, for two reasons. First, most scientists did not believe that DNA was chemically complex enough to be the genetic material, especially given the much greater chemical complexity of proteins. Second, and perhaps more important, bacterial genetics was a new field of study—it was not yet clear that bacteria even had genes.

Viral replication experiments confirmed that DNA is the genetic material

The questions about bacteria and other simple organisms were soon resolved, as researchers identified genes and mutations. Bacteria and viruses seemed to undergo genetic processes

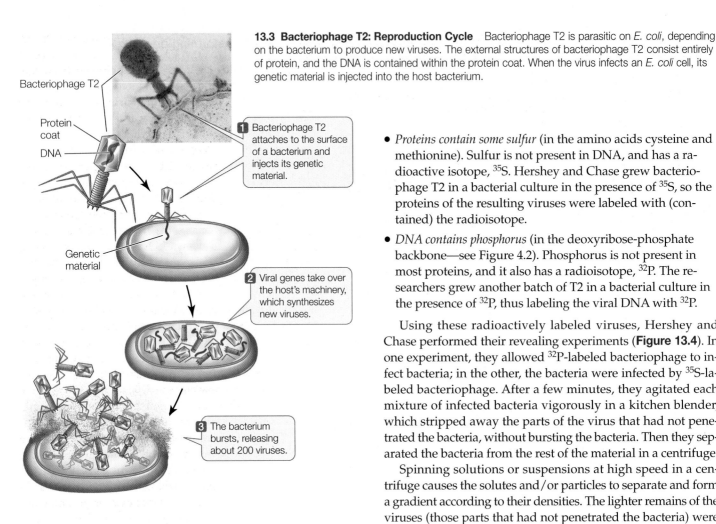

Bacteriophage T2

Protein coat

DNA

Genetic material

13.3 Bacteriophage T2: Reproduction Cycle Bacteriophage T2 is parasitic on *E. coli*, depending on the bacterium to produce new viruses. The external structures of bacteriophage T2 consist entirely of protein, and the DNA is contained within the protein coat. When the virus infects an *E. coli* cell, its genetic material is injected into the host bacterium.

1 Bacteriophage T2 attaches to the surface of a bacterium and injects its genetic material.

2 Viral genes take over the host's machinery, which synthesizes new viruses.

3 The bacterium bursts, releasing about 200 viruses.

similar to those in fruit flies and pea plants. Experiments with these relatively simple organisms were designed to discover the nature of the genetic material.

In 1952, Alfred Hershey and Martha Chase of the Carnegie Laboratory of Genetics published a paper that had a much greater immediate impact than Avery's 1944 paper. The Hershey–Chase experiment, which sought to determine whether DNA or protein was the genetic material, was carried out with a virus that infects bacteria. This virus, called bacteriophage T2, consists of little more than a DNA core packed inside a protein coat (**Figure 13.3**). Thus the virus is made of the two materials that were, at the time, the leading candidates for the genetic material.

When bacteriophage T2 attacks a bacterium, part (but not all) of the virus enters the bacterial cell. About 20 minutes later, the cell bursts, releasing dozens of particles that are virtually identical to the infecting virus particle. Clearly the virus is somehow able to replicate itself inside the bacterium. Hershey and Chase deduced that the entry of some viral component affects the genetic program of the host bacterial cell, transforming it into a bacteriophage factory. They set out to determine which part of the virus—protein or DNA—enters the bacterial cell. To trace the two components of the virus over its life cycle, Hershey and Chase labeled each component with a specific radioisotope:

- *Proteins contain some sulfur* (in the amino acids cysteine and methionine). Sulfur is not present in DNA, and has a radioactive isotope, ^{35}S. Hershey and Chase grew bacteriophage T2 in a bacterial culture in the presence of ^{35}S, so the proteins of the resulting viruses were labeled with (contained) the radioisotope.

- *DNA contains phosphorus* (in the deoxyribose-phosphate backbone—see Figure 4.2). Phosphorus is not present in most proteins, and it also has a radioisotope, ^{32}P. The researchers grew another batch of T2 in a bacterial culture in the presence of ^{32}P, thus labeling the viral DNA with ^{32}P.

Using these radioactively labeled viruses, Hershey and Chase performed their revealing experiments (**Figure 13.4**). In one experiment, they allowed ^{32}P-labeled bacteriophage to infect bacteria; in the other, the bacteria were infected by ^{35}S-labeled bacteriophage. After a few minutes, they agitated each mixture of infected bacteria vigorously in a kitchen blender, which stripped away the parts of the virus that had not penetrated the bacteria, without bursting the bacteria. Then they separated the bacteria from the rest of the material in a centrifuge.

Spinning solutions or suspensions at high speed in a centrifuge causes the solutes and/or particles to separate and form a gradient according to their densities. The lighter remains of the viruses (those parts that had not penetrated the bacteria) were captured in the "supernatant" fluid, while the heavier bacterial cells segregated into a "pellet" in the bottom of the centrifuge tube. The scientists found that the supernatant fluid contained most of the ^{35}S (and thus the viral protein), while most of the ^{32}P (and thus the viral DNA) had stayed with the bacteria. These results suggested that it was DNA that had been transferred into the bacteria, and that DNA was the compound responsible for redirecting the genetic program of the bacterial cell.

Hershey and Chase performed similar but longer-term experiments, allowing the progeny (offspring) generation of viruses to grow. The resulting viruses contained almost no ^{35}S and none of the parental viral protein. They did, however, contain about one-third of the original ^{32}P—and thus, presumably, one-third of the original DNA. Because DNA was carried over in the viruses from generation to generation but protein was not, the logical conclusion was that the hereditary information was contained in the DNA.

Eukaryotic cells can also be genetically transformed by DNA

With the publication of the evidence for DNA as the genetic material in bacteria and viruses, the question arose as to whether DNA was also the genetic material in complex eukaryotes. Some dubious experimental results were reported. For example, a

INVESTIGATING LIFE

13.4 The Hershey–Chase Experiment

When bacterial cells were infected with radioactively labeled T2 bacteriophage, only labeled DNA was found in the bacteria. After centrifuging the culture to make the bacteria form a pellet, the labeled protein remained in the supernatant. This showed that DNA, not protein, is the genetic material.

HYPOTHESIS Either component of a bacteriophage—DNA or protein—might be the hereditary material that enters a bacterial cell to direct the assembly of new viruses.

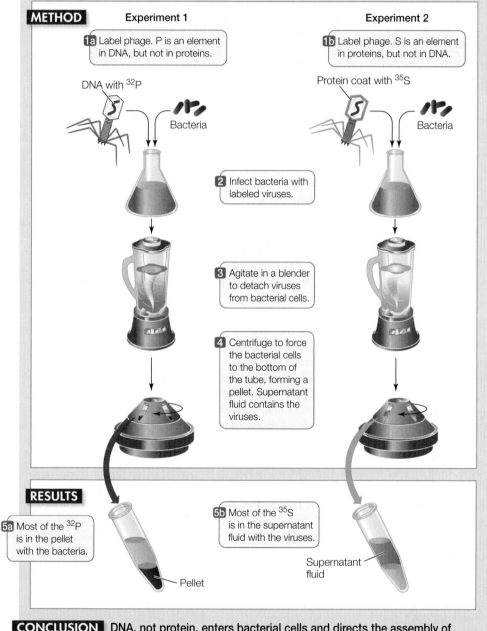

METHOD

Experiment 1

1a Label phage. P is an element in DNA, but not in proteins.

DNA with ^{32}P

Bacteria

Experiment 2

1b Label phage. S is an element in proteins, but not in DNA.

Protein coat with ^{35}S

Bacteria

2 Infect bacteria with labeled viruses.

3 Agitate in a blender to detach viruses from bacterial cells.

4 Centrifuge to force the bacterial cells to the bottom of the tube, forming a pellet. Supernatant fluid contains the viruses.

RESULTS

5a Most of the ^{32}P is in the pellet with the bacteria.

Pellet

5b Most of the ^{35}S is in the supernatant fluid with the viruses.

Supernatant fluid

CONCLUSION DNA, not protein, enters bacterial cells and directs the assembly of new viruses.

Go to **yourBioPortal.com** for original citations, discussions, and relevant links for all INVESTIGATING LIFE figures.

white duck was injected with DNA from a brown duck and the recipient was reported to turn brown. In another example, flatworms were fed DNA from worms that had learned a simple task, and the recipient worms were reported to immediately get smarter. However, no one could duplicate these results. This episode underscores a central aspect of experimental biology: that published research should be repeated with the same results before the conclusions can be considered valid.

It would be impossible for a large molecule such as DNA to avoid hydrolysis into nucleotides in the digestive system, let alone get into all the cells of the body, after being ingested by an animal. However, genetic transformation of eukaryotic cells by DNA (called **transfection**) can be demonstrated. The key is to use a **genetic marker**, a gene whose presence in the recipient cell confers an observable phenotype. In the experiments with pneumococcus, these phenotypes were the smooth polysaccharide capsule and virulence. In eukaryotes, researchers usually use a nutritional or antibiotic resistance marker gene that permits the growth of transformed recipient cells but not of nontransformed cells. For example, thymidine kinase is an enzyme needed to make use of thymidine in the synthesis of deoxythymidine triphosphate (dTTP), one of the four deoxyribonucleoside triphosphates used in the synthesis of DNA. Mammalian cells that lack the gene for thymidine kinase cannot grow in a medium that contains thymidine as the only source for dTTP synthesis. When DNA containing the marker gene encoding thymidine kinase is added to a culture of mammalian cells lacking this gene, some cells will grow in the thymidine medium, demonstrating that they have been transfected with the gene (**Figure 13.5**). Any cell can be transfected in this way, even an egg cell. In this case, a whole new genetically transformed organism can result; such an organism is referred to as *transgenic*. Transformation in eukaryotes is the final line of evidence for DNA as the genetic material.

INVESTIGATING LIFE

13.5 Transfection in Eukaryotic Cells

The use of a marker gene shows that mammalian cells can be genetically transformed by DNA. Usually, the marker gene is carried by a larger molecule (a virus or a small chromosome).

HYPOTHESIS DNA can transform eukaryotic cells.

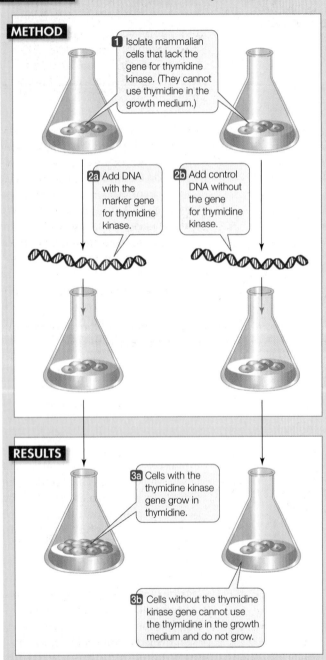

METHOD

1 Isolate mammalian cells that lack the gene for thymidine kinase. (They cannot use thymidine in the growth medium.)

2a Add DNA with the marker gene for thymidine kinase.

2b Add control DNA without the gene for thymidine kinase.

RESULTS

3a Cells with the thymidine kinase gene grow in thymidine.

3b Cells without the thymidine kinase gene cannot use the thymidine in the growth medium and do not grow.

CONCLUSION The cells were transformed by DNA.

Go to **yourBioPortal.com** for original citations, discussions, and relevant links for all INVESTIGATING LIFE figures.

13.1 RECAP

Experiments on bacteria and on viruses demonstrated that DNA is the genetic material.

- At the time of Griffith's experiments in the 1920s, what circumstantial evidence suggested to scientists that DNA might be the genetic material? **See p. 267**

- Why were the experiments of Avery, MacLeod, and McCarty definitive evidence that DNA was the genetic material? **See p. 269 and Figure 13.2**

- What attributes of bacteriophage T2 were key to the Hershey–Chase experiments demonstrating that DNA is the genetic material? **See p. 270 and Figure 13.4**

As soon as scientists became convinced that the genetic material was DNA, they began efforts to learn its precise three-dimensional chemical structure. The chemical makeup of DNA, as a polymer made up of nucleotide monomers, had been known for several decades. In determining the structure of DNA, scientists hoped to find the answers to two questions: (1) how is DNA replicated between cell divisions, and (2) how does it direct the synthesis of specific proteins? They were eventually able to answer both questions.

13.2 What Is the Structure of DNA?

The structure of DNA was deciphered only after many types of experimental evidence were considered together in a theoretical framework. The most crucial evidence was obtained using X-ray crystallography. Some chemical substances, when they are isolated and purified, can be made to form crystals. The positions of atoms in a crystallized substance can be inferred from the diffraction pattern of X rays passing through the substance (**Figure 13.6A**). The structure of DNA would not have been characterized without the crystallographs prepared in the early 1950s by the English chemist Rosalind Franklin (**Figure 13.6B**). Franklin's work, in turn, depended on the success of the English biophysicist Maurice Wilkins, who prepared samples containing very uniformly oriented DNA fibers. These DNA samples were far better for diffraction than previous ones, and the crystallographs Franklin prepared from them suggested a spiral or helical molecule.

The chemical composition of DNA was known

The chemical composition of DNA also provided important clues to its structure. Biochemists knew that DNA was a polymer of nucleotides. Each nucleotide consists of a molecule of the sugar deoxyribose, a phosphate group, and a nitrogen-containing base (see Figures 4.1 and 4.2). The only differences among the four nucleotides of DNA are their nitrogenous bases: the purines **adenine** (**A**) and **guanine** (**G**), and the pyrimidines **cytosine** (**C**) and **thymine** (**T**).

(A)

(B)

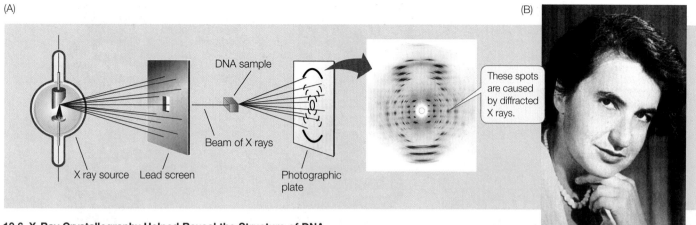

13.6 X-Ray Crystallography Helped Reveal the Structure of DNA (A) The positions of atoms in a crystallized chemical substance can be inferred by the pattern of diffraction of X rays passed through it. The pattern of DNA is both highly regular and repetitive. (B) Rosalind Franklin's crystallography helped scientists to visualize the helical structure of the DNA molecule.

In 1950, biochemist Erwin Chargaff at Columbia University reported some observations of major importance. He and his colleagues found that DNA from many different species—and from different sources within a single organism—exhibits certain regularities. In almost all DNA, the following rule holds: The amount of adenine equals the amount of thymine (A = T), and the amount of guanine equals the amount of cytosine (G = C) (**Figure 13.7**). As a result, the total abundance of purines (A + G) equals the total abundance of pyrimidines (T + C). The structure of DNA could not have been worked out without this observation, now known as Chargaff's rule, yet its significance was overlooked for at least three years.

Watson and Crick described the double helix

The solution to the structure of DNA was finally achieved through model building: the assembly of three-dimensional representations of possible molecular structures using known relative molecular dimensions and known bond angles. This technique was originally applied to molecular structural studies by the American biochemist Linus Pauling. The English physicist Francis Crick and the American geneticist James D. Watson (**Figure 13.8A**), who were both then at the Cavendish Laboratory of Cambridge University, used model building to solve the structure of DNA.

Watson and Crick attempted to combine all that had been learned so far about DNA structure into a single coherent model. Rosalind Franklin's crystallography results (see Figure 13.6) convinced Watson and Crick that the DNA molecule must be **helical** (cylindrically spiral). Density measurements and previous model building results suggested that there are two polynucleotide chains in the molecule. Modeling studies also showed that the strands run in opposite directions, that is, they are **antiparallel**; that two strands would not fit together in the model if they were parallel.

How are the nucleotides oriented in these chains? Watson and Crick suggested that:

- The nucleotide bases are on the interior of the two strands, with a sugar-phosphate backbone on the outside:

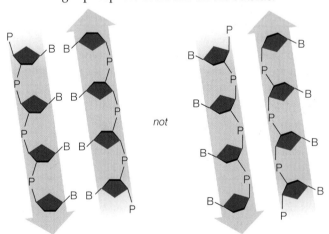

- To satisfy Chargaff's rule (purines = pyrimidines), a purine on one strand is always paired with a pyrimidine on the opposite strand. These **base pairs** (A-T and G-C) have the same width down the double helix, a uniformity shown by x-ray diffraction.

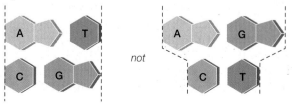

13.7 Chargaff's Rule In DNA, the total abundance of purines is equal to the total abundance of pyrimidines.

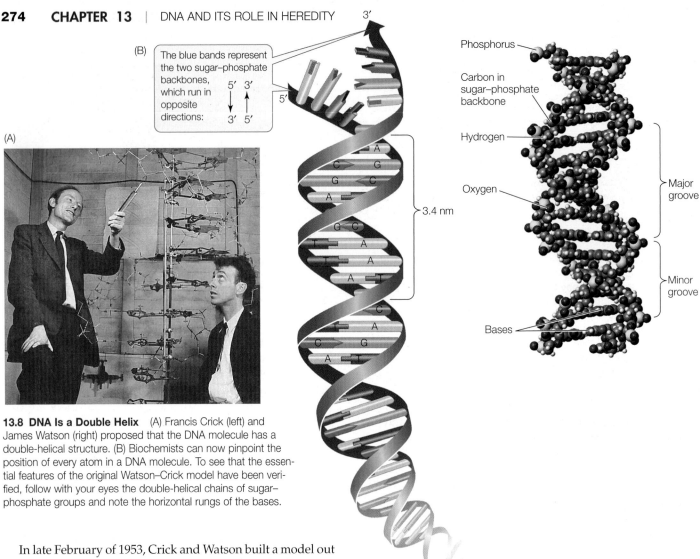

(B)

The blue bands represent the two sugar–phosphate backbones, which run in opposite directions:

5′ 3′
↓   ↑
3′ 5′

3′

5′

3.4 nm

Phosphorus

Carbon in sugar–phosphate backbone

Hydrogen

Oxygen

Major groove

Minor groove

Bases

13.8 DNA Is a Double Helix   (A) Francis Crick (left) and James Watson (right) proposed that the DNA molecule has a double-helical structure. (B) Biochemists can now pinpoint the position of every atom in a DNA molecule. To see that the essential features of the original Watson–Crick model have been verified, follow with your eyes the double-helical chains of sugar–phosphate groups and note the horizontal rungs of the bases.

In late February of 1953, Crick and Watson built a model out of tin that established the general structure of DNA. This structure explained all the known chemical properties of DNA, and it opened the door to understanding its biological functions. There have been minor amendments to that first published structure, but its principal features remain unchanged.

Four key features define DNA structure

Four features summarize the molecular architecture of the DNA molecule (see **Figure 13.8B**):

- It is a *double-stranded helix* of uniform diameter.
- It is *right-handed*. (Hold your right hand with the thumb pointing up. Imagine the curve of the helix following the direction of your fingers as it winds upward and you have the idea.)
- It is *antiparallel* (the two strands run in opposite directions).
- The outer edges of the nitrogenous bases are *exposed* in the major and minor grooves. These grooves exist because the backbones of the two strands are closer together on one side of the double helix (forming the minor groove) than on the other side (forming the major groove).

THE HELIX   The sugar–phosphate "backbones" of the polynucleotide chains coil around the outside of the helix, and the ni-

trogenous bases point toward the center. The two chains are held together by hydrogen bonding between specifically paired bases (**Figure 13.9**). Consistent with Chargaff's rule,

- Adenine (A) pairs with thymine (T) by forming two hydrogen bonds.
- Guanine (G) pairs with cytosine (C) by forming three hydrogen bonds.

Every base pair consists of one purine (A or G) and one pyrimidine (T or C). This pattern is known as **complementary base pairing**.

Because the A-T and G-C pairs are of equal length, they fit into a fixed distance between the two chains (like rungs on a ladder), and the diameter of the helix is thus uniform. The base pairs are flat, and their stacking in the center of the molecule is stabilized by hydrophobic interactions (see Section 2.2), contributing to the overall stability of the double helix.

ANTIPARALLEL STRANDS   What does it mean to say that the two DNA strands are *antiparallel*? The direction of each strand is determined by examining the bonds between the alternating phosphate and sugar groups that make up the backbone of each strand. Look closely at the five-carbon sugar (deoxyribose)

13.9 Base Pairing in DNA Is Complementary The purines (A and G) pair with the pyrimidines (T and C, respectively) to form base pairs that are equal in size and resemble the rungs on a ladder whose sides are formed by the sugar–phosphate backbones. The deoxyribose sugar (left) is where the 3' and 5' carbons are located. The two strands are antiparallel.

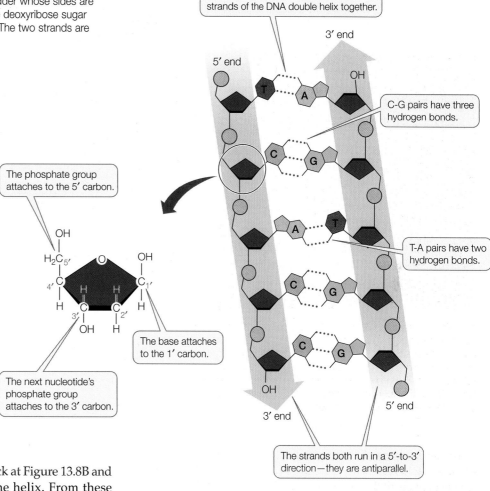

The phosphate group attaches to the 5' carbon.

The next nucleotide's phosphate group attaches to the 3' carbon.

The base attaches to the 1' carbon.

Pairs of complementary bases form hydrogen bonds that hold the two strands of the DNA double helix together.

C-G pairs have three hydrogen bonds.

T-A pairs have two hydrogen bonds.

The strands both run in a 5'-to-3' direction—they are antiparallel.

molecule in Figure 13.9. The number followed by a prime (') designates the position of a carbon atom in the sugar. In the sugar–phosphate backbone of DNA, the phosphate groups are connected to the 3' carbon of one deoxyribose molecule and the 5' carbon of the next, linking successive sugars together.

Thus the two ends of a polynucleotide chain differ. At one end of a chain is a free (not connected to another nucleotide) 5' phosphate group (—OPO$_3^-$); this is called the **5' end**. At the other end is a free 3' hydroxyl group (—OH); this is called the **3' end**. In a DNA double helix, the 5' end of one strand is paired with the 3' end of the other strand, and vice versa. In other words, if you drew an arrow for each strand running from 5' to 3', the arrows would point in opposite directions.

BASE EXPOSURE IN THE GROOVES Look back at Figure 13.8B and note the major and minor grooves in the helix. From these grooves, the exposed outer edges of the flat, hydrogen-bonded base pairs are accessible for additional hydrogen bonding. As seen in Figure 13.9, two hydrogen bonds join each A-T base pair, while three hydrogen bonds join each G-C base pair. Hydrogen-bonding opportunities also exist at an unpaired C=O group in T and an "N" in A. The G-C base pair offers additional hydrogen bonding possibilities as well. Thus the surfaces of the A-T and G-C base pairs are chemically distinct, allowing other molecules, such as proteins, to recognize specific base pair sequences and bind to them. Access to the exposed base-pair sequences in the major and minor grooves is the key to protein–DNA interactions, which are necessary for the replication and expression of the genetic information in DNA.

The double-helical structure of DNA is essential to its function

The genetic material performs four important functions, and the DNA structure proposed by Watson and Crick was elegantly suited to three of them.

● *The genetic material stores an organism's genetic information.* With its millions of nucleotides, the base sequence of a DNA molecule can encode and store an enormous amount

of information. Variations in DNA sequences can account for species and individual differences. DNA fits this role nicely.

● *The genetic material is susceptible to mutations* (permanent changes) *in the information it encodes.* For DNA, mutations might be simple changes in the linear sequence of base pairs.

● *The genetic material is precisely replicated in the cell division cycle.* Replication could be accomplished by complementary base pairing, A with T and G with C. In the original publication of their findings in 1953, Watson and Crick coyly pointed out, "It has not escaped our notice that the specific pairing we have postulated immediately suggests a possible copying mechanism for the genetic material."

● *The genetic material (the coded information in DNA) is expressed as the phenotype.* This function is not obvious in the structure of DNA. However, as we will see in the next chapter, the nucleotide sequence of DNA is copied into RNA, which uses the coded information to specify a linear sequence of amino acids—a protein. The folded forms of proteins determine many of the phenotypes of an organism.

13.2 RECAP

DNA is a double helix made up of two antiparallel polynucleotide chains. The two chains are joined by hydrogen bonds between the nucleotide bases, which pair specifically: A with T, and G with C. Chemical groups on the bases that are exposed in the grooves of the helix are available for hydrogen bonding with other molecules, such as proteins. These molecules can recognize specific sequences of nucleotide bases.

- Describe the evidence that Watson and Crick used to come up with the double helix model for DNA. **See p. 273**

- How does the double-helical structure of DNA relate to its function? **See p. 275**

Once the structure of DNA was understood, it was possible to investigate how DNA replicates itself. Let's examine the experiments that taught us how this elegant process works.

13.3 How Is DNA Replicated?

The mechanism of DNA replication that suggested itself to Watson and Crick was soon confirmed. First, experiments showed that DNA could be replicated in a test tube containing simple substrates and an enzyme. Then a truly classic experiment showed that each of the two strands of the double helix can serve as a template for a new strand of DNA.

——————— **yourBioPortal.com** ———————

GO TO **Animated Tutorial 13.1** • DNA Replication, Part 1: Replication of a Chromosome and DNA Polymerization

Three modes of DNA replication appeared possible

The prediction that the DNA molecule contains the information needed for its own replication was confirmed by the work of Arthur Kornberg, then at Washington University in St. Louis. He showed that new DNA molecules with the same base composition as the original molecules could be synthesized in a test tube containing the following substances:

- The substrates were the deoxyribonucleoside triphosphates dATP, dCTP, dGTP, and dTTP.

- A **DNA polymerase** enzyme catalyzed the reaction.

- DNA served as a **template** to guide the incoming nucleotides.

- The reaction also contained salts and a pH buffer, to create an appropriate chemical environment for the DNA polymerase to function.

Recall that a nucleoside is a nitrogen base attached to a sugar. The four deoxyribonucleoside triphosphates (dNTPs) each consist of a nitrogen base attached to deoxyribose, which in turn is attached to three phosphate groups. When a dNTP is added

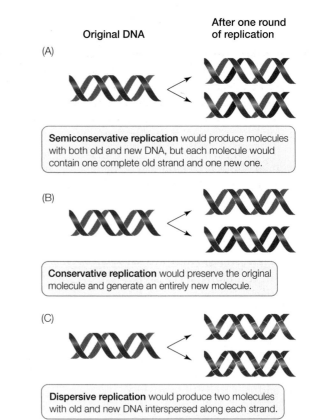

Original DNA — After one round of replication

(A)

Semiconservative replication would produce molecules with both old and new DNA, but each molecule would contain one complete old strand and one new one.

(B)

Conservative replication would preserve the original molecule and generate an entirely new molecule.

(C)

Dispersive replication would produce two molecules with old and new DNA interspersed along each strand.

13.10 Three Models for DNA Replication In each model, the original DNA is shown in blue and the newly synthesized DNA is in red.

to a DNA strand during DNA synthesis, the two terminal phosphates are removed, resulting in a monophosphate nucleotide.

The next challenge was to determine which of three possible replication patterns occurs during DNA replication:

- *Semiconservative replication*, in which each parent strand serves as a template for a new strand, and the two new DNA molecules each have one old and one new strand (**Figure 13.10A**)

- *Conservative replication*, in which the original double helix serves as a template for, but does not contribute to, a new double helix (**Figure 13.10B**)

- *Dispersive replication*, in which fragments of the original DNA molecule serve as templates for assembling two new molecules, each containing old and new parts, perhaps at random (**Figure 13.10C**)

Watson and Crick's original paper suggested that DNA replication was semiconservative, but Kornberg's experiment did not provide a basis for choosing among these three models.

An elegant experiment demonstrated that DNA replication is semiconservative

The work of Matthew Meselson and Franklin Stahl convinced the scientific community that DNA is reproduced by **semiconservative replication**. Working at the California Institute of Technology, Meselson and Stahl devised a simple way to distinguish between old parent strands of DNA and newly copied ones: *density labeling*.

The key to their experiment was the use of a "heavy" isotope of nitrogen. Heavy nitrogen (^{15}N) is a rare, nonradioactive isotope that makes molecules containing it denser than chemically identical molecules containing the common isotope, ^{14}N. Meselson, Stahl, and Jerome Vinograd grew two cultures of the bacterium *Escherichia coli* for many generations:

• One culture was grown in a medium whose nitrogen source (ammonium chloride, NH_4Cl) was made with ^{15}N instead of ^{14}N. As a result, all the DNA in the bacteria was "heavy."

─── **yourBioPortal.com** ───
GO TO Animated Tutorial 13.2 • The Meselson–Stahl Experiment

• Another culture was grown in a medium containing ^{14}N, and all the DNA in these bacteria was "light."

When DNA extracts from the two cultures were combined and centrifuged, two separate bands formed, showing that this method could be used to distinguish between DNA samples of slightly different densities.

Next, the researchers grew another *E. coli* culture on ^{15}N medium, then transferred it to normal ^{14}N medium and allowed the bacteria to continue growing (**Figure 13.11**). Under the conditions they used, *E. coli* cells replicate their DNA and divide

INVESTIGATING LIFE

13.11 The Meselson–Stahl Experiment
A centrifuge was used to separate DNA molecules labeled with isotopes of different densities. This experiment revealed a pattern that supports the semiconservative model of DNA replication.

HYPOTHESIS DNA replicates semiconservatively.

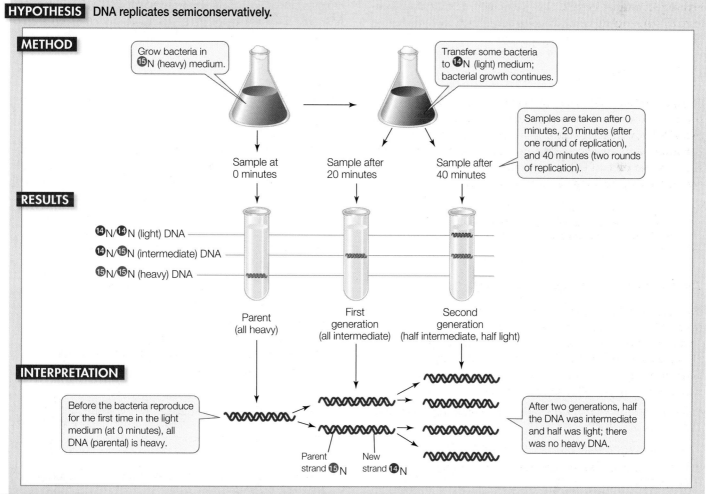

CONCLUSION This pattern could only have been observed if each DNA molecule contains a template strand from the parental DNA; thus DNA replication is semiconservative.

FURTHER INVESTIGATION: If you continued this experiment for two more generations (as Meselson and Stahl actually did), what would be the composition (in terms of density) of the fourth generation DNA?

Go to **yourBioPortal.com** for original citations, discussions, and relevant links for all INVESTIGATING LIFE figures.

every 20 minutes. Meselson and Stahl collected some of the bacteria after each division and extracted DNA from the samples. They found that the density gradient was different in each bacterial generation:

- At the time of the transfer to the ^{14}N medium, the DNA was uniformly labeled with ^{15}N, and hence formed a single band corresponding with dense DNA.

- After one generation in the ^{14}N medium, when the DNA had been duplicated once, all the DNA was of intermediate density.

- After two generations, there were two equally large DNA bands: one of low density and one of intermediate density.

- In samples from subsequent generations, the proportion of low-density DNA increased steadily.

The results of this experiment can be explained only by the semiconservative model of DNA replication. In the first round of DNA replication in the ^{14}N medium, the strands of the double helix—both heavy with ^{15}N—separated. Each strand then acted as the template for a second strand, which contained only ^{14}N and hence was less dense. Each double helix then consisted of one ^{15}N strand and one ^{14}N strand, and was of intermediate density. In the second replication, the ^{14}N-containing strands directed the synthesis of partners with ^{14}N, creating low-density DNA, and the ^{15}N strands formed new ^{14}N partners.

The crucial observation demonstrating the semiconservative model was that intermediate-density DNA (^{15}N–^{14}N) appeared in the first generation and continued to appear in subsequent generations. With the other models, the results would have been quite different (see Figure 13.10):

- If conservative replication had occurred, the first generation would have had both high-density DNA (^{15}N–^{15}N) and low-density DNA (^{14}N–^{14}N), but no intermediate-density DNA.

- If dispersive replication had occurred, the density of the new DNA would have been intermediate, but DNA of this density would not continue to appear in subsequent generations.

Some scientists consider the Meselson–Stahl experiment to be one of the most elegant experiments ever performed by biologists, and it is an excellent example of the scientific method. It began with three hypotheses—the three models of DNA replication—and was designed so that the results could differentiate between them.

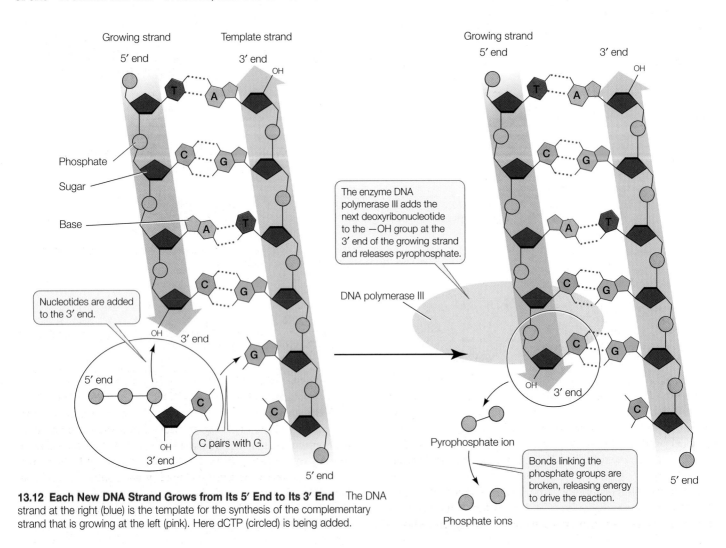

13.12 Each New DNA Strand Grows from Its 5′ End to Its 3′ End   The DNA strand at the right (blue) is the template for the synthesis of the complementary strand that is growing at the left (pink). Here dCTP (circled) is being added.

There are two steps in DNA replication

Semiconservative DNA replication in the cell involves a number of different enzymes and other proteins. It takes place in two general steps:

- The DNA double helix is unwound to separate the two template strands and make them available for new base pairing.

- As new nucleotides form complementary base pairs with template DNA, they are covalently linked together by phosphodiester bonds, forming a polymer whose base sequence is complementary to the bases in the template strand.

A key observation is that *nucleotides are added to the growing new strand at the 3′ end*—the end at which the DNA strand has a free hydroxyl (—OH) group on the 3′ carbon of its terminal deoxyribose (**Figure 13.12**). One of the three phosphate groups in a dNTP is attached to the 5′ position of the sugar. The bonds linking the other two phosphate groups to the dNTP are broken, resulting in a monophosphate nucleotide, and releasing energy for the reaction.

DNA polymerases add nucleotides to the growing chain

DNA is replicated through the interaction of the template strand with a huge protein complex called the **replication complex**, which contains at least four proteins, including DNA polymerase. All chromosomes have at least one region called the **origin of replication** (*ori*), to which the replication complex binds. Binding occurs when proteins in the complex recognize a specific DNA sequence within the origin of replication.

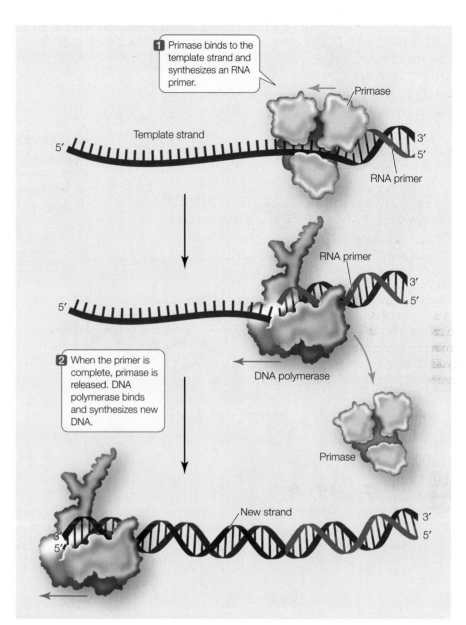

1 Primase binds to the template strand and synthesizes an RNA primer.

Primase

Template strand

5′

3′
5′

RNA primer

RNA primer

3′
5′

2 When the primer is complete, primase is released. DNA polymerase binds and synthesizes new DNA.

DNA polymerase

Primase

New strand

3′
5′

3′
5′

13.13 DNA Forms with a Primer DNA polymerases require a primer—a "starter" strand of DNA or RNA to which they can add new nucleotides.

DNA REPLICATION BEGINS WITH A PRIMER A DNA polymerase elongates a polynucleotide strand by covalently linking new nucleotides to a previously existing strand. However, it cannot start this process without a short "starter" strand, called a **primer**. In DNA replication, the primer is usually a short single strand of RNA (**Figure 13.13**) but in some organisms it is DNA. This RNA primer strand is complementary to the DNA template, and is synthesized one nucleotide at a time by an enzyme called a **primase**. The DNA polymerase then adds nucleotides to the 3′ end of the primer and continues until the replication of that section of DNA has been completed. Then the RNA primer is degraded, DNA is added in its place, and the resulting DNA fragments are connected by the action of

other enzymes. When DNA replication is complete, each new strand consists only of DNA.

DNA POLYMERASES ARE LARGE DNA polymerases are much larger than their substrates, the dNTPs, and the template DNA, which is very thin (**Figure 13.14A**). Molecular models of the enzyme–substrate–template complex from bacteria show that the enzyme is shaped like an open right hand with a palm, a thumb, and fingers (**Figure 13.14B**). The palm holds the active site of the enzyme and brings together each substrate and the template. The finger regions rotate inward and have precise shapes that can recognize the different shapes of the four nucleotide bases.

(A)

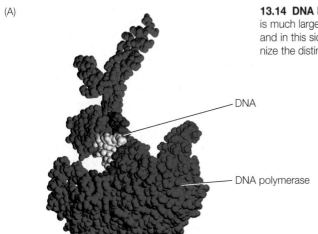

13.14 DNA Polymerase Binds to the Template Strand (A) The DNA polymerase enzyme (blue) is much larger than the DNA molecule (red and white). (B) DNA polymerase is shaped like a hand, and in this side-on view, its "fingers" can be seen curling around the DNA. These "fingers" can recognize the distinctive shapes of the four bases.

DNA

DNA polymerase

CELLS CONTAIN SEVERAL DIFFERENT DNA POLYMERASES Most cells contain more than one kind of DNA polymerase, but only one of them is responsible for chromosomal DNA replication. The others are involved in primer removal and DNA repair. Fifteen DNA polymerases have been identified in humans; the ones catalyzing most replication are DNA polymerases δ (delta) and ε (epsilon). In the bacterium *E. coli* there are five DNA polymerases; the one responsible for replication is DNA polymerase III.

Many other proteins assist with DNA polymerization

Various other proteins play roles in other replication tasks; some of these are shown in **Figure 13.15**. The first event at the origin of replication is the localized unwinding and separation (denaturation) of the DNA strands. As we saw in Chapter 4, there are several forces that hold the two strands together, including hydrogen bonding and the hydrophobic interactions of the bases. An enzyme called **DNA helicase** uses energy from ATP hydrolysis to unwind and separate the strands, and spe-

(B)

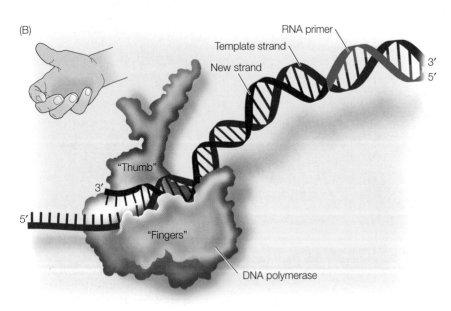

RNA primer

Template strand

New strand

3'
5'

"Thumb"

3'

5'

"Fingers"

DNA polymerase

13.15 Many Proteins Collaborate in the Replication Complex Several proteins in addition to DNA polymerase are involved in DNA replication. The two molecules of DNA polymerase shown here are actually part of the same complex.

yourBioPortal.com
GO TO Web Activity 13.1 • The Replication Complex

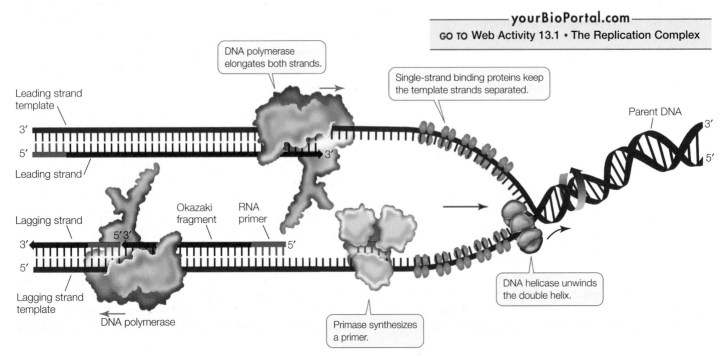

DNA polymerase elongates both strands.

Single-strand binding proteins keep the template strands separated.

Parent DNA

Leading strand template

3'
5'

Leading strand

3'

Lagging strand

3'
5'

5' 3'

Okazaki fragment

RNA primer

5'

Lagging strand template

DNA polymerase

3'
5'

DNA helicase unwinds the double helix.

Primase synthesizes a primer.

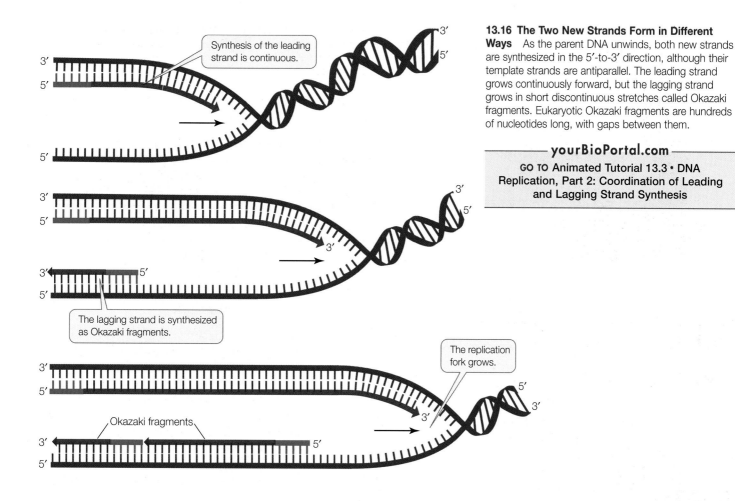

Synthesis of the leading strand is continuous.

The lagging strand is synthesized as Okazaki fragments.

The replication fork grows.

Okazaki fragments

13.16 The Two New Strands Form in Different Ways As the parent DNA unwinds, both new strands are synthesized in the 5'-to-3' direction, although their template strands are antiparallel. The leading strand grows continuously forward, but the lagging strand grows in short discontinuous stretches called Okazaki fragments. Eukaryotic Okazaki fragments are hundreds of nucleotides long, with gaps between them.

─── **yourBioPortal.com** ───

GO TO Animated Tutorial 13.3 • DNA Replication, Part 2: Coordination of Leading and Lagging Strand Synthesis

cial proteins called **single-strand binding proteins** bind to the unwound strands to keep them from reassociating into a double helix. This process makes each of the two template strands available for complementary base pairing.

THE TWO DNA STRANDS GROW DIFFERENTLY As Figure 13.15 shows, the DNA at the **replication fork**—the site(s) where DNA unwinds to expose the bases so that they can act as templates—opens up like a zipper in one direction. Study **Figure 13.16** and try to imagine what is happening over a short period of time. Remember that the two DNA strands are antiparallel; that is, the 3' end of one strand is paired with the 5' end of the other.

- One newly replicating strand (the **leading strand**) is oriented so that it can grow continuously at its 3' end as the fork opens up.

- The other new strand (the **lagging strand**) is oriented so that as the fork opens up, its exposed 3' end gets farther and farther away from the fork, and an unreplicated gap is formed. This gap would get bigger and bigger if there were not a special mechanism to overcome this problem.

Synthesis of the lagging strand requires the synthesis of relatively small, discontinuous stretches of sequence (100 to 200 nucleotides in eukaryotes; 1,000 to 2,000 nucleotides in prokaryotes). These discontinuous stretches are synthesized just as the

leading strand is, by the addition of new nucleotides one at a time to the 3' end of the new strand, but the synthesis of this new strand moves in the direction opposite to that in which the replication fork is moving. These stretches of new DNA are called **Okazaki fragments** (after their discoverer, the Japanese biochemist Reiji Okazaki). While the leading strand grows continuously "forward," the lagging strand grows in shorter, "backward" stretches with gaps between them.

A single primer is needed for synthesis of the leading strand, but each Okazaki fragment requires its own primer to be synthesized by the primase. In bacteria, DNA polymerase III then synthesizes an Okazaki fragment by adding nucleotides to one primer until it reaches the primer of the previous fragment. At this point, DNA polymerase I (discovered by Arthur Kornberg) removes the old primer and replaces it with DNA. Left behind is a tiny nick—the final phosphodiester linkage between the adjacent Okazaki fragments is missing. The enzyme **DNA ligase** catalyzes the formation of that bond, linking the fragments and making the lagging strand whole (**Figure 13.17**).

Working together, DNA helicase, the two DNA polymerases, primase, DNA ligase, and the other proteins of the replication complex do the job of DNA synthesis with a speed and accuracy that are almost unimaginable. In *E. coli*, the replication complex makes new DNA at a rate in excess of 1,000 base pairs per second, committing errors in fewer than one base in a million.

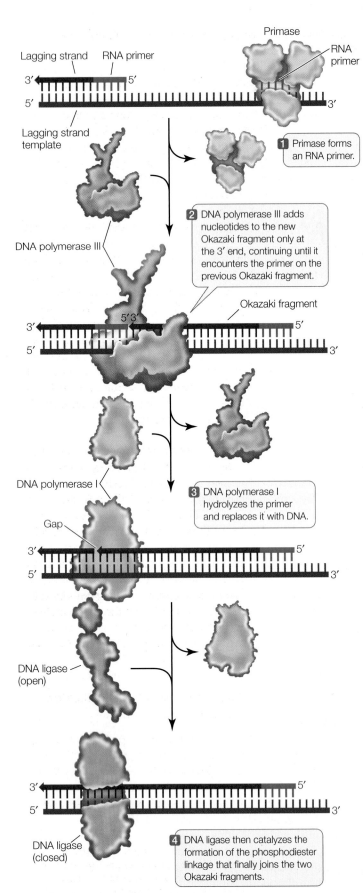

13.17 The Lagging Strand Story In bacteria, DNA polymerase I and DNA ligase cooperate with DNA polymerase III to complete the complex task of synthesizing the lagging strand.

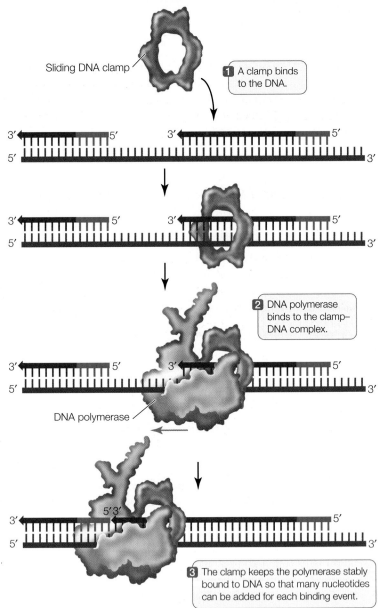

13.18 A Sliding DNA Clamp Increases the Efficiency of DNA Polymerization The clamp increases the efficiency of polymerization by keeping the enzyme bound to the substrate, so the enzyme does not have to repeatedly bind to template and substrate.

A SLIDING CLAMP INCREASES THE RATE OF DNA REPLICATION How do DNA polymerases work so fast? We saw in Section 8.3 that an enzyme catalyzes a chemical reaction:

substrate binds to enzyme → one product is formed → enzyme is released → cycle repeats

DNA replication would not proceed as rapidly as it does if it went through such a cycle for each nucleotide. Instead, DNA polymerases are **processive**—that is, they catalyze many polymerizations each time they bind to a DNA molecule:

substrates bind to one enzyme → many products are formed → enzyme is released → cycle repeats

The newly replicated strand is stabilized by a **sliding DNA clamp**, which is shaped like a screw cap on a bottle (**Figure 13.18**). This protein has multiple identical subunits assembled into a doughnut shape. The doughnut's "hole" is just large enough to encircle the DNA double helix, along with a single layer of water molecules for lubrication. The clamp binds to the DNA polymerase–DNA complex, keeping the enzyme and the DNA associated tightly with each other. If the clamp is absent, DNA polymerase dissociates from DNA after 20–100 polymerizations. With the clamp, it can polymerize up to 50,000 nucleotides before it detaches.

PCNA IS THE MAESTRO OF THE REPLICATION FORK In mammals, the sliding clamp was first recognized in rapidly dividing cells and is called **proliferating cell nuclear antigen** (**PCNA**). PCNA does more than just keep the DNA polymerase bound to the DNA; it also helps to orient the polymerase for binding to the substrates. Furthermore, PCNA has binding sites for many other proteins, including chromosome structural proteins, DNA ligase, DNA methylation enzymes (see Section 16.4) and enzymes involved in DNA repair (see below). It also removes the prereplication complex from *ori*, ensuring that replication only happens once per cell cycle. For all that it does, PCNA has been called the "maestro of the replication fork."

DNA IS THREADED THROUGH A REPLICATION COMPLEX Until recently, DNA replication was always depicted to look like a locomotive (the replication complex) moving along a railroad track (the DNA). While this does occur in some organisms, most commonly in eukaryotes the replication complex seems to be stationary, attached to chromatin structures, and it is the DNA that moves, essentially threading through the complex as single strands and emerging as double strands).

SMALL CIRCULAR CHROMOSOMES REPLICATE FROM A SINGLE ORIGIN Small circular chromosomes, such as those of bacteria (consisting of 1–4 million base pairs), have a single origin of replication. Two replication forks form at this *ori*, and as the DNA moves through the replication complex, the replication forks extend around the circle (**Figure 13.19A**). Two interlocking circular DNA molecules are formed, and they are separated by an enzyme called **DNA topoisomerase**. As we mentioned above, DNA polymerases are very fast. In *E. coli*, replication can be as fast as 1,000 bases per second, and it takes 20–40 minutes to replicate the bacterium's 4.7 million base pairs.

LARGE LINEAR CHROMOSOMES HAVE MANY ORIGINS Human DNA polymerases are slower than those of *E. coli*, and can replicate DNA at a rate of about 50 bases per second. Human chromosomes are much larger than those of bacteria (about 80 million base pairs) and linear. Large linear chromosomes such as those of humans contain hundreds of origins of replication. Numerous replication complexes bind to these sites at the same time and catalyze simultaneous replication. Thus there are many replication forks in eukaryotic DNA (**Figure 13.19B**).

Telomeres are not fully replicated and are prone to repair

As we have just seen, replication of the lagging strand occurs by the addition of Okazaki fragments to RNA primers. When the terminal RNA primer is removed, no DNA can be synthesized to replace it be-

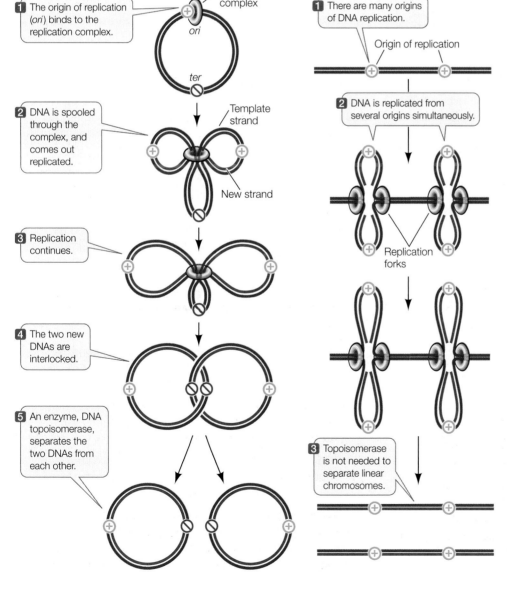

(A) Circular chromosome

Replication complex

1 The origin of replication (*ori*) binds to the replication complex.

ori

ter

2 DNA is spooled through the complex, and comes out replicated.

Template strand

New strand

3 Replication continues.

4 The two new DNAs are interlocked.

5 An enzyme, DNA topoisomerase, separates the two DNAs from each other.

(B) Linear chromosome

1 There are many origins of DNA replication.

Origin of replication

2 DNA is replicated from several origins simultaneously.

Replication forks

3 Topoisomerase is not needed to separate linear chromosomes.

13.19 Replication of Small Circular and Large Linear Chromosomes (A) Small circular chromosomes, typical of prokaryotes, have a single origin (*ori*) and terminus (*ter*) of replication. (B) Larger linear chromosomes, typical of nuclear DNA in eukaryotes, have many origins of replication.

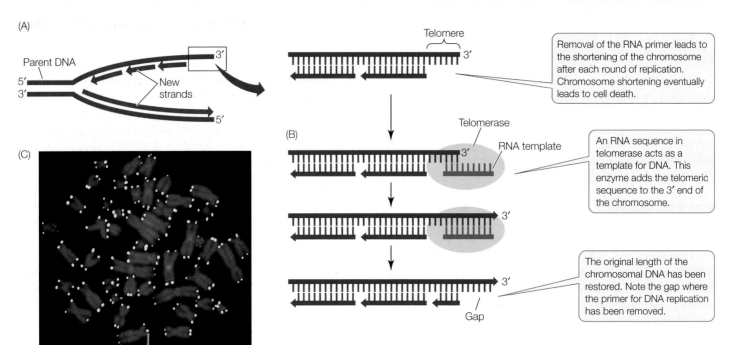

13.20 Telomeres and Telomerase (A) Removal of the RNA primer at the 3′ end of the template for the lagging strand leaves a region of DNA—the telomere—unreplicated. (B) In continuously dividing cells, the enzyme telomerase binds to the 3′ end and extends the lagging strand of DNA, so the chromosome does not get shorter. (C) Bright fluorescent staining marks the telomeric regions on these blue-stained human chromosomes.

The callout boxes in the figure read:

Removal of the RNA primer leads to the shortening of the chromosome after each round of replication. Chromosome shortening eventually leads to cell death.

An RNA sequence in telomerase acts as a template for DNA. This enzyme adds the telomeric sequence to the 3′ end of the chromosome.

The original length of the chromosomal DNA has been restored. Note the gap where the primer for DNA replication has been removed.

cause there is no 3′ end to extend (**Figure 13.20A**). So the new chromosome has a bit of single-stranded DNA at each end. This situation activates a mechanism for cutting off the single-stranded region, along with some of the intact double-stranded DNA. Thus the chromosome becomes slightly shorter with each cell division.

There is another, more serious problem at the ends of chromosomes, and that is simply that they are ends! In Section 11.2 we described checkpoints in the cell cycle for the integrity of DNA. At one of the checkpoints, the DNA is examined for DNA breaks (due to radiation, etc.) and DNA repair is initiated if breaks are found. This involves joining the breaks via a combination of DNA synthesis and DNA ligase activity. This system might recognize the ends of chromosomes as breaks and join two chromosomes together. This would create havoc with genomic integrity.

In many eukaryotes, there are repetitive sequences at the ends of chromosomes called **telomeres**. In humans, the telomere sequence is TTAGGG, and it is repeated about 2,500 times. These repeats bind special proteins that prevent the DNA repair system from recognizing the ends as breaks. In addition, the repeats may form loops that have a similar protective role. So the telomere acts like the plastic tip of shoelaces to prevent fraying.

Each human chromosome can lose 50–200 base pairs of telomeric DNA after each round of DNA replication and cell division. After 20–30 divisions, the chromosomes are unable to participate in cell division, and the cell dies. This phenomenon explains, in part, why many cell lineages do not last the entire lifetime of the organism: their telomeres are lost. Yet continuously dividing cells, such as bone marrow stem cells and gamete-producing cells, maintain their telomeric DNA. An enzyme, appropriately called **telomerase**, catalyzes the addition of

any lost telomeric sequences in these cells (**Figure 13.20B**). Telomerase contains an RNA sequence that acts as a template for the telomeric DNA repeat sequence.

Telomerase is expressed in more than 90 percent of human cancers, and may be an important factor in the ability of cancer cells to divide continuously. Since most normal cells do not have this ability, telomerase is an attractive target for drugs designed to attack tumors specifically.

There is also interest in telomerase and aging. When a gene expressing high levels of telomerase is added to human cells in culture, their telomeres do not shorten. Instead of living 20–30 cell generations and then dying, the cells become immortal. It remains to be seen how this finding relates to the aging of a whole organism.

13.3 RECAP

Meselson and Stahl showed that DNA replication is semiconservative: each parent DNA strand serves as a template for a new strand. A complex of proteins, most notably DNA polymerases, is involved in replication. New DNA is polymerized in one direction only, and since the two strands are antiparallel, one strand is made continuously and the other is synthesized in short Okazaki fragments that are eventually joined.

- How did the Meselson–Stahl experiment differentiate between the three models for DNA replication? See pp. 276–278 and Figures 13.10 and 13.11

- What are the five enzymes needed for DNA replication and what are their roles? See pp. 279–283 and Figures 13.13–13.17

- How is the leading strand of DNA replicated continuously while the lagging strand must be replicated in fragments? See p. 281 and Figure 13.16

The complex process of DNA replication is amazingly accurate, but it is not perfect. What happens when things go wrong?

13.4 How Are Errors in DNA Repaired?

DNA must be accurately replicated and faithfully maintained. The price of failure can be great; the accurate transmission of genetic information is essential for the functioning and even the life of a single cell or multicellular organism. Yet the replication of DNA is not perfectly accurate, and the DNA of nondividing cells is subject to damage by natural chemical alterations and by environmental agents. In the face of these threats, how has life gone on so long?

DNA repair mechanisms help to preserve life. DNA polymerases initially make significant numbers of mistakes in assembling polynucleotide strands. Without DNA repair, the observed error rate of one for every 10^5 bases replicated would result in about 60,000 mutations every time a human cell divided. Fortunately, our cells can repair damaged nucleotides and DNA replication errors, so that very few errors end up in the replicated DNA. Cells have at least three DNA repair mechanisms at their disposal:

- A **proofreading** mechanism corrects errors in replication as DNA polymerase makes them.

- A **mismatch repair** mechanism scans DNA immediately after it has been replicated and corrects any base-pairing mismatches.

- An **excision repair** mechanism removes abnormal bases that have formed because of chemical damage and replaces them with functional bases.

Most DNA polymerases perform a **proofreading** function each time they introduce a new nucleotide into a growing DNA strand (**Figure 13.21A**). When a DNA polymerase recognizes a mispairing of bases, it removes the improperly introduced nucleotide and tries again. (Other proteins in the replication complex also play roles in proofreading.) The error rate for this process is only about 1 in 10,000 repaired base pairs, and it lowers the overall error rate for replication to about one error in every 10^{10} bases replicated.

After the DNA has been replicated, a second set of proteins surveys the newly replicated molecule and looks for mismatched base pairs that were missed in proofreading (**Figure**

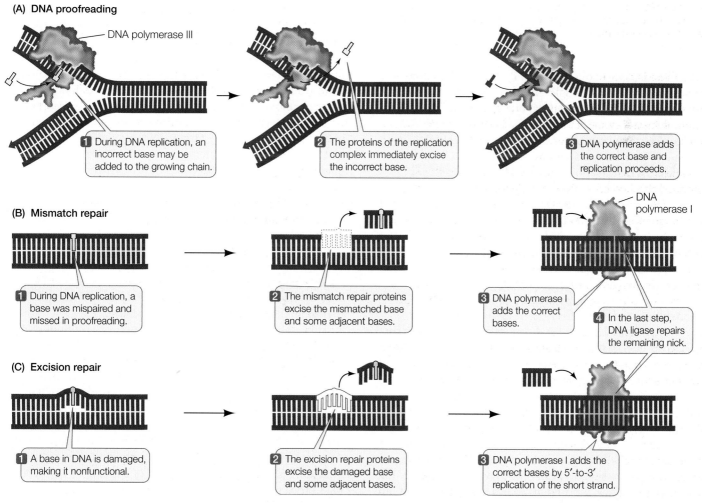

(A) DNA proofreading

DNA polymerase III

1 During DNA replication, an incorrect base may be added to the growing chain.

2 The proteins of the replication complex immediately excise the incorrect base.

3 DNA polymerase adds the correct base and replication proceeds.

(B) Mismatch repair

DNA polymerase I

1 During DNA replication, a base was mispaired and missed in proofreading.

2 The mismatch repair proteins excise the mismatched base and some adjacent bases.

3 DNA polymerase I adds the correct bases.

4 In the last step, DNA ligase repairs the remaining nick.

(C) Excision repair

1 A base in DNA is damaged, making it nonfunctional.

2 The excision repair proteins excise the damaged base and some adjacent bases.

3 DNA polymerase I adds the correct bases by 5'-to-3' replication of the short strand.

13.21 DNA Repair Mechanisms The proteins of the replication complex function in DNA repair mechanisms, reducing the rate of errors in the replicated DNA. Another mechanism (excision repair) repairs damage to existing DNA molecules.

13.21B). For example, this **mismatch repair** mechanism might detect an A-C base pair instead of an A-T pair. But how does the repair mechanism "know" whether the A-C pair should be repaired by removing the C and replacing it with T or by removing the A and replacing it with G?

The mismatch repair mechanism can detect the "wrong" base because a DNA strand is chemically modified some time after replication. In prokaryotes, methyl groups ($—CH_3$) are added to some adenines. In eukaryotes, cytosine bases are methylated. Immediately after replication, methylation has not yet occurred on the newly replicated strand, so the new strand is "marked" (distinguished by being unmethylated) as the one in which errors should be corrected.

When mismatch repair fails, DNA sequences are altered. One form of colon cancer arises in part from a failure of mismatch repair.

DNA molecules can also be damaged during the life of a cell (for example, when it is in G_1). High-energy radiation, chemicals from the environment, and random spontaneous chemical reactions can all damage DNA. **Excision repair** mechanisms deal with these kinds of damage (**Figure 13.21C**). Individuals who suffer from a condition known as xeroderma pigmentosum lack an excision repair mechanism that normally corrects the damage caused by ultraviolet radiation. They can develop skin cancers after even a brief exposure to sunlight.

13.4 RECAP

DNA replication is not perfect; in addition, DNA may be naturally altered or damaged. Repair mechanisms exist that detect and repair mismatched or damaged DNA.

- Explain the roles of DNA proofreading, mismatch repair, and excision repair. See Figure 13.21

Understanding how DNA is replicated and repaired has allowed scientists to develop techniques for studying genes. We'll look at just one of those techniques next.

13.5 How Does the Polymerase Chain Reaction Amplify DNA?

The principles underlying DNA replication in cells have been used to develop an important laboratory technique that has been vital in analyzing genes and genomes. This technique allows researchers to make multiple copies of short DNA sequences.

The polymerase chain reaction makes multiple copies of DNA sequences

In order to study DNA and perform genetic manipulations, it is necessary to make multiple copies of a DNA sequence. This is necessary because the amount of DNA isolated from a biological sample is often too small to work with. The **polymerase chain reaction** (**PCR**) technique essentially automates this replication process by copying a short region of DNA many times in a test tube. This process is referred to as *DNA amplification*.

The PCR reaction mixture contains:

- a sample of double stranded DNA from a biological sample, to act as the template,
- two short, artificially synthesized primers that are complementary to the ends of the sequence to be amplified,
- the four dNTPs (dATP, dTTP, dCTP and dGTP),
- a DNA polymerase that can tolerate high temperatures without becoming degraded, and
- salts and a buffer to maintain a near-neutral pH.

The PCR amplification is a cyclic process in which a sequence of steps is repeated over and over again (**Figure 13.22**):

- The first step involves heating the reaction to near boiling point, to separate (*denature*) the two strands of the DNA template.
- The reaction is then cooled to allow the primers to bind (or *anneal*) to the template strands.
- Next, the reaction is warmed to an optimum temperature for the DNA polymerase to catalyze the production of the complementary new strands.

A single cycle takes a few minutes to produce two copies of the target DNA sequence, leaving the new DNA in the double-stranded state. Repeating the cycle many times leads to an exponential increase in the number of copies of the DNA sequence.

The PCR technique requires that the base sequences at the 3' end of each strand of the target DNA sequence be known, so that complementary primers, usually 15–30 bases long, can be made in the laboratory. Because of the uniqueness of DNA sequences, a pair of primers this length will usually bind to only a single region of DNA in an organism's genome. This specificity, despite the incredible diversity of DNA sequences, is a key to the power of PCR.

One initial problem with PCR was its temperature requirements. To denature the DNA, it must be heated to more than 90°C—a temperature that destroys most DNA polymerases. The PCR technique would not be practical if new polymerase had to be added after denaturation in each cycle.

This problem was solved by nature: in the hot springs at Yellowstone National Park, as well as in other high-temperature locations, there lives a bacterium called, appropriately, *Thermus aquaticus* ("hot water"). The means by which this organism survives temperatures of up to 95°C was investigated by Thomas Brock and his colleagues at the University of Wisconsin, Madison. They discovered that *T. aquaticus* has an entire metabolic machinery that is heat-resistant, including a DNA polymerase that does not denature at these high temperatures.

Scientists pondering the problem of copying DNA by PCR read Brock's basic research articles and got a clever idea: why not use *T. aquaticus* DNA polymerase in the PCR technique? It

TOOLS FOR INVESTIGATING LIFE

13.22 The Polymerase Chain Reaction
The steps in this cyclic process are repeated many times to produce multiple identical copies of a DNA fragment. This makes enough DNA for chemical analysis and genetic manipulations.

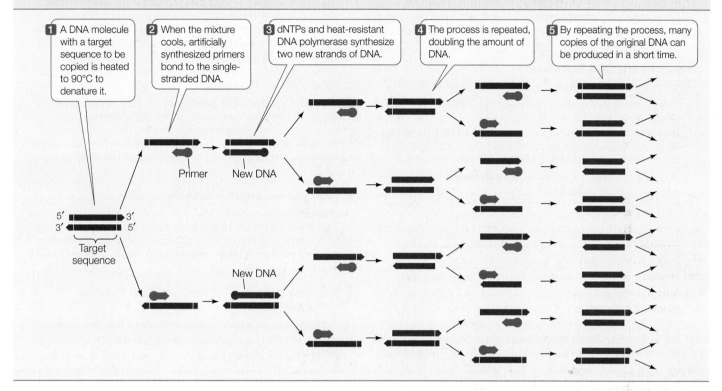

1 A DNA molecule with a target sequence to be copied is heated to 90°C to denature it.

2 When the mixture cools, artificially synthesized primers bond to the single-stranded DNA.

3 dNTPs and heat-resistant DNA polymerase synthesize two new strands of DNA.

4 The process is repeated, doubling the amount of DNA.

5 By repeating the process, many copies of the original DNA can be produced in a short time.

Primer   New DNA

New DNA

5′
3′
3′
5′

Target sequence

could withstand the 90°C denaturation temperature and would not have to be added during each cycle. The idea worked, and it earned biochemist Kary Mullis a Nobel prize. PCR has had an enormous impact on genetic research. Some of its most striking applications will be described in Chapters 15–18. These applications range from amplifying DNA in order to identify an individual person or organism, to detection of diseases.

13.5 RECAP

Knowledge of the mechanisms of DNA replication led to the development of a technique for making multiple copies of DNA sequences.

- What is the role of primers in PCR? **See pp. 286 and Figure 13.22**

CHAPTER SUMMARY

13.1 What Is the Evidence that the Gene Is DNA?

- Griffith's experiments in the 1920s demonstrated that some substance in cells—then called a **transforming principle**—can cause heritable changes in other cells. **Review Figure 13.1**

- The location and quantity of DNA in the cell suggested that DNA might be the genetic material. Avery, MacLeod, and McCarty isolated the transforming principle from bacteria and identified it as DNA. **Review Figure 13.2**

- The Hershey–Chase experiment established conclusively that DNA (and not protein) is the genetic material, by tracing the

DNA of radioactively labeled viruses, with which they infected bacterial cells. **Review Figure 13.4**

- Genetic transformation of eukaryotic cells is called **transfection**. Transformation and transfection can be studied with the aid of a **marker** gene that confers a known and observable phenotype. **Review Figure 13.5**

13.2 What Is the Structure of DNA?

- Chargaff's rule states that the amount of **adenine** in DNA is equal to the amount of **thymine**, and that the amount of **guanine** is equal to the amount of **cytosine**; thus the total

- abundance of purines (A + G) equals the total abundance of pyrimidines (T + C).
- X-ray crystallography showed that the DNA molecule is **helical**. Watson and Crick proposed that DNA is a double-stranded helix in which the strands are **antiparallel**. Review Figure 13.8
- **Complementary base pairing** between A and T and between G and C accounts for Chargaff's rule. The bases are held together by hydrogen bonding. Review Figure 13.9

13.3 How Is DNA Replicated?

SEE ANIMATED TUTORIAL 13.1

- Meselson and Stahl showed the replication of DNA to be **semiconservative**. Each parent strand acts as a **template** for the synthesis of a new strand; thus the two replicated DNA molecules each contain one parent strand and one newly synthesized strand. Review Figure 13.11, **ANIMATED TUTORIAL 13.2**
- In DNA replication, the enzyme **DNA polymerase** catalyzes the addition of nucleotides to the 3′ end of each strand. Which nucleotides are added is determined by complementary base pairing with the template strand. Review Figure 13.12
- The **replication complex** is a huge protein complex that attaches to the chromosome at the **origin of replication** (*ori*).
- Replication proceeds from the origin of replication on both strands in the 5′-to-3′ direction, forming a **replication fork**.
- **Primase** catalyzes the synthesis of a short RNA **primer** to which nucleotides are added by DNA polymerase. Review Figure 13.13
- Many proteins assist in DNA replication. **DNA helicase** separates the strands, and **single-strand binding proteins** keep the strands from reassociating. Review Figure 13.13, WEB ACTIVITY 13.1

- The **leading strand** is synthesized continuously and the **lagging strand** in pieces called **Okazaki fragments**. The fragments are joined together by **DNA ligase**. Review Figures 13.16 and 13.17, **ANIMATED TUTORIAL 13.3**
- The speed with which DNA polymerization proceeds is attributed to the **processive** nature of DNA polymerases, which can catalyze many polymerizations at a time. A **sliding DNA clamp** helps ensure the stability of this process. Review Figure 13.18
- In prokaryotes, two interlocking circular DNA molecules are formed; they are separated by an enzyme called **DNA topoisomerase**. Review Figure 13.19
- In eukaryotes, DNA replication leaves a short, unreplicated sequence, the **telomere**, at the 3′ end of the chromosome. Unless the enzyme **telomerase** is present, the sequence is removed. After multiple cell cycles, the telomeres shorten, leading to chromosome instability and cell death. Review Figure 13.20

13.4 How Are Errors in DNA repaired?

- DNA polymerases make about one error in 10^5 bases replicated. DNA is also subject to natural alterations and chemical damage. DNA can be repaired by three different mechanisms: **proofreading**, **mismatch repair**, and **excision repair**. Review Figure 13.21

13.5 How Does the Polymerase Chain Reaction Amplify DNA?

- The **polymerase chain reaction** technique uses DNA polymerase to make multiple copies of DNA in the laboratory. Review Figure 13.22

SELF-QUIZ

1. Griffith's studies of *Streptococcus pneumoniae*
   a. showed that DNA is the genetic material of bacteria.
   b. showed that DNA is the genetic material of bacteriophage.
   c. demonstrated the phenomenon of bacterial transformation.
   d. proved that prokaryotes reproduce sexually.
   e. proved that protein is not the genetic material.

2. In the Hershey–Chase experiment,
   a. DNA from parent bacteriophage appeared in progeny bacteriophage.
   b. most of the phage DNA never entered the bacteria.
   c. more than three-fourths of the phage protein appeared in progeny phage.
   d. DNA was labeled with radioactive sulfur.
   e. DNA formed the coat of the bacteriophage.

3. Which statement about complementary base pairing is *not* true?
   a. Complementary base pairing plays a role in DNA replication.
   b. In DNA, T pairs with A.
   c. Purines pair with purines, and pyrimidines pair with pyrimidines.
   d. In DNA, C pairs with G.
   e. The base pairs are of equal length.

4. In semiconservative replication of DNA,
   a. the original double helix remains intact and a new double helix forms.
   b. the strands of the double helix separate and act as templates for new strands.
   c. polymerization is catalyzed by RNA polymerase.
   d. polymerization is catalyzed by a double-helical enzyme.
   e. DNA is synthesized from amino acids.

5. Which of the following does *not* occur during DNA replication?
   a. Unwinding of the parent double helix
   b. Formation of short pieces that are connected by DNA ligase
   c. Complementary base pairing
   d. Use of a primer
   e. Polymerization in the 3′-to-5′ direction

6. The primer used for DNA replication
   a. is a short strand of RNA added to the 3′ end.
   b. is needed only once on a leading strand.
   c. remains on the DNA after replication.
   d. ensures that there will be a free 5′ end to which nucleotides can be added.
   e. is added to only one of the two template strands.

7. One strand of DNA has the sequence 5'-ATTCCG-3' The complementary strand for this is
   a. 5'-TAAGGC-3'
   b. 5'-ATTCCG-3'
   c. 5'-ACCTTA-3'
   d. 5'-CGGAAT-3'
   e. 5'-GCCTTA-3'

8. The role of DNA ligase in DNA replication is to
   a. add more nucleotides to the growing strand one at a time.
   b. open up the two DNA strands to expose template strands.
   c. ligate base to sugar to phosphate in a nucleotide.
   d. bond Okazaki fragments to one another.
   e. remove incorrectly paired bases.

9. The polymerase chain reaction
   a. is a method for sequencing DNA.
   b. is used to transcribe specific genes.
   c. amplifies specific DNA sequences.
   d. does not require DNA replication primers.
   e. uses a DNA polymerase that denatures at 55°C.

10. What is the correct order for the following events in excision repair of DNA? (1) DNA polymerase I adds correct bases by 5' to 3' replication; (2) damaged bases are recognized; (3) DNA ligase seals the new strand to existing DNA; (4) part of a single strand is excised.
    a. 1, 2, 3, 4
    b. 2, 1, 3, 4
    c. 2, 4, 1, 3
    d. 3, 4, 2, 1
    e. 4, 2, 3, 1

FOR DISCUSSION

1. Suppose that Meselson and Stahl had continued their experiment on DNA replication for another ten bacterial generations. Would there still have been any ^{14}N–^{15}N hybrid DNA present? Would it still have appeared in the centrifuge tube? Explain.

2. If DNA replication were conservative rather than semiconservative, what results would Meselson and Stahl have observed? Draw a diagram of the results using the conventions of Figure 13.10.

3. Using the following information, calculate the number of origins of DNA replication on a human chromosome: DNA polymerase adds nucleotides at 3,000 base pairs per minute in one direction; replication is bidirectional; S phase lasts 300 minutes; there are 120 million base pairs per chromosome. In a typical chromosome 3 μm long, how many origins are there per μm?

4. The drug dideoxycytidine, used to treat certain viral infections, is a nucleotide made with 2',3'-dideoxyribose. This sugar lacks —OH groups at both the 2' and the 3' positions. Explain why this drug stops the growth of a DNA chain when added to DNA.

ADDITIONAL INVESTIGATION

Outline a series of experiments using radioactive isotopes (such as ^{32}P and ^{35}S) to show that it is bacterial DNA and not bacterial protein that enters the host cell and is responsible for bacterial transformation.

WORKING WITH DATA (GO TO yourBioPortal.com)

The Hershey-Chase Experiment   The experiments in which labeled bacteriophage were used to infect host *E. coli* cells were key evidence for the identification of DNA as the gene (Figure 13.4). In this exercise, you will analyze the data that Hershey and Chase obtained, as well as important control experiments that ruled out protein and pointed to DNA as the gene.

The Meselson-Stahl Experiment   Because of its elegant simplicity, this experiment has been called one of the most beautiful in the history of biology (Figure 13.11). In this real-world exercise, you will examine the experimental protocol and make calculations based on the actual centrifuge photographs that the experimenters obtained.

14 From DNA to Protein: Gene Expression

An unexpected wedding gift

The wedding and honeymoon began spectacularly. Andrew Speaker, an Atlanta lawyer, and law student Sarah Cooksey began their honeymoon in Rome. Days later, they got shocking news from the U.S. Centers for Disease Control and Prevention: Andrew had drug-resistant tuberculosis (TB) and would have to be quarantined to prevent him from spreading the disease to others.

Several months before, Speaker had gone to see his physician, complaining of a sore rib. The doctor ordered an X-ray and saw some fluid in Speaker's lungs. Suspicious, the physician sent samples of the fluid to a lab, which confirmed the diagnosis of TB. Moreover, the TB appeared to be resistant to several drugs.

Before the nineteenth-century German microbiologist Robert Koch identified the bacterium *Mycobacterium tuberculosis* as its causative agent, TB was known as consumption. What started as a bloody cough with fever, chills, and night sweats would usually progress to other organs, including the nervous system. Death was almost inevitable. Today tuberculosis cases are fairly rare in the United States and Europe. Worldwide, however, there are more than 8 million cases and 1.6 million deaths annually, and TB remains the scourge it was in the nineteenth century. Speaker probably picked up the bacterium on his travels and it hid in his tissues (possibly for years) before flaring up at the time of his wedding.

Two drugs are used as the first approach to treating TB. One of them, isoniazid, is activated inside the bacterial cell, and the activated form blocks an enzyme essential for the assembly of the bacterial cell walls. Without functional cell walls, new bacterial cells cannot survive. The second drug, rifampin, binds to a part of the enzyme RNA polymerase that is necessary for gene expression. Without the appropriate expression of its genes, the bacterium soon dies.

In both cases, the targets of the antibiotics are proteins, each encoded by a gene (a sequence of DNA). Mutations in these genes can lead to altered amino acid sequences in the proteins, so that they no longer have three-dimensional structures that bind to the antibiotics. Unfortunately, these altered genes can be transferred from one bacterium to another, so a single *M. tuberculosis* strain can evolve to have both mutations and be resistant to both antibiotics. That is what happened in the case of Andrew Speaker.

He made his way to Denver for treatment, this time with a third antibiotic, kanamycin, which binds to the bacterial ribosome. The ribosome is the cell's protein synthesis factory, and is also essential for gene expression. Finally, the treatment was successful.

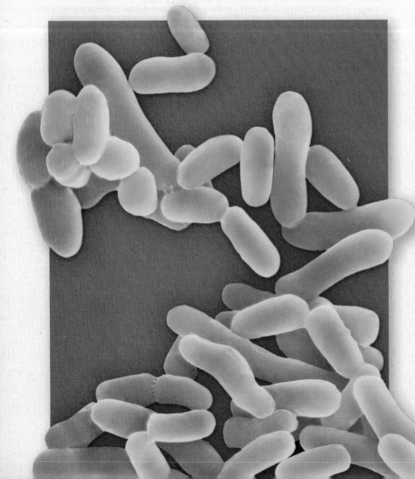

Mycobacterium tuberculosis The causative agent of TB can be killed with antibiotics, but resistance sometimes occurs.

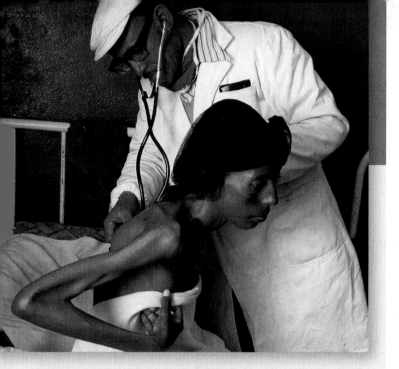

Tuberculosis Is a World Health Issue Drug-resistant TB has become a major medical problem throughout the world. Here a doctor examines a patient in Ethiopia, which ranks high among the world's nations in number of TB cases.

Proteins are the major products of gene expression. Some proteins play vital structural roles in cells, and others act as enzymes, which are essential for most aspects of phenotypic expression. So, when protein synthesis is inhibited, cells cannot survive. This is what happened to the TB-causing bacteria when Speaker was treated with kanamycin. But genes can mutate, and the alleles that result may encode proteins that have altered surfaces. The mutant alleles in resistant TB encoded proteins that would no longer bind antibiotics.

IN THIS CHAPTER we will describe how genes are expressed as proteins, first discussing the evidence for the relationship between genes and proteins. We will then describe how the DNA sequence of a gene is copied (transcribed) into a sequence of RNA, and how the RNA sequence is translated to make a polypeptide with a defined sequence of amino acids. We will discuss some of the modifications to proteins that occur after they are made by the ribosomes. Following Mendel's definition of the gene as a physical entity, scientists characterized the gene as DNA (see Chapter 13). In this chapter we see how the gene gets expressed as a phenotype at the molecular level.

14.1 What Is the Evidence that Genes Code for Proteins?

In Chapter 12, we defined genes as sequences of DNA and learned that genes are expressed as physical characteristics known as the phenotype. Here, we show that in most cases, genes code for proteins, and it is the proteins that determine the phenotype. What is the evidence for this?

The molecular basis of phenotypes was actually discovered before it was known that DNA was the genetic material. Scientists had studied the chemical differences between individuals carrying wild-type and mutant alleles in organisms as diverse as humans and bread molds. They found that the major phenotypic differences resulted from differences in specific proteins.

Observations in humans led to the proposal that genes determine enzymes

The identification of a gene product as a protein began with a mutation. In the early twentieth century, the English physician Archibald Garrod saw a number of children with a rare disease. One symptom of the disease was that the urine turned dark brown in air. This was especially noticeable on the infants' diapers. The disease was given the descriptive name alkaptonuria ("black urine").

Garrod noticed that the disease was most common in children whose parents were first cousins. Mendelian genetics had just been "rediscovered," and Garrod realized that because first cousins share alleles (can you calculate what fraction?), the children of first cousins might inherit a rare mutant allele from both parents. He proposed that alkaptonuria was a phenotype caused by a recessive, mutant allele.

Garrod took the analysis one step further. He identified the biochemical abnormality in the affected children. He isolated from them an unusual substance, homogentisic acid, which accumulated in blood, joints (where it crystallized and caused severe pain), and urine (where it turned black). The chemical structure of homogentisic acid is similar to that of the amino acid tyrosine:

Homogentisic acid   Tyrosine

Enzymes as biological catalysts had just been discovered. Garrod proposed that homogentisic acid was a breakdown product of tyrosine. Normally, homogentisic acid is converted to a harmless product. According to Garrod, there is a normal (wild-type) human allele that determines the synthesis of an enzyme that catalyzes this conversion:

normal allele
↓
active enzyme
↓
tyrosine → homogentisic acid → harmless product

When the allele has been mutated, the enzyme is inactive and homogentisic acid accumulates instead:

mutant allele
↓
inactive enzyme
↓
tyrosine → homogentisic acid ⊣ (HA accumulates)

Therefore, Garrod correlated *one gene to one enzyme* and coined the term "inborn error of metabolism" to describe this genetically determined biochemical disease. But his hypothesis needed direct confirmation by the identification of the specific enzyme and specific gene mutation involved. This did not occur until the enzyme, homogentisic acid oxidase, was described as active in healthy people and inactive in alkaptonuria patients in 1958, and the specific DNA mutation was described in 1996.

To directly relate genes and enzymes, biologists first turned to simpler organisms that could be manipulated in the laboratory.

Experiments on bread mold established that genes determine enzymes

As they work to explain the principles that underlie life, biologists often turn to organisms that they can manipulate experimentally. It wasn't possible to follow up on Garrod's hypothesis relating genes to enzymes in humans, because it is unethical to perform genetics experiments on people. Instead, biologists use *model organisms* that are easy to grow in the laboratory or greenhouse, and use them to develop principles of genetics that can then be applied more generally to other organisms. You have seen some of these model organisms in previous chapters:

- Pea plants (*Pisum sativum*) were used by Mendel in his genetics experiments.

- Fruit flies (*Drosophila*) were used by Morgan in his genetics experiments.

- *E. coli* was used by Meselson and Stahl to study DNA replication.

To this list we now add the common bread mold, *Neurospora crassa*. *Neurospora* is a type of fungus known as an ascomycete (see Chapter 30). This mold is haploid for most of its life, so that there are no dominant or recessive alleles: all alleles are expressed phenotypically and are not masked by a heterozygous condition. *Neurospora* is simple to culture and grows well in the laboratory. In the 1940s, George W. Beadle and Edward L. Tatum at Stanford University undertook studies to chemically define the phenotypes in *Neurospora*.

Beadle and Tatum knew about the roles of enzymes in biochemistry when they began their work, and like Garrod, they hypothesized that the expression of a specific gene results in the activity of a specific enzyme. Now, they set out to *test this hypothesis directly*. They grew *Neurospora* on a minimal nutritional medium containing sucrose, minerals, and a vitamin. Using this medium, the enzymes of wild-type *Neurospora* could catalyze all the metabolic reactions needed to make all the chemical constituents of their cells, including amino acids and proteins. These wild-type strains are called *prototrophs* ("original eaters"). From these wild type strains, they were able to produce and isolate distinct mutant strains that showed specific biochemical deficiencies.

Mutations provide a powerful way to determine cause and effect in biology. Nowhere has this been so evident as in the elucidation of biochemical pathways. Such pathways consist of sequential events (chemical reactions) in which each event is dependent on the occurrence of the preceding event. The general reasoning is as follows:

- *Observation*. Condition (1) occurs and condition (2) occurs; that is, (1) and (2) are *correlated*.

- *Hypothesis*. Condition (1) results in condition (2); that is, (1) *causes* (2).

In biochemical genetics, this can be stated as follows:

- *Observation*. A particular gene (*a*) is present and a particular reaction catalyzed by a particular enzyme (A) occurs; the two are correlated.

- *Hypothesis*. The gene (*a*) encodes (causes the synthesis of) the specific enzyme (A).

- *Test of hypothesis*. A mutant gene (*a'*) encodes a nonfunctional enzyme (A') and the reaction does not occur.

Beadle and Tatum treated wild-type *Neurospora* with X rays, which act as a **mutagen** (something known to cause mutations—inherited genotypic changes). When they tested the treated molds, they found that some mutant strains could no longer grow on the minimal medium, but grew only if they were supplied with additional nutrients, such as amino acids. The scientists hypothesized that these *auxotrophs* ("increased eaters") must have suffered mutations in genes that encoded the enzymes used to synthesize the nutrients that they now needed to obtain from their environment.

For each auxotrophic mutant strain, Beadle and Tatum were able to find a single compound that, when added to the minimal medium, supported the growth of that strain. These results suggested that mutations have simple effects, and that each mutation causes a defect in only one enzyme in a metabolic pathway. These conclusions confirmed Garrod's **one-gene, one-enzyme hypothesis** (**Figure 14.1**).

One group of auxotrophs, for example, could grow only if the minimal medium was supplemented with the amino acid arginine. These strains were designated *arg* mutants. Beadle and Tatum found several different *arg* mutant strains. They proposed two alternative hypotheses to explain why these different genetic strains had the same phenotype:

- The different *arg* mutants could have mutations in the same gene, as is the case for some eye color mutations in fruit flies. In this case, the gene might code for an enzyme involved in arginine synthesis.
- The different *arg* mutants could have mutations in different genes, each coding for a separate function that leads to arginine production. These independent functions might be different enzymes along the same biochemical pathway.

Some of the *arg* mutant strains fell into one of these two categories, and some into the other. Genetic crosses showed that some of the mutations were at the same chromosomal locus, and were different alleles of the same gene. Other mutations were at different loci, or on different chromosomes, and so were not alleles of the same gene. Beadle and Tatum concluded that these different genes participated in governing a single biosynthetic pathway—in this case, the pathway leading to arginine synthesis. Next, they set out to elucidate each step in this pathway (see the Interpretation in Figure 14.1).

By growing different *arg* mutants in the presence of various compounds suspected to be intermediates in the biosynthetic pathway for arginine, Beadle and Tatum were able to classify each mutation as affecting one enzyme or another, and to order the compounds along the pathway. Then they broke open the wild-type and mutant cells and examined them for enzyme activities. The results confirmed their hypothesis: each mutant strain was indeed missing a single enzyme activity in the pathway. In general, gene expression controls metabolism.

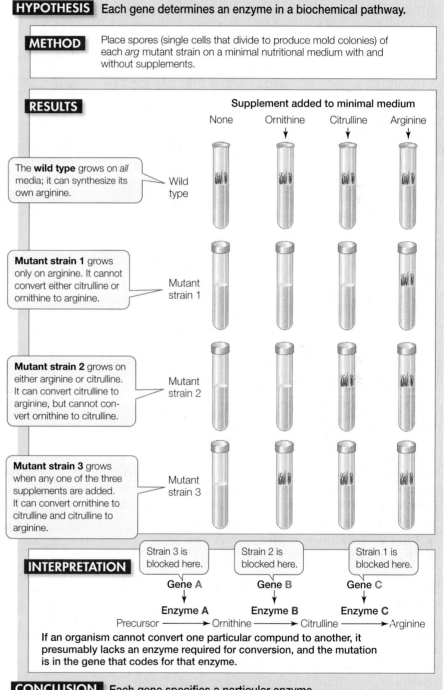

INVESTIGATING LIFE

14.1 One Gene, One Enzyme

Beadle and Tatum had several mutant strains of *Neurospora* that could not make arginine (*arg*). Several compounds are needed for arginine synthesis. By testing these compounds in the growth media for the mutant strains, the researchers deduced that each mutant strain was deficient in one enzyme along a biochemical pathway.

HYPOTHESIS Each gene determines an enzyme in a biochemical pathway.

METHOD Place spores (single cells that divide to produce mold colonies) of each *arg* mutant strain on a minimal nutritional medium with and without supplements.

RESULTS

Supplement added to minimal medium

None   Ornithine   Citrulline   Arginine

The **wild type** grows on *all* media; it can synthesize its own arginine. — Wild type

Mutant strain 1 grows only on arginine. It cannot convert either citrulline or ornithine to arginine. — Mutant strain 1

Mutant strain 2 grows on either arginine or citrulline. It can convert citrulline to arginine, but cannot convert ornithine to citrulline. — Mutant strain 2

Mutant strain 3 grows when any one of the three supplements are added. It can convert ornithine to citrulline and citrulline to arginine. — Mutant strain 3

INTERPRETATION

Strain 3 is blocked here. | Strain 2 is blocked here. | Strain 1 is blocked here.

Gene A → Enzyme A | Gene B → Enzyme B | Gene C → Enzyme C

Precursor ——→ Ornithine ——→ Citrulline ——→ Arginine

If an organism cannot convert one particular compund to another, it presumably lacks an enzyme required for conversion, and the mutation is in the gene that codes for that enzyme.

CONCLUSION Each gene specifies a particular enzyme.

FURTHER INVESTIGATION: If a diploid *Neurospora* spore were made from two haploid cells, one with mutant 3 and the other with mutant 2, what would be its phenotype?

Go to **yourBioPortal.com** for original citations, discussions, and relevant links for all INVESTIGATING LIFE figures.

One gene determines one polypeptide

The gene–enzyme relationship has undergone several modifications in light of our current knowledge of molecular biology. Many proteins, including many enzymes, are composed of more than one polypeptide chain, or subunit (that is, they have a quaternary structure; see Section 3.2). Look at the illustration of hemoglobin in Figure 3.10. This protein has four polypeptides—two α and two β subunits, and the different subunits are encoded by separate genes. Thus it is more correct to speak of a **one-gene, one-polypeptide relationship**.

So far we have seen that in terms of protein synthesis, the *function of a gene is to inform the production of a single, specific polypeptide*. But not all genes code for polypeptides. As we will see below and in Chapter 16, there are many DNA sequences that code for RNA molecules that are not translated into polypeptides but instead have other functions.

14.1 RECAP

Beadle and Tatum's studies of mutations in bread molds led to our understanding of the one-gene, one-polypeptide relationship. In most cases, the function of a gene is to code for a specific polypeptide.

- What is a model organism, and why is *Neurospora* a good model for studying biochemical genetics? **See p. 292**

- How were Beadle and Tatum's experiments on *Neurospora* set up to determine the order of steps in a biochemical pathway? **See pp. 292–293 and Figure 14.1**

- Explain the distinction between the phrases "one-gene, one-enzyme" and "one-gene, one-polypeptide." **See pp. 293–294**

Now that we have established the one-gene, one-polypeptide relationship, how does it work? That is, how is the information encoded in DNA used to produce a particular polypeptide?

14.2 How Does Information Flow from Genes to Proteins?

Much of the biochemical genetics in the middle of the twentieth century was directed at revealing the relationship between genes and protein synthesis. As we discussed in Section 14.1, the expression of a specific gene usually results in the synthesis of a specific polypeptide. The process of gene expression was outlined in Section 4.1. To review, this process occurs in two major steps:

- During **transcription**, the information in a DNA sequence (a gene) is copied into a complementary RNA sequence.

- During **translation**, this RNA sequence is used to create the amino acid sequence of a polypeptide.

In 1958, Francis Crick described this process as "the **central dogma** of molecular biology."

RNA differs from DNA and plays a vital role in gene expression

RNA (ribonucleic acid) is a key intermediary between a DNA sequence and a polypeptide. RNA is an informational polynucleotide similar to DNA (see Figure 4.2), but it differs from DNA in three ways:

- RNA generally consists of only one polynucleotide strand.

- The sugar molecule found in RNA is ribose, rather than the deoxyribose found in DNA.

- Although three of the nitrogenous bases (adenine, guanine, and cytosine) in RNA are identical to those in DNA, the fourth base in RNA is **uracil (U)**, which is similar to thymine but lacks the methyl (—CH_3) group.

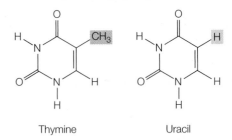

Thymine   Uracil

The bases in RNA can pair with those in a single strand of DNA. This pairing obeys the same complementary base-pairing rules as in DNA, except that adenine pairs with uracil instead of thymine. Single-stranded RNA can fold into complex shapes by internal base pairing, as seen below.

Three types of RNA participate in protein synthesis:

- **Messenger RNA (mRNA)** carries a copy of a gene sequence in DNA to the site of protein synthesis at the ribosome.

- **Transfer RNA (tRNA)** carries amino acids to the ribosome for assembly into polypeptides.

- **Ribosomal RNA (rRNA)** catalyzes peptide bond formation and provides a structural framework for the ribosome.

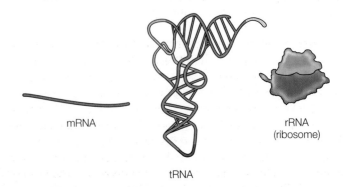

mRNA

rRNA
(ribosome)

tRNA

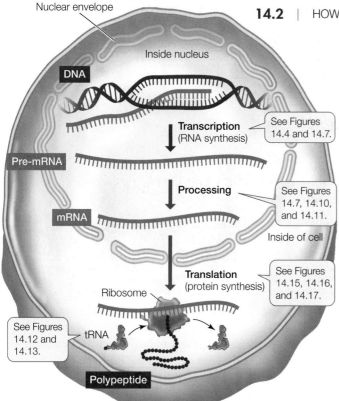

Nuclear envelope

Inside nucleus

DNA

Transcription
(RNA synthesis)

See Figures
14.4 and 14.7.

Pre-mRNA

Processing

See Figures
14.7, 14.10,
and 14.11.

mRNA

Inside of cell

Translation
(protein synthesis)

See Figures
14.15, 14.16,
and 14.17.

Ribosome

See Figures
14.12 and
14.13.

tRNA

Polypeptide

14.2 From Gene to Protein This diagram summarizes the processes of gene expression in eukaryotes.

yourBioPortal.com

GO TO **Web Activity 14.1 • Eukaryotic Gene Expression**

Two hypotheses were proposed to explain information flow from DNA to protein

The central dogma suggested that information flows from DNA to RNA to protein, and not in the reverse direction. It raised two questions:

• How does genetic information get from the nucleus of a eukaryotic cell to the cytoplasm? (As Section 5.3 explains, most of the DNA of a eukaryotic cell is confined to the nucleus, but proteins are synthesized in the cytoplasm.)

• What is the relationship between a specific nucleotide sequence in DNA and a specific amino acid sequence in a protein?

To answer these questions, Crick proposed two hypotheses, the messenger hypothesis and the adapter hypothesis.

THE MESSENGER HYPOTHESIS AND TRANSCRIPTION Crick and his colleagues proposed that an RNA molecule forms as a complementary copy of one DNA strand in a gene. This messenger RNA, or mRNA, then travels from the nucleus to the cytoplasm, where it serves as an informational sequence of **codons**. Each codon consists of three consecutive nucleotides, and different codons encode particular amino acids. Thus the mRNA sequence determines the ordered sequence of amino acids in a polypeptide chain, which is built by the ribosome. The process by which RNA forms is called transcription (**Figure 14.2**).

This hypothesis has been tested repeatedly, and the result is always the same: each DNA sequence that encodes a protein is transcribed as a sequence of mRNA. Today it is routine in thousands of laboratories around the world to test for gene expression by examining the mRNA copy of the gene, which is often called the **transcript**. There is no longer any question that Crick's model was correct.

THE ADAPTER HYPOTHESIS AND TRANSLATION To answer the question of how a DNA sequence gets transformed into the specific amino acid sequence of a polypeptide, Crick proposed the adapter hypothesis: that there must be an adapter molecule that can both bind a specific amino acid and recognize a specific sequence of nucleotides. He proposed that this recognition function occurs because the adapter molecule contains an **anticodon** complementary to the codon in the mRNA. He envisioned such adapters as molecules with two regions, one serving the binding function and the other serving the recognition function.

In due course, such adapter molecules were found: they are known as transfer RNA, or tRNA. Each tRNA recognizes a specific codon in the mRNA and simultaneously carries the specific amino acid corresponding to that codon. Thus, the tRNAs together can translate the language of DNA into the language of proteins. The tRNA adapters, carrying bound amino acids, line up on the mRNA sequence so that the amino acids are in the proper sequence for a growing polypeptide chain—in the process of *translation* (see Figure 14.2). Once again, actual observations of the expression of thousands of genes in all types of organisms have confirmed the hypothesis that tRNA acts as the intermediary between the nucleotide sequence information in mRNA and the amino acid sequence in a protein.

We can summarize the main features of the central dogma, the messenger hypothesis, and the adapter hypothesis as follows: a given gene is transcribed to produce an mRNA molecule that is complementary to one of the DNA strands, and then the tRNA molecules translate the sequence of codons in the mRNA into a sequence of linked amino acids, to form a polypeptide.

RNA viruses are exceptions to the central dogma

Certain viruses present exceptions to the central dogma. As we saw in Section 13.1, a virus is a non-cellular infectious particle that reproduces inside cells. Many viruses, such as the tobacco mosaic virus, influenza viruses, and poliovirus, have *RNA* rather than DNA as their genetic material. With its nucleotide sequence, RNA could potentially act as an information carrier and be expressed as a protein. But if RNA is usually single-stranded, how do these viruses replicate? They generally solve this problem by transcribing from RNA to RNA, making an RNA strand that is complementary to their genomes. This "opposite" strand is then used to make multiple copies of the viral genome by transcription:

RNA ⇌ RNA → Polypeptide

Human immunodeficiency viruses (HIV) and certain rare tumor viruses also have RNA as their genomes, but do not replicate by transcribing from RNA to RNA. Instead, after infecting a host cell, such a virus makes a DNA copy of its genome, which becomes incorporated into the host's genome. The virus relies on the host cell's transcription machinery to make more RNA. This RNA can be either translated to produce viral proteins, or incorporated as the viral genome into new viral particles. Synthesis of DNA from RNA is called **reverse transcription**, and not surprisingly, such viruses are called **retroviruses**.

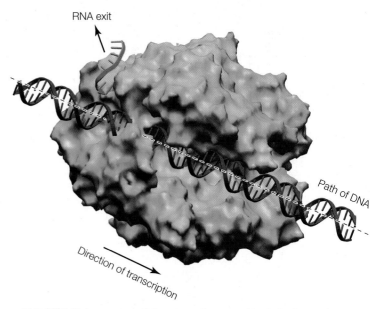

14.3 RNA Polymerase   This enzyme from yeast is similar to most other RNA polymerases. Note the size relationship between enzyme and DNA. See Figure 14.4 for details.

14.2 RECAP

The central dogma of molecular biology states that the DNA code is used to produce RNA and the RNA sequence determines the sequence of amino acids in a polypeptide. Transcription is the process by which a DNA sequence is copied into mRNA. Translation is the process by which this information is converted into polypeptide chain. Transfer RNAs recognize the genetic information in messenger RNA and bring the appropriate amino acids into position in a growing polypeptide chain.

- What is the central dogma of molecular biology? **See p. 294**

- What are the roles of mRNA and tRNA in gene expression? **See p. 294 and Figure 14.2**

The central dogma is indeed central to gene expression in all organisms. Understanding its details is essential for understanding how organisms function at the molecular level, and this understanding is key to the application of biology to human welfare, in areas such as agriculture and medicine. Much of the remainder of this book will in one way or another involve DNA and proteins. Let's begin by describing how the information in DNA is transcribed to produce RNA.

14.3 How Is the Information Content in DNA Transcribed to Produce RNA?

In normal prokaryotic and eukaryotic cells, RNA synthesis is directed by DNA. Transcription—the formation of a specific RNA sequence from a specific DNA sequence—requires several components:

- A DNA template for complementary base pairing; one of the two strands of DNA

- The appropriate nucleoside triphosphates (ATP, GTP, CTP, and UTP) to act as substrates

- An RNA polymerase enzyme

Not only mRNA is produced by transcription. The same process is responsible for the synthesis of tRNA and ribosomal RNA (rRNA), whose important roles in protein synthesis will be described below. Like polypeptides, these RNAs are encoded by specific genes. In addition, as we will see in Chapter 16,

many small RNAs, called microRNAs, are transcribed. These molecules stay in the nucleus, where they play roles in stimulating or inhibiting gene expression.

RNA polymerases share common features

RNA polymerases from both prokaryotes and eukaryotes catalyze the synthesis of RNA from the DNA template. There is only one kind of RNA polymerase in bacteria, while there are several kinds in eukaryotes; however, they all share a common structure (**Figure 14.3**). Like DNA polymerases, RNA polymerases are *processive*; that is, a single enzyme–template binding event results in the polymerization of hundreds of RNA bases. But unlike DNA polymerases, RNA polymerases *do not require a primer* and *do not have a proofreading function*.

Transcription occurs in three steps

Transcription can be divided into three distinct processes: initiation, elongation, and termination. You can follow these processes in **Figure 14.4**.

INITIATION   Transcription begins with initiation, which requires a **promoter**, a special sequence of DNA to which the RNA polymerase binds very tightly (see Figure 14.4A). Eukaryotic genes generally have one promoter each, while in prokaryotes and viruses, several genes often share one promoter. Promoters are important control sequences that "tell" the RNA polymerase two things:

- Where to start transcription

- Which strand of DNA to transcribe

A promoter, which is a specific sequence in the DNA that reads in a particular direction, orients the RNA polymerase and thus "aims" it at the correct strand to use as a template. Promoters

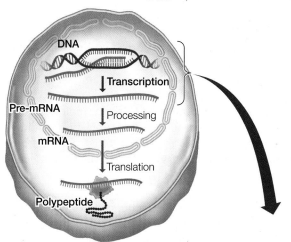

14.4 DNA Is Transcribed to Form RNA DNA is partially unwound by RNA polymerase to serve as a template for RNA synthesis. The RNA transcript is formed and then peels away, allowing the DNA that has already been transcribed to rewind into a double helix. Three distinct processes—initiation, elongation, and termination—constitute DNA transcription. RNA polymerase is much larger in reality than indicated here, covering about 50 base pairs.

yourBioPortal.com

GO TO Animated Tutorial 14.1 • Transcription

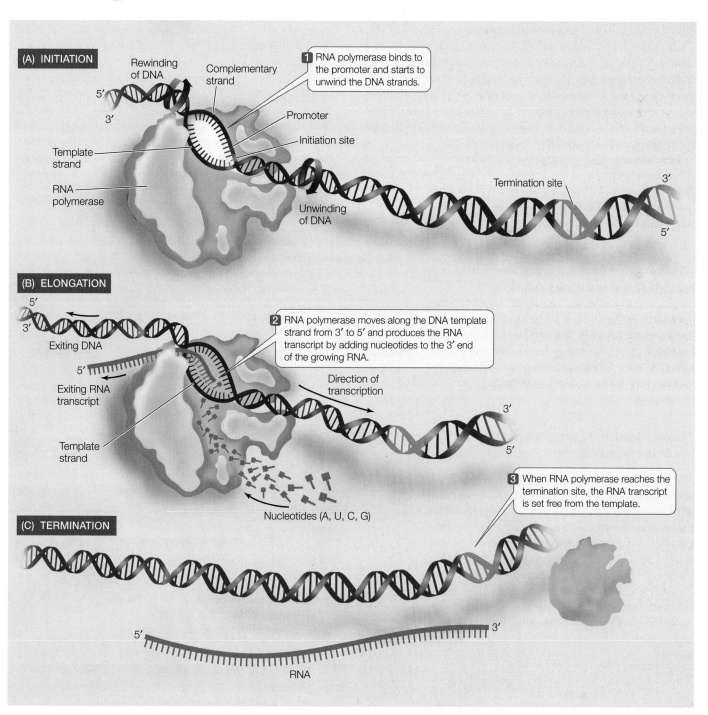

(A) INITIATION

Rewinding of DNA

Complementary strand

1 RNA polymerase binds to the promoter and starts to unwind the DNA strands.

Promoter

Initiation site

Template strand

RNA polymerase

Termination site

Unwinding of DNA

(B) ELONGATION

Exiting DNA

Exiting RNA transcript

2 RNA polymerase moves along the DNA template strand from 3′ to 5′ and produces the RNA transcript by adding nucleotides to the 3′ end of the growing RNA.

Direction of transcription

Template strand

3 When RNA polymerase reaches the termination site, the RNA transcript is set free from the template.

Nucleotides (A, U, C, G)

(C) TERMINATION

RNA

function somewhat like the capital letter at the beginning of a sentence, indicating how the sequence of words should be read. Part of each promoter is the **initiation site**, where transcription begins. Groups of nucleotides lying "upstream" from the initiation site (5' on the non-template strand, and 3' on the template strand) help the RNA polymerase bind.

Although every gene has a promoter, not all promoters are identical. Some promoters are more effective at transcription initiation than others. Furthermore, there are differences between transcription initiation in prokaryotes and in eukaryotes. But despite these variations, the basic mechanisms of initiation are the same throughout the living world.

ELONGATION Once RNA polymerase has bound to the promoter, it begins the process of **elongation** (see Figure 14.4B). RNA polymerase unwinds the DNA about 10 base pairs at a time and reads the template strand in the 3'-to-5' direction. Like DNA polymerase, RNA polymerase adds new nucleotides to the 3' end of the growing strand, but does not require a primer to get this process started. The new RNA elongates from the first base, which forms its 5' end, to its 3' end. The RNA transcript is thus antiparallel to the DNA template strand.

Because RNA polymerases do not proofread, transcription errors occur at a rate of one for every 10^4 to 10^5 bases. Because many copies of RNA are made, however, and because they often have only a relatively short life span, these errors are not as potentially harmful as mutations in DNA.

TERMINATION Just as initiation sites in the DNA template strand specify the starting point for transcription, particular base sequences specify its **termination** (see Figure 14.4C). The mechanisms of termination are complex and of more than one kind. For some genes, the newly formed transcript falls away from the DNA template and the RNA polymerase. For others, a helper protein pulls the transcript away.

The information for protein synthesis lies in the genetic code

The **genetic code** relates genes (DNA) to mRNA and mRNA to the amino acids that make up proteins. The genetic code specifies which amino acids will be used to build a protein. You can think of the genetic information in an mRNA molecule as a series of sequential, nonoverlapping three-letter "words." Each sequence of three nucleotide bases (the three "letters") along the mRNA polynucleotide chain specifies a particular amino acid. Each three-letter "word" is called a **codon**. Each

codon is complementary to the corresponding triplet of bases in the DNA molecule from which it was transcribed. The genetic code relates codons to their specific amino acids.

CHARACTERISTICS OF THE CODE Molecular biologists "broke" the genetic code in the early 1960s. The problem they addressed was perplexing: how could more than 20 "code words" be written with an "alphabet" consisting of only four "letters"? In other words, how could four bases (A, U, G, and C) code for 20 different amino acids?

A triplet code, based on three-letter codons, was considered likely. Since there are only four letters (A, G, C, and U), a one-letter code clearly could not unambiguously encode 20 amino acids; it could encode only four of them. A two-letter code could have only $4 \times 4 = 16$ unambiguous codons—still not enough. But a triplet code could have $4 \times 4 \times 4 = 64$ codons, more than enough to encode the 20 amino acids.

Marshall W. Nirenberg and J. H. Matthaei, at the U.S. National Institutes of Health, made the first decoding breakthrough in 1961 when they realized that they could use a simple artificial polynucleotide instead of a complex natural mRNA as a messenger. They could then identify the polypeptide that the artificial messenger encoded. This led to the identification of the first three codons (**Figure 14.5**).

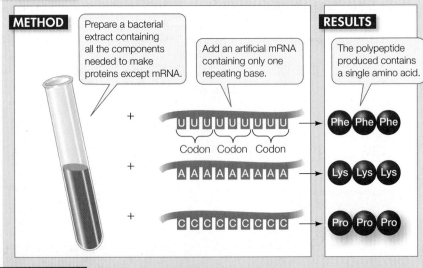

INVESTIGATING LIFE

14.5 Deciphering the Genetic Code
Nirenberg and Matthaei used a test-tube protein synthesis system to determine the amino acids specified by synthetic mRNAs of known codon compositions.

HYPOTHESIS A triplet codon based on three-base codons specifies amino acids.

METHOD Prepare a bacterial extract containing all the components needed to make proteins except mRNA.

Add an artificial mRNA containing only one repeating base.

RESULTS The polypeptide produced contains a single amino acid.

U U U U U U U U U
Codon Codon Codon → Phe Phe Phe

A A A A A A A A A → Lys Lys Lys

C C C C C C C C C → Pro Pro Pro

CONCLUSION UUU is an mRNA codon for phenylalanine.
AAA is an mRNA codon for lysine.
CCC is an mRNA codon for proline.

FURTHER INVESTIGATION: What would be the result if the artificial mRNA were poly-G?

yourBioPortal.com
GO TO Animated Tutorial 14.2 • Deciphering the Genetic Code

Go to **yourBioPortal.com** for original citations, discussions, and relevant links for all INVESTIGATING LIFE figures.

Other scientists later found that simple artificial mRNAs only three nucleotides long—each amounting to one codon—could bind to a ribosome, and that the resulting complex could then bind to the corresponding tRNA with its specific amino acid. Thus, for example, a simple UUU mRNA caused the tRNA carrying phenyl-alanine to bind to the ribosome. After this discovery, the complete deciphering of the genetic code was relatively simple. To discover which amino acid a codon represented, the scientists simply re-peated the experiment using a sample of artificial mRNA for that codon, and observed which amino acid became bound to it.

The complete genetic code is shown in **Figure 14.6**. Notice that there are many more codons than there are different amino acids in proteins. All possible combinations of the four avail-able "letters" (the bases) give 64 (4^3) different three-letter codons, yet these codons determine only 20 amino acids. AUG, which codes for methionine, is also the **start codon**, the initia-tion signal for translation. Three of the codons (UAA, UAG, UGA) are **stop codons**, or termination signals for translation. When the translation machinery reaches one of these codons, translation stops, and the polypeptide is released from the trans-lation complex.

What happens if a stop codon isn't there? In humans a se-vere anemic condition, α-thalassemia, results from various mu-tations in either of the two genes that encode the α-polypeptide chain of hemoglobin. In one of these mutant alleles, the stop codon UAA has been converted to GAA, a codon for glutamine. The next stop codon doesn't occur until much further along the mRNA, resulting in a protein molecule with larger, defective α subunits.

THE GENETIC CODE IS REDUNDANT BUT NOT AMBIGUOUS The 60 codons that are not start or stop codons are far more than enough to code for the other 19 amino acids—and indeed, for almost all amino acids, there is more than one codon. Thus we say that the genetic code is redundant (or degenerate). For ex-

ample, leucine is represented by six different codons (see Fig-ure 14.6). Only methionine and tryptophan are represented by just one codon each.

A *redundant* code should not be confused with an *ambiguous* code. If the code were ambiguous, a single codon could specify either of two (or more) different amino acids, and there would be doubt about which amino acid should be incorporated into a growing polypeptide chain. Redundancy in the code simply means that there is more than one clear way to say, "Put leucine here." The genetic code is not ambiguous: a given amino acid may be encoded by more than one codon, but a codon can code for only one amino acid.

THE GENETIC CODE IS (NEARLY) UNIVERSAL The same genetic code is used by all the species on our planet. Thus the code must be an ancient one that has been maintained intact throughout the evolution of living organisms. Exceptions are known: within mi-tochondria and chloroplasts, the code differs slightly from that in prokaryotes and in the nuclei of eukaryotic cells; and in one group of protists, UAA and UAG code for glutamine rather than functioning as stop codons. The significance of these differences is not yet clear. What is clear is that the exceptions are few.

The common genetic code means that there is also a common language for evolution. Natural selection acts on phenotypic variations that result from genetic variation. The genetic code probably originated early in the evolution of life. As we saw in Chapter 4, simulation experiments indicate the plausibility of individual nucleotides and nucleotide polymers arising spon-taneously on the primeval Earth. The common code also has profound implications for genetic engineering, as we will see in Chapter 18, since it means that the code for a human gene is the same as that for a bacterial gene. It is therefore impressive, but not surprising, that a human gene can be expressed in *E. coli* via laboratory manipulations, since these cells speak the same "molecular language."

The codons in Figure 14.6 are mRNA codons. The base sequence of the DNA strand that is tran-scribed to produce the mRNA is complementary and antiparallel to these codons. Thus, for example,

- 3'-AAA-5' in the template DNA strand corre-sponds to phenylalanine (which is encoded by the mRNA codon 5'-UUU-3')

Second letter

	U	C	A	G	
U	UUU UUC Phenyl-alanine / UUA UUG Leucine	UCU UCC UCA UCG Serine	UAU UAC Tyrosine / UAA Stop codon UAG Stop codon	UGU UGC Cysteine / UGA Stop codon UGG Tryptophan	U C A G
C	CUU CUC CUA CUG Leucine	CCU CCC CCA CCG Proline	CAU CAC Histidine / CAA CAG Glutamine	CGU CGC CGA CGG Arginine	U C A G
A	AUU AUC AUA Isoleucine / AUG Methionine; start codon	ACU ACC ACA ACG Threonine	AAU AAC Asparagine / AAA AAG Lysine	AGU AGC Serine / AGA AGG Arginine	U C A G
G	GUU GUC GUA GUG Valine	GCU GCC GCA GCG Alanine	GAU GAC Aspartic acid / GAA GAG Glutamic acid	GGU GGC GGA GGG Glycine	U C A G

First letter (left side) / Third letter (right side)

14.6 The Genetic Code Genetic information is encod-ed in mRNA in three-letter units—codons—made up of nucleoside monophosphates with the bases uracil (U), cytosine (C), adenine (A), and guanine (G) and is read in a 5' to 3' direction on mRNA. To decode a codon, find its first letter in the left column, then read across the top to its second letter, then read down the right column to its third letter. The amino acid the codon specifies is given in the corresponding row. For example, AUG codes for methionine, and GUA codes for valine.

— **yourBioPortal.com** —
GO TO **Web Activity 14.2** • The Genetic Code

- 3′-ACC-5′ in the template DNA corresponds to tryptophan (which is encoded by the mRNA codon 5′-UGG-3′)

The non-template strand has the same sequence as the mRNA (but with T's instead of U's), and is often referred to as the "coding strand." By convention, DNA sequences are usually shown beginning with the 5′ end of the coding sequence.

The features of transcription that we have described were first elucidated in model prokaryotes, such as *E. coli*. Biologists then used the same methods to analyze this process in eukaryotes, and, although the basics are the same, there are some notable (and important) differences. We now turn to eukaryotic gene expression.

14.4 How Is Eukaryotic DNA Transcribed and the RNA Processed?

Since the genetic code is the same, you might expect the process of gene expression to be the same in eukaryotes as it is in prokaryotes. And basically it is. However, there are significant differences in gene structure between prokaryotes and eukaryotes, that is, there are differences in the organization of the nucleotide sequences in the genes. In addition, in eukaryotes but not prokaryotes, a nucleus separates transcription and translation (**Table 14.1**). Let's look at the distinctive *eukaryotic* process of transcription.

Eukaryotic genes have noncoding sequences

A diagram of the structure and transcription of a typical eukaryotic gene is shown in **Figure 14.7**. In prokaryotes, several adjacent genes sometimes share one promoter; however, in eukaryotes, each gene has its own promoter, which usually precedes the coding region. Unlike the prokaryotic RNA polymerase, a eukaryotic RNA polymerase does not recognize the promoter sequence by itself, but requires help from other molecules, as we'll see in more detail in Chapter 16. At the other end of the gene, downstream from the coding region, is a DNA sequence appropriately called the **terminator**, which signals the end of transcription.

Eukaryotic genes may also contain noncoding base sequences, called **introns** (*int*ervening *regi*ons). One or more introns may be interspersed with the coding sequences, which are called **exons** (*ex*pressed *regi*ons). Both introns and exons appear in the primary mRNA transcript, called **pre-mRNA**, but the introns are removed by the time the mature mRNA—the mRNA that will be translated—leaves the nucleus. Pre-mRNA processing involves cutting introns out of the pre-mRNA transcript and splicing together the remaining exon transcripts (see Figure 14.7). If this seems surprising, you are in good company. For scientists who were familiar with prokaryotic genes and gene expression, the discovery of introns in eukaryotic genes was entirely unexpected.

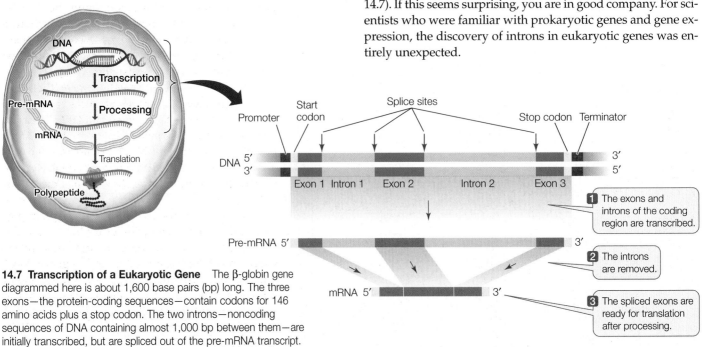

14.7 Transcription of a Eukaryotic Gene The β-globin gene diagrammed here is about 1,600 base pairs (bp) long. The three exons—the protein-coding sequences—contain codons for 146 amino acids plus a stop codon. The two introns—noncoding sequences of DNA containing almost 1,000 bp between them—are initially transcribed, but are spliced out of the pre-mRNA transcript.

1 The exons and introns of the coding region are transcribed.

2 The introns are removed.

3 The spliced exons are ready for translation after processing.

TABLE 14.1		
Differences between Prokaryotic and Eukaryotic Gene Expression		
CHARACTERISTIC	PROKARYOTES	EUKARYOTES
Transcription and translation occurrence	At the same time in the cytoplasm	Transcription in the nucleus, then translation in the cytoplasm
Gene structure	DNA sequence is read in the same order as the amino acid sequence	Noncoding introns within coding sequence
Modification of mRNA after initial transcription but before translation	None	Introns spliced out; 5′ cap and 3′ poly A added

How can we locate introns within a eukaryotic gene? One way is by **nucleic acid hybridization**, the method that originally revealed the existence of introns. This method, outlined in **Figure 14.8**, has been crucial for studying the relationship between eukaryotic genes and their transcripts. It involves two steps:

• The target DNA is denatured by heat to break the hydrogen bonds between the base pairs and separate the two strands.

• A single-stranded nucleic acid from another source (called a **probe**) is incubated with the denatured DNA. If the probe has a base sequence complementary to the target DNA, a probe–target double helix forms by hydrogen bonding between the bases. Because the two strands are from different sources, the resulting double-stranded molecule is called a hybrid.

Biologists used nucleic acid hybridization to examine the β-globin gene, which encodes one of the globin polypeptides that make up hemoglobin. Follow the experiment in **Figure 14.9** carefully as we describe what they did and what happened.

The researchers first denatured DNA containing the β-globin gene by heating it slowly, then added previously isolated, mature β-globin mRNA. They were able to view the hybridized molecules using electron microscopy. As expected, the mRNA bound to the DNA by complementary base pairing. The researchers expected to obtain a linear (1:1) matchup of the mRNA to the coding DNA. That expectation was only partly met: there were indeed stretches of RNA–DNA hybrid, but some looped structures were also visible. These loops were not expected, and initially the scientists thought that something must be wrong with the experimental procedure. However, when they repeated the experiment, they got the same results, and when they did it with other genes and mRNAs, loops again appeared. The loops turned out to be the introns, stretches of DNA that did not have complementary base sequences on the mature mRNA.

When pre-mRNA was used instead of mature mRNA to hybridize to the DNA, there was complete hybridization, revealing that the introns were indeed part of the pre-mRNA transcript. Somewhere on the path from primary transcript (pre-mRNA) to mature mRNA, the introns had been removed, and the exons had been spliced together. We will examine this splicing process in the next section.

Introns *interrupt, but do not scramble*, the DNA sequence of a gene. The base sequences of the exons in the template strand, if joined and taken in order, form a continuous sequence that is complementary to that of the mature mRNA. In some cases, the separated exons encode different functional regions, or **domains**, of the protein. For example, the globin polypeptides that make up hemoglobin each have two domains: one for binding to a nonprotein pigment called heme, and another for binding to the other globin subunits. These two domains are encoded by different exons in the globin genes. Most (but not all) eukaryotic genes contain introns, and in rare cases, introns are also found in prokaryotes. The largest human gene encodes a muscle protein called titin; it has 363 exons, which together code for 38,138 amino acids.

TOOLS FOR INVESTIGATING LIFE

14.8 Nucleic Acid Hybridization
Base pairing permits the detection of a sequence that is complementary to the probe.

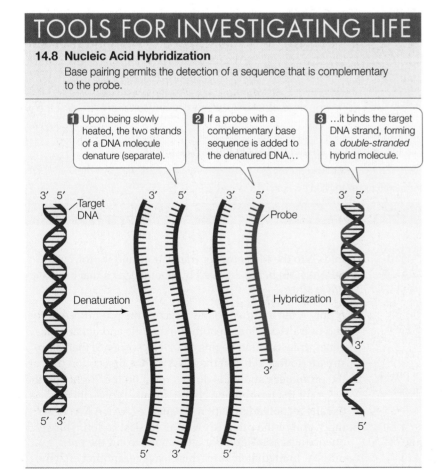

1 Upon being slowly heated, the two strands of a DNA molecule denature (separate).

2 If a probe with a complementary base sequence is added to the denatured DNA…

3 …it binds the target DNA strand, forming a *double-stranded* hybrid molecule.

INVESTIGATING LIFE

14.9 Demonstrating the Existence of Introns

When an mRNA transcript of the ß-globin gene was hybridized with the double-stranded DNA of that gene, the introns in the DNA "looped out." This demonstrated that the coding region of a eukaryotic gene can contain noncoding DNA that is not present in the mature mRNA transcript.

HYPOTHESIS Some regions within the coding sequence of a gene do not end up in its mRNA.

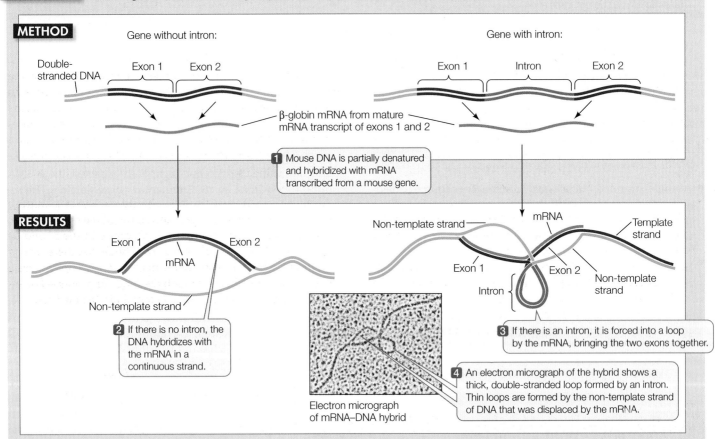

METHOD

Gene without intron:

Double-stranded DNA · Exon 1 · Exon 2

ß-globin mRNA from mature mRNA transcript of exons 1 and 2

Gene with intron:

Exon 1 · Intron · Exon 2

1 Mouse DNA is partially denatured and hybridized with mRNA transcribed from a mouse gene.

RESULTS

Exon 1 · Exon 2 · mRNA

Non-template strand

2 If there is no intron, the DNA hybridizes with the mRNA in a continuous strand.

Non-template strand — mRNA — Template strand

Exon 1 · Exon 2 · Non-template strand · Intron

3 If there is an intron, it is forced into a loop by the mRNA, bringing the two exons together.

4 An electron micrograph of the hybrid shows a thick, double-stranded loop formed by an intron. Thin loops are formed by the non-template strand of DNA that was displaced by the mRNA.

Electron micrograph of mRNA–DNA hybrid

CONCLUSION The DNA contains noncoding regions within the genes that are not present in the mature mRNA.

FURTHER INVESTIGATION: Draw the result assuming that there were three exons and two introns.

Go to **yourBioPortal.com** for original citations, discussions, and relevant links for all INVESTIGATING LIFE figures.

Eukaryotic gene transcripts are processed before translation

The primary transcript of a eukaryotic gene is modified in several ways before it leaves the nucleus: both ends of the pre-mRNA are modified, and the introns are removed.

MODIFICATION AT BOTH ENDS Two steps in the processing of pre-mRNA take place in the nucleus, one at each end of the molecule (**Figure 14.10**):

• A **G cap** is added to the 5′ end of the pre-mRNA as it is transcribed. The G cap is a chemically modified molecule of guanosine triphosphate (GTP). It facilitates the binding of

mRNA to the ribosome for translation, and it protects the mRNA from being digested by ribonucleases that break down RNAs.

• A **poly A tail** is added to the 3′ end of the pre-mRNA at the end of transcription. In both prokaryotic and eukaryotic genes, transcription begins at a DNA sequence that is upstream (to the "left" on the DNA) of the first codon (i.e., at the promoter), and ends downstream (to the "right" on the DNA) of the termination codon. In eukaryotes, there is usually a "polyadenylation" sequence (AAUAAA) near the 3′ end of the pre-mRNA, after the last codon. This sequence acts as a signal for an enzyme to cut the pre-mRNA. Immediately after this cleavage, another enzyme

14.10 Processing the Ends of Eukaryotic Pre-mRNA Modifications at each end of the pre-mRNA transcript—the G cap and the poly A tail—are important for mRNA function.

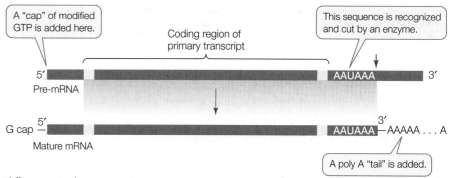

A "cap" of modified GTP is added here.

Coding region of primary transcript

This sequence is recognized and cut by an enzyme.

5′ —————— AAUAAA —— 3′
Pre-mRNA

G cap — 5′ ————— AAUAAA—AAAAA . . . A 3′
Mature mRNA

A poly A "tail" is added.

adds 100 to 300 adenine nucleotides (a "poly A" sequence) to the 3′ end of the pre-mRNA. This "tail" may assist in the export of the mRNA from the nucleus and is important for mRNA stability.

SPLICING TO REMOVE INTRONS The next step in the processing of eukaryotic pre-mRNA within the nucleus is removal of the introns. If these RNA sequences were not removed, a very different amino acid sequence, and possibly a nonfunctional protein, would result. A process called **RNA splicing** removes the introns and splices the exons together.

As soon as the pre-mRNA is transcribed, several **small nuclear ribonucleoprotein particles (snRNPs)** bind at each end. There are several types of these RNA–protein particles in the nucleus.

At the boundaries between introns and exons are **consensus sequences**—short stretches of DNA that appear, with little variation ("consensus"), in many different genes. The RNA in one of the snRNPs has a stretch of bases complementary to the consensus sequence at the 5′ exon–intron boundary, and it binds to the pre-mRNA by complementary base pairing. Another snRNP binds to the pre-mRNA near the 3′ intron–exon boundary (**Figure 14.11**).

Next, using energy from adenosine triphosphate (ATP), proteins are added to form a large RNA–protein complex called a **spliceosome**. This complex cuts the pre-mRNA, releases the introns, and joins the ends of the exons together to produce mature mRNA.

Molecular studies of human genetic diseases have provided insights into intron consensus sequences and splicing machinery. For example, people with the genetic disease β-thalassemia, like those with α-thalassemia discussed earlier in the chapter, have a defect in the production of one of the

hemoglobin subunits. These people suffer from severe anemia because they have an inadequate supply of red blood cells. In some cases, the genetic mutation that causes the disease occurs at an intron consensus sequence in the β-globin gene. Consequently, β-globin pre-mRNA cannot be spliced correctly, and nonfunctional β-globin mRNA is made. This finding offers another example of how biologists can use mutations to elucidate cause-and-effect relationships.

After processing is completed in the nucleus, the mature mRNA moves out into the cytoplasm through the nuclear pores. In the nucleus, a protein called TAP binds to the 5′ end of processed mRNA. This protein in turn binds to others, which are recognized by a receptor at the nuclear pore. Together, these proteins lead the mRNA through the pore. Unprocessed or incompletely processed pre-mRNAs remain in the nucleus.

yourBioPortal.com
GO TO Animated Tutorial 14.3 • RNA Splicing

14.11 The Spliceosome: An RNA Splicing Machine The binding of snRNPs to consensus sequences bordering the introns on the pre-mRNA results in a series of proteins binding and forming a large complex called a spliceosome. This structure determines the exact position of each cut in the pre-mRNA with great precision.

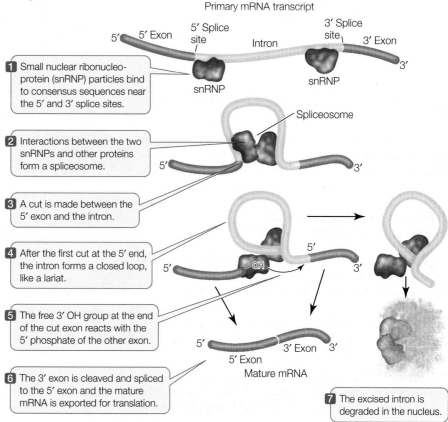

Primary mRNA transcript

5′ Exon   5′ Splice site   Intron   3′ Splice site   3′ Exon

snRNP   snRNP

1 Small nuclear ribonucleoprotein (snRNP) particles bind to consensus sequences near the 5′ and 3′ splice sites.

2 Interactions between the two snRNPs and other proteins form a spliceosome.

Spliceosome

3 A cut is made between the 5′ exon and the intron.

4 After the first cut at the 5′ end, the intron forms a closed loop, like a lariat.

5 The free 3′ OH group at the end of the cut exon reacts with the 5′ phosphate of the other exon.

6 The 3′ exon is cleaved and spliced to the 5′ exon and the mature mRNA is exported for translation.

5′ Exon   3′ Exon
Mature mRNA

7 The excised intron is degraded in the nucleus.

Transcription and post-transcriptional events produce an mRNA that is ready to be translated into a sequence of amino acids in a polypeptide. We turn now to the events of translation.

14.5 How Is RNA Translated into Proteins?

As Crick's adapter hypothesis proposed, the translation of mRNA into proteins requires a molecule that links the information contained in mRNA codons with specific amino acids in proteins. That function is performed by transfer RNA (tRNA). Two key events must take place to ensure that the protein made is the one specified by the mRNA:

- The tRNAs must read mRNA codons correctly.

- The tRNAs must deliver the amino acids that correspond to each mRNA codon.

Once the tRNAs "decode" the mRNA and deliver the appropriate amino acids, components of the ribosome catalyze the formation of peptide bonds between amino acids. We now turn to these two steps.

─── **yourBioPortal.com** ───
GO TO Animated Tutorial 14.4 • Protein Synthesis

Transfer RNAs carry specific amino acids and bind to specific codons

A codon in mRNA and an amino acid in a protein are related by way of an adapter—a specific tRNA with an attached amino acid. For each of the 20 amino acids, there is at least one specific type ("species") of tRNA molecule. The tRNA molecule has three functions:

- It binds to a particular amino acid. When it is carrying an amino acid, the tRNA is said to be "charged."

- It associates with mRNA.

- It interacts with ribosomes.

The tRNA molecular structure relates clearly to all of these functions. The molecule has about 75 to 80 nucleotides. It has a conformation (a three-dimensional shape) that is maintained by complementary base pairing (hydrogen bonding) between bases within its own sequence (**Figure 14.12**).

The conformation of a tRNA molecule is exquisitely suited for its interaction with specific binding sites on ribosomes. In addition, at the 3′ end of every tRNA molecule is its amino acid attachment site: a site to which its specific amino acid binds co-

14.12 Transfer RNA The stem and loop structure of a tRNA molecule is well suited to its functions: binding to amino acids, associating with mRNA molecules, and interacting with ribosomes.

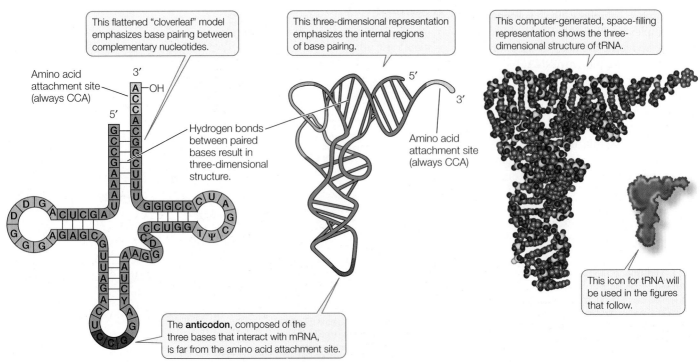

This flattened "cloverleaf" model emphasizes base pairing between complementary nucleotides.

This three-dimensional representation emphasizes the internal regions of base pairing.

This computer-generated, space-filling representation shows the three-dimensional structure of tRNA.

Amino acid attachment site (always CCA)

Hydrogen bonds between paired bases result in three-dimensional structure.

Amino acid attachment site (always CCA)

This icon for tRNA will be used in the figures that follow.

The **anticodon**, composed of the three bases that interact with mRNA, is far from the amino acid attachment site.

valently. At about the midpoint of the tRNA sequence is a group of three bases, called the anticodon, which is the site of complementary base pairing (via hydrogen bonding) with the codon on the mRNA. Thus, each tRNA species has a unique anticodon that corresponds to the amino acid it carries. When the tRNA and the mRNA come into contact on the surface of the ribosome, the codon and anticodon are antiparallel, permitting hydrogen bonding to occur between the complementary bases. As an example of this process, consider the amino acid arginine:

- The template strand DNA sequence that codes for arginine is 3'-GCC-5', which is transcribed, by complementary base pairing, to produce the mRNA codon 5'-CGG-3'
- That mRNA codon binds by complementary base pairing to a tRNA with the anticodon 3'-GCC-5', which is charged with arginine.

Recall that 61 different codons encode the 20 amino acids in proteins (see Figure 14.6). Does this mean that the cell must produce 61 different tRNA species, each with a different anticodon? No. The cell gets by with about two-thirds of that number of tRNA species because the specificity for the base at the 3' end of the codon (and the 5' end of the anticodon) is not always strictly observed. This phenomenon, called *wobble*, allows the alanine codons GCA, GCC, and GCU, for example, all to be recognized by the same tRNA. Wobble is allowed in some matches but not in others; of most importance, it does not allow the genetic code to be ambiguous. That is, each mRNA codon binds to just one tRNA species, carrying a specific amino acid.

Activating enzymes link the right tRNAs and amino acids

The charging of each tRNA with its correct amino acid is achieved by a family of activating enzymes, known more formally as aminoacyl-tRNA synthases (**Figure 14.13**). Each activating enzyme is specific for one amino acid and for its corresponding tRNA. The enzyme has a three-part active site that recognizes three molecules: a specific amino acid, ATP, and a specific tRNA. Since tRNA has a complex three-dimensional structure, the activating enzyme recognizes a specific tRNA with a very low error rate. Remarkably, the error rate for amino acid recognition is also low, on the order of one in 1,000. Because the activating enzymes are so highly specific, the process of tRNA charging is sometimes called the *second genetic code*. Follow the events of activation in Figure 14.13.

A clever experiment by Seymour Benzer and his colleagues at Purdue University demonstrated the importance of specificity in the attachment of tRNA to its amino acid. In their laboratory, the amino acid cysteine, already properly attached to its tRNA, was chemically modified to become a different amino acid, alanine. Which component—the amino acid or the tRNA— would be recognized when this hybrid charged tRNA was put

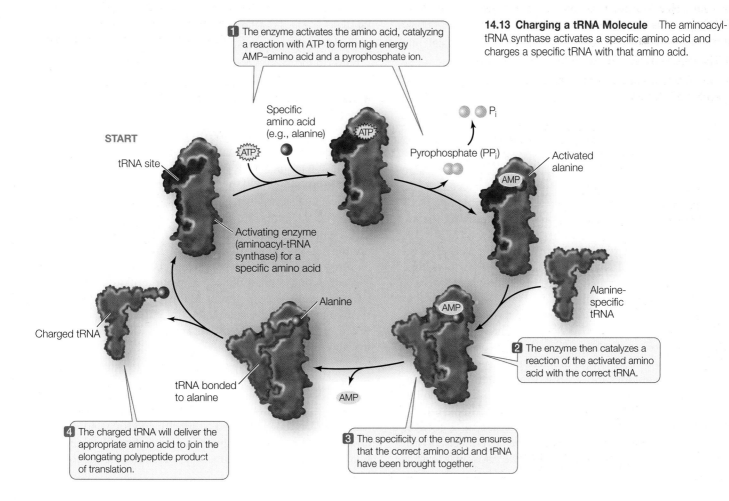

14.13 Charging a tRNA Molecule The aminoacyl-tRNA synthase activates a specific amino acid and charges a specific tRNA with that amino acid.

1 The enzyme activates the amino acid, catalyzing a reaction with ATP to form high energy AMP–amino acid and a pyrophosphate ion.

2 The enzyme then catalyzes a reaction of the activated amino acid with the correct tRNA.

3 The specificity of the enzyme ensures that the correct amino acid and tRNA have been brought together.

4 The charged tRNA will deliver the appropriate amino acid to join the elongating polypeptide product of translation.

START

tRNA site

Specific amino acid (e.g., alanine)

ATP

Pyrophosphate (PP_i)

P_i

Activated alanine

AMP

Activating enzyme (aminoacyl-tRNA synthase) for a specific amino acid

Alanine-specific tRNA

Alanine

AMP

Charged tRNA

tRNA bonded to alanine

AMP

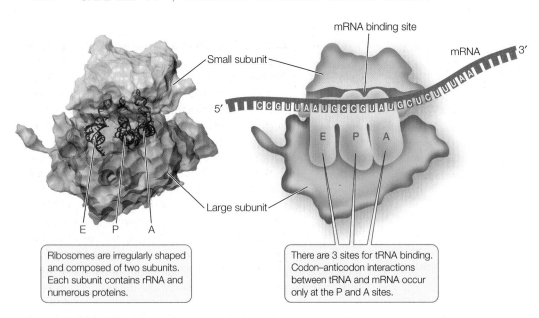

mRNA binding site

Small subunit

mRNA   3′

5′   CGGUUAAUGCCGUAUGCUCUUUUAAI

E   P   A

Large subunit

E   P   A

14.14 Ribosome Structure Each ribosome consists of a large and a small subunit. The subunits remain separate when they are not in use for protein synthesis.

Ribosomes are irregularly shaped and composed of two subunits. Each subunit contains rRNA and numerous proteins.

There are 3 sites for tRNA binding. Codon–anticodon interactions between tRNA and mRNA occur only at the P and A sites.

into a protein-synthesizing system? The answer was the tRNA. Everywhere in the synthesized protein where cysteine was supposed to be, alanine appeared instead. The cysteine-specific tRNA had delivered its cargo (alanine) to every mRNA codon for cysteine. This experiment showed that the protein synthesis machinery recognizes the anticodon of the charged tRNA, not the amino acid attached to it. If activating enzymes in nature did what Benzer did in the laboratory and charged tRNAs with the wrong amino acids, those amino acids would be inserted into proteins at inappropriate places, leading to alterations in protein shape and function and endangering cell life.

The ribosome is the workbench for translation

The **ribosome** is the molecular workbench where the task of translation is accomplished. Its structure enables it to hold mRNA and charged tRNAs in the right positions, thus allowing a polypeptide chain to be assembled efficiently. A given ribosome does not specifically produce just one kind of protein. A ribosome can use any mRNA and all species of charged tRNAs, and thus can be used to make many different polypeptide products. Ribosomes can be used over and over again, and there are thousands of them in a typical cell.

Although ribosomes are small in contrast to other cellular organelles, their mass of several million daltons makes them large in comparison with charged tRNAs. Each ribosome consists of two subunits, a large one and a small one (**Figure 14.14**). In eukaryotes, the large subunit consists of three different molecules of ribosomal RNA (rRNA) and 49 different protein molecules, arranged in a precise pattern. The small subunit consists of one rRNA molecule and 33 different protein molecules.

These two subunits and several dozen other molecules interact non-covalently, like a jigsaw puzzle. In fact, when hydrophobic interactions between the proteins and RNAs are disrupted, the ribosome falls apart. If the disrupting agent is removed, the complex structure self-assembles perfectly! When not active in the translation of mRNA, the ribosomes exist as two separate subunits.

The ribosomes of prokaryotes are somewhat smaller than those of eukaryotes, and their ribosomal proteins and RNAs are different. Mitochondria and chloroplasts also contain ribosomes, some of which are similar to those of prokaryotes (see Chapter 5).

On the large subunit of the ribosome there are three sites to which a tRNA can bind, and these are designated A, P, and E (see Figure 14.14). The mRNA and ribosome move in relation to one another, and as they do so, a charged tRNA traverses these three sites in order:

- The *A (amino acid) site* is where the charged tRNA anticodon binds to the mRNA codon, thus lining up the correct amino acid to be added to the growing polypeptide chain.

- The *P (polypeptide) site* is where the tRNA adds its amino acid to the polypeptide chain.

- The *E (exit) site* is where the tRNA, having given up its amino acid, resides before being released from the ribosome and going back to the cytosol to pick up another amino acid and begin the process again.

The ribosome has a *fidelity function* that ensures that the mRNA–tRNA interactions are accurate; that is, that a charged tRNA with the correct anticodon (e.g., 3′-UAC-5′) binds to the appropriate codon in mRNA (e.g., 5′-AUG-3′). When proper binding occurs, hydrogen bonds form between the base pairs. The rRNA of the small ribosomal subunit plays a role in validating the three-base-pair match. If hydrogen bonds have not formed between all three base pairs, the tRNA must be the wrong one for that mRNA codon, and that tRNA is ejected from the ribosome.

Translation takes place in three steps

Translation is the process by which the information in mRNA (derived from DNA) is used to specify and link a specific sequence of amino acids, producing a polypeptide. Like transcription, translation occurs in three steps: initiation, elongation, and termination.

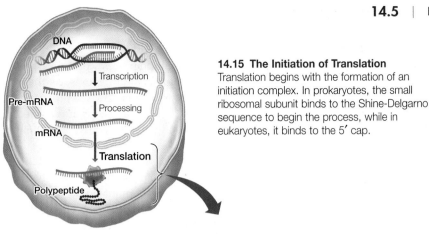

14.15 The Initiation of Translation
Translation begins with the formation of an initiation complex. In prokaryotes, the small ribosomal subunit binds to the Shine-Delgarno sequence to begin the process, while in eukaryotes, it binds to the 5′ cap.

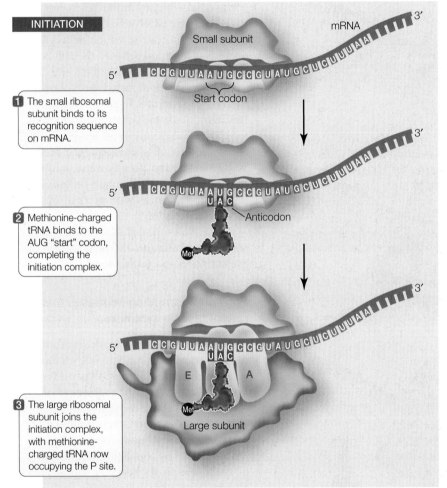

INITIATION

1 The small ribosomal subunit binds to its recognition sequence on mRNA.

2 Methionine-charged tRNA binds to the AUG "start" codon, completing the initiation complex.

3 The large ribosomal subunit joins the initiation complex, with methionine-charged tRNA now occupying the P site.

binds to this start codon by complementary base pairing to complete the initiation complex. Thus the first amino acid in a polypeptide chain is always methionine. However, not all mature proteins have methionine as their N-terminal amino acid. In many cases, the initiator methionine is removed by an enzyme after translation.

After the methionine-charged tRNA has bound to the mRNA, the large subunit of the ribosome joins the complex. The methionine-charged tRNA now lies in the P site of the ribosome, and the A site is aligned with the second mRNA codon. These ingredients—mRNA, two ribosomal subunits, and methionine-charged tRNA—are put together properly by a group of proteins called *initiation factors*.

ELONGATION A charged tRNA whose anticodon is complementary to the second codon of the mRNA now enters the open A site of the large ribosomal subunit. The large subunit then catalyzes two reactions:

- It breaks the bond between the tRNA and its amino acid in the P site.

- It catalyzes the formation of a peptide bond between that amino acid and the one attached to the tRNA in the A site.

Because the large ribosomal subunit performs these two actions, it is said to have **peptidyl transferase** activity. In this way, methionine (the amino acid in the P site) becomes the N terminus of the new protein. The second amino acid is now bound to methionine, but remains attached to its tRNA at the A site.

How does the large ribosomal subunit catalyze this binding? Harry Noller and his colleagues at the University of California at Santa Cruz did a series of experiments and found that:

- If they removed almost all of the proteins from the large subunit, it still catalyzed peptide bond formation.

- If the rRNA was destroyed, so was peptidyl transferase activity.

Thus *rRNA is the catalyst*. The purification and crystallization of ribosomes has allowed scientists to examine their structure in detail, and the catalytic role of rRNA in peptidyl transferase activity has been confirmed. This supports the hypothesis that RNA, and catalytic RNA in particular, evolved before DNA (see Section 4.3).

After the first tRNA releases its methionine, it moves to the E site and is then dissociated from the ribosome, returning to the cytosol to become charged with another methionine. The second tRNA, now bearing a dipeptide (a two-amino-acid chain), is shifted to the P site as the ribosome moves one codon along the mRNA in the 5′-to-3′ direction.

INITIATION The translation of mRNA begins with the formation of an **initiation complex**, which consists of a charged tRNA and a small ribosomal subunit, both bound to the mRNA (**Figure 14.15**).

In prokaryotes, the rRNA of the small ribosomal subunit first binds to a complementary ribosome binding site (known as the Shine–Dalgarno sequence) on the mRNA. This sequence is upstream of the actual start codon, but lines up the start codon so that is will be adjacent to the P site of the large subunit.

Eukaryotes do this somewhat differently: the small ribosomal subunit binds to the 5′cap on the mRNA. After binding, the small subunit moves along the mRNA until it reaches the start codon.

Recall that the mRNA start codon in the genetic code is AUG (see Figure 14.6). The anticodon of a methionine-charged tRNA

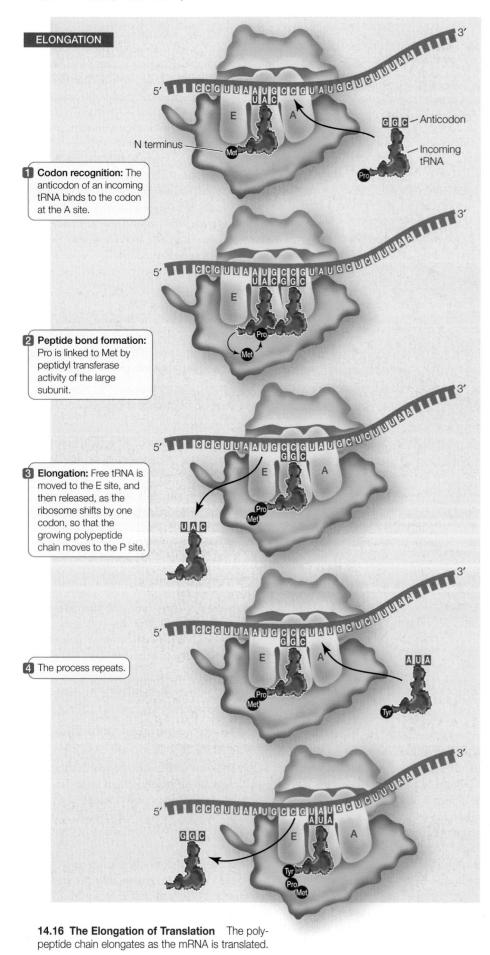

ELONGATION

1 **Codon recognition:** The anticodon of an incoming tRNA binds to the codon at the A site.

2 **Peptide bond formation:** Pro is linked to Met by peptidyl transferase activity of the large subunit.

3 **Elongation:** Free tRNA is moved to the E site, and then released, as the ribosome shifts by one codon, so that the growing polypeptide chain moves to the P site.

4 The process repeats.

14.16 The Elongation of Translation The polypeptide chain elongates as the mRNA is translated.

The elongation process continues, and the polypeptide chain grows, as these steps are repeated. Follow the process in **Figure 14.16**. All these steps are assisted by ribosomal proteins called *elongation factors.*

TERMINATION The elongation cycle ends, and translation is terminated, when a stop codon—UAA, UAG, or UGA—enters the A site (**Figure 14.17**). These codons do not correspond with any amino acids, nor do they bind any tRNAs. Rather, they bind a *protein release factor*, which allows hydrolysis of the bond between the polypeptide chain and the tRNA in the P site.

The newly completed polypeptide thereupon separates from the ribosome. Its C terminus is the last amino acid to join the chain. Its N terminus, at least initially, is methionine, as a consequence of the AUG start codon. In its amino acid sequence, it contains information specifying its conformation, as well as its ultimate cellular destination.

Table 14.2 summarizes the nucleic acid signals for initiation and termination of transcription and translation.

Polysome formation increases the rate of protein synthesis

Several ribosomes can work simultaneously at translating a single mRNA molecule, producing multiple polypeptides at the same time. As soon as the first ribosome has moved far enough from the site of translation initiation, a second initiation complex can form, then a third, and so on. An assemblage consisting of a strand of mRNA with its beadlike ribosomes and their growing polypeptide chains is called a **polyribosome**, or **polysome** (**Figure 14.18**). Cells that are actively synthesizing proteins contain large numbers of polysomes and few free ribosomes or ribosomal subunits.

A polysome is like a cafeteria line in which patrons follow one another, adding items to their trays. At any moment, the person at the start has a little food (a newly initiated protein); the person at the end has a complete meal (a completed protein). However, in the polysome cafeteria, everyone gets the same meal: many copies of the same protein are made from a single mRNA.

14.17 The Termination of Translation Translation terminates when the A site of the ribosome encounters a stop codon on the mRNA.

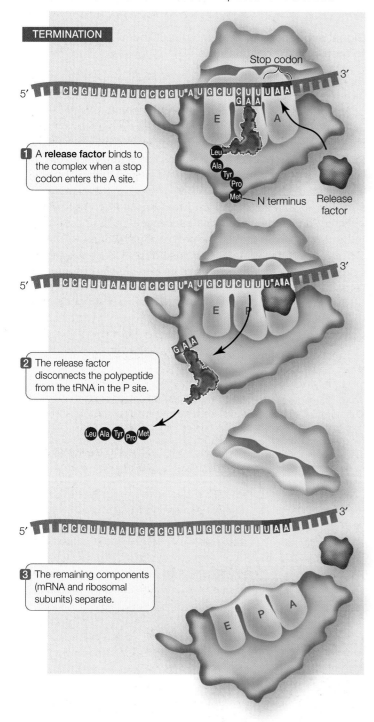

TERMINATION

1 A **release factor** binds to the complex when a stop codon enters the A site.

2 The release factor disconnects the polypeptide from the tRNA in the P site.

3 The remaining components (mRNA and ribosomal subunits) separate.

(A)

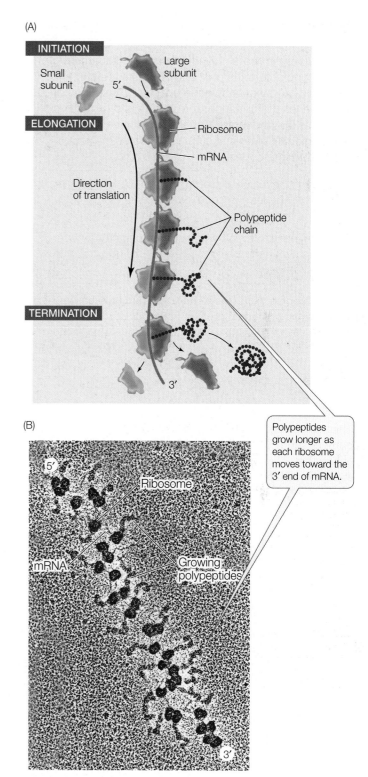

(B)

Polypeptides grow longer as each ribosome moves toward the 3′ end of mRNA.

14.18 A Polysome (A) A polysome consists of multiple ribosomes and their growing polypeptide chains moving along an mRNA molecule. (B) An electron micrograph of a polysome.

TABLE 14.2

Signals that Start and Stop Transcription and Translation

	TRANSCRIPTION	TRANSLATION
Initiation	Promoter DNA	AUG start codon in the mRNA
Termination	Terminator DNA	UAA, UAG, or UGA in the mRNA

14.5 RECAP

A key step in protein synthesis is the attachment of an amino acid to its proper tRNA. This attachment is carried out by an activating enzyme. Translation of the genetic information from mRNA into protein occurs at the ribosome. Multiple ribosomes may act on a single mRNA to make multiple copies of the protein that it encodes.

- How is an amino acid attached to a specific tRNA, and why is the term "second genetic code" associated with this process? **See pp. 304–305 and Figure 14.13**

- Describe the events of initiation, elongation, and termination of translation. **See pp. 306–308 and Figures 14.15–14.17**

The polypeptide chain that is released from the ribosome is not necessarily a functional protein. Let's look at some of the posttranslational changes that can affect the fate and function of a polypeptide.

14.6 What Happens to Polypeptides after Translation?

The site of a polypeptide's function may be far away from its point of synthesis in the cytoplasm. This is especially true for eukaryotes. The polypeptide may be moved into an organelle, or even out of the cell. In addition, polypeptides are often modified by the addition of new chemical groups that have func-

tional significance. In this section we examine these *posttranslational* aspects of protein synthesis.

Signal sequences in proteins direct them to their cellular destinations

As a polypeptide chain emerges from the ribosome, it folds into its three-dimensional shape. As we described in Section 3.2, the polypeptide's conformation is determined by the sequence of amino acids that make it up. Properties such as the polarity and charge of the R groups in the amino acids determine how they interact with each other in the folded molecule. Ultimately, a polypeptide's conformation allows it to interact with other molecules in the cell, such as a substrate or another polypeptide. In addition to this structural information, the newly formed polypeptide can contain a **signal sequence**—an "address label" indicating where in the cell the polypeptide belongs.

Protein synthesis always begins on free ribosomes in the cytoplasm. But as a polypeptide chain is made, the information contained in its amino acid sequence gives it one of two sets of further instructions (**Figure 14.19**):

- *"Complete translation and be released to an organelle, or remain in the cytosol."* Some proteins contain signal sequences that direct them to the nucleus, mitochondria, plastids, or per-

14.19 Destinations for Newly Translated Polypeptides in a Eukaryotic Cell   Signal sequences on newly synthesized polypeptides bind to specific receptor proteins on the outer membranes of the organelles to which they are "addressed." Once the protein has bound to it, the receptor forms a channel in the membrane, and the protein enters the organelle.

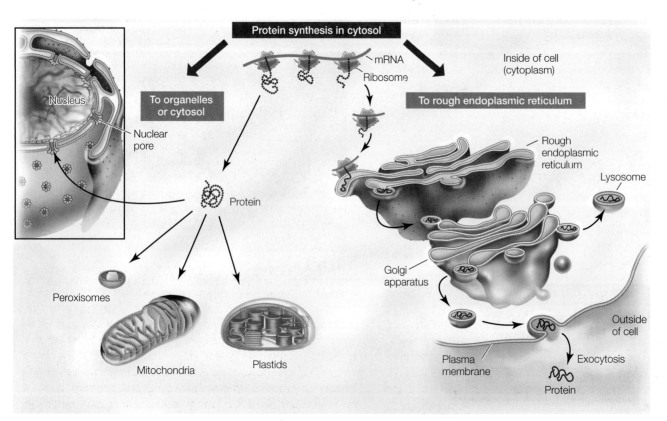

oxisomes. If they lack a signal sequence, they remain in the cytosol by default.

- *"Stop translation, go to the endoplasmic reticulum, and finish synthesis there."* Other proteins contain a signal sequence that directs them to the endoplasmic reticulum (ER) before translation is complete. Such proteins may be retained in the lumen (the inside) of the ER, or be sent to the Golgi apparatus. From there, they may be sent to the lysosomes or the plasma membrane. Alternatively if they lack such specific instructions, they may be secreted from the cell via vesicles that fuse with the plasma membrane.

DESTINATION: NUCLEUS, MITOCHONDRION, OR CHLOROPLAST After translation, some folded polypeptides have a short exposed sequence of amino acids that acts like a postal "zip code," directing them to an organelle. These signal (or localization) sequences are either at the N terminus or in the interior of the amino acid chain. For example, the following sequence directs a protein to the nucleus:

—Pro—Pro—Lys—Lys—Lys—Arg—Lys—Val—

A nuclear localization sequence would occur in histone proteins associated with nuclear DNA, but not in citric acid cycle enzymes, which are addressed to the mitochondria. Signal sequences for a particular organelle vary, so not all the polypeptides destined for the nucleus have the same signal sequence.

How do we know that the amino acid sequence shown above is the signal? To investigate this question, Stephen Dilworth and colleagues at the University of Cambridge injected cells with nuclear and cytoplasmic proteins (**Figure 14.20**). The experiments involved the nuclear protein nucleoplasmin, the cytoplasmic protein pyruvate kinase, the nuclear localization signal (see above) and a "mix-and-match" procedure. For example, the putative nuclear signal was removed from nucleoplasmin, which normally carries it, or attached to pyruvate kinase, which does not normally carry it. The result was that it did not matter where in the cell the protein normally resided. If it had the signal, it went to the nucleus and if it did not have the signal, it stayed in the cytoplasm.

A signal sequence binds to a specific receptor protein, appropriately called a **docking protein**, on the outer membrane of the appropriate organelle. Once the signal sequence has bound to it, the docking protein forms a channel in the membrane, allowing the signal-bearing protein to pass through the membrane and enter the organelle. In this process, the protein is usually unfolded by a chaperonin protein (see Figure 3.12) so that it can pass through the channel; then it refolds into its normal conformation.

DESTINATION: ENDOPLASMIC RETICULUM If a specific hydrophobic sequence of 15–30 amino acids occurs at the N terminus of an elongated polypeptide chain, the polypeptide is sent initially to the ER. Some proteins are retained in the ER, but most move on to the Golgi, where they can be modified for eventual transport to the lysosomes, the plasma membrane, or out of the cell. In the cytoplasm, before translation is finished and while the polypeptide is still attached to a ribosome, this signal sequence

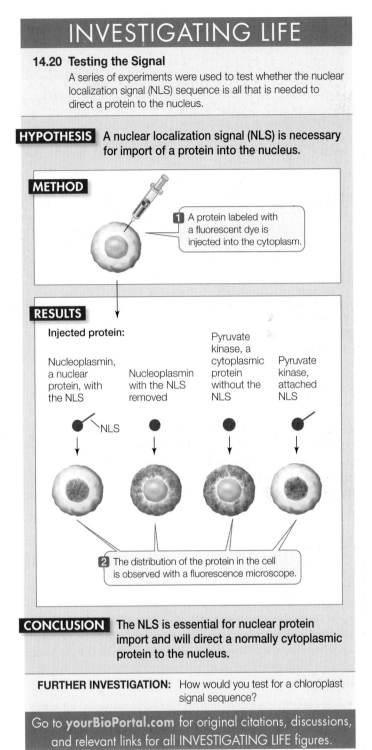

INVESTIGATING LIFE

14.20 Testing the Signal

A series of experiments were used to test whether the nuclear localization signal (NLS) sequence is all that is needed to direct a protein to the nucleus.

HYPOTHESIS A nuclear localization signal (NLS) is necessary for import of a protein into the nucleus.

METHOD

1 A protein labeled with a fluorescent dye is injected into the cytoplasm.

RESULTS

Injected protein:

Nucleoplasmin, a nuclear protein, with the NLS

Nucleoplasmin with the NLS removed

Pyruvate kinase, a cytoplasmic protein without the NLS

Pyruvate kinase, attached NLS

NLS

2 The distribution of the protein in the cell is observed with a fluorescence microscope.

CONCLUSION The NLS is essential for nuclear protein import and will direct a normally cytoplasmic protein to the nucleus.

FURTHER INVESTIGATION: How would you test for a chloroplast signal sequence?

Go to **yourBioPortal.com** for original citations, discussions, and relevant links for all INVESTIGATING LIFE figures.

binds to a **signal recognition particle** composed of protein and RNA (**Figure 14.21**). This binding blocks further protein synthesis until the ribosome becomes attached to a specific receptor protein in the membrane of the rough ER. Once again, the receptor protein is converted into a channel, through which the growing polypeptide passes. After the formation of the channel, protein synthesis resumes, and the chain grows longer until its sequence is completed. The elongating polypeptide may be retained in the ER membrane itself, or it may enter the

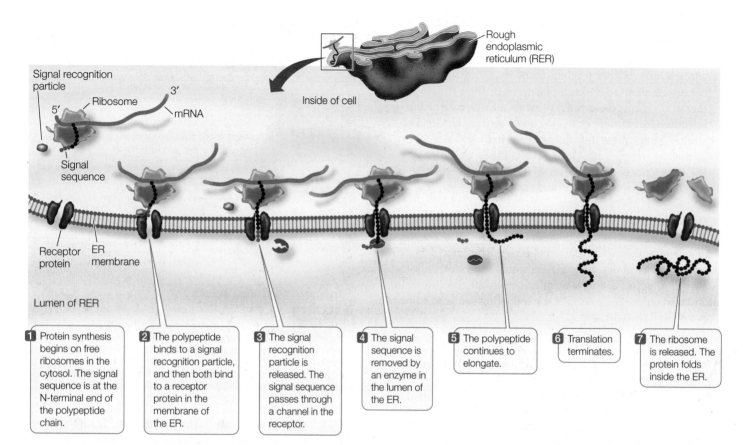

Rough endoplasmic reticulum (RER)

Inside of cell

Signal recognition particle

Ribosome

5′   3′

mRNA

Signal sequence

Receptor protein

ER membrane

Lumen of RER

1 Protein synthesis begins on free ribosomes in the cytosol. The signal sequence is at the N-terminal end of the polypeptide chain.

2 The polypeptide binds to a signal recognition particle, and then both bind to a receptor protein in the membrane of the ER.

3 The signal recognition particle is released. The signal sequence passes through a channel in the receptor.

4 The signal sequence is removed by an enzyme in the lumen of the ER.

5 The polypeptide continues to elongate.

6 Translation terminates.

7 The ribosome is released. The protein folds inside the ER.

14.21 A Signal Sequence Moves a Polypeptide into the ER   When a certain signal sequence of amino acids is present at the beginning of a polypeptide chain, the polypeptide will be taken into the endoplasmic reticulum (ER). The finished protein is thus segregated from the cytosol.

interior space—the lumen—of the ER. In either case, an enzyme in the lumen of the ER removes the signal sequence from the polypeptide chain.

If the finished protein enters the ER lumen, it can be transported to its appropriate location—to other cellular compartments or to the outside of the cell—via the ER and the Golgi apparatus, without mixing with other molecules in the cytoplasm.

After removal of the terminal signal sequence in the lumen of the ER, additional signals are needed to direct the protein to its destination. These signals are of two kinds:

- Some are sequences of amino acids that allow the protein's retention within the ER.

- Others are sugars, which are added in the Golgi apparatus. The resulting *glycoproteins* end up either at the plasma membrane or in a lysosome (or plant vacuole), or are secreted, depending on which sugars are added.

Proteins with no additional signals pass from the ER through the Golgi apparatus and are secreted from the cell.

The importance of signals is shown by Inclusion-cell (I-cell) disease, an inherited disease that causes death in early childhood. People with this disease have a mutation in the gene encoding a Golgi enzyme that adds targeting sugars to proteins destined for the lysosomes. As a result, enzymes that are essential for the hydrolysis of various macromolecules cannot reach the lysosomes, where they are normally active. The macromolecules accumulate in the lysosomes, and this lack of cellular recycling has drastic effects, resulting in early death.

Many proteins are modified after translation

Most mature proteins are not identical to the polypeptide chains that are translated from mRNA on the ribosomes. Instead, most polypeptides are modified in any of a number of ways after translation (**Figure 14.22**). These modifications are essential to the final functioning of the protein.

- **Proteolysis** is the cutting of a polypeptide chain. Cleavage of the signal sequence from the growing polypeptide chain in the ER is an example of proteolysis; the protein might move back out of the ER through the membrane channel if the signal sequence were not cut off. Some proteins are actually made from *polyproteins* (long polypeptides) that are cut into final products by enzymes called *proteases*. These protease enzymes are essential to some viruses, including human immunodeficiency virus (HIV), because the large viral polyprotein cannot fold properly unless it is cut. Certain drugs used to treat acquired immune deficiency syndrome (AIDS) work by inhibiting the HIV protease, thereby preventing the formation of proteins needed for viral reproduction.

- **Glycosylation** is the addition of sugars to proteins to form glycoproteins. In both the ER and the Golgi apparatus, resident enzymes catalyze the addition of various sugars or short sugar chains to certain amino acid R groups on pro-

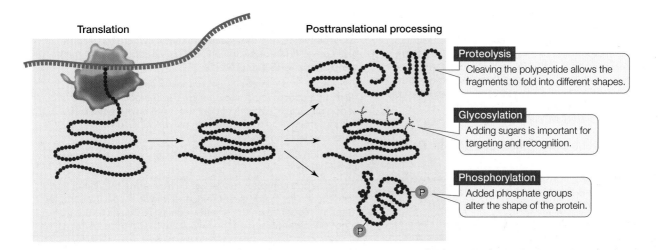

Translation | Posttranslational processing

Proteolysis
Cleaving the polypeptide allows the fragments to fold into different shapes.

Glycosylation
Adding sugars is important for targeting and recognition.

Phosphorylation
Added phosphate groups alter the shape of the protein.

14.22 Posttranslational Modifications of Proteins Most polypeptides must be modified after translation in order to become functional proteins.

teins. One such type of "sugar coating" is essential for directing proteins to lysosomes, as mentioned above. Other types are important in the conformation of proteins and their recognition functions at the cell surface. Other attached sugars help to stabilize extracellular proteins, or proteins stored in vacuoles in plant seeds.

● **Phosphorylation** is the addition of phosphate groups to proteins, and is catalyzed by *protein kinases*. The charged phosphate groups change the conformation of a protein, often exposing the active site of an enzyme or the binding site for another protein. We have seen the role of phosphorylation in cell signaling (see Chapter 7).

14.6 RECAP

Signal sequences in polypeptides direct them to their appropriate destinations inside or outside the cell. Many polypeptides are modified after translation.

● How do signal sequences determine where a protein will go after it is made? **See pp. 310–312 and Figure 14.21**

● What are some ways in which posttranslational modifications alter protein structure and function? **See pp. 312–313 and Figure 14.22**

All of the processes we have just described result in a functional protein, but only if the amino acid sequence of that protein is correct. If the sequence is not correct, cellular dysfunction may result. Changes in the DNA—mutations—are a major source of errors in amino acid sequences. This is the subject of the next chapter.

CHAPTER SUMMARY

14.1 What Is the Evidence that Genes Code for Proteins?

● Beadle and Tatum's experiments on metabolic enzymes in the bread mold *Neurospora* led to the **one-gene, one-enzyme hypothesis**. We now know that there is a **one-gene, one-polypeptide relationship**. Review Figure 14.1

14.2 How Does Information Flow from Genes to Proteins?

● The **central dogma** of molecular biology states that DNA encodes RNA, and RNA encodes proteins. Proteins do not encode proteins, RNA, or DNA.

● The process by which the information in DNA is copied to RNA is called **transcription**. The process by which a protein is built from the information in RNA is called **translation**. Review Figure 14.2, **WEB ACTIVITY 14.1**

● Certain RNA viruses are exceptions to the central dogma. These **retroviruses** synthesize DNA from RNA in a process called **reverse transcription**.

● The product of transcription is **messenger RNA (mRNA)**. **Transfer RNA (tRNA)** molecules are adapters that translate the

genetic information in the mRNA into a corresponding sequence of amino acids to produce a polypeptide.

14.3 How Is the Information Content in DNA Transcribed to Produce RNA?

● In a given gene, only one of the two strands of DNA (the **template strand**) acts as a template for transcription. **RNA polymerase** is the catalyst for transcription.

● RNA transcription from DNA proceeds in three steps: **initiation**, **elongation**, and **termination**. Review Figure 14.4, **ANIMATED TUTORIAL 14.1**

● Initiation requires a **promoter**, to which RNA polymerase binds. Part of each promoter is the **initiation site**, where transcription begins.

● Elongation of the RNA molecule proceeds from the 5′ to 3′ end.

● Particular base sequences specify termination, at which point transcription ends and the RNA transcript separates from the DNA template.

- The **genetic code** is a "language" of triplets of mRNA nucleotide bases (**codons**) corresponding to 20 specific amino acids; there are **start** and **stop codons** as well. The code is redundant (an amino acid may be represented by more than one codon), but not ambiguous (no single codon represents more than one amino acid). Review Figures 14.5 and 14.6, **ANIMATED TUTORIAL 14.2 AND WEB ACTIVITY 14.2**

14.4 How Is Eukaryotic DNA Transcribed and the RNA Processed?

- Unlike prokaryotes, where transcription and translation occur in the cytoplasm and are coupled, in eukaryotes transcription occurs in the nucleus and translation occurs later in the cytoplasm.

- The initial transcript of a eukaryotic protein-coding gene is modified with a **5′ cap** and a **3′ poly A sequence**. Review Figure 14.10

- Eukaryotic genes contain **introns**, which are noncoding sequences within the transcribed regions of genes.

- Pre-mRNA contains the introns. They are removed in the nucleus via **mRNA splicing** by the **small nuclear ribonucleoprotein particles**. Then the mRNA passes through the nuclear pore into the cytoplasm, where it is translated on the surfaces of **ribosomes**. Review Figure 14.11, **ANIMATED TUTORIAL 14.3**

14.5 How Is RNA Translated into Proteins?
SEE ANIMATED TUTORIAL 14.4

- During translation, amino acids are linked together in the order specified by the codons in the mRNA. This task is achieved by tRNAs, which bind to (are charged with) specific amino acids.

- Each tRNA species has an amino acid attachment site as well as an **anticodon** complementary to a specific mRNA codon. A specific activating enzyme charges each tRNA with its specific amino acid. Review Figures 14.12 and 14.13

- The **ribosome** is the molecular workbench where translation takes place. It has one large and one small subunit, both made of **ribosomal RNA** and proteins.

- Three sites on the large subunit of the ribosome interact with tRNA anticodons. The **A site** is where the charged tRNA anticodon binds to the mRNA codon; the **P site** is where the tRNA adds its amino acid to the growing polypeptide chain; and the **E site** is where the tRNA is released.

- Translation occurs in three steps: **initiation**, **elongation**, and **termination**.

- The **initiation complex** consists of tRNA bearing the first amino acid, the small ribosomal subunit, and mRNA. A specific complementary sequence on the small subunit rRNA binds to the transcription initiation site on the mRNA. Review Figure 14.15

- The growing polypeptide chain is elongated by the formation of peptide bonds between amino acids, catalyzed by the rRNA. Review Figure 14.16

- When a stop codon reaches the A site, it terminates translation by binding a release factor. Review Figure 14.17

- In a **polysome**, more than one ribosome moves along a strand of mRNA at one time. Review Figure 14.18

14.6 What Happens to Polypeptides after Translation?

- **Signal sequences** of amino acids direct polypeptides to their cellular destinations. Review Figure 14.19

- Destinations in the cytoplasm include organelles, which proteins enter after being recognized and bound by surface receptors called **docking proteins**.

- Proteins "addressed" to the ER bind to a **signal recognition particle**. Review Figure 14.21

- Posttranslational modifications of polypeptides include **proteolysis**, in which a polypeptide is cut into smaller fragments; **glycosylation**, in which sugars are added; and **phosphorylation**, in which phosphate groups are added. Review Figure 14.22

SELF-QUIZ

1. Which of the following is *not* a difference between RNA and DNA?
   a. RNA has uracil; DNA has thymine.
   b. RNA has ribose; DNA has deoxyribose.
   c. RNA has five bases; DNA has four.
   d. RNA is a single polynucleotide strand; DNA is a double strand.
   e. RNA molecules are smaller than human chromosomal DNA molecules.

2. Normally, *Neurospora* can synthesize all 20 amino acids. A certain strain of this mold cannot grow in minimal nutritional medium, but grows only when the amino acid leucine is added to the medium. This strain
   a. is dependent on leucine for energy.
   b. has a mutation affecting a biochemical pathway leading to the synthesis of carbohydrates.
   c. has a mutation affecting the biochemical pathways leading to the synthesis of all 20 amino acids.
   d. has a mutation affecting the biochemical pathway leading to the synthesis of leucine.
   e. has a mutation affecting the biochemical pathways leading to the syntheses of 19 of the 20 amino acids.

3. An mRNA has the sequence 5′-AUGAAAUCCUAG-3′. What is the template DNA strand for this sequence?
   a. 5′-TACTTTAGGATC-3′
   b. 5′-ATGAAATCCTAG-3′
   c. 5′-GATCCTAAAGTA-3′
   d. 5′-TACAAATCCTAG-3′
   e. 5′-CTAGGATTTCAT-3′

4. The adapters that allow translation of the four-letter nucleic acid language into the 20-letter protein language are called
   a. aminoacyl-tRNA synthase.
   b. transfer RNAs.
   c. ribosomal RNAs.
   d. messenger RNAs.
   e. ribosomes.

5. Which of the following does *not* occur after eukaryotic mRNA is transcribed?
   a. Binding of RNA polymerase to the promoter
   b. Capping of the 5′ end
   c. Addition of a poly A tail to the 3′ end
   d. Splicing out of the introns
   e. Transport to the cytosol

6. Transcription
   a. produces only mRNA.
   b. requires ribosomes.
   c. requires tRNAs.
   d. produces RNA growing from the 5′ end to the 3′ end.
   e. takes place only in eukaryotes.
7. Which statement about translation is *not* true?
   a. Translation is RNA-directed polypeptide synthesis.
   b. An mRNA molecule can be translated by only one ribosome at a time.
   c. The same genetic code operates in almost all organisms and organelles.
   d. Any ribosome can be used in the translation of any mRNA.
   e. There are both start and stop codons.
8. Which statement about RNA is *not* true?
   a. Transfer RNA functions in translation.
   b. Ribosomal RNA functions in translation.
   c. RNAs are produced by transcription.
   d. Messenger RNAs are produced on ribosomes.
   e. DNA codes for mRNA, tRNA, and rRNA.
9. The genetic code
   a. is different for prokaryotes and eukaryotes.
   b. has changed during the course of recent evolution.
   c. has 64 codons that code for amino acids.
   d. has more than one codon for many amino acids
   e. is ambiguous.
10. Which statement about RNA splicing is *not* true?
    a. It removes introns.
    b. It is performed by small nuclear ribonucleoprotein particles (snRNPs).
    c. It removes the introns at the ribosome.
    d. It is usually directed by consensus sequences.
    e. It shortens the RNA molecule.

FOR DISCUSSION

1. In rats, a gene 1,440 base pairs (bp) long codes for an enzyme made up of 192 amino acids. Discuss this apparent discrepancy. How long would the initial and final mRNA transcripts be?

2. Har Gobind Khorana at the University of Wisconsin synthesized artificial mRNAs such as poly CA (CACA …) and poly CAA (CAACAACAA …). He found that poly CA codes for a polypeptide consisting of alternating threonine (Thr) and histidine (His) residues. There are two possible codons in poly CA, CAC and ACA. One of these must encode histidine and the other threonine—but which is which? The answer comes from results with poly CAA, which produces three different polypeptides: poly Thr, poly Gln (glutamine), and poly Asn (asparagine). (An artificial mRNA can be read, inefficiently, beginning at any point in the chain; there is no specific initiation signal. Thus poly CAA can be read as a polymer of CAA, of ACA, or of AAC.) Compare the results of the poly CA and poly CAA experiments, and determine which codon corresponds with threonine and which with histidine.

3. Look back at Question 2. Using the genetic code in Figure 14.6 as a guide, deduce what results Khorana would have obtained had he used poly UG and poly UGG as artificial mRNAs. In fact, very few such artificial mRNAs would have given useful results. For an example of what could happen, consider poly CG and poly CGG. If poly CG were the mRNA, a mixed polypeptide of arginine and alanine (Arg–Ala–Ala–Arg …) would be obtained; poly CGG would give three polypeptides: poly Arg, poly Ala, and poly Gly (glycine). Can any codons be determined from only these data? Explain.

4. Errors in transcription occur about 100,000 times as often as errors in DNA replication. Why can this high rate be tolerated in RNA synthesis but not in DNA synthesis?

ADDITIONAL INVESTIGATION

Beadle and Tatum's experiments showed that a biochemical pathway could be deduced from mutant strains. In bacteria, the biosynthesis of the amino acid tryptophan (T) from the precursor chorismate (C) involves four intermediate chemical compounds, which we will call D, E, F, and G. Here are the phenotypes of various mutant strains. Each strain has a mutation in a gene for a different enzyme; + means growth with the indicated compound added to the medium, and 0 means no growth. Based on these data, order the compounds (C, D, E, F, G, and T) and enzymes (1, 2, 3, 4, and 5) in a biochemical pathway.

	Addition to medium					
Mutant strain	C	D	E	F	G	T
1	0	0	0	0	+	+
2	0	+	+	0	+	+
3	0	+	0	0	+	+
4	0	+	+	+	+	+
5	0	0	0	0	0	+

WORKING WITH DATA (GO TO yourBioPortal.com)

Deciphering the Genetic Code The identification of the first mRNA codons associated with specific amino acids was a landmark in molecular biology (Figure 14.5). In this hands-on exercise, you will learn about the experimental protocol that Nirenberg and Matthei followed, using artificial mRNA, and analyze the results they obtained.

15

Gene Mutation and Molecular Medicine

Baby 81

The tsunami of December 26, 2004, struck the coastal town of Kalmunai, Sri Lanka, with such force that 4-month-old Abilass Jeyarajah was torn from his mother's arms and swept away. Hours later, while his parents desperately searched the devastated town, their tiny son washed up on the beach a kilometer away, alive. A local schoolteacher found him and brought him to the hospital—the eighty-first patient admitted that day. The hospital was overwhelmed with 1,000 dead bodies, many of them children. Since Abilass was alive and healthy, he was dubbed "Baby 81, the miracle baby" and became an instant celebrity among the staff as they went about their grim duties of caring for the injured and dying.

Meanwhile, the parents kept looking. Two days later, they met the schoolteacher, who told them about the baby he had found. Rushing to the hospital, the Jeyarajahs were elated to find their son, but were in for a rude shock. Eight other couples who had also lost infants were claiming Baby 81 as theirs. The baby remained in the hospital while the case went to court.

Judge M. P. Mohaideen faced a situation not unlike one faced by King Solomon 3,000 years ago, who was asked to decide which of two women was the mother of an infant. Solomon's method of determining parentage is told in a famous biblical passage—he ordered the baby cut in two, and the real mother indicated that she would rather give the baby away than have the baby killed. The Sri Lankan judge had a different method: he called in molecular biologists.

With 6 billion base pairs of DNA packaged in 46 chromosomes, each one of us is unique. Although our protein-coding sequences are similar (after all, our phenotypes are similar), only 1.5 percent of the DNA in the human genome actually codes for proteins. The eukaryotic genome contains many repeated sequences, and the repeat frequencies may differ between individuals, offering one way to differentiate one individual from another. A base pair at a particular site may also vary between individuals, due to DNA replication errors or random muta-

After the Tsunami In December of 2004, a tsunami originating in the Indian Ocean struck a broad region that encompassed many nations in Southeast Asia. The result was an unprecedented humanitarian disaster that left almost a quarter of a million people dead and many more homeless.

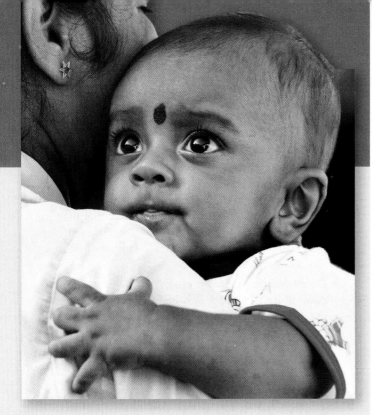

Baby 81 Abilass Jeyarajah survived the tsunami and was reunited with his parents by court order after DNA testing proved that he is indeed their son.

tions. Both of these types of differences are mutations, defined as inherited changes in DNA.

It is now possible to analyze these differences in DNA sequences (amplified by PCR) to identify people, in a process called DNA fingerprinting. The most common DNA fingerprinting technique used today involves the detection of variations in repeat sequences at different loci throughout the genome. When DNA samples from the nine sets of contesting parents were analyzed and compared with a sample from Baby 81, only one pair of parents carried sequences that were the same as those of the baby. On February 14, 2005, the judge ruled that the Jeyarajahs were the biological parents, and Baby 81 got his real name and parents back.

IN THIS CHAPTER we will discuss the nature and detection of mutations at the molecular and chromosomal levels. We will describe how abnormal proteins can cause human genetic diseases, and how these diseases and the alleles that produce them can be detected. Finally, we'll see how this knowledge of mutations has been applied in the development of new treatments.

15.1 What Are Mutations?

In Chapter 12, we described mutations as inherited changes in genes, and we saw that different alleles may produce different phenotypes (short pea plants versus tall, for example). Now that we understand the chemical nature of genes and how they are expressed as phenotypes (in particular, proteins) we can return to the concept of mutations for a more specific definition. We can now state that mutations are changes in the nucleotide sequence of DNA that are passed on from one cell, or organism, to another.

As an example of just one cause of mutations, recall from Chapter 13 that DNA polymerases make errors. Repair systems such as proofreading are in place to correct them. But some errors escape being corrected and are passed on to the daughter cells.

Mutations in multicellular organisms can be divided into two types:

- **Somatic mutations** are those that occur in somatic (body) cells. These mutations are passed on to the daughter cells during mitosis, and to the offspring of those cells in turn, but are not passed on to sexually produced offspring. For example, a mutation in a single human skin cell could result in a patch of skin cells that all have the same mutation, but it would not be passed on to the person's children.

- **Germ line mutations** are those that occur in the cells of the germ line—the specialized cells that give rise to gametes. A gamete with the mutation passes it on to a new organism at fertilization.

In either case, the mutations may or may not have phenotypic effects.

Mutations have different phenotypic effects

Phenotypically, we can understand mutations in terms of their effects on proteins and their function (**Figure 15.1**).

- **Silent mutations** do not affect protein function. They can be mutations in noncoding DNA, such as the repeat sequences that were used to identify Baby 81 in the opening story of this chapter. Or they can be in the coding portion of DNA but not have any effect on the protein.

- **Loss of function mutations** affect protein function. These mutations may lead to nonfunctional proteins that no longer work as structural proteins or enzymes. They almost

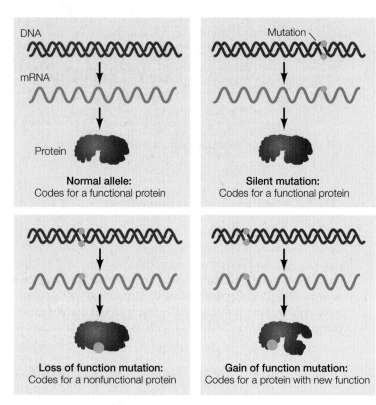

15.1 Mutation and Phenotype   Mutations may or may not affect the protein phenotype.

always show recessive inheritance in a diploid organism, because the presence of one wild-type allele will usually result in sufficient functional protein for the cell. For example, the familiar wrinkled-seed allele in pea plants, originally studied by Mendel (see Figure 12.3), is due to a mutation in the gene *SBE1* (starch *b*ranching *e*nzyme). Normally the protein made by this gene catalyzes the branching of starch as seeds develop. In the mutant, the SBE1 protein is not functional and that leads to osmotic changes, causing the wrinkled appearance.

- A **gain of function mutation** leads to a protein with an altered function. This kind of mutation usually shows dominant inheritance, because the presence of the wild-type allele does not prevent the mutant allele from functioning. This is common in cancer. For example, a receptor for a growth factor normally requires binding of the growth factor (the ligand) to activate the cell division cycle. Some cancers are caused by mutations in genes coding for these receptors such that they no longer require stimulation by their particular ligands. The mutant receptors are "always on," leading to the unrestrained cell proliferation that is characteristic of cancer cells.

- **Conditional mutations** cause their phenotypes only under certain *restrictive* conditions. They are not detectable under other, *permissive* conditions. Many conditional mutants are temperature-sensitive; that is, they show the altered phenotype only at a certain temperature (recall the rabbit in Figure 12.11). The mutant allele in such an organism may code

for an enzyme with an unstable tertiary structure that is altered at the restrictive temperature.

All mutations are alterations in the nucleotide sequence of DNA. At the molecular level, we can divide mutations into two categories:

- A **point mutation** results from the gain, loss, or substitution of a single nucleotide. After DNA replication, the altered nucleotide becomes a mutant base pair. If a point mutation occurs within a gene (rather than in a noncoding DNA sequence), then one allele of that gene (usually dominant) becomes another allele (usually recessive).

- **Chromosomal mutations** are more extensive than point mutations. They may change the position or orientation of a DNA segment without actually removing any genetic information, or they may cause a segment of DNA to be duplicated or irretrievably lost.

Point mutations change single nucleotides

Point mutations result from the addition or subtraction of a nucleotide base, or the substitution of one base for another. Point mutations can arise due to errors in DNA replication that are not corrected during proofreading, or they may be caused by environmental **mutagens** (substances that cause mutations, such as radiation or certain chemicals).

Point mutations in the coding regions of DNA usually result in changes in the mRNA, but changes in the mRNA may or may not result in changes in the protein. Silent mutations by definition have no effect on the protein. Missense and nonsense mutations result in changes in the protein, some of them drastic (**Figure 15.2**).

SILENT MUTATIONS   Silent mutations have no effect on amino acid sequences. This is because they are often found in noncoding DNA. Also, because of the redundancy of the genetic code, a base substitution in a coding region will not always cause a change in the amino acid sequence when the altered mRNA is translated. Silent mutations are quite common, and they result in genetic diversity that is not expressed as phenotypic differences.

MISSENSE MUTATIONS   Some base substitutions change the genetic code such that one amino acid substitutes for another in a protein. These changes are called **missense mutations**. A specific example of a missense mutation is the one that causes sickle-cell disease, a serious heritable blood disorder. The disease occurs in people who carry two copies of the sickle allele of the gene for human β-globin (a subunit of hemoglobin, the protein in human blood that carries oxygen). The sickle allele differs from the normal allele by one base pair, resulting in a polypeptide that differs by one amino acid from the normal protein. Individuals who are homozygous for this recessive allele have defective, sickle-shaped red blood cells (**Figure 15.3**).

A missense mutation may result in a defective protein, but often it has no effect on the protein's function. For example, a

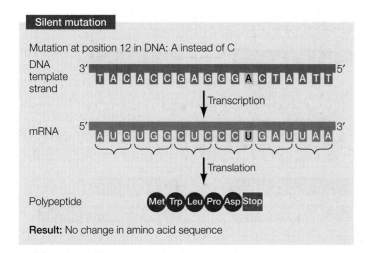

Silent mutation

Mutation at position 12 in DNA: A instead of C

DNA template strand 3′ T A C A C C G A G G G A C T A A T T 5′

↓ Transcription

mRNA 5′ A U G U G G C U C C C U G A U U A A 3′

↓ Translation

Polypeptide Met Trp Leu Pro Asp Stop

Result: No change in amino acid sequence

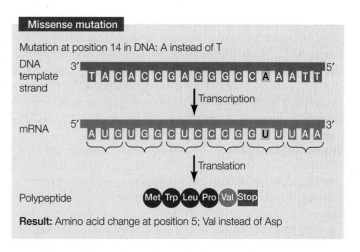

Missense mutation

Mutation at position 14 in DNA: A instead of T

DNA template strand 3′ T A C A C C G A G G G C C A A T T 5′

↓ Transcription

mRNA 5′ A U G U G G C U C C C G G U U A A 3′

↓ Translation

Polypeptide Met Trp Leu Pro Val Stop

Result: Amino acid change at position 5; Val instead of Asp

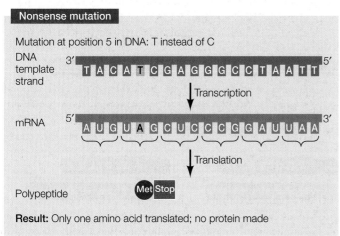

Nonsense mutation

Mutation at position 5 in DNA: T instead of C

DNA template strand 3′ T A C A T C G A G G G C C T A A T T 5′

↓ Transcription

mRNA 5′ A U G U A G C U C C C G G A U U A A 3′

↓ Translation

Polypeptide Met Stop

Result: Only one amino acid translated; no protein made

15.2 Point Mutations When they occur in the coding regions of proteins, single-base pair changes can cause missense, nonsense, or frameshift mutations. Some of these mutations are silent, while others affect the protein's amino acid sequence.

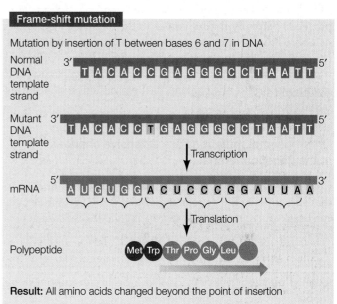

Frame-shift mutation

Mutation by insertion of T between bases 6 and 7 in DNA

Normal DNA template strand 3′ T A C A C C G A G G G C C T A A T T 5′

Mutant DNA template strand 3′ T A C A C C T G A G G G C C T A A T T 5′

↓ Transcription

mRNA 5′ A U G U G G A C U C C C G G A U U A A 3′

↓ Translation

Polypeptide Met Trp Thr Pro Gly Leu

Result: All amino acids changed beyond the point of insertion

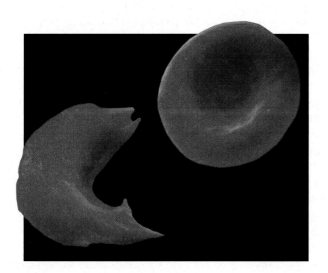

15.3 Sickled and Normal Red Blood Cells The misshapen red blood cell on the left is caused by a missense mutation and an incorrect amino acid in one of the two polypeptides of hemoglobin.

hydrophilic amino acid may be substituted for another hydrophilic amino acid, so that the shape of the protein is unchanged. Or a missense mutation might reduce the functional efficiency of a protein rather than completely inactivating it. Therefore, individuals homozygous for a missense mutation in a protein essential for life may survive if enough of the protein's function is retained.

In some cases, a gain of function missense mutation occurs. An example is a mutation in the human *TP53* gene, which codes for a tumor suppressor; that is, the TP53 protein normally functions to inhibit the cell cycle. Certain mutations of the *TP53* gene cause this protein to no longer inhibit cell division, but to promote it and prevent programmed cell death. So a TP53 protein mutated in this way has a gain of oncogenic (cancer-causing) function.

NONSENSE MUTATIONS A **nonsense mutation** involves a base substitution that causes a stop codon (for translation) to form somewhere in the mRNA. A nonsense mutation results in a shortened protein, since translation does not proceed beyond the point where the mutation occurred. For example, a common mutation causing thalassemia (another blood disorder affecting hemoglobin) in Mediterranean populations is a nonsense mutation that drastically shortens the α-globin subunit. Shortened proteins are usually not functional; however, if the nonsense mutation occurs near the 3′ end of the gene, it may have no effect on function.

FRAME-SHIFT MUTATIONS Not all point mutations are base substitutions. Single or double bases may be inserted into or deleted from DNA. Such mutations in coding sequences are known as **frame-shift mutations** because they interfere with the translation of the genetic message by throwing it out of register. Think again of codons as three-letter words, each corresponding to a particular amino acid. Translation proceeds codon by codon; if a base is added to the mRNA or subtracted from it, translation proceeds perfectly until it comes to the one-base insertion or deletion. From that point on, the three-letter words in the genetic message are one letter out of register. In other words, such mutations shift the "reading frame" of the message. Frame-shift mutations almost always lead to the production of nonfunctional proteins.

Chromosomal mutations are extensive changes in the genetic material

Changes in single nucleotides are not the most dramatic changes that can occur in the genetic material. Whole DNA molecules can break and rejoin, grossly disrupting the sequence of genetic information. There are four types of such chromosomal mutations: *deletions*, *duplications*, *inversions*, and *translocations*. These mutations can be caused by severe damage to chromosomes resulting from mutagens or by drastic errors in chromosome replication.

- **Deletions** result from the removal of part of the genetic material (**Figure 15.4A**). Like frame-shift point mutations, their consequences can be severe unless they affect noncoding DNA or unnecessary genes, or are masked by the presence of normal alleles of the deleted genes in the same cell. It is easy to imagine one mechanism that could produce deletions: a DNA molecule might break at two points and the two end pieces might rejoin, leaving out the DNA between the breaks.

- **Duplications** can be produced at the same time as deletions (**Figure 15.4B**). A duplication would arise if homologous chromosomes broke at different positions and then reconnected to the wrong partners. One of the two chromosomes produced by this mechanism would lack a segment of DNA (it would have a deletion), and the other would have two copies (a duplication) of the segment that was deleted from the first chromosome.

- **Inversions** can also result from breaking and rejoining of chromosomes. A segment of DNA may be removed and reinserted into the same location in the chromosome, but "flipped" end over end so that it runs in the opposite direction (**Figure 15.4C**). If the break site includes part of a DNA

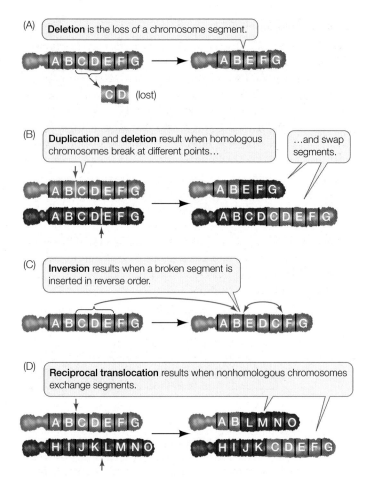

15.4 Chromosomal Mutations Chromosomes may break during replication, and parts of chromosomes may then rejoin incorrectly. The letters on these chromosome illustrations represent large segments of the chromosomes. Each segment may include anywhere from zero to hundreds or thousands of genes.

segment that codes for a protein, the resulting protein will be drastically altered and almost certainly nonfunctional.

- **Translocations** result when a segment of a chromosome breaks off and is inserted into a different chromosome. Translocations may involve reciprocal exchanges of chromosome segments, as in **Figure 15.4D**. Translocations often lead to duplications and deletions and may result in sterility if normal chromosome pairing in meiosis cannot occur.

Mutations can be spontaneous or induced

It is useful to distinguish two types of mutations in terms of their causes:

- **Spontaneous mutations** are permanent changes in the genetic material that occur without any outside influence. In other words, they occur simply because cellular processes are imperfect.

- **Induced mutations** occur when some agent from outside the cell—a mutagen—causes a permanent change in DNA.

Spontaneous mutations may occur by several mechanisms:

- *The four nucleotide bases of DNA can have different structures.* Each can exist in two different forms (called *tautomers*), one of which is common and one rare. When a base temporarily forms its rare tautomer, it can pair with the wrong base. For example, C normally pairs with G, but if C is in its rare tautomer at the time of DNA replication, it pairs with (and DNA polymerase will insert) an A. The result is a point mutation: G → A (**Figure 15.5A and C**).

- *Bases in DNA may change because of a chemical reaction*—for example, loss of an amino group in cytosine (a reaction called *deamination*). If this occurs in a DNA molecule, the error will usually be repaired. However, since the repair mechanism is not perfect, the altered nucleotide will sometimes remain during replication. Then, DNA polymerase will add an A (which base-pairs with U) instead of G (which normally pairs with C).

- *DNA polymerase can make errors in replication* (see Section 13.4)—for example, inserting a T opposite a G. Most of these errors are repaired by the proofreading function of the replication complex, but some errors escape detection and become permanent.

- *Meiosis is not perfect.* Nondisjunction—failure of homologous chromosomes to separate during meiosis—can occur, leading to one too many or one too few chromosomes (aneuploidy; see Figure 11.21). Random chromosome breakage and rejoining can produce deletions, duplications, inversions, or translocations.

Induced mutations result from alterations of DNA by mutagens:

- *Some chemicals can alter the nucleotide bases.* For example, nitrous acid (HNO_2) and similar molecules can react with cytosine and convert it to uracil by deamination. More specifically, they convert an amino group on the cytosine ($—NH_2$) into a keto group ($—C=O$). This alteration has the same result as spontaneous deamination: instead of a G, DNA polymerase inserts an A (**Figure 15.5B and C**).

- *Some chemicals add groups to the bases.* For instance, benzopyrene, a component of cigarette smoke, adds a large chemical group to guanine, making it unavailable for base pairing. When DNA polymerase reaches such a modified guanine, it inserts any one of the four bases; of course, three-fourths of the time the inserted base will not be cytosine, and a mutation results.

- *Radiation damages the genetic material.* Radiation can damage DNA in two ways. First, ionizing radiation (including X rays, gamma rays, and radiation from unstable isotopes) produces highly reactive chemicals called *free radicals*. Free radicals can change bases in DNA to forms that are not recognized by DNA polymerase. Ionizing radiation can also

15.5 Spontaneous and Induced Mutations (A) All four nitrogenous bases in DNA exist in both a prevalent (common) form and a rare form. When a base spontaneously forms its rare tautomer, it can pair with a different base. (B) Mutagens such as nitrous acid can induce changes in the bases. (C) The results of both spontaneous and induced mutations are permanent changes in the DNA sequence following replication.

(A) A spontaneous mutation

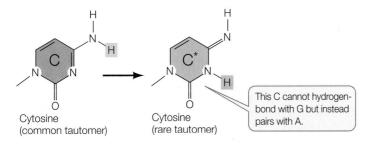

(B) An induced mutation

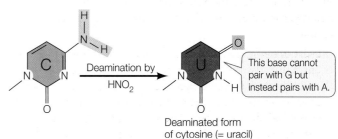

(C) The consequences of either mutation

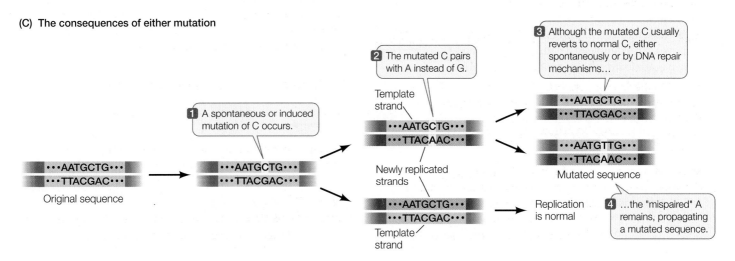

break the sugar–phosphate backbone of DNA, causing chromosomal abnormalities. Second, ultraviolet radiation (from the sun or a tanning lamp) can damage DNA in another way. It is absorbed by thymine, causing it to form covalent bonds with adjacent bases. This, too, plays havoc with DNA replication by distorting the double helix.

Some base pairs are more vulnerable than others to mutation

DNA sequencing has revealed that mutations occur most often at certain base pairs. These "hot spots" are often located where cytosine has been methylated to 5-methylcytosine.

As we discussed above, unmethylated cytosine can lose its amino group, either spontaneously or because of a chemical mutagen, to form uracil (see Figure 15.5B). This type of error is usually detected by the cell and repaired. The DNA repair mechanism recognizes uracil as inappropriate for DNA (since uracil occurs only in RNA) and replaces it with cytosine.

However, when 5-methylcytosine loses its amino group, the product is thymine, a natural base for DNA. The DNA repair mechanism ignores this thymine (**Figure 15.6**). During replication, however, the mismatch repair mechanism recognizes that G-T is a mismatched pair, although it cannot tell which base was incorrectly inserted into the sequence. So half of the time it matches a new C to the G, but the other half of the time it matches a new A to the T, resulting in a mutation. It is not surprising that 5-methylcytosine residues are hot spots for mutation.

Mutagens can be natural or artificial

Many people associate mutagens with materials made by humans, but just as there are many human-made chemicals that cause mutations, there are also many mutagenic substances that occur naturally. Plants (and to a lesser extent animals) make thousands of small molecules that they use for their own purposes, such as defense against pathogens (see Chapter 39). Some of these are mutagenic and potentially carcinogenic. Examples of human-made mutagens are nitrites, which are used to preserve meats. Once in mammals, nitrites get converted by the smooth endoplasmic reticulum (ER) to nitrosamines, which are strongly mutagenic because they cause deamination of cytosine (see above). An example of a naturally occurring mutagen is aflatoxin, which is made by the mold *Aspergillus*. When mammals ingest the mold, the aflatoxin is converted by the ER into a product that, like benzopyrene from cigarette smoke, binds to guanine; this also causes mutations.

Radiation can also be human-made or natural. Some of the isotopes made in nuclear reactors and nuclear bomb explosions are certainly harmful. For example, extensive studies have shown increased mutations in the survivors of the atom bombs dropped on Japan in 1945. You probably know that natural ultraviolet radiation in sunlight also causes mutations, in this case by affecting thymine and, to a lesser extent, other bases in DNA.

Biochemists have estimated how much DNA damage occurs in the human genome under normal circumstances: among the genome's 2.3 billion base pairs there are about 16,000 DNA-damaging events per cell per day, of which 80 percent are repaired.

Mutations have both benefits and costs

What is the overall effect of mutation? For an organism, there are benefits and costs.

- *Mutations are the raw material of evolution.* Without mutation, there would be no evolution. As we will see in Part Seven of this book, mutation alone does not drive evolution, but it provides the genetic diversity that makes natural selection possible. This diversity can be beneficial in two ways. First, a mutation in somatic cells may benefit the organism immediately. Second, a mutation in germ line cells may have no immediate selective advantage to the organism but may cause a phenotypic change in offspring. If the environment changes in a later generation, that mutation may be advantageous and thus selected for under these conditions.

- *Germ line and somatic mutations can be harmful.* Mutations in germ line cells that get carried to the next generation are often deleterious, especially if the offspring are homozygous for a harmful recessive allele. In their extreme form, such mutations produce phenotypes that are lethal. Lethal mutations can kill an organism during early development, or the organism may die before maturity and reproduction.

In Chapter 11 we described how genetic changes in somatic cells can lead to cancer. Typically these are mutations in oncogenes (the "gas pedal") that result in the stimulation of cell division, or mutations in tumor suppressor genes (the "brakes") that result in a lack of inhibition of cell division. These muta-

15.6 5-Methylcytosine in DNA Is a "Hot Spot" for Mutations   If cytosine has been methylated to 5-methylcytosine, the mutation is unlikely to be repaired and a C-G base pair is replaced with a T-A pair.

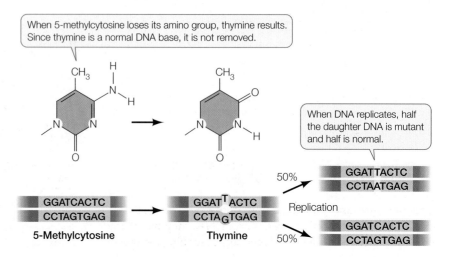

When 5-methylcytosine loses its amino group, thymine results. Since thymine is a normal DNA base, it is not removed.

When DNA replicates, half the daughter DNA is mutant and half is normal.

GGATCACTC
CCTAGTGAG
5-Methylcytosine

GGAT^{T}ACTC
CCTA$_G$TGAG
Thymine

Replication

50%   GGATTACTC / CCTAATGAG

50%   GGATCACTC / CCTAGTGAG

tions can occur by either spontaneous or induced mutagenesis. While spontaneous mutagenesis is not in our control, we can certainly try to avoid mutagenic substances and radiation. Not surprisingly, many things that cause cancer (carcinogens) are also mutagens. A good example is benzopyrene (discussed above), which is found in coal tar, car exhaust fumes, and charbroiled foods, as well as in cigarette smoke.

A major environmental issue is the effect of both human-made and natural mutagens on public health. Identifying mutagens to which people are exposed, and estimating their risk for both mutagenesis and carcinogenesis, is a major public policy goal. Here are two recent examples:

- The Montreal Protocol is the only international environmental agreement signed and adhered to by all nations. It bans chlorofluorocarbons and other substances that cause depletion of the ozone layer in the upper atmosphere of Earth. Such depletion can result in increased ultraviolet radiation reaching Earth's surface. This would cause more somatic mutations which lead to skin cancer.

- Bans on cigarette smoking have rapidly spread throughout the world. Cigarette smoking causes cancer due to increased exposure of somatic cells in the lungs and throat to benzopyrene and other carcinogens.

15.1 RECAP

Mutations are alterations in the nucleotide sequence of DNA. They may be changes in single nucleotides or extensive rearrangements of chromosomes. If they occur in somatic cells, they will be passed on to daughter cells; if they occur in germ line cells, they will be passed on to offspring.

- What are the various kinds of point mutations? **See pp. 318–320 and Figure 15.2**

- What distinguishes the various kinds of chromosomal mutations: deletions, duplications, inversions, and translocations? **See p. 320 and Figure 15.4**

- Explain the difference between spontaneous and induced mutagenesis. Give an example of each. **See pp. 320–322 and Figure 15.5**

- Why do many mutations involve G-C base pairs? **See p. 322 and Figure 15.6**

We have seen that there are many different ways in which DNA can be altered, in terms of both the types of changes and the mechanisms by which they occur. We turn now to the ways that biologists detect mutations in DNA.

15.2 How Are DNA Molecules and Mutations Analyzed?

Once biologists understood the connections between phenotype and proteins, and between genes and DNA, they were faced with the important challenge of precisely describing the specific DNA changes that lead to specific protein changes—an area of research called *molecular genetics*. To begin this work, biologists needed tools to analyze DNA molecules for mutations. In this section we will see how some of the numerous naturally occurring enzymes that cleave DNA have now become one of the most important tools used in molecular genetics laboratories.

Restriction enzymes cleave DNA at specific sequences

All organisms, including bacteria, must have ways of dealing with their enemies. As we saw in Section 13.1, bacteria are attacked by viruses called bacteriophage. These viruses inject their genetic material into the host cell and turn it into a virus-producing factory, eventually killing the cell. Some bacteria defend themselves against such invasions by producing **restriction enzymes** (also known as *restriction endonucleases*), which cut double-stranded DNA molecules—such as those injected by bacteriophage—into smaller, noninfectious fragments (**Figure 15.7**). These enzymes break the bonds of the DNA backbone between the 3' hydroxyl group of one nucleotide and the 5' phosphate group of the next nucleotide. This cutting process is called **restriction digestion**.

There are many such restriction enzymes, each of which cleaves DNA at a specific sequence of bases called a **recognition sequence** or a **restriction site**. Most recognition sequences are 4–6 base pairs long. The sequence is recognized through the principles of protein–DNA interactions (see Section 13.2). That is, the base pairs inside the DNA double helix vary slightly in shape, so that a particular short sequence of base pairs will fit a specific three-dimensional structure on an enzyme.

Why doesn't a restriction enzyme cut the DNA of the bacterial cell that makes it? One way that the cell protects itself is by modifying the restriction sites on its own DNA. Specific modifying enzymes called *methylases* add methyl ($-CH_3$) groups to certain bases at the restriction sites on the host's DNA after it has been replicated. The methylation of the host's bases makes

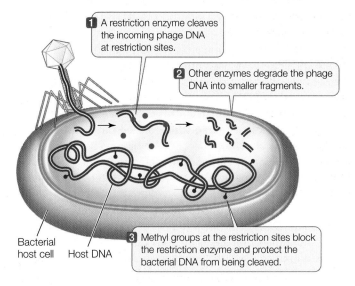

1 A restriction enzyme cleaves the incoming phage DNA at restriction sites.

2 Other enzymes degrade the phage DNA into smaller fragments.

Bacterial host cell   Host DNA

3 Methyl groups at the restriction sites block the restriction enzyme and protect the bacterial DNA from being cleaved.

15.7 Bacteria Fight Invading Viruses by Making Restriction Enzymes

the recognition sequence unrecognizable to the restriction enzyme. But unmethylated phage DNA is efficiently recognized and cleaved.

Bacterial restriction enzymes can be isolated from the cells that make them and used as biochemical reagents in the laboratory to give information about the nucleotide sequences of DNA molecules from other organisms. If DNA from any organism is incubated in a test tube with a restriction enzyme (along with buffers and salts that help the enzyme to function), that DNA will be cut wherever the restriction site occurs. A specific sequence of bases defines each restriction site. For example, the enzyme *Eco*RI (named after its source, a strain of the bacterium *E. coli*) cuts DNA only where it encounters the following paired sequence in the DNA double helix:

$$5' \ldots \text{GAATTC} \ldots 3'$$
$$3' \ldots \text{CTTAAG} \ldots 5'$$

Note that this sequence is palindromic, like the word "mom," in that the opposite strands have the same sequences when they are read from their 5′ ends. The *Eco*RI enzyme has two identical active sites on its two subunits, which cleave the two strands simultaneously between the G and the A of each strand.

The *Eco*RI recognition sequence occurs, on average, about once in every 4,000 base pairs in a typical prokaryotic genome, or about once per four prokaryotic genes. So *Eco*RI can chop a large piece of DNA into smaller pieces containing, on average, just a few genes. Using *Eco*RI in the laboratory to cut small genomes, such as those of viruses that have tens of thousands of base pairs, may result in a few fragments. For a huge eukaryotic chromosome with tens of millions of base pairs, a very large number of fragments will be created.

Of course, "on average" does not mean that the enzyme cuts all stretches of DNA at regular intervals. For example, the *Eco*RI recognition sequence does not occur even once in the 40,000 base pairs of the T7 phage genome—a fact that is crucial to the survival of this virus, since its host is *E. coli*. Fortunately for *E. coli*, the *Eco*RI recognition sequence does appear in the DNA of other bacteriophage.

Hundreds of restriction enzymes (all with unique recognition sequences) have been purified from various microorganisms. In the laboratory, different restriction enzymes can be used to cut samples of DNA from the same source. Thus restriction enzymes can be used to cut a sample of DNA in many different, specific places. The fragments formed can be used to create a physical map of the intact DNA molecule. Before DNA sequencing technology became automated and widely available, this was the principal way that DNA from different organisms was mapped and characterized.

Restriction enzyme digestion is used very widely today to manipulate DNA in the laboratory, and to identify and analyze point mutations. Fragments of DNA from different organisms can be amplified using the polymerase chain reaction (PCR; see Section 13.5). Even between closely related individuals, these amplified fragments often contain variations in DNA sequences (due in most cases to silent mutations). If these variations affect restriction sites, then digestion of the fragments with restriction enzymes can be used to distinguish between the samples. Restriction enzymes are also used to cut DNA for use in genetic engineering experiments, and in many other types of experiments aimed at understanding how organisms function at the molecular level.

Gel electrophoresis separates DNA fragments

After a laboratory sample of DNA has been cut with a restriction enzyme, the DNA is in fragments, which must be separated to identify (map) where the cuts were made. Because the recognition sequence does not occur at regular intervals, the fragments are not all the same size, and this property provides a way to separate them from one another. Separating the fragments is necessary to determine the number and molecular sizes (in base pairs) of the fragments produced, or to identify and purify an individual fragment for further analysis or for use in an experiment.

A convenient way to separate or purify DNA fragments is by **gel electrophoresis**. Samples containing the fragments are placed in wells at one end of a semisolid gel (usually made of agarose or polyacrylamide polymers) and an electric field is applied to the gel (**Figure 15.8**). Because of its phosphate groups, DNA is negatively charged at neutral pH; therefore, because opposite charges attract, the DNA fragments move through the gel toward the positive end of the field. Because the spaces between the polymers of the gel are small, small DNA molecules can move through the gel faster than larger ones. Thus, DNA fragments of different sizes separate from one another and can be detected with a dye. This gives us three types of information:

- *The number of fragments.* The number of fragments produced by digestion of a DNA sample with a given restriction enzyme depends on how many times that enzyme's recognition sequence occurs in the sample. Thus gel electrophoresis can provide some information about the presence of specific DNA sequences in the DNA sample.

- *The sizes of the fragments.* DNA fragments of known size are often placed in one well of the gel to provide a standard for comparison. This tells us how large the DNA fragments in the other wells are. By comparing the fragment sizes obtained with two or more restriction enzymes, the locations of their recognition sites relative to one another can be worked out (mapped).

- *The relative abundance of a fragment.* In many experiments, the investigator is interested in how much DNA is present. The relative *intensity* of a band produced by a specific fragment can indicate the amount of that fragment.

After separation on a gel, a fragment with a specific DNA sequence can be revealed with a single-stranded DNA probe (as we will see later in this chapter; see Figure 15.16). The gel region containing the desired fragment (in size or sequence) can be cut out as a lump of gel, and the pure DNA fragment can then be removed from the gel by diffusion into a small volume of water. This fragment can then be analyzed in terms of sequence or amplified and used experimentally.

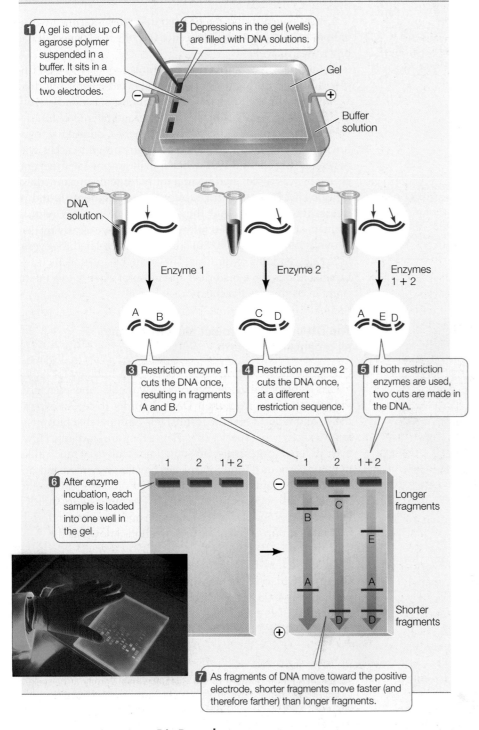

TOOLS FOR INVESTIGATING LIFE

15.8 Separating Fragments of DNA by Gel Electrophoresis

A mixture of DNA fragments is placed in a gel and an electric field is applied across the gel. The negatively charged DNA moves toward the positive end of the field, with smaller molecules moving faster than larger ones. After minutes to hours for separation, the electric power is shut off and the separated fragments can be analyzed.

1 A gel is made up of agarose polymer suspended in a buffer. It sits in a chamber between two electrodes.

2 Depressions in the gel (wells) are filled with DNA solutions.

Gel

Buffer solution

DNA solution

Enzyme 1   Enzyme 2   Enzymes 1 + 2

A   B   C   D   A   E D

3 Restriction enzyme 1 cuts the DNA once, resulting in fragments A and B.

4 Restriction enzyme 2 cuts the DNA once, at a different restriction sequence.

5 If both restriction enzymes are used, two cuts are made in the DNA.

6 After enzyme incubation, each sample is loaded into one well in the gel.

1   2   1 + 2   1   2   1 + 2

Longer fragments

B   C   E   A   A   D   D

Shorter fragments

7 As fragments of DNA move toward the positive electrode, shorter fragments move faster (and therefore farther) than longer fragments.

yourBioPortal.com
GO TO Animated Tutorial 15.1 • Separating Fragments of DNA by Gel Electrophoresis

DNA fingerprinting uses restriction analysis and electrophoresis

The two methods we have just described—restriction digestion to cut DNA into fragments and gel electrophoresis to separate them by size—are techniques used in **DNA fingerprinting**, which identifies individuals based on their DNA profiles. DNA fingerprinting works best with sequences that are highly polymorphic—that is, sequences that have multiple alleles (due to many point mutations during the evolution of the organism) and are therefore likely to be different in different individuals. Two types of polymorphisms are especially informative:

- **Single nucleotide polymorphisms** (**SNPs**; pronounced "snips") are inherited variations involving a single nucleotide base (so SNPs are point mutations). These polymorphisms have been mapped for many organisms. If one parent is homozygous for A at a certain point on the genome, and the other parent has a G at that point, the offspring will be heterozygous: one chromosome will have A at that point and the other will have G. If a SNP occurs in a restriction enzyme recognition site, such that one variant is recognized by the enzyme and the other isn't, then individuals can be distinguished from one another by amplifying a DNA fragment containing that site from a sample of total DNA isolated from each individual. The fragments are then cut with the restriction enzyme and analyzed by gel electrophoresis.

- **Short tandem repeats** (**STRs**) are short, repetitive DNA sequences that occur side by side on the chromosomes, usually in the noncoding regions. These repeat patterns, which contain 1–5 base pairs, are also inherited. For example, at a particular locus on chromosome 15 there may be an STR of "AGG." An individual may inherit an allele with six copies of the repeat (AGGAGGAGGAGGAGGAGG) from her mother and an allele with two copies (AGGAGG) from her father. Again, PCR is used to amplify DNA fragments containing these repeat sequences, and the fragments are distinguished by gel electrophoresis (**Figure 15.9**).

The method of DNA fingerprinting used most commonly today involves STR analysis. When several different STR loci, each with numerous alleles, are analyzed, an individual's unique pat-

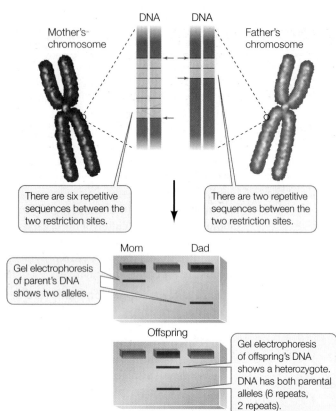

Mother's chromosome — DNA DNA — Father's chromosome

There are six repetitive sequences between the two restriction sites.

There are two repetitive sequences between the two restriction sites.

Mom   Dad

Gel electrophoresis of parent's DNA shows two alleles.

Offspring

Gel electrophoresis of offspring's DNA shows a heterozygote. DNA has both parental alleles (6 repeats, 2 repeats).

15.9 DNA Fingerprinting with Short Tandem Repeats A particular STR locus can be analyzed to determine the number of repeat sequences that were inherited by an individual from each parent. The two alleles can be identified in an electrophoresis gel on the basis of their sizes. When several STR loci are analyzed, the pattern can constitute a definitive identification of an individual.

tern becomes apparent. The Federal Bureau of Investigation in the United States uses 13 STR loci in its Combined DNA Index System (CODIS) database.

DNA fingerprinting can be used in forensics (crime investigation) to help prove the innocence or guilt of a suspect. It has other uses, as well.

A fascinating example demonstrates the use of DNA fingerprinting in the analysis of historical events. Three hundred years of rule by the Romanov dynasty in Russia ended on July 16, 1918, when Tsar Nicholas II, his wife, and their five children were executed by a firing squad during the Communist revolution. A report that the bodies had been burned to ashes was never questioned until 1991, when a shallow grave with several skeletons was discovered several miles from the presumed execution site. DNA fingerprinting of bone fragments found in this grave indicated that they came from an older man, a woman, and three female children, who all were clearly related to one another (**Figure 15.10**) and were also related to several living descendants of the Tsar. The accuracy and specificity of these methods gave historical and cultural closure to a major event in the twentieth century.

The DNA barcode project aims to identify all organisms on Earth

One of the most exciting aspects of DNA technology for biologists is its potential to identify species, varieties, and even individual organisms from their DNA. In order to repeat experiments and report scientific results, it is essential that biologists know exactly what species or varieties they are studying. However, different organisms can sometimes look very much alike in nature. About 1.7 million species have been named and described, but about ten times that number probably have yet to be identified. A proposal to use DNA technology to identify known species and detect the unknown ones has been endorsed by a large group of scientific organizations known as the Consortium for the Barcode of Life (CBOL).

Evolutionary biologist Paul Hebert at the University of Guelph in Ontario, Canada was walking down the aisle of a supermarket in 1998 when he noticed the barcodes on all the packaged foods. This gave him an idea to identify each species with a "DNA barcode" that is based on a short sequence from a sin-

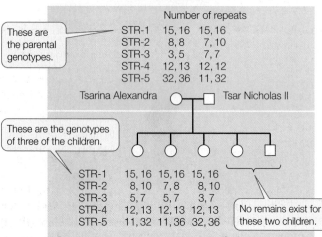

	Number of repeats	
STR-1	15,16	15,16
STR-2	8,8	7,10
STR-3	3,5	7,7
STR-4	12,13	12,12
STR-5	32,36	11,32

These are the parental genotypes.

Tsarina Alexandra   Tsar Nicholas II

These are the genotypes of three of the children.

STR-1	15,16	15,16	15,16
STR-2	8,10	7,8	8,10
STR-3	5,7	5,7	3,7
STR-4	12,13	12,13	12,13
STR-5	11,32	11,36	32,36

No remains exist for these two children.

15.10 DNA Fingerprinting of the Russian Royal Family The skeletal remains of Tsar Nicholas II, his wife Alexandra, and three of their children were found in 1991 and subjected to DNA fingerprinting. Five STRs were tested. The results can be interpreted by looking at the inheritance of alleles from each parent in the children. In STR-2, for example, the parents had genotypes 8,8 (homozygous) and 7,10 (heterozygous). The three children all inherited type 8 from Alexandra and either type 7 or type 10 from Nicholas.

15.11 A DNA Barcode A 650- to 750-base-pair region of the cytochrome oxidase gene can be amplified by PCR from any organism and then sequenced. This knowledge is used to make a bar code in which each of the four DNA bases is represented as a different color. Such a species barcode permits accurate and rapid identification of a particular species for experimental, ecological or evolutionary studies.

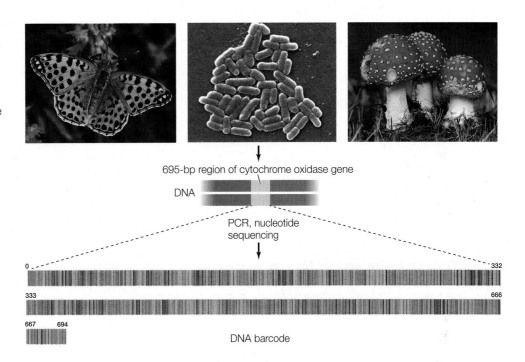

695-bp region of cytochrome oxidase gene

DNA

PCR, nucleotide sequencing

DNA barcode

gle gene. The gene he chose is the cytochrome oxidase gene, a component of the respiratory chain that is present in most organisms. Because this gene mutates readily, there are many allelic differences between species. A fragment of 650–750 base pairs in this gene is being sequenced for all organisms, and so far sufficient variation has been detected to make it diagnostic for each species (**Figure 15.11**).

Once the DNA of the targeted gene fragment has been sequenced for all known species, a simple device for conducting field analyses can be developed. The barcode project has the potential to advance biological research on evolution, to track species diversity in ecologically significant areas, to help identify new species, and even to detect undesirable microbes or bioterrorism agents.

15.2 RECAP

Large DNA molecules can be cut into smaller pieces by restriction digestion and then sorted by gel electrophoresis. PCR is used to amplify sequences of interest from complex samples. These techniques are used in DNA fingerprinting to analyze DNA polymorphisms for the purpose of identifying individuals. Scientists hope to identify all species using DNA analysis.

- How does a restriction enzyme recognize a restriction site on DNA? **See p. 323**
- How does gel electrophoresis separate DNA fragments? **See p. 324 and Figure 15.8**
- What are STRs and how are they used to identify individuals? **See pp. 325–326**

We have seen that molecular methods can be used to identify individuals because of mutations in their DNA. Many of the STRs and SNPs used in these analyses do not occur in protein-coding regions, and so probably do not affect the phenotype. Nevertheless, they are mutations—inherited changes in the DNA. We now turn to mutations that affect phenotype, using humans as our model organism.

15.3 How Do Defective Proteins Lead to Diseases?

The biochemistry that relates genotype (DNA) and phenotype (proteins) has been most completely described for model organisms, such as the prokaryote *E. coli* and the eukaryotes yeast and *Drosophila*. While the details vary, there is great similarity in the fundamental processes among these forms of life. These similarities have permitted the application of knowledge and methods discovered using these model organisms to the study of human biochemical genetics. Of particular interest are the effects of mutations on human phenotypes, sometimes leading to diseases.

Genetic mutations may make proteins dysfunctional

Genetic mutations are often expressed phenotypically as proteins that differ from normal (wild-type) proteins. Abnormalities in enzymes, receptor proteins, transport proteins, structural proteins, and most of the other functional classes of proteins have all been implicated in genetic diseases.

DYSFUNCTIONAL ENZYMES In 1934, the urine of two mentally retarded young siblings was found to contain phenylpyruvic acid, an unusual by-product of the metabolism of the amino acid phenylalanine. It was not until two decades later, however, that the complex clinical phenotype of the disease that afflicted these children, called *phenylketonuria* (*PKU*), was traced back to its molecular cause. The disease resulted from an abnormality in a single enzyme, phenylalanine hydroxylase (**Figure 15.12**). This enzyme normally catalyzes the conversion of dietary phenylalanine to tyrosine, but it was not active in the livers of PKU patients. Lack of this conversion led to excess phenylalanine in the blood and explained the accumulation of phenylpyruvic acid. Later, the amino acid sequence of phenylalanine hydroxylase (PAH) in normal people was compared with the amino acid sequences

1 The enzyme that converts phenylalanine to tyrosine is nonfunctional.

2 Because conversion to tyrosine is blocked, phenylalanine and phenylpyruvic acid accumulate.

15.12 One Gene, One Enzyme Phenylketonuria is caused by an abnormality in a specific enzyme that metabolizes the amino acid phenylalanine. Knowing the molecular causes of such single-gene, single-enzyme metabolic diseases can aid researchers in developing screening tests as well as treatments.

ABNORMAL HEMOGLOBIN The first human genetic disease known to be caused by an amino acid sequence abnormality was sickle-cell disease. This blood disorder most often afflicts people whose ancestors came from the tropics or from the Mediterranean. About 1 in 655 African-Americans are homozygous for the sickle allele and have the disease. The abnormal allele produces abnormal hemoglobin that results in sickle-shaped red blood cells (see Figure 15.3). These cells tend to block narrow blood capillaries, especially when the oxygen concentration of the blood is low. The result is tissue damage and eventually death by organ failure.

Recall that human hemoglobin is a protein with quaternary structure, containing four globin subunits—two α chains and two β chains—as well as the pigment heme (see Figure 3.10). In sickle-cell disease, one of the 146 amino acids in the β-globin chain is abnormal: at position 6, the normal glutamic acid has been replaced by valine. This replacement changes the charge of the protein (glutamic acid is negatively charged and valine is neutral), causing it to form long, needle-like aggregates in the red blood cells. The phenotypic result is anemia, a deficiency of normal red blood cells and an impaired ability of the blood to carry oxygen.

Because hemoglobin is easy to isolate and study, its variations in the human population have been extensively documented (**Figure 15.13**). Hundreds of single amino acid alterations in β-globin have been reported. For example, at the same position that is mutated in sickle-cell disease (resulting in hemoglobin S), the normal glutamic acid may be replaced by lysine, causing hemoglobin C disease. In this case, the resulting anemia is usually not severe. Many alterations of hemoglobin do not affect the protein's function. That is fortunate, because about 5 percent of all humans are carriers for one of these variants.

There are hundreds of inherited diseases in humans in which the primary phenotypes are caused by specific mutations leading to protein abnormalities. Some of the more common examples are listed in **Table 15.2**. These mutations can be domi-

from individuals with PKU. Many people with PKU had tryptophan instead of arginine at position 408 of this long polypeptide chain of 452 amino acids (**Table 15.1**).

The exact cause of mental retardation in PKU remains elusive, although, as we will see later in this chapter, it can be prevented. We can, however, understand why most people with PKU have light skin and hair color. The pigment melanin, which is responsible for dark skin and hair, is made from tyrosine, which people with PKU cannot synthesize adequately.

Hundreds of human genetic diseases that result from enzyme abnormalities have been discovered, many of which lead to mental retardation and premature death. Most of these diseases are rare; PKU, for example, shows up in one out of every 12,000 newborns. But these diseases are just the tip of the mutation iceberg. Some mutations result in amino acid changes that have no obvious clinical effects. In fact, amino acid differences among individuals have been detected in at least 30 percent of all human proteins whose sequences are known. Thus polymorphism does not necessarily mean disease. There can be numerous alleles of a gene, some producing proteins that function normally, while others produce variants that cause disease—as we will now see for hemoglobin.

TABLE 15.1
Two Common Mutations That Cause Phenylketonuria

	NORMAL CODON 408	MUTANT CODON 408 (20% OF PKU CASES)	NORMAL CODON 280	MUTANT CODON 280 (2% OF PKU CASES)
Length of PAH protein	452 amino acids	452 amino acids	452 amino acids	452 amino acids
DNA at codon	xxCGGxx xxGCCxx	xxTGGxx xxACCxx	xxGAAxx xxCTTxx	xxAAAxx xxTTTxx
mRNA at codon	xxCGGxx	xxUGGxx	xxGAAxx	xxAAAxx
Amino acid at codon	Arginine	Tryptophan	Glutamic acid	Lysine
Active PAH enzyme?	Yes	No	Yes	No

	TABLE 15.2		
	Some Human Genetic Diseases		
DISEASE NAME	**INHERITANCE PATTERN; FREQUENCY**	**GENE MUTATED; PROTEIN PRODUCT**	**CLINICAL PHENOTYPE**
Familial hypercholesterolemia	Autosomal codominant; 1 in 500 heterozygous	*LDLR*; low-density lipoprotein receptor	High blood cholesterol, heart disease
Cystic fibrosis	Autosomal recessive; 1 in 4000	*CFTR*; chloride ion channel in membrane	Immune, digestive, and respiratory illness
Duchenne muscular dystrophy	Sex-linked recessive; 1 in 3500 males	*DMD*; the muscle membrane protein dystrophin	Muscle weakness
Hemophilia A	Sex-linked recessive; 1 in 5000 males	*HEMA*; factor VIII blood clotting protein	Inability to clot blood after injury, hemorrhage

nant, codominant, or recessive, and some are sex-linked. Before we examine how these diseases can be analyzed at the molecular level, we turn briefly to a fascinating exception to the association between genes and proteins.

Prion diseases are disorders of protein conformation

Transmissible spongiform encephalopathies (TSEs) are degenerative brain diseases that occur in many mammals, including humans. The brain gradually develops holes, making it look like a sponge. Scrapie is a TSE that has been known for 250 years. It causes affected sheep and goats to show the abnormal behavior of rubbing ("scraping") the wool off their bodies (as well as causing more severe neurological problems). In the 1980s, a TSE that appeared in cows in Britain was traced to the cows having eaten products from sheep that had scrapie. These cows would shake and rub their bodies against fences, and their staggering led farmers to dub them "mad cows." In the 1990s, some people who ate beef from these cows got a human version of the disease, dubbed "mad cow disease" by the media. Those with the disease eventually died.

At first, viruses were suspected to cause TSEs. But when Tikva Alper at Hammersmith Hospital, London, treated infectious extracts with high doses of ultraviolet light to inactivate nucleic acids, they still caused TSEs. She proposed that the causative agent was a protein, not a virus. Later, Stanley Prusiner at the University of California purified the protein responsible and showed it to be free of DNA or RNA. He called it a *proteinaceous infective particle*, or **prion**. This is a violation of the central dogma of molecular biology (DNA → RNA → protein; see Chapter 14), because in this case the protein was "doing it all." There was no genetic material involved.

This is a rare case of a mutant phenotype without a mutant gene. Normal brain cells contain a membrane protein called PrP^c. A protein with the *same amino acid sequence* is present in TSE-affected brain tissues, but that protein, called PrP^{sc}, has a different three-dimensional shape (**Figure 15.14**). Thus TSEs are not caused by a mutated gene (the primary structures of the two proteins are the same), but are somehow caused by an alteration in protein conformation. The altered three-dimensional structure of the protein has profound effects on its function in the cell. PrP^{sc} is insoluble, and it piles up as fibers in brain tissue, causing cell death.

How can the exposure of a normal cell to material containing PrP^{sc} result in a TSE? The abnormal PrP^{sc} protein seems to induce a conformational change in the normal PrP^c protein so that it, too, becomes abnormal. Just how this conversion occurs, and how it causes a TSE, is unclear.

To try to understand how TSEs develop, scientists are asking "What is the *normal* role of the prion protein?" Recently, it was shown that in the brain the prion protein blocks a key enzyme in the synthesis of a protein called β-amyloid. This is the protein that accumulates in the brains of patients with Alzheimer's disease. People with early-onset Alzheimer's (age 40) have less PrP^c in their brains than people who age normally. So the PrP^c pro-

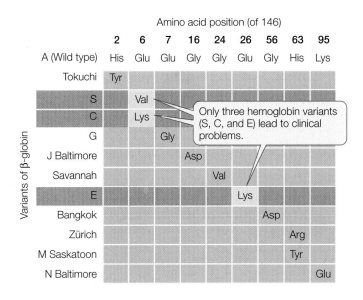

15.13 Hemoglobin Polymorphism Each of these mutant alleles codes for a protein with a single amino acid change in the 146-amino acid chain of β-globin. Only three of the hundreds of known variants of β-globin, shown on the left, are known to lead to clinical abnormalities. "S" is the sickle-cell anemia allele.

15.14 Prion Diseases are Disorders of Protein Conformation A normal membrane protein in brain cells (PrP^C, left) can be converted to the disease-causing form (PrP^SC, right), which has a different three-dimensional structure.

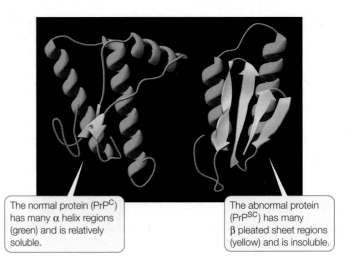

The normal protein (PrP^C) has many α helix regions (green) and is relatively soluble.

The abnormal protein (PrP^SC) has many β pleated sheet regions (yellow) and is insoluble.

tein appears to play a role in protecting against Alzheimer's disease. Other functions for this protein are also being discovered, but the mechanism by which it appears to spread TSE disease is not yet understood.

Prions are an unusual phenomenon in human disease. The vast majority of inherited diseases are caused by mutations in genes that reduce the levels of their protein products, or make the proteins dysfunctional. But the expression of these genes, like that of all genes, is influenced by the environment.

Most diseases are caused by multiple genes and environment

The human diseases for which clinical phenotypes can be traced to a single altered protein and its altered gene may number in the thousands. Taken together, these diseases have a frequency of about one percent in the human population.

Far more common, however, are diseases that are **multifactorial**; that is, diseases that are caused by the interactions of many genes and proteins with one or more factors in the environment. When studying genetics, we tend to call individuals either normal (wild type) or abnormal (mutant); however, in reality every individual contains thousands or millions of genetic variations that arose through mutations. Our susceptibility to disease is often determined by complex interactions between these genotypes and factors in the environment, such as the foods we eat or the pathogens we encounter. For example, a complex set of genotypes determine who among us can eat a high-fat diet and not experience a heart attack, or will succumb to disease when exposed to infectious bacteria. Estimates suggest that up to 60 percent of all people are affected by diseases that are *genetically influenced*. Identifying these genetic influences is a major task of molecular medicine and human genome sequencing.

15.3 RECAP

Many genetic mutations are expressed as nonfunctional enzymes, structural proteins, or membrane proteins. Human genetic diseases may be inherited in dominant, codominant, or recessive patterns, and they may be sex-linked.

- Describe an example of an abnormal protein in humans that results from a genetic mutation and causes a disease. **See pp. 327–328**

- Describe an example of an abnormal protein in humans that results from a genetic mutation and does *not* cause a disease. **See p. 328**

- How is the brain cell membrane protein PrP^c related to diseases caused by prions? **See p. 329**

The abnormal proteins that cause disease result (with the exception of TSEs) from genetic mutations. We now turn to the identification of such mutations, an important task for molecular medicine.

15.4 What DNA Changes Lead to Genetic Diseases?

We have seen for diseases such as PKU and sickle-cell anemia that the clinical phenotype of inherited diseases could be traced to individual proteins, and that the genes could then be identified. With the advent of new ways to identify DNA variations, a new pattern of human genetic analysis has emerged. In these cases, the clinical phenotype is first related to a DNA variation, and then the protein involved is identified. This pattern of discovery is called **reverse genetics**, because it proceeds in the opposite direction to genetic analyses done before the mid-1980s. For example, in sickle-cell anemia, the protein abnormality in hemoglobin was described first (a single amino acid change), and then the gene for β-globin was isolated and the DNA mutation was pinpointed.

clinical phenotype → protein phenotype → gene

On the other hand, for cystic fibrosis (see Table 15.2), a mutant version of the gene *CFTR* was isolated first, and then the protein was characterized:

clinical phenotype → gene → protein phenotype

Whichever approach is used, final identification of the protein(s) involved in a disease is important in designing specific therapies.

Genetic markers can point the way to important genes

To identify a mutant gene by reverse genetics, close linkage to a marker sequence is used. To understand this linkage, imagine an astronaut looking down from space, trying to find her son on a park bench on Chicago's North Shore. The astronaut first picks out reference points—landmarks that will lead her to the park. She recognizes the shape of North America, then moves to Lake Michigan, then the Willis Tower, and so on. Once she has zeroed in on the North Shore Park, she can use advanced optical instruments to find her son. The reference points for gene isolation are the **genetic markers**.

- Knowledge of at least *two mutations* is needed. One mutation determines the disease phenotype and the other mutation is a closely linked "marker mutation" that does not affect the disease phenotype but is easy to identify. In early genetic studies, markers that produced visible phenotypes were used to follow the inheritance of important traits. Today, single nucleotide polymorphisms (SNPs) or STRs are usually used.

- *Genetic linkage* is the co-inheritance of the marker and the disease-causing allele. If they are always together, they must be close together on the chromosome.

A key requirement for a genetic marker is that it has allelic polymorphisms (differences in sequence) that are identifiable by current methods of rapid DNA analysis. As we saw in Section 15.2, an STR can have varying numbers of a short repeat sequence, and thus there are multiple alleles of these markers. We also saw in Section 15.2 that restriction enzymes can be used to identify SNPs, provided the SNP occurs within a restriction site. Restriction enzymes can also be used to identify mutations such as insertions or deletions if the affected sequences contain restriction sites. We will examine in more detail the use of restriction enzymes to identify genetic polymorphisms, and other SNP markers, before returning to the discussion of human genes and their abnormalities.

RESTRICTION FRAGMENT LENGTH POLYMORPHISMS

As Section 15.2 describes, restriction enzymes cut DNA molecules at specific recognition sequences. On a particular human chromosome, a given restriction enzyme may make thousands of cuts, producing many DNA fragments. The enzyme *Eco*RI, for example, cuts DNA at

$$5'\ldots \text{GAATTC} \ldots 3'$$

Suppose this recognition sequence exists in a certain stretch of human chromosome 7. The restriction enzyme will cut this stretch once and make two fragments of DNA. Now suppose that, in some people, this sequence contains a SNP and is mutated as follows:

$$5'\ldots \text{GAGTTC} \ldots 3'$$

This sequence will not be recognized by the restriction enzyme; thus it will remain intact and yield one larger fragment of DNA.

Differences in DNA sequences due to mutations in restriction sites are called **restriction fragment length polymorphisms (RFLPs) (Figure 15.15)**. They can be easily visualized as bands on an electrophoresis gel. An RFLP band pattern is inherited in a Mendelian fashion and can be followed through a pedigree. Thousands of such markers have been described for humans and many other organisms.

Before the advent of PCR technology, the only way to analyze RFLPs was by digesting total genomic DNA samples with restriction enzymes. These samples contain thousands of DNA fragments of various sizes. In order to visualize a particular fragment, the DNA from the gel is transferred (blotted) onto a nylon membrane, denatured to separate the double-stranded molecules, and mixed with a single-stranded DNA fragment (a *probe*) containing at least part of the sequence within the RFLP fragment of interest (**Figure 15.16**). The probe hybridizes (by base pairing) with the DNA band containing the RFLP. Because the probe is "la-

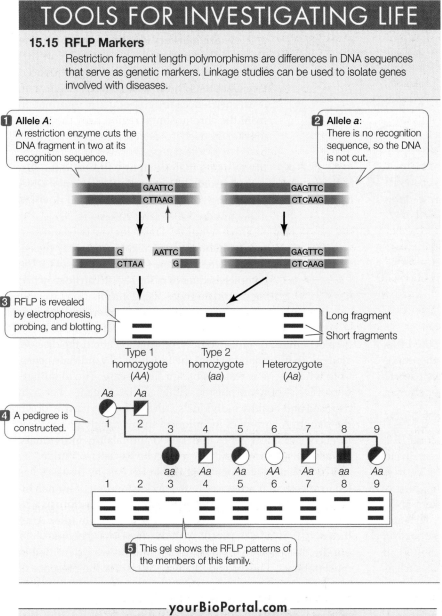

TOOLS FOR INVESTIGATING LIFE

15.15 RFLP Markers
Restriction fragment length polymorphisms are differences in DNA sequences that serve as genetic markers. Linkage studies can be used to isolate genes involved with diseases.

1 Allele *A*:
A restriction enzyme cuts the DNA fragment in two at its recognition sequence.

2 Allele *a*:
There is no recognition sequence, so the DNA is not cut.

GAATTC
CTTAAG

GAGTTC
CTCAAG

G AATTC
CTTAA G

GAGTTC
CTCAAG

3 RFLP is revealed by electrophoresis, probing, and blotting.

Long fragment
Short fragments

Type 1 homozygote (AA)
Type 2 homozygote (aa)
Heterozygote (Aa)

4 A pedigree is constructed.

Aa Aa
1   2

3 4 5 6 7 8 9

aa Aa Aa AA Aa aa Aa
1   2   3   4   5   6   7   8   9

5 This gel shows the RFLP patterns of the members of this family.

TOOLS FOR INVESTIGATING LIFE

15.16 Analyzing DNA Fragments by DNA Gel Blotting
A probe can be used to locate a specific DNA fragment on an electrophoresis gel.

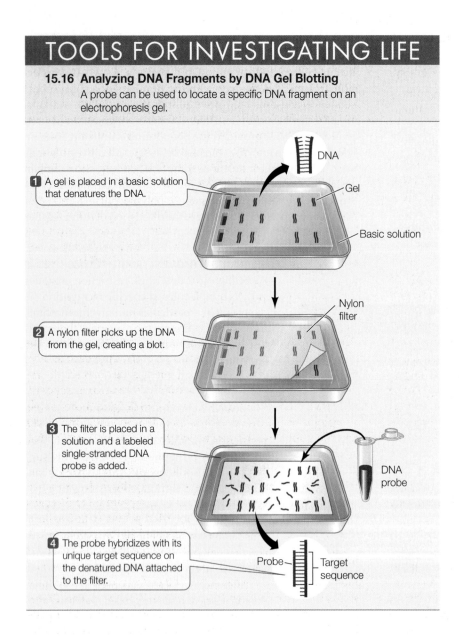

DNA

1 A gel is placed in a basic solution that denatures the DNA.

Gel

Basic solution

2 A nylon filter picks up the DNA from the gel, creating a blot.

Nylon filter

3 The filter is placed in a solution and a labeled single-stranded DNA probe is added.

DNA probe

4 The probe hybridizes with its unique target sequence on the denatured DNA attached to the filter.

Probe — Target sequence

established observation that if two genes are located near each other on the same chromosome, they are usually passed on together from parent to offspring. The same holds true for any pair of genetic markers. The idea is to find markers that are close by on the chromosome to get to the gene of interest.

To narrow down the location of a gene, a scientist must find a marker and a gene that are *always inherited together*. To do this, family medical histories are taken and pedigrees are constructed. If a genetic marker and a genetic disease are inherited together, then they must be near each other on the same chromosome. Unfortunately, "near each other" still might be as much as several million base pairs apart. The process of locating a gene is thus similar to that of the astronaut looking for her son: the first landmarks lead to only an approximate location.

How are markers identified that are more closely linked to the gene of interest? Now that the human genome has been fully sequenced (see Chapter 17), the task is much easier than it was just a decade ago. Sequence information from the chromosomal region near the linked marker is used to develop additional SNP-based or STR markers. These are tested to identify markers that are tightly linked to the disease phenotype. Eventually, the region of DNA containing the gene can be narrowed down to a few hundred thousand base pairs.

Once a linked DNA region is identified, many methods are available to identify the actual gene responsible for a genetic disease. The complete sequence of the region can be searched for candidate genes, using information available from databases of genome sequences. With luck a scientist can make an educated guess, based

beled" with a radioactive isotope or a chemical tag, the DNA fragment containing the RFLP can be seen among the thousands of other fragments on the blot. This technology was used to create "RFLP maps" of the human genome and of genomes of many other organisms.

SINGLE NUCLEOTIDE POLYMORPHISMS As noted in Section 15.2, single nucleotide polymorphisms (SNPs) are widespread in eukaryotic genomes. There is roughly one SNP for every 1,330 base pairs in the human genome. Not all SNPs occur within restriction sites, but those that don't can still be used as markers. SNPs can be detected by direct sequence comparisons, or by PCR amplification using primers that contain one version of the SNP, so that only one allele will be amplified efficiently. SNPs can also be detected using sophisticated chemical methods such as mass spectrometry (see Section 17.5).

Genetic markers such as STRs, RFLPs, and SNPs can be used as landmarks to find genes of interest if the genes also have alleles that are polymorphic. The key to this method is the well-

on biochemical or physiological information about the disease, along with information about the functions of candidate genes, as to which gene is responsible for the disease. The identification of DNA polymorphisms within candidate genes, between diseased and healthy individuals, can also help to narrow down the search. A variety of techniques, such as analyzing mRNA levels of candidate genes in diseased and healthy individuals, are used to confirm that the correct gene has been identified.

The isolation of genes responsible for genetic diseases has led to spectacular advances in the understanding of human biology. For example, the gene responsible for cystic fibrosis was first identified by its close association with a SNP marker. After the gene sequences of people with the disease and people without the disease were compared, a mutation was identified in most patients. This provided a way to test for the presence of that mutation in people, by extracting DNA from cells or tissues that could be easily sampled. Moreover, knowing the gene sequence led to the identification of the protein it codes for and a characterization of the abnormal protein. Treatments that specifically target this protein are now being devised. Research on the protein in normal people has led to an understanding of its role in the body. So reverse genetics can lead to diagnosis, treatment, and biological understanding.

Disease-causing mutations may involve any number of base pairs

Disease-causing mutations may involve a single base pair (as we saw in the case of hemophilia), a long stretch of DNA (as in cases of Duchenne muscular dystrophy, which we will discuss shortly), multiple segments of DNA (as in fragile-X syndrome), or even entire chromosomes (as we saw with Down syndrome in Section 11.5).

POINT MUTATIONS   There are many examples of point mutations in human genetic diseases. In some cases, all of the people with the disease have the same genetic mutation. This is the case with sickle-cell anemia, where a single base pair change in the β-globin gene causes a single amino acid change, which leads to the abnormal protein and phenotype. This is not the situation with most other genetic diseases. For example, over 500 different mutations in the *PAH* (phenylalanine hydroxylase) gene have been discovered in different patients with phenylketonuria (PKU; see Table 15.1). This makes sense if you think about the three-dimensional structure of an enzyme protein and the many amino acid changes that could affect its activity.

LARGE DELETIONS   Larger mutations may involve many base pairs of DNA. For example, deletions in the X chromosome that include the gene for the protein dystrophin result in Duchenne muscular dystrophy. Dystrophin is important in organizing the structure of muscles, and people who have only the abnormal form have severe muscle weakness. Sometimes only part of the dystrophin gene is missing, leading to an incomplete but partly functional protein and a mild form of the disease. In other cases, however, deletions span the entire sequence of the gene, so that the protein is missing entirely, resulting in a severe form of the disease. In yet other cases, deletions involve millions of base pairs and cover not only the dystrophin gene but adjacent genes as well; the result may be several diseases simultaneously.

CHROMOSOMAL ABNORMALITIES   Chromosomal abnormalities also cause human diseases. Such abnormalities include the gain

or loss of one or more chromosomes (aneuploidy) (see Figure 11.21), loss of a piece of chromosome (deletion), and the transfer of a piece of one chromosome to another chromosome (translocation) (see Figure 15.4). About one newborn in 200 has a chromosomal abnormality. While some of these abnormalities are inherited as preexisting aberrations from one or both parents, others are the result of meiotic events, such as nondisjunction, that occurred during the formation of gametes in one of the parents.

One common cause of mental retardation is *fragile-X syndrome* (**Figure 15.17**). About one male in 1,500 and one female in 2,000 are affected. These people have a constriction near the tip of the X chromosome. Although the basic pattern of inheritance is that of an X-linked recessive trait, there are departures from this pattern. Not all people with the fragile-X chromosomal abnormality are mentally retarded, as we will see.

Expanding triplet repeats demonstrate the fragility of some human genes

About one-fifth of all males that have the fragile-X chromosomal abnormality are phenotypically normal, as are most of their daughters. But many of those daughters' sons are mentally retarded. In a family in which the fragile-X syndrome appears, later generations tend to show earlier onset and more severe symptoms of the disease. It is almost as if the abnormal allele itself is changing—and getting worse. And that's exactly what is happening.

The gene responsible for fragile-X syndrome (*FMR1*) contains a repeated triplet, CGG, at a certain point in the promoter region. In normal people, this triplet is repeated 6 to 54 times (the average is 29). In mentally retarded people with fragile-X syndrome, the CGG sequence is repeated 200 to 2,000 times.

Males carrying a moderate number of repeats (55–200) show no symptoms and are called premutated. These repeats become more numerous as the daughters of these men pass the chromosome on to their children (**Figure 15.18**). With more than 200 repeats, increased methylation of the cytosines in the CGG triplets is likely, accompanied by transcriptional inactivation of the *FMR1* gene. The normal role of the protein product of this gene is to bind to mRNAs involved in neuron function and regulate their translation at the ribosome. When the FMR1 protein is not made in adequate amounts, these mRNAs are not properly translated, and nerve cells die. Their loss often results in mental retardation.

This phenomenon of **expanding triplet repeats** has been found in over a dozen other diseases, such as myotonic dystrophy (involving repeated CTG triplets) and Huntington's disease (in which CAG is repeated). Such repeats, which may be found within a protein-coding region or outside it, appear to be present in many other genes without causing harm. How the repeats expand is not known; one theory is that DNA polymerase may slip after copying a repeat and then fall back to copy it again.

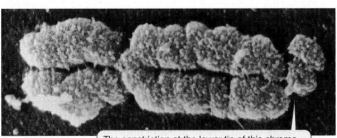

The constriction at the lower tip of this chromosome is the location of the fragile-X abnormality.

15.17 A Fragile-X Chromosome at Metaphase   The chromosomal abnormality associated with fragile-X syndrome shows up under the microscope as a constriction in the chromosome. This occurs during preparation of the chromosome for microscopy.

15.18 The CGG Repeats in the FMR1 Gene Expand with Each Generation The genetic defect in fragile-X syndrome is caused by 200 or more repeats of the CGG triplet.

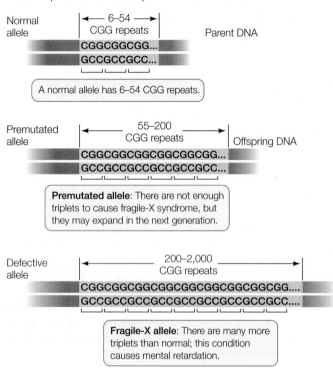

15.4 RECAP

Genes involved in disease can be identified by first detecting the abnormal DNA sequence and then the protein that the wild-type allele encodes. Unusual features such as expanding triplet repeats have been detected in the human genome.

- How can a gene be identified before its protein product is known? **See pp. 330–331 and Figure 15.15**

- How do expanding repeats cause genetic diseases? **See p. 333 and Fig. 15.18**

The determination of the precise molecular phenotypes and genotypes of various human genetic diseases has made it possible to diagnose these diseases even before symptoms first appear. Let's take a detailed look at some of these genetic screening techniques.

15.5 How Is Genetic Screening Used to Detect Diseases?

Genetic screening is the use of a test to identify people who have, are predisposed to, or are carriers of a genetic disease. It can be done at many times of life and used for many purposes.

- *Prenatal screening* can be used to identify an embryo or fetus with a disease so that medical intervention can be applied or decisions can be made about whether or not to continue the pregnancy.

- *Newborn babies* can be screened so that proper medical intervention can be initiated quickly for those babies who need it.

- *Asymptomatic people* who have a relative with a genetic disease can be screened to determine whether they are carriers of the disease or are likely to develop the disease themselves.

Genetic screening can be done at the level of either the phenotype or the genotype.

Screening for disease phenotypes involves analysis of proteins

At the level of the phenotype, genetic screening involves examining a protein relevant to the phenotype for abnormal structure or function. Since many proteins are enzymes, low enzyme activity is strongly suggestive of a mutation, as we saw in Section 15.1. Perhaps the best example of this kind of protein screening is a test for phenylketonuria (PKU), which has made it possible to identify the disease in newborns, so that treatment of the disease can be started. It is very likely that you were screened for PKU.

Initially, babies born with PKU have a normal phenotype because excess phenylalanine in their blood before birth diffuses across the placenta to the mother's circulatory system. Since the mother is almost always heterozygous, and therefore has adequate phenylalanine hydroxylase activity, her body metabolizes the excess phenylalanine from the fetus. After birth, however, the baby begins to consume protein-rich food (milk) and to break down some of his or her own proteins. Phenylalanine enters the baby's blood and accumulates. After a few days, the phenylalanine level in the baby's blood may be ten times higher than normal. Within days, the developing brain is damaged, and untreated children with PKU become severely mentally retarded. If detected early, PKU can be treated with a special diet low in phenylalanine to avoid the brain damage that would otherwise result. Thus, early detection is imperative.

Newborn screening for PKU and other diseases began in 1963 with the development of a simple, rapid test for the presence of excess phenylalanine in blood serum (**Figure 15.19**). This method uses dried blood spots from newborn babies and can be automated so that a screening laboratory can process many samples in a day.

Screening using newborn babies' blood is now done for up to 25 genetic diseases. Some are rare, such as maple syrup urine disease, which occurs once in 185,000 births. This disease is caused by a defect in an enzyme that metabolizes certain amino acids, and results in sweet-smelling urine and severe brain damage. Other genetic diseases are more common, such as congenital hypothyroidism, which occurs about once in 4,000 births, and causes reduced growth and mental retardation due to low levels of thyroid hormone. With early intervention, many of these infants can be successfully treated. So it is not surprising that newborn screening is legally mandatory in many countries, including the United States and Canada.

15.19 Genetic Screening of Newborns for Phenylketonuria A blood test is used to screen newborns for phenylketonuria. Small samples of blood are taken from a newborn's heel. The samples are placed in a machine that measures the phenylalanine concentration in the blood. Early detection means that the symptoms of the condition can be prevented by putting the baby on a therapeutic diet.

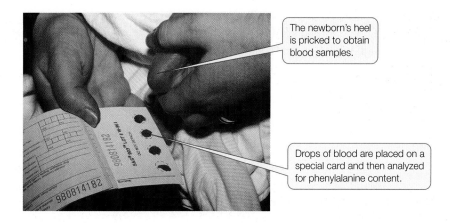

The newborn's heel is pricked to obtain blood samples.

Drops of blood are placed on a special card and then analyzed for phenylalanine content.

DNA testing is the most accurate way to detect abnormal genes

The level of phenylalanine in the blood is an indirect measure of phenylalanine hydroxylase activity in the liver. But how can we screen for genetic diseases that are not detectable by blood tests? What if blood is difficult to obtain, as it is in a fetus? How are genetic abnormalities in heterozygotes, who express the normal protein at some level, identified?

DNA testing is the direct analysis of DNA for a mutation, and it offers the most direct and accurate way of detecting an abnormal allele. Now that the mutations responsible for many human diseases have been identified, any cell in the body can be examined at any time of life for mutations. With the amplification power of PCR, only one or a few cells are needed for testing. These methods work best for diseases caused by only one or a few different mutations.

Consider, for example, two parents who are both heterozygous for the cystic fibrosis allele, who have had a child with the disease, and want a normal child. If treated with the appropriate hormones, the mother can be induced to "superovulate," releasing several eggs. An egg can be injected with a single sperm from her husband and the resulting zygote allowed to divide to the 8-cell stage. If one of these embryonic cells is removed, it can be tested for the presence of the cystic fibrosis allele. If the test is negative, the remaining 7-cell embryo can be implanted in the mother's womb where with luck, it will develop normally.

Such *preimplantation screening* is performed only rarely. More typical are analyses of fetal cells after normal fertilization and implantation in the womb. Fetal cells can be analyzed at about the tenth week of pregnancy by chorionic villus sampling, or during the thirteenth to seventeenth weeks by amniocentesis. In either case, only a few fetal cells are necessary to perform DNA testing.

DNA testing can also be performed with newborns. The blood samples used for screening for PKU and other disorders contain enough of the baby's blood cells to permit DNA analysis using PCR-based techniques. Screening tests using DNA analysis are now being used for sickle-cell disease and cystic fibrosis; similar tests for other diseases will surely follow.

Of the numerous methods of DNA testing available, two are the most widespread. We will describe their use to detect the mutation in the β-globin gene that results in sickle-cell disease.

SCREENING FOR ALLELE-SPECIFIC CLEAVAGE DIFFERENCES The first method uses RFLP analysis, as we described earlier. There is a difference between the normal and the sickle allele of the β-globin gene, with respect to a restriction enzyme recognition sequence. Around the sixth codon in the normal gene is the sequence

5'... CCTGAGGAG... 3'

This sequence is recognized by the restriction enzyme *Mst*II, which will cleave DNA at

5'... CCTNAGGAG... 3'

where N is any base. In the sickle allele, the DNA sequence is

5'... CCTGTGGAG... 3'

The point mutation at codon 6 makes this sequence unrecognizable by *Mst*II. The sequence surrounding the mutant site can be amplified by PCR and digested with *Mst*II. Gel electrophoresis is used to distinguish between PCR products derived from the normal allele, which are cut by the enzyme, and products from the sickle allele, which are not cut (**Figure 15.20**).

This *allele-specific cleavage* method of DNA testing works only if a restriction enzyme exists that can recognize the sequence of either the normal or the mutant allele.

SCREENING BY ALLELE-SPECIFIC OLIGONUCLEOTIDE HYBRIDIZATION The *allele-specific oligonucleotide hybridization* method uses short synthetic DNA strands called *oligonucleotide probes* that will hybridize with denatured PCR products from either the normal or the mutant allele. Usually, an oligonucleotide probe of at least a dozen bases is needed to form a stable double helix with the target DNA. If the probe is radioactively or fluorescently labeled, hybridization can be readily detected (**Figure 15.21**).

Detection of a mutation by either DNA screening method can be used for diagnosis of a genetic disease, so that appropriate treatment can begin. In addition, identification provides a person with important information about his or her genome.

TOOLS FOR INVESTIGATING LIFE

15.20 DNA Testing by Allele-Specific Cleavage

Allele-specific cleavage can be used to detect mutations such as the one that causes sickle-cell disease.

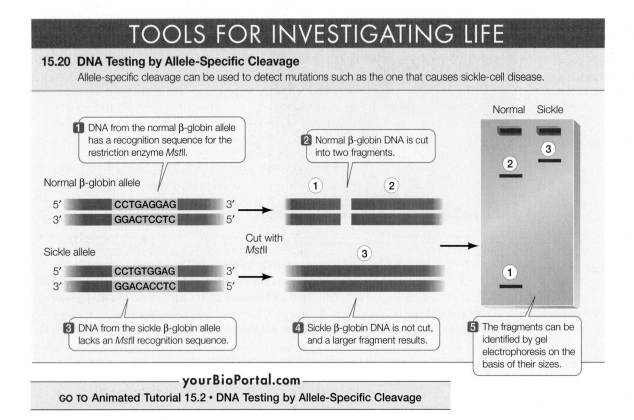

1 DNA from the normal β-globin allele has a recognition sequence for the restriction enzyme *Mst*II.

2 Normal β-globin DNA is cut into two fragments.

Normal β-globin allele

5′ ▭CCTGAGGAG▭ 3′
3′ ▭GGACTCCTC▭ 5′

Sickle allele

5′ ▭CCTGTGGAG▭ 3′
3′ ▭GGACACCTC▭ 5′

Cut with *Mst*II

3 DNA from the sickle β-globin allele lacks an *Mst*II recognition sequence.

4 Sickle β-globin DNA is not cut, and a larger fragment results.

5 The fragments can be identified by gel electrophoresis on the basis of their sizes.

Normal   Sickle

yourBioPortal.com
GO TO Animated Tutorial 15.2 • DNA Testing by Allele-Specific Cleavage

TOOLS FOR INVESTIGATING LIFE

15.21 DNA Testing by Allele-Specific Oligonucleotide Hybridization

Testing of this family reveals that three of them are heterozygous carriers of the sickle allele. The first child, however, has inherited two normal alleles and is neither affected by the disease nor a carrier.

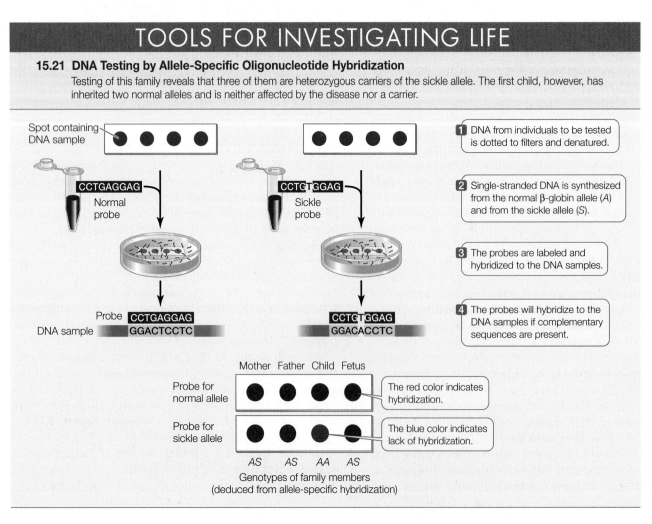

Spot containing DNA sample

1 DNA from individuals to be tested is dotted to filters and denatured.

CCTGAGGAG
Normal probe

CCTGTGGAG
Sickle probe

2 Single-stranded DNA is synthesized from the normal β-globin allele (*A*) and from the sickle allele (*S*).

3 The probes are labeled and hybridized to the DNA samples.

4 The probes will hybridize to the DNA samples if complementary sequences are present.

Probe   CCTGAGGAG
DNA sample   GGACTCCTC

CCTGTGGAG
GGACACCTC

Mother   Father   Child   Fetus

Probe for normal allele

The red color indicates hybridization.

Probe for sickle allele

The blue color indicates lack of hybridization.

AS   *AS*   *AA*   *AS*

Genotypes of family members
(deduced from allele-specific hybridization)

Ongoing research has resulted in the development of increasingly accurate diagnostic tests and a better understanding of various genetic diseases at the molecular level. This knowledge is now being applied to the development of new treatments for genetic diseases. In the next section we will survey various approaches to treatment, including modifications of the mutant phenotype and gene therapy, in which the normal version of a mutant gene is supplied.

15.6 How Are Genetic Diseases Treated?

Most treatments for genetic diseases simply try to alleviate the patient's symptoms. But to effectively treat these diseases—whether they affect all cells, as in inherited disorders such as PKU, or only somatic cells, as in cancer—physicians must be able to diagnose the disease accurately, understand how the disease works at the molecular level, and intervene early, before the disease ravages or kills the individual. There are two main approaches to treating genetic diseases: modifying the disease phenotype, or replacing the defective gene.

Genetic diseases can be treated by modifying the phenotype

Altering the phenotype of a genetic disease so that it no longer harms an individual is commonly done in one of three ways: by restricting the substrate of a deficient enzyme, by inhibiting a harmful metabolic reaction, or by supplying a missing protein product (**Figure 15.22**).

RESTRICTING THE SUBSTRATE Restricting the substrate of a deficient enzyme is the approach taken when a newborn is diagnosed with PKU. In this case, the deficient enzyme is phenylalanine hydroxylase, and the substrate is phenylalanine. The infant's inability to break down phenylalanine in food leads to a buildup of the substrate, which causes the clinical symptoms. So the infant is immediately put on a special diet that contains only enough phenylalanine for immediate use. Lofenelac, a milk-based product that is low in phenylalanine, is fed to these

infants just like formula. Later, certain fruits, vegetables, cereals, and noodles low in phenylalanine can be added to the diet. Meat, fish, eggs, dairy products, and bread, which contain high amounts of phenylalanine, must be avoided, especially during childhood, when brain development is most rapid. The artificial sweetener aspartame must also be avoided because it is made of two amino acids, one of which is phenylalanine.

People with PKU are generally advised to stay on a low-phenylalanine diet for life. Although maintaining these dietary restrictions may be difficult, it is effective. Numerous follow-up studies since newborn screening was initiated have shown that people with PKU who stay on the diet are no different from the rest of the population in terms of mental ability. This is an impressive achievement in public health, given the severity of mental retardation in untreated patients.

METABOLIC INHIBITORS In Section 11.7, we described how drugs that are inhibitors of various cell cycle processes are used to treat cancer. Drugs are also used to treat the symptoms of many genetic diseases. As biologists have gained insight into the molec-

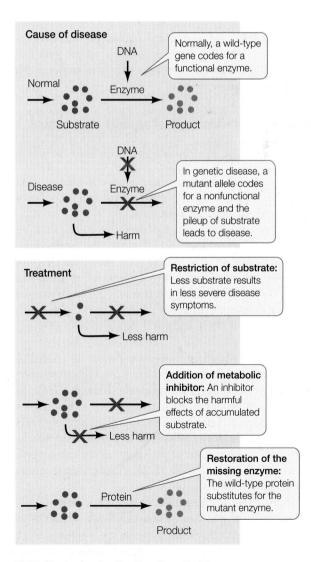

15.22 Strategies for Treating Genetic Diseases

ular characteristics of these diseases and the specific proteins involved, a more specific approach to treatment is taking shape. This is called *molecular medicine.*

An example of this approach is the treatment of chronic myelogenous leukemia. In this cancer, certain white blood cells undergo a gain-of-function mutation, making a totally new protein that is not made in any other cells. This new protein was isolated and a drug was made that specifically targets and inactivates the protein, thereby preventing the proliferation of the cancerous cells. The result has been greatly improved survival in these patients.

SUPPLYING THE MISSING PROTEIN An obvious way to treat a disease phenotype in which a functional protein is missing is to supply that protein. This approach is the basis of treatment for hemophilia, in which the missing blood factor VIII is supplied to the patient. At first this protein was obtained from blood and was sometimes contaminated with viruses or other pathogens. Now, however, the production of human clotting proteins by recombinant DNA technology (see Chapter 18) has made it possible to provide the protein in a much purer form.

Unfortunately, the phenotypes of many diseases caused by genetic mutations are very complex. In these cases, simple interventions like those we have just described do not work. Indeed, a recent survey of 351 diseases caused by single-gene mutations showed that current therapies increased patients' life spans by only 15 percent.

Gene therapy offers the hope of specific treatments

Clearly, if a cell lacks a functional allele, it would be optimal to provide that allele. This is the aim of gene therapy. Diseases ranging from rare inherited disorders caused by single-gene mutations to cancer are under intensive investigation, in an effort to develop gene therapy treatments.

The object of **gene therapy** is to insert a new gene that will be expressed in the host. The new DNA must be attached to a promoter that will be active in human cells. The physicians who are developing such treatments are confronted by numerous challenges. They must find an effective way for the new gene to be taken up by the patient's cells, for the gene to be precisely inserted into the host DNA, and for the gene to be expressed.

Which human cells should be the targets of gene therapy? The best approach would be to replace the nonfunctional allele with a functional one in every cell of the body. But delivery of a gene to every cell poses a formidable challenge. Until recently, attempts at gene therapy have used *ex vivo* techniques. That is, physicians have taken cells from the patient's body, added the new gene to those cells in the laboratory, and then returned the cells to the patient in the hope that the correct gene product would be made (**Figure 15.23**). A successful example demonstrates this technique.

Adenosine deaminase is needed for the maturation of white blood cells, and a genetic disease results when a person is homozygous for a mutant allele for this enzyme. People without this enzyme have severe immune system deficiencies. The wild-type gene for adenosine deaminase has been isolated and in-

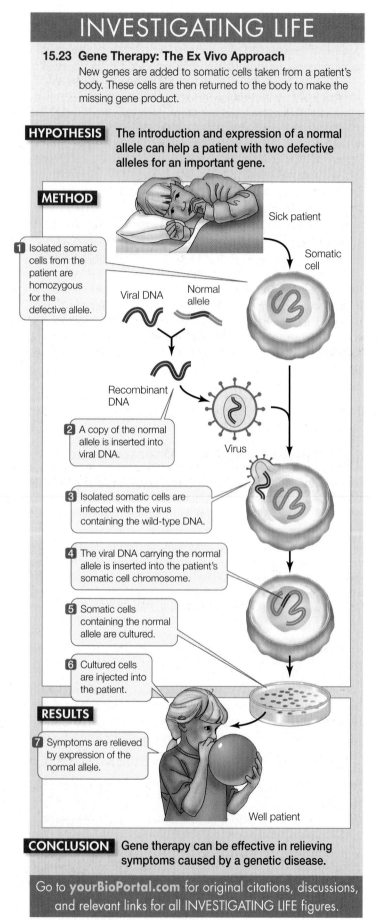

INVESTIGATING LIFE

15.23 Gene Therapy: The Ex Vivo Approach
New genes are added to somatic cells taken from a patient's body. These cells are then returned to the body to make the missing gene product.

HYPOTHESIS The introduction and expression of a normal allele can help a patient with two defective alleles for an important gene.

METHOD

Sick patient

1 Isolated somatic cells from the patient are homozygous for the defective allele.

Somatic cell

Viral DNA   Normal allele

Recombinant DNA

2 A copy of the normal allele is inserted into viral DNA.

Virus

3 Isolated somatic cells are infected with the virus containing the wild-type DNA.

4 The viral DNA carrying the normal allele is inserted into the patient's somatic cell chromosome.

5 Somatic cells containing the normal allele are cultured.

6 Cultured cells are injected into the patient.

RESULTS

7 Symptoms are relieved by expression of the normal allele.

Well patient

CONCLUSION Gene therapy can be effective in relieving symptoms caused by a genetic disease.

Go to **yourBioPortal.com** for original citations, discussions, and relevant links for all INVESTIGATING LIFE figures.

serted into a virus that can carry the gene into white blood cells of a patient lacking the enzyme. The recombinant virus lacks the genes for reproduction inside cells, but retains the genes coding for cell uptake and insertion into the host DNA. The recombinant virus was added to white blood cells from a patient that had inherited the mutant form of adenosine deaminase. The wild-type adenosine deaminase gene became inserted into the cells' chromosomes, along with viral DNA. When these transformed white blood cells with the wild-type gene were put back into the patient, the cells made adenosine deaminase and the patient's condition improved.

The other approach to gene therapy is to insert the gene directly into cells in the body of the patient. This in vivo approach is being attempted for various types of cancer. Lung cancer cells, for example, are accessible to such treatment if the DNA is given as an aerosol through the respiratory system. Several thousand patients, over half of them with cancer, have undergone this treatment. In preliminary clinical trials, people are given the therapy to see whether it has any toxicity and whether the new gene is actually incorporated into the patients' genomes. In more ambitious trials, larger numbers of patients receive the therapy with the hope that their disease will disappear, or at least improve.

15.6 RECAP

Treatment of a human genetic disease may involve an attempt to modify the abnormal phenotype by restricting the substrate of a deficient enzyme, inhibiting a harmful metabolic reaction, or supplying a missing protein. On the other hand, gene therapy aims to address a genetic defect by inserting a normal allele into a patient's cells.

- How do metabolic inhibitors used in chemotherapy function in treating cancer? See pp. 337–338 and Figure 5.22

- How does ex vivo gene therapy work? Can you give an example? See p. 338 and Figure 15.23

In this chapter, we dealt with mutations in general, focusing on DNA changes that affect phenotypes through specific protein products. But there is much more to molecular genetics than genes and proteins. Determining which genes will be expressed when and where is a major function of the genome. In Chapter 16 we turn to gene regulation.

CHAPTER SUMMARY

15.1 What Are Mutations?

- **Mutations** are heritable changes in DNA. **Somatic mutations** are passed on to daughter cells, but only **germ line mutations** are passed on to sexually produced offspring.

- **Point mutations** result from alterations in single base pairs of DNA. **Silent mutations** can occur in noncoding DNA or in coding regions of genes and do not affect the amino acid sequences of proteins. **Missense**, **nonsense**, and **frame-shift** mutations all cause changes in protein sequences. Review Figure 15.2

- Chromosomal mutations (**deletions**, **duplications**, **inversions**, or **translocations**) involve large regions of chromosomes. Review Figure 15.4

- **Spontaneous mutations** occur because of instabilities in DNA or chromosomes. **Induced mutations** occur when a mutagen damages DNA. Review Figure 15.5

- Mutations can occur in hot spots where cytosine has been methylated to 5-methylcytosine. Review Figure 15.6

- Mutations, although often detrimental to an individual organism, are the raw material of evolution.

15.2 How Are DNA Molecules and Mutations Analyzed?

- **Restriction enzymes**, which are made by microorganisms as a defense against viruses, bind to and cut DNA at specific **recognition sequences** (also called **restriction sites**). These enzymes can be used to produce small fragments of DNA for study, a technique known as restriction digestion. Review Figure 15.7

- DNA fragments can be separated by size using **gel electrophoresis**. Review Figure 15.8, ANIMATED TUTORIAL 15.1

- **DNA fingerprinting** is used to distinguish between specific individuals, or to reveal which individuals are most closely related

to one another. It involves the detection of DNA polymorphisms, including **single nucleotide polymorphisms** (**SNPs**) and **short tandem repeats** (**STRs**). Review Figure 15.9

- The goal of the DNA barcoding project is to sequence a single region of DNA in all species for identification purposes.

15.3 How Do Defective Proteins Lead to Diseases?

- Abnormalities in nearly all classes of proteins, including enzymes, transport proteins, receptor proteins, and structural proteins, have been implicated in genetic diseases.

- While a single amino acid difference can be the cause of disease, amino acid variations have been detected in many functional proteins. Review Figure 15.13

- Transmissible spongiform encephalopathies (TSEs) are degenerative brain diseases that can be transmitted from one animal to another by consumption of infected tissues. The infective agent is a **prion**, a protein with an abnormal conformation.

- **Multifactorial** diseases are caused by the interactions of many genes and proteins with the environment. They are much more common than diseases caused by mutations in a single gene.

- Predictable patterns of inheritance are associated with some human genetic diseases. Autosomal recessive, autosomal dominant, and sex-linked patterns are common.

15.4 What DNA Changes Lead to Genetic Diseases?

- It is possible to isolate both the mutant genes and the abnormal proteins responsible for human diseases. Review Figure 15.15, WEB ACTIVITY 15.1

- The effects of fragile-X syndrome worsen with each generation. This pattern is the result of an **expanding triplet repeat**. Review Figure 15.18

15.5 How Is Genetic Screening Used to Detect Human Diseases?

- **Genetic screening** is used to detect human genetic diseases, alleles predisposing people to those diseases, or carriers of those diseases.

- Genetic screening can be done by looking for abnormal protein expression. **Review Figure 15.19**

- **DNA testing** is the direct identification of mutant alleles. Any cell can be tested at any time in the life cycle.

- The two predominant methods of DNA testing are the allele-specific cleavage method and allele-specific oligonucleotide hybridization method. **Review Figures 15.20 and 15.21, ANIMATED TUTORIAL 15.2**

15.6 How Are Genetic Diseases Treated?

- There are three ways to modify the phenotype of a genetic disease: restrict the substrate of a deficient enzyme, inhibit a harmful metabolic reaction, or supply a missing protein. **Review Figure 15.22**

- Cancer is treated with metabolic inhibitors.

- In **gene therapy**, a mutant gene is replaced with a normal gene. Both ex vivo and in vivo therapies are being developed. **Review Figure 15.23**

SELF-QUIZ

1. Phenylketonuria is an example of a genetic disease in which
   a. a single enzyme is not functional.
   b. inheritance is sex-linked.
   c. two parents without the disease cannot have a child with the disease.
   d. mental retardation always occurs, regardless of treatment.
   e. a transport protein does not work properly.

2. Mutations of the gene for β-globin
   a. are usually lethal.
   b. occur only at amino acid position 6.
   c. number in the hundreds.
   d. always result in sickling of red blood cells.
   e. can always be detected by gel electrophoresis.

3. Multifactorial (complex) diseases
   a. are less common than single-gene diseases.
   b. involve the interaction of many genes with the environment.
   c. affect less than 1 percent of humans.
   d. involve the interactions of several mRNAs.
   e. are exemplified by sickle-cell disease.

4. In fragile-X syndrome,
   a. females are affected more severely than males.
   b. a short sequence of DNA is repeated many times to create the fragile site.
   c. both the X and Y chromosomes tend to break when prepared for microscopy.
   d. all people who carry the gene that causes the syndrome are mentally retarded.
   e. the basic pattern of inheritance is autosomal dominant.

5. Most genetic diseases are rare because
   a. each person is unlikely to be a carrier for harmful alleles.
   b. genetic diseases are usually sex-linked and so uncommon in females.
   c. genetic diseases are always dominant.
   d. two parents probably do not carry the same recessive alleles.
   e. mutation rates in humans are low.

6. Mutational "hot spots" in human DNA
   a. always occur in genes that are transcribed.
   b. are common at cytosines that have been modified to 5-methylcytosine.
   c. involve long stretches of nucleotides.
   d. occur only where there are long repeats.
   e. are very rare in genes that code for proteins.

7. Newborn genetic screening for PKU
   a. is very expensive.
   b. detects phenylketones in urine.
   c. has not led to the prevention of mental retardation resulting from this disorder.
   d. should be done during the second or third day of an infant's life.
   e. uses bacterial growth to detect excess phenylketones in blood.

8. Genetic diagnosis by DNA testing
   a. detects only mutant and not normal alleles.
   b. can be done only on eggs or sperm.
   c. involves hybridization to rRNA.
   d. often utilizes restriction enzymes and a polymorphic site.
   e. cannot be done with PCR.

9. Which of the following is *not* a way to treat a genetic disease?
   a. Inhibiting a harmful biochemical reaction
   b. Adding the wild-type allele to cells expressing the mutation
   c. Restricting the substrate of a harmful biochemical reaction
   d. Replacing a mutant allele with the wild-type allele in the fertilized egg
   e. Supplying a wild-type protein that is missing due to mutation

10. Current treatments for genetic diseases include all of the following *except*
    a. restricting a dietary substrate.
    b. replacing the mutant gene in all cells.
    c. alleviating the patient's symptoms.
    d. inhibiting a harmful metabolic reaction.
    e. supplying a protein that is missing.

FOR DISCUSSION

1. In the past, it was common for people with phenylketonuria (PKU) who were placed on a low-phenylalanine diet after birth to be allowed to return to a normal diet during their teenage years. Although the levels of phenylalanine in their blood were high, their brains were thought to be beyond the stage when they could be harmed. If a woman with PKU becomes pregnant, however, a problem arises. Typically, the fetus is heterozygous, but is unable, at early stages of development, to metabolize the high levels of phenylalanine that arrive from the mother's blood. Why is the fetus likely to be heterozygous? What do you think would happen to the fetus during this "maternal PKU" situation? What would be your advice to a woman with PKU who wants to have a child?

2. Cystic fibrosis is an autosomal recessive disease in which thick mucus is produced in the lungs and airways. The gene responsible for this disease encodes a protein composed of 1,480 amino acids. In most patients with cystic fibrosis, the protein has 1,479 amino acids: a phenylalanine is missing at position 508. A baby is born with cystic fibrosis. He has an older brother who is not affected. How would you test the DNA of the older brother to determine whether he is a carrier for cystic fibrosis? How would you design a gene therapy protocol to "cure" the cells in the younger brother's lungs and airways?

3. A number of efforts are under way to identify human genetic polymorphisms that correlate with multifactorial diseases such as diabetes, heart disease, and cancer. What would be the uses of such information? What concerns do you think are being raised about this kind of genetic testing?

ADDITIONAL INVESTIGATION

Tay-Sachs disease is caused by a recessively inherited mutation in the gene coding for the enzyme hexosaminidase A (HexA), which normally breaks down a lipid called GM2 ganglioside. Accumulation of this lipid in the brain leads to progressive deterioration of the nervous system and death, usually by age 4. HexA activity in blood serum is 0–6 percent in homozygous recessives and 7–35 percent in heterozygous carriers, compared to non-carriers (100 percent). The most common mutation in the *HexA* gene is an insertion of four base pairs, which presumably leads to a premature stop codon. How would you do genetic screening for carriers of this disease by enzyme testing and by DNA testing? What are the advantages of DNA testing? How would you investigate the premature stop codon hypothesis?

———— WORKING WITH DATA (GO TO yourBioPortal.com) ————

Gene Therapy: The Ex Vivo Approach   In this exercise, you use the original research paper to examine the protocol used to treat two patients with gene therapy for adenosine deaminase deficiency (Figure 15.23). You will examine the kinds of evidence used to detect the wild-type gene in the cells of these patients, and will analyze the results in terms of immune system cell function.

Alcoholism and the control of gene expression

Many people drink alcoholic beverages but relatively few of them become addicted (alcoholic). When they do, the results are often disastrous, both socially and physiologically. Alcoholism often disrupts relationships with family, friends, and colleagues. Lost productivity leads to economic costs estimated at over $100 billion per year in the U.S. alone. Physiologically, alcoholism is characterized by a compulsion to consume alcohol, tolerance (increasing doses are needed for the same effect), and dependence (abrupt cessation of consumption leads to severe withdrawal symptoms). In most of these people, alcohol acts not just to provide pleasant sensations (positive reinforcement) but also to alleviate unpleasant ones such as anxiety (negative reinforcement).

Why do only some people become alcoholic? Alcoholism is a complex behavioral disease. Psychologists sometimes speak of "addictive personalities," and genetic studies indicate there may be inherited factors. It would help both alcoholics and those who treat them if we understood the differences in brain chemistry between alcoholic and nonalcoholic individuals. But we can't do the necessary experiments on humans; instead, animal models are used to study alcoholism at the molecular level. James Murphy at Indiana University has bred a strain of rats, called P rats, that prefer alcohol when given the choice of alcohol-containing or alcohol-free water. These rats show many of the symptoms of true addiction, including compulsive drinking, tolerance, and withdrawal. In effect, they are a genetic strain of alcoholic animals.

People often drink alcoholic beverages to relieve anxiety, and there are clear links between anxiety disorders and alcoholism. Like many of their human counterparts, the P rats appear more anxious than wild-type rats, spending more time in a closed rather than an open environment. Drinking alcohol alters this behavior and seems to relieve their anxiety.

There may also be a link between the transcription factor CREB and alcohol consumption. CREB (or *cyclic AMP response element binding protein*) is especially abundant in the brain and regulates the expression of hundreds of genes that are important in metabolism. CREB becomes activated when it is phosphorylated by the enzyme protein kinase A, which in turn is activated by the second messenger cyclic AMP. In an effort to understand the molecular basis of alcoholism and anxiety, neuroscientist Subhash Pandey and his colleagues at the University of Illinois compared CREB levels in the brains of P rats and wild-type rats.

Alcoholism Huge social and economic costs are associated with alcohol abuse. Scientists are trying to understand its molecular basis.

An Explanation for Alcoholism?   The transcription factor, CREB, binds to DNA and activates promoters of genes involved in addictive behaviors in alcoholism.

They found that P rats have inherently lower levels of CREB in certain parts of the brain. When these rats consumed alcohol, the total levels of CREB did not increase, but the levels of phosphorylated CREB did. It is the phosphorylated version of CREB that binds to DNA and regulates gene transcription.

The prospect that CREB, a transcription factor that regulates gene expression, is a key element in the genetic propensity for alcoholism is important because it begins to explain the molecular nature of a complex behavioral disease. Such understanding may permit more effective treatment of alcohol abuse or its prevention. Equally important to our purpose here, it underscores the importance of the regulation of gene expression in biological processes.

IN THIS CHAPTER we will focus on the control of gene expression in many types of organisms. We begin with the simplest systems, viruses, which undertake an ordered series of molecular events when they infect a host cell. Then we turn to prokaryotes, which respond to changes in their environment with coordinated changes in gene expression. In eukaryotes, similar principles are used to regulate gene expression, but with added levels of complexity. Finally, we turn to the regulation of gene expression by modification of the genome—the field of epigenetics.

16.1 How Do Viruses Regulate Their Gene Expression?

"A virus is a piece of bad news wrapped in protein." This quote from immunologist Sir Peter Medawar is certainly true for the cells that viruses infect. As we describe in Chapter 13, a virus injects its genetic material into a host cell and turns that cell into a virus factory (see Figure 13.3). Viral life cycles are very efficient. Perhaps the record is held by poliovirus: a single poliovirus infecting a mammalian cell can produce over 100,000 new virus particles!

Unlike organisms, **viruses** are *acellular*; that is, they are not cells, do not consist of cells, and do not carry out many of the processes characteristic of life. Most virus particles, called **virions**, are composed of only nucleic acid and a few proteins. Viruses do not carry out two of the basic functions of cellular life: they do not regulate the transport of substances into and out of themselves by membranes, and they do not perform metabolic functions involved with taking in nutrients, refashioning them, and expelling wastes. But they can reproduce in systems that do perform these metabolic functions—namely, living cells. By studying the relatively simple viral reproductive cycle, biologists have discovered principles of gene expression and its regulation that apply to cellular systems that may be much more complex.

As we describe in Chapter 14, gene expression begins at the *promoter,* where RNA polymerase binds to initiate transcription. In a genome with many genes, not all promoters are active at a given time—there is *selective gene transcription.* The "decision" regarding which genes to activate involves two types of regulatory proteins that bind to DNA: repressor proteins and activator proteins. In both cases, these proteins bind to the promoter to regulate the gene (**Figure 16.1**):

● In **negative regulation**, the gene is normally transcribed. Binding of a repressor protein prevents transcription.

● In **positive regulation**, the gene is normally not transcribed. An activator protein binds to stimulate transcription.

You will see these mechanisms, or combinations of them, as we examine regulation in viruses, prokaryotes, and eukaryotes.

Bacteriophage undergo a lytic cycle

The Hershey–Chase experiment (see Figure 13.4) involved the typical viral reproductive cycle, the **lytic cycle**, so named because the infected host cell lyses (bursts), releasing progeny viruses. Once a virus has injected its nucleic acid into a cell, that

(A) Negative regulation

DNA
Repressor binding site
5′
3′
→ Transcription

DNA
5′
3′
⊣ No transcription

Binding of repressor protein blocks transcription.

(B) Positive regulation

DNA
Activator binding site
5′
3′
⊣ No transcription

DNA
5′
3′
→ Transcription

Binding of activator protein stimulates transcription.

16.1 Positive and Negative Regulation Proteins regulate gene expression by binding to DNA and preventing or allowing RNA polymerase to bind DNA at the promotor region to control transcription.

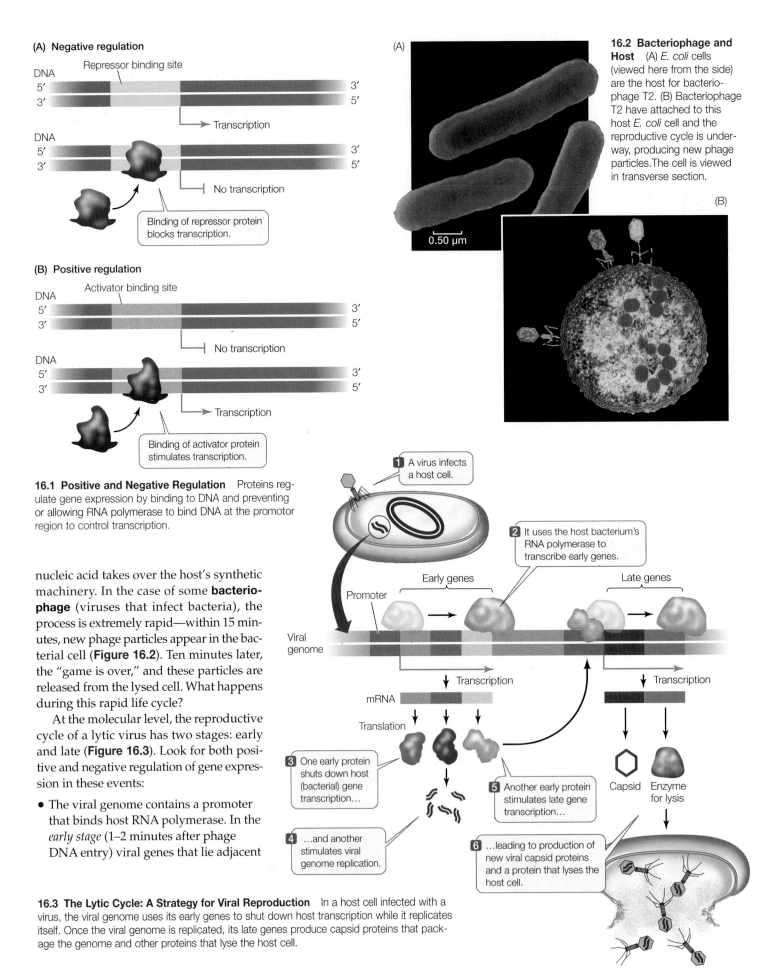

(A)

16.2 Bacteriophage and Host (A) *E. coli* cells (viewed here from the side) are the host for bacteriophage T2. (B) Bacteriophage T2 have attached to this host *E. coli* cell and the reproductive cycle is underway, producing new phage particles. The cell is viewed in transverse section.

(B)

0.50 μm

1 A virus infects a host cell.

2 It uses the host bacterium's RNA polymerase to transcribe early genes.

Promoter

Early genes

Late genes

Viral genome

Transcription

mRNA

Transcription

Translation

3 One early protein shuts down host (bacterial) gene transcription…

4 …and another stimulates viral genome replication.

5 Another early protein stimulates late gene transcription…

6 …leading to production of new viral capsid proteins and a protein that lyses the host cell.

Capsid

Enzyme for lysis

nucleic acid takes over the host's synthetic machinery. In the case of some **bacteriophage** (viruses that infect bacteria), the process is extremely rapid—within 15 minutes, new phage particles appear in the bacterial cell (**Figure 16.2**). Ten minutes later, the "game is over," and these particles are released from the lysed cell. What happens during this rapid life cycle?

At the molecular level, the reproductive cycle of a lytic virus has two stages: early and late (**Figure 16.3**). Look for both positive and negative regulation of gene expression in these events:

- The viral genome contains a promoter that binds host RNA polymerase. In the *early stage* (1–2 minutes after phage DNA entry) viral genes that lie adjacent

16.3 The Lytic Cycle: A Strategy for Viral Reproduction In a host cell infected with a virus, the viral genome uses its early genes to shut down host transcription while it replicates itself. Once the viral genome is replicated, its late genes produce capsid proteins that package the genome and other proteins that lyse the host cell.

to this promoter are transcribed. These early genes often encode proteins that shut down host transcription and stimulate viral genome replication and transcription of viral late genes. Three minutes after DNA entry, viral nuclease enzymes digest the host's chromosome, providing nucleotides for the synthesis of viral genomes.

- In the *late stage*, viral late genes are transcribed; they encode the viral capsid proteins and enzymes that lyse the host cell to release the new virions. This begins 9 minutes after DNA entry and 6 minutes before the first new phage particles appear.

The whole process—from binding and infection to release of new phage—takes about half an hour. During this period, the sequence of transcriptional events is carefully controlled to produce complete, infective virons.

Some bacteriophage can carry bacterial genes from one cell to another

During the lytic cycle some bacteriophage package their DNA in **capsids** (outer shells). In rare cases, a bacterial DNA fragment is inserted into a capsid instead of, or along with, the phage DNA. When such a virion infects another bacterium, the bacterial DNA is injected into the new host cell, a mechanism of gene transfer called **transduction**. The viral infection does not produce new viruses. Instead, the incoming DNA fragment can recombine with the host chromosome, replacing host genes with genes from the virus's former host. The recipeint cell survives under these conditions because there is no virus replication.

Some bacteriophage can undergo a lysogenic cycle

Like all nucleic acid genomes, those of viruses can mutate and evolve by natural selection. Some viruses have evolved an advantageous process called **lysogeny** that postpones the lytic cycle. In lysogeny, the viral DNA becomes integrated into the host DNA and becomes a **prophage** (**Figure 16.4**). As the host cell divides, the viral DNA gets replicated along with that of the host. The prophage can remain inactive within the bacterial genome for thousands of generations, producing many copies of the original viral DNA.

However, if the host cell is not growing well, the virus "cuts its losses." It immediately switches to a lytic cycle, in which the prophage excises itself from the host chromosome and reproduces. In other words, the virus is able to enhance its chances of multiplication and survival by inserting its DNA into the host chromosome, where it sits as a silent partner until conditions are right for lysis.

16.4 The Lytic and Lysogenic Cycles of Bacteriophage In the lytic cycle, infection of a bacterium by viral DNA leads directly to the multiplication of the virus and lysis of the host cell. In the lysogenic cycle, an inactive prophage is integrated into the host DNA where it is replicated during the bacterial life cycle.

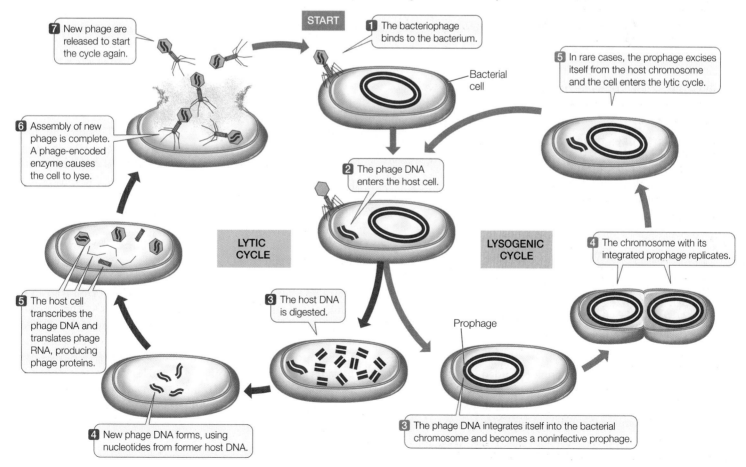

16.5 Control of Bacteriophage λ Lysis and Lysogeny Two regulatory proteins, Cro and cI, compete to control expression of one another and genes for viral lysis and lysogeny.

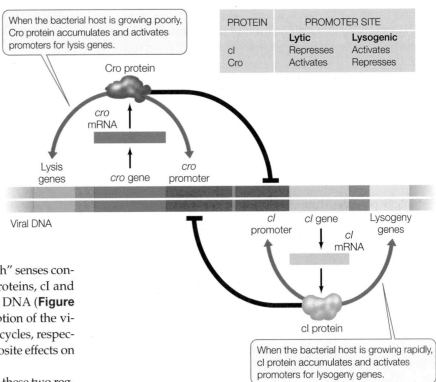

When the bacterial host is growing poorly, Cro protein accumulates and activates promoters for lysis genes.

Cro protein

PROTEIN	PROMOTER SITE	
	Lytic	**Lysogenic**
cI	Represses	Activates
Cro	Activates	Represses

cro mRNA

Lysis genes

cro gene

cro promoter

Viral DNA

cI promoter

cI gene

Lysogeny genes

cI mRNA

cI protein

When the bacterial host is growing rapidly, cI protein accumulates and activates promoters for lysogeny genes.

Uncovering the regulation of gene expression that underlies the lysis/lysogeny switch was a major achievement of molecular biologists. Here we present just an outline of the process to give you an idea of the positive and negative regulatory mechanisms involved (**Figure 16.5**). The model virus *bacteriophage λ* (lambda) has been used extensively to study the lysogenic mechanism.

How does the phage "know" when to switch to the lytic cycle? A kind of "genetic switch" senses conditions within the host. Two viral regulatory proteins, cI and Cro, compete for two promoters on the phage DNA (**Figure 16.5**). These two promoters control the transcription of the viral genes involved in the lytic and the lysogenic cycles, respectively, and the two regulatory proteins have opposite effects on each promoter.

Phage infection is essentially a "race" between these two regulatory proteins. In a rapidly growing *E. coli* host cell, Cro synthesis is low, so cI "wins," and the phage enters a lysogenic cycle. If the host cell is growing slowly, Cro synthesis is higher, and the genes involved in lysis are activated. The two regulatory proteins are made very early in phage infection, and each binds to a specific DNA sequence.

The reproductive cycle of bacteriophage λ is a paradigm for our understanding of viral life cycles in general. This relatively simple system has served as a model to help us understand how the complicated reproductive cycles of other viruses, including HIV, are controlled.

Eukaryotic viruses have complex regulatory mechanisms

Many eukaryotes are susceptible to infections by various kinds of viruses: RNA and DNA viruses, as well as retroviruses (see also Section 26.6).

- *DNA viruses*. Many viral particles contain double-stranded DNA. However, some contain single-stranded DNA, and a complementary strand is made after the viral genome has been injected into the host cell. Like some bacteriophage, DNA viruses that infect eukaryotes are capable of undergoing both lytic and lysogenic life cycles. Examples include the herpes viruses and papillomaviruses (which cause warts).

- *RNA viruses*. Some viral genomes are made up of RNA that is usually, but not always, single-stranded. The RNA is translated by the host's machinery to produce viral proteins, some of which are involved in replication of the RNA genome. The influenza virus has an RNA genome.

- *Retroviruses*. The retroviral genome is RNA, and the **retrovirus** encodes a protein that makes a DNA strand that is complementary to the RNA. The DNA is integrated into the host chromosome and acts as a template for both mRNA and new viral genomes. Human immunodeficiency virus (HIV) is the retrovirus that causes acquired immune deficiency syndrome (AIDS).

REGULATING HIV GENES As an example of viral genome regulation, we will consider the reproductive cycle of HIV (**Figure 16.6**). HIV is an **enveloped virus**; it is enclosed within a phospholipid membrane derived from its host cell. Proteins in the membrane are involved in infection of new host cells, which HIV enters by direct fusion of the viral envelope with the host plasma membrane.

As indicated above, a distinctive feature of the retroviral life cycle is RNA-directed DNA synthesis. This process is catalyzed by the viral enzyme **reverse transcriptase**, which uses the RNA template to produce a complementary DNA (cDNA) strand, while at the same time degrading the viral RNA. The reverse transcriptase also makes a complementary copy of the cDNA, and it is the double-stranded cDNA that gets integrated into the host's chromosome. The integrated DNA is referred to as the **provirus** and, like the prophage, it contains promoters that are recognized by the host cell transcription apparatus. Both the reverse transcriptase and the integrase are needed for the very early stages of infection and are carried inside the HIV virion.

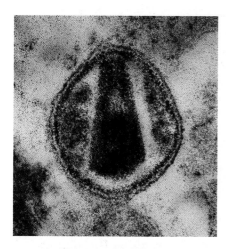

16.6 The Reproductive Cycle of HIV
This retrovirus enters a host cell via fusion of its envelope with the host's plasma membrane. Reverse transcription of retroviral RNA then produces a DNA provirus—a molecule of complementary DNA that inserts itself into the host's genome.

The provirus resides permanently in the host chromosome and is occasionally activated to produce new virions. When this happens, the provirus is transcribed as mRNA, which is then translated by the host cell's protein-synthesizing machinery.

Under normal circumstances, the host cell regulates viral gene expression using proteins that may have originated as a defense mechanism against invaders. Host proteins bind to viral mRNA as it is being made and causes RNA polymerase to fall off the viral DNA, thereby terminating transcription. However, HIV can counteract this regulation with a virus-encoded protein called tat (*trans*activator of *t*ranscription), which binds to the terminator proteins and blocks their action. This *antitermination* allows viral gene transcription and the rest of the viral reproductive cycle to proceed (**Figure 16.7**).

Almost every step in the complex reproductive cycle of HIV is, in principle, a potential target for drugs to treat AIDS.

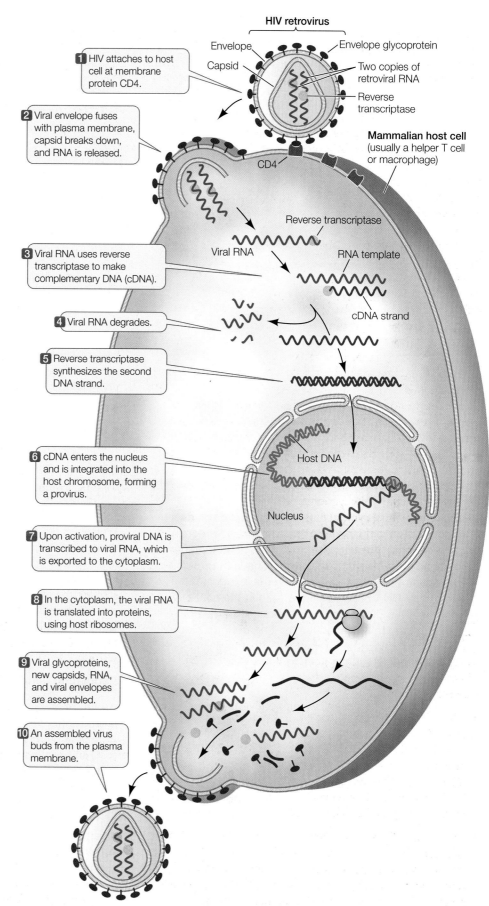

HIV retrovirus

Envelope
Envelope glycoprotein
Capsid
Two copies of retroviral RNA
Reverse transcriptase

1 HIV attaches to host cell at membrane protein CD4.

2 Viral envelope fuses with plasma membrane, capsid breaks down, and RNA is released.

CD4

Mammalian host cell (usually a helper T cell or macrophage)

Reverse transcriptase
Viral RNA
RNA template

3 Viral RNA uses reverse transcriptase to make complementary DNA (cDNA).

cDNA strand

4 Viral RNA degrades.

5 Reverse transcriptase synthesizes the second DNA strand.

Host DNA

6 cDNA enters the nucleus and is integrated into the host chromosome, forming a provirus.

Nucleus

7 Upon activation, proviral DNA is transcribed to viral RNA, which is exported to the cytoplasm.

8 In the cytoplasm, the viral RNA is translated into proteins, using host ribosomes.

9 Viral glycoproteins, new capsids, RNA, and viral envelopes are assembled.

10 An assembled virus buds from the plasma membrane.

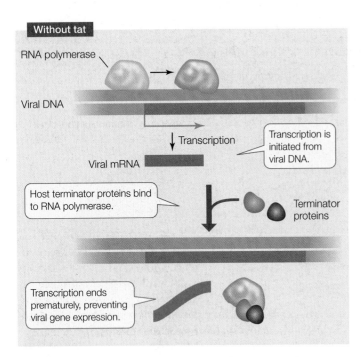

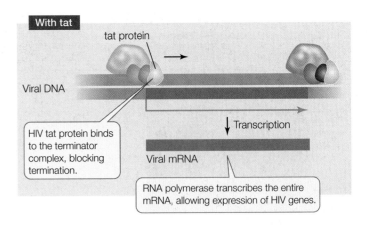

16.7 Regulation of Transcription by HIV The tat protein acts as an antiterminator, allowing transcription of the HIV genome.

16.1 RECAP

Viruses are not cells. They consist of nucleic acids and a few proteins, and require a host cell to reproduce. In the lytic cycle, the viral genome directs the host cell to generate new virions along with proteins that cause the host cell to lyse. In the lysogenic cycle, viral DNA becomes integrated in the host's genome. This DNA is multiplied along with the host cells but may remain inactive for long periods. Special viral proteins that interact with host and viral DNA sequences are the keys to the regulation of viral gene expression.

- What is the difference between positive and negative regulation of gene expression? **See Figure 16.1**

- What are the lytic and lysogenic cycles of bacteriophage? **See p. 345 and Figure 16.4**

- Describe positive and negative regulation of gene expression in bacteriophage and HIV life cycles. **See pp. 346–347 and Figures 16.5 and 16.7**

The environment surrounding prokaryotic cells can change abruptly, requiring rapid responses by the cell. We now turn to these responses, which often involve, as in viruses, the positive and negative regulation of gene expression by proteins binding to DNA.

16.2 How Is Gene Expression Regulated in Prokaryotes?

Prokaryotes conserve energy and resources by making certain proteins only when they are needed. The protein content of a bacterium can change rapidly when conditions warrant. There are several ways in which a prokaryotic cell can shut off the supply of an unneeded protein. The cell can:

- downregulate the transcription of mRNA for that protein;

- hydrolyze the mRNA after it is made, thereby preventing translation;

- prevent translation of the mRNA at the ribosome;

- hydrolyze the protein after it is made; or

- inhibit the function of the protein.

Whichever mechanism is used, it must be both responsive to environmental signals and efficient. The earlier the cell intervenes in the process of protein synthesis, the less energy it wastes. Selective blocking of transcription is far more efficient than transcribing the gene, translating the message, and then degrading or inhibiting the protein. While all five mechanisms for regulating protein levels are found in nature, prokaryotes generally use the most efficient one: transcriptional regulation.

Regulating gene transcription conserves energy

As a normal inhabitant of the human intestine, *E. coli* must be able to adjust to sudden changes in its chemical environment. Its host may present it with one foodstuff one hour (e.g., glucose) and another the next (e.g., lactose). Such changes in nutrients present the bacterium with a metabolic challenge. Glucose is its preferred energy source, and is the easiest sugar to metabolize, but not all of its host's foods contain an abundant supply of glucose. For example, the bacterium may suddenly be deluged with milk, whose main sugar is lactose. Lactose is a β-galactoside—a disaccharide containing galactose β-linked to glucose (see Section 3.3). Three proteins are involved in the initial uptake and metabolism of lactose by *E. coli*:

- *β-galactoside permease* is a carrier protein in the bacterial plasma membrane that moves the sugar into the cell.

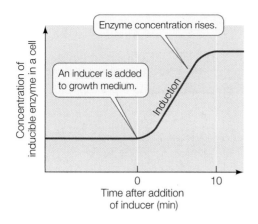

Enzyme concentration rises.

An inducer is added to growth medium.

Induction

Concentration of inducible enzyme in a cell

Time after addition of inducer (min)

0   10

16.8 An Inducer Stimulates the Expression of a Gene for an Enzyme It is most efficient for a cell to produce an enzyme only when it is needed. Some enzymes are induced by the presence of the substance they act upon (for example, β-galactosidase is induced by the presence of lactose).

- *β-galactosidase* is an enzyme that hydrolyses lactose to glucose and galactose.
- *β-galactoside transacetylase* transfers acetyl groups from acetyl CoA to certain β-galactosides. Its role in the metabolism of lactose is not clear.

When *E. coli* is grown on a medium that contains glucose but no lactose or other β-galactosides, the levels of these three proteins are extremely low—the cell does not waste energy and materials making the unneeded enzymes. But if the environment changes such that lactose is the predominant sugar available and very little glucose is present, the bacterium promptly begins making all three enzymes. There are only two molecules of β-galactosidase present in an *E. coli* cell when glucose is present in the medium. But when glucose is absent, the presence of lactose can induce the synthesis of 3,000 molecules of β-galactosidase per cell!

If lactose is removed from *E. coli's* environment, synthesis of the three enzymes stops almost immediately. The enzyme molecules already present do not disappear; they are merely diluted during subsequent cell divisions until their concentration falls to the original low level within each bacterium.

Compounds that, like lactose, stimulate the synthesis of a protein are called **inducers** (**Figure 16.8**). The proteins that are

produced are called **inducible proteins**, whereas proteins that are made all the time at a constant rate are called **constitutive proteins**. (Think of the constitution of a country, a document that does not change under normal circumstances.)

We have now seen two basic ways of regulating the rate of a metabolic pathway. In Section 8.5 we described allosteric regulation of enzyme activity (the rate of enzyme-catalyzed reactions); this mechanism allows rapid fine-tuning of metabolism. Regulation of protein synthesis—that is, regulation of the concentration of enzymes—is slower, but results in greater savings of energy and resources. Protein synthesis is a highly endergonic process, since assembling mRNA, charging tRNA, and moving the ribosomes along mRNA all require the hydrolysis of ATP. **Figure 16.9** compares these two modes of regulation.

Operons are units of transcriptional regulation in prokaryotes

The genes that encode the three enzymes for processing lactose in *E. coli* are **structural genes**; they specify the primary structure (the amino acid sequence) of a protein molecule. Structural genes are genes that can be transcribed into mRNA.

The three structural genes involved in the metabolism of lactose lie adjacent to one another on the *E. coli* chromosome. This arrangement is no coincidence: the genes share a single promoter, and their DNA is transcribed into a single, continuous molecule of mRNA. Because this particular mRNA governs the synthesis of all three lactose-metabolizing enzymes, either all or none of these enzymes are made, depending on whether their common message—their mRNA—is present in the cell.

A cluster of genes with a single promoter is called an **operon**, and the operon that encodes the three lactose-metabolizing enzymes in *E. coli* is called the *lac operon*. The *lac* operon promoter can be very efficient (the maximum rate of mRNA synthesis can be high) but mRNA synthesis can be shut down when the enzymes are not needed. This example of negative regulation was elegantly worked out by Nobel Prize winners François Jacob and Jacques Monod.

In addition to the promoter, an operon has other regulatory sequences that are not transcribed. A typical operon consists of a promoter, an operator, and two or more

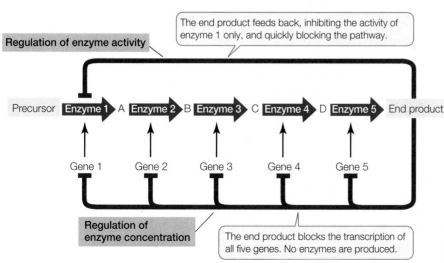

Regulation of enzyme activity

The end product feeds back, inhibiting the activity of enzyme 1 only, and quickly blocking the pathway.

Precursor → Enzyme 1 → A → Enzyme 2 → B → Enzyme 3 → C → Enzyme 4 → D → Enzyme 5 → End product

Gene 1   Gene 2   Gene 3   Gene 4   Gene 5

Regulation of enzyme concentration

The end product blocks the transcription of all five genes. No enzymes are produced.

16.9 Two Ways to Regulate a Metabolic Pathway Feedback from the end product of a metabolic pathway can block enzyme activity (allosteric regulation), or it can stop the transcription of genes that code for the enzymes in the pathway (transcriptional regulation).

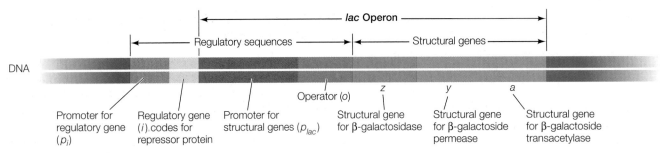

16.10 The *lac* Operon of *E. coli*   The *lac* operon of *E. coli* is a segment of DNA that includes a promoter, an operator, and the three structural genes that code for lactose-metabolizing enzymes.

structural genes (**Figure 16.10**). The **operator** is a short stretch of DNA that lies between the promoter and the structural genes. It can bind very tightly with regulatory proteins that either activate or repress transcription. There are numerous mechanisms to control the transcription of operons; here we will focus on three examples:

- An inducible operon regulated by a repressor protein

- A repressible operon regulated by a repressor protein

- An operon regulated by an activator protein

Operator–repressor interactions control transcription in the *lac* and *trp* operons

The *lac* operon contains a promoter, to which RNA polymerase binds to initiate transcription, and an operator, to which a **repressor** protein can bind. When the repressor is bound, transcription of the operon is blocked.

The repressor protein has two binding sites: one for the operator and the other for the inducer (allolactose). Binding with the inducer changes the shape of the repressor protein. This change in three-dimensional structure (conformation) prevents the repressor from binding to the operator (**Figure 16.11**). As a result, RNA polymerase can bind to the promoter and start transcribing the structural genes of the *lac* operon.

Study Figure 16.11 for the features of this negative control. You will notice that:

- in the absence of inducer, the operon is turned off;

- control is exerted by a regulatory protein—the repressor—that turns the operon off;

- the inducer, when present, binds to and changes the shape of the repressor so that it no longer binds to the operator, turning the operon on;

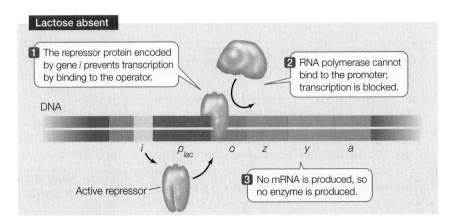

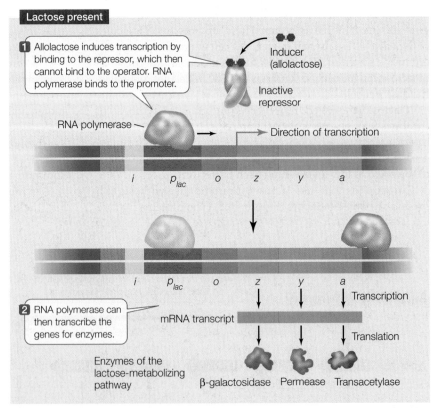

16.11 The *lac* Operon: An Inducible System   Allolactose (the inducer), a disaccharide that forms from lactose in bacterial cells, leads to synthesis of the enzymes in the lactose-metabolizing pathway by binding to the repressor protein and preventing its binding to the operator.

---

yourBioPortal.com

GO TO Animated Tutorial 16.1 • The *lac* Operon

- the **regulatory gene** produces a protein whose sole function is to regulate expression of the other genes; and

- certain DNA sequences (operators and promoters) do not code for proteins, but are binding sites for regulatory or other proteins.

In contrast to the inducible system of the *lac* operon, other operons in *E. coli* are repressible; that is, they are repressed when molecules called **co-repressors** bind to their repressors. This binding causes the repressor to change shape and bind to the operator, thereby inhibiting transcription.

An example is the operon whose structural genes catalyze the synthesis of the amino acid tryptophan:

5 enzyme-catalyzed reactions

precursor molecules $\;\rightarrow\;\;\rightarrow\;\;\rightarrow\;\;\rightarrow\;\;\rightarrow\;$ tryptophan

When tryptophan is present in the cell in adequate concentrations, it is advantageous to stop making the enzymes for tryptophan synthesis. To do this, the cell uses a repressor that binds to an operator upstream of the genes of the *trp* operon. But the repressor of the *trp* operon is not normally bound to the operator; it only binds when its shape is changed by binding to tryptophan, the co-repressor. To summarize the differences between these two types of operons:

- In *inducible* systems, the substrate of a metabolic pathway (the inducer) interacts with a regulatory protein (the repressor), rendering the repressor incapable of binding to the operator and thus allowing transcription.

- In *repressible* systems, the product of a metabolic pathway (the co-repressor) binds to a regulatory protein, which is then able to bind to the operator and block transcription.

─── **yourBioPortal.com** ───
GO TO Animated Tutorial 16.2 • The *trp* Operon

In general, inducible systems control catabolic pathways (which are turned on only when the substrate is available), whereas repressible systems control anabolic pathways (which are turned on until the concentration of the product becomes excessive). In both of the systems described here, the regulatory protein is a repressor that functions by binding to the operator. Next we will consider an example of positive control involving an activator.

16.12 Catabolite Repression Regulates the *lac* Operon The promoter for the *lac* operon does not function efficiently in the absence of cAMP, as occurs when glucose levels are high. High glucose levels thus repress the enzymes that metabolize lactose.

Protein synthesis can be controlled by increasing promoter efficiency

The examples described in the previous section are termed negative control because transcription is *decreased* in the presence of a repressor protein. *E. coli* can also use positive control to *increase* transcription through the presence of an **activator** protein. For an example we return to the *lac* operon, where the relative levels of glucose and lactose determine the amount of transcription. When lactose is present and glucose is low, the *lac* operon is activated by binding of a protein called cAMP receptor protein (CRP) to the *lac* operon promoter. CRP is an activator of transcription, because its binding results in more efficient binding of RNA polymerase to the promoter, and thus increased transcription of the structural genes (**Figure 16.12**).

In the presence of abundant glucose, CRP does not bind to the promoter and so the efficiency of transcription of the *lac* operon is reduced. This is an example of **catabolite repression**, a system of gene regulation in which the presence of the preferred energy source represses other catabolic pathways. The signaling pathway that controls catabolite repression of the *lac* operon involves the second messenger cAMP (see Section 7.3). The mechanisms controlling positive and negative regulation of the *lac* operon are summarized in **Table 16.1**.

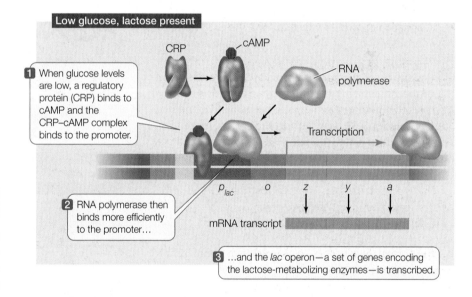

Low glucose, lactose present

CRP   cAMP

RNA polymerase

1 When glucose levels are low, a regulatory protein (CRP) binds to cAMP and the CRP–cAMP complex binds to the promoter.

Transcription

2 RNA polymerase then binds more efficiently to the promoter…

p_{lac} o   z   y   a

mRNA transcript

3 …and the *lac* operon—a set of genes encoding the lactose-metabolizing enzymes—is transcribed.

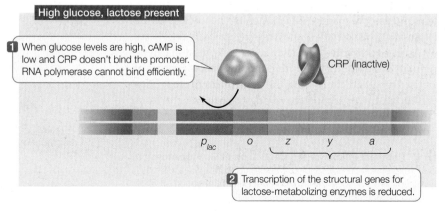

High glucose, lactose present

1 When glucose levels are high, cAMP is low and CRP doesn't bind the promoter. RNA polymerase cannot bind efficiently.

CRP (inactive)

p_{lac} o   z   y   a

2 Transcription of the structural genes for lactose-metabolizing enzymes is reduced.

Positive and Negative Regulation in the *lac* Operon[a]

GLUCOSE	cAMP LEVELS	RNA POLYMERASE BINDING TO PROMOTER	LACTOSE	LAC REPRESSOR	TRANSCRIPTION OF *lac* GENES?	LACTOSE USED BY CELLS?
Present	Low	Absent	Absent	Active and bound to operator	No	No
Present	Low	Present, not efficient	Present	Inactive and not bound to operator	Low level	No
Absent	High	Present, very efficient	Present	Inactive and not bound to operator	High level	Yes
Absent	High	Absent	Absent	Active and bound to operator	No	No

[a]Negative regulators are in red type.

16.2 RECAP

Gene expression in prokaryotes is most commonly regulated through control of transcription. An operon consists of a set of closely linked structural genes and the DNA sequences (promoter and operator) that control their transcription. Operons can be regulated by both negative and positive controls.

- Describe the molecular conditions at the *lac* operon promoter in the presence versus absence of lactose. **See Figure 16.11**

- What are the key differences between an inducible system and a repressible system? **See p. 351**

- What are the differences between positive and negative control of transcription? **See p. 351 and Table 16.1**

Studies of viruses and bacteria provide a basic understanding of mechanisms that regulate gene expression and of the roles of regulatory proteins in both positive and negative regulation. We now turn to the control of gene expression in eukaryotes. You will see both negative and positive control of transcription, as well as posttranscriptional mechanisms of regulation.

16.3 How Is Eukaryotic Gene Transcription Regulated?

For the normal development of an organism from fertilized egg to adult, and for each cell to acquire and maintain its proper specialized function, certain proteins must be made at just the right times and in just the right cells; these proteins must not be made at other times in other cells. Thus the expression of eukaryotic genes must be precisely regulated.

As in prokaryotes, eukaryotic gene expression can be regulated at a number of different points in the process of transcribing and translating the gene into a protein (**Figure 16.13**). In this section we will describe the mechanisms

16.13 Potential Points for the Regulation of Gene Expression
Gene expression can be regulated before transcription (1), during transcription (2, 3), after transcription but before translation (4, 5), at translation (6), or after translation (7).

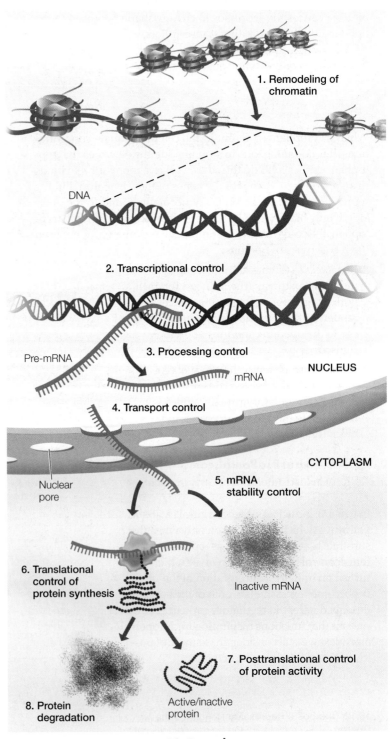

1. Remodeling of chromatin

DNA

2. Transcriptional control

Pre-mRNA

3. Processing control

mRNA

NUCLEUS

4. Transport control

Nuclear pore

CYTOPLASM

5. mRNA stability control

6. Translational control of protein synthesis

Inactive mRNA

7. Posttranslational control of protein activity

8. Protein degradation

Active/inactive protein

TABLE 16.2
Transcription in Prokaryotes and Eukaryotes

	PROKARYOTES	EUKARYOTES
Locations of functionally related genes	Often clustered in operons	Often distant from one another with separate promoters
RNA polymerases	One	Three: I transcribes rRNA II transcribes mRNA III transcribes tRNA and small RNAs
Promoters and other regulatory sequences	Few	Many
Initiation of transcription	Binding of RNA polymerase to promoter	Binding of many proteins, including RNA polymerase

that result in the selective transcription of specific genes. The mechanisms for regulating gene expression in eukaryotes have similar themes to those of prokaryotes. Both types of cells use DNA–protein interactions and negative and positive control. However, there are many differences, some of them dictated by the presence of a nucleus, which physically separates transcription and translation (**Table 16.2**).

Transcription factors act at eukaryotic promoters

As in prokaryotes, a promoter in eukaryotes is a sequence of DNA near the 5′ end of the coding region of a gene where RNA polymerase binds and initiates transcription. There are typically two important sequences in a promoter: One is the **recognition sequence**—the sequence recognized by RNA polymerase. The second, closer to the transcription initiation site, is the **TATA box** (so called because it is rich in AT base pairs), where DNA begins to denature so that the template strand can be exposed.

Eukaryotic RNA polymerase II cannot simply bind to the promoter and initiate transcription. Rather, it does so only after various regulatory proteins, called **transcription factors**, have assembled on the chromosome (**Figure 16.14**). First, the protein TFIID ("TF" stands for transcription factor) binds to the TATA box. Binding of TFIID changes both its own shape and that of the DNA, presenting a new surface that attracts the binding of other transcription factors to form a transcription complex. RNA polymerase II binds only after several other proteins have bound to this complex.

Some regulatory DNA sequences, such as the TATA box, are common to the promoters of many eukaryotic genes and are recognized by transcription factors that are found in all the cells of an organism. Other sequences found in promoters are specific

16.14 The Initiation of Transcription in Eukaryotes Apart from TFIID, which binds to the TATA box, each transcription factor in this transcription complex has binding sites only for the other proteins in the complex, and does not bind directly to DNA. B, E, F, and H are transcription factors.

──────── **yourBioPortal.com** ────────
GO TO Animated Tutorial 16.3 • Initiation of Transcription

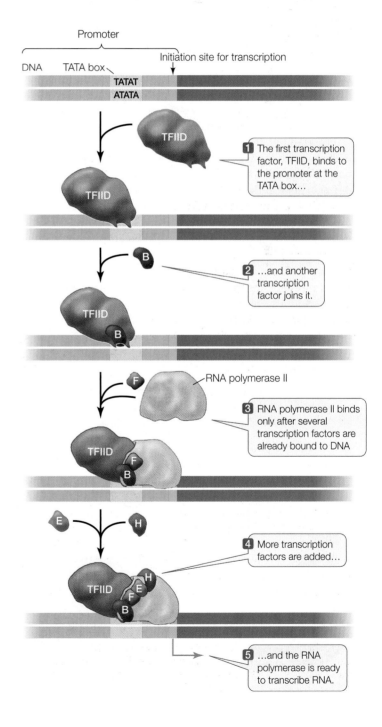

Promoter

DNA   TATA box   Initiation site for transcription

TATAT
ATATA

TFIID

TFIID

1 The first transcription factor, TFIID, binds to the promoter at the TATA box…

B

2 …and another transcription factor joins it.

TFIID
B

F   RNA polymerase II

3 RNA polymerase II binds only after several transcription factors are already bound to DNA

TFIID
F
B

E   H

4 More transcription factors are added…

TFIID   H
E
F
B

5 …and the RNA polymerase is ready to transcribe RNA.

to only a few genes and are recognized by transcription factors found only in certain types of cells. These specific transcription factors play an important role in cell differentiation, the structural and functional specialization of cells during development.

Other proteins can recognize and bind to DNA sequences and regulate transcription

In addition to the promoter, there are other short sequences (elements) of DNA that bind regulatory proteins, which in turn interact with RNA polymerase to regulate the rate of transcription (**Figure 16.15**). Some of these DNA elements are positive regulators (termed enhancers, which bind activator proteins) and others are negative (silencers, which bind repressor proteins). Some occur near the promoter and others as far as 20,000 base pairs away. One example of a transcription factor is CREB, which you read about in the opening essay of this chapter. When the activators and/or repressors (collectively termed transcription factors) bind to these elements, they interact with the RNA polymerase complex, causing DNA to bend. Often many such binding proteins are involved, and *the combination of factors present determines the rate of transcription.*

For example, the immature red blood cells in bone marrow make large amounts of β-globin. At least thirteen different transcription factors are involved in regulating transcription of the β-globin gene in these cells. Not all of these factors are present or active in other cells, such as the immature white blood cells produced by the same bone marrow. As a result the β-globin gene is not transcribed in those cells. So although the same genes are present in all cells, the fate of the cell is determined by which of its genes are expressed. How do transcription factors recognize specific DNA sequences?

Specific protein–DNA interactions underlie binding

As we have seen, transcription factors with specific DNA binding domains are involved in the activation and inactivation of specific genes. There are four common structural themes in the protein domains that bind to DNA. These themes, or **structural motifs**, consist of different combinations of structural elements (protein conformations) and may include special components such as zinc. The four common structural motifs in DNA binding domains are: helix-turn-helix, leucine zipper, zinc finger, and helix-loop-helix (**Figure 16.16**).

Let's look at how one of these motifs works. As pointed out in Section 13.2, the complementary bases in DNA not only form hydrogen bonds with each other, but also can form additional hydrogen bonds with proteins, particularly at points exposed in the major and minor grooves. In this way, an intact DNA double helix can be recognized by a protein motif whose structure:

- fits into the major or minor groove;
- has amino acids that can project into the interior of the double helix; and
- has amino acids that can form hydrogen bonds with the interior bases.

The helix-turn-helix motif, in which two α-helices are connected via a non-helical turn, fits these three criteria. The interior-facing "recognition" helix is the one whose amino acids interact with the bases inside the DNA. The exterior-facing helix sits on the sugar–phosphate backbone, ensuring that the interior helix is presented to the bases in the correct configuration. Many repressor proteins have this helix-turn-helix motif in their structure.

Repressors can inhibit transcription in several different ways. They can prevent the binding of transcriptional activators to DNA, or they can interact with other DNA binding proteins to decrease the rate of transcription.

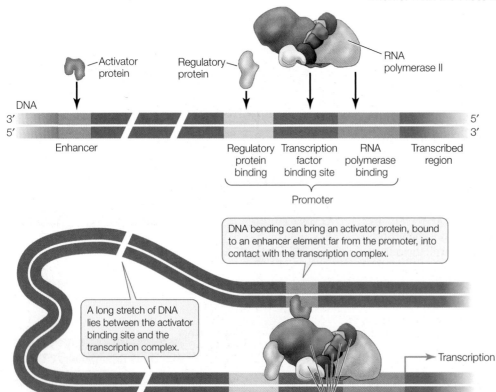

16.15 Transcription Factors, Repressors, and Activators The actions of many proteins determine whether and where RNA polymerase II will transcribe DNA.

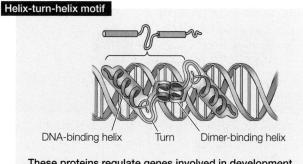

Helix-turn-helix motif

DNA-binding helix   Turn   Dimer-binding helix

These proteins regulate genes involved in development.

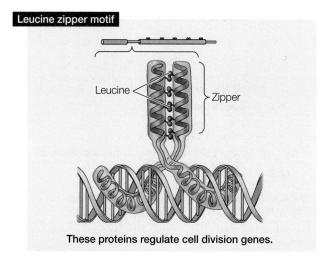

Leucine zipper motif

Leucine   Zipper

These proteins regulate cell division genes.

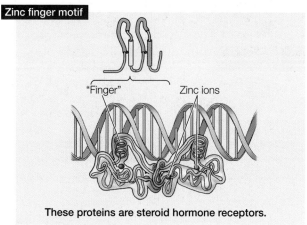

Zinc finger motif

"Finger"   Zinc ions

These proteins are steroid hormone receptors.

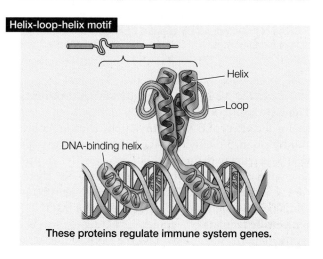

Helix-loop-helix motif

Helix

Loop

DNA-binding helix

These proteins regulate immune system genes.

16.16 Protein–DNA Interactions The DNA-binding domains of most regulatory proteins contain one of four structural motifs.

The expression of sets of genes can be coordinately regulated by transcription factors

How do eukaryotic cells coordinate the regulation of several genes whose transcription must be turned on at the same time? Prokaryotes solve this problem by arranging multiple genes in an operon that is controlled by a single promoter. But most eukaryotic genes have their own separate promoters, and genes that are coordinately regulated may be far apart. In these cases, the expression of genes can be coordinated if they share regulatory sequences that bind the same transcription factors.

This type of coordination is used by organisms to respond to stress—for example, by plants in response to drought. Under conditions of drought stress, a plant must simultaneously synthesize a number of proteins whose genes are scattered throughout the genome. The synthesis of these proteins comprises the stress response. To coordinate expression, each of these genes has a specific regulatory sequence near its promoter called the *stress response element* (*SRE*). A transcription factor binds to this element and stimulates mRNA synthesis (**Figure 16.17**). The stress re-

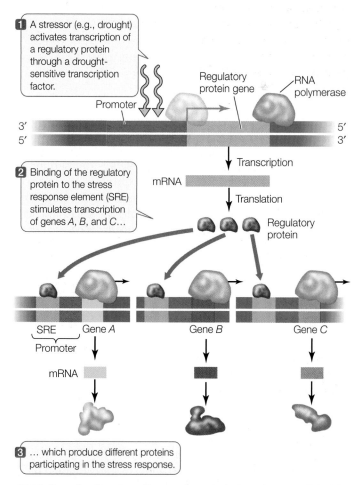

1 A stressor (e.g., drought) activates transcription of a regulatory protein through a drought-sensitive transcription factor.

Promoter   Regulatory protein gene   RNA polymerase

3′   5′
5′   3′

Transcription

2 Binding of the regulatory protein to the stress response element (SRE) stimulates transcription of genes *A*, *B*, and *C*...

mRNA

Translation

Regulatory protein

SRE   Gene *A*   Gene *B*   Gene *C*
Promoter

mRNA

3 ... which produce different proteins participating in the stress response.

16.17 Coordinating Gene Expression A single environmental signal, such as drought stress, causes the synthesis of a transcriptional regulatory protein that acts on many genes.

sponse proteins not only help the plant conserve water, but also protect the plant against excess salt in the soil and freezing. This finding has considerable importance for agriculture because crops are often grown under less than optimal conditions.

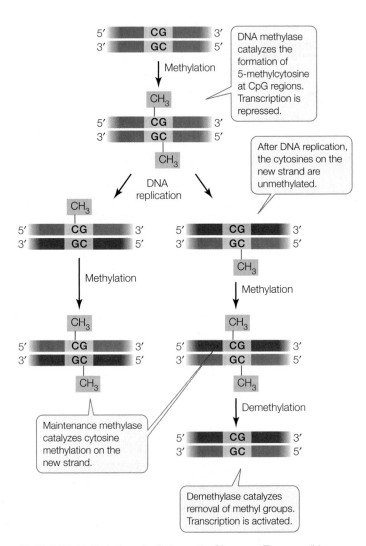

16.18 DNA Methylation: An Epigenetic Change The reversible formation of 5-methylcytosine in DNA can alter the rate of transcription.

16.3 RECAP

A number of transcription factors must bind to a eukaryotic promoter before RNA polymerase will bind to it and begin transcription. This provides a number of ways to increase or decrease transcription.

- Describe some of the different ways in which transcription factors regulate gene transcription. **See pp. 353–354 and Figure 16.15**

- How can more than one gene be regulated at the same time? **See p. 355 and Figure 16.17**

The mechanisms for control of gene expression that we have discussed so far involve direct interactions between proteins and specific DNA elements. If the sequences of the DNA elements are altered, then transcription of the gene will be affected. However, there are other mechanisms for controlling gene expression that do not depend on specific DNA sequences. We will discuss these mechanisms in the next section.

16.4 How Do Epigenetic Changes Regulate Gene Expression?

In the mid-twentieth century, the great developmental biologist Conrad Hal Waddington coined the term "epigenetics" and defined it as "that branch of biology which studies the causal interactions between genes and their products which bring the phenotype into being." Today **epigenetics** is defined more specifically, referring to changes in the expression of a gene or set of genes that occur without changing the DNA sequence. These changes are reversible, but sometimes are stable and heritable. They include two processes: DNA **methylation** and chromosomal protein alterations.

DNA methylation occurs at promoters and silences transcription

Depending on the organism, from 1 to 5 percent of cytosine residues in the DNA are chemically modified by the addition of a methyl group (—CH$_3$) to the 5'-carbon, to form 5-methylcytosine (**Figure 16.18**). This covalent addition is catalyzed by the enzyme **DNA methyltransferase** and, in mammals, usually occurs in C residues that are adjacent to G residues. DNA regions rich in these doublets are called **CpG islands**, and are especially abundant in promoters.

This covalent change in DNA is heritable: when DNA is replicated, a **maintenance methylase** catalyzes the formation of 5-methylcytosine in the new DNA strand. However, the pattern of cytosine methylation can also be altered, because methylation is reversible: a third enzyme, appropriately called **demethy-**

lase, catalyzes the removal of the methyl group from cytosine (see Figure 16.18).

What is the effect of DNA methylation? During replication and transcription, 5-methylcytosine behaves just like plain cytosine: it base pairs with guanine. But extra methyl groups in a promoter attract proteins that bind methylated DNA. These proteins are generally involved in the repression of gene transcription; thus heavily methylated genes tend to be inactive. This form of genetic regulation is epigenetic because it affects gene expression patterns without altering the DNA sequence.

DNA methylation is important in development from egg to embryo. For example, when a mammalian sperm enters an egg, many genes in first the male and then the female genome become demethylated. Thus many genes that are usually inactive are expressed during early development. As the embryo develops and its cells become more specialized, genes whose products are not needed in particular cell types become methylated. These methylated genes are "silenced"; their transcription is repressed. However, unusual or abnormal events can sometimes turn silent genes back on.

For example, DNA methylation may play roles in the genesis of some cancers. In cancer cells, oncogenes get activated and promote cell division, and tumor suppressor genes (that normally inhibit cell division) are turned off (see Chapter 11). This misregulation can occur when the promoters of oncogenes become demethylated while those of tumor suppressor genes become methylated. This is the case in colorectal cancer.

Histone protein modifications affect transcription

Another mechanism for epigenetic gene regulation is the alteration of chromatin structure, or *chromatin remodeling*. DNA is packaged with histone proteins into nucleosomes, which can make DNA physically inaccessible to RNA polymerase and the rest of the transcription apparatus. Each histone protein has a "tail" of approximately 20 amino acids at its N terminus that sticks out of the compact structure and contains certain positively charged amino acids (notably lysine). Enzymes called histone acetyltransferases can add acetyl groups to these positively charged amino acids, thus changing their charges:

Lysine in histone   Acetyl-CoA   Acetyl-lysine

Ordinarily, there is strong electrostatic attraction between the positively charged histone proteins and DNA, which is negatively charged because of its phosphate groups. Reducing the positive charges of the histone tails reduces the affinity of the histones for DNA, opening up the compact nucleosome. Additional chromatin remodeling proteins can bind to the loosened nucleosome–DNA complex, opening up the DNA for gene expression (**Figure 16.19**). Histone acetyltransferases can thus activate transcription.

Another kind of chromatin remodeling protein, histone deacetylase, can remove the acetyl groups from histones and

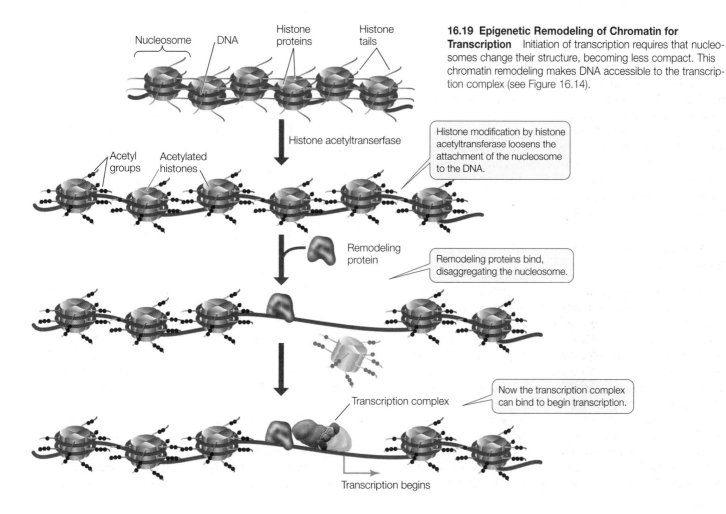

16.19 Epigenetic Remodeling of Chromatin for Transcription Initiation of transcription requires that nucleosomes change their structure, becoming less compact. This chromatin remodeling makes DNA accessible to the transcription complex (see Figure 16.14).

Histone modification by histone acetyltransferase loosens the attachment of the nucleosome to the DNA.

Remodeling proteins bind, disaggregating the nucleosome.

Now the transcription complex can bind to begin transcription.

Nucleosome   DNA   Histone proteins   Histone tails

Histone acetyltranserfase

Acetyl groups   Acetylated histones

Remodeling protein

Transcription complex

Transcription begins

thereby repress transcription. Histone deacetylases are targets for drug development to treat some forms of cancer. As noted above, certain genes block cell division in normal specialized tissues. In some cancers these genes are less active than normal, and the histones near them show excessive levels of deacetylation. Theoretically, a drug acting as a histone deacetylase inhibitor could tip the balance toward acetylation and this might activate genes that normally inhibit cell division.

Other types of histone modification can affect gene activation and repression. For example, histone methylation is associated with gene inactivation and histone phosphorylation also affects gene expression, the specific effect depending on which amino acid is modified. All of these effects are reversible and so the activity of a eukaryotic gene may be determined by very complex patterns of histone modification. David Allis of the Rockefeller University in New York City has dubbed this epigenetic system the "histone code."

Epigenetic changes induced by the environment can be inherited

Despite that fact that they are reversible, many epigenetic changes such as DNA methylation and histone modification can permanently alter gene expression patterns in a cell. If the cell is a germ line cell that forms gametes, the epigenetic changes can be passed on to the next generation. But what determines these epigenetic changes? A clue comes from a recent study of monozygotic twins.

Monozygotic twins come from a single fertilized egg that divides to produce two separate cells; each of these goes on to develop a separate individual. Twin brothers or sisters thus have identical genomes. But are they identical in their *epigenomes*? A comparison of DNA in hundreds of such twin pairs shows that in tissues of three-year-olds, the DNA methylation patterns are virtually the same. But by age 50, by which time the twins have usually been living apart for decades, in different environments, the patterns are quite different. This indicates that the *environment plays an important role in epigenetic modifications* and, therefore, in the regulation of genes that these modifications affect.

What factors in the environment lead to epigenetic changes? One might be stress: when mice are put in a stressful situation, genes that are involved in important brain pathways become heavily methylated (and transcriptionally inactive). Treatment of the stressed mice with an antidepressant drug "hits the undo button," reversing these changes. Transcription factors such as CREB that mediate addiction (see the opening story of this chapter) are involved with histone acetylation, which leads to subsequent gene activation. The sperm of men with psychosis have different methylation patterns than sperm from nonpsychotic men. This last observation is especially provocative, as it suggests that epigenetic patterns, some of which may have formed during life, can be passed on to the next generation. This means that some phenotypic characteristics acquired during the lifetime of an organism might be heritable, contrary to biologists' long-held views. The idea that epigenetic changes can be inherited remains controversial.

DNA methylation can result in genomic imprinting

In mammals specific patterns of methylation develop for each sex during gamete formation. This happens in two stages: first, the existing methyl groups are removed from the 5-methylcytosines by a demethylase, and then a DNA methylase adds methyl groups to the appropriate cytosines. When the gametes form they carry this new pattern of methylation (epigenetic information).

The DNA methylation pattern in male gametes (sperm) differs from that in female gametes (eggs) at about 200 genes in the mammalian genome. That is, a given gene in this group may be methylated in eggs but unmethylated in sperm (**Figure 6.20**). In this case the offspring would inherit a maternal gene that is transcriptionally inactive (methylated) and a paternal gene that is transcriptionally active (demethylated). This is called **genomic imprinting**.

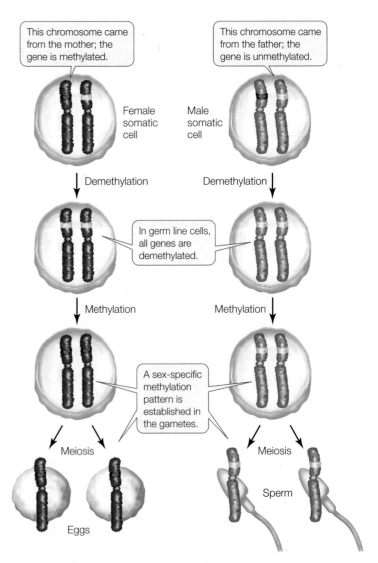

This chromosome came from the mother; the gene is methylated.

This chromosome came from the father; the gene is unmethylated.

Female somatic cell

Male somatic cell

Demethylation

Demethylation

In germ line cells, all genes are demethylated.

Methylation

Methylation

A sex-specific methylation pattern is established in the gametes.

Meiosis

Meiosis

Sperm

Eggs

16.20 Genomic Imprinting For some genes, epigenetic DNA methylation differs in male and female gametes. As a result, an individual might inherit an allele from the female parent that is transcriptionally silenced; but the same allele from the male parent would be expressed.

An example of imprinting is found in a region on human chromosome 15 called 15q11. This region is imprinted differently during the formation of male and female gametes, and offspring normally inherit both the paternally and maternally derived patterns. In rare cases, there is a chromosome deletion in one of the gametes, and the newborn baby inherits just the male or the female imprinting pattern in this particular chromosome region. If the male pattern is the only one present (female region deleted), the baby develops Angelman Syndrome, characterized by epilepsy, tremors, and constant smiling. If the female pattern is the only one present (male region deleted), the baby develops a quite different phenotype called Prader-Willi syndrome, marked by muscle weakness and obesity. Note that the *gene sequences are the same* in both cases: it is *the epigenetic patterns that are different.*

Imprinting of specific genes occurs primarily in mammals and flowering plants. Most imprinted genes are involved with embryonic development. An embryo must have both the paternally and maternally imprinted gene patterns to develop properly. In fact, attempts to make an embryo that has chromosomes from only one sex (for example, by chemically treating an egg cell to double its chromosomes) usually fail. So imprinting has an important lesson for genetics: *males and females may be the same genetically (except for the X and Y chromosomes), but they differ epigenetically.*

Global chromosome changes involve DNA methylation

Like single genes, large regions of chromosomes or even entire chromosomes can have distinct patterns of DNA methylation. Under a microscope, two kinds of chromatin can be distinguished in the stained interphase nucleus: *euchromatin* and *heterochromatin*. The euchromatin appears diffuse and stains lightly; it contains the DNA that is transcribed into mRNA. Heterochromatin is condensed and stains darkly; any genes it contains are generally not transcribed.

Perhaps the most dramatic example of heterochromatin is the inactive X chromosome of mammals. A normal female mammal has two X chromosomes; a normal male has an X and a Y (see Section 12.4). The X and Y chromosomes probably arose from a pair of autosomes (non–sex chromosomes) about 300 million years ago. Over time, mutations in the Y chromosome resulted in maleness-determining genes, and the Y chromosome gradually lost most of the genes it once shared with its X homolog. As a result, females and males differ greatly in the "dosage" of X-linked genes. Each female cell has two copies of each gene on the X chromosome, and therefore has the potential to produce twice as much of each protein product. Nevertheless, for 75 percent of the genes on the X chromosome, transcription is generally the same in males and in females. How does this happen?

Mary Lyon, Liane Russell, and Ernest Beutler independently hypothesized in 1961 that one of the X chromosomes in each cell of a female is, to a significant extent, transcriptionally inactivated early in embryonic development. They proposed that one copy of X becomes inactive in each embryonic cell, and the same X remains inactive in all that cell's descendants. Several lines of evidence have since confirmed this hypothesis.

In a given embryonic cell, the "choice" of which X in the pair to inactivate is random. Recall that one X in a female comes from her father and one from her mother. Thus, in one embryonic cell the paternal X might be the one remaining transcriptionally active, but in a neighboring cell the maternal X might be active.

The inactivated X chromosome does not vanish, but is identifiable within the nucleus. During interphase a single, stainable nuclear body called a Barr body (after its discoverer, Murray Barr) can be seen in cells of human females under the light microscope (**Figure 16.21A**). This clump of heterochromatin, which is not present in normal males, is the inactivated X chromosome, and it consists of heavily methylated DNA. A female with the normal two X chromosomes will have one Barr body, while a rare female with three Xs will have two, and an XXXX female will have three. Males that are XXY will have one. These observations suggest that the interphase cells of each person, male

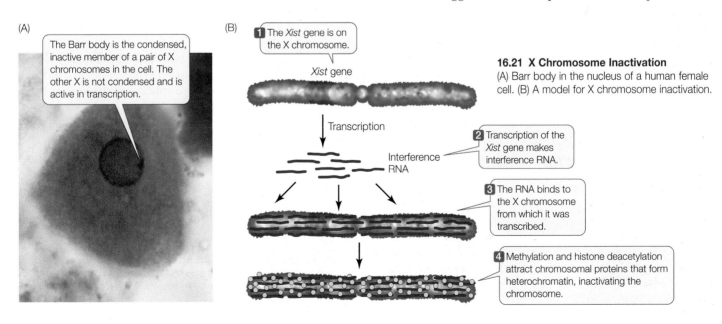

(A) The Barr body is the condensed, inactive member of a pair of X chromosomes in the cell. The other X is not condensed and is active in transcription.

(B)
1 The *Xist* gene is on the X chromosome.

Xist gene

Transcription

Interference RNA

2 Transcription of the *Xist* gene makes interference RNA.

3 The RNA binds to the X chromosome from which it was transcribed.

4 Methylation and histone deacetylation attract chromosomal proteins that form heterochromatin, inactivating the chromosome.

16.21 X Chromosome Inactivation
(A) Barr body in the nucleus of a human female cell. (B) A model for X chromosome inactivation.

or female, have a single active X chromosome, and thus a constant dosage of expressed X chromosome genes.

Condensation of the inactive X chromosome makes its DNA sequences physically unavailable to the transcriptional machinery. Most of the genes of the inactive X are heavily methylated. However, one gene, *Xist* (for *X inactivation-specific transcript*), is only lightly methylated and is transcriptionally active. On the active X chromosome, *Xist* is heavily methylated and not transcribed. The RNA transcribed from *Xist* binds to the X chromosome from which it is transcribed, and this binding leads to a spreading of inactivation along the chromosome. The *Xist* RNA transcript is an example of **interference RNA (Figure 16.21B)**.

16.4 RECAP

Epigenetics describes stable changes in gene expression that do not involve changes in DNA sequences. These changes involve modifications of DNA (cytosine methylation) or of histone proteins bound to DNA. Epigenetic changes can be affected by the environment, and can also result in genome imprinting, in which expression of some genes depends on their parental origin.

- How are DNA methylation patterns established and how do they affect gene expression? **See p. 356 and Figure 16.18**

- Explain how histone modifications affect transcription. **See pp. 357–358 and Figure 16.19**

- Why and how does X chromosome inactivation occur? **See p. 359**

Gene expression involves transcription and then translation. So far we have described how eukaryotic gene expression is regulated at the transcriptional level. But as Figure 16.13 shows, there are many points at which regulation can occur after the initial gene transcript is made.

16.5 How Is Eukaryotic Gene Expression Regulated After Transcription?

Eukaryotic gene expression can be regulated both in the nucleus prior to mRNA export, and after the mRNA leaves the nucleus. Posttranscriptional control mechanisms can involve alternative splicing of pre-mRNA, microRNAs, repressors of translation, or regulation of protein breakdown in the proteasome.

Different mRNAs can be made from the same gene by alternative splicing

Most primary mRNA transcripts contain several introns (see Figure 14.7). We have seen how the splicing mechanism recognizes the boundaries between exons and introns. What would happen if the β-globin pre-mRNA, which has two introns, were spliced from the start of the first intron to the end of the second? The middle exon would be spliced out along with the two introns. An entirely new protein (certainly not a β-globin) would be made, and the functions of normal β-globin would be lost. Such **alternative splicing** can be a deliberate mechanism for generating a family of different proteins with different activities and functions from a single gene (**Figure 16.22**).

Before the human genome was sequenced, most scientists estimated that they would find between 80,000 and 150,000 protein-coding genes. You can imagine their surprise when the actual sequence revealed only about 24,000 genes! In fact, there are many more human mRNAs than there are human genes, and most of this variation comes from alternative splicing. Indeed, recent surveys show that about half of all human genes are alternatively spliced. Alternative splicing may be a key to the differences in levels of complexity among organisms. For example, although humans and chimpanzees have similar-sized genomes, there is more alternative splicing in the human brain than in the brain of a chimpanzee.

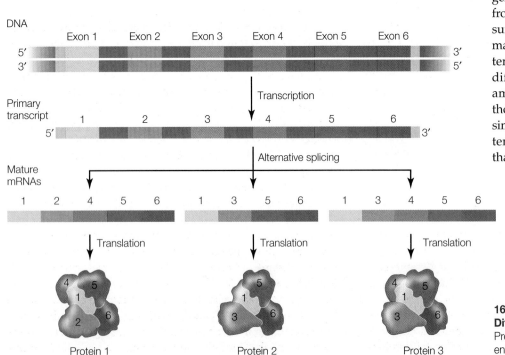

16.22 Alternative Splicing Results in Different Mature mRNAs and Proteins Pre-mRNA can be spliced differently in different tissues, resulting in different proteins.

MicroRNAs are important regulators of gene expression

As we discuss in the next chapter, less than 5 percent of the genome in most plants and animals codes for proteins. Some of the genome encodes ribosomal RNA and transfer RNAs, but until recently biologists thought that the rest of the genome was not transcribed; some even called it "junk." Recent investigations, however, have shown that some of these noncoding regions are transcribed. The noncoding RNAs are often very small and therefore difficult to detect. These tiny RNA molecules are called **microRNA** (miRNA).

The first miRNA sequences were found in the worm *Caenorhabditis elegans*. This model organism, which has been studied extensively by developmental biologists, goes through several larval stages. Victor Ambros at the University of Massachusetts found mutations in two genes that had different effects on progress through these stages:

- *lin-14* mutations (named for abnormal cell *lin*eage) caused the larvae to skip the first stage and go straight to the second stage. Thus the gene's normal role is to facilitate events of the first larval stage.

- *lin-4* mutations caused certain cells in later larval stages to repeat a pattern of development normally shown in the first larval stage. It was as if the cells were stuck in that stage. So the normal role of this gene is to *negatively regulate lin-14*, turning its expression off so the cells can progress to the next stage.

Not surprisingly, further investigation showed that *lin-14* encodes a transcription factor that affects the transcription of genes involved in larval cell progression. It was originally expected that *lin-4,* the negative regulator, would encode a protein that downregulates genes activated by the lin-14 protein. But this turned out to be incorrect. Instead, *lin-4* encodes a 22-base miRNA that inhibits *lin-14* expression posttranscriptionally by binding to its mRNA.

Several hundred miRNAs have now been described in many eukaryotes. Each one is about 22 bases long and usually has dozens of mRNA targets. These miRNAs are transcribed as longer precursors that are then cleaved through a series of steps to double-stranded miRNAs. A protein complex guides the miRNA to its target mRNA, where translation is inhibited and the mRNA is degraded (**Figure 16.23**). The remarkable conservation of the miRNA gene silencing mechanism in eukaryotes indicates that it is evolutionarily ancient and biologically important.

Translation of mRNA can be regulated

Is the amount of a protein in a cell determined by the amount of its mRNA? Recently, scientists examined the relationship between mRNA abundance and protein abundance in yeast cells. For about a third of the many genes surveyed, there was a clear correlation between mRNA and protein: more of one led to more of the other. But for two-thirds of the proteins, there was no apparent relationship between the two: sometimes there was lots of mRNA and little or no protein, or lots of protein and lit-

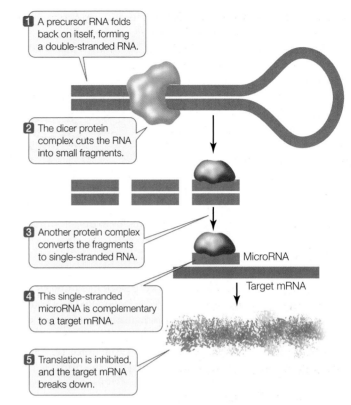

1 A precursor RNA folds back on itself, forming a double-stranded RNA.

2 The dicer protein complex cuts the RNA into small fragments.

3 Another protein complex converts the fragments to single-stranded RNA.

MicroRNA

Target mRNA

4 This single-stranded microRNA is complementary to a target mRNA.

5 Translation is inhibited, and the target mRNA breaks down.

16.23 mRNA Inhibition by MicroRNAs MicroRNAs result in inhibition of translation and in breakdown of the target mRNA.

tle mRNA. The concentrations of these proteins must therefore be determined by factors acting after the mRNA is made. Cells do this in two major ways: by blocking the translation of mRNA, or by altering how long newly synthesized proteins persist in the cell (protein longevity).

REGULATION OF TRANSLATION There are three known ways in which the translation of mRNA can be regulated. One way, as we saw in the previous section, is to inhibit translation with miRNAs. A second way involves modification of the guanosine triphosphate cap on the 5′ end of the mRNA (see Section 14.4). An mRNA that is capped with an unmodified GTP molecule is not translated. For example, stored mRNAs in the egg cells of the tobacco hornworm moth are capped with unmodified GTP molecules and are not translated. After the egg is fertilized, however, the caps are modified, allowing the mRNA to be translated to produce the proteins needed for early embryonic development.

In another system, repressor proteins directly block translation. For example, in mammalian cells the protein ferritin binds free iron ions (Fe^{2+}). When iron is present in excess, ferritin synthesis rises dramatically, but the amount of ferritin mRNA remains constant, indicating that the increase in ferritin synthesis is due to an increased rate of mRNA translation. Indeed, when the iron level in the cell is low, a translational repressor protein binds to ferritin mRNA and prevents its translation by blocking its attachment to a ribosome. When the iron level rises, some of the excess Fe^{2+} ions bind to the repressor and alter its three-

16.24 A Proteasome Breaks Down Proteins
Proteins targeted for degradation are bound to ubiquitin, which then binds the targeted protein to a proteasome. The proteasome is a complex structure where proteins are digested by several powerful proteases.

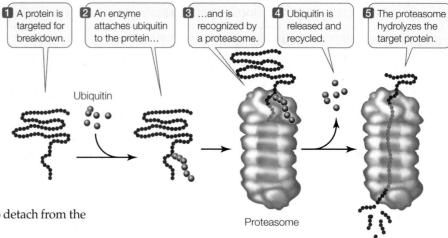

1 A protein is targeted for breakdown.

2 An enzyme attaches ubiquitin to the protein...

3 ...and is recognized by a proteasome.

4 Ubiquitin is released and recycled.

5 The proteasome hydrolyzes the target protein.

Ubiquitin

Proteasome

dimensional structure, causing the repressor to detach from the mRNA and allowing translation to proceed.

REGULATION OF PROTEIN LONGEVITY   The protein content of any cell at a given time is a function of both protein synthesis and protein degradation. Certain proteins can be targeted for destruction in a chain of events that begins when an enzyme attaches a 76-amino acid protein called **ubiquitin** (so named because it is ubiquitous, or widespread) to a lysine residue of the protein to be destroyed. Other ubiquitins then attach to the primary one, forming a polyubiquitin chain. The protein–polyubiquitin complex then binds to a huge protein complex called a **proteasome** (from *protease* and *soma*, body) (**Figure 16.24**). Upon entering the proteasome, the polyubiquitin is removed and ATP energy is used to unfold the target protein. Three different proteases then digest the protein into small peptides and amino acids.

You may recall from Section 11.2 that cyclins are proteins that regulate the activities of key enzymes at specific points in the cell cycle. Cyclins must be broken down at just the right time, and this is done by proteasomes. Viruses can hijack this system. For example, some strains of the human papillomavirus target p53 protein and retinoblastoma protein, which normally inhibit the cell cycle, for proteasomal degradation, resulting in unregulated cell division (cancer).

16.5 RECAP

One of the most important means of posttranscriptional regulation is alternative RNA splicing, which allows more than one protein to be made from a single gene. The stability of mRNA in the cytoplasm can also be regulated. MicroRNAs, mRNA modifications, and translational repressors can prevent mRNA translation. Proteins in the cell can be targeted for breakdown by ubiquitin and then hydrolyzed in proteasomes.

- How can a single pre-mRNA sequence encode several different proteins? **See p. 360 and Figure 16.22**

- How do miRNAs regulate gene expression? **See p. 361 and Figure 16.23**

- Explain the role of the proteasome. **See p. 362 and Figure 16.24**

CHAPTER SUMMARY

16.1 How Do Viruses Regulate Their Gene Expression?

- **Viruses** are not cells, and rely on host cells to reproduce.

- The basic unit of a virus is a **virion**, which consists of a nucleic acid genome (DNA or RNA) and a protein coat, called a **capsid**.

- **Bacteriophage** are viruses that infect bacteria.

- Viruses undergo a **lytic cycle**, which causes the host cell to burst, releasing new virions.

- Some viruses have promoters that bind host RNA polymerase, which they use to transcribe their own genes and proteins. **Review Figure 16.3**

- Rarely, a phage will transfer bacterial genes to a new host in the process of **transduction**.

- Some viruses can also undergo a **lysogenic cycle**, in which a molecule of their DNA, called a **prophage**, is inserted into the host chromosome, where it replicates for generations. **Review Figure 16.4**

- The cellular environment determines whether a phage undergoes a lytic or a lysogenic cycle. Regulatory proteins that compete for promoters on phage DNA control the switch between the two life cycles. **Review Figure 16.5**

- A **retrovirus** uses reverse transcriptase to generate a cDNA **provirus** from its RNA genome. The provirus is incorporated into the host's DNA and can be activated to produce new virions. **Review Figure 16.6**

16.2 How Is Gene Expression Regulated in Prokaryotes?

- Some proteins are synthesized only when they are needed. Proteins that are made only in the presence of a particular compound—an **inducer**—are **inducible** proteins. Proteins that are made at a constant rate regardless of conditions are **constitutive** proteins.

- An **operon** consists of a promoter, an **operator**, and two or more **structural genes**. Promoters and operators do not code for proteins, but serve as binding sites for regulatory proteins. Review Figure 16.10

- **Regulatory genes** code for regulatory proteins, such as **repressors**. When a repressor binds to an operator, transcription of the structural gene is inhibited. Review Figure 16.11, **ANIMATED TUTORIALS 16.1 AND 16.2**

- The *lac* operon is an example of an inducible system, in which the presence of an inducer (lactose) keeps the repressor from binding the operator, allowing the transcription of structural genes for lactose metabolism.

- Transcription can be enhanced by the binding of an **activator** protein to the promoter. Review Figure 16.12

- **Catabolite repression** is the inhibition of a catabolic pathway for one energy source by a different, preferred energy source.

16.3 How Is Eukaryotic Gene Transcription Regulated?

- Eukaryotic gene expression is regulated both during and after transcription. Review Figure 16.13, **WEB ACTIVITY 16.1**

- **Transcription factors** and other proteins bind to DNA and affect the rate of initiation of transcription at the promoter. Review Figure 16.14 and 16.15, **ANIMATED TUTORIAL 16.3**

- The interactions of these proteins with DNA are highly specific and depend on protein domains and DNA sequences.

- Genes at distant locations from one another can be coordinately regulated by common transcription factors and promoter elements. Review Figure 16.17

16.4 How Do Epigenetic Changes Regulate Gene Expression?

- **Epigenetics** refers to changes in gene expression that do not involve changes in DNA sequences.

- **Methylation** of cytosine residues generally inhibits transcription. Review Figure 16.18

- Modifications of histone proteins in nucleosomes make transcription either easier or more difficult. Review Figure 16.19

- Epigenetic changes can occur because of the environment.

- DNA methylation can explain **genome imprinting**, where the expression of a gene depends on its parental origin. Review Figure 16.20

16.5 How Is Eukaryotic Gene Expression Regulated After Transcription?

- **Alternative splicing** of pre-mRNA can produce different proteins. Review Figure 16.22

- **MicroRNAs** are small RNAs that do not code for proteins, but regulate the translation and longevity of mRNA. Review Figure 16.23

- The translation of mRNA to proteins can be regulated by translational repressors.

- The **proteasome** can break down proteins, thus affecting protein longevity. Review Figure 16.24

SEE WEB ACTIVITY 16.2 for a concept review of this chapter.

SELF-QUIZ

1. Which of the following statements about the *lac* operon is *not* true?
   a. When lactose binds to the repressor, the repressor can no longer bind to the operator.
   b. When lactose binds to the operator, transcription is stimulated.
   c. When the repressor binds to the operator, transcription is inhibited.
   d. When lactose binds to the repressor, the shape of the repressor is changed.
   e. The repressor has binding sites for both DNA and lactose.

2. Which of the following is *not* a type of viral reproduction?
   a. DNA virus in a lytic cycle
   b. DNA virus in a lysogenic cycle
   c. DNA virus (single-stranded) with a double-stranded DNA intermediate
   d. RNA virus with reverse transcription to make cDNA
   e. RNA virus acting as tRNA

3. In the lysogenic cycle of bacteriophage λ,
   a. a repressor, cI, blocks the lytic cycle.
   b. the bacteriophage carries DNA between bacterial cells.
   c. both early and late phage genes are transcribed.
   d. the viral genome is made into RNA, which stays in the host cell.
   e. many new viruses are made immediately, regardless of host health.

4. An operon is
   a. a molecule that can turn genes on and off.
   b. an inducer bound to a repressor.
   c. a series of regulatory sequences controlling transcription of protein-coding genes.

   d. any long sequence of DNA.
   e. a promoter, an operator, and a group of linked structural genes.

5. Which of the following is true of both positive and negative gene regulation?
   a. They reduce the rate of transcription of certain genes.
   b. They involve regulatory proteins (or RNA) binding to DNA.
   c. They involve transcription of all genes in the genome.
   d. They are not both active in the same organism or virus.
   e. They act away from the promoter.

6. In DNA, 5-methylcytosine
   a. forms a base pair with adenine.
   b. is not recognized by DNA polymerase.
   c. is related to transcriptional silencing of genes.
   d. does not occur at promoters.
   e. is an irreversible modification of cytosine.

7. Which statement about selective gene transcription in eukaryotes is *not* true?
   a. Regulatory proteins can bind at a site on DNA distant from the promoter.
   b. Transcription requires transcription factors.
   c. Genes are usually transcribed as groups called operons.
   d. Both positive and negative regulation occur.
   e. Many proteins bind at the promoter.

8. Control of gene expression in eukaryotes includes all of the following except
   a. alternative RNA splicing.
   b. binding of proteins to DNA.
   c. transcription factors.
   d. stabilization of mRNA by miRNA.
   e. DNA methylation.

9. The promoter in the *lac* operon is
   a. the region that binds the repressor.
   b. the region that binds RNA polymerase.
   c. the gene that codes for the repressor.
   d. a structural gene.
   e. an operon.

10. Epigenetic changes
    a. can involve DNA methylation.
    b. are due to nonhistone protein acetylation.
    c. are due to changes in the genetic code.
    d. are an example of positive control of translation.
    e. are never reversible.

FOR DISCUSSION

1. Compare the life cycles of a lysogenic bacteriophage and HIV (Figures 16.4 and 16.6) with respect to:
   a. how the virus enters the cell.
   b. how the virion is released in the cell.
   c. how the viral genome is replicated.
   d. how new viruses are produced.

2. Compare promoters adjacent to early and late genes in the bacteriophage lytic cycle.

3. The repressor protein that acts on the *lac* operon of *E. coli* is encoded by a regulatory gene. The repressor is made in small quantities and at a constant rate. Would you surmise that the promoter for this repressor protein is efficient or inefficient? Is synthesis of the repressor constitutive, or is it under environmental control?

4. A protein-coding gene in a eukaryote has three introns. How many different proteins could be made by alternative splicing of the pre-mRNA from this gene?

ADDITIONAL INVESTIGATION

In colorectal cancer, tumor suppressor genes are not active. This is an important factor resulting in uncontrolled cell division. Two possible explanations for the inactive genes are: a mutation in the coding region, resulting in an inactive protein, or epigenetic silencing at the promoter of the gene, resulting in reduced transcription. How would you investigate these two possibilities?

17

Genomes

The dog genome

Canis lupus familiaris, the dog, was domesticated by humans from the gray wolf thousands of years ago. While there are many kinds of wolves, they all look more or less the same. Not so with "man's best friend." The American Kennel Club recognizes about 155 different breeds. Dog breeds not only look different, they vary greatly in size. For example, an adult Chihuahua weighs just 1.5 kg, while a Scottish deerhound weighs 70 kg. No other mammal shows such large phenotypic variation, and biologists are curious about how this occurs. Also, there are hundreds of genetic diseases in dogs, and many of these diseases have counterparts in humans. To find out about the genes behind the phenotypic variation, and to elucidate the relationships between genes and diseases, the Dog Genome Project began in the late 1990s. Since then the sequences of several dog genomes have been published.

Two dogs—a boxer and a poodle—were the first to have their entire genomes sequenced. The dog genome contains 2.8 billion base pairs of DNA in 39 pairs of chromosomes. There are 19,000 protein-coding genes, most of them with close counterparts in other mammals, including humans. The whole genome sequence made it easy to create a map of genetic markers—specific nucleotides or short sequences of DNA at particular locations on the genome that differ between individual dogs and/or breeds.

Genetic markers are used to map the locations of (and thus identify) genes that control particular traits. For example, Dr. Elaine Ostrander and her colleagues at the National Institutes of Health studied Portuguese water dogs to identify genes that control size. Taking samples of cells for DNA isolation was relatively easy: a cotton swab was swept over the inside of the cheek. As Dr. Ostrander said, the dogs "didn't care, especially if they thought they were going to get a treat or if there was a tennis ball in our other hand." It turned out that the gene for *i*nsulin-like *g*rowth *f*actor 1 (IGF-1) is important in determining size: large breeds have an allele that codes for an active IGF-1 and small breeds have a different allele that codes for a less active IGF-1.

Another gene important to phenotypic variation was found in whippets, sleek dogs that run fast and are often raced. A mutation in the gene for myostatin, a protein that inhibits overdevelopment of muscles, results in a

Variation in Dogs The Chihuahua (bottom) and the Brazilian mastiff (top) are the same species, *Canis lupus familiaris*, and yet show great variation in size. Genome sequencing has revealed insights into how size is controlled by genes.

Genetic Bully   These dogs are both whippets, but the muscle-bound dog (right) has a mutation in a gene that limits muscle buildup.

whippet that is more muscular and runs faster. Myostatin is important in human muscles as well.

Inevitably, some scientists have set up companies to test dogs for genetic variations, using DNA supplied by anxious owners and breeders. Some traditional breeders frown on this practice, but others say it will improve the breeds and give more joy (and prestige) to owners. So the issues surrounding the Dog Genome Project are not very different from ones arising from the Human Genome Project.

Powerful methods have been developed to analyze DNA sequences, and the resulting information is accumulating at a rapid rate. Comparisons of sequenced genomes are providing new insights into evolutionary relationships and confirming old ones. We are in a new era of biology.

IN THIS CHAPTER we look at genomes. First we look at how large molecules of DNA are cut and sequenced, and what kinds of information these genome sequences provide. Then we turn to the results of ongoing sequencing efforts in both prokaryotes and eukaryotes. We next consider the human genome and some of the real and potential uses of human genome information. Finally, we will describe the emerging fields of proteomics and metabolomics, which attempt to give a complete inventory of a cell's proteins and metabolic activity.

17.1 How Are Genomes Sequenced?

As you saw in the opening story on dogs, one reason for sequencing genomes is to compare different organisms. Another is to identify changes in the genome that result in disease. In 1986, the Nobel laureate Renato Dulbecco and others proposed that the world scientific community be mobilized to undertake the sequencing of the entire human genome. One challenge discussed at the time was to detect DNA damage in people who had survived the atomic bomb attacks and been exposed to radiation in Japan during World War II. But in order to detect changes in the human genome, scientists first needed to know its normal sequence.

The result was the publicly funded **Human Genome Project**, an enormous undertaking that was successfully completed in 2003. This effort was aided and complemented by privately funded groups. The project benefited from the development of many new methods that were first used in the sequencing of smaller genomes—those of prokaryotes and simple eukaryotes.

Two approaches were used to sequence the human genome

Many prokaryotes have a single chromosome, while eukaryotes have several to many. Because of their differing sizes, chromosomes can be separated from one another, identified, and experimentally manipulated. It might seem that the most straightforward approach to sequencing a chromosome would be to start at one end and simply sequence the entire DNA molecule. However, this approach is not practical since only about 700 base pairs can be sequenced at a time using current methods. Prokaryotic chromosomes contain 1–4 million base pairs and human chromosome 1 contains 246 million base pairs.

To sequence an entire genome, chromosomal DNA must be cut into short fragments about 500 base pairs long, which are separated and sequenced. For the haploid human genome, which has about 3.3 billion base pairs, there are more than 6 million such fragments. When all of the fragments have been sequenced, the problem becomes how to put these millions of sequences together. This task can be accomplished using larger, *overlapping fragments*.

Let's illustrate this process using a single, 10 base-pair (bp) DNA molecule. (This is a double-stranded molecule, but for convenience we show only the sequence of the noncoding strand.) The molecule is cut three ways. The first cut generates the fragments:

TG, ATG, and CCTAC

The second cut of the same molecule generates the fragments:

AT, GCC, and TACTG

The third cut results in:

CTG, CTA, and ATGC

Can you put the fragments into the correct order? (The answer is ATGCCTACTG.) Of course, the problem of ordering 6 million fragments, each about 500 bp long, is more of a challenge! The field of **bioinformatics** was developed to analyze DNA sequences using complex mathematics and computer programs.

Until recently, two broad approaches were used to analyze DNA fragments for alignment: hierarchical sequencing and shotgun sequencing. These were developed for the Human Genome Project, but have been applied to other organisms as well.

HIERARCHICAL SEQUENCING The publicly funded human genome sequencing team developed a method known as **hierarchical sequencing**. The first step was to systematically identify short marker sequences along the chromosomes, ensuring that every fragment of DNA to be sequenced would contain a marker (**Figure 17.1A**). Genetic markers can be short tandem repeats (STRs), single nucleotide polymorphisms (SNPs), or the recognition sites for *restriction enzymes*, which recognize and cut DNA at specific sequences (see Chapter 15).

Some restriction enzymes recognize sequences of 4–6 base pairs and generate many fragments from a large DNA molecule. For example, the enzyme *Sau*3A cuts DNA every time it encounters GATC. Other restriction enzymes recognize sequences of 8–12 base pairs (*Not*I cuts at GCGGCCGC, for example) and generate far fewer, but much larger, fragments.

In hierarchical sequencing, genomic DNA is cut up into a set of relatively large (55,000 to 2 million bp) fragments. If different enzymes are used in separate digests, the fragments will overlap so that some fragments share particular markers. Each fragment is inserted into a bacterial plasmid to create a **bacterial artificial chromosome** (**BAC**), which is then inserted into bacteria. Each bacterium gets just one plasmid with its fragment of (for example) the human genome and is allowed to grow into a colony containing millions of genetically identical bacteria (called a *clone*). Clones differ from one another in that each has a different fragment from the human genome. A collection of clones, containing many different fragments of a genome, is called a **genomic library**.

The DNA from each clone is then extracted and cut into smaller overlapping pieces, which in turn are cloned, purified, and sequenced. The overlapping parts of the sequences allow researchers (with the aid of computers) to align them to create the complete sequence of the BAC clone. The genetic markers on each BAC clone are used to arrange the larger fragments in the proper order along the chromosome map. This method works, but it is slow. An alternative approach, shotgun sequencing, makes far greater use of use of computers to align the sequences.

---

yourBioPortal.com
GO TO Animated Tutorial 17.1 • Sequencing the Genome

TOOLS FOR INVESTIGATING LIFE

17.1 Sequencing Genomes Involves Fragment Overlaps

Short fragments of the whole genome can be sequenced, but then the fragments must be correctly aligned. Historically two approaches were used. Both involved the use of bacterial clones to separate and amplify individual DNA fragments.

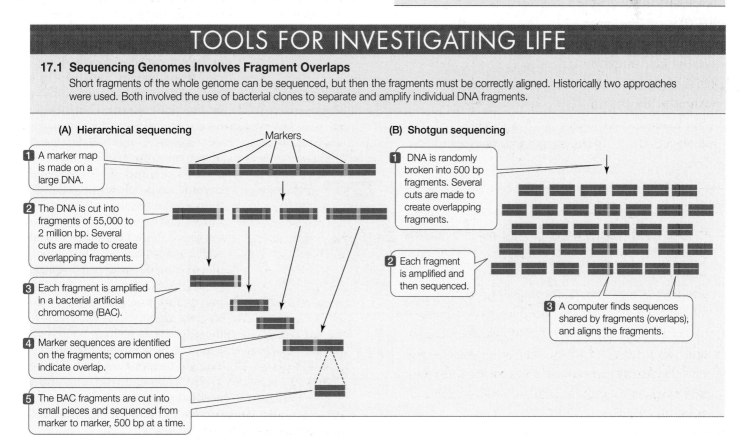

(A) Hierarchical sequencing

Markers

1. A marker map is made on a large DNA.

2. The DNA is cut into fragments of 55,000 to 2 million bp. Several cuts are made to create overlapping fragments.

3. Each fragment is amplified in a bacterial artificial chromosome (BAC).

4. Marker sequences are identified on the fragments; common ones indicate overlap.

5. The BAC fragments are cut into small pieces and sequenced from marker to marker, 500 bp at a time.

(B) Shotgun sequencing

1. DNA is randomly broken into 500 bp fragments. Several cuts are made to create overlapping fragments.

2. Each fragment is amplified and then sequenced.

3. A computer finds sequences shared by fragments (overlaps), and aligns the fragments.

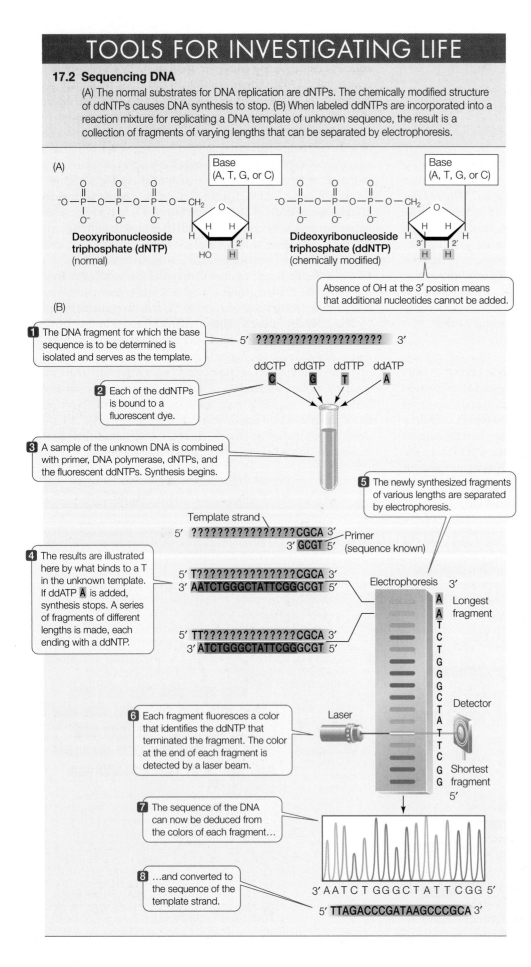

TOOLS FOR INVESTIGATING LIFE

17.2 Sequencing DNA

(A) The normal substrates for DNA replication are dNTPs. The chemically modified structure of ddNTPs causes DNA synthesis to stop. (B) When labeled ddNTPs are incorporated into a reaction mixture for replicating a DNA template of unknown sequence, the result is a collection of fragments of varying lengths that can be separated by electrophoresis.

(A)

Deoxyribonucleoside triphosphate (dNTP) (normal)

Dideoxyribonucleoside triphosphate (ddNTP) (chemically modified)

Absence of OH at the 3' position means that additional nucleotides cannot be added.

(B)

1 The DNA fragment for which the base sequence is to be determined is isolated and serves as the template.

5' ??????????????????? 3'

ddCTP ddGTP ddTTP ddATP
C G T A

2 Each of the ddNTPs is bound to a fluorescent dye.

3 A sample of the unknown DNA is combined with primer, DNA polymerase, dNTPs, and the fluorescent ddNTPs. Synthesis begins.

5 The newly synthesized fragments of various lengths are separated by electrophoresis.

Template strand
5' ??????????????CGCA 3'
3' GCGT 5' Primer (sequence known)

4 The results are illustrated here by what binds to a T in the unknown template. If ddATP A is added, synthesis stops. A series of fragments of different lengths is made, each ending with a ddNTP.

5' T??????????????CGCA 3'
3' AATCTGGGCTATTCGGGCGT 5'

5' TT??????????????CGCA 3'
3' ATCTGGGCTATTCGGGCGT 5'

Electrophoresis 3'
A Longest fragment
A
T
C
T
G
G
G
C
T
A
T
T
C
G Shortest fragment
G
5'

6 Each fragment fluoresces a color that identifies the ddNTP that terminated the fragment. The color at the end of each fragment is detected by a laser beam.

Laser Detector

7 The sequence of the DNA can now be deduced from the colors of each fragment...

8 ...and converted to the sequence of the template strand.

3' AATCT GGGCT ATT CGG 5'

5' TTAGACCCGATAAGCCCGCA 3'

SHOTGUN SEQUENCING Instead of mapping the genome and creating a BAC library, the **shotgun sequencing** method involves directly cutting genomic DNA into smaller, overlapping fragments that are cloned and sequenced. Powerful computers align the fragments by finding sequence homologies in the overlapping regions (**Figure 17.1B**). As sequencing technologies and computers have improved, the shotgun approach has become much faster and cheaper than the hierarchical approach.

As a demonstration, researchers used this method to sequence a 1.8 million-base-pair prokaryotic genome in just a few months. Next came larger genomes. The entire 180 million-base-pair fruit fly genome was sequenced by the shotgun method in little over a year. This success proved that the shotgun method might work for the much larger human genome, and in fact it was used to sequence the human genome rapidly relative to the hierarchical method.

The nucleotide sequence of DNA can be determined

How are the individual DNA fragments generated by the hierarchical or shotgun methods sequenced? Current techniques are variations of a method developed in the late 1970s by Frederick Sanger. This method uses chemically modified nucleosides that were originally developed to stop cell division in cancer. As we discuss in Chapter 13, deoxyribonucleoside triphosphates (dNTPs) are the normal substrates for DNA replication, and contain the sugar deoxyribose. If that sugar is replaced with 2,3-dideoxyribose, the resulting dideoxyribonucleoside triphosphate (ddNTP) will still be added by DNA polymerase to a growing polynucleotide chain. However, because the ddNTP has no hydroxyl group (—OH) at the 3' position, the next nucleotide cannot be added (**Figure 17.2A**). Thus synthesis stops at the position where ddNTP has been incorporated into the growing end of a DNA strand.

To determine the sequence of a DNA fragment (usually no more than 700 base pairs long), it is isolated and mixed with

- DNA polymerase
- A short primer appropriate for the DNA sequence
- The four dNTPs (dATP, dGTP, dCTP, and dTTP)
- Small amounts of the four ddNTPs, each bonded to a differently colored fluorescent "tag"

In the first step of the reaction, the DNA is heated to denature it (separate it into single strands). Only one of these strands will act as a template for sequencing—the one to which the primer binds. DNA replication proceeds, and the test tube soon contains a mixture of the original DNA strands and shorter, new complementary strands. The new strands, each ending with a fluorescent ddNTP, are of varying lengths. For example, each time a T is reached on the template strand, DNA polymerase adds either a dATP or a ddATP to the growing complementary strand. If dATP is added, the strand continues to grow. If ddATP is added, growth stops (**Figure 17.2B**).

After DNA replication has been allowed to proceed for a while, the new DNA fragments are denatured and the single-stranded fragments separated by electrophoresis (see Figure 15.8), which sorts the DNA fragments by length. During the electrophoresis run, the fragments pass through a laser beam that excites the fluorescent tags, and the distinctive color of light emitted by each ddNTP is detected. The color indicates which ddNTP is at the end of each strand. A computer processes this information and prints out the DNA sequence of the fragment (see Figure 17.2B).

The delivery of chemical reagents by automated machines, coupled with automated analysis, has made DNA sequencing faster than ever. Huge laboratories often have 80 sequencing machines operating at once, each of which can sequence and analyze up to 70,000 bp in a typical 4-hour run. This may be fast enough for a prokaryotic genome with 1.5 million base pairs (20 runs), but when it comes to routine sequencing of larger genomes (like the 3.3 billion-base-pair human genome), even more speed is needed.

High-throughput sequencing has been developed for large genomes

The first decade of the new millennium has seen rapid development of **high-throughput sequencing** methods—fast, cheap ways to sequence and analyze large genomes. A variety of different approaches are being used. They generally involve the amplification of DNA templates by the polymerase chain reaction (PCR; see Section 13.5), and the physical binding of template DNA to a solid surface or to tiny beads called microbeads. These techniques are often referred to as *massively parallel DNA sequencing*, because thousands or millions of sequencing reactions are run at once to greatly speed up the process. One such high-throughput method is illustrated in **Figure 17.3**. In one 7-hour run, these machines can sequence 50,000,000 base pairs of DNA! How does it work?

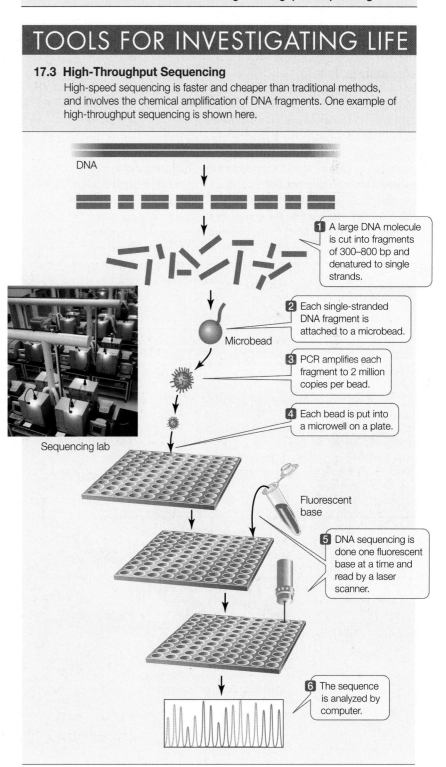

──── **yourBioPortal.com** ────
GO TO Animated Tutorial 17.2 • High-Throughput Sequencing

TOOLS FOR INVESTIGATING LIFE

17.3 High-Throughput Sequencing
High-speed sequencing is faster and cheaper than traditional methods, and involves the chemical amplification of DNA fragments. One example of high-throughput sequencing is shown here.

DNA

1 A large DNA molecule is cut into fragments of 300–800 bp and denatured to single strands.

2 Each single-stranded DNA fragment is attached to a microbead.

Microbead

3 PCR amplifies each fragment to 2 million copies per bead.

4 Each bead is put into a microwell on a plate.

Sequencing lab

Fluorescent base

5 DNA sequencing is done one fluorescent base at a time and read by a laser scanner.

6 The sequence is analyzed by computer.

For massively parallel sequencing using microbeads, the genomic DNA is first cut into 300- to 800-base-pair fragments. The fragments are denatured to single strands and attached to tiny beads that are less than 20 μm in diameter, one DNA fragment (template) per bead. PCR is used to create several million identical copies of the fragment on each bead. Then each bead is loaded into a tiny (40 μm diameter) well in a multi-well plate, and the sequencing begins.

The automated sequencer adds a reaction mix like the one described above, but containing only one of four fluorescently labeled dNTPs. That nucleotide will become incorporated as the first nucleotide in a complementary strand only in wells where the first nucleotide in the template strand can base-pair with it. For example, if the first nucleotide on the template in well #1 has base T, then a fluorescent nucleotide with base A will bind to that well. Next, the reaction mix is removed and a scanner captures an image of the plate, indicating which wells contain the fluorescent nucleotide. This process is repeated with a different labeled nucleotide. The machine cycles through many repeats using all four dNTPs, and records which wells gain new nucleotides after each cycle. A computer then identifies the sequence of nucleotides that were gained by each well, and aligns the fragments to provide the complete sequence of the genome.

This method was used to sequence the genome of James Watson, codiscoverer of the DNA double helix. It took less than two months and cost less than $1 million. Sequencing methods are being continually refined to increase speed and accuracy and decrease costs.

Genome sequences yield several kinds of information

New genome sequences are published more and more frequently, creating a torrent of biological information (**Figure 17.4**). In general, biologists use sequence information to identify:

- *Open reading frames*, the coding regions of genes. For protein-coding genes, these regions can be recognized by the start and stop codons for translation, and by intron consensus sequences that indicate the locations of introns.

- *Amino acid sequences* of proteins, which can be deduced from the DNA sequences of open reading frames by applying the genetic code (see Figure 14.6).

- *Regulatory sequences*, such as promoters and terminators for transcription.

- *RNA genes*, including rRNA, tRNA, and small nuclear RNA (snRNA) genes.

- *Other noncoding sequences* that can be classified into various categories including centromeric and telomeric regions, nuclear matrix attachment regions, transposons, and repetitive sequences such as short tandem repeats.

Sequence information is also used for *comparative genomics*, the comparison of a newly sequenced genome (or parts thereof) with sequences from other organisms. This can give information about the functions of sequences, and can be used to trace evolutionary relationships among different organisms.

17.4 The Genomic Book of Life Genome sequences contain many features, some of which are summarized in this overview. Sifting through all the information contained in a genome sequence can help us understand how an organism functions and what its evolutionary history might be.

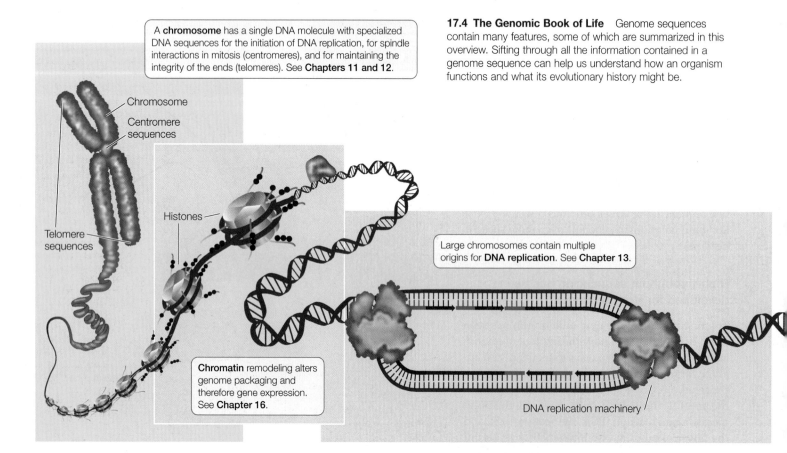

A **chromosome** has a single DNA molecule with specialized DNA sequences for the initiation of DNA replication, for spindle interactions in mitosis (centromeres), and for maintaining the integrity of the ends (telomeres). See **Chapters 11 and 12**.

Chromosome

Centromere sequences

Histones

Telomere sequences

Chromatin remodeling alters genome packaging and therefore gene expression. See **Chapter 16**.

Large chromosomes contain multiple origins for **DNA replication**. See **Chapter 13**.

DNA replication machinery

17.1 RECAP

The sequencing of genomes required the development of ways to cut large chromosomes into fragments, sequence the fragments, and then line them up on the chromosome. Two ways to do this are hierarchical sequencing and shotgun sequencing. Today new procedures are being developed that require automation and powerful computers. Actual DNA sequencing involves labeled nucleotides that are detected at the ends of growing polynucleotide chains.

- What are the hierarchical and shotgun approaches to genome analysis? **See pp. 367–368 and Figure 17.1**

- What is the dideoxy method for DNA sequencing? **See pp. 368–369 and Figure 17.2**

- Explain how high-throughput sequencing methods work. **See pp. 369–370 and Figure 17.3**

- How are open reading frames recognized in a genomic sequence? What kind of information can be derived from an open reading frame? **See p. 370**

We now turn to the first organisms whose sequences were determined, prokaryotes, and the information these sequences provided.

17.2 What Have We Learned from Sequencing Prokaryotic Genomes?

When DNA sequencing became possible in the late 1970s, the first life forms to be sequenced were the simplest viruses with their relatively small genomes. The sequences quickly provided new information on how these viruses infect their hosts and reproduce. But the manual sequencing techniques used on viruses were not up to the task of studying the genomes of prokaryotes and eukaryotes. The newer, automated sequencing techniques we just described made such studies possible. We now have genome sequences for many prokaryotes, to the great benefit of microbiology and medicine.

The sequencing of prokaryotic genomes led to new genomics disciplines

In 1995 a team led by Craig Venter and Hamilton Smith determined the first complete genomic sequence of a free-living cellular organism, the bacterium *Haemophilus influenzae*. Many more prokaryotic sequences have followed, revealing not only how prokaryotes apportion their genes to perform different cellular functions, but also how their specialized functions are carried out. Soon we may even be able to ask the provocative question of what the minimal requirements of a living cell might be.

FUNCTIONAL GENOMICS **Functional genomics** is the biological discipline that assigns functions to the products of genes. This

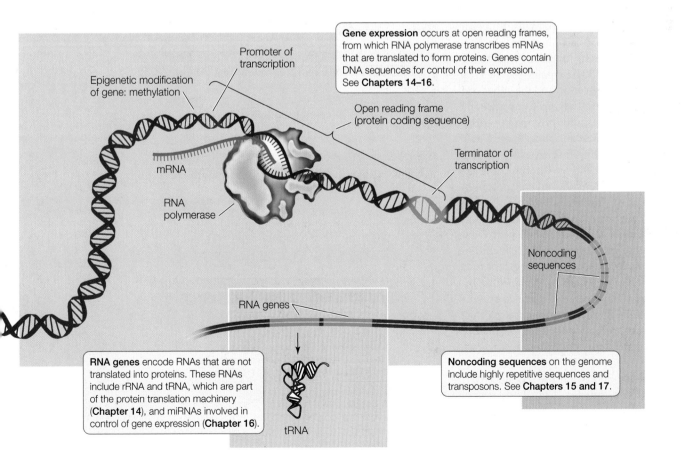

Gene expression occurs at open reading frames, from which RNA polymerase transcribes mRNAs that are translated to form proteins. Genes contain DNA sequences for control of their expression. See **Chapters 14–16**.

Epigenetic modification of gene: methylation

Promoter of transcription

Open reading frame (protein coding sequence)

mRNA

RNA polymerase

Terminator of transcription

Noncoding sequences

RNA genes

tRNA

RNA genes encode RNAs that are not translated into proteins. These RNAs include rRNA and tRNA, which are part of the protein translation machinery (**Chapter 14**), and miRNAs involved in control of gene expression (**Chapter 16**).

Noncoding sequences on the genome include highly repetitive sequences and transposons. See **Chapters 15 and 17**.

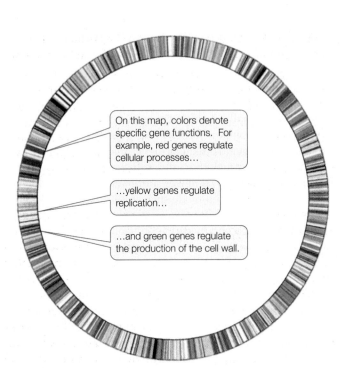

On this map, colors denote specific gene functions. For example, red genes regulate cellular processes...

...yellow genes regulate replication...

...and green genes regulate the production of the cell wall.

17.5 Functional Organization of the Genome of *H. influenzae* The entire DNA sequence has 1,830,137 base pairs. Different colors reflect different classes of gene function.

field, less than 15 years old, is now a major occupation of biologists. Let's see how funtional genomics methods were applied to the bacterium *H. influenzae* once its sequence was known.

The only host for *H. influenzae* is humans. It lives in the upper respiratory tract and can cause ear infections or, more seriously, meningitis in children. Its single circular chromosome has 1,830,138 base pairs (**Figure 17.5**). In addition to its origin of replication and the genes coding for rRNAs and tRNAs, this bacterial chromosome has 1,738 open reading frames with promoters nearby.

When this sequence was first announced, only 1,007 (58 percent) of the open reading frames coded for proteins with known functions. The remaining 42 percent coded for proteins whose functions were unknown. Since then scientists have identified many of these proteins' roles. For example, they found genes for enzymes of glycolysis, fermentation, and electron transport. Other gene sequences code for membrane proteins, including those involved in active transport. An important finding was that highly infective strains of *H. influenzae*, but not noninfective strains, have genes for surface proteins that attach the bacterium to the human respiratory tract. These surface proteins are now a focus of research on possible treatments for *H. influenzae* infections.

COMPARATIVE GENOMICS Soon after the sequence of *H. influenzae* was announced, smaller (*Mycoplasma genitalium*; 580,073 base pairs) and larger (*E. coli*; 4,639,221 base pairs) prokaryotic sequences were completed. Thus began a new era in biology, that of **comparative genomics**,

which compares genome sequences from different organisms. Scientists can identify genes that are present in one bacterium and missing in another, allowing them to relate these genes to bacterial function.

M. genitalium, for example, lacks the enzymes needed to synthesize amino acids, which *E. coli* and *H. influenzae* both possess. This finding reveals that *M. genitalium* must obtain all its amino acids from its environment (usually the human urogenital tract). Furthermore, *E. coli* has 55 regulatory genes coding for transcriptional activators and 58 for repressors; *M. genitalium* only has 3 genes for activators. What do such findings tell us about an organism's lifestyle? For example, is the biochemical flexibility of *M. genitalium* limited by its relative lack of control over gene expression?

Some sequences of DNA can move about the genome

Genome sequencing allowed scientists to study more broadly a class of DNA sequences that had been discovered by geneticists decades earlier. Segments of DNA called **transposable elements** can move from place to place in the genome and can even be inserted into another piece of DNA in the same cell (e.g., a plasmid). A transposable element might be at one location in the genome of one *E. coli* strain, and at a different location in another strain. The insertion of this movable DNA sequence from elsewhere in the genome into the middle of a protein-coding gene disrupts that gene (**Figure 17.6A**). Any mRNA expressed from the disrupted gene will have the extra sequence and the

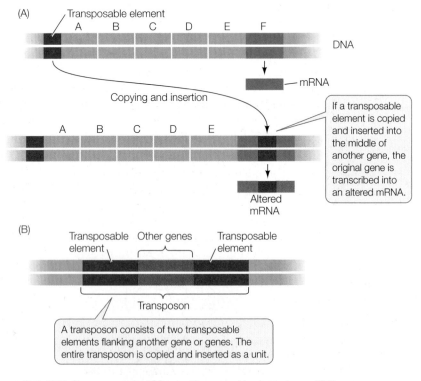

If a transposable element is copied and inserted into the middle of another gene, the original gene is transcribed into an altered mRNA.

A transposon consists of two transposable elements flanking another gene or genes. The entire transposon is copied and inserted as a unit.

17.6 DNA Sequences that Move Transposable elements are DNA sequences that move from one location to another. (A) In one method of transposition, the DNA sequence is replicated and the copy inserts elsewhere in the genome. (B) Transposons contain transposable elements and other genes.

protein will be abnormal. So transposable elements can produce significant phenotypic effects by inactivating genes.

Transposable elements are often short sequences of 1,000–2,000 base pairs, and are found at many sites in prokaryotic genomes. The mechanisms that allow them to move vary. For example, a transposable element may be replicated, and then the copy inserted into another site in the genome. Or the element might splice out of one location and move to another location.

Longer transposable elements (up to 5,000 bp) carry additional genes and are called **transposons (Figure 17.6B)**. Sometimes these DNA regions contain a gene for antibiotic resistance.

The sequencing of prokaryotic and viral genomes has many potential benefits

Prokaryotic genome sequencing promises to provide insights into microorganisms that cause human diseases. Genome sequencing has revealed unknown genes and proteins that can be targeted for isolation and functional study. Such studies are revealing new methods to combat pathogens and their infections. Sequencing has also revealed surprising relationships between some pathogenic organisms, suggesting that genes may be transferred between different strains.

- *Chlamydia trachomatis* causes the most common sexually transmitted disease in the United States. Because it is an intracellular parasite, it has been very hard to study. Among its 900 genes are several for ATP synthesis—something scientists used to think this bacterium could not accomplish on its own.

- *Rickettsia prowazekii* causes typhus; it is carried by lice and infects people bitten by the lice. Of its 634 genes, 6 encode proteins that are essential for virulence. These virulence proteins are being used to develop vaccines.

- *Mycobacterium tuberculosis* causes tuberculosis. It has a relatively large genome, coding for 4,000 proteins. Over 250 of these are used to metabolize lipids, so this may be the main way that this bacterium gets its energy. Some of its genes code for previously unidentified cell surface proteins; these proteins are targets for potential vaccines.

- *Streptomyces coelicolor* and its close relatives are the source for the genes for two-thirds of all naturally occuring antibiotics currently in clinical use. These antibiotics include streptomycin, tetracycline, and erythromycin. The genome sequence of *S. coelicolor* reveals 22 clusters of genes responsible for antibiotic production, of which only four were previously known. This finding may lead to new antibiotics to combat pathogens that have evolved resistance to conventional antibiotics.

- *E. coli* strain O157:H7 causes illness (sometimes severe) in at least 70,000 people a year in the United States. Its genome has 5,416 genes, of which 1,387 are different from those in the familiar (and harmless) laboratory strains of this bacterium. Many of these unique genes are also present in other pathogenic bacteria, such as *Salmonella* and *Shigella*. This finding suggests that there is extensive genetic ex-

change among these species, and that "superbugs" that share genes for antibiotic resistance may be on the horizon.

- *Severe acute respiratory syndrome (SARS)* was first detected in southern China in 2002 and rapidly spread in 2003. There is no effective treatment and 10 percent of infected people die. Isolation of the causative agent, a virus, and the rapid sequencing of its genome revealed several novel proteins that are possible targets for antiviral drugs or vaccines. Research is underway on both fronts, since another outbreak is anticipated.

Genome sequencing also provides insights into organisms involved in global ecological cycles (see Chapter 58). In addition to the well-known carbon dioxide, another important gas contributing to the atmospheric "greenhouse effect" and global warming is methane (CH_4; see Figure 2.7). Some bacteria, such as *Methanococcus*, produce methane in the stomachs of cows. Others, such as *Methylococcus*, remove methane from the air and use it as an energy source. The genomes of both of these bacteria have been sequenced. Understanding the genes involved in methane production and oxidation may help us to slow the progress of global warming.

Metagenomics allows us to describe new organisms and ecosystems

If you take a microbiology laboratory course you will learn how to identify various prokaryotes on the basis of their growth in lab cultures. For example, staphylococci are a group of bacteria that infect skin and nasal passages. When grown on a special medium called blood agar they form round, raised colonies. Microorganisms can also be identified by their nutritional requirements or the conditions under which they will grow (for example, aerobic versus anaerobic). Such culture methods have been the mainstay of microbial identification for over a century and are still useful and important. However, scientists can now use PCR and modern DNA analysis techniques to analyze microbes *without* culturing them in the laboratory.

In 1985, Norman Pace, then at Indiana University, came up with the idea of isolating DNA directly from environmental samples. He used PCR to amplify specific sequences from the samples to determine whether particular microbes were present. The PCR products were sequenced to explore their diversity. The term **metagenomics** was coined to describe this approach of analyzing genes without isolating the intact organism. It is now possible to perform shotgun sequencing with samples from almost any environment. The sequences can be used to detect the presence of known microbes and pathogens, and perhaps even the presence of heretofore unidentified organisms (**Figure 17.7**). For example:

- Shotgun sequencing of DNA from 200 liters of seawater indicated that it contained 5,000 different viruses and 2,000 different bacteria, many of which had not been described previously.

- One kilogram of marine sediment contained a million different viruses, most of them new.

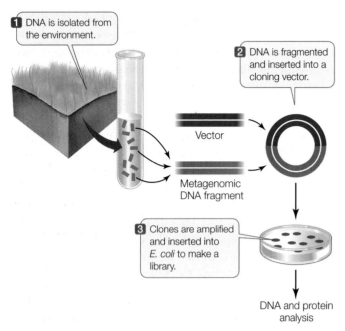

1 DNA is isolated from the environment.

2 DNA is fragmented and inserted into a cloning vector.

Vector

Metagenomic DNA fragment

3 Clones are amplified and inserted into *E. coli* to make a library.

DNA and protein analysis

17.7 Metagenomics Microbial DNA extracted from the environment can be amplified and analyzed. This has led to the description of many new genes and species.

- Water runoff from a mine contained many new species of prokaryotes thriving in this apparently inhospitable environment. Some of these organisms exhibited metabolic pathways that were previously unknown to biologists. These organisms and their capabilities may be useful in cleaning up pollutants from the water.

These and other discoveries are truly extraordinary and potentially very important. It is estimated that 90 percent of the microbial world has been invisible to biologists and is only now being revealed by metagenomics. Entirely new ecosystems of bacteria and viruses are being discovered in which, for example, one species produces a molecule that another metabolizes. It is hard to overemphasize the importance of such an increase in our knowledge of the hidden world of microbes. This knowledge will help us to understand natural ecological processes, and has the potential to help us find better ways to manage environmental catastrophes such as oil spills, or remove toxic heavy metals from soil.

Will defining the genes required for cellular life lead to artificial life?

When the genomes of prokaryotes and eukaryotes are compared, a striking conclusion arises: certain genes are present in all organisms (universal genes). There are also some (nearly) universal gene segments that are present in many genes in many organisms; for example, the sequence that codes for an ATP binding site. These findings suggest that there is some ancient, minimal set of DNA sequences common to all cells. One way to identify these sequences is to look for them in computer analyses of sequenced genomes.

Another way to define the minimal genome is to take an organism with a simple genome and deliberately mutate one gene at a time to see what happens. *M. genitalium* has one of the smallest known genomes—only 482 protein-coding genes. Even so, some of its genes are dispensable under some circumstances. For example, it has genes for metabolizing both glucose and fructose, but it can survive in the laboratory on a medium containing only one of these sugars.

What about other genes? Researchers have addressed this question with experiments involving the use of transposons as mutagens. When transposons in the bacterium are activated, they insert themselves into genes at random, mutating and inactivating them (**Figure 17.8**). The mutated bacteria are tested for growth and survival, and DNA from interesting mutants is sequenced to find out which genes contain transposons.

The astonishing result of these studies is that *M. genitalium* can survive in the laboratory with a minimal genome of only 382 functional genes! Is this really all it takes to make a viable organism? Experiments are underway to make a synthetic genome based on that of *M. genitalium*, and then insert it into an empty bacterial cell. If the cell starts transcribing mRNA and making proteins—is in fact viable—it may turn out to be the first life created by humans.

In addition to the technical feat of creating artificial life, this technique could have important applications. New microbes could be made with entirely new abilities, such as degrading oil spills, making synthetic fibers, reducing tooth decay, or converting cellulose to ethanol for use as fuel. On the other hand, fears of the misuse or mishandling of this knowledge are not unfounded. For example, it might also be possible to develop synthetic bacteria harmful to people, animals or plants, and use them as agents of biological warfare or bioterrorism. The "genomics genie" is, for better or worse, already out of the bottle. Hopefully human societies will use it to their benefit.

17.2 RECAP

DNA sequencing is used to study the genomes of prokaryotes that are important to humans and to ecosystems. Functional genomics uses gene sequences to determine the functions of the gene products. Comparative genomics compares gene sequences from different organisms to help identify their functions and evolutionary relationships. Transposable elements and transposons move from one place to another in the genome.

- Give some examples of prokaryotic genomes that have been sequenced. What have the sequences shown? See pp. 371–373

- What is metagenomics and how is it used? See pp. 373–374 and Figure 17.7

- How are selective inactivation studies being used to determine the minimal genome? See p. 374 and Figure 17.8

INVESTIGATING LIFE

17.8 Using Transposon Mutagenesis to Determine the Minimal Genome

Mycoplasma genitalium has the smallest number of genes of any prokaryote. But are all of its genes essential to life? By inactivating the genes one by one, scientists determined which of them are essential for the cell's survival. This research may lead to the construction of artificial cells with customized genomes, designed to perform functions such as degrading oil and making plastics.

HYPOTHESIS Only some of the genes in a bacterial genome are essential for cell survival.

METHOD

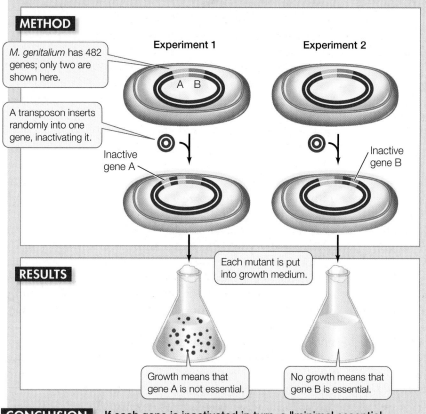

M. genitalium has 482 genes; only two are shown here.

Experiment 1   Experiment 2

A transposon inserts randomly into one gene, inactivating it.

Inactive gene A

Inactive gene B

Each mutant is put into growth medium.

RESULTS

Growth means that gene A is not essential.

No growth means that gene B is essential.

CONCLUSION If each gene is inactivated in turn, a "minimal essential genome" can be determined.

Go to **yourBioPortal.com** for original citations, discussions, and relevant links for all INVESTIGATING LIFE figures.

Advances in DNA sequencing and analysis have led to the rapid sequencing of eukaryotic genomes. We now turn to the results of these analyses.

17.3 What Have We Learned from Sequencing Eukaryotic Genomes?

As genomes have been sequenced and described, a number of major differences have emerged between eukaryotic and prokaryotic genomes (**Table 17.1**). Key differences include:

- *Eukaryotic genomes are larger than those of prokaryotes*, and they have more protein-coding genes. This difference is not surprising, given that multicellular organisms have many cell types with specific functions. Many proteins are needed to do those specialized jobs. A typical virus contains enough DNA to code for only a few proteins—about 10,000 base pairs (bp). As we saw above, the simplest prokaryote, *Mycoplasma*, has several hundred protein-coding genes in a genome of 0.5 million bp. A rice plant, in contrast, has 37,544 genes.

- *Eukaryotic genomes have more regulatory sequences*—and many more regulatory proteins—than prokaryotic genomes. The greater complexity of eukaryotes requires much more regulation, which is evident in the many points of control associated with the expression of eukaryotic genes (see Figure 16.13).

- *Much of eukaryotic DNA is noncoding*. Distributed throughout many eukaryotic genomes are various kinds of DNA sequences that are not transcribed into mRNA, most notably introns and gene control sequences. As we discuss in Chapter 16, some noncoding sequences are transcribed into microRNAs. In addition, eukaryotic genomes contain various kinds of repeated sequences. These features are rare in prokaryotes.

- *Eukaryotes have multiple chromosomes*. The genomic "encyclopedia" of a eukaryote is separated into multiple "volumes." Each chromosome must have, at a minimum, three defining DNA sequences that we have described in previous chapters: an origin of replication (*ori*) that is recognized by the DNA replication machinery; a centromere region that holds the replicated chromosomes together before mitosis; and a telomeric sequence at each end of the chromosome that maintains chromosome integrity.

Model organisms reveal many characteristics of eukaryotic genomes

Most of the lessons learned from eukaryotic genomes have come from several simple model organisms that have been studied extensively: the yeast *Saccharomyces cerevisiae*, the nematode (roundworm) *Caenorhabditis elegans*, the fruit fly *Drosophila melanogaster*, and—representing plants—the thale cress, *Arabidopsis thaliana*. Model organisms have been chosen because they are relatively easy to grow and study in a laboratory, their genetics are well studied, and they exhibit characteristics that represent a larger group of organisms.

YEAST: THE BASIC EUKARYOTIC MODEL Yeasts are single-celled eukaryotes. Like most eukaryotes, they have membrane-enclosed organelles, such as the nucleus and endoplasmic reticulum, and a life cycle that alternates between haploid and diploid generations (see Figure 11.15).

TABLE 17.1
Representative Sequenced Genomes

ORGANISM	HAPLOID GENOME SIZE (Mb)	NUMBER OF GENES	PROTEIN-CODING SEQUENCE
Bacteria			
M. genitalium	0.58	485	88%
H. influenzae	1.8	1,738	89%
E. coli	4.6	4,377	88%
Yeasts			
S. cerevisiae	12.5	5,770	70%
S. pombe	12.5	4,929	60%
Plants			
A. thaliana	115	28,000	25%
Rice	390	37,544	12%
Animals			
C. elegans	100	19,427	25%
D. melanogaster	123	13,379	13%
Pufferfish	342	27,918	10%
Chicken	1,130	25,000	3%
Human	3,300	24,000	1.2%

Mb = millions of base pairs

TABLE 17.2
Comparison of the Genomes of *E. coli* and Yeast

	E. COLI	YEAST
Genome length (base pairs)	4,640,000	12,068,000
Number of protein-coding genes	4,290	5,770
Proteins with roles in:		
Metabolism	650	650
Energy production/storage	240	175
Membrane transport	280	250
DNA replication/repair/ recombination	120	175
Transcription	230	400
Translation	180	350
Protein targeting/secretion	35	430
Cell structure	180	250

While the prokaryote *E. coli* has a single circular chromosome with about 4.6 million bp and 4,290 protein-coding genes, budding yeast (*Saccharomyces cerevisiae*) has 16 linear chromosomes and a haploid content of more than 12.5 million bp, with 5,770 protein-coding genes. Gene inactivation studies similar to those carried out for *M. genitalium* (see Figure 17.7) indicate that fewer than 20 percent of these genes are essential to survival.

The most striking difference between the yeast genome and that of *E. coli* is in the number of genes for targeting proteins to organelles (**Table 17.2**). Both of these single-celled organisms appear to use about the same numbers of genes to perform the basic functions of cell survival. It is the compartmentalization of the eukaryotic yeast cell into organelles that requires it to have many more genes. This finding is direct, quantitative confirmation of something we have known for a century: the eukaryotic cell is structurally more complex than the prokaryotic cell.

THE NEMATODE: UNDERSTANDING EUKARYOTIC DEVELOPMENT In 1965 Sydney Brenner, fresh from being part of the team that first isolated mRNA, looked for a simple organism in which to study multicellularity. He settled on *Caenorhabditis elegans*, a millimeter-long nematode (roundworm) that normally lives in the soil. It can also live in the laboratory, where it has become a favorite

model organism of developmental biologists (see Section 19.4). The nematode has a transparent body that develops over 3 days from a fertilized egg to an adult worm made up of nearly 1,000 cells. In spite of its small number of cells, the nematode has a nervous system, digests food, reproduces sexually, and ages. So it is not surprising that an intense effort was made to sequence the genome of this model organism.

The *C. elegans* genome (100 million bp) is eight times larger than that of yeast and has 3.5 times as many protein-coding genes (19,427). Gene inactivation studies have shown that the worm can survive in laboratory cultures with only 10 percent of these genes. So the "minimum genome" of a worm is about twice the size of that of yeast, which in turn is four times the size of the minimum genome for *Mycoplasma*. What do these extra genes do?

All cells must have genes for survival, growth, and division. In addition, the cells of multicellular organisms must have genes for holding cells together to form tissues, for cell differentiation, and for intercellular communication. Looking at **Table 17.3**, you will recognize functions that we discussed in earlier chapters,

TABLE 17.3
C. elegans Genes Essential to Multicellularity

FUNCTION	PROTEIN/DOMAIN	NUMBER OF GENES
Transcription control	Zinc finger; homeobox	540
RNA processing	RNA binding domains	100
Nerve impulse transmission	Gated ion channels	80
Tissue formation	Collagens	170
Cell interactions	Extracellular domains; glycotransferases	330
Cell–cell signaling	G protein-linked receptors; protein kinases; protein phosphatases	1,290

including gene regulation (see Chapter 16) and cell communication (see Chapter 7).

DROSOPHILA MELANOGASTER: RELATING GENETICS TO GENOMICS

The fruit fly *Drosophila melanogaster* is a famous model organism. Studies of fruit fly genetics resulted in the formulation of many basic principles of genetics (see Section 12.4). Over 2,500 mutations of *D. melanogaster* had been described by the 1990s when genome sequencing began, and this fact alone was a good reason for sequencing the fruit fly's DNA. The fruit fly is a much larger organism than *C. elegans*, both in size (it has 10 times more cells) and complexity, and it undergoes complicated developmental transformations from egg to larva to pupa to adult.

Not surprisingly, the fly's genome (about 123 million bp) is larger than that of *C. elegans*. But as we mentioned earlier, genome size does not necessarily correlate with the number of genes encoded. In this case, the larger fruit fly genome contains fewer genes (13,379) than the smaller nematode genome. **Figure 17.9** summarizes the functions of the *Drosophila* genes that have been characterized so far; this distribution is typical of complex eukaryotes.

ARABIDOPSIS: STUDYING THE GENOMES OF PLANTS

About 250,000 species of flowering plants dominate the land and fresh water. But in the context of the history of life, the flowering plants are fairly young, having evolved only about 200 million years ago. The genomes of some plants are huge—for example, the genome of corn is about 3 billion bp, and that of wheat is 16 billion bp. So although we are naturally most interested in the genomes of plants we use as food and fiber, it is not surprising that scientists first chose to sequence a simpler flowering plant.

Arabidopsis thaliana, thale cress, is a member of the mustard family and has long been a favorite model organism of plant biologists. It is small (hundreds could grow and reproduce in the space occupied by this page) and easy to manipulate, and has a relatively small (115 million bp) genome.

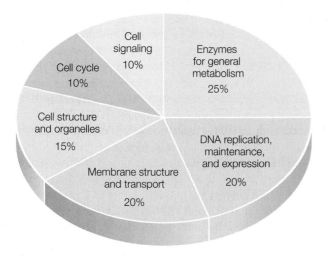

17.9 Functions of the Eukaryotic Genome The distribution of gene functions in *Drosophila melanogaster* shows a pattern that is typical of many complex organisms.

TABLE 17.4	
Arabidopsis Genes Unique to Plants	
FUNCTION	NUMBER OF GENES
Cell wall and growth	42
Water channels	300
Photosynthesis	139
Defense and metabolism	94

The *Arabidopsis* genome has about 28,000 protein-coding genes but, remarkably, many of these genes are duplicates and probably originated by chromosomal rearrangements. When these duplicate genes are subtracted from the total, about 15,000 unique genes are left—similar to the gene numbers found in fruit flies and nematodes. Indeed, many of the genes found in these animals have homologs (genes with very similar sequences) in *Arabidopsis* and other plants, suggesting that plants and animals have a common ancestor.

But *Arabidopsis* has some genes that distinguish it as a plant (**Table 17.4**). These include genes involved in photosynthesis, in the transport of water into the root and throughout the plant, in the assembly of the cell wall, in the uptake and metabolism of inorganic substances from the environment, and in the synthesis of specific molecules used for defense against microbes and herbivores (organisms that eat plants). These plant defense molecules may be a major reason why the number of protein-coding genes in plants is higher than in animals. Plants cannot escape their enemies or other adverse conditions as animals can, and so they must cope with the situation where they are. So they make tens of thousands of molecules to fight their enemies and adapt to the environment (see Chapter 39).

These plant-specific genes are also found in the genomes of other plants, including rice, the first major crop plant whose sequence has been determined. Rice (*Oryza sativa*) is the world's most important crop; it is a staple in the diet of 3 billion people. The larger genome in rice has a set of genes remarkably similar to that of *Arabidopsis*. More recently the genome of the poplar tree, *Populus trichocarpa*, was sequenced to gain insight into the potential for this rapidly growing tree to be used as a source of fixed carbon for making fuel. A comparison of the three genomes shows many genes in common, comprising the *basic plant genome* (**Figure 17.10**).

Eukaryotes have gene families

About half of all eukaryotic protein-coding genes exist as only one copy in the haploid genome (two copies in somatic cells). The rest are present in multiple copies, which arose from gene duplications. Over evolutionary time, different copies of genes have undergone separate mutations, giving rise to groups of closely related genes called **gene families**. Some gene families, such as those encoding the globin proteins that make up hemoglobin, contain only a few members; other families, such as the genes encoding the immunoglobulins that make up antibodies, have hundreds of members. In the human genome,

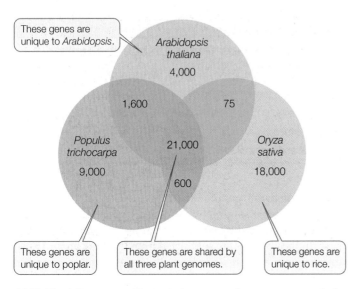

17.10 Plant Genomes Three plant genomes share a common set of approximately 21,000 genes that appear to comprise the "minimal" plant genome.

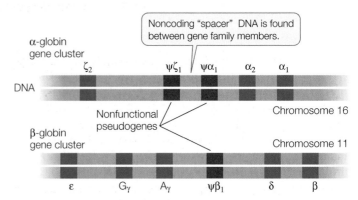

17.11 The Globin Gene Family The α-globin and β-globin clusters of the human globin gene family are located on different chromosomes. The genes of each cluster are separated by noncoding "spacer" DNA. The nonfunctional pseudogenes are indicated by the Greek letter psi (ψ). The γ gene has two variants, A_γ and G_γ.

there are 24,000 protein-coding genes, but 16,000 distinct gene families. So only one-third of the human genes are unique.

The DNA sequences in a gene family are usually different from one another. As long as at least one member encodes a functional protein, the other members may mutate in ways that change the functions of the proteins they encode. For evolution, the availability of multiple copies of a gene allows for selection of mutations that provide advantages under certain circumstances. If a mutated gene is useful, it may be selected for in succeeding generations. If the mutated gene is a total loss, the functional copy is still there to carry out its role.

The gene family encoding the globins is a good example of the gene families found in vertebrates. These proteins are found in hemoglobin and myoglobin (an oxygen-binding protein present in muscle). The globin genes all arose long ago from a single common ancestral gene. In humans, there are three functional members of the α-globin cluster and five in the β-globin cluster (**Figure 17.11**). In adults, each hemoglobin molecule is a tetramer containing two identical α-globin subunits, two identical β-globin subunits, and four heme pigments (see Figure 3.10).

During human development, different members of the globin gene cluster are expressed at different times and in different tissues. This differential gene expression has great physiological significance. For example, hemoglobin containing γ-globin, a subunit found in the hemoglobin of the human fetus, binds O_2 more tightly than adult hemoglobin does. This specialized form of hemoglobin ensures that in the placenta, O_2 will be transferred from the mother's blood to the developing fetus's blood. Just before birth the liver stops synthesizing fetal hemoglobin and the bone marrow cells take over, making the adult forms (2α and 2β). Thus hemoglobins with different binding affinities for O_2 are provided at different stages of human development.

In addition to genes that encode proteins, many gene families include nonfunctional **pseudogenes**, which are designated with the Greek letter psi (Ψ) (see Figure 17.11). These pseudo-

genes result from mutations that cause a loss of function rather than an enhanced or new function. The DNA sequence of a pseudogene may not differ greatly from that of other family members. It may simply lack a promoter, for example, and thus fail to be transcribed. Or it may lack a recognition site needed for the removal of an intron, so that the transcript it makes is not correctly processed into a useful mature mRNA. In some gene families pseudogenes outnumber functional genes. Because some members of the family are functional, there appears to be little selection pressure for the deletion of pseudogenes.

Eukaryotic genomes contain many repetitive sequences

Eukaryotic genomes contain numerous repetitive DNA sequences that do not code for polypeptides. These include highly repetitive sequences, moderately repetitive sequences, and transposons.

Highly repetitive sequences are short (less than 100 bp) sequences that are repeated thousands of times in tandem (side-by-side) arrangements in the genome. They are not transcribed. Their proportion in eukaryotic genomes varies, from 10 percent in humans to about half the genome in some species of fruit flies. Often they are associated with heterochromatin, the densely packed, transcriptionally inactive part of the genome. Other highly repetitive sequences are scattered around the genome. For example, short tandem repeats (STRs) of 1–5 bp can be repeated up to 100 times at a particular chromosomal location. The copy number of an STR at a particular location varies between individuals and is inherited. In Chapter 15 we describe how STRs can be used in the identification of individuals (DNA fingerprinting).

Moderately repetitive sequences are repeated 10–1000 times in the eukaryotic genome. These sequences include the genes that are transcribed to produce tRNAs and rRNAs, which are used in protein synthesis. The cell makes tRNAs and rRNAs constantly, but even at the maximum rate of transcription, single copies of the tRNA and rRNA genes would be inadequate to supply the large amounts of these molecules needed by most cells. Thus the genome has multiple copies of these genes.

In mammals, four different rRNA molecules make up the ribosome: the 18S, 5.8S, 28S, and 5S rRNAs. (The S stands for Svedberg unit, which is a measure of size.) The 18S, 5.8S, and

28S rRNAs are transcribed together as a single precursor RNA molecule (**Figure 17.12**). As a result of several posttranscriptional steps, the precursor is cut into the final three rRNA products, and the noncoding "spacer" RNA is discarded. The sequence encoding these RNAs is moderately repetitive in humans: a total of 280 copies of the sequence are located in clusters on five different chromosomes.

TRANSPOSONS Apart from the RNA genes, most moderately repetitive sequences are not stably integrated into the genome. Instead, these sequences can move from place to place, and are thus called transposable elements or transposons. Prokaryotes also have transposons (see Figure 17.6). Transposons make up over 40 percent of the human genome and about 50 percent of the maize genome, although the percentage is smaller (3–10 percent) in many other eukaryotes.

There are four main types of transposons in eukaryotes:

1. *SINEs* (short *in*terspersed *e*lements) are up to 500 bp long and are transcribed but not translated. There are about 1.5 million of them scattered over the human genome, making up about 15 percent of the total DNA content. A single type, the 300-bp Alu element, accounts for 11 percent of the human genome; it is present in a million copies.

2. *LINEs* (long *in*terspersed *e*lements) are up to 7,000 bp long, and some are transcribed and translated into proteins. They constitute about 17 percent of the human genome.

SINEs and LINEs move about the genome in a distinctive way: they are transcribed into RNA, which then acts as a template for new DNA. The new DNA becomes inserted at a new location in the genome. This "copy and paste" mechanism results in two copies of the transposon: one at the original location and the other at a new location.

3. *Retrotransposons* also make RNA copies of themselves when they move about the genome. Some of them encode proteins needed for their own transposition, and others do not. SINEs and LINEs are types of retrotransposons. Non-SINE, non-LINE retrotransposons constitute about 8 percent of the human genome.

4. *DNA transposons* do not use RNA intermediates. Like some prokaryotic transposable elements, they are excised from the original location and become inserted at a new location without being replicated.

What role do these moving sequences play in the cell? The best answer so far seems to be that transposons are simply cellular parasites that can be replicated. The insertion of a transposon at a new location can have important consequences. For example, the insertion of a transposon into the coding region of a gene results in a mutation (see Figure 17.8). This phenomenon accounts for a few rare forms of several genetic diseases in humans, including hemophilia and muscular dystrophy. If the insertion of a transposon takes place in the germ line, a gamete with a new mutation results. If the insertion takes place in a somatic cell, cancer may result.

Sometimes an adjacent gene can be replicated along with a transposon, resulting in a gene duplication. A transposon can carry a gene, or a part of it, to a new location in the genome, shuffling the genetic material and creating new genes. Clearly, transposition stirs the genetic pot in the eukaryotic genome and thus contributes to genetic variation.

Section 5.5 describes the theory of *endosymbiosis*, which proposes that chloroplasts and mitochondria are the descendants of once free-living prokaryotes. Transposons may have played a role in endosymbiosis. In living eukaryotes the chloroplasts and mitochondria contain some DNA, but the nucleus contains most of the genes

(A)

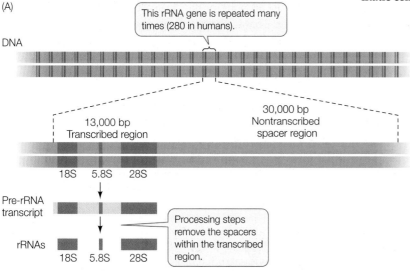

This rRNA gene is repeated many times (280 in humans).

DNA

13,000 bp Transcribed region

30,000 bp Nontranscribed spacer region

18S   5.8S   28S

Pre-rRNA transcript

Processing steps remove the spacers within the transcribed region.

rRNAs
18S   5.8S   28S

(B)

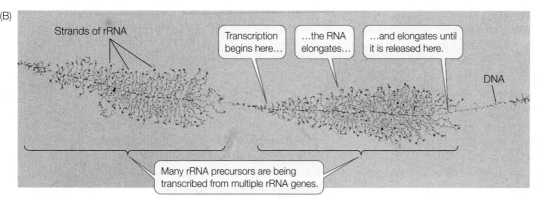

Strands of rRNA

Transcription begins here...

...the RNA elongates...

...and elongates until it is released here.

DNA

Many rRNA precursors are being transcribed from multiple rRNA genes.

17.12 A Moderately Repetitive Sequence Codes for rRNA (A) This rRNA gene, along with its nontranscribed spacer region, is repeated 280 times in the human genome, with clusters on five chromosomes. (B) This electron micrograph shows transcription of multiple rRNA genes.

17.13 Sequences in the Eukaryotic Genome There are many types of DNA sequences. Some are transcribed, and some of those sequences are translated.

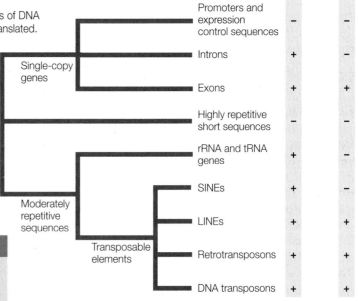

that encode the organelles' proteins. If the organelles were once independent, they must originally have contained all of those genes. How did the genes move to the nucleus? They may have done so by DNA transpositions between organelles and the nucleus, which still occur today. The DNA that remains in the organelles may be the remnants of more complete prokaryotic genomes.

See **Figure 17.13** for a summary of the various types of sequences in the human genome.

17.3 RECAP

The sequencing of the genomes of model organisms demonstrated common features of the eukaryotic genome, including the presence of repetitive sequences and transposons. Some eukaryotic genes are in families, which may include members that are mutated and nonfunctional. Some sequences are transcribed, but others are not.

- What are the major differences between prokaryotic and eukaryotic genomes? **See p. 375**

- Describe one function of genes found in *C. elegans* that has no counterpart in the genome of yeast. **See p. 376 and Table 17.3**

- What is the evolutionary role of eukaryotic gene families? **See p. 377**

- Why are there multiple copies of sequences coding for rRNA in the mammalian genome? **See p. 378**

- What effects can transposons have on a genome? **See p. 379**

The analysis of eukaryotic genomes has resulted in an enormous amount of useful information, as we have seen. In the next section we look more closely at the human genome.

17.4 What Are the Characteristics of the Human Genome?

By the start of 2005 the first human genome sequences were completed, two years ahead of schedule and well under budget. The published sequences, one produced by the publically funded Human Genome Project, and the other by a private company, were haploid genomes that were composites of several people. Since 2005, the diploid genomes of several individuals have been sequenced and published.

The human genome sequence held some surprises

The following are just some of the interesting facts that we have learned about the human genome:

- Of the 3.3 billion base pairs in the haploid human genome, fewer than 2 percent (about 24,000 genes) make up protein-coding regions. This was a surprise. Before sequencing began, humans were estimated, based on the diversity of their proteins, to have 80,000–150,000 genes. The actual number of genes—not many more than in a fruit fly—means that posttranscriptional mechanisms (such as alternative splicing) must account for the observed number of proteins in humans. That is, the average human gene must code for several different proteins.

- The average gene has 27,000 base pairs. Gene sizes vary greatly, from about 1,000 to 2.4 million base pairs. Variation in gene size is to be expected given that human proteins (and RNAs) vary in size, from 100 to about 5,000 amino acids per polypeptide chain.

- Virtually all human genes have many introns.

- Over 50 percent of the genome is made up of transposons and other highly repetitive sequences. Repetitive sequences near genes are GC-rich, while those farther away from genes are AT-rich.

- Most of the genome (about 97 percent) is the same in all people. Despite this apparent homogeneity, there are, of course, many individual differences. Scientists have mapped over 7 million single nucleotide polymorphisms (SNPs) in humans.

- Genes are not evenly distributed over the genome. Chromosome 19 is packed densely with genes, while chromosome 8 has long stretches without coding regions. The Y chromosome has the fewest genes (231), while chromosome 1 has the most (2,968).

Comparisons between sequenced genomes from prokaryotes and eukaryotes have revealed some of the evolutionary relationships between genes. Some genes are present in both prokaryotes and eukaryotes; others are only in eukaryotes; still others are only in animals, or only in vertebrates (**Figure 17.14**).

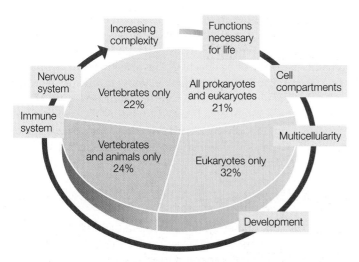

17.14 Evolution of the Genome A comparison of the human and other genomes has revealed how genes with new functions have been added over the course of evolution. Each percentage number refers to genes in the human genome. Thus, 21 percent of human genes have homologs in prokaryotes and other eukaryotes, 32 percent of human genes occur only in other eukaryotes, and so on.

More comparative genomics is possible now that the genomes of two other primates, the chimpanzee and the rhesus macaque, have been sequenced. The chimpanzee is evolutionarily close to humans, and shares 95 percent of the human genome sequence. The more distantly related rhesus macaque shares 91 percent of the human sequence. The search is on for a set of human genes that differ from the other primates and "make humans human."

Human genomics has potential benefits in medicine

Complex phenotypes are determined not by single genes, but by multiple genes interacting with the environment. The single-allele models of phenylketonuria and sickle-cell anemia (see Chapter 15) do not apply to such common disorders as diabetes, heart disease, and Alzheimer's disease. To understand the genetic bases of these diseases, biologists are now using rapid genotyping technologies to create "haplotype maps," which are used to identify SNPs (pronounced "snips") that are linked to genes involved in disease.

HAPLOTYPE MAPPING The SNPs that differ between individuals are not inherited as independent alleles. Rather, *a set of SNPs that are present on a segment of chromosome are usually inherited as a unit*. This linked piece of a chromosome is called a **haplotype**. You can think of the haplotype as a sentence and the SNP as a word in the sentence. Analyses of haplotypes in humans from all over the world have shown that there are at most 500,000 common variations.

GENOTYPING TECHNOLOGY AND PERSONAL GENOMICS New technologies are continually being developed to analyze thousands or millions of SNPs in the genomes of individuals. Such technologies include rapid sequencing methods and DNA microarrays that depend on DNA hybridization to identify specific SNPs. For example, a microarray of 500,000 SNPs has been used to analyze thousands of people to find out which SNPs are associated with specific diseases. The aim is to *correlate the SNP-defined haplotype with a disease state*. The amount of data is prodigious: 500,000 SNPs, thousands of people, thousands of medical records. With so much natural variation, statistical measures of association between a haplotype and a disease need to be very rigorous.

These association tests have revealed particular haplotypes or alleles that are associated with modestly increased risks for such diseases as breast cancer, diabetes, arthritis, obesity, and coronary heart disease (**Figure 17.15** and **Table 17.5**). Private companies will now scan a human genome for these variants—and the price for this service keeps getting lower. However, at this point it is unclear what a person without symptoms should do with the information, since multiple genes, environmental influences, and epigenetic effects all contribute to the development of these diseases.

Of course, the best way to analyze a person's genome is by actually sequencing it. Until recently, this was prohibitively expensive. As we mentioned earlier, DNA pioneer James Watson's genome cost over $1 million, certainly too much for a typical person or insurance company to afford in the context of health care. But with advances in sequencing technologies the cost is decreasing rapidly. One new method automatically sequences protein-coding exons only, for example. Once the cost of genome sequencing is within an affordable range, SNP testing will be superseded.

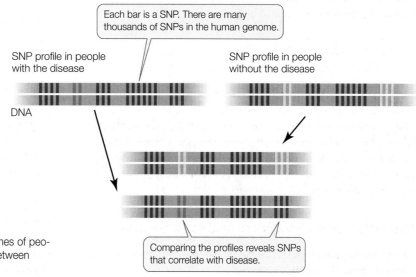

17.15 SNP Genotyping and Disease Scanning the genomes of people with and without particular diseases reveals correlations between SNPs and complex diseases.

TABLE 17.5

SNP Human Genome Scans and Diseases

DISEASE	LOCATION OF SNP (CHROMOSOME NUMBER)	% INCREASED RISK	
		HETEROZYGOTES	HOMOZYGOTES
Breast cancer	8	20	63
Coronary heart disease	9	20	56
Heart attack	9	25	64
Obesity	16	32	67
Diabetes	10	65	277
Prostate cancer	8	26	58

PHARMACOGENOMICS Genetic variation can affect how an individual responds to a particular drug. For example, a drug may be chemically modified in the liver to make it more or less active. Consider an enzyme that catalyzes the following reaction:

active drug → less active drug

A mutation in the gene that encodes this enzyme may make the enzyme less active. For a given dose of the drug, a person with the mutation would have more active drug in the bloodstream than a person without the mutation. So the effective dose of the drug would be lower in these people.

Now consider a different case, in which the liver enzyme is needed to make the drug active:

inactive drug → active drug

A person carrying a mutation in the gene encoding this liver enzyme would not be affected by the drug, since the activating enzyme is not present.

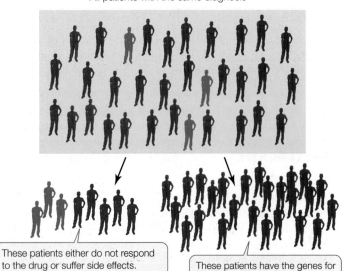

All patients with the same diagnosis

These patients either do not respond to the drug or suffer side effects. They need an alternative drug or dose.

These patients have the genes for an effective response to the drug.

17.16 Pharmacogenomics Correlations between genotypes and responses to drugs will help physicians develop personalized medical care.

The study of how an individual's genome affects his or her response to drugs or other outside agents is called **pharmacogenomics**. This type of analysis makes it possible to predict whether a drug will be effective. The objective is to *personalize drug treatment* so that a physician can know in advance whether an individual will benefit from a particular drug (**Figure 17.16**). This approach might also be used to reduce the incidence of adverse drug reactions by identifying individuals that will metabolize a drug slowly, which can lead to a dangerously high level of the drug in the body.

17.4 RECAP

The haploid human genome has 3.3 billion base pairs, but less than 2 percent of the genome codes for proteins. Most human genes are subject to alternative splicing; this may account for the fact that there are more proteins than genes. SNP mapping to find correlations with disease and drug susceptibility holds promise for personalized medicine.

- What are some of the major characteristics of the human genome? See p. 380

- How does SNP mapping work in personalized medicine? See pp. 381–382 and Figures 17.15 and 17.16

Genome sequencing has had great success in advancing biological understanding. High-throughput technologies are now being applied to other components of the cell: proteins and metabolites. We now turn to the results of these studies.

17.5 What Do the New Disciplines Proteomics and Metabolomics Reveal?

"The human genome is the book of life." Statements like this were common at the time the human genome sequence was first revealed. They reflect "genetic determinism," that a person's phenotype is determined by his or her genotype. But is an organism just a product of gene expression? We know that it is not. The proteins and small molecules present in any cell at a given point in time reflect not just gene expression but modifications by the intracellular and extracellular environment. Two new fields have emerged to complement genomics and take a more complete snapshot of a cell and organism—proteomics and metabolomics.

The proteome is more complex than the genome

As mentioned above, many genes encode more than a single protein (**Figure 17.17A**). Alternative splicing leads to different combinations of exons in the mature mRNAs transcribed from a single gene (see Figure 16.22). Posttranslational modifications also increase the number of proteins that can be derived from one gene (see Figure 14.22). The **proteome** is the sum total of the proteins produced by an organism, and it is more complex than its genome.

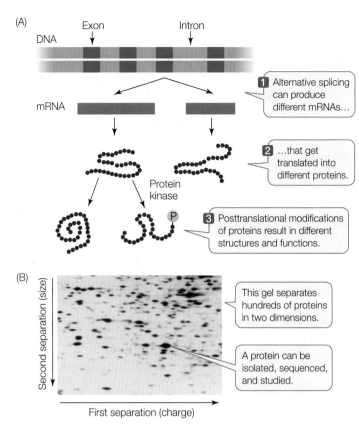

17.17 Proteomics (A) A single gene can code for multiple proteins. (B) A cell's proteins can be separated on the basis of charge and size by two-dimensional gel electrophoresis. The two separations can distinguish most proteins from one another.

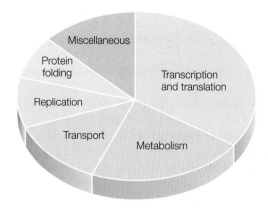

17.18 Proteins of the Eukaryotic Proteome About 1,300 proteins are common to all eukaryotes and fall into these categories. Although their amino acid sequences may differ to a limited extent, they perform the same essential functions in all eukaryotes.

Two methods are commonly used to analyze the proteome:

- Because of their unique amino acid compositions (primary structures), most proteins have unique combinations of electric charge and size. On the basis of these two properties, they can be separated by two-dimensional gel electrophoresis. Thus isolated, individual proteins can be analyzed, sequenced, and studied (**Figure 17.17B**).

- Mass spectrometry uses electromagnets to identify proteins by the masses of their atoms and displays them as peaks on a graph.

The ultimate aim of proteomics is just as ambitious as that of genomics. While genomics seeks to describe the genome and its expression, proteomics seeks to identify and characterize all of the expressed proteins.

Comparisons of the proteomes of humans and other eukaryotic organisms have revealed a common set of proteins that can be categorized into groups with similar amino acid sequences and similar functions. Forty-six percent of the yeast proteome, 43 percent of the worm proteome, and 61 percent of the fly proteome are shared by the human proteome. Functional analyses indicate that this set of 1,300 proteins provide the basic metabolic functions of a eukaryotic cell, such as glycolysis, the citric acid cycle, membrane transport, protein synthesis, DNA replication, and so on. (**Figure 17.18**).

Of course, these are not the only human proteins. There are many more, which presumably distinguish us as *human* eukaryotic organisms. As we have mentioned before, proteins have different functional regions called domains (for example, a domain for binding a substrate, or a domain for spanning a membrane). While a particular organism may have many unique proteins, those proteins are often just unique combinations of domains that exist in other organisms. *This reshuffling of the genetic deck is a key to evolution.*

Metabolomics is the study of chemical phenotype

Studying genes and proteins gives a limited picture of what is going on in a cell. But as we have seen, both gene function and protein function are affected by the internal and external environments of the cell. Many proteins are enzymes and their activities affect the concentrations of their substrates and products. So as the proteome changes, so will the abundances of these often-small molecules, called metabolites. The **metabolome** is the quantitative description of all of the *small molecules* in a cell or organism. These include:

- *Primary metabolites* involved in normal processes, such as intermediates in pathways like glycolysis. This category also includes hormones and other signaling molecules.

- *Secondary metabolites*, which are often unique to particular organisms or groups of organisms. They are often involved in special responses to the environment. Examples are antibiotics made by microbes, and the many chemicals made by plants that are used in defense against pathogens and herbivores.

Not surprisingly, measuring metabolites involves sophisticated analytical instruments. If you have studied organic or analytical chemistry, you may be familiar with gas chromatography and high-performance liquid chromatography, which separate molecules, and mass spectrometry and nuclear magnetic resonance spectroscopy, which are used to identify them. These measurements result in "chemical snapshots" of cells or organisms, which can be related to physiological states.

There has been some progress in defining the human metabolome. A database created by David Wishart and colleagues at the University of Alberta contains over 6,500 metabolite entries. The challenge now is to relate levels of these substances to physiology. For example, you probably know high levels of glucose in the blood are associated with diabetes. But what about early stages of heart disease? There may be a pattern of metabolites that is diagnostic of this disease. This could aid in early diagnosis and treatment.

Plant biologists are far ahead of medical researchers in the study of metabolomics. Over the years, tens of thousands of secondary metabolites have been identified in plants, many of them made in response to environmental challenges. Some of these are discussed in Chapter 39. The metabolome of the model organism *Arabidopsis thaliana* is being described, and will give insight into how a plant copes with stresses such as drought or pathogen attack. This knowledge could be helpful in optimizing plant growth for agriculture.

17.5 RECAP

The proteome is the total of all proteins produced by an organism. There are more proteins than genes in the genome. The metabolome is the total content of small molecules such as intermediates in metabolism, hormones, and secondary metabolites. The proteome and the metabolome can be analyzed using chemical methods that separate and identify molecules.

- How is the proteome analyzed? **See p. 383 and Figure 17.17**

- Explain the differences between genome, protoeome, and metabolome.

CHAPTER SUMMARY

17.1 How Are Genomes Sequenced?

- The sequencing of genomes required the development of ways to cut large chromosomes into fragments, sequence each of the fragments, and then line them up on the chromosome. **Review Figure 17.1, ANIMATED TUTORIAL 17.1**

- **Hierarchical sequencing** involves mapping the genome with genetic markers, cutting the genome into smaller pieces and sequencing them, then lining up the sequences using the markers.

- **Shotgun sequencing** involves directly cutting the genome into overlapping fragments, sequencing them, and using a computer to line up the sequences.

- DNA sequencing technologies involve labeled nucleotides that terminate the growing polynucleotide chain. **Review Figure 17.2**

- Rapid, automated methods for **high-throughput sequencing** are being developed. **Review Figure 17.3, ANIMATED TUTORIAL 17.2**

17.2 What Have We Learned from Sequencing Prokaryotic Genomes?

- DNA sequencing is used to study the genomes of prokaryotes that are important to humans and ecosystems.

- **Functional genomics** aims to determine the functions of gene products. **Comparative genomics** involves comparisons of genes and genomes from different organisms to identify common features and functions.

- **Transposable elements** and **transposons** can move about the genome. **Review Figure 17.6**

- **Metagenomics** is the identification of DNA sequences without first isolating, growing and identifying the organisms present in an environmental sample. Many of these sequences are from prokaryotes that were heretofore unknown to biologists. **Review Figure 17.7**

- Transposon mutagenesis can be used to inactivate genes one by one. Then the organism can be tested for survival. In this way, a minimal genome of less than 350 genes was identified for the bacterium *Mycoplasma genitalium*. **Review Figure 17.8**

17.3 What Have We Learned from Sequencing Eukaryotic Genomes?

- Genome sequences from model organisms have demonstrated some common features of the eukaryotic genome. In addition, there are specialized genes for cellular compartmentation, development, and features unique to plants. **Review Tables 17.1–17.4 and Figures 17.9 and 17.10**

- Some eukaryotic genes exist as members of **gene families**. Proteins may be made from these closely related genes at different times and in different tissues. Some members of gene families may be nonfunctional **pseudogenes**.

- Repeated sequences are present in the eukaryotic genome.

- **Moderately repeated sequences** include those coding for rRNA. **Review Figure 17.12**

17.4 What Are the Characteristics of the Human Genome?

- The haploid human genome has 3.3 billion base pairs.

- Only 2 percent of the genome codes for proteins; the rest consists of repeated sequences and noncoding DNA.

- Virtually all human genes have introns, and alternative splicing leads to the production of more than one protein per gene.

- SNP genotyping correlates variations in the genome with diseases or drug sensitivity. It may lead to personalized medicine. **Review Figure 17.15**

- **Pharmacogenomics** is the analysis of genetics as applied to drug metabolism.

17.5 What Do the New Disciplines of Proteomics and Metabolomics Reveal?

- The **proteome** is the total protein content of an organism.

- There are more proteins than protein-coding genes in the genome.

- The proteome can be analyzed using chemical methods that separate and identify proteins. These include two-dimensional electrophoresis and mass spectrometry. **See Figure 17.17**

- The **metabolome** is the total content of small molecules, such as intermediates in metabolism, hormones, and secondary metabolites.

SEE WEB ACTIVITY 17.1 for a concept review of this chapter.

SELF-QUIZ

1. Eukaryotic protein-coding genes differ from their prokaryotic counterparts in that eukaryotic genes
   a. are double-stranded.
   b. are present in only a single copy.
   c. contain introns.
   d. have promoters.
   e. are transcribed into mRNA.

2. A comparison of the genomes of yeast and bacteria shows that only yeast has many genes for
   a. energy metabolism.
   b. cell wall synthesis.
   c. intracellular protein targeting.
   d. DNA-binding proteins.
   e. RNA polymerase.

3. The genomes of the fruit fly and the nematode are similar to that of yeast, except that the former organisms have many genes for
   a. intercellular signaling.
   b. synthesis of polysaccharides.
   c. cell cycle regulation.
   d. intracellular protein targeting.
   e. transposable elements.

4. The minimum genome of *Mycoplasma genitalium*
   a. has 100 genes.
   b. has been used to create new species.
   c. has an RNA genome.
   d. is larger than the genome of *E. coli*.
   e. was derived by transposon mutagenesis.

5. Which is *not* true of metagenomics?
   a. It has been done with bacteria.
   b. It is done on rRNA sequences.
   c. It has revealed many new metabolic capacities.
   d. It involves extracting DNA from the environment.
   e. It cannot be done on seawater.

6. Transposons
   a. always use RNA for replication.
   b. are approximately 50 bp long.
   c. are made up of either DNA or RNA.
   d. do not contain genes coding for proteins.
   e. make up about 40 percent of the human genome.

7. Vertebrate gene families
   a. have mostly inactive genes.
   b. include the globins.
   c. are not produced by gene duplications.
   d. increase the number of unique genes in the genome.
   e. are not transcribed.

8. The DNA sequences that code for eukaryotic rRNA
   a. are transcribed only at the ribosome.
   b. are repeated hundreds of times.
   c. contain all the genes clustered directly beside one another.
   d. are on only one human chromosome.
   e. are identical to the sequences that code for miRNA.

9. The human genome
   a. contains very few repeated sequences.
   b. has 3.3 billion base pairs.
   c. was sequenced by hierarchical sequencing only.
   d. has genes evenly distributed along chromosomes.
   e. has few genes with introns.

10. Which of the following about genome sequencing is true?
    a. In hierarchical sequencing, but not high-throughput sequencing, DNA is amplified in BAC vectors.
    b. In hierarchical sequencing, a genetic map is made after the DNA is sequenced.
    c. Shotgun sequencing is considerably slower than hierarchical sequencing.
    d. The human genome was first sequenced by high-throughput methods.
    e. DNA sequence determination by chain termination is the basis of shotgun sequencing only.

FOR DISCUSSION

1. In rats, a protein-coding gene 1,440 bp long codes for an enzyme made up of 192 amino acids. Discuss this apparent discrepancy. How long would the initial and final mRNA transcripts be?

2. The genomes of rice, wheat, and corn are similar to one another and to that of *Arabidopsis* in many ways. Discuss how these plants might nevertheless have very different proteins.

3. Why are the proteome and the metabolome more complex than the genome?

ADDITIONAL INVESTIGATION

It is the year 2025. You are taking care of a patient who is concerned about having an early stage of kidney cancer. His mother died from this disease.

a. Assume that the SNPs linked to genes involved in the development of this type of cancer have been identified. How would you determine if this man has a genetic predisposition for developing kidney cancer? Explain how you would do the analysis.

b. How might you develop a metabolomic profile for kidney cancer and then use it to determine whether your patient has kidney cancer?

c. If the patient was diagnosed with cancer by the methods in (a) and (b), how would you use pharmacogenomics to choose the right medications to treat the tumor in this patient?

18 Recombinant DNA and Biotechnology

Pollution fighters

In the summer of 1990, soldiers from Iraq invaded neighboring Kuwait. The reason was oil: the Iraqis were angry because Kuwait was pumping too much of it, keeping prices low. Six months later, a United Nations–sponsored coalition army from more than 30 countries drove the Iraqis out of Kuwait and back to their homeland. For Kuwait, the Gulf War was a success, but it left an environmental disaster. As they fled, the Iraqi soldiers set fire to more than 700 oil wells. It took over six months to put the fires out, and in the meantime an astounding 250 million gallons of crude oil were released into the desert. Twenty years later, much of the oil remains as a gooey coating, severely affecting the organisms that live there.

The government of Kuwait is using a variety of processes to get rid of the contaminating oil. Among them is the addition of bacteria that break down and consume the oil, utilizing the hydrocarbons in it as an energy source for growth. This process—using an organism to remove a pollutant—is called bioremediation. The Kuwait episode is not the first major use of bacteria for bioremediation. In 1989, the oil tanker *Exxon Valdez* ran aground near the Alaskan shore, releasing 11 million gallons of crude oil along 500 miles of shoreline. Physical methods such as skimming the water were used to remove more than half of the oil. Nitrogen and phosphorus salts were then sprayed on the oily rocks to stimulate the growth of oil-consuming bacteria already there, and other bacteria were added as part of the recovery effort. The oil gradually disappeared.

Some species of bacteria, because of their genetic capacity to produce unusual enzymes and biochemical pathways, thrive on all sorts of nutrients besides the usual glucose, including pollutants. Scientists have discovered these organisms simply by mixing polluted soil with water and seeing what grows. Many of the genes coding for enzymes involved in breaking down crude oil are carried on small chromosomes called plasmids. In 1971, Ananda Chakrabarty at the General Electric Research Center in New York used genetic crosses to develop a single strain of the bacterium *Pseudomonas* with multiple plasmids carrying genes for the breakdown of various hydrocarbons in oil. He and his company applied for a patent to legally protect their discovery and profit from it. In a landmark case, the U.S. Supreme Court ruled in 1980 that "a live, human-made microorganism is

The Spoils of War Massive oil spills occurred in Kuwait during the 1991 Gulf War.

Using Biotechnology to Clean Up the Environment
Ananda Chakrabarty received the first patent for a genetically modified organism, a bacterium that breaks down crude oil.

patentable" under the U.S. Constitution. Since then other bacteria have been patented that remove toxic metals such as mercury and copper from soils. In these cases, the bacteria use metabolic pathways to convert the metals to biologically inert forms.

The Supreme Court ruling came at a time when new laboratory methods were being developed to insert specific DNA sequences into organisms by recombinant DNA technology. Since then, an entirely new biotechnology industry has sprung up, its activities legally protected. The resulting flood of patents for DNA sequences and genetically modified organisms continues to this day.

IN THIS CHAPTER we will describe some of the techniques that are used to manipulate DNA. First, we will describe how DNA molecules are cut into smaller fragments and how these fragments are spliced together to create recombinant DNA. This will lead to a discussion of how recombinant DNA is introduced into suitable host cells. After describing some other ways to manipulate DNA, we will show how scientists have applied these methods to create a new biotechnology industry.

18.1 What Is Recombinant DNA?

You are familiar with *restriction endonucleases* (restriction enzymes), which occur naturally in bacteria and are used in the laboratory to cut DNA into fragments (see Chapter 15). Our focus in Chapter 15 was on the use of these enzymes for detecting mutations. In this chapter we examine how they are used, along with other enzymes, to construct recombinant DNA.

During the late 1960s, scientists discovered other enzymes that act on DNA. One of these is **DNA ligase**, which catalyzes the joining of DNA fragments. This is the enzyme that joins Okazaki fragments during DNA replication (see Section 13.3). Once they had isolated restriction enzymes and DNA ligase, scientists could use these enzymes to cut DNA into fragments and then splice them together in new combinations. Stanley Cohen and Herbert Boyer did just that in 1973. They used restriction enzymes to cut sequences from two *E. coli* plasmids (small chromosomal DNAs—see Figure 12.27) containing different antibiotic resistance genes. Then they used DNA ligase to join the fragments together. The resulting plasmid, when inserted into new *E. coli* cells, gave those cells resistance to both antibiotics (**Figure 18.1**). The era of **recombinant DNA**—a DNA molecule made in the laboratory that is derived from at least two genetic sources—was born.

Hundreds of different restriction enzymes are now available. They recognize *palindromic* DNA sequences—sequences that read the same way in both directions. For example, you can read the DNA recognition sequence for the restriction enzyme *Eco*R1 from 5′ to 3′ as GAATTC on both strands:

$$5'.......GAATTC......3'$$
$$3'.......CTTAAG......5'$$

Some restriction enzymes cut the DNA straight through the middle of the palindrome, generating "blunt-ended" fragments. Others, such as *Eco*RI, make staggered cuts—they cut one strand of the double helix several bases away from where they cut the other (**Figure 18.2**). After *Eco*RI makes its two cuts in the complementary strands, the ends of the strands are held together only by the hydrogen bonds between four base pairs. These hydrogen bonds are too weak to persist at warm temperatures (above room temperature), so the DNA separates into fragments when it is warmed. As a result, each fragment carries a single-stranded "overhang" at the location of each cut. These overhangs are called **sticky ends** because they have specific base sequences that can bind by base pairing with complementary sticky ends.

INVESTIGATING LIFE

18.1 Recombinant DNA

With the discovery of restriction enzymes and DNA ligase, it became possible to combine DNA fragments from different sources in the laboratory. But would such "recombinant DNA" be functional when inserted into a living cell? The results of this experiment completely changed the scope of genetic research, increasing our knowledge of gene structure and function, and ushered in the new field of biotechnology.

HYPOTHESIS Biologically functional recombinant chromosomes can be made in the laboratory.

METHOD

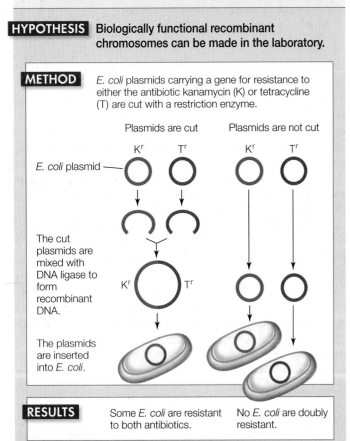

E. coli plasmids carrying a gene for resistance to either the antibiotic kanamycin (K) or tetracycline (T) are cut with a restriction enzyme.

Plasmids are cut   Plasmids are not cut

K^r T^r K^r T^r

E. coli plasmid

The cut plasmids are mixed with DNA ligase to form recombinant DNA.

K^r T^r

The plasmids are inserted into *E. coli*.

RESULTS

Some *E. coli* are resistant to both antibiotics.

No *E. coli* are doubly resistant.

CONCLUSION Two DNA fragments with different genes can be joined to make a recombinant DNA molecule, and the resulting DNA is functional.

FURTHER INVESTIGATION: Only one cell in 10,000 took up the plasmid in the experiment. The spontaneous mutation rate to T^r or K^r is one cell in 10^6. How would you distinguish between genetic transformation and spontaneous mutation in this experiment?

Go to **yourBioPortal.com** for original citations, discussions, and relevant links for all INVESTIGATING LIFE figures.

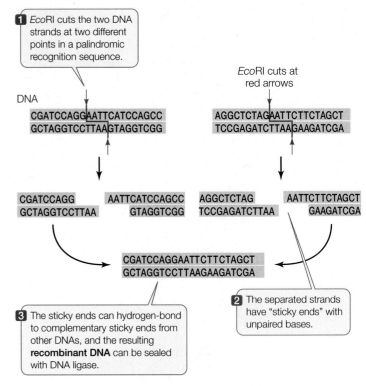

1 *Eco*RI cuts the two DNA strands at two different points in a palindromic recognition sequence.

*Eco*RI cuts at red arrows

DNA

CGATCCAGGAATTCATCCAGCC
GCTAGGTCCTTAAGTAGGTCGG

AGGCTCTAGAATTCTTCTAGCT
TCCGAGATCTTAAGAAGATCGA

CGATCCAGG   AATTCATCCAGCC
GCTAGGTCCTTAA   GTAGGTCGG

AGGCTCTAG   AATTCTTCTAGCT
TCCGAGATCTTAA   GAAGATCGA

CGATCCAGGAATTCTTCTAGCT
GCTAGGTCCTTAAGAAGATCGA

3 The sticky ends can hydrogen-bond to complementary sticky ends from other DNAs, and the resulting **recombinant DNA** can be sealed with DNA ligase.

2 The separated strands have "sticky ends" with unpaired bases.

18.2 Cutting, Splicing, and Joining DNA Some restriction enzymes (*Eco*RI is shown here) make staggered cuts in DNA. *Eco*RI can be used to cut two different DNA molecules (blue and orange). The exposed bases can hydrogen bond with complementary exposed bases on other DNA fragments, forming recombinant DNA. DNA ligase stabilizes the recombinant molecule by forming covalent bonds in the DNA backbone.

joined to a fragment from another source, such as a bacterium. Initially the fragments are held together by weak hydrogen bonds, but then the enzyme ligase catalyzes the formation of covalent bonds between adjacent nucleotides at the ends of the fragments, joining them to form a single, larger molecule.

With these tools—restriction enzymes and DNA ligase—scientists can cut and rejoin different DNA molecules from any and all sources, including artificially synthesized DNA sequences.

18.1 RECAP

DNA fragments from different sources can be linked together to make recombinant DNA.

● How did Cohen and Boyer make the first recombinant DNA? **See Figure 18.1**

● How does a staggered cut in DNA create a "sticky end"? **See p. 387 and Figure 18.2**

After a DNA molecule has been cut with a restriction enzyme, complementary sticky ends can form hydrogen bonds with one another. The original ends may rejoin, or two different fragments with complementary sticky ends may join. Indeed, a fragment from one source, such as a human, can be

Recombinant DNA has no biological significance until it is inserted inside a living cell, which can replicate and transcribe the transplanted genetic information. How can recombinant DNA made in the laboratory be inserted and expressed in living cells?

18.2 How Are New Genes Inserted into Cells?

One goal of recombinant DNA technology is to **clone**—that is, to produce many identical copies of—a particular gene. Cloning might be done for analysis, to produce a protein product in quantity, or as a step toward creating an organism with a new phenotype. Recombinant DNA is cloned by inserting it into host cells in a process known as **transformation** (or **transfection** if the host cells are derived from an animal). A host cell or organism that contains recombinant DNA is referred to as a **transgenic** cell or organism. Later in this chapter we will encounter many examples of transgenic cells and organisms, including yeast, mice, wheat plants, and even cows.

Various methods are used to create transgenic cells. Generally, these methods are inefficient in that only a few of the cells that are exposed to the recombinant DNA actually become transformed with it. In order to grow only the transgenic cells, **selectable marker** genes, such as genes that confer resistance to antibiotics, are often included as part of the recombinant DNA molecule. Antibiotic resistance genes were the markers used in Cohen and Boyer's experiment (see Figure 18.1).

Genes can be inserted into prokaryotic or eukaryotic cells

The initial successes with recombinant DNA technology were achieved using bacteria as hosts. As we have seen in preceding chapters, bacterial cells are easily grown and manipulated in the laboratory. Much of their molecular biology is known, especially for certain well-studied bacteria such as *E. coli*. Furthermore, bacteria contain plasmids, which are easily manipulated to carry recombinant DNA into the cell.

In some important ways, however, bacteria are not ideal organisms for studying and expressing eukaryotic genes. Consider how differently the processes of transcription and translation proceed in prokaryotes and eukaryotes, and recall that DNA often contains the signals for these specific functions (see Chapter 14). Furthermore, scientists often want to study how genes function in multicellular eukaryotic organisms rather than in cells grown in cultures. Or they might want to create a crop plant or farm animal with a new phenotype for use in agriculture. For these reasons, scientists have developed methods to transform or transfect eukaryotic cells.

Yeasts such as *Saccharomyces* are commonly used as eukaryotic hosts for recombinant DNA studies. The advantages of using yeasts include rapid cell division (a life cycle completed in 2–8 hours), ease of growth in the laboratory, and a relatively small genome size (about 12 million base pairs and 6,000 genes). In addition, yeasts have most of the characteristics of other eukaryotes, except for those characteristics involved in multicellularity.

Plant cells can also be used as hosts. One property that makes plant cells good hosts is the ability to make stem cells (unspecialized, totipotent cells; see Chapter 5 opener) from mature plant tissues. When these unspecialized plant cells are isolated and grown in culture, they can be transformed with recombinant DNA. These transgenic cells can be studied in culture, or manip-ulated to form entire new plants. There are also methods for making whole transgenic plants without going through the cell culture step. These methods result in plants that carry the recombinant DNA in all their cells, including the germ line cells.

If biologists want to study expression of human or animal genes, for example for medical purposes, they use cultured animal cells as hosts. Whole transgenic animals can also be created.

Recombinant DNA enters host cells in a variety of ways

Methods for inserting DNA into host cells vary. The cells may be chemically treated to make their outer membranes more permeable, and then mixed with the DNA so that it can diffuse into the cells. Another approach is called *electroporation*; a short electric shock is used to create temporary pores in the membranes, through which the DNA can enter. Viruses can be altered so that they carry recombinant DNA into cells. Plants are often transformed using a bacterium that has evolved mechanisms to transfer its DNA into cells and then insert the DNA into a plant chromosome. Transgenic animals can be produced by injecting recombinant DNA into the nuclei of fertilized eggs. There are even "gene guns," which "shoot" the host cells with tiny particles carrying the DNA.

The challenge of inserting new DNA into a cell lies not just in getting it into the host cell, but in getting it to replicate as the host cell divides. DNA polymerase does not bind to just any sequence. If the new DNA is to be replicated, it must become part of a segment of DNA that contains an origin of replication. Such a DNA molecule is called a **replicon**, or replication unit.

There are two general ways in which the newly introduced DNA can become part of a replicon:

- It may be inserted into a host chromosome. Although the site of insertion is usually random, this is nevertheless a common method of integrating new genes into host cells.

- It can enter the host cell as part of a carrier DNA sequence, called a **vector**, that already has an origin of replication.

Several types of vectors are used to get DNA into cells. Once inside the cells, some vectors replicate independently, while others incorporate all or part of their DNA into the host chromosomes.

PLASMIDS AS VECTORS As you learned in Chapter 12, plasmids are small chromosomes that exist in prokaryotic cells in addition to the main chromosomes. Yeast cells can also harbor plasmids. A number of characteristics make plasmids useful as transformation vectors:

- They are relatively small (an *E. coli* plasmid has 2,000–6,000 base pairs) and therefore easy to manipulate in the laboratory.

- A plasmid will usually have one or more restriction enzyme recognition sequences that each occur only once in the plasmid sequence. These sites make it easy to insert additional DNA into the plasmid before it is used to transform host cells.

- Many plasmids contain genes that confer resistance to antibiotics, which can serve as selectable markers.

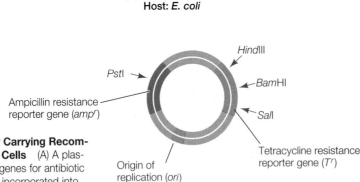

(A) Plasmid pBR322
Host: *E. coli*

*Hind*III
*Pst*I
*Bam*HI
*Sal*I
Ampicillin resistance
reporter gene (*amp*^r)
Tetracycline resistance
reporter gene (*T*^r)
Origin of
replication (*ori*)

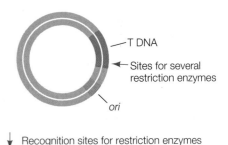

(B) Ti plasmid
Hosts: *Agrobacterium tumefaciens* (plasmid)
and infected plants (T DNA)

T DNA
Sites for several
restriction enzymes
ori

↓ Recognition sites for restriction enzymes

18.3 Vectors for Carrying Recombinant DNA into Cells (A) A plasmid with reporter genes for antibiotic resistance can be incorporated into an *E. coli* cell. (B) The Ti plasmid, isolated from the bacterium *Agrobacterium tumefaciens*, is used to insert DNA into many types of plants.

- Plasmids have a bacterial origin of replication (*ori*) and can replicate independently of the host chromosome. It is not uncommon for a bacterial cell to contain hundreds of copies of a recombinant plasmid. For this reason, the power of bacterial transformation to amplify a gene is extraordinary. A one-liter culture of bacteria harboring the human β-globin gene in a typical plasmid has as many copies of that gene as the sum total of all the cells in a typical adult human being (10^{14}). A typical bacterial plasmid is shown in **Figure 18.3A**.

The plasmids used as vectors in the laboratory have been extensively altered by recombinant DNA technology to include convenient features: multiple cloning sites with 20 or more unique restriction enzyme sites for cloning purposes; origins of replication for a variety of host cells; and various kinds of reporter genes and selectable marker genes.

VIRUSES AS VECTORS Constraints on plasmid replication limit the size of the new DNA that can be inserted into a plasmid to about 10,000 base pairs. Although many prokaryotic genes may be smaller than this, most eukaryotic genes—with their introns and extensive flanking sequences—are bigger. A vector that accommodates larger DNA inserts is needed for these genes.

Both prokaryotic and eukaryotic viruses are often used as vectors for eukaryotic DNA. Bacteriophage λ, which infects *E. coli*, has a DNA genome of about 45,000 base pairs. If the genes that cause the host cell to die and lyse—about 20,000 base pairs—are eliminated, the virus can still attach to a host cell and inject its DNA. The deleted 20,000 base pairs can be replaced with DNA from another organism. Because viruses infect cells naturally, they offer a great advantage over plasmids, which often require artificial means to coax them to enter host cells. As we saw in Section 15.6, viruses are important vectors in human gene therapy.

PLASMID VECTORS FOR PLANTS An important vector for carrying new DNA into many types of plants is a plasmid found in *Agrobacterium tumefaciens*. This bacterium lives in the soil, infects plants, and causes a disease called crown gall, which is

characterized by the presence of growths, or tumors, in the plant. *A. tumefaciens* contains a plasmid called Ti (for tumor-inducing) (**Figure 18.3B**). The Ti plasmid carries genes that allow the bacterium to infect plant cells and then insert a region of its DNA called the T DNA into the chromosomes of infected cells. The T DNA contains genes that cause the growth of tumors and the production of specific sugars that the bacterium uses as sources of energy. Scientists have exploited this remarkable natural "genetic engineer" to insert foreign DNA into the genomes of plants.

When used as a vector for plant transformation, the tumor-inducing and sugar-producing genes on the T DNA are removed and replaced with foreign DNA. The altered Ti plasmids are first used to transform *Agrobacterium* cells from which the original Ti plasmids have been removed. Then the *Agrobacterium* cells are used to infect plant cells. Whole plants can be regenerated from transgenic cells or, in the case of the model plant *Arabidopsis* (see Section 17.3), the *Agrobacterium* can be used to directly infect germ line cells of whole plants.

Reporter genes identify host cells containing recombinant DNA

Even when a population of host cells interacts with an appropriate vector, only a small proportion of the cells actually take up the vector. Furthermore, the process of making recombinant DNA is far from perfect. After a ligation reaction, not all the vector copies contain the foreign DNA. How can we identify or select the host cells that contain that sequence?

Selectable markers such as antibiotic resistance genes can be used to select cells containing those genes. Only cells carrying the antibiotic resistance gene can grow in the presence of that antibiotic. If a vector carrying genes for resistance to two different antibiotics is used, one antibiotic can be used to selectively grow cells carrying the vector. If the other antibiotic resistance gene is inactivated by the insertion of foreign DNA, then cells carrying copies of the vector with the inserted DNA can be identified by their sensitivity to that antibiotic (**Figure 18.4**). Since the uptake of recombinant DNA is a rare event (only about 1 cell in 10,000 takes up a plasmid in such experiments), it is vital to be able to select the small number of cells harboring the recombinant DNA.

TOOLS FOR INVESTIGATING LIFE

18.4 Marking Recombinant DNA by Inactivating a Gene

Selectable marker (reporter) genes are used by scientists to select for bacteria that have taken up a plasmid. A second reporter gene allows for the identification of bacteria harboring the recombinant plasmid. The host bacteria in this experiment could display any of the three phenotypes indicated in the table.

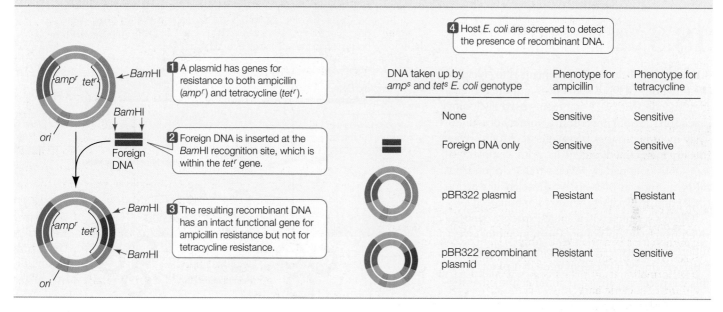

DNA taken up by *amp^s* and *tet^s E. coli* genotype	Phenotype for ampicillin	Phenotype for tetracycline
None	Sensitive	Sensitive
Foreign DNA only	Sensitive	Sensitive
pBR322 plasmid	Resistant	Resistant
pBR322 recombinant plasmid	Resistant	Sensitive

1 A plasmid has genes for resistance to both ampicillin (*amp^r*) and tetracycline (*tet^r*).

2 Foreign DNA is inserted at the *Bam*HI recognition site, which is within the *tet^r* gene.

3 The resulting recombinant DNA has an intact functional gene for ampicillin resistance but not for tetracycline resistance.

4 Host *E. coli* are screened to detect the presence of recombinant DNA.

Selectable markers are one type of **reporter gene**, which is any gene whose expression is easily observed. Other reporter genes code for proteins that can be detected visually. For example:

- The β-galactosidase (*lacZ*) gene in the *E. coli lac* operon (see Figure 16.10) codes for an enzyme that can convert the substrate X-Gal into a bright blue product. Many plasmids contain the *lacZ* gene with a multiple cloning site within its sequence. Bacterial colonies containing the plasmid (which also includes an antibiotic resistance gene) are selected on a solid medium containing the antibiotic. X-Gal is also included in the medium, so that bacterial colonies containing the recombinant DNA inserted into the *lacZ* gene produce white, rather than blue, colonies.

- Green fluorescent protein, which normally occurs in the jellyfish *Aequopora victoriana*, emits visible light when exposed to ultraviolet light. The gene for this protein has been isolated and incorporated into vectors. It is now widely used as a reporter gene (**Figure 18.5**).

Such reporters are not just used to select and identify cells carrying recombinant DNA. They can be attached to promoters in order to study how the promoters function under different conditions or in different tissues of a transgenic multicellular organism. They can also be attached to other proteins, to study how and where those proteins become localized within eukaryotic cells.

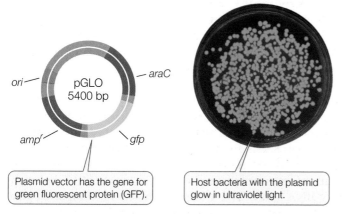

Plasmid vector has the gene for green fluorescent protein (GFP).

Host bacteria with the plasmid glow in ultraviolet light.

18.5 Green Fluorescent Protein as a Reporter The presence of a plasmid with the gene for green fluorescent protein is readily apparent in transgenic cells because they glow under ultraviolet light. This allows the identification of cells carrying a plasmid without the use of selection on antibiotics. That is, no cells are killed during the selection process.

18.2 RECAP

Recombinant DNA can be cloned by using a vector to insert it into a suitable host cell. The vector often has a selectable marker or other reporter gene that gives the host cell a phenotype by which transgenic cells can be identified.

- List the characteristics of a plasmid that make it suitable for introducing new DNA into a host cell. See pp. 389–390

- How are cells harboring a vector that carries recombinant DNA selected? See p. 390 and Figure 18.4

We have described how DNA can be cut or amplified, inserted into a vector, and introduced into host cells. We have also seen how host cells carrying recombinant DNA can be identified. Now let's consider where the genes or DNA fragments used in these procedures come from.

18.3 What Sources of DNA Are Used in Cloning?

A major goal of cloning experiments is to elucidate the functions of DNA sequences and the proteins they encode. The DNA fragments used in cloning procedures are obtained from a number of sources. They include random fragments of chromosomes that are maintained as gene libraries, complementary DNA obtained by reverse transcription from mRNA, products of the polymerase chain reaction (PCR), and artificially synthesized or mutated DNA.

Often a scientist will want to express a gene derived from one kind of organism in another, very different organism—for example, a human gene in a bacterium, or a bacterial gene in a plant. To do this it is necessary to use a promoter and other regulatory sequences from the host organism: a bacterial promoter will not function in a plant cell, for example. The coding region of the gene of interest is inserted between a promoter and a transcription termination sequence derived from the host organism, or from one that uses similar mechanisms for gene regulation.

Libraries provide collections of DNA fragments

In Chapter 17 we introduced the concept of a **genomic library**: a collection of DNA fragments that together comprise the genome of an organism. Now we provide details on how a genomic or other gene library is generated and used.

Restriction enzymes or other means, such as mechanical shearing, can be used to break chromosomes into smaller pieces. These smaller DNA fragments still constitute a genome (**Figure 18.6A**), but the information is now in many smaller "volumes." Each fragment is inserted into a vector, which is then taken up by a host cell. Proliferation of a single transformed cell produces a colony of recombinant cells, each of which harbors many copies of the same fragment of DNA.

When plasmids are used as vectors, about 700,000 separate fragments are required to make a library of the human genome. By using bacteriophage λ, which can carry four times as much DNA as a plasmid, the number of "volumes" in the library can be reduced to about 160,000. Although this seems like a large number, a single petri plate can hold thousands of phage colonies, or plaques, and is easily screened for the presence of a particular DNA sequence by hybridization to an appropriate DNA probe.

TOOLS FOR INVESTIGATING LIFE

18.6 Constructing Libraries

Intact genomic DNA is too large to be introduced into host cells. A genomic library can be made by breaking the DNA into small fragments, incorporating the fragments into a vector, and then transforming host cells with the recombinant vectors. Each colony of cells contains many copies of a small part of the genome. Similarly, there are many mRNAs in a cell. These can be copied into cDNAs and a library made from them. The DNA in these colonies can then be isolated for analysis.

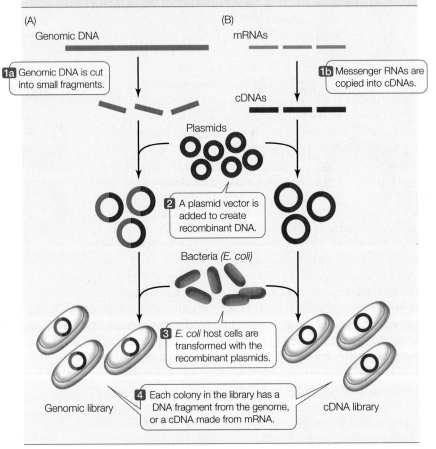

cDNA libraries are constructed from mRNA transcripts

A much smaller DNA library—one that includes only the genes transcribed in a particular tissue—can be made from **complementary DNA**, or **cDNA** (**Figure 18.6B**). This involves isolating mRNA from cells, then making cDNA copies of that mRNA by complementary base pairing. An enzyme, reverse transcriptase, catalyzes this reaction.

A collection of cDNAs from a particular tissue at a particular time in the life cycle of an organism is called a **cDNA library**. Messenger RNAs do not last long in the cytoplasm and are often present in small amounts, so a cDNA library is a "snapshot" that preserves the transcription pattern of the cell. Complementary DNA libraries have been invaluable for comparing gene expression in different tissues at different stages of development. For example, their use has shown that up to one-third of all the genes of an animal are expressed only during development. Complementary DNA is also a good starting point for cloning eukaryotic genes (because the clones contain only the coding sequences of the genes) and genes that are expressed in only a few cell types.

Synthetic DNA can be made by PCR or by organic chemistry

In Chapter 13 we describe the polymerase chain reaction (PCR), a method of amplifying DNA in a test tube. PCR can begin with as little as 10^{-12} g of DNA (a picogram). Any fragment of DNA can be amplified by PCR as long as appropriate primers are available. You will recall that DNA replication (by PCR or any other system) requires not just a template on which DNA polymerase adds complementary nucleotides, but also a short oligonucleotide primer where replication begins (see Figure 13.22). If the appropriate primers (two are needed—one for each strand of DNA) are added to denatured DNA, more than two billion copies of the DNA region between the primers can be produced in just a few hours. This amplified DNA can then be inserted into plasmids to create recombinant DNA and cloned in host cells.

The *artificial synthesis* of DNA by organic chemistry is now fully automated, and a special service laboratory can make short- to medium-length sequences overnight for any number of investigators. Synthetic oligonucleotides (single-stranded DNA fragments of up to 40 bp) are used as primers in PCR reactions. These primers can be designed to create short new sequences at the ends of the PCR products. This might be done to create a mutation in a recombinant gene, or to add restriction enzyme sites at the ends of the PCR product to aid in cloning.

Longer synthetic sequences can be pieced together to construct an artificial gene. If we know the amino acid sequence of the desired protein product, we can use the genetic code to figure out the corresponding DNA sequence. As mentioned above, other sequences must be added, such as the promoter and transcription termination sequences. Appropriate selection of the codon for a given amino acid is another important consideration: many amino acids are encoded by more than one codon (see Figure 14.6), and host organisms vary in their use of synonymous codons.

DNA mutations can be created in the laboratory

Mutations that occur in nature have been important in demonstrating cause-and-effect relationships in biology. However, mutations in nature are rare events. Recombinant DNA technology allows us to ask "what if" questions by creating mutations artificially. Because synthetic DNA can be made with any desired sequence, it can be manipulated to create specific mutations, the consequences of which can be observed when the mutant DNA is expressed in host cells. These mutagenesis techniques have revealed many cause-and-effect relationships.

For example, consider the experiment illustrated in Figure 14.20. Researchers hypothesized that a nuclear localization signal (NLS) sequence of amino acids is necessary for targeting a protein to the nucleus after it is made at the ribosome. The researchers used recombinant DNA technology to synthesize genes encoding proteins with and without the sequence, which

were then used to transform cells. Without the NLS, newly synthesized proteins did not enter the nucleus. Knowing this, the researchers then asked, "Are certain amino acids more functionally important to the NLS than others?" In follow-up experiments, they made a series of mutated genes to test whether certain amino acids were needed at certain locations in the NLS. They found that changing the amino acids at the very beginning or very end of the NLS, but not the middle, abolished its function. This led to a fuller description of the binding of the NLS to its nuclear receptor. Without the ability to generate specific mutations, these experiments would not have been possible.

18.3 RECAP

DNA for cloning can be obtained from genomic libraries, cDNA made from mRNA, or artificially synthesized DNA fragments. Gene function can be investigated by intentionally introducing mutations into natural or synthetic genes and organisms.

- How are genomic DNA and cDNA libraries made and used? **See p. 392 and Figure 18.6**

- Explain how recombinant DNA and mutagenesis are used to test "what if" questions in biology. **See p. 393**

We've explored the various sources of DNA that can be used to make recombinant DNA molecules and the ways the resulting molecules can be used to study the functions of genes and proteins. We now turn to some additional tools that are available for studying DNA.

18.4 What Other Tools Are Used to Study DNA Function?

Sections 13.5 and 17.1 describe PCR and DNA sequencing, two important techniques arising from our understanding of DNA replication. In this section we will examine three additional techniques for studying DNA, including homologous recombination to inactivate genes, antisense and RNAi to block gene expression, and DNA microarrays to analyze large numbers of nucleotide sequences.

Genes can be inactivated by homologous recombination

One way to study a gene or protein in order to understand its function is to inactivate the gene so that it is not transcribed and translated into a functional protein. Such a manipulation is called a **knockout** experiment. In plants, transposons or T DNA insertions can be used to create thousands of knockout mutants, and then the mutants are screened to identify those with altered phenotypes. For example, the mutants can be screened for those that are susceptible to a particular disease. This is an important way to identify genes that are involved in processes such as resistance to disease or other environmental stresses, such as drought and temperature extremes.

A technique called **homologous recombination** is a much more targeted way to produce knockout mutants. In this case, the gene of interest has already been identified, and recombinant DNA technology is used to specifically inactivate that gene. Mice are frequently used in such knockout experiments (**Figure 18.7**). The normal allele of the mouse gene to be tested is inserted into a plasmid. Restriction enzymes are then used to insert a fragment containing a reporter gene into the middle of the normal gene. This addition of extra DNA plays havoc with the targeted gene's transcription and translation; a functional mRNA is seldom made from a gene whose sequence has been thus interrupted.

Once the recombinant plasmid has been made, it is used to transfect mouse embryonic stem cells. (A **stem cell** is an unspecialized cell that divides and differentiates into specialized cells.) Much of the targeted gene is still present in the plasmid (although in two separated regions), and these sequences tend to line up with their homologous sequences in the normal allele on the mouse chromosome. Sometimes recombination occurs, and the plasmid's inactive allele is "swapped" with the functional allele in the host cell. The inactive allele is inserted permanently into the host cell's genome and the normal allele is lost (because the plasmid cannot replicate in mouse cells). The active reporter gene in the insert is used to select those stem cells carrying the inactivated gene.

A transfected stem cell is now transplanted into an early mouse embryo. If the mouse that develops from this embryo has the mutant gene in its germ line cells, its progeny will have the knockout gene in every cell in their bodies. Such mice are inbred to create *knockout mice* carrying the inactivated gene in homozygous form. The mutant mouse can then be observed for phenotypic changes, to find clues about the function of the targeted gene in the normal (wild-type) animal. The knockout technique has been important in assessing the roles of many genes, and has been especially valuable in studying human genetic diseases. Many of these diseases, such as phenylketonuria, have knockout mouse *models*—mouse strains that suffer from an analogous disease—produced by homologous recombination. These models can be used to study a disease and to test potential treatments. Mario Capecchi, Martin Evans, and Oliver Smithies shared the Nobel Prize for developing the knockout mouse technique.

Complementary RNA can prevent the expression of specific genes

Another way to study the expression of a specific gene is to block the translation of its mRNA. This is an example of scientists imitating nature. As described in Section 16.5, gene expression is sometimes controlled by the production of double-stranded RNA molecules, which are cut up and unwound to produce short, single-stranded RNA molecules (microRNAs) that are complementary to specific mRNA sequences (see Figure 16.23). Such a complementary molecule is called **antisense RNA** because it binds by base pairing to the "sense" bases on the mRNA. The resulting partially double-stranded RNA hybrid inhibits translation of the mRNA, and the hybrid tends to be broken down rapidly in the cytoplasm. Although the gene continues to be transcribed, translation does not take place. After determining the sequence of a gene and its mRNA in the lab-

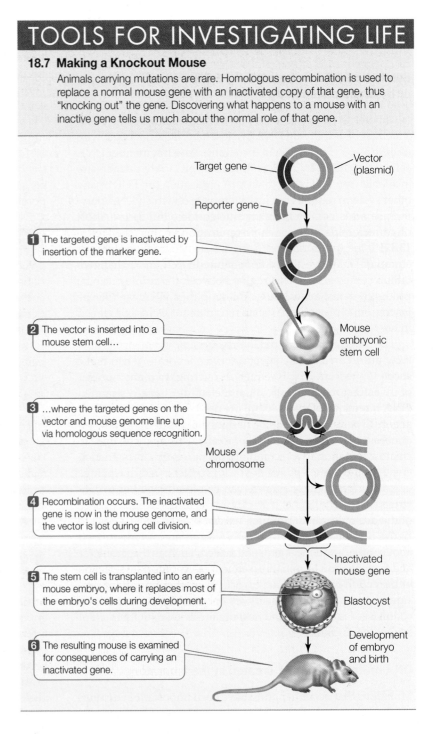

TOOLS FOR INVESTIGATING LIFE

18.7 Making a Knockout Mouse

Animals carrying mutations are rare. Homologous recombination is used to replace a normal mouse gene with an inactivated copy of that gene, thus "knocking out" the gene. Discovering what happens to a mouse with an inactive gene tells us much about the normal role of that gene.

Target gene

Vector (plasmid)

Reporter gene

1 The targeted gene is inactivated by insertion of the marker gene.

2 The vector is inserted into a mouse stem cell...

Mouse embryonic stem cell

3 ...where the targeted genes on the vector and mouse genome line up via homologous sequence recognition.

Mouse chromosome

4 Recombination occurs. The inactivated gene is now in the mouse genome, and the vector is lost during cell division.

Inactivated mouse gene

5 The stem cell is transplanted into an early mouse embryo, where it replaces most of the embryo's cells during development.

Blastocyst

6 The resulting mouse is examined for consequences of carrying an inactivated gene.

Development of embryo and birth

18.8 Using Antisense RNA and siRNA to Block Translation of mRNA Once a gene's sequence is known, the synthesis of its protein can be prevented by making either an antisense RNA (left) or a small interfering RNA (siRNA, right) that is complementary to its mRNA.

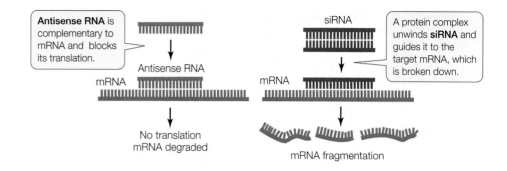

oratory, scientists can make a specific, single-stranded antisense RNA and add it to a cell to prevent translation of that gene's mRNA (**Figure 18.8, left**).

Several antisense drugs are being developed to reduce the expression of genes involved with cancer. For example, the gene *bcl2* codes for a protein that blocks apoptosis, and in some forms of cancer *bcl2* is activated inappropriately through mutation. These cells fail to undergo apoptosis, continue to divide, and form a tumor. Treatment with oblimersen, an antisense RNA that binds to *bcl2* mRNA, prevents production of the protein, and leads to apoptosis of tumor cells and shrinkage of the tumor.

A related technique to antisense RNA takes advantage of **RNA interference (RNAi)**, a rare, natural mechanism for inhibiting mRNA translation. In a process similar to that involved in processing microRNAs, a short (about 20 nucleotides) double-stranded RNA is unwound to single strands by a protein complex that guides this RNA to a complementary region on mRNA. The protein complex catalyzes the breakdown of the targeted mRNA. RNAi was not discovered until the late 1990s, but since then scientists have synthesized double-stranded siRNAs to inhibit the expression of known genes (**Figure 18.8, right**). Because these double-stranded siRNAs are more stable than antisense RNAs, the use of siRNAs is the preferred approach for blocking translation. Macular degeneration is an eye disease that results in near-blindness when blood vessels proliferate in the eye. The signaling molecule that stimulates vessel proliferation is a growth factor. An RNAi-based therapy is being developed to target this growth factor's mRNA and shows promise in stopping and even reversing the progress of the disease.

Although medical applications for RNAi are still at the experimental stage, antisense RNA and RNAi have been widely used to test cause-and-effect relationships in biological research. Another powerful research tool with great potential for medicine is the gene chip, or DNA microarray.

DNA microarrays can reveal RNA expression patterns

The emerging science of genomics has to face two major quantitative realities. First, there are very large numbers of genes in eukaryotic genomes. Second, the pattern of gene expression in different tissues at different times is quite distinctive. For example, a cell from a skin cancer at its early stage may have a unique mRNA "fingerprint" that differs from that of both normal skin cells and the cells of a more advanced skin cancer.

To find such patterns, scientists could isolate mRNA from a cell and test it by hybridization with each gene in the genome, one gene at a time. But that would involve many steps and take a very long time. It is far simpler to do these hybridizations all in one step. This is possible with **DNA microarray** technology, which provides large arrays of sequences for hybridization experiments.

The development of DNA arrays ("gene chips") was inspired by methods used for decades by the semiconductor industry. A silicon microchip consists of an array of microscopic electric circuits etched onto a tiny silicon base, called a chip. In the same way, a series of DNA sequences can be attached to a glass slide in a precise order (**Figure 18.9**). The slide is divided into a grid of microscopic spots, or "wells." Each spot contains thousands of copies of a particular oligonucleotide of 20 or more bases. A computer controls the addition of these oligonucleotide sequences in a predetermined pattern. Each oligonucleotide can hybridize with only one DNA or RNA sequence, and thus is a unique identifier of a gene. Many thousands of different oligonucleotides can be placed on a single microarray.

As we mention in Section 17.4, DNA microarrays can be used to identify specific single nucleotide polymorphisms or other mutations in genomic DNA samples. Or they can be used to analyze RNA from different tissues or cells to identify which genes are expressed in those cell types. If mRNA is to be analyzed, it is usually incubated with reverse transcriptase to make cDNA (see Figure 18.6B). Fluorescent dyes are used to tag the cDNAs from different samples with different colors (usually red and green; see Figure 18.9). The cDNAs are used to probe the DNA on the microarray. Complementary sequences that form hybrids with the DNA on the microarray can be located using a sensitive scanner that detects the fluorescent light.

A clinical use of DNA chips was developed by Laura van 't Veer and her colleagues at the Netherlands Cancer Institute (**Figure 18.10**). Most women with breast cancer are treated with surgery to remove the tumor, and then treated with radiation soon afterward to kill cancer cells that the surgery may have missed. But a few cancer cells may still survive in some patients, and these eventually form tumors in the breast or elsewhere in the body. The challenge for physicians is to develop criteria to identify patients with surviving cancer cells so that they can be treated aggressively with tumor-killing chemotherapy. The scientists in van 't Veer's group followed the medical histories of breast cancer patients to identify those patients whose cancer recurred. They then used a DNA microarray to examine the expression of about 1,000 genes in these patients' original tu-

TOOLS FOR INVESTIGATING LIFE

18.9 DNA on a Chip

Large arrays of DNA sequences can be used to identify specific sequences in a sample of DNA or RNA by hybridization. For example, thousands of known, synthetic DNA sequences can be attached to a glass slide in an organized grid pattern. This can be hybridized with cDNA samples derived from two different tissues to find out what genes are being expressed in the tissues.

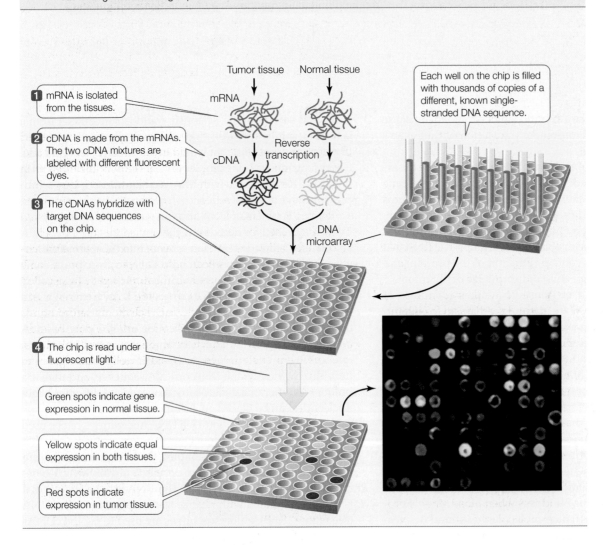

1 mRNA is isolated from the tissues.

2 cDNA is made from the mRNAs. The two cDNA mixtures are labeled with different fluorescent dyes.

3 The cDNAs hybridize with target DNA sequences on the chip.

Each well on the chip is filled with thousands of copies of a different, known single-stranded DNA sequence.

4 The chip is read under fluorescent light.

Green spots indicate gene expression in normal tissue.

Yellow spots indicate equal expression in both tissues.

Red spots indicate expression in tumor tissue.

Tumor tissue   Normal tissue

mRNA

cDNA

Reverse transcription

DNA microarray

---

yourBioPortal.com

GO TO Animated Tutorial 18.1 • DNA Chip Technology

mors (which had been stored after their surgical removal) relative to normal tissue. They found 70 genes whose expression differed dramatically between tumors from patients whose cancers recurred and tumors from patients whose cancers did not recur. From this information, the Dutch group was able to identify what is called a *gene expression signature*. This expression pattern is useful in clinical decision-making: patients with a good prognosis can avoid unnecessary chemotherapy, while those with a poor prognosis can receive aggressive treatment.

18.10 Using DNA Arrays for Medical Diagnosis

The pattern of expression of 70 genes in tumor tissues indicates whether breast cancer is likely to recur. Actual arrays have more dots than shown here.

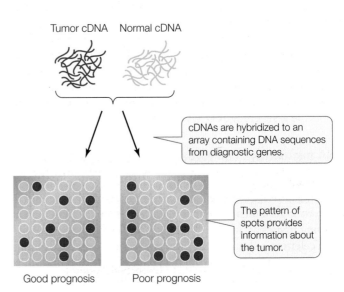

Tumor cDNA   Normal cDNA

cDNAs are hybridized to an array containing DNA sequences from diagnostic genes.

The pattern of spots provides information about the tumor.

Good prognosis   Poor prognosis

Now that you've seen how DNA can be fragmented, recombined, manipulated, and put back into living organisms, let's see some examples of how these techniques are used to make useful products.

18.5 What Is Biotechnology?

Biotechnology is the use of cells or whole living organisms to produce materials useful to people, such as foods, medicines, and chemicals. People have been doing this for a very long time. For example, the use of yeasts to brew beer and wine dates back at least 8,000 years, and the use of bacterial cultures to make cheese and yogurt is a technique many centuries old. For a long time people were not aware of the molecular basis of each of these biochemical transformations.

About 100 years ago, thanks largely to Louis Pasteur's work, it became clear that specific bacteria, yeasts, and other microbes could be used as biological converters to make certain products. Alexander Fleming's discovery that the mold *Penicillium* makes the antibiotic penicillin led to the large-scale commercial culture of microbes to produce antibiotics as well as other useful chemicals. Today, microbes are grown in vast quantities to make much of the industrial-grade alcohol, glycerol, butyric acid, and citric acid that are used by themselves or as starting materials in the manufacture of other products.

Nevertheless the commercial harvesting of proteins, including hormones and enzymes, was limited by the (often) minuscule amounts that could be extracted from organisms that produce them naturally. Yields were low, and purification was difficult and costly. Gene cloning has changed all this. The ability to insert almost any gene into bacteria or yeasts, along with methods to induce the gene to make its product in large amounts and export it from the cells, has turned these microbes into versatile factories for important products. Today there is interest in producing nutritional supplements and pharmaceuticals in whole transgenic animals and harvesting them in large quantities, for example from the milk of cows or the eggs of chickens. Key to this boom in biotechnology has been the development of specialized vectors that not only carry genes into cells, but also make those cells express them at high levels.

Expression vectors can turn cells into protein factories

If a eukaryotic gene is inserted into a typical plasmid and used to transform *E. coli*, little if any of the gene product will be made. Other key prokaryotic DNA sequences must be included with the gene. A bacterial promoter, a signal for transcription termination, and a special sequence that is necessary for ribosome binding on the mRNA must all be included in the transformation vector if the gene is to be expressed in the bacterial cell.

To solve this kind of problem, scientists make **expression vectors** that have all the characteristics of typical vectors, as well as the extra sequences needed for the foreign gene (also called a *transgene*) to be expressed in the host cell. For bacterial hosts, these additional sequences include the elements named above (**Figure 18.11**); for eukaryotes, they include the poly A–addition sequence, transcription factor binding sites, and enhancers. An expression vector can be designed to deliver transgenes to

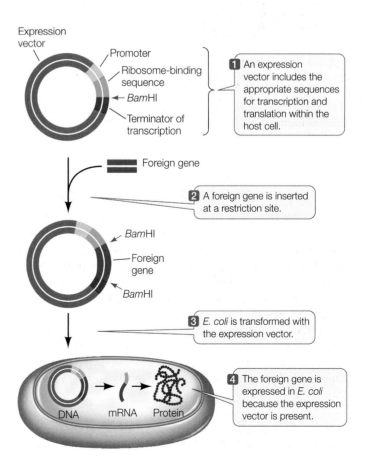

18.11 Expression of a Transgene in a Host Cell Produces Large Amounts of its Protein Product To be expressed in *E. coli*, a gene derived from a eukaryote requires bacterial sequences for transcription initiation (promoter), transcription termination, and ribosome binding. Expression vectors contain these additional sequences, enabling the eukaryotic protein to be synthesized in the prokaryotic cell.

yourBioPortal.com

GO TO Web Activity 18.1 • Expression Vectors

any class of prokaryotic or eukaryotic host and may include additional features:

- An *inducible promoter*, which responds to a specific signal, can be included. For example, a promoter that responds to hormonal stimulation can be used so that the transgene will be expressed at high levels when the hormone is added.

- A *tissue-specific promoter*, which is expressed only in a certain tissue at a certain time, can be used if localized expression is desired. For example, many seed proteins are expressed only in the plant embryo. Coupling a transgene to a seed-specific promoter will allow it to be expressed only in seeds.

- *Signal sequences* can be added so that the gene product is directed to an appropriate destination. For example, when a protein is made by yeast or bacterial cells in a liquid medium, it is economical to include a signal directing the protein to be secreted into the extracellular medium for easier recovery.

18.5 RECAP

Expression vectors maximize the expression of transgenes inserted into host cells.

- How do expression vectors work? See pp. 397–398 and Figure 18.11

This chapter has introduced many of the methods that are used in biotechnology. Let's turn now to the ways biotechnology is being applied to meet some specific human needs.

18.6 How Is Biotechnology Changing Medicine, Agriculture, and the Environment?

Huge potential for improvements in health, agriculture, and the environment derive from recent developments in biotechnology. We now have the ability to make virtually any protein by recombinant DNA technology and to insert transgenes into many kinds of host cells. With these revolutionary developments in biological capability, concerns have been raised about ethics and safety. We now turn to the promises and problems of biotechnology that uses DNA manipulation.

Medically useful proteins can be made by biotechnology

Many medically useful products are being made by biotechnology (**Table 18.1**), and hundreds more are in various stages of development. The manufacture of *tissue plasminogen activator* (*TPA*) provides a good illustration of a medical application of biotechnology.

When a wound begins bleeding, a blood clot soon forms to stop the flow. Later, as the wound heals, the clot dissolves. How does the blood perform these conflicting functions at the right times? Mammalian blood contains an enzyme called *plasminogen*. When activated, it becomes *plasmin* and catalyzes the dissolution of the clotting proteins. The conversion of plasminogen to plasmin is catalyzed by the enzyme TPA, which is produced by cells lining the blood vessels:

$$\text{plasminogen} \xrightarrow{\text{TPA}} \text{plasmin}$$
$$\text{(inactive)} \qquad\qquad \text{(active)}$$

Heart attacks and strokes can be caused by blood clots that form in major blood vessels leading to the heart or the brain, respectively. During the 1970s, a bacterial enzyme called *streptokinase* was found to stimulate the dissolution of clots in some patients. Treatment with this enzyme saved lives, but being a foreign protein, it triggered the body's immune system to react against it. More important, the drug sometimes prevented clotting throughout the entire circulatory system, sometimes leading to a dangerous situation in which blood could not clot where needed.

When TPA was discovered, it had many advantages: it bound specifically to clots, and it did not provoke an immune reaction. But the amounts of TPA that could be harvested from human tissues were tiny, certainly not enough to inject at the site of a clot in the emergency room.

Recombinant DNA technology solved this problem. TPA mRNA was isolated and used to make cDNA, which was then

---

TABLE 18.1
Some Medically Useful Products of Biotechnology

PRODUCT	USE
Colony-stimulating factor	Stimulates production of white blood cells in patients with cancer and AIDS
Erythropoietin	Prevents anemia in patients undergoing kidney dialysis and cancer therapy
Factor VIII	Replaces clotting factor missing in patients with hemophilia A
Growth hormone	Replaces missing hormone in people of short stature
Insulin	Stimulates glucose uptake from blood in people with insulin-dependent (Type I) diabetes
Platelet-derived growth factor	Stimulates wound healing
Tissue plasminogen activator	Dissolves blood clots after heart attacks and strokes
Vaccine proteins: Hepatitis B, herpes, influenza, Lyme disease, meningitis, pertussis, etc.	Prevent and treat infectious diseases

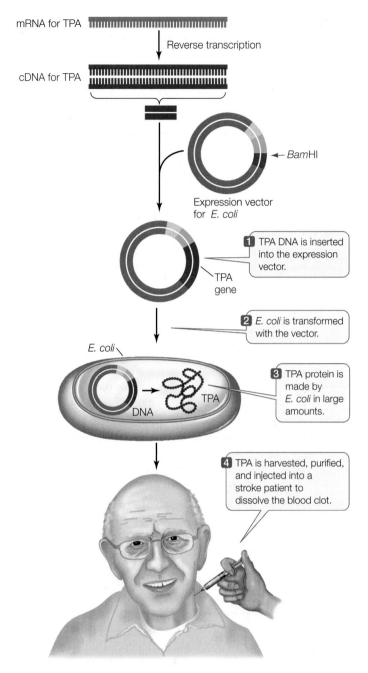

18.12 Tissue Plasminogen Activator: From Protein to Gene to Drug
TPA is a naturally occurring human protein that dissolves blood clots. It is used to treat patients suffering from blood clotting in heart attacks or strokes, and is manufactured using recombinant DNA technology.

inserted into an expression vector and used to transform *E. coli* (**Figure 18.12**). The transgenic bacteria made the protein in quantity, and it soon became available commercially. This drug has had considerable success in dissolving blood clots in people experiencing strokes and heart attacks.

Another way of making medically useful products in large amounts is **pharming**: the production of pharmaceuticals in farm animals or plants. For example, a gene encoding a useful protein might be placed next to the promoter of the gene that encodes lactoglobulin, an abundant milk protein. Transgenic animals car-

rying this recombinant DNA will secrete large amounts of the foreign protein into their milk. These natural "bioreactors" can produce abundant supplies of the protein, which can be separated easily from the other components of the milk (**Figure 18.13**).

Human growth hormone is a protein made in the pituitary gland in the brain and has many effects, especially in growing children (see Chapter 41). Children with growth hormone deficiency have short stature as well as other abnormalities. In the past they were treated with protein isolated from the pituitary glands of dead people, but the supply was too limited to meet demand. Recombinant DNA technology was used to coax bacteria to make the protein, but the cost of treatment was high ($30,000 a year). In 2004, a team led by Daniel Salamone at the

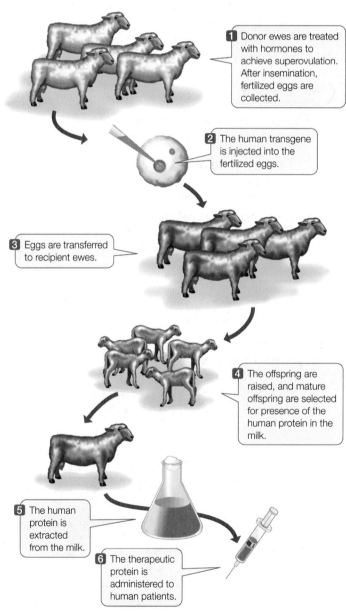

18.13 Pharming An expression vector carrying a desired gene can be put into an animal egg, which is implanted into a surrogate mother. The transgenic offspring produce the new protein in their milk. The milk is easily harvested and the protein isolated, purified, and made clinically available to patients.

University of Buenos Aires made a transgenic cow that secretes human growth hormone in her milk. The yield is prodigious: only 15 such cows are needed to meet the needs worldwide of children suffering from this type of dwarfism.

DNA manipulation is changing agriculture

The cultivation of plants and the husbanding of animals provide the world's oldest examples of biotechnology, dating back more than 10,000 years. Over the centuries, people have adapted crops and farm animals to their needs. Through selective breeding of these organisms, desirable characteristics such as large seeds, high fat content in milk, or resistance to disease have been imparted and improved.

Until recently, the most common way to improve crop plants and farm animals was to identify individuals with desirable phenotypes that existed as a result of natural variation. Through many deliberate crosses, the genes responsible for the desirable trait could be introduced into a widely used variety or breed of that organism.

Despite some spectacular successes, such as the breeding of high-yielding varieties of wheat, rice, and hybrid corn, such deliberate crossing can be a hit-or-miss affair. Many desirable traits are controlled by multiple genes, and it is hard to predict the results of a cross or to maintain a prized combination as a pure-breeding variety year after year. In sexual reproduction, combinations of desirable genes are quickly separated by meiosis. Furthermore, traditional breeding takes a long time: many plants and animals take years to reach maturity and then can reproduce only once or twice a year—a far cry from the rapid reproduction of bacteria.

Modern recombinant DNA technology has several advantages over traditional methods of breeding (**Figure 18.14**):

- *The ability to identify specific genes.* The development of genetic markers allows breeders to select for specific desirable genes, making the breeding process more precise and rapid.
- *The ability to introduce any gene from any organism into a plant or animal species.* This ability, combined with mutagenesis techniques, vastly expands the range of possible new traits.
- *The ability to generate new organisms quickly.* Manipulating cells in the laboratory and regenerating a whole plant by cloning is much faster than traditional breeding.

Consequently, recombinant DNA technology has found many applications in agriculture (**Table 18.2**). We will describe a few examples to demonstrate the approaches that plant scientists have used to improve crop plants.

PLANTS THAT MAKE THEIR OWN INSECTICIDES   Plants are subject to infections by viruses, bacteria, and fungi, but probably the most important crop pests are herbivorous insects. From the locusts of biblical (and modern) times to the cotton boll weevil, insects have continually eaten the crops people grow.

The development of insecticides has improved the situation somewhat, but insecticides have their own problems. Many, including the organophosphates, are relatively nonspecific and kill beneficial insects in the broader ecosystem as well as crop pests. Some even have

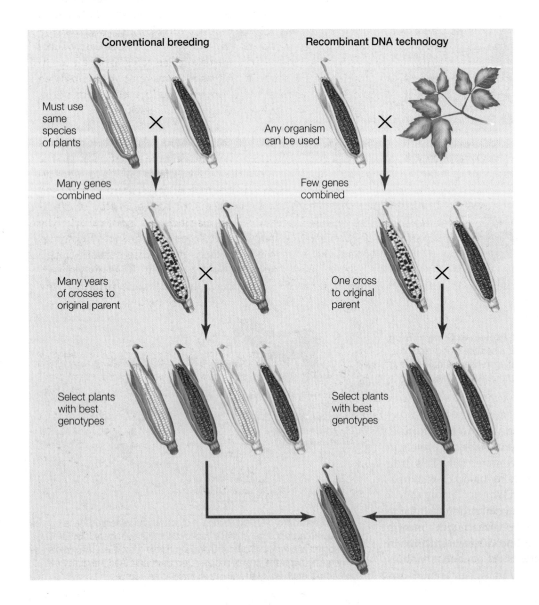

18.14 Genetically Modified Plants versus Conventional Plant Breeding Plant biotechnology offers many potential advantages over conventional breeding.

TABLE 18.2	
Agricultural Applications of Biotechnology under Development	
PROBLEM	TECHNOLOGY/GENES
Improving the environmental adaptations of plants	Genes for drought tolerance, salt tolerance
Improving nutritional traits	High-lysine seeds; β-carotene in rice
Improving crops after harvest	Delay of fruit ripening; sweeter vegetables
Using plants as bioreactors	Plastics, oils, and drugs produced in plants

toxic effects on other groups of organisms, including people. What's more, many insecticides persist in the environment for a long time.

Some bacteria protect themselves by producing proteins that can kill insects. For example, the bacterium *Bacillus thuringiensis* produces a protein that is toxic to the insect larvae that prey on it. The toxicity of this protein is 80,000 times greater than that of a typical commercial insecticide. When a hapless larva eats the bacteria, the toxin becomes activated and binds specifically to the insect's gut, producing holes and killing the insect. Dried preparations of *B. thuringiensis* have been sold for decades as safe insecticides that break down rapidly in the environment. But the biodegradation of these preparations is their limitation, because it means that the dried bacteria must be applied repeatedly during the growing season.

A more permanent approach is to have the crop plants themselves make the toxin, and this is exactly what plant scientists have done. The toxin gene from *B. thuringiensis* has been isolated, cloned, and extensively modified by the addition of a plant promoter and other regulatory sequences. Transgenic corn, cotton, soybeans, tomatoes, and other crops are now being grown successfully with this added gene. Pesticide usage by farmers growing these transgenic crops is greatly reduced.

CROPS THAT ARE RESISTANT TO HERBICIDES Herbivorous insects are not the only threat to agriculture. Weeds may grow in fields and compete with crop plants for water and soil nutrients. Glyphosate is a widely used and effective herbicide, or weed killer, that works only on plants. It inhibits an enzyme system in the chloroplast that is involved in the synthesis of amino acids. Glyphosate is a broad-spectrum herbicide that kills most weeds, but unfortunately it also kills crop plants. One solution to this problem is to use it to rid a field of weeds before the crop plants start to grow. But, as any gardener knows, when the crop begins to grow, the weeds reappear. If the crop were not affected by the herbicide, the herbicide could be applied to the field at any time.

Scientists have used expression vectors to make plants that synthesize a different form of the target enzyme for glyphosate that is unaffected by the herbicide. The gene for this enzyme has been inserted into corn, cotton, and soybean plants, making them resistant to glyphosate. This technology has expanded rapidly and a large proportion of cotton and soybean plants now carry this gene.

GRAINS WITH IMPROVED NUTRITIONAL CHARACTERISTICS To remain healthy, humans must consume adequate amounts of β-carotene, which the body converts into vitamin A (see Figure 3.21). About 400 million people worldwide suffer from vitamin A deficiency, which makes them susceptible to infections and blindness. One reason is that rice grains, which do not contain β-carotene, make up a large part of their diets. Other parts of the rice plant, and indeed many plants and other organisms, contain enzymes for the biochemical pathway that leads to β-carotene production.

Plant biologists Ingo Potrykus and Peter Beyer isolated one of the genes for the β-carotene pathway from the bacterium *Erwinia uredovora* and another from daffodil plants. They added a promoter and other signals for expression in the developing rice grain, and then transformed rice plants with the two genes. The resulting rice plants produce grains that look yellow because of their high β-carotene content. A newer variety with a corn gene replacing the one from daffodils makes even more β-carotene and is golden in color (**Figure 18.15**). A daily intake of about 150 grams of this cooked rice can supply all the β-carotene a person needs. This new transgenic strain has been crossed with strains adapted for various local environments, in the hope of improving the diets of millions of people.

CROPS THAT ADAPT TO THE ENVIRONMENT Agriculture depends on ecological management—tailoring the environment to the needs of crop plants and animals. A farm field is an unnatural, human-designed system that must be carefully managed to maintain optimal conditions for crop growth. For example, excessive irrigation can cause increases in soil salinity. The Fertile Crescent, the region between the Tigris and Euphrates rivers in the Middle East where agriculture probably originated 10,000 years ago, is no longer fertile. It is now a desert, largely because the soil has a high salt concentration. Few plants can grow on salty soils, partly because of osmotic ef-

Wild type   Golden rice 1   Golden rice 2

18.15 Transgenic Rice Rich in β-Carotene Right and middle: The grains from these transgenic rice strains are colored because they make the pigment β-carotene, which is converted to vitamin A in the human body. Left: Normal rice grains do not contain β-carotene.

fects that result in wilting, and partly because excess salt ions are toxic to plant cells.

Some plants can tolerate salty soils because they have a protein that transports Na$^+$ ions out of the cytoplasm and into the vacuole, where the ions can accumulate without harming plant growth (see Section 5.3 for a description of the plant vacuole). In many salt-intolerant plants, including *Arabidopsis thaliana*, the gene for this protein exists but is inactive. Recombinant DNA technology has allowed scientists to create active versions of this gene, and to use it to transform crop plants such as rapeseed, wheat, and tomatoes. When this gene was added to tomato plants, they grew in water that was four times as salty as the typical lethal level (**Figure 18.16**). This finding raises the prospect of growing useful crops on what were previously unproductive soils.

This example illustrates what could become a fundamental shift in the relationship between crop plants and the environment. *Instead of manipulating the environment to suit the plant, biotechnology may allow us to adapt the plant to the environment.* As a result, some of the negative effects of agriculture, such as water pollution, could be lessened.

Biotechnology can be used for environmental cleanup

The thousands of species of bacteria have many unique enzymes and biochemical pathways. Bacteria are nature's recyclers, thriving on many types of nutrients—including what humans refer to as wastes. **Bioremediation** is the use by humans of other organisms to remove contaminants from the environment. Two well-known uses of bacteria for bioremediation are composting and wastewater treatment.

Composting involves the use of bacteria to break down large molecules, including carbon-rich polymers and proteins in waste products such as wood chips, paper, straw, and kitchen scraps. For example, some species of bacteria make cellulase, an enzyme that hydrolyzes cellulose. Bacteria are used in *wastewater treatment* to break down human wastes, paper products, and household chemicals.

Transgenic organisms can also be used to clean up environmental contaminants. As we saw at the opening of this chapter, bacteria are being used to help clean up oil spills. As another example, plants that have been modified to take up heavy metals are being explored as a way to remediate contaminated soils, such as mine tailings (see Section 39.4).

There is public concern about biotechnology

Concerns have been raised about the safety and wisdom of genetically modifying crops and other organisms. These concerns are centered on three claims:

- Genetic manipulation is an unnatural interference with nature.
- Genetically altered foods are unsafe to eat.
- Genetically altered crop plants are dangerous to the environment.

Advocates of biotechnology tend to agree with the first claim. However, they point out that all crops are unnatural in the sense that they come from artificially bred plants growing in a manipulated environment (a farmer's field). Recombinant DNA technology just adds another level of sophistication to these technologies.

To counter the concern about whether genetically engineered crops are safe for human consumption, biotechnology advocates point out that only single genes are added and that these genes are specific for plant function. For example, the *B. thuringiensis* toxin produced by transgenic plants has no effect on people. However, as plant biotechnology moves from adding genes that improve plant growth to adding genes that affect human nutrition, such concerns will become more pressing.

Various negative environmental impacts have been envisaged. There is concern about the possible "escape" of transgenes from crops to other species. If the gene for herbicide resistance, for example, were inadvertently transferred from a crop plant to a closely related weed, that weed could thrive in her-

(A)

(B)

18.16 Salt-Tolerant Tomato Plants
Transgenic plants containing a gene for salt tolerance thrive in salty water (A), while plants without the transgene die (B). This technology may allow crops to be grown on salty soils.

bicide-treated areas. Another negative impact would be the development of new super-weeds from transgenic crops. For example, a drought tolerant crop plant might spread into, and upset the ecology of, a desert. Or beneficial insects could eat plant materials containing *B. thuringiensis* toxin and die. Transgenic plants undergo extensive field-testing before they are approved for use, but the complexity of the biological world makes it impossible to predict all potential environmental effects of transgenic organisms. In fact, some spreading of transgenes has been detected. Because of the potential benefits of agricultural biotechnology (see Table 18.2), scientists believe that it is wise to proceed with caution.

18.6 RECAP

Biotechnology has been used to produce medicines and to develop transgenic plants with improved agricultural and nutritional characteristics.

- What are the advantages of using biotechnology for plant breeding compared with traditional methods? **See Figure 18.14**

- What are some of the concerns that people might have about agricultural biotechnology? **See pp. 402–403**

CHAPTER SUMMARY

18.1 What Is Recombinant DNA?

- **Recombinant DNA** is formed by the combination of two DNA sequences from different sources. **Review Figure 18.1**

- Many **restriction enzymes** make staggered cuts in the two strands of DNA, creating fragments that have **sticky ends** with unpaired bases.

- DNA fragments with sticky ends can be used to create recombinant DNA. DNA molecules from different sources can be cut with the same restriction enzyme and spliced together using **DNA ligase**. **Review Figure 18.2**

18.2 How Are New Genes Inserted into Cells?

- One goal of recombinant DNA technology is to **clone** a particular gene, either for analysis or to produce its protein product in quantity.

- Bacteria, yeasts, and cultured plant and animal cells are commonly used as hosts for recombinant DNA. The insertion of foreign DNA into host cells is called **transformation** or **transfection** (for animal cells). Transformed or transfected cells are called **transgenic** cells.

- Various methods are used to get recombinant DNA into cells. These include chemical or electrical treatment of the cells, the use of viral vectors, and injection. *Agrobacterium tumefaciens* is often used to insert DNA into plant cells.

- To identify host cells that have taken up a foreign gene, the inserted sequence can be tagged with one or more **reporter genes**, which are genetic markers with easily identifiable phenotypes. **Selectable markers** allow for the selective growth of transgenic cells.

- Replication of the foreign gene in the host cell requires that it become part of a segment of DNA that contains a **replicon** (origin and terminus of replication).

- **Vectors** are DNA sequences that can carry new DNA into host cells. Plasmids and viruses are commonly used as vectors. **Review Figure 18.3**

18.3 What Sources of DNA Are Used in Cloning?

- DNA fragments from a genome can be inserted into host cells to create a **genomic library**. **Review Figure 18.6A**

- The mRNAs produced in a certain tissue at a certain time can be extracted and used to create **complementary DNA (cDNA)** by reverse transcription. **Review Figure 18.6B**

- PCR products can be used for cloning.

- Synthetic DNA containing any desired sequence can be made and mutated in the laboratory.

18.4 What Other Tools Are Used to Study DNA Function?

- Homologous recombination can be used to **knock out** a gene in a living organism. **Review Figure 18.7**

- Gene silencing techniques can be used to inactivate the mRNA transcript of a gene, which may provide clues to the gene's function. Artificially created **antisense RNA** or **siRNA** can be added to a cell to prevent translation of a specific mRNA. **Review Figure 18.8**

- **DNA microarray** technology permits the screening of thousands of cDNA sequences at the same time. **Review Figure 18.9, ANIMATED TUTORIAL 18.1**

18.5 What Is Biotechnology?

- **Biotechnology** is the use of living cells to produce materials useful to people. Recombinant DNA technology has resulted in a boom in biotechnology.

- **Expression vectors** allow a transgene to be expressed in a host cell. **Review Figure 18.11, WEB ACTIVITY 18.1**

18.6 How Is Biotechnology Changing Medicine, Agriculture, and the Environment?

- Recombinant DNA techniques have been used to make medically useful proteins. **Review Figure 18.12**

- **Pharming** is the use of transgenic plants or animals to produce pharmaceuticals. **Review Figure 18.13**

- Because recombinant DNA technology has several advantages over traditional agricultural biotechnology, it is being extensively applied to agriculture. **Review Figure 18.14**

- Transgenic crop plants can be adapted to their environments, rather than vice versa.

- **Bioremediation** is the use of organisms, which are often genetically modified, to improve the environment by breaking down pollutants.

- There is public concern about the application of recombinant DNA technology to food production.

SELF-QUIZ

1. Restriction enzymes
   a. play no role in bacteria.
   b. cleave DNA at highly specific recognition sequences.
   c. are inserted into bacteria by bacteriophage.
   d. are made only by eukaryotic cells.
   e. add methyl groups to specific DNA sequences.

2. Which of the following is used as a reporter gene in recombinant DNA work with bacteria as host cells?
   a. rRNA
   b. Green fluorescent protein
   c. Antibiotic sensitivity
   d. Ability to make ornithine
   e. Vitamin synthesis

3. From the list below, select the sequence of steps for inserting a piece of foreign DNA into a plasmid vector, introducing the plasmid into bacteria, and verifying that the plasmid and the foreign gene are present:

   (1) Transform host cells.

   (2) Select for the lack of plasmid reporter gene 1 function.

   (3) Select for the plasmid reporter gene 2 function.

   (4) Digest vector and foreign DNA with a restriction enzyme, which inactivates plasmid reporter gene 1.

   (5) Ligate the digested plasmid together with the foreign DNA.
   a. 4, 5, 1, 3, 2
   b. 4, 5, 1, 2, 3
   c. 1, 3, 4, 2, 5
   d. 3, 2, 1, 4, 5
   e. 1, 3, 2, 5, 4

4. Possession of which feature is *not* desirable in a vector for gene cloning?
   a. An origin of DNA replication
   b. Genetic markers for the presence of the vector
   c. Many recognition sequences for the restriction enzyme to be used
   d. One recognition sequence each for one or more different restriction enzymes
   e. Genes other than the target for transfection

5. RNA interference (RNAi) inhibits
   a. DNA replication.
   b. neither transcription nor translation of specific genes.
   c. recognition of the promoter by RNA polymerase.
   d. transcription of all genes.
   e. translation of specific mRNAs.

6. Complementary DNA (cDNA)
   a. is produced from ribonucleoside triphosphates.
   b. is produced by reverse transcription.
   c. is the "other strand" of single-stranded DNA in a virus.
   d. requires no template for its synthesis.
   e. cannot be placed into a vector because it has the opposite base sequence of the vector DNA.

7. In a genomic library of frog DNA in *E. coli* bacteria,
   a. all bacterial cells have the same sequences of frog DNA.
   b. all bacterial cells have different sequences of frog DNA.
   c. each bacterial cell has a random fragment of frog DNA.
   d. each bacterial cell has many fragments of frog DNA.
   e. the frog DNA is transcribed into mRNA in the bacterial cells.

8. An expression vector requires all of the following except
   a. genes for ribosomal RNA.
   b. a reporter gene.
   c. a promoter of transcription.
   d. an origin of DNA replication.
   e. restriction enzyme recognition sequences.

9. "Pharming" is a term that describes
   a. the use of animals in transgenic research.
   b. plants making genetically altered foods.
   c. synthesis of recombinant drugs by bacteria.
   d. large-scale production of cloned animals.
   e. synthesis of a drug by a transgenic plant or animal.

10. Which of the following could *not* be used to test whether expression of a particular gene is necessary for a particular biological function?
    a. RNAi
    b. Knockout technology
    c. Antisense
    d. Mutant tRNA
    e. Transposon mutagenesis

FOR DISCUSSION

1. Compare PCR (see Section 13.5) and cloning as methods to amplify a gene. What are the requirements, benefits, and drawbacks of each method?

2. As specifically as you can, outline the steps you would take to (a) insert and express the gene for a new, nutritious seed protein in wheat, and (b) insert and express a gene for a human enzyme in sheep's milk.

3. Compare traditional genetic methods with molecular methods for producing genetically altered plants. For each case, describe (a) sources of new genes; (b) numbers of genes transferred; and (c) how long the process takes.

ADDITIONAL INVESTIGATION

Green fluorescent protein (GFP) from a jellyfish can be incorporated into a vector as a reporter gene to signal the presence of the vector in a host cell (see Figure 18.5). How would you alter the technique in Figure 18.4 to substitute GFP for one (or both) of the antibiotic resistance markers?

WORKING WITH DATA (GO TO yourBioPortal.com)

Recombinant DNA In 1973, Stanley Cohen and Herbert Boyer pioneered the field of recombinant DNA technology when they demonstrated that biologically functional recombinant bacterial plasmids can be constructed in the laboratory (Figure 18.1). In this exercise, you will examine their original research article and calculations from their data that show that recombinant DNA was made.

On track with stem cells

In horse racing, bettors speak of the "future book" odds on a horse's chances in an upcoming race. On the morning after winning a race in 2005, the future book odds for Greg's Gold did not look good—he was limping because of a shredded tendon in his right front leg. A tendon is like a rubber band connecting muscles and bones, and tendons in the legs store energy when an animal runs. Typically, a damaged tendon is allowed to heal naturally, but scar tissue makes it less flexible, and a horse cannot run as fast as it did before injuring a tendon. So it looked as if Greg's Gold might have to retire from racing.

Greg's Gold's trainer, David Hofmans, decided to try a new therapy. A veterinarian removed a small amount of adipose (fatty) tissue from the horse's hindquarters and sent it to a cell biology laboratory. There, the tissue was treated with enzymes to digest the extracellular molecules that held the cells together. Several cell populations were obtained, among them mesenchymal stem cells.

Stem cells are actively dividing, unspecialized cells that have the potential to produce different cell types depending on the signals they receive from the body. Mesenchymal stem cells are able to differentiate into various kinds of connective tissue, including bone, cartilage, blood vessels, tendons, and muscle.

Two days after the tissue was taken, Greg's Gold's veterinarian received the stem cells back from the lab and injected them into the site of the damaged tendon. After several months, the tendon healed with little scar tissue, and Greg's Gold's trainer returned him to the racetrack. Greg's Gold raced for almost two more years, winning over $1 million in purse money before being retired.

The mesenchymal stem cell treatment has been used successfully on several thousand horses, and on dogs with arthritis. Most stem cell therapies for humans are still at the experimental stage, particularly in the United States, where controversy over the use of embryonic stem cells has slowed the progress of research and the adoption of therapeutic techniques. But in Japan, women undergoing reconstructive surgery after the removal of breast cancer have had more favorable outcomes when treated with their own mesenchymal stem cells. Bone marrow transplantation is one form of stem cell therapy that has been used successfully for more than thirty years in the United States, to treat patients with cancers such as leukemia and lymphoma.

Greg's Gold Fat stem cells helped repair damage to his tendons and he was able to race—and win—again.

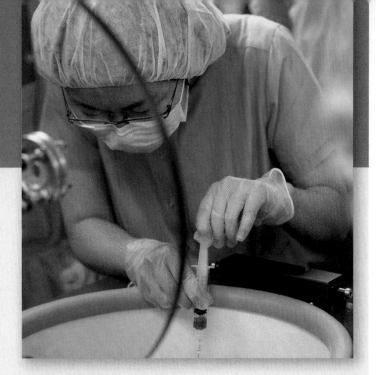

Fat as a Source of Stem Cells This centrifuge separates dense fatty tissues from the lighter stem cells. Stem cells from fat have been found to be capable of differentiating into several specialized cell types.

The processes by which an unspecialized stem cell proliferates and forms specialized cells and tissues with distinctive appearances and functions are similar to the developmental processes that occur in the embryo. Much of our knowledge of developmental biology has come from studies on model organisms such as the fruit fly *Drosophila melanogaster*, the nematode worm *Caenorhabditis elegans*, zebrafish, the mouse, and the small flowering plant *Arabidopsis thaliana*. Eukaryotes share many similar genes, and the cellular and molecular principles underlying their development also turn out to be similar. Thus discoveries from one organism can aid us in understanding other organisms, including ourselves.

IN THIS CHAPTER we begin by describing how almost every cell in a multicellular organism contains all of the genes present in the zygote that gave rise to that organism. Then we explain how cellular changes during development result from the differential expression of those genes. Finally, we show how the various mechanisms of transcriptional control and chemical signaling that are discussed in previous chapters work together to produce a complex organism.

19.1 What Are the Processes of Development?

Development is the process by which a multicellular organism, beginning with a single cell, goes through a series of changes, taking on the successive forms that characterize its life cycle (**Figure 19.1**). After the egg is fertilized, it is called a zygote, and in the earliest stages of development a plant or animal is called an **embryo**. Sometimes the embryo is contained within a protective structure such as a seed coat, an eggshell, or a uterus. An embryo does not photosynthesize or feed itself. Instead, it obtains its food from its mother either directly (via the placenta) or indirectly (by way of nutrients stored in a seed or egg). A series of embryonic stages precedes the birth of the new, independent organism. Many organisms continue to develop throughout their life cycle, with development ceasing only with death.

Development involves distinct but overlapping processes

The developmental changes an organism undergoes as it progresses from an embryo to mature adulthood involve four processes:

- **Determination** sets the developmental *fate* of a cell—what type of cell it will become—even before any characteristics of that cell type are observable. For example, the mesenchymal stem cells described in the opening story look unspecialized, but their fate to become connective tissue cells has already been determined.

- **Differentiation** is the process by which different types of cells arise, leading to cells with specific structures and functions. For example, mesenchymal stem cells differentiate to become muscle, fat, tendon, or other connective tissue cells.

- **Morphogenesis** (Greek for "origin of form") is the organization and spatial distribution of differentiated cells into the multicellular body and its organs.

- **Growth** is the increase in size of the body and its organs by cell division and cell expansion.

Determination and differentiation occur largely because of differential gene expression. The cells that arise from repeated mitoses in the early embryo may look the same superficially, but they soon begin to differ in terms of which of the thousands of genes in the genome are expressed.

Morphogenesis involves differential gene expression and the interplay of signals between cells. Morphogenesis can occur in several ways:

ANIMAL DEVELOPMENT

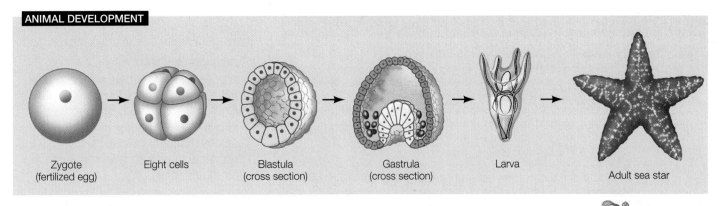

Zygote (fertilized egg) → Eight cells → Blastula (cross section) → Gastrula (cross section) → Larva → Adult sea star

PLANT DEVELOPMENT

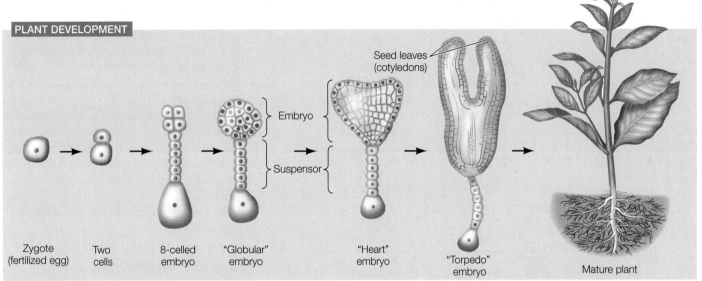

Zygote (fertilized egg) → Two cells → 8-celled embryo → "Globular" embryo → "Heart" embryo → "Torpedo" embryo → Mature plant

Seed leaves (cotyledons)

Embryo

Suspensor

19.1 From Fertilized Egg to Adult The stages of development from zygote to maturity are shown for an animal and for a plant. The blastula is a hollow sphere of cells; the gastrula has three cell layers.

— yourBioPortal.com —
GO TO Web Activity 19.1 • Stages of Development

- *Cell division* is important in both plants and animals.
- *Cell expansion* is especially important in plant development, where a cell's position and shape are constrained by the cell wall.
- *Cell movements* are very important in animal morphogenesis (see Section 44.2)
- *Apoptosis* (programmed cell death) is essential in organ development.

Growth can occur by an increase in the number of cells or by the enlargement of existing cells. Growth continues throughout the individual's life in some organisms, but reaches a more or less stable end point in others.

Cell fates become progressively more restricted during development

During development, each undifferentiated cell will become part of a particular type of tissue—this is referred to as the **cell fate** of that undifferentiated cell. A cell's fate is a function of both differential gene expression and morphogenesis. The role of morphogenesis in determining cell fate was revealed in experiments in which undifferentiated cells were removed from specific locations in early embryos and grafted into new positions on other embryos. The cells were marked with stains so that their development into adult structures could be traced. Such experiments on amphibian embryos indicated that the fates of early embryonic cells are not irrevocably determined, but depend on the cells' environment and stage of development (**Figure 19.2**). In this example, the cells that would have become skin tissue if left in place became brain or notochord tissues, depending on the locations of the grafts.

But as development proceeds from zygote to mature organism, the developmental potential of cells becomes more restricted. For example, if tissue is removed from the brain area of a later-stage frog embryo, it will become brain tissue, even if transplanted to a part of an early-stage embryo that is destined to become another structure.

As we will discuss in this chapter, cell fate determination is influenced by changes in gene expression as well as the extracellular environment. Determination is not something that is visible under the microscope—cells do not change their appearance when they become determined. Determination is followed by differentiation—the actual changes in biochemistry, structure, and function that result in cells of different types. *Determination is a commitment; the final realization of that commitment is differentiation.*

INVESTIGATING LIFE

19.2 Developmental Potential in Early Frog Embryos

In an early embryo, the cells look alike. But marking experiments suggested that the fates of these cells were determined early in development. Was the fate of a cell irrevocable or did it still retain the ability to become a different cell type? To answer this question, biologists transplanted cells from one location in one embryo to a different location in a second embryo. The cells took on the fate of cells at the new location. Therefore, cells in the early embryo retained the ability to form other cell types if placed in the right environment.

HYPOTHESIS The fate of the cells in an early amphibian embryo is irrevocably determined.

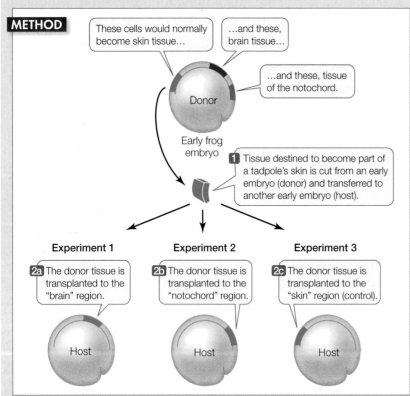

METHOD

These cells would normally become skin tissue...

...and these, brain tissue...

...and these, tissue of the notochord.

Donor

Early frog embryo

1 Tissue destined to become part of a tadpole's skin is cut from an early embryo (donor) and transferred to another early embryo (host).

Experiment 1

2a The donor tissue is transplanted to the "brain" region.

Host

Experiment 2

2b The donor tissue is transplanted to the "notochord" region.

Host

Experiment 3

2c The donor tissue is transplanted to the "skin" region (control).

Host

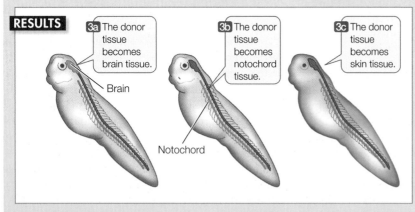

RESULTS

3a The donor tissue becomes brain tissue.

Brain

3b The donor tissue becomes notochord tissue.

Notochord

3c The donor tissue becomes skin tissue.

CONCLUSION The hypothesis is rejected. Cell fates in the early embryo are not determined, but can change depending on the environment.

FURTHER INVESTIGATION: What would happen if tissue from an adult were transplanted into an early embryo?

Go to **yourBioPortal.com** for original citations, discussions, and relevant links for all INVESTIGATING LIFE figures.

19.1 RECAP

Development takes place via the processes of determination, differentiation, morphogenesis, and growth. Cells in the very early embryo have not yet had their fates determined; as development proceeds, their potential fates become more and more restricted.

- What are the four processes of development? **See p. 405**

- Explain what the experiment in Figure 19.2 told us about how cell fates become determined. **See p. 407**

Is a mesophyll cell in a plant leaf or a liver cell in a human being irrevocably committed to that specialization? Under the right experimental circumstances, differentiation is reversible in some cells. The next section describes how the genomes of some cells can be induced to express *different* sets of genes used in differentiation.

19.2 Is Cell Differentiation Irreversible?

A zygote has the ability to give rise to every type of cell in the adult body; in other words, it is **totipotent** (*toti*, "all"; *potent*, "capable"). Its genome contains instructions for all of the structures and functions that will arise throughout the life cycle of the organism. Later in development, the cellular descendants of the zygote lose their totipotency and become determined. These determined cells then differentiate into specialized cells. The human liver cell and the leaf mesophyll cell generally retain their differentiated forms and functions throughout their lives. But this does not necessarily mean that they have irrevocably lost their totipotency. Most of the differentiated cells of an animal or plant have nuclei containing the entire genome of the organism and therefore have the genetic capacity for totipotency. We explore here several examples of how this capacity has been demonstrated experimentally.

Plant cells can be totipotent

A carrot root cell normally faces a dark future. It cannot photosynthesize and generally does not give rise to new carrot plants. However, in 1958 Frederick Steward at Cornell University showed that if he isolated cells from a carrot root and maintained them in a suitable nutrient medium, he could induce them to dedifferentiate—to lose their differentiated characteristics. The cells could divide and give rise to masses of undifferentiated cells called *calli* (singular *callus*), which could be maintained in culture indefinitely. But, if they were provided with the right chemical cues, the cells could develop into embryos and eventually into complete new plants (**Figure 19.3**). Since the new plants were genetically identical to the cells from which they came, they were clones of the original carrot plant.

The ability to clone an entire carrot plant from a differentiated root cell indicated that the cell contained the entire carrot genome, and that under the right conditions, the cell and its descendants could express the appropriate genes in the right sequence to form a new plant. Many types of cells from other plant species show similar behavior in the laboratory. This ability to generate a whole plant from a single cell has been invaluable in agriculture and forestry. For example, trees from planted forests are used in making paper, lumber, and other products. To replace the trees reliably, forestry companies regenerate new trees from the leaves of selected trees with desirable traits. The characteristics of these clones are more uniform and predictable than those of trees grown from seeds.

Nuclear transfer allows the cloning of animals

Animal somatic cells cannot be manipulated as easily as plant cells can. However, experiments such as the one shown in Figure 19.2 have demonstrated the totipotency of early embryonic cells from animals. In humans, this totipotency permits both genetic screening (see Section 15.5) and certain assisted reproductive technologies (see Section 43.4). A human embryo can be isolated in the laboratory and one or a few cells removed and examined to determine whether a certain genetic condition is present. Due to their totipotency, the remaining cells can develop into a complete embryo, which can be implanted into the mother's uterus, where it develops into a normal fetus and infant.

Until recently, it was not possible to induce a cell from a fully developed animal to dedifferentiate and then redifferentiate into another cell type. However, nuclear transfer experiments have shown that the genetic information from an animal cell can be used to create cloned animals. Robert Briggs and Thomas King performed the first such experiments in the 1950s using frog embryos. First they removed the nucleus from an unfertilized egg, forming an *enucleated* egg. Then, with a very fine glass needle, they punctured a cell from an early embryo and drew up part of its contents, including the nucleus, which they injected into the enucleated egg. They stimulated the eggs to divide, and many went on to form embryos, and eventually frogs, that were clones from the original implanted nucleus. These experiments led to two important conclusions:

- No information is lost from the nuclei of cells as they pass through the early stages of embryonic development. This fundamental principle of developmental biology is known as **genomic equivalence**.

- The cytoplasmic environment around a cell nucleus can modify its fate.

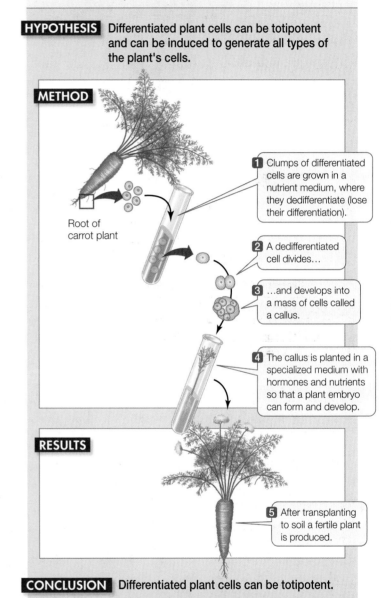

INVESTIGATING LIFE

19.3 Cloning a Plant

When cells were removed from a plant and put into a medium with nutrients and hormones, they lost many of their specialized features—in other words, they dedifferentiated. Did these cells retain the ability to differentiate again? Frederick Steward found that a cultured carrot cell did indeed retain the ability to develop into an embryo and a new plant.

HYPOTHESIS Differentiated plant cells can be totipotent and can be induced to generate all types of the plant's cells.

METHOD

Root of carrot plant

1 Clumps of differentiated cells are grown in a nutrient medium, where they dedifferentiate (lose their differentiation).

2 A dedifferentiated cell divides...

3 ...and develops into a mass of cells called a callus.

4 The callus is planted in a specialized medium with hormones and nutrients so that a plant embryo can form and develop.

RESULTS

5 After transplanting to soil a fertile plant is produced.

CONCLUSION Differentiated plant cells can be totipotent.

Go to **yourBioPortal.com** for original citations, discussions, and relevant links for all INVESTIGATING LIFE figures.

In 1996, Ian Wilmut and his colleagues in Scotland cloned the first mammal by the cell fusion method. To produce donor cells suitable for nuclear transfer, they took differentiated cells from a ewe's udder and starved them of nutrients for a week, halting the cells in the G1 phase of the cell cycle. One of these cells was fused with an enucleated egg from a different breed

TOOLS FOR INVESTIGATING LIFE

19.4 Cloning a Mammal

The experimental procedure described here produced the first cloned mammal, a Dorset sheep named Dolly (shown on the left in the photo). As an adult, Dolly mated and subsequently gave birth to a normal offspring (the lamb on the right), thus proving the genetic viability of cloned mammals.

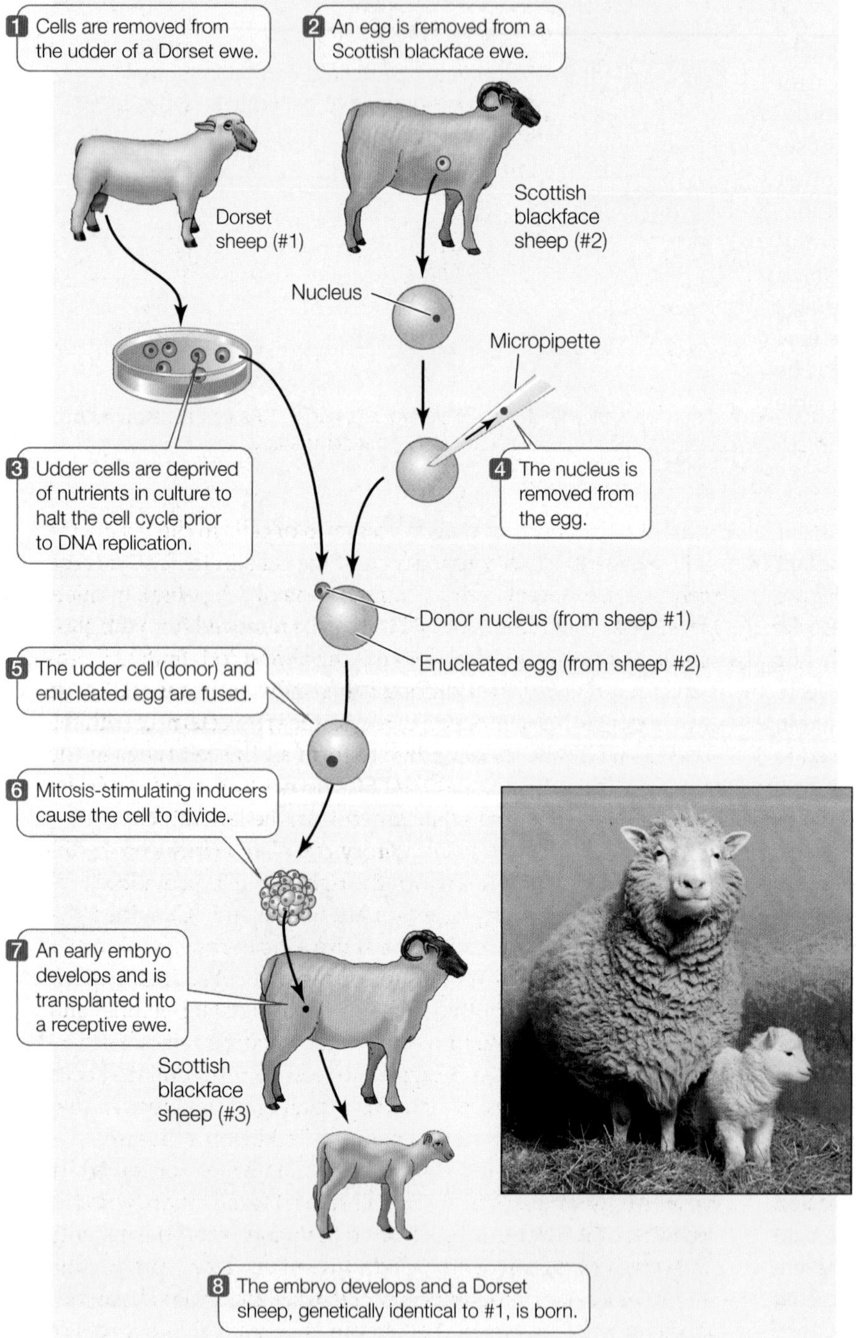

1 Cells are removed from the udder of a Dorset ewe.

2 An egg is removed from a Scottish blackface ewe.

Dorset sheep (#1)

Scottish blackface sheep (#2)

Nucleus

Micropipette

3 Udder cells are deprived of nutrients in culture to halt the cell cycle prior to DNA replication.

4 The nucleus is removed from the egg.

Donor nucleus (from sheep #1)

Enucleated egg (from sheep #2)

5 The udder cell (donor) and enucleated egg are fused.

6 Mitosis-stimulating inducers cause the cell to divide.

7 An early embryo develops and is transplanted into a receptive ewe.

Scottish blackface sheep (#3)

8 The embryo develops and a Dorset sheep, genetically identical to #1, is born.

of ewe. Signals from the egg's cytoplasm stimulated the donor nucleus to enter S phase, and the rest of the cell cycle proceeded normally. After several cell divisions, the resulting early embryo was transplanted into the womb of a surrogate mother (**Figure 19.4**).

Out of 277 successful attempts to fuse adult cells with enucleated eggs, one lamb survived to be born; she was named Dolly, and she became world-famous overnight. DNA analyses confirmed that Dolly's nuclear genes were identical to those of the ewe from whose udder the donor nucleus had been obtained. Dolly grew to adulthood, mated, and produced offspring in the normal manner, thus proving her status as a fully functioning adult animal.

Many other animal species, including cats, dogs, horses, pigs, rabbits, and mice have since been cloned by nuclear transfer. The cloning of animals has practical uses and has given us important information about developmental biology. There are several reasons to clone animals:

- *Expansion of the numbers of valuable animals*: One goal of Wilmut's experiments was to develop a method of cloning transgenic animals carrying genes with therapeutic properties. For example, a cow that was genetically engineered to make human growth hormone in milk has been cloned to produce two more cows that do the same thing. Only 15 such cows would supply the world's need for this medication, which is used to treat short stature due to growth hormone deficiency.

- *Preservation of endangered species*: The banteng, a relative of the cow, was the first endangered animal to be cloned, using a cow enucleated egg and a cow surrogate mother. Cloning may be the only way to save endangered species with low rates of natural reproduction, such as the giant panda.

- *Preservation of pets*: Many people get great personal benefit from pets, and the death of a pet can be devastating. Companies have been set up to clone cats and dogs from cells provided by their owners. Of course, the behavioral characteristics of the beloved pet, which are certainly derived in part from the environment, may not be the same in the cloned pet as in its genetic parent.

Multipotent stem cells differentiate in response to environmental signals

In plants, the growing regions at the tips of the roots and stems contain *meristems*, which are clusters of undifferentiated, rapidly dividing **stem cells**. These cells can give rise to the specialized cell types that make up the various parts of roots and stems. In general, plants have far fewer (15–20) broad cell types than animals (as many as 200).

In mammals, stem cells are found in adult tissues that need frequent cell replacement, such as the skin, the inner lining of the intestine, and the bone marrow, where blood and other types of cells are formed. Canadian cell biologists Ernest McCulloch and James Till discovered mammalian stem cells in the early 1960s when they injected bone marrow cells into adult mice. They noticed that the recipient mice developed small clumps of tissue in the spleen. When they looked more carefully at the clumps, they found that each was composed of undifferentiated stem cells. Before this, stem cells were believed to be present only in animal embryos.

As they divide, stem cells produce daughter cells that differentiate to replace dead cells and maintain the tissues. These adult stem cells in animals are not totipotent, because their ability to differentiate is limited to a relatively few cell types. In other words, they are **multipotent**. For example, there are two types of multipotent stem cells in bone marrow. One type (called hematopoietic stem cells) produces the various kinds of red and white blood cells, while the other type (mesenchymal stem cells) produces the cells that make bone and surrounding *connective* tissues, such as muscle.

The differentiation of multipotent stem cells is "on demand." The blood cells that differentiate in the bone marrow do so in response to specific signals known as growth factors. This is the basis of an important cancer therapy called *hematopoietic stem cell transplantation* (HSCP) (**Figure 19.5**). Because some treatments that kill cancer cells also kill other dividing cells, bone marrow stem cells in patients will die if exposed to these treatments. To circumvent this problem, stem cells are removed from the patient's blood and given growth factors to increase their numbers in the laboratory. The cells are stored during treatment, and then added back to populate the depleted bone marrow when treatment is over. The stored stem cells retain their ability to differentiate in the bone marrow environment. By allowing the use of high doses of treatment to kill tumors, bone marrow transplantation saves thousands of lives each year.

Adjacent cells can also influence stem cell differentiation. We saw this in the opening story of this chapter, in which stem cells from fat differentiated to form cells of the tendon. Bone marrow stem cells that can form muscle will do so if implanted into the heart. Such stem cell transplantation for heart repair has been demonstrated in animals and even in people who had heart attacks, in experiments that used the stem cells to repair a damaged heart. Multipotent stem cells have been found in many organs and tissues, and their use in treating diseases is under intensive investigation.

Pluripotent stem cells can be obtained in two ways

As stated earlier, totipotent stem cells that can form an entire new animal are found only in very early embryos. In both mice and humans, the earliest embryonic stage before differentiation occurs is called a *blastocyst* (see Figure 44.4). Although they

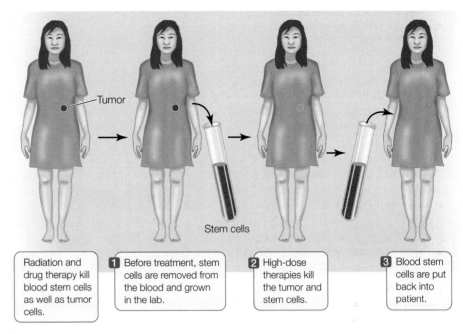

Tumor

Stem cells

	Radiation and drug therapy kill blood stem cells as well as tumor cells.
1	Before treatment, stem cells are removed from the blood and grown in the lab.
2	High-dose therapies kill the tumor and stem cells.
3	Blood stem cells are put back into patient.

19.5 Stem Cell Transplantation Multipotent blood stem cells can be used in hematopoietic stem cell transplantation, to replace stem cells destroyed by cancer therapy.

cannot form an entire embryo, a group of cells in the blastocyst still retains the ability to form all of the cells in the body: these cells are **pluripotent** ("pluri," many; "potent," capable). In mice, these **embryonic stem cells** (**ESCs**) can be removed from the blastocyst and grown in laboratory culture almost indefinitely if provided with the right conditions. When cultured mouse ESCs are injected back into a mouse blastocyst, the stem cells mix with the resident cells and differentiate to form all the cell types in the mouse. This indicates that the ESCs do not lose any of their developmental potential while growing in the laboratory.

ESCs growing in the laboratory can also be induced to differentiate in a particular way if the right signal is provided (**Figure 19.6A**). For example, treatment of mouse ESCs with a derivative of vitamin A causes them to form neurons, while other growth factors induce them to form blood cells. Such experiments demonstrate both the cells' developmental potential and the roles of environmental signals. This finding raises the possibility of using ESC cultures as sources of differentiated cells to repair specific tissues, such as a damaged pancreas in diabetes, or a brain that malfunctions in Parkinson's disease.

ESCs can be harvested from human embryos conceived by *in vitro* ("under glass"—in the laboratory) fertilization, with the consent of the donors. Since more than one embryo is usually conceived in this procedure, embryos not used for reproduction might be available for embryonic stem cell isolation. These cells could then be grown in the laboratory and used as sources of tissues for transplantation into patients with tissue damage. There are two problems with this approach:

- Some people object to the destruction of human embryos for this purpose.

- The stem cells, and tissues derived from them, would provoke an immune response in a recipient (see Chapter 42).

19.6 Two Ways to Obtain Pluripotent Stem Cells   Pluripotent stem cells can be obtained either from human embryos (A) or by adding highly expressed genes to skin cells to transform them into stem cells (B).

yourBioPortal.com

GO TO Animated Tutorial 19.1 • Embryonic Stem Cells

Shinya Yamanaka and coworkers at Kyoto University in Japan have developed another way to produce pluripotent stem cells that gets around these two problems (**Figure 19.6B**). Instead of extracting ESCs from blastocysts, they make **induced pluripotent stem cells (iPS cells)** from skin cells. They developed this method systematically:

1. First, they used gene chips to compare the genes expressed in ESCs with nonstem cells (see Figure 18.9). They found several genes that were uniquely expressed at high levels in ESCs. These genes were believed to be essential to the undifferentiated state and function of stem cells.

2. Next, they isolated the genes and inserted them into a vector for genetic transformation of skin cells (see Section 18.5). They found that the skin cells now expressed the newly added genes at high levels.

3. Finally, they showed that the iPS cells were pluripotent and could be induced to differentiate into many tissues.

Because the iPS cells can be made from skin cells of the individual who is to be treated, an immune response may be avoided. Such cells have already been used for cell therapy in animals for diseases similar to human Parkinson's disease (a brain disorder), diabetes, and sickle cell anemia. Human uses are sure to follow.

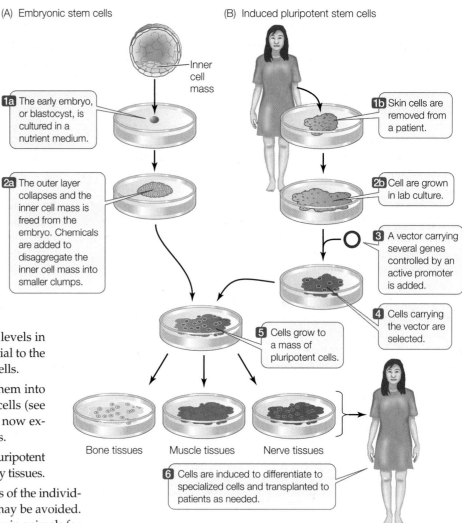

(A) Embryonic stem cells

(B) Induced pluripotent stem cells

1a The early embryo, or blastocyst, is cultured in a nutrient medium.

Inner cell mass

1b Skin cells are removed from a patient.

2a The outer layer collapses and the inner cell mass is freed from the embryo. Chemicals are added to disaggregate the inner cell mass into smaller clumps.

2b Cell are grown in lab culture.

3 A vector carrying several genes controlled by an active promoter is added.

4 Cells carrying the vector are selected.

5 Cells grow to a mass of pluripotent cells.

Bone tissues   Muscle tissues   Nerve tissues

6 Cells are induced to differentiate to specialized cells and transplanted to patients as needed.

19.2 RECAP

Even differentiated cells retain their ability to differentiate into other cell types, given appropriate chemical signals. This has made cloning and stem cell technologies possible.

- Describe the differences between totipotent, pluripotent, and multipotent cells. **See pp. 408–411**

- How are stem cells found in adult body tissues different from embryonic stem cells? **See pp. 410–411**

- What are the two ways to produce pluripotent stem cells? **See pp. 411–412 and Figure19.6**

Cloning experiments and observations of stem cells have shown that most differentiated cells in an organism share the same genes. But not all genes are expressed in every cell. What turns gene expression on and off as cells differentiate? In the next section we explore several of the mechanisms controlling the changes in gene expression that lead to cell differentiation.

19.3 What Is the Role of Gene Expression in Cell Differentiation?

Although every cell contains all the genes needed to produce every protein encoded by its genome, each cell expresses only selected genes. For example, certain cells in hair follicles produce keratin, the protein that makes up hair, while other cell types in the body do not. What determines whether a cell will produce keratin? Chapter 16 describes a number of ways in which cells regulate gene expression and protein production—by controlling transcription, translation, and posttranslational protein modifications. But the mechanisms that control gene expression resulting in cell differentiation generally work at the level of transcription.

Differential gene transcription is a hallmark of cell differentiation

The gene for β-globin, one of the protein components of hemoglobin, is expressed in red blood cells as they form in the bone marrow of mammals. That this same gene is also present—but unexpressed—in neurons in the brain (which do not make hemoglobin) can be demonstrated by nucleic acid hybridization. Recall that in nucleic acid hybridization, a probe made of single-stranded DNA or RNA of known sequence is added to denatured DNA to reveal complementary coding regions on the DNA template strand (see Figure 15.16). A probe for the β-globin gene can be applied to DNA from brain cells and immature red blood cells (recall that mature mammalian red blood cells lose their nuclei during development). In both cases, the probe finds its complement, showing that the β-globin gene is present in both types of cells. On the other hand, if the β-globin probe is applied to mRNA, rather than DNA, from the two cell types, it finds β-globin mRNA only in the red blood cells, not in the brain cells. This result shows that the gene is expressed in only one of the two cell types.

What leads to this differential gene expression? One well-studied example of cell differentiation is the conversion of undifferentiated muscle precursor cells into cells that are destined to form muscle (**Figure 19.7**). In the vertebrate embryo these cells come from a layer called the *mesoderm* (see Section 44.2). A key event in the commitment of these cells to become muscle is that they stop dividing. Indeed, in many parts of the embryo, *cell division and cell differentiation are mutually exclusive.* Cell signaling activates the gene for a transcription factor called **MyoD**

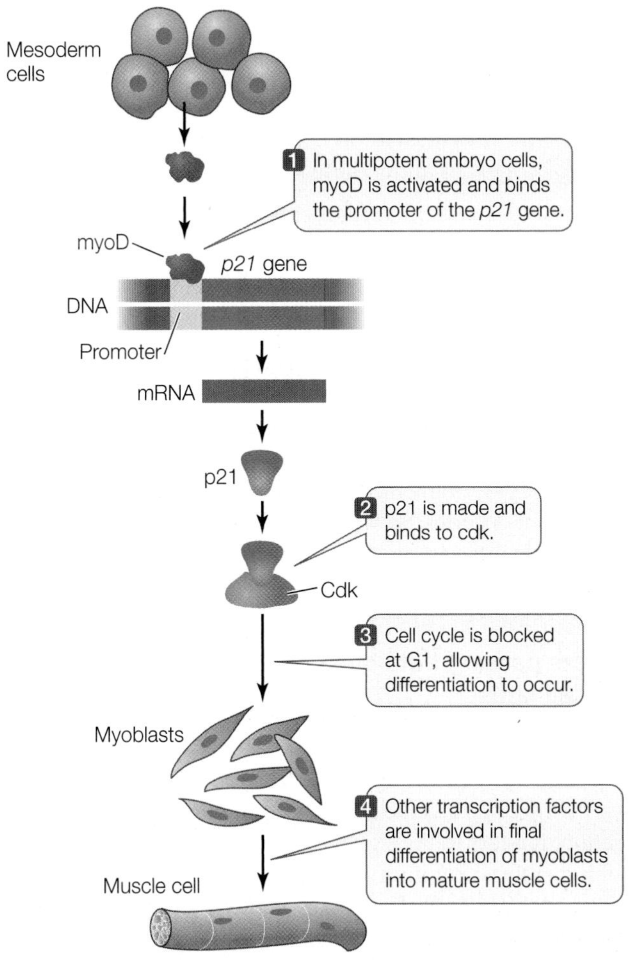
Mesoderm cells
1 In multipotent embryo cells, myoD is activated and binds the promoter of the *p21* gene.
myoD
p21 gene
DNA
Promoter
mRNA
p21
2 p21 is made and binds to cdk.
Cdk
3 Cell cycle is blocked at G1, allowing differentiation to occur.
Myoblasts
4 Other transcription factors are involved in final differentiation of myoblasts into mature muscle cells.
Muscle cell

19.7 Transcription and Differentiation in the Formation of Muscle Cells Activation of a transcription factor, MyoD, is important in muscle cell differentiation.

(*myo*blast-determining gene). Recall that transcription factors are DNA binding proteins that regulate the expression of specific genes. In this case, MyoD activates the gene for p21, an inhibitor of cyclin-dependent kinases that normally stimulate the cell cycle at G1 (see Figure 11.6). Expression of the *p21* gene causes the cell cycle to stop, and other transcription factors then enter the picture so that differentiation can begin. Interestingly, *myoD* is also activated in stem cells that are present in adult muscle, indicating a role of this transcription factor in repair of muscle as it gets damaged and worn out.

Genes such as *myoD* that direct the most fundamental decisions in development (often by regulating other genes on other chromosomes) usually encode transcription factors. In some cases, a single transcription factor can cause a cell to differentiate in a certain way. In others, complex interactions between genes and proteins determine a sequence of transcriptional events that leads to differential gene expression.

19.3 RECAP

Differentiation involves selective gene expression, controlled at the level of transcription by transcription factors.

- What techniques could you use to identify genes expressed during cell differentiation? **See pp. 412–413**
- What is the role of transcription factors in controlling differentiation? **See p. 413 and Figure 19.7**

Cell differentiation involves extensive transcriptional regulation of genes. But what causes a cell to express one set of genes, and not some other set? In other words, how is a cell's fate determined?

19.4 How Is Cell Fate Determined?

The fertilized egg undergoes many cell divisions to produce the many differentiated cells in the body (such as liver, muscle, and nerve cells). How can one cell produce so many different cell types? There are two ways that this occurs:

- **Cytoplasmic segregation** (unequal cytokinesis). A factor within an egg, zygote, or precursor cell may be unequally distributed in the cytoplasm. After cell division, the factor ends up in some daughter cells or regions of cells, but not others.
- **Induction** (cell-to-cell communication). A factor is actively produced and secreted by certain cells to induce other cells to become determined.

Cytoplasmic segregation can determine polarity and cell fate

Some differences in gene expression patterns are the result of *cytoplasmic* differences between cells. One such cytoplasmic difference is the emergence of distinct "top" and "bottom" ends of an organism or structure; such a difference is called **polarity**.

INVESTIGATING LIFE

19.8 Asymmetry in the Early Sea Urchin Embryo

As an embryo develops, cells become determined and their ultimate fate gets more and more narrowly defined. The cells of an eight-celled sea urchin embryo look identical and so might be expected to have the same developmental potential. But do they? Hans Driesch separated different parts of this tiny embryo from one another, to examine their developmental potentials. His experiments showed that even at the eight-cell stage, cell fate determination is underway.

HYPOTHESIS Different regions in the fertilized egg and the embryo have different developmental fates.

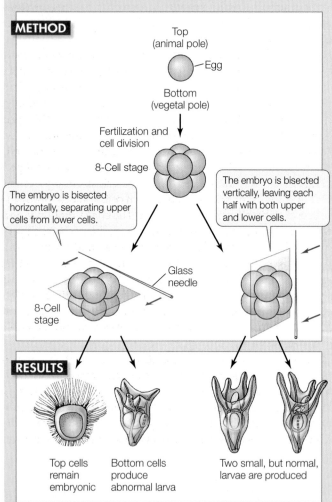

METHOD

Top (animal pole)

Egg

Bottom (vegetal pole)

Fertilization and cell division

8-Cell stage

The embryo is bisected horizontally, separating upper cells from lower cells.

The embryo is bisected vertically, leaving each half with both upper and lower cells.

Glass needle

8-Cell stage

RESULTS

Top cells remain embryonic

Bottom cells produce abnormal larva

Two small, but normal, larvae are produced

CONCLUSION The upper and lower halves of a sea urchin embryo differ in their developmental potential.

Go to **yourBioPortal.com** for original citations, discussions, and relevant links for all INVESTIGATING LIFE figures.

yourBioPortal.com

GO TO Animated Tutorial 19.2 • Early Asymmetry in the Embryo

Many examples of polarity are observed as development proceeds. Our heads are distinct from our rear ends, and the distal (far) ends of our arms and legs (wrists, ankles, fingers, toes) differ from the proximal (near) ends (shoulders and hips). Polarity may develop early; even within the fertilized egg, the yolk and other factors are often distributed asymmetrically. During early development in animals, polarity is specified by an *animal pole* at the top of the zygote and a *vegetal pole* at the bottom.

In the early twentieth century Hans Driesch at the Marine Biological Station in Naples, Italy, demonstrated the effects of cytoplasmic segregation on development (**Figure 19.8**). Very early development in sea urchins occurs by rapid, equal mitotic divisions of the fertilized egg; there is no increase in size at this stage, just a partitioning of the cells. If an eight-cell embryo is carefully separated vertically into two four-celled halves, both halves develop into normal (albeit small) larvae. But if an eight-cell embryo is cut horizontally, the top half does not develop at all, while the bottom half develops into a small, abnormal larva.

Clearly, then, there must be at least one factor essential for development that is segregated in the vegetal half of the sea urchin egg, such that the bottom cells of the 8-cell embryo get this essential factor and the top cells do not. Experiments have established that certain materials, called **cytoplasmic determinants**, are distributed unequally in the egg cytoplasm. Cytoplasmic determinants play roles in directing the embryonic development of many organisms (**Figure 19.9**). What are these determinants and what accounts for their unequal distribution?

The cytoskeleton contributes to the asymmetric distribution of cytoplasmic determinants in the egg. Recall from Section 5.3 that an important function of the microtubules and microfilaments in the cytoskeleton is to help move materials in the cell. Two properties allow these structures to accomplish this:

• Microtubules and microfilaments have polarity—they grow by adding subunits to the plus end.

• Cytoskeletal elements can bind specific proteins, which can be used in the transport of mRNA.

For example, in the sea urchin egg, a protein binds to both the growing (+) end of a microfilament and to an mRNA encoding a cytoplasmic determinant. As the microfilament grows toward one end of the cell, it carries the mRNA along with it. The asymmetrical distribution of the mRNA leads to a similar distribution of the protein it encodes. So what were once unspecified cytoplasmic determinants can now be defined in terms of cellular structures, mRNAs, and proteins.

Inducers passing from one cell to another can determine cell fates

The term "induction" has different meanings in different contexts. In biology it can be used broadly to refer to the initiation of, or cause of, a change or process. But in the context of cellular differentiation, it refers to the signaling events by which cells in a developing embryo communicate and influence one another's developmental fate. Induction involves chemical signals and signal transduction mechanisms. We will describe two examples of this form of induction: one in the developing vertebrate eye, and the other in a developing reproductive structure of the nematode *Caenorhabditis elegans*.

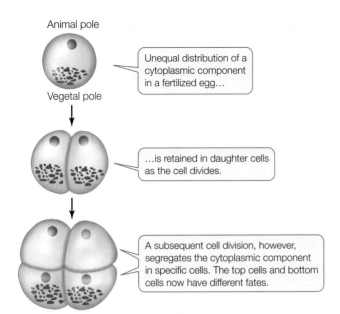

19.9 The Principle of Cytoplasmic Segregation The unequal distribution of some component in the cytoplasm of a cell may determine the fates of its descendants.

LENS DIFFERENTIATION IN THE VERTEBRATE EYE The development of the lens in the vertebrate eye is a classic example of induction. In a frog embryo, the developing forebrain bulges out at both sides to form the *optic vesicles*, which expand until they come into contact with the cells at the surface of the head (**Figure 19.10**). The surface tissue in the region of contact thickens, forming a *lens placode*—tissue that will ultimately form the lens. The lens placode bends inward, folds over on itself, and ultimately detaches from the surface tissue to produce a structure that will develop into the lens. If the growing optic vesicle is cut away before it contacts the surface cells, no lens forms. Placing an impermeable barrier between the optic vesicle and the surface cells also prevents the lens from forming. These observations suggest that the surface tissue begins to develop into a lens when it receives a signal from the optic vesicle. Such a signal is termed an **inducer**.

Inducers trigger sequences of gene expression in the responding cells. How cells switch on different sets of genes that govern development and direct the formation of body plans is of great interest to developmental and evolutionary biologists. They use model organisms to investigate the major principles governing these processes.

VULVAL DIFFERENTIATION IN THE NEMATODE The tiny nematode *Caenorhabditis elegans* is a favorite model organism for studying development. Its genome was one of the first eukaryotic genomes to be sequenced (see Section 17.3). It develops from fertilized egg to larva in only about 8 hours, and the worm reaches the adult stage in just 3.5 days. The process is easily observed using a low-magnification dissecting microscope because the body covering is transparent (**Figure 19.11A**).

The adult nematode is *hermaphroditic*, containing both male and female reproductive organs. It lays eggs through a pore called the *vulva* on the ventral (lower) surface. During development, a single cell, called the *anchor cell*, induces the vulva to form from six cells on the worm's ventral surface. In this case, there are two molecular signals, the *primary inducer* and the *secondary inducer* (or *lateral signal*). Each of the six ventral cells has three possible fates: it may become a primary vulval precursor cell, a secondary vulval precursor cell, or simply become part of the worm's skin—an epidermal cell. You can follow the sequence of events in **Figure 19.11B**. The concentration gradient of the primary inducer, LIN-3, is key: the anchor cell produces LIN-3, which diffuses out of the cell and forms a concentration gradient with respect to adjacent cells. Cells that receive the most LIN-3 become vulval precursor cells; cells slightly farther from the anchor cell receive less LIN-3 and become epidermal cells. Induction involves the activation or inactivation of specific genes through signal transduction cascades in the responding cells (**Figure 19.12**).

Nematode development illustrates the important observation that *much of development is controlled by molecular switches that allow a cell to proceed down one of two alternative tracks*. One challenge for developmental biologists is to find these switches and determine how they work. The primary inducer, LIN-3, released by the *C. elegans* anchor cell is a growth factor homologous to a ver-

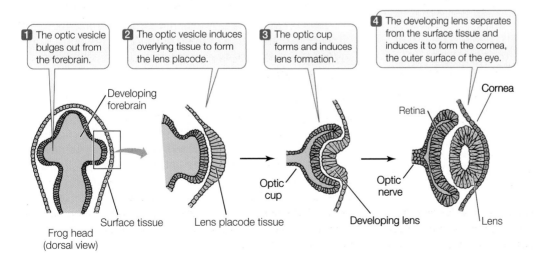

1 The optic vesicle bulges out from the forebrain.

2 The optic vesicle induces overlying tissue to form the lens placode.

3 The optic cup forms and induces lens formation.

4 The developing lens separates from the surface tissue and induces it to form the cornea, the outer surface of the eye.

Developing forebrain

Surface tissue

Frog head (dorsal view)

Lens placode tissue

Optic cup

Developing lens

Retina

Optic nerve

Cornea

Lens

19.10 Embryonic Inducers in Vertebrate Eye Development The eye of a frog develops as different cells induce changes in neighboring cells.

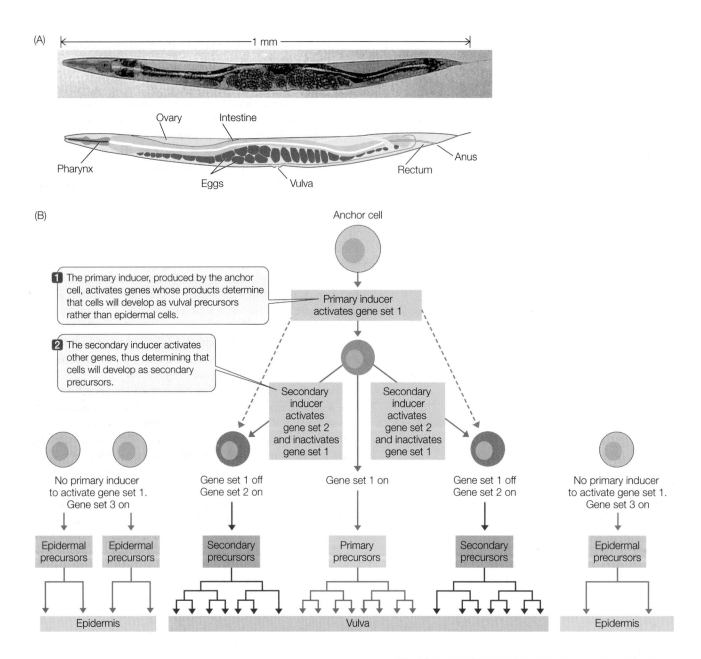

19.11 Induction during Vulval Development in *Caenorhabditis elegans* (A) In the nematode *C. elegans*, it has been possible to follow all of the cell divisions from the fertilized egg to the 959 cells found in the fully developed adult. (B) During vulval development, a molecule secreted by the anchor cell (the LIN-3 protein) acts as the primary inducer. The primary precursor cell (the one that received the highest concentration of LIN-3) then secretes a secondary inducer (the lateral signal) that acts on its neighbors. The gene expression patterns triggered by these molecular switches determine cell fates.

tebrate growth factor called EGF (*epidermal growth factor*). LIN-3 binds to a receptor on the surfaces of vulval precursor cells, setting in motion a signal transduction cascade involving the Ras protein and MAP kinases (see Figure 7.12). This results in increased transcription of the genes involved in the differentiation of vulval cells.

19.4 RECAP

Cellular differentiation involves cytoplasmic segregation and induction. Cytoplasmic segregation is the unequal distribution of gene products in the egg, zygote, or early embryo. Induction occurs when one cell or tissue sends a chemical signal to another.

- How does cytoplasmic segregation result in polarity in a fertilized egg, and how does polarity affect cell differentiation? **See pp. 413–414 and Figure 19.9**

- Describe an example of how induction influences tissue formation in the vertebrate eye. **See p. 416 and Figure 19.10**

- How do inducer molecules interact with transcription factors to produce differentiated cells? **See p. 416 and Figure 19.12**

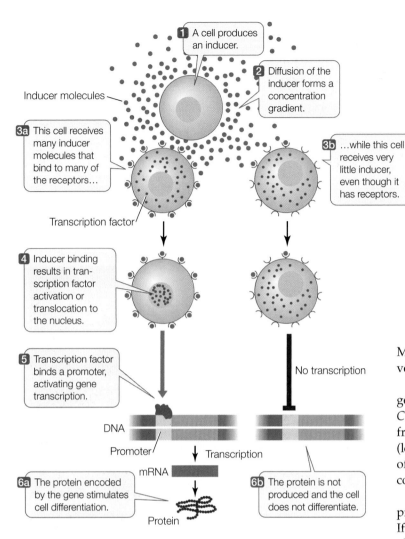

1 A cell produces an inducer.

2 Diffusion of the inducer forms a concentration gradient.

Inducer molecules

3a This cell receives many inducer molecules that bind to many of the receptors...

3b ...while this cell receives very little inducer, even though it has receptors.

Transcription factor

4 Inducer binding results in transcription factor activation or translocation to the nucleus.

5 Transcription factor binds a promoter, activating gene transcription.

No transcription

DNA

Promoter

Transcription

mRNA

6a The protein encoded by the gene stimulates cell differentiation.

6b The protein is not produced and the cell does not differentiate.

Protein

19.12 Induction The concentration of an inducer directly affects the degree to which a transcription factor is activated. The inducer acts by binding to a receptor on the target cell. This binding is followed by signal transduction involving transcription factor activation or translocation from the cytoplasm to the nucleus. In the nucleus it acts to stimulate the expression of genes involved in cell differentiation.

We have seen that cytoplasmic segregation and induction lead to cell differentiation, and have seen two examples of how these processes lead to organ formation in developing multicellular organisms. We now take a closer look at how gene expression affects differentiation and development.

19.5 How Does Gene Expression Determine Pattern Formation?

Pattern formation is the process that results in the spatial organization of a tissue or organism. It is inextricably linked to morphogenesis, the creation of body form. You might expect morphogenesis to involve a lot of cell division, followed by differentiation—and it does. But what you might not expect is the amount of programmed cell death—apoptosis—that occurs during morphogenesis.

Multiple genes interact to determine developmental programmed cell death

We noted in Section 11.6 that apoptosis is a programmed series of events that leads to cell death. Apoptosis is an integral part of the normal development and life of an organism. For example, in an early human embryo, the hands and feet look like tiny paddles: the tissues that will become fingers and toes are linked by connective tissue. Between days 41 and 56 of development, the cells between the digits die, freeing the individual fingers and toes:

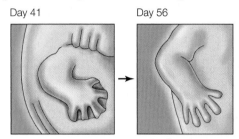

Day 41   Day 56

Many cells and structures form and then disappear during development, in processes involving apoptosis.

Model organisms have been very useful in studying the genes involved in apoptosis. For example, the nematode worm *C. elegans* produces precisely 1,090 somatic cells as it develops from a fertilized egg into an adult, but 131 of those cells die (leaving 959 cells in the adult worm). The sequential expression of two genes called *ced-4* and *ced-3* (for *c*ell *d*eath) appears to control this programmed cell death (**Figure 19.13**).

In the nematode nervous system, 302 neurons come from 405 precursors; thus 103 neural precursor cells undergo apoptosis. If the protein encoded by either *ced-3* or *ced-4* is nonfunctional, *all* 405 cells form neurons, resulting in abnormal brain develop-

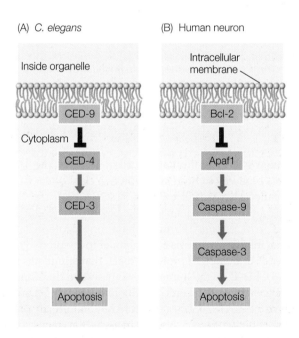

(A) *C. elegans*

(B) Human neuron

Inside organelle

Intracellular membrane

CED-9

Bcl-2

Cytoplasm

CED-4

Apaf1

CED-3

Caspase-9

Caspase-3

Apoptosis

Apoptosis

19.13 Pathways for Apoptosis In the worm *C. elegans* (A) and humans (B) similar pathways for apoptosis are controlled by genes with similar sequences and functions.

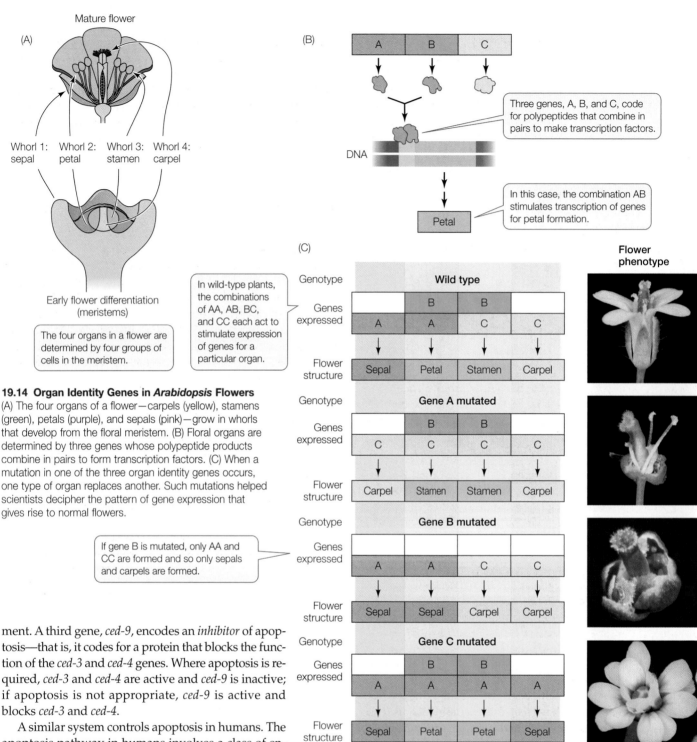

Mature flower

(A)

Whorl 1: sepal Whorl 2: petal Whorl 3: stamen Whorl 4: carpel

Early flower differentiation (meristems)

The four organs in a flower are determined by four groups of cells in the meristem.

(B)

| A | B | C |

Three genes, A, B, and C, code for polypeptides that combine in pairs to make transcription factors.

DNA

In this case, the combination AB stimulates transcription of genes for petal formation.

Petal

(C)

In wild-type plants, the combinations of AA, AB, BC, and CC each act to stimulate expression of genes for a particular organ.

If gene B is mutated, only AA and CC are formed and so only sepals and carpels are formed.

Flower phenotype

Wild type

Genotype				
Genes expressed		B	B	
	A	A	C	C
Flower structure	Sepal	Petal	Stamen	Carpel

Gene A mutated

Genotype				
Genes expressed		B	B	
	C	C	C	C
Flower structure	Carpel	Stamen	Stamen	Carpel

Gene B mutated

Genotype				
Genes expressed				
	A	A	C	C
Flower structure	Sepal	Sepal	Carpel	Carpel

Gene C mutated

Genotype				
Genes expressed		B	B	
	A	A	A	A
Flower structure	Sepal	Petal	Petal	Sepal

Whorl 1 Whorl 2 Whorl 3 Whorl 4

19.14 Organ Identity Genes in *Arabidopsis* Flowers
(A) The four organs of a flower—carpels (yellow), stamens (green), petals (purple), and sepals (pink)—grow in whorls that develop from the floral meristem. (B) Floral organs are determined by three genes whose polypeptide products combine in pairs to form transcription factors. (C) When a mutation in one of the three organ identity genes occurs, one type of organ replaces another. Such mutations helped scientists decipher the pattern of gene expression that gives rise to normal flowers.

ment. A third gene, *ced-9*, encodes an *inhibitor* of apoptosis—that is, it codes for a protein that blocks the function of the *ced-3* and *ced-4* genes. Where apoptosis is required, *ced-3* and *ced-4* are active and *ced-9* is inactive; if apoptosis is not appropriate, *ced-9* is active and blocks *ced-3* and *ced-4*.

A similar system controls apoptosis in humans. The apoptosis pathway in humans involves a class of enzymes called caspases (see Figure 11.22), which are similar in amino acid sequence to the protein encoded by *ced-3* in *C. elegans*. Humans have one protein, Bcl-2, that inhibits apoptosis and is similar to the product of *ced-9*, and another protein, Apaf1, that stimulates apoptosis like CED-4. As the human nervous system develops, half of the neurons that are formed undergo apoptosis. So humans and nematodes, two species separated by more than 600 million years of evolutionary history, have similar genes controlling programmed cell death (see Figure 19.13). The commonality of this pathway indicates its importance: mutations are harmful and evolution selects against them.

Plants have organ identity genes

Like animals, plants have organs—for example, leaves and roots. Many plants form flowers, and many flowers are composed of four types of organs: sepals, petals, stamens (male reproductive organs), and carpels (female reproductive organs). These floral organs occur in concentric *whorls*, with groups of each organ type encircling a central axis. The sepals are on the outside and the carpels are on the inside (**Figure 19.14A**).

In the model plant *Arabidopsis thaliana* (thale cress), flowers develop in a radial pattern around the shoot apex as it develops and elongates. The whorls develop from a *meristem* of about 700 undifferentiated cells arranged in a dome, which is at the growing point on the stem. How is the identity of a particular whorl determined? A group of genes called **organ identity genes** encode proteins that act in combination to produce specific whorl features (**Figure 19.14B and C**):

- Genes in class A are expressed in whorls 1 and 2 (which form sepals and petals, respectively).
- Genes in class B are expressed in whorls 2 and 3 (which form petals and stamens).
- Genes in class C are expressed in whorls 3 and 4 (which form stamens and carpels).

Two lines of experimental evidence support this model of organ identity gene function:

- *Loss-of-function mutations*: for example, a mutation in a class A gene results in no sepals or petals. In any organism, the replacement of one organ for another is called *homeosis*, and this type of mutation is a **homeotic mutation** (see Figure 19.14C).
- *Gain-of-function mutations*: for example, a promoter for a class C gene can be artificially coupled to a class A gene. In this case, the class A gene is expressed in all four whorls, resulting in only sepals and petals.

Genes in classes A, B, and C code for transcription factors that are active as dimers, that is, proteins with two polypeptide subunits. Gene regulation in these cases is *combinatorial*—that is, the composition of the dimer determines which genes will be activated. For example, a dimer made up of two class A monomers activates transcription of the genes that make sepals; a dimer made up of A and B monomers results in petals, and so forth. A common feature of the A, B, and C proteins, as well as many other plant transcription factors, is a DNA-binding domain called the **MADS box**. These proteins also have domains that can bind to other proteins in a *transcription initiation complex*. As we discuss in Chapter 16, transcription initiation in eukaryotes is controlled by a complex of proteins that interact with DNA and other proteins at the promoter. The MADS box proteins participate in this complex to control the expression of specific genes.

Some familiar ornamental plants have mutations in floral organ identity genes. For example, many rose varieties have mutations in a C gene, resulting in multiple rows of petals instead of the single set of five petals found in wild roses. An understanding of the molecular basis of floral organ identity may have practical uses. Many of the foods that make up the human diet come from fruits and seeds, which form from parts of the carpel—the female reproductive organ of the flower. Genetically modifying plants to produce more carpels could increase the amount of fruit or grain a crop produces. A genetic system similar to the one described here for *Arabidopsis* controls floral organ formation in rice, humanity's most widely consumed plant. Appropriate mutations in these genes might lead to more grain produced per plant.

Transcription of the floral organ identity genes is controlled by other gene products, including the LEAFY protein. Plants with loss-of-function mutations in the *LEAFY* gene make flowering stems instead of flowers, with increased numbers of modified leaves called bracts. The wild-type LEAFY protein acts as a transcription factor, stimulating expression of the class A, B, and C genes so that they produce flowers. This finding, too, has practical applications. It usually takes 6–20 years for a citrus tree to produce flowers and fruits. Scientists have made transgenic orange trees expressing the *LEAFY* gene coupled to a strongly expressed promoter. These trees flower and fruit years earlier than normal trees.

Morphogen gradients provide positional information

During development, the key cellular question, "What am I (or what will I be)?" is often answered in part by "Where am I?" Think of the cells in the developing nematode, which develop into different parts of the vulva depending on their positions relative to the anchor cell. This spatial "sense" is called **positional information**. Positional information often comes in the form of an inducer called a **morphogen**, which diffuses from one group of cells to surrounding cells, setting up a concentration gradient. There are two requirements for a signal to be considered a morphogen:

- *It must directly affect target cells, rather than triggering a secondary signal that affects target cells.*
- *Different concentrations of the signal must cause different effects.*

Developmental biologist Lewis Wolpert uses the "French flag model" to explain morphogens (**Figure 19.15**). This model can be applied to the differentiation of the vulva in *C. elegans* (see Figure 19.11) and to the development of vertebrate limbs.

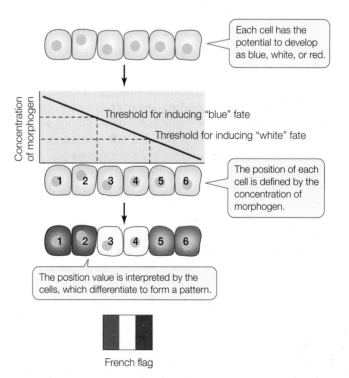

Each cell has the potential to develop as blue, white, or red.

Concentration of morphogen

Threshold for inducing "blue" fate

Threshold for inducing "white" fate

1 2 3 4 5 6

The position of each cell is defined by the concentration of morphogen.

1 2 3 4 5 6

The position value is interpreted by the cells, which differentiate to form a pattern.

French flag

19.15 The French Flag Model In the "French flag" model, a concentration gradient of a diffusible morphogen signals each cell to specify its position.

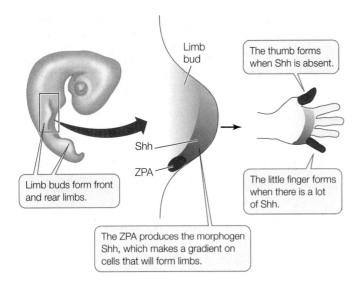

Limb bud

The thumb forms when Shh is absent.

Limb buds form front and rear limbs.

Shh

ZPA

The little finger forms when there is a lot of Shh.

The ZPA produces the morphogen Shh, which makes a gradient on cells that will form limbs.

19.16 Specification of the Vertebrate Limb and the French Flag Model The zone of polarizing activity (ZPA) in the limb bud of the embryo secretes the morphogen *Sonic hedgehog* (Shh). Cells in the bud form different digits depending on the concentration of Shh.

- First, cells in the mother that are adjacent to the maturing egg release products that set up anterior–posterior and dorsal–ventral axes in the egg.

- Next, a series of gene products in the embryo successively define the position of each cell in a segment relative to these axes. For example, a cell might first be defined as being in the head rather than in the abdomen in the anterior–posterior axis; then it might be defined as being on the ventral (top) side of the head.

- Finally, a set of genes called *Hox genes* control the ultimate identity of each body part; for example, determining that the cells at a particular position in the head will make mouthparts.

The genes involved in each of these steps code for transcription factors, which in turn control the synthesis of other transcription factors acting on the next set of genes. This cascade of events may remind you of a signal transduction cascade (see Section 7.3), only in this case it is a *cascade of events that occurs over time and location*, rather than abruptly and in a single cell. The genes finally expressed are the ones familiar to you: they code for protein kinases, receptors, and other proteins that carry out the functions of the cell.

The description of these events in fruit fly development is one of the great achievements in modern biology. It gave biologists a deep understanding of how the events that specify cell identity unfold. We will only skim the surface of the process here, but keep in mind the basic principle of a transcriptional cascade. As we will see in Chapter 20, the fruit fly has been a true model organism in this case, because these findings have informed research on other organisms, including mammals.

Experimental genetics was used to elucidate the events leading to cell fate determination in *Drosophila*:

- First, developmental mutations were identified. For example, a mutant strain might produce larvae with two heads or no segments.

- Then, the mutant was compared with wild-type flies, and the gene responsible for the developmental mistake, and its protein product (if appropriate), were isolated.

- Finally, experiments with the gene (making transgenic flies) and protein (injecting the protein into an egg or into an embryo) were done to confirm the proposed developmental pathway.

Together, these approaches revealed a sequential pattern of gene expression that results in the determination of each segment within 24 hours after fertilization. Several classes of genes are involved.

The vertebrate limb develops from a paddle-shaped *limb bud* (**Figure 19.16**). The cells that develop into different digits must receive positional information; if they do not, the limb will be totally disorganized (imagine a hand with only thumbs or only little fingers). A group of cells at the posterior base of the limb bud, just where it joins the body wall, is called the *zone of polarizing activity* (ZPA). The cells of the ZPA secrete a morphogen called *Sonic hedgehog* (Shh), which forms a gradient that determines the posterior–anterior (little finger to thumb) axis of the developing limb. The cells getting the highest dose of Shh form the little finger; those getting the lowest dose develop into the thumb. Recall the French flag model when considering the gradient of Shh.

A cascade of transcription factors establishes body segmentation in the fruit fly

Perhaps the best-studied example of how gene expression affects cell fate in response to morphogens is body segmentation in the fruit fly *Drosophila melanogaster*. The body segments of this model organism are clearly different from one another. The adult fly has an anterior *head* (composed of several fused segments), three different *thoracic* segments, and eight *abdominal* segments at the posterior end. Each segment develops into different body parts: for example, antennae and eyes develop from head segments, wings from the thorax, and so on.

The life cycle of *Drosophila* from fertilized egg to adult takes about 2 weeks at room temperature. The egg hatches into a larva, which then forms a pupa, which finally is transformed into the adult fly. By the time a larva appears—about 24 hours after fertilization—there are recognizable segments. The thoracic and abdominal segments all look similar, but *the fates of the cells to become different adult segments is already determined*. The determination events in the first 24 hours will be our focus here.

Several types of genes are expressed sequentially in the embryo to define these segments:

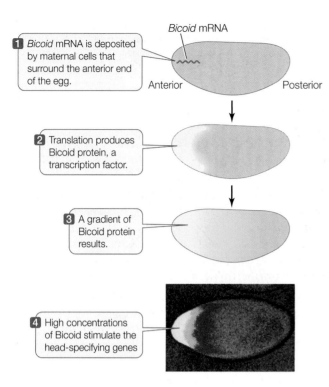

1. *Bicoid* mRNA is deposited by maternal cells that surround the anterior end of the egg.

Bicoid mRNA

Anterior   Posterior

2. Translation produces Bicoid protein, a transcription factor.

3. A gradient of Bicoid protein results.

4. High concentrations of Bicoid stimulate the head-specifying genes

19.17 Bicoid Protein Provides Positional Information The anterior–posterior axis of *Drosophila* arises from the gradient of a morphogen encoded by *Bicoid*, a maternal effect gene. Bicoid protein is also a transcription factor, which activates a gene to specify that the anterior region will become the head of the fly. Other maternal effect genes in the posterior region of the embryo inhibit Bicoid, thus limiting its activity in that region.

How did biologists elucidate these pathways? Let's look more closely at the experimental approaches used in this case.

- Females that are homozygous for a particular *bicoid* mutation produce larvae with no head and no thorax; thus the Bicoid protein must be needed for the anterior structures to develop.

- If the eggs of these *bicoid* mutants are injected at the anterior end with cytoplasm from the anterior region of a wild-type egg, the injected eggs develop into normal larvae. This experiment also shows that the Bicoid protein is involved in the development of anterior structures.

- If cytoplasm from the anterior region of a wild-type egg is injected into the posterior region of another egg, anterior structures develop there. The degree of induction depends on how much cytoplasm is injected.

- Eggs from homozygous *nanos* mutant females develop into larvae with missing abdominal segments.

- If cytoplasm from the posterior region of a wild-type egg is injected into the posterior region of a *nanos* mutant egg, it will develop normally.

These and other experiments led scientists to understand the cascade of events that determine cell fates.

The events involving *Bicoid*, *Nanos*, and *Hunchback* begin before fertilization and continue after it, during the multinucleate stage, which lasts a few hours. At this stage, the embryo looks like a bunch of indistinguishable nuclei under the light microscope. But the cell fates have already begun to be determined. After the anterior and posterior ends have been established, the next step in pattern formation is the determination of segment number and locations.

MATERNAL EFFECT GENES Like the eggs and early embryos of sea urchins, *Drosophila* eggs and larvae are characterized by unevenly distributed cytoplasmic determinants (see Figure 19.9). These molecular determinants, which include both mRNAs and proteins, are the products of specific **maternal effect genes**. These genes are transcribed in the cells of the mother's ovary that surround what will be the anterior portion of the egg. The transcription products are passed to the egg by cytoplasmic bridges. Two maternal effect genes, called *Bicoid* and *Nanos*, help determine the anterior–posterior axis of the egg. (The dorsal–ventral [back–belly] axis is determined by other maternal effect genes that will not be described here.)

The mRNAs for *Bicoid* and *Nanos* diffuse from the mother's cells into what will be the anterior end of the egg. The *Bicoid* mRNA is translated to produce Bicoid protein, which diffuses away from the anterior end, establishing a gradient in the egg cytoplasm (**Figure 19.17**). Where it is present in sufficient concentration, Bicoid acts as a transcription factor to stimulate the transcription of the *Hunchback* gene in the early embryo. A gradient of the Hunchback protein establishes the head, or anterior, region.

Meanwhile, the egg's cytoskeleton transports the *Nanos* mRNA from the anterior end of the egg, where it was deposited, to the posterior end, where it is translated. The Nanos protein forms a gradient with the highest concentration at the posterior end. At that end, the Nanos protein inhibits the translation of *Hunchback* mRNA. Thus, the action of both Bicoid and Nanos establish the Hunchback gradient, which determines the anterior and posterior ends of the embryo.

SEGMENTATION GENES The number, boundaries, and polarity of the *Drosophila* larval segments are determined by proteins encoded by the **segmentation genes**. These genes are expressed when there are about 6,000 nuclei in the embryo (about three hours after fertilization). Three classes of segmentation genes act one after the other to regulate finer and finer details of the segmentation pattern:

- **Gap genes** organize broad areas along the anterior–posterior axis. Mutations in gap genes result in gaps in the body plan—the omission of several consecutive larval segments.

- **Pair rule genes** divide the embryo into units of two segments each. Mutations in pair rule genes result in embryos missing every other segment.

19.18 A Gene Cascade Controls Pattern Formation in the *Drosophila* Embryo (A) Maternal effect genes (see Figure 19.17) induce gap, pair rule, and segment polarity genes—collectively referred to as segmentation genes. (B) Two gap genes, *Hunchback* (orange) and *Krüppel* (green) overlap; both genes are transcribed in the yellow area. (C) The pair rule gene *Fushi tarazu* is transcribed in the dark blue areas. (D) The segment polarity gene *Engrailed* (bright green) is seen here at a slightly more advanced stage than is depicted in (A). By the end of this cascade, a group of nuclei at the anterior of the embryo, for example, is determined to become the first head segment in the adult fly.

---

yourBioPortal.com

GO TO Animated Tutorial 19.3 •
Pattern Formation in the *Drosophila* Embryo

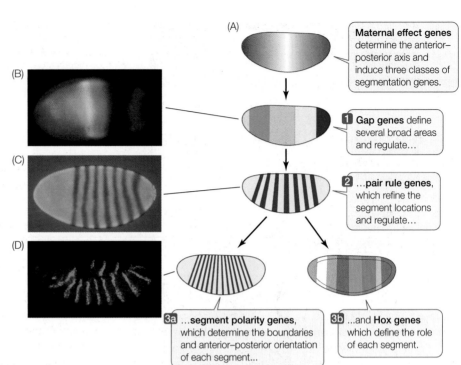

Maternal effect genes determine the anterior–posterior axis and induce three classes of segmentation genes.

1 Gap genes define several broad areas and regulate…

2 …**pair rule genes**, which refine the segment locations and regulate…

3a …**segment polarity genes**, which determine the boundaries and anterior–posterior orientation of each segment…

3b …and **Hox genes** which define the role of each segment.

- **Segment polarity genes** determine the boundaries and anterior–posterior organization of the individual segments. Mutations in segment polarity genes can result in segments in which posterior structures are replaced by reversed (mirror-image) anterior structures.

The expression of these genes is sequential (**Figure 19.18**). The maternal effect protein Bicoid, which begins the cascade, acts as a morphogen and transcription factor to stimulate the expression of genes such as *Hunchback* that set up the anterior–posterior axis. As a result, a nucleus in the early embryo "knows" where it is. The Hunchback protein stimulates gap gene transcription, the products of the gap genes activate pair rule genes, and the pair rule gene products activate segment polarity genes. By the end of this cascade, nuclei throughout the embryo "know" which segment they will be part of in the adult fly.

The next set of genes in the cascade determines the form and function of each segment.

HOX GENES Hox genes encode a family of transcription factors that are expressed in different combinations along the length of the embryo, and help determine cell fate within each segment. Hox gene expression tells the cells of a segment in the head to make eyes, those of a segment in the thorax to make wings, and so on. The *Drosophila* Hox genes occur in two clusters on chromosome 3, in the same order as the segments whose function they determine (**Figure 19.19**). By the time the fruit fly larva hatches, its segments are completely determined. Hox genes are homeotic genes that are shared by all animals, and they are functionally analogous to the organ identity genes of plants. However, they differ from plant homeotic genes in DNA sequence and encoded

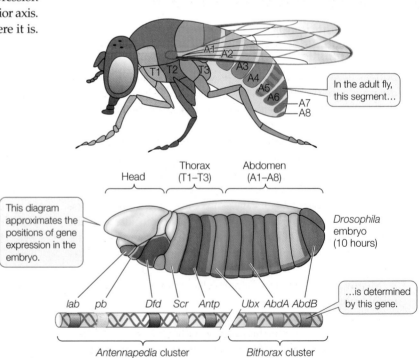

In the adult fly, this segment…

This diagram approximates the positions of gene expression in the embryo.

Head   Thorax (T1–T3)   Abdomen (A1–A8)

Drosophila embryo (10 hours)

…is determined by this gene.

lab pb   Dfd Scr Antp   Ubx AbdA AbdB

Antennapedia cluster   *Bithorax* cluster

19.19 Hox Genes in *Drosophila* Determine Segment Identity Two clusters of Hox genes on chromosome 3 (center) determine segment function in the adult fly (top). These genes are expressed in the embryo (bottom) long before the structures of the segments actually appear.

(A)

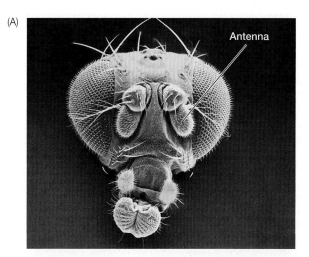

Antenna

(B)

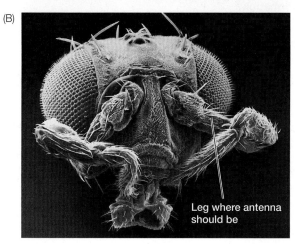

Leg where antenna
should be

19.20 A Homeotic Mutation in *Drosophila* Mutations of the Hox genes cause body parts to form on inappropriate segments. (A) A wild-type fruit fly. (B) An *antennapedia* mutant fruit fly. Mutations such as this reveal the normal role of the *Antennapedia* gene in determining segment function.

esized that all of the Hox genes might have come from the duplication of a single gene in an ancestral, unsegmented organism. Since Lewis put forward this hypothesis, molecular research methods became available to test it.

Hox genes encode transcription factors

Molecular biologists confirmed Lewis's hypothesis using nucleic acid hybridization. Several scientists found that probes for a sequence found in the *Bithorax* cluster could bind to other sequences in both the *Bithorax* and *Antennapedia* clusters. In other words, this DNA sequence is common to all the Hox genes in both clusters. It is also found in several of the segmentation genes, as well as other genes that encode transcription factors.

This 180-base-pair DNA sequence is called the **homeobox**. It encodes a 60-amino acid sequence called the *homeodomain*. The homeodomain recognizes and binds to a specific DNA sequence in the promoters of its target genes. However, this recognition is usually not sufficient to allow the transcription factor to bind fully to a promoter and turn the target gene on or off. Other transcription factors are also involved.

The Hox genes are found in animals with an anterior–posterior axis, where they play a role in development similar to that played by MADS box genes in plants. But the homeobox is found in many different transcription factors, including some from plants. The evolutionary significance of these common pathways for development will be discussed in the next chapter.

protein structure. This is not surprising, given that the last common ancestor of plants and animals was unicellular, and therefore multicellularity evolved independently for plants and animals.

In *Drosophila*, the maternal effect genes, segmentation genes, and Hox genes interact to "build" a larva step by step, beginning with the unfertilized egg. How do we know that the Hox genes determine segment identity? An important clue came from bizarre homeotic mutations observed in *Drosophila*. The *antennapedia* mutation causes legs to grow on the head in place of antennae (**Figure 19.20**), and the *bithorax* mutation causes an extra pair of wings to grow in a thoracic segment where wings do not normally occur (see Figure 20.3). Edward Lewis at Caltech found that *antennapedia* and *bithorax* mutations resulted from changes in Hox genes.

The first cluster of Hox genes—the *Antennapedia* cluster—specifies anterior segments, starting with genes for the different head segments and ending with thoracic segments. The second cluster (*Bithorax*) contains three genes. It begins with a gene specifying the last thoracic segment, followed by a gene for the anterior abdominal segments, and ends with a gene for the posterior abdominal segments (see Figure 19.19). Lewis hypoth-

---

19.5 RECAP

A cascade of transcription factors governs pattern formation and the subsequent development of animal and plant organs. Often these transcription factors create or respond to morphogen gradients. In plants, cell fate is often determined by MADS box genes, and in animal embryos, cell fate is determined in part by Hox genes.

- How is apoptosis crucial in shaping the developing embryo? **See p. 417**

- How do organ identity genes act in *Arabidopsis*? **See pp. 418–419 and Figure 19.14**

- List the key attributes of a morphogen. How does the Bicoid protein fit this definition? **See pp. 420–421 and Figure 19.17**

- How is segment identity established in the *Drosophila* embryo? **Review pp. 421–422 and Figure 19.19**

CHAPTER SUMMARY

19.1 What Are the Processes of Development?

- A multicellular organism begins its development as an embryo. A series of embryonic stages precedes the birth of an independent organism. **Review Figure 19.1, WEB ACTIVITY 19.1**

- The processes of development are **determination, differentiation, morphogenesis**, and **growth**.

- **Differential gene expression** is responsible for the differences between cell types. **Cell fate** is determined by environmental factors, such as the cell's position in the embryo, as well as by intracellular influences. **Review Figure 19.2**

- Determination is followed by differentiation, the actual changes in biochemistry, structure, and function that result in cells of different types. Determination is a commitment; differentiation is the realization of that commitment.

19.2 Is Cell Differentiation Irreversible?

- The zygote is **totipotent**; it is capable of forming every type of cell in the adult body.

- The ability to create **clones** from differentiated cells demonstrates the principle of **genomic equivalence**. **Review Figures 19.3 and 19.4**

- **Stem cells** produce daughter cells that differentiate when provided with appropriate intercellular signals. Some **multipotent** stem cells in the adult body can differentiate into a limited number of cell types to replace dead cells and maintain tissues. **Review Figure 19.5**

- Embryonic stem cells are **pluripotent** and can be cultured in the laboratory. Under suitable environmental conditions, these cells can differentiate into any tissue type. **Induced pluripotent stem cells** have similar characteristics. This has led to technologies to replace cells or tissues damaged by injury or disease. **Review Figure 19.6, ANIMATED TUTORIAL 19.1**

19.3 What Is the Role of Gene Expression in Cell Differentiation?

- Differential gene expression results in cell differentiation. Transcription factors are especially important in regulating gene expression during differentiation.

- Complex interactions of many genes and their products are responsible for differentiation during development. **Review Figure 19.7**

19.4 How Is Cell Fate Determined?

- **Cytoplasmic segregation**—the unequal distribution of **cytoplasmic determinants** in the egg, zygote, or early embryo—can establish **polarity** and lead to cell fate determination. **Review Figures 19.8 and 19.9, ANIMATED TUTORIAL 19.2**

- **Induction** is a process by which embryonic animal tissues direct the development of neighboring cells and tissues by secreting chemical signals, called **inducers**. **Review Figure 19.10**

- The induction of the vulva in the nematode *Caenorhabditis elegans* offers an example of how inducers act as molecular switches to direct a cell down one of two differentiation paths. **Review Figures 19.11 and 19.12**

19.5 How Does Gene Expression Determine Pattern Formation?

- **Pattern formation** is the process that results in the spatial organization of a tissue or organism.

- During development, selective elimination of cells by apoptosis results from the expression of specific genes. **Review Figure 19.13**

- Sepals, petals, stamens, and carpels form in plants as a result of combinatorial interactions between transcription factors encoded by **organ identity genes**. **Review Figure 19.14**

- The transcription factors encoded by floral organ identity genes contain an amino acid sequence called the **MADS box** that can bind to DNA.

- Both plants and animals use **positional information** as a basis for pattern formation. Positional information usually comes in the form of a signal called a **morphogen**. Different concentrations of the morphogen cause different effects. **See Figures 19.15 and 19.16**

- In the fruit fly *D. melanogaster*, a cascade of transcriptional activation sets up the axes of the embryo, the development of the segments, and finally the determination of cell fate in each segment. The cascade involves the sequential expression of maternal effect genes, gap genes, pair rule genes, segment polarity genes, and Hox genes. **Review Figures 19.18 and 19.19, ANIMATED TUTORIAL 19.3**

- Hox genes help to determine cell fate in the embryos of all animals. The **homeobox** is a DNA sequence found in Hox genes and other genes that code for transcription factors. The sequence of amino acids encoded by the homeobox is called the homeodomain.

SELF-QUIZ

1. Which statement about determination is true?
   a. Differentiation precedes determination.
   b. All cells are determined after two cell divisions in most organisms.
   c. A determined cell will keep its determination no matter where it is placed in an embryo.
   d. A cell changes its appearance when it becomes determined.
   e. A differentiated cell has the same pattern of transcription as a determined cell.

2. Cloning experiments on sheep, frogs, and mice have shown that
   a. nuclei of adult cells are totipotent.
   b. nuclei of embryonic cells can be totipotent.
   c. nuclei of differentiated cells have different genes than zygote nuclei have.
   d. differentiation is fully reversible in all cells of a frog.
   e. differentiation involves permanent changes in the genome.

3. The term "induction" describes a process in which a cell or cells
   a. influence the development of another group of cells.
   b. trigger cell movements in an embryo.
   c. stimulate the transcription of their own genes.
   d. organize the egg cytoplasm before fertilization.
   e. inhibit the movement of the embryo.

4. Stem cells from adult animals
   a. are always totipotent.
   b. divide when provided with external signals.
   c. are not present in bone marrow.
   d. are present in an embryo but not an adult.
   e. can be turned into differentiated cells with only a few genes.

5. Which statement about cytoplasmic determinants in *Drosophila* is *not* true?
   a. They specify the dorsal–ventral and anterior–posterior axes of the embryo.
   b. Their positions in the embryo are determined by cytoskeletal action.
   c. Some are products of specific genes in the mother fruit fly.
   d. They do not produce gradients.
   e. They have been studied by the transfer of cytoplasm from egg to egg.

6. In fruit flies, the following genes are used to determine segment polarity: (k) gap genes; (l) Hox genes; (m) maternal effect genes; (n) pair rule genes. In what order are these genes expressed during development?
   a. k, l, m, n
   b. l, k, n, m
   c. m, k, n, l
   d. n, k, m, l
   e. n, m, k, l

7. Which statement about induction is *not* true?
   a. One group of cells induces adjacent cells to develop in a certain way.
   b. It triggers a sequence of gene expression in target cells.
   c. Single cells cannot form an inducer.
   d. A tissue may be induced as well as make an inducer.
   e. The chemical identification of specific inducers has not been achieved.

8. In the process of pattern formation in the *Drosophila* embryo,
   a. the first steps are specified by Hox genes.
   b. mutations in pair rule genes result in embryos missing every other segment.
   c. mutations in gap genes result in the insertion of extra segments.
   d. segment polarity genes determine the dorsal–ventral axes of segments.
   e. all segments develop the same organs.

9. Homeotic mutations
   a. are often severe and result in structures at inappropriate places.
   b. cause subtle changes in the forms of larvae or adults.
   c. occur only in prokaryotes.
   d. do not affect the animal's DNA.
   e. are confined to the zone of polarizing activity.

10. Which statement about the homeobox is *not* true?
    a. It is transcribed and translated.
    b. It is found only in animals.
    c. Proteins containing the homeodomain bind to DNA.
    d. It is a sequence of DNA shared by more than one gene.
    e. It occurs in Hox genes.

FOR DISCUSSION

1. Molecular biologists can attach genes to active promoters and insert them into cells (see Section 18.5). What would happen if the following were inserted and overexpressed? Explain your answers.
   a. *ced-9* in embryonic neuron precursors of *C. elegans*
   b. *MyoD* in undifferentiated myoblasts
   c. the gene for Sonic hedgehog in a chick limb bud
   d. *Nanos* at the anterior end of the *Drosophila* embryo

2. A powerful method to test for the function of a gene in development is to generate a "knockout" organism, in which the gene in question is inactivated (see Section 18.4). What do you think would happen in each of the following cases?
   a. a knocked-out *ced-9* in *C. elegans*
   b. a knocked-out *Nanos* in *Drosophila*

3. If you wanted a rose flower with only petals, what kind of homeotic mutation would you seek in the rose genome?

4. During development, an animal cell's potential for differentiation becomes ever more limited. In the normal course of events, most cells in the adult animal have the potential to be only one or a few cell types. On the basis of what you have learned in this chapter, discuss possible mechanisms for the progressive limitation of the cell's potential.

5. How were biologists able to obtain such a complete accounting of all the cells in *C. elegans*? What major conclusions came from these studies?

ADDITIONAL INVESTIGATION

Cloning involves considerable reprogramming of gene expression in a differentiated cell so that it acts like an egg cell. How would you investigate this reprogramming?

WORKING WITH DATA (GO TO yourBioPortal.com)

Cloning a Mammal In this hands-on exercise, you will examine the experimental protocol used by Wilmut and colleagues to clone Dolly the sheep (Figure 19.4). You will see the data on the efficiency of this process, as well as the genetic evidence that Dolly was indeed a clone.

20 Development and Evolutionary Change

The eyes have it

Eyes are not essential for survival; many animals and all plants get by just fine without them. However, almost all animals *do* have eyes or some type of light-sensing organs, and having eyes can confer a selective advantage.

About a dozen different kinds of eyes are found among the different animals, including the camera-like eyes of humans and the compound eyes of insects, with their thousands of individual units. In trying to understand the origin of this variety, scientists—starting with Charles Darwin—proposed that eyes evolved independently many times in different animal groups, and that each improvement in the ability of eyes to gather light and form images conferred a selective advantage on their possessor.

A remarkable discovery in the 1990s may have overturned this long-held dogma about the evolution of eyes. Years earlier a mutant fruit fly without eyes was found,

and the gene involved—appropriately called *eyeless*—was mapped onto one of its chromosomes. This mutant fly remained a laboratory curiosity until 1994, when the Swiss developmental biologists Rebecca Quiring and Walter Gehring began looking for transcription factors that might be involved in fly development. The gene for one of the proteins they identified mapped to the *eyeless* locus. Thus, the product of the *eyeless* gene is a transcription factor that controls the formation of the eye. Quiring and Gehring demonstrated this by making recombinant DNA constructs that allowed the *eyeless* gene to be expressed in various embryonic tissues of transgenic flies. These experiments resulted in adult flies with extra eyes on various body parts—including on the legs, under the wings, and on the antennae—depending on where the *eyeless* gene was expressed in the embryos.

But the big surprise came when the scientists performed a database search and found that the *eyeless* gene sequence was quite similar to that of *Pax6*, a gene in mice that, when mutated, leads to the development of very small eyes. Could the very different eyes of flies and mice just be variations on a common developmental theme? To test for functional similarity between the insect and mammalian genes, Gehring and colleagues repeated their gene expression experiments using the mouse *Pax6* gene instead of the fly *eyeless* gene. Once again, eyes developed at various sites on the transgenic flies. So a gene whose expression normally leads to the development of a mammalian "camera" eye now led to the development of an insect's "compound"

Eye of the Fly Unlike the single-lensed eyes of vertebrates, the compound eyes of flies and other insects are composed of thousands of individual lenses, or ommatidia.

A Mouse Gene Can Produce a Fly's Eye When the mouse *Pax6* eye-specifying gene was implanted in the part of the fruit fly embryo that normally produces a limb, ommatidia emerged in place of a leg.

eye—a very different eye type. Thus a single transcription factor appears to function as a molecular switch that turns on eye development. Although eyes evolved many times during animal evolution, all of them depend on the same gene. The special features of the many different eyes in diverse animals all evolved from a common developmental process.

The discovery that the same genes govern development in a wide variety of animals led to the rapid growth of the discipline of evolutionary developmental biology, often known as "evo-devo." Evolutionary developmental biologists compare the genes that regulate development in many different multicellular organisms to understand how a single gene can do so many different things.

IN THIS CHAPTER we show that the genes controlling pattern formation, which we introduced in Chapter 19, are shared by a diverse array of organisms. We next describe how changes can occur in some parts of an organism without causing undesirable changes in other parts. We see how a common set of genes can produce a great variety of body forms. We then turn to the ways some organisms can modulate their development by responding to signals from their environment. Finally, we examine how developmental processes constrain evolution.

20.1 What Is Evo-Devo?

The modern study of evolution and development is called **evolutionary developmental biology**, or **evo-devo**. Its ideas have come from studies of the molecular mechanisms that underlie the development of morphology, and how the genes controlling these mechanisms have evolved. The principles of evo-devo are:

- Many groups of animals and plants, even distantly related ones, share similar molecular mechanisms for morphogenesis and pattern formation. As we saw in the opening essay, some genes that are experimentally swapped from one organism to another can retain similar functions in the new organisms. These mechanisms can be thought of as "toolkits," in the same sense that a few tools in a carpenter's toolkit can be used to build many different structures.

- The molecular pathways that determine different developmental processes, such as anterior–posterior polarity and organ formation in animals, operate independently from one another. This is called **modularity**.

- Changes in the location and timing of expression of particular genes are important in the evolution of new body forms and structures.

- Development produces morphology, and much of morphological evolution occurs by modifications of existing development genes and pathways, rather than the introduction of radically new developmental mechanisms.

- Mechanisms of development have often evolved to be responsive to environmental conditions.

Biologists have long known that the morphological differences between species are due to differences in their genomes. But we have also discussed how the genomes of different species—including distantly related ones—share numerous similar regulatory and coding sequences (see Section 17.3). When developmental biologists began to describe the events of differentiation, morphogenesis, and pattern formation at the molecular level, they found common regulatory genes and pathways in organisms that don't appear similar at all, such as fruit flies and mice.

Developmental genes in distantly related organisms are similar

In the opening story of this chapter, we describe how a single developmental switch turns on the production of eyes in two widely divergent species—fruit flies and mice—that are only

Mouse *Pax*6 gene:

DNA: GTATCCAACGGTTGTGTGAGTAAAATTCTGGGCAGGTATTACGAGACTGGCTCCATCAGA

Amino acids: V S N G C V S K I L G R Y Y E T G S I R

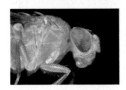

Fly *eyeless* gene:

77%: GTATCAAATGGATGTGTGAGCAAAATTCTCGGGAGGTATTATGAAACAGGAAGCATACGA

100%: V S N G C V S K I L G R Y Y E T G S I R

Shark eye control gene:

85%: GTGTCCAACGGTTCTGTCAGTAAAATCCTGGGCAGATACTATGAAACAGGATCCATCAGA

100%: V S N G C V S K I L G R Y Y E T G S I R

Squid eye control gene:

78%: GTCTCCAACGGCTGCGTTAGCAAGATTCTCGGACGGTACTATGAGACGGGCTCCATAAGA

100%: V S N G C V S K I L G R Y Y E T G S I R

20.1 DNA Sequence Similarity in Eye Development Genes Genes controlling eye development contain regions that are highly conserved, even among species with very different eyes. These sequences, from a conserved region of the *Pax6* gene and its homologs in other species, are similar at the DNA level (top sequence in each pair) and identical at the amino acid level (bottom sequence). The percentages beside the sequences represent the percent match with the corresponding DNA and protein sequences in the mouse.

distantly related by evolution. The genes that control this switch, *eyeless* in fruit flies and *Pax6* in mice, contain sequences that are highly conserved in these species and in other animals (**Figure 20.1**). As described in Section 22.1, biologists infer from these similarities that the genes are *homologous*, meaning that they evolved from a gene present in a common ancestor of mice and fruit flies. In recent years, thousands of genes have been found that are homologous across distantly related species.

An even more dramatic example of homology in genes that control development, because it involves a whole set of genes, is the Hox gene cluster. These genes provide positional information and control pattern formation in early *Drosophila* embryos (see Figure 19.19). When scientists looked for similar sequences in the mouse and human genomes, the results were amazing. The Hox genes had homologs in mammals, and what is more, the genes were arranged in similar clusters in the genomes of mammals and fruit flies, and were expressed in similar patterns in their embryos (**Figure 20.2**). Over the millions of years that have elapsed since the common ancestor of these animals, the genes in question have mostly been main-

tained, suggesting that their functions were favored over many different conditions.

These and other examples have lead evo-devo biologists to the idea that certain developmental mechanisms, controlled by specific DNA sequences, have been conserved over long periods during the evolution of multicellular organisms. These sequences comprise the **genetic toolkit**, which has been modified and reshuffled over the course of evolution to produce the remarkable diversity of plants, animals, and other organisms in the world today.

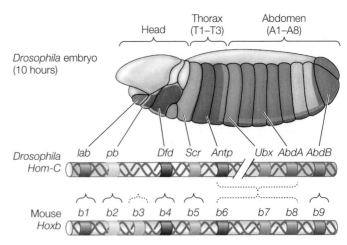

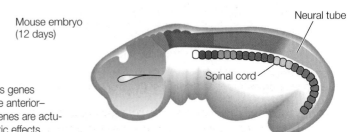

20.2 Regulatory Genes Show Similar Expression Patterns Homologous genes encoding similar transcription factors are expressed in similar patterns along the anterior–posterior axes of both insects and vertebrates. The mouse (and human) Hox genes are actually present in multiple copies; this prevents a single mutation from having drastic effects.

Many developmental mutations in fruit flies that result in striking abnormalities (e.g., a head segment that forms a leg; see Figure 19.20) affect only a single structure, segment or region. The rest of the embryo is often unaffected. How is this possible?

20.2 How Can Mutations With Large Effects Change Only One Part of the Body?

In Chapter 19 we describe how development involves interactions between gene products, which determine a sequence of transcriptional events leading to differential gene expression. On the other hand, the study of homeotic mutations revealed that embryos, like adults, are made up of **developmental modules**—functional entities encompassing genes and various signaling pathways that determine physical structures such as body segments and legs.

The form of each module in an organism may be changed independently of the other modules because some developmental genes exert their effects on only a single module. For example, the form of a developing animal's heart can change independently of changes in its limbs, because some of the genes that govern heart formation do not affect limb formation, and vice versa. If this were not true, a mutation in any developmental gene might result in an adult with multiple, widely different deformities. Such an adult would have difficulty functioning well in any environment.

---
yourBioPortal.com

GO TO Animated Tutorial 20.1 • Modularity

Genetic switches govern how the genetic toolkit is used

Different structures can evolve within a single organism using a common set of genetic instructions because there are mecha-

nisms called **genetic switches** that control how the genetic toolkit is used. These mechanisms involve promoters and the transcription factors that bind them. The signal cascades that converge on and operate these switches determine when and where genes will be turned on and off. Multiple switches control each gene by influencing its expression at different times and in different places. In this way, elements of the genetic toolkit can be involved in multiple developmental processes while still allowing individual modules to develop independently.

Genetic switches integrate positional information in the developing embryo and play key roles in determining the developmental pathways of different modules. For example, each Hox gene codes for a transcription factor that is expressed in a particular segment or appendage of the developing fruit fly. The pattern and functioning of each segment depend on the unique Hox gene or combination of Hox genes that are expressed in the segment.

Consider the formation of fruit fly wings. *Drosophila* has three thoracic segments, the first of which bears no wings. The second segment bears the large forewings, and the third segment bears small hindwings, called *halteres*, that function as balancing organs. Hox proteins are not expressed in forewing cells, but all hind wing cells express the Hox gene *Ultrabithorax* (*Ubx*) because a set of genetic switches activates the *Ubx* gene in the third thoracic segment. Ubx turns off genes that promote the formation of the veins and other structures of the forewing, and it turns on genes that promote the formation of hind wing features (**Figure 20.3**). In butterflies, on the other hand, Ubx influences target genes so that wings develop in the third-segment cells, so full hind wings develop. Therefore, a simple genetic change results in a major morphological difference in the wings of flies and butterflies.

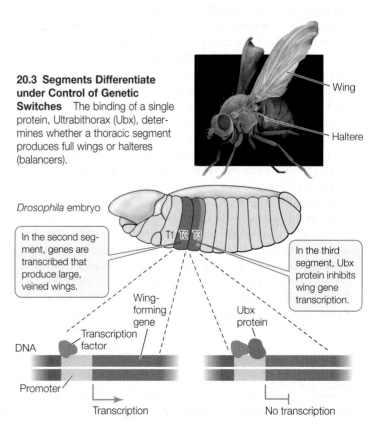

20.3 Segments Differentiate under Control of Genetic Switches The binding of a single protein, Ultrabithorax (Ubx), determines whether a thoracic segment produces full wings or halteres (balancers).

Wing

Haltere

Drosophila embryo

In the second segment, genes are transcribed that produce large, veined wings.

T1 T2 T3

In the third segment, Ubx protein inhibits wing gene transcription.

Wing-forming gene

Ubx protein

Transcription factor

DNA

Promoter

Transcription

No transcription

(A) *B. rostratus*
(terrestrial)

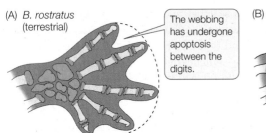

The webbing has undergone apoptosis between the digits.

(B) *B. occidentalis*
(arboreal)

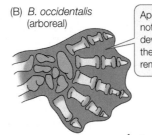

Apoptosis did not occur during development, so the webbing remains.

20.4 Heterochrony Resulted in the Evolution of a Tree-Climbing Salamander (A) The foot of an adult *B. rostratus*, a terrestrial salamander. (B) The foot of *B. occidentalis*, a closely related salamander, does not lose its webbing. This species uses the suction of its webbed feet in an arboreal lifestyle.

Modularity allows for differences in the timing and spatial pattern of gene expression

Modularity allows the relative *timing* of different developmental processes to shift independently of one another, in a process called **heterochrony**. That is, the genes regulating the development of one module (say, the eyes of vertebrates) may be expressed at different developmental stages in different species.

Salamanders of the genus *Bolitoglossa* illustrate how heterochrony can result in major morphological changes. Salamander embryos have webbing between their toes, but in most species of salamanders a particular gene triggers apoptosis in the webbing as the salamanders develop. The resulting independent digits allow the adult salamander to walk more easily than if it had webbed feet. This is the case with *Bolitoglossa rostratus*, a species that lives on the forest floor (**Figure 20.4A**). But in arboreal species such as *Bolitoglossa occidentalis* (**Figure 20.4B**), this gene is not expressed and apoptosis does not occur. The feet of *B. occidentalis* are webbed throughout life, acting like suction cups so the animal can adhere to vertical surfaces such as tree trunks. Thus a simple change in gene expression led to a major morphological change and allowed a new lifestyle.

The evolution of the giraffe's neck provides another example of heterochrony. As in virtually all mammals (with the exception of manatees and sloths) there are seven vertebrae in the neck of the giraffe. So the giraffe did not get a longer neck by adding vertebrae. Instead the cervical (neck) vertebrae of the giraffe are much longer than those of other mammals (**Figure 20.5**). Bones grow due to the proliferation of cartilage-producing cells called chondrocytes. Bone growth is stopped by a signal that results in death of the chondrocytes and calcification of the bone matrix. In giraffes this signaling process is delayed in the cervical vertebrae, with the result that these vertebrae grow longer. Thus, the evolution of longer necks acted through *changes in the timing of expression* of the genes that control bone formation.

Differences in the *spatial expression pattern* of a developmental gene can also result in evolutionary change. Foot webbing in salamanders is determined by the temporal expression of a developmental gene, but foot webbing in ducks and chickens is affected by alterations in the spatial expression of a gene. The feet of all bird embryos have webs of skin that connect their toes. This webbing is retained in adult ducks (and other aquatic birds) but not in adult chickens (and other non-aquatic birds). The loss of webbing is due to a signaling protein called bone morphogenetic protein 4 (BMP4) that instructs the cells in the webbing to undergo apoptosis. The death of these cells destroys the webbing between the toes.

Embryonic duck and chicken hindlimbs both express the *BMP4* gene in the webbing between the toes, but they differ in expression of a gene called *Gremlin*, which encodes a BMP *inhibitor* protein (**Figure 20.6**). In ducks, but not chickens, the *Gremlin* gene is expressed in the webbing cells. The Gremlin protein inhibits the BMP4 protein from signaling for apoptosis, and the result is a webbed foot. If chicken hindlimbs are experimentally exposed to Gremlin during development, apoptosis does not occur, and ducklike webbed feet form on the chicken (**Figure 20.7**).

(A) Giraffe

(B) Human

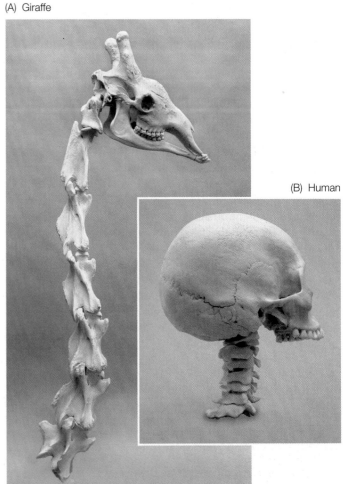

20.5 Heterochrony in the Development of a Longer Neck There are seven vertebrae in the neck of the giraffe (left) and human (right; not to scale). But the vertebrae of the giraffe are much longer (25 cm compared to 1.5 cm) because during development, growth continues for a longer period of time. This timing difference is called heterochrony.

Chick hindlimb   Duck hindlimb

Purple dye marks the presence of BMP4 proteins.

Chick limbs do not produce Gremlin (a BMP4 inhibitor) in the webbing.

Duck limbs produce Gremlin in the webbing (arrows).

Red dye shows the pattern of cell death (apoptosis).

No apoptosis occurs.

In the chicken, webbing undergoes apoptosis, resulting in the separated toes of the adult.

Webbing in the adult duck's foot remains intact.

20.6 Changes in Gremlin Expression Correlate with Changes in Hindlimb Structure The left column of photos shows the development of a chicken's foot; the right column shows foot development in a duck. Gremlin protein in the webbing of the duck foot inhibits BMP4 signaling, thus preventing the embryonic webbing from undergoing apoptosis.

INVESTIGATING LIFE

20.7 Changing the Form of an Appendage

Ducks have webbed feet and chickens do not—a major difference in the adaptations of these species. Webbing is initially present in the chick embryo, but undergoes apoptosis that is stimulated by the protein BMP4. In ducks another protein, Gremlin, binds to BMP4 and inhibits it, preventing apoptosis and resulting in webbed feet. J. J. Hurle and colleagues at the Universidad de Cantabria in Spain asked what would happen if Gremlin were put onto a developing chick foot. They hypothesized that apoptosis would be inhibited, and it was: the chick developed webbed feet. Thus, a single developmental switch controls foot shape—an important adaptation to the environment.

HYPOTHESIS Adding Gremlin protein (a BMP4 inhibitor) to a developing chicken foot will transform it into a ducklike foot.

METHOD Chip a small window in chick egg shell and carefully add Gremlin-secreting beads to the webbing of embryonic chicken hindlimbs. Add beads that do not contain Gremlin to other hindlimbs (controls). Close the eggs and observe limb development.

RESULTS In the hindlimbs in which Gremlin was secreted, the webbing does not undergo apoptosis, and the hindlimb resembles that of a duck. The control hindlimbs develop the normal chicken form.

Control   Gremlin added

 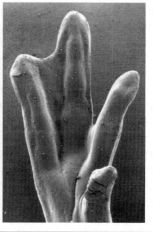

CONCLUSION Differences in *Gremlin* gene expression cause differences in morphology, allowing duck hindlimbs to retain their webbing.

Go to **yourBioPortal.com** for original citations, discussions, and relevant links for all INVESTIGATING LIFE figures.

20.2 RECAP

Embryos and adult organisms are made up of self-contained units called modules. The form of each module may change independently because some developmental genes exert their effects on only a single module.

- How do genetic switches control the way a gene is used? **See p. 429 and Figure 20.3**

- Explain how heterochrony can result in evolutionary change. **See p. 430 and Figures 20.4 and 20.5**

Genetic manipulations and studies of pattern formation within embryos have shown that the same signals can control development of different structures in an individual organism. For example, the protein BMP4 promotes apoptosis between developing digits in feet, and then is involved later in the formation of bone. These studies suggest that the processes that generate multiple structures *within* an organism might also explain how different structures develop in different species.

20.3 How Can Differences among Species Evolve?

Can the processes that allow different structures to develop in different regions of an embryo also explain major morphological differences among species? Apparently they can. The genetic switches that determine where and when genes will be expressed appear to underlie both the transformation of an individual from egg to adult and the many major differences in body form that exist among species. Arthropods provide good examples of how morphological differences among species can evolve through mutations in the genes that regulate the differentiation of segments.

The arthropods (which include crustaceans, centipedes, spiders, and insects) are segmented, with head, thoracic, and abdominal segments. In centipedes, both thoracic and abdominal segments form legs; but in insects, only the thoracic segments do. Arthropods express a gene called *Distal-less* (*Dll*) that causes legs to form from segments. What shuts down *Dll* expression in insect abdominal segments? The product of the Hox gene *Ubx* is produced in arthropod abdominal segments. But it has very different effects in different organisms. In centipedes, the Ubx protein apparently activates expression of the *Dll* gene to promote the formation of legs. But during the evolution of insects, a change in the *Ubx* gene sequence resulted in a modified Ubx protein that *represses Dll* expression in abdominal segments, so leg formation is inhibited. A phylogenetic tree of arthropods shows that this change in *Ubx* occurred in the ancestor of insects, at the same time that abdominal legs were lost (**Figure 20.8**).

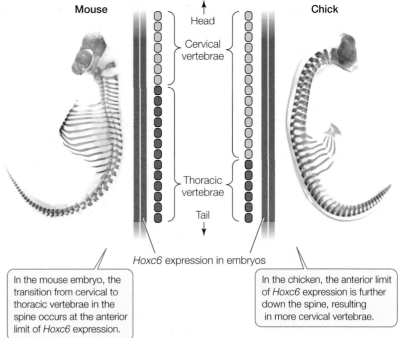

Hoxc6 expression in embryos

In the mouse embryo, the transition from cervical to thoracic vertebrae in the spine occurs at the anterior limit of *Hoxc6* expression.

In the chicken, the anterior limit of *Hoxc6* expression is further down the spine, resulting in more cervical vertebrae.

20.9 Changes in Gene Expression and Evolution of the Spine Differences in the pattern of *Hoxc6* expression result in a different boundary between the cervical and thoracic vertebrae in mice and chicks.

In vertebrates, a similar process governs the development of differences in segments of the vertebral column. The vertebral column consists of a set of anterior-to-posterior regions: the cervical (neck), thoracic (chest), lumbar (back), sacral, and caudal (tail) regions. The spatial pattern of Hox gene expression governs the transition from one region to another (**Figure 20.9**). For example, the anterior limit of expression of *Hoxc6* always falls at the boundary between the cervical and thoracic vertebrae of mice and chickens, even though these animals have different numbers of cervical and thoracic vertebrae. The anterior-most

20.8 A Mutation in a Hox Gene Changed the Number of Legs in Insects In the insect lineage (blue box) of the arthropods, a change to the *Ubx* gene resulted in a protein that inhibits the *Dll* gene, which is required for legs to form. Because insects express this modified *Ubx* gene in their abdominal segments, no legs grow from these segments. Other arthropods, such as centipedes, do grow legs from their abdominal segments.

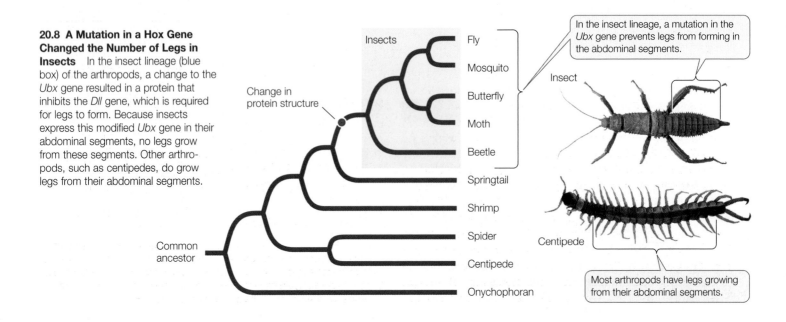

segment that expresses *Hoxc6* is the segment where the fore-limbs will develop. Thus, genetic changes that expanded or contracted the expression domains of different Hox genes resulted in changes in the characteristic numbers of different vertebrae during evolution.

20.3 RECAP

Changes in the genetic switches that determine where and when genes will be expressed underlie the evolution of differences in form among species.

- Why do insects, unlike other arthropods, lack abdominal limbs? **See p. 432 and Figure 20.8**

- How can the evolution of the spinal column be explained by changes in a developmental gene? **See p. 432 and Figure 20.9**

So far in this chapter, we have focused on how modular genetic signaling cascades control the development of an organism and how changes in genetic switches can produce differences between species. You may have the impression that all of these processes unfold from the genetic information contained in the fertilized egg, but that is not the case. Information from the environment can influence the genetic signaling cascades and thereby alter the form of the organism.

20.4 How Does the Environment Modulate Development?

The environment an individual lives in may differ from the one its parents lived in. Some environmental signals can produce developmental changes in an organism. If such changes result in higher reproductive fitness, they will be favored by natural selection. The ability of an organism to modify its development in response to environmental conditions is called **developmental plasticity** or *phenotypic plasticity*. It means that *a single genotype has the capacity to produce two or more different phenotypes.*

Temperature can determine sex

In Chapter 12 we discuss the genetic mechanisms that determine sex. In mammals there are two sex chromosomes; XX individuals are female and XY individuals are male. But in some reptiles, sex is determined not by genetic differences between individuals, but rather by the temperature at which the eggs are incubated—a remarkable case of developmental plasticity.

Research in the laboratory of David Crews at the University of Texas has shown that if eggs of the red-eared slider turtle are incubated at temperatures below 28.6°C, they will all become males, whereas if the eggs are incubated above 29.4°C, they will all become females. In the less than 1°C range between these two temperatures, a mix of males and females will hatch from the eggs (**Figure 20.10**). In other species with temperature-dependent sex determination, the incubation temperatures that produce males and females may differ from those that produce males and females in the red-eared slider. These different temperature dependencies indicate that the effects of incubation temperature can vary among species. But how does temperature control this developmental plasticity?

In vertebrates, the development of male and female organs in the embryo is controlled by the actions of sex steroid hormones. This is the case whether the organism's sex determination is controlled genetically or by temperature. Sex steroid biosynthesis in both males and females begins with cholesterol, which goes through many chemical reactions to produce the male sex steroids (androgens) and the female sex steroids (estrogens). In this biosynthetic sequence, the step that produces the first androgen—testosterone—precedes the step that produces estrogens; therefore, both males and females produce testosterone.

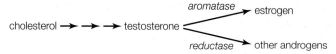

In animals with temperature-controlled sex determination, incubation temperature influences sex development by controlling the expression of the enzyme aromatase, which converts testosterone to estrogen. If aromatase is abundantly expressed, estrogens are dominant and female organs develop. If aromatase is not expressed, testosterone is dominant, and male organs develop. Applying estrogen to eggs results in the development of females, even at the male-inducing temperature.

What is the evolutionary advantage of this sex determination mechanism? It might be that incubation temperature influences the growth rate of the embryo and the time of hatching.

20.10 Hot Females, Cool Males Whether the embryo of a red-eared slider turtle develops into a male or a female depends on the temperature at which the egg is incubated. Higher temperatures produce only females and lower temperatures produce only males. This is apparently due to temperature sensitivity in the synthesis of sex hormones.

In species in which males compete for territories and for females, a larger body size would be a benefit for males, but not necessarily for females. Depending on availability of food in the environment, earlier hatching may have a positive or a negative effect on growth rate. For these kinds of reasons, incubation temperature may have a differential effect on the reproductive successes of males and females in a population.

One interesting experiment clearly demonstrated the fitness value of temperature-determined sex (**Figure 20.11**). At the University of Sydney, Daniel Warner and Richard Shine used hormones to manipulate the eggs of a lizard called the jacky dragon (*Amphibolurus muricatus*) to produce males at temperatures that would normally result in females. In this species, females are produced at all incubation temperatures, but males are only produced at incubation temperatures between 27°C and 30°C. The hormonal manipulations allowed the investigators to obtain both males and females from three different incubation temperatures—low, medium, and high—and to compare their subsequent growth characteristics and reproductive success.

The young lizards were released into outdoor enclosures and allowed to behave naturally for the next three years. The males incubated at the medium temperature had higher reproductive success over the three-year study. The reproductive successes of females from the low and medium incubation temperatures did not differ, but were higher for the high-temperature female group. These data support the hypothesis that the incubation temperature differentially affects reproductive success in males and females, and provides an explanation for why there could be selection for temperature-dependent sex determination in some species.

Organisms use information that predicts future conditions

In many cases of phenotypic plasticity, the adaptation of the different body forms to different but predictable environments is quite obvious. An excellent example is the moth *Nemoria arizonaria*, which produces two generations each year. Caterpillars

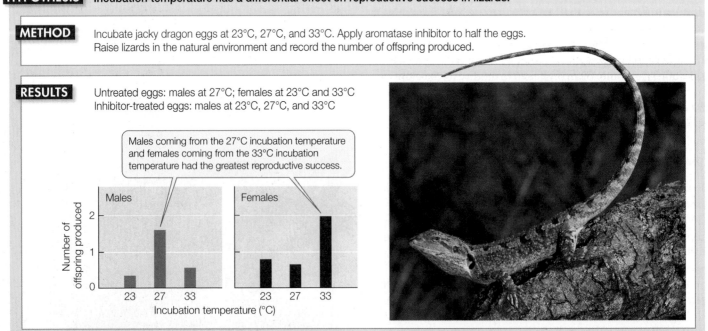

INVESTIGATING LIFE

20.11 Temperature-Dependent Sex Determination is Associated with Sex-Specific Fitness Differences

In some reptiles, sex is determined by the incubation temperature of the developing embryo. This led to the hypothesis that male-inducing temperatures during development result in males with higher reproductive fitness. Warner and Shine tested this hypothesis by using a drug to block estrogen synthesis, so that males developed instead of females at high and low temperatures. These males had much lower reproductive fitness than males that developed at the normal male-inducing temperature (which is intermediate). In contrast, females showed highest fitness when they developed from eggs incubated at higher temperatures.

HYPOTHESIS Incubation temperature has a differential effect on reproductive success in lizards.

METHOD Incubate jacky dragon eggs at 23°C, 27°C, and 33°C. Apply aromatase inhibitor to half the eggs. Raise lizards in the natural environment and record the number of offspring produced.

RESULTS Untreated eggs: males at 27°C; females at 23°C and 33°C
Inhibitor-treated eggs: males at 23°C, 27°C, and 33°C

Males coming from the 27°C incubation temperature and females coming from the 33°C incubation temperature had the greatest reproductive success.

CONCLUSION Incubation temperature during development has a differential effect on fitness of male and female jacky dragons, and thus could be a selective pressure leading to the pattern of temperature determination of sex in this species.

Go to **yourBioPortal.com** for original citations, discussions, and relevant links for all INVESTIGATING LIFE figures.

that hatch from eggs in spring feed on oak tree flowers (catkins). These caterpillars complete their development and transform into adult moths in summer. The summer moths lay their eggs on oak leaves, and the caterpillars that hatch eat the leaves. When these caterpillars transform into adult moths, they lay eggs that over-winter and hatch in the spring when the catkins are once again in bloom. Both types of caterpillars are camouflaged in the environments in which they feed. The body form of the spring caterpillars resembles the catkins (**Figure 20.12A**), and the body form of the summer caterpillars resembles small oak branches (**Figure 20.12C**). At the time of hatching, the young caterpillars all look similar, but their diets trigger developmental changes that result in the differences in appearance. The ability to avoid predation by phenotypic plasticity increases evolutionary fitness.

A variety of environmental signals influence development

In addition to temperature and diet, there are other environmental signals that initiate developmental changes. A ubiquitous and dependable source of environmental information is light, which provides predictive information about seasonal changes. Outside of the equatorial region, lengthening days herald spring and summer while shortening days indicate oncoming fall and winter. Many insects use day length to enter or exit a period of developmental or reproductive arrest called **diapause**, which enables them to better survive harsh conditions. Deer, moose, and elk use day length to time the development and the dropping of antlers, and many organisms use day length to optimize the timing of reproduction. As we discuss in Chapter 38, many plants initiate reproduction in response to the length of the night (an absence of light) and others respond to certain wavelengths of light with developmental changes.

You may wonder why we are mentioning processes like antler growth and seasonal reproduction in a chapter on development. Development encompasses more than the events that occur before an organism reaches maturity. Development includes changes in body form and function that can occur throughout the life of the organism.

Plants provide a particularly clear example of this. Redwood trees that are thousands of years old still have undifferentiated tissues called meristems that produce new differentiated tissues for the tree—stems, leaves, reproductive structures, and so on—throughout its life. These developmental processes are not a simple read-out of a genetic program; they are adjusted to optimize plant form in the environment in which the plant grows. Light, which plants need for photosynthesis, is an important environmental signal in plant development. Dim light stimulates the elongation of stem cells, so that plants growing in the shade become tall and spindly (**Figure 20.13**). This developmental plasticity is adaptive because a spindly plant is more likely to reach a patch of brighter light than a plant that remains compact. In bright light a plant does not need to grow tall, and can put its energy into growing leaves.

---
yourBioPortal.com
GO TO **Web Activity 20.1** • Plant Development

Natural selection can act on any genes or signaling pathways with important developmental functions that can influence reproductive success. Antler growth cycles involve the turning on and off of genes controlling bone growth. Seasonal breeding involves turning on and off the same genes that were involved in sex development and maturation. The evolution of development extends to all stages of life.

(A) Catkins

(B) Summer adult

(C) *Nemoria arizonaria* caterpillars

Spring: on catkin

Summer: on branch

20.12 Spring and Summer Forms of a Caterpillar (A) Spring caterpillars of the moth *N. arizonaria* resemble the oak catkins on which they feed. They develop into adults (B), which lay eggs on oak leaves. The summer caterpillars (C) of the same species resemble oak twigs.

In the dark, stem cells elongate and leaf growth is reduced.

In the light, stem cells are shorter and leaves are bigger.

20.13 Light Seekers The bean plants on the left were grown under low light levels. The plant's cells have elongated in response to the low light, and the plants have become spindly. The control plants on the right were grown under normal light conditions.

20.4 RECAP

Developmental plasticity enables developing organisms to adjust their forms to fit the environments in which they live. Organisms respond to environmental signals that are accurate predictors of future conditions. Development continues throughout life, and can result in adaptive changes in the forms and functions of adult organisms.

- Describe several examples of how an organism's phenotype can be a response to environmental signals. See pp. 433–435 and Figures 20.10 and 20.12

- How would you determine whether or not an environmental effect on development is adaptive? See pp. 434–435 and Figure 20.11

Appropriate responses to new environmental conditions are likely to evolve over time, but what are the limits of such evolution? Do developmental genes dictate what structures and forms are possible?

20.5 How Do Developmental Genes Constrain Evolution?

Four decades ago, the French geneticist François Jacob made the analogy that evolution works like a tinker, assembling new structures by *combining and modifying the available materials*, and not like an engineer, who is free to develop dramatically different designs (say, a jet engine to replace a propeller-driven engine). We have seen that the evolution of morphology has not been governed by the appearance of radically new genes, but by modifications of existing genes and their regulatory pathways. Thus, developmental genes and their expression constrain evolution in two major ways:

- Nearly all evolutionary innovations are modifications of previously existing structures.

- The genes that control development are highly conserved; that is, the regulatory genes themselves change slowly over the course of evolution.

Evolution proceeds by changing what's already there

The features of organisms almost always evolve from preexisting features in their ancestors. New "wing genes" did not suddenly appear in insects, birds, and bats; instead, wings arose as modifications of existing structures. Wings evolved independently in insects and vertebrates—once in insects, and in three independent instances among the vertebrates (**Figure 20.14**). *In vertebrates, the wings are modified limbs.*

Like limbs, wings have a common structure: a humerus that articulates with the body; two longer bones, the radius and ulna, that project away from the humerus; and then metacarpals and phalanges (digits). During development these bones have different lengths and weights in different organisms.

Developmental controls also influence how organisms lose structures. The ancestors of present-day snakes lost their forelimbs as a result of changes in the segmental expression of Hox genes. The snake lineage subsequently lost its hindlimbs by the loss of expression of the *Sonic hedgehog* gene in the limb bud tissue. But some snake species such as boas and pythons still have rudimentary pelvic bones and upper leg bones.

Conserved developmental genes can lead to parallel evolution

The nucleotide sequences of many of the genes that govern development have been highly conserved throughout the evolution of multicellular organisms—in other words, these genes exist in similar form across a broad spectrum of species (see Figure 20.2).

The existence of highly conserved developmental genes makes it likely that similar traits will evolve repeatedly, especially among closely related species—a process called **parallel phenotypic evolution**. A good example of this process is provided by a small fish, the three-spined stickleback (*Gasterosteus aculeatus*).

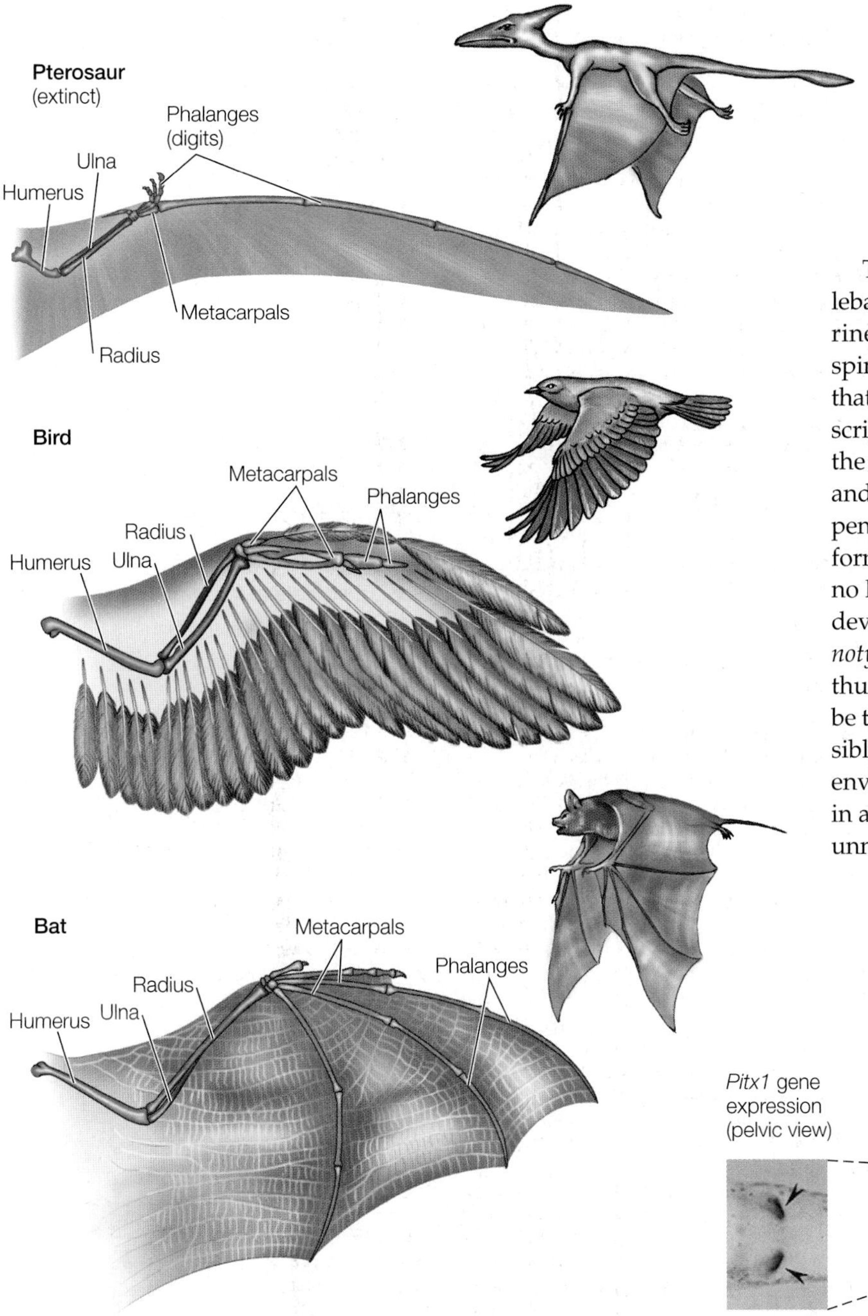

Pterosaur
(extinct)

Phalanges
(digits)

Ulna

Humerus

Metacarpals

Radius

20.14 Wings Evolved Three Times in Vertebrates
The wings of pterosaurs (the earliest flying vertebrates, which lived from 265 to 220 million years ago), birds, and bats are all modified forelimbs constructed from the same skeletal components. However, the components have different forms in the different groups of vertebrates.

Bird

Metacarpals

Phalanges

Radius

Humerus   Ulna

Bat

Metacarpals

Phalanges

Radius

Humerus   Ulna

The difference between marine and freshwater sticklebacks is not induced by environmental conditions. Marine species that are reared in fresh water still grow spines. Not surprisingly, the difference is due to a gene that affects development. The *Pitx1* gene codes for a transcription factor that is normally expressed in regions of the developing embryo that form the head, trunk, tail, and pelvis of the marine stickleback. However, in independent populations from Japan, British Columbia, California, and Iceland, the gene has evolved such that it is no longer expressed in the pelvis, and the spines do not develop. *This same gene has evolved to produce similar phenotypic changes in several independent populations*, and is thus a good example of parallel evolution. What could be the common selective mechanism in these cases? Possibly, the decreased predation pressure in the freshwater environment allows for increased reproductive success in animals that invest less energy in the development of unnecessary protective structures.

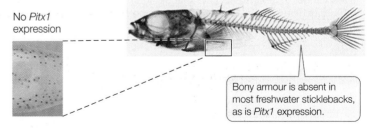

Bony plates and pronounced spines characterize marine sticklebacks.

Dorsal spines

Pitx1 gene
expression
(pelvic view)

Pelvic spine

No *Pitx1*
expression

Bony armour is absent in most freshwater sticklebacks, as is *Pitx1* expression.

Sticklebacks are widely distributed across the Atlantic and Pacific Oceans and are also found in many freshwater lakes. Marine populations of this species spend most of their lives at sea, but return to fresh water to breed. Members of freshwater populations live in lakes and never journey to salt water. Genetic evidence shows that freshwater populations have arisen independently from marine populations many times, most recently at the end of the last ice age. Marine sticklebacks have several structures that protect them from predators: well-developed pelvic bones with pelvic spines, and bony plates. In the freshwater populations descended from them, this body armor is greatly reduced, and dorsal and pelvic spines are much shorter or even lacking (**Figure 20.15**).

20.15 Parallel Phenotypic Evolution in Sticklebacks A developmental gene, *Pitx1*, encodes a transcription factor that stimulates the production of plates and spines. This gene is active in marine sticklebacks, but mutated and inactive in various freshwater populations of the fish. The fact that this mutation is found in geographically distant and isolated freshwater populations is evidence for parallel evolution.

20.5 RECAP

Developmental controls constrain evolution because nearly all evolutionary innovations are modifications of previously existing structures. The conservation of many genes makes it likely that similar traits will evolve repeatedly.

- How have diverse body forms evolved by means of modifications in the functioning of existing genes? **See p. 436 and Figures 20.14**

- Explain how the differences between marine and freshwater sticklebacks exemplify parallel evolution via changes in gene regulation. **See p. 437 and Figure 20.15**

During the course of evolution, many novel traits have arisen, but failed to persist beyond a single generation. Part Six of this book examines the processes of evolution—the powerful forces that influence the survival and reproductive success of various life forms. We will examine how different adaptations become prevalent in different environments, resulting in the extraordinary diversity of life on Earth today.

CHAPTER SUMMARY

20.1 What Is Evo-Devo?

- **Evolutionary developmental biology** (**evo-devo**) is the modern study of the evolutionary aspects of development, and it focuses on molecular mechanisms.

- Changes in development underlie evolutionary changes in morphology that produce major differences in body forms.

- Similarities in the basic mechanisms of development between widely divergent organisms reflect common ancestry. **Review Figure 20.1**

- Evolutionary diversity is produced using a modest number of regulatory genes.

- The transcription factors and chemical signals that govern pattern formation in the bodies of multicellular organisms, and the genes that encode them, can be thought of as a **genetic toolkit**.

- Regulatory genes have been highly conserved during evolution. **Review Figure 20.2**

20.2 How Can Mutations with Large Effects Change Only One Part of the Body?

SEE ANIMATED TUTORIAL 20.1

- The bodies of developing and mature organisms are organized into self-contained units called **developmental modules** that can be modified independently. Modularity allows the timing of different developmental processes to shift independently, in a process called **heterochrony**. **Review Figure 20.4 and 20.5**

- Alterations in the spatial expression patterns of regulatory genes can also result in evolutionary changes. **Review Figure 20.6**

20.3 How Can Differences among Species Evolve?

- Changes in **genetic switches** that determine where and when a set of genes will be expressed underlie both the transformation

of an individual from egg to adult and the evolution of differences among species.

- Morphological changes in species can evolve through mutations in the genes that regulate the differentiation of body segments. **Review Figure 20.8**

20.4 How Does the Environment Modulate Development?

SEE WEB ACTIVITY 20.1

- The ability of an organism to modify its development in response to environmental conditions is called **developmental plasticity**.

- In many species of reptiles, sex development is determined by incubation temperature, which acts through genes that control the production, modification, and action of sex hormones. **Review Figure 20.10**

- The adaptive significance of developmental plasticity is not always obvious, but experiments can test for effects on reproductive success. **Review Figure 20.11**

- Some environmental cues, such as those that anticipate seasons, are highly regular and can reliably drive seasonal adaptations in body form and function. **Review Figure 20.12**

- Environmental cues that trigger developmental change are diverse and can act at any stage of the life of an organism.

20.5 How Do Developmental Genes Constrain Evolution?

- Virtually all evolutionary innovations are modifications of preexisting structures. **Review Figure 20.14**

- Because many genes that govern development have been highly conserved, similar traits are likely to evolve repeatedly, especially among closely related species. This process is called **parallel phenotypic evolution**. **Review Figure 20.15**

SELF-QUIZ

1. Which of the following is *not* one of the principles of evolutionary developmental biology (evo-devo)?
   a. Animal groups share similar molecular mechanisms for morphogenesis.
   b. Changes in the timing of gene expression are important in the evolution of new structures.
   c. Evolution of development is not responsive to the environment.

   d. Changes in the locations of gene expression in the embryo can lead to new structures.
   e. Evolution occurs by modification of existing developmental genes and pathways.

2. The developmental control pathway that results in polarity and pattern formation in the head–abdomen axis in *Drosophila*
   a. has a similar gene sequence and chromosome order in the mouse.

b. arose only in insects during evolution.

c. determines only the organs that arise in head segments.

d. involves only gene products made by the embryo.

e. arose through new genes that had not existed before in any form.

3. Which of the following is *not* true of genetic switches?

a. They control how a genetic toolkit is used.

b. They integrate positional information in an embryo.

c. A single switch controls each gene.

d. They allow different structures to develop within an individual organism.

e. They determine when and where a gene is turned on or off.

4. Ducks have webbed feet and chickens do not because

a. ducks need webbed feet to swim, whereas terrestrial chickens do not.

b. both duck and chicken embryos express *BMP4* in the webbing between the toes, but the *Gremlin* gene is expressed in the webbing cells only in ducks.

c. both duck and chicken embryos express *BMP4* in the webbing between the toes, but the *Gremlin* gene is expressed in the webbing cells only in chickens.

d. only duck embryos express *BMP4* in the webbing between the toes.

e. only chicken embryos express *BMP4* in the webbing between the toes.

5. Modularity is important for development because it

a. guarantees that all units of a developing embryo will change in a coordinated way.

b. coordinates the establishment of the anterior–posterior axis of the developing embryo.

c. allows changes in developmental genes to change one part of the body without affecting other parts.

d. guarantees that the timing of gene expression is the same in all parts of a developing embryo.

e. allows organisms to be built up one module at a time.

6. Organisms often respond to environmental signals that accurately predict future conditions by

a. stopping development until the signal changes.

b. altering their development to adapt to the future environment.

c. altering their development such that the resulting adult can produce offspring adapted to the future environment.

d. producing new mutants.

e. developing normally because the predicted conditions may not last long.

7. The process whereby changes in the timing of developmental events can change the form of an organism is called

a. heterochrony.

b. developmental plasticity.

c. adaptation.

d. modularity.

e. mutation.

8. Which of the following is true about temperature determination of sex in some reptiles?

a. It ensures that males and females are produced at different seasons of the year.

b. It ensures that males are faster than females.

c. It acts through the inactivation of the male sex chromosome.

d. There is no evidence that it has evolved because of effects on reproductive success.

e. Temperature effects are due to modifications of concentrations and actions of sex steroids.

9. Which of the following examples of evolutionary change do *not* involve Hox genes?

a. Difference in numbers of legs between bees and centipedes.

b. Difference in number of cervical vertebrae between a goose and a giraffe.

c. Loss of forelegs in snakes.

d. Loss of webbing in the feet of chickens.

e. Location of legs and antennae in *Drosophila.*

10. Parallel phenotypic evolution is likely to occur because

a. closely related organisms typically face similar problems.

b. the conservation of regulatory genes during evolution means that similar traits are likely to evolve repeatedly.

c. many different phenotypes can be produced by a given genotype.

d. phenotypic plasticity, which generates parallel phenotypic evolution, is widespread.

e. evolutionary biologists have looked especially hard to find evidence of it.

FOR DISCUSSION

1. What environmental influences on development would probably be missed if investigations were confined to unicellular organisms such as bacteria and single-celled eukaryotes?

2. If evolutionary innovations can result from rather simple changes in the timing of expression of a few genes, why have such innovations arisen relatively infrequently during evolution?

3. François Jacob stated that evolution was more like tinkering than engineering. Does the observation that developmental genes have changed little over evolutionary time support his assertion? Why?

4. Despite their major differences, plants and animals share many of the genes that regulate development. What are the implications of this observation for the ways in which humans can respond to the adverse effects of the many substances we release into the environment that cause developmental abnormalities in plants and animals? What kinds of substances are most likely to have such effects? Why?

ADDITIONAL INVESTIGATION

Figure 20.7 describes an experiment in which the protein Gremlin, which inhibits expression of the *BMP4* gene, was introduced into the foot of a developing chicken. What results would you expect from introducing Gremlin into other parts of a developing chicken? Why? Into what other body parts would it be most informative to introduce Gremlin? If you were particularly interested in parallel phenotypic evolution, what other organisms might you use in these experiments?

Appendix A: The Tree of Life

Phylogeny is the organizing principle of modern biological taxonomy. A guiding principle of modern phylogeny is *monophyly*. A monophyletic group is considered to be one that contains an ancestral lineage and *all* of its descendants. Any such group can be extracted from a phylogenetic tree with a single cut.

The tree shown here provides a guide to the relationships among the major groups of extant (living) organisms in the tree of life as we have presented them throughout this book. The position of the branching "splits" indicates the relative branching order of the lineages of life, but the time scale is not meant to be uniform. In addition, the groups appearing at the branch tips do not necessarily carry equal phylogenetic "weight." For example, the ginkgo [63] is indeed at the apex of its lineage; this gymnosperm group consists of a single living species. In contrast, a phylogeny of the eudicots [55] could continue on from this point to fill many more trees the size of this one.

The glossary entries that follow are informal descriptions of some major features of the organisms described in Part Seven of this book. Each entry gives the group's common name, followed by the formal scientific name of the group (in parentheses). Numbers in square brackets reference the location of the respective groups on the tree.

It is sometimes convenient to use an informal name to refer to a collection of organisms that are not monophyletic but nonetheless all share (or all lack) some common attribute. We call these "convenience terms"; such groups are indicated in these entries by quotation marks, and we do not give them formal scientific names. Examples include "prokaryotes," "protists," and "algae." Note that these groups cannot be removed with a single cut; they represent a collection of distantly related groups that appear in different parts of the tree. We also use quotation marks here to designate two groups of fungi that are not believed to be monophyletic.

An interactive version of this tree, with links to much greater detail (such as photos, distribution maps, species lists, and identification keys), can be found at yourBioPortal.com.

– A –

acorn worms (*Enteropneusta*) Benthic marine hemichordates [120] with an acorn-shaped proboscis, a short collar (neck), and a long trunk.

"algae" A convenience term encompassing various distantly related groups of aquatic, photosynthetic chromalveolates [5] and certain members of the Plantae [8].

alveolates (*Alveolata*) [7] Unicellular eukaryotes with a layer of flattened vesicles (alveoli) supporting the plasma membrane. Major alveolate groups include the dinoflagellates [52], apicomplexans [53], and ciliates [54].

amborella (*Amborella*) [60] An understory shrub or small tree found in New Caledonia. Thought to be the sister-group of the remaining living angiosperms [14].

ambulacrarians (*Ambulacraria*) [30] The echinoderms [119] and hemichordates [120].

amniotes (*Amniota*) [37] Mammals, reptiles, and their extinct close relatives. Characterized by many adaptations to terrestrial life, including an amniotic egg (with a unique set of membranes—the amnion, chorion, and allantois), a water-repellant epidermis (with epidermal scales, hair, or feathers), and, in males, a penis that allows internal fertilization.

amoebozoans (*Amoebozoa*) [85] A group of eukaryotes [4] that use lobe-shaped pseudopods for locomotion and to engulf food. Major amoebozoan groups include the loboseans, plasmodial slime molds, and cellular slime molds.

amphibians (*Amphibia*) [129] Tetrapods [36] with glandular skin that lacks epidermal scales, feathers, or hair. Many amphibian species undergo a complete metamorphosis from an aquatic larval form to a terrestrial adult form, although direct development is also common. Major amphibian groups include frogs and toads (anurans), salamanders, and caecilians.

amphipods (*Amphipoda*) Small crustaceans [117] that are abundant in many marine and freshwater habitats. They are important herbivores, scavengers, and micropredators, and are an important food source for many aquatic organisms.

angiosperms (*Anthophyta* or *Magnoliophyta*) [14] The flowering plants. Major angiosperm groups include the monocots, eudicots, and magnoliids.

animals (*Animalia* or *Metazoa*) [21] Multicellular heterotrophic eukaryotes. The majority of animals are bilaterians [24]. Other groups of animals include the cnidarians [98], ctenophores [97], placozoans [96], and sponges [22]. The closest living relatives of the animals are the choanoflagellates [92].

annelids (*Annelida*) [105] Segmented worms, including earthworms, leeches, and polychaetes. One of the major groups of lophotrochozoans [26].

anthozoans (*Anthozoa*) One of the major groups of cnidarians [98]. Includes the sea anemones, sea pens, and corals.

anurans (*Anura*) Comprising the frogs and toads, this is the largest group of living amphibians [129]. They are tail-less, with a shortened vertebral column and elongate hind legs modified for jumping. Many species have an aquatic larval form known as a tadpole.

apicomplexans (*Apicomplexa*) [53] Parasitic alveolates [7] characterized by the possession of an apical complex at some stage in the life cycle.

arachnids (*Arachnida*) Chelicerates [115] with a body divided into two parts: a cephalothorax that bears six pairs of appendages (four pairs of which are usually used as legs) and an abdomen that bears the genital opening. Familiar arachnids include spiders, scorpions, mites and ticks, and harvestmen.

arbuscular mycorrhizal fungi (*Glomeromycota*) [88] A group of fungi [19] that associate with plant roots in a close symbiotic relationship.

archaeans (*Archaea*) [3] Unicellular organisms lacking a nucleus and lacking peptidoglycan in the cell wall. Once grouped with the bacteria, archaeans possess distinctive membrane lipids.

archosaurs (*Archosauria*) [39] A group of reptiles [38] that includes dinosaurs and crocodilians [134]. Most dinosaur groups became extinct at the end of the Cretaceous; birds [133] are the only surviving dinosaurs.

arrow worms (*Chaetognatha*) [107] Small planktonic or benthic predatory marine worms with fins and a pair of hooked, prey-grasping spines on each side of the head.

arthropods (*Arthropoda*) The largest group of ecdysozoans [27]. Arthropods are characterized by a stiff exoskeleton, segmented bodies, and jointed appendages. Includes the chelicerates [115], myriapods [116], crustaceans [117], and hexapods (insects and their relatives) [118].

ascidians (*Ascidiacea*) "Sea squirts"; the largest group of urochordates [122]. Also known as tunicates, they are sessile (as adults), marine, saclike filter feeders.

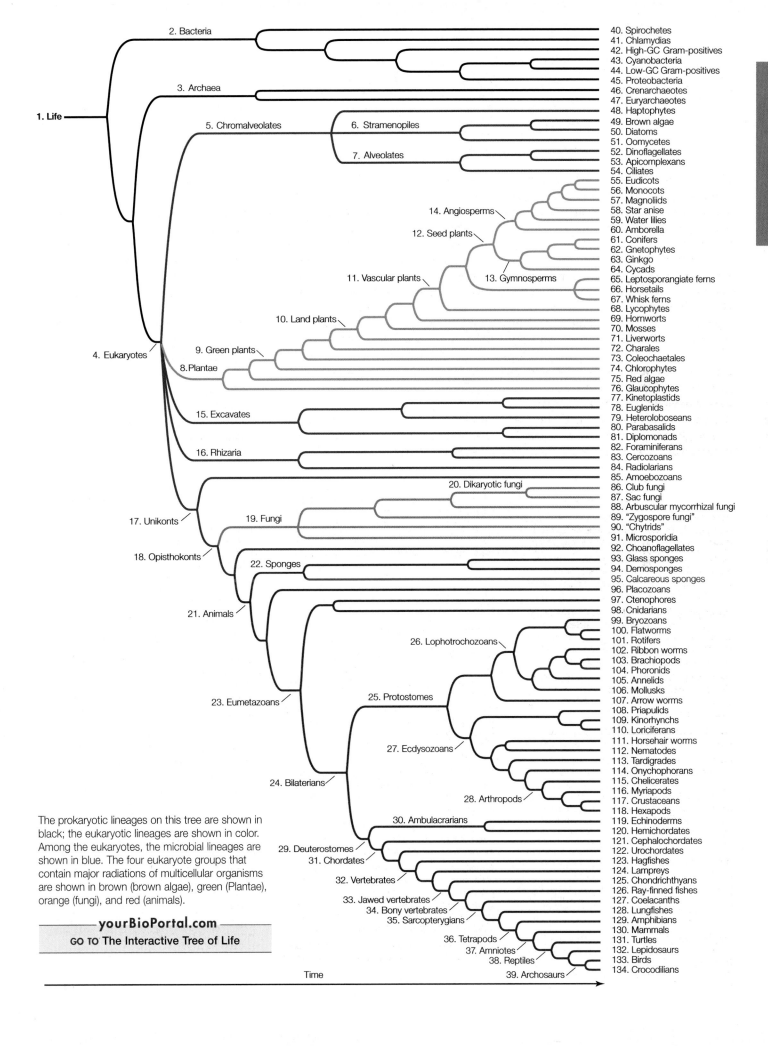

The prokaryotic lineages on this tree are shown in black; the eukaryotic lineages are shown in color. Among the eukaryotes, the microbial lineages are shown in blue. The four eukaryote groups that contain major radiations of multicellular organisms are shown in brown (brown algae), green (Plantae), orange (fungi), and red (animals).

yourBioPortal.com
GO TO The Interactive Tree of Life

– B –

bacteria (*Eubacteria*) [2]   Unicellular organisms lacking a nucleus, possessing distinctive ribosomes and initiator tRNA, and generally containing peptidoglycan in the cell wall. Different bacterial groups are distinguished primarily on nucleotide sequence data.

barnacles (*Cirripedia*)   Crustaceans [117] that undergo two metamorphoses—first from a feeding planktonic larva to a nonfeeding swimming larva, and then to a sessile adult that forms a "shell" composed of four to eight plates cemented to a hard substrate.

bilaterians (*Bilateria*) [24]   Those animal groups characterized by bilateral symmetry and three distinct tissue types (endoderm, ectoderm, and mesoderm). Includes the protostomes [25] and deuterostomes [29].

birds (*Aves*) [133]   Feathered, flying (or secondarily flightless) tetrapods [36].

bivalves (*Bivalvia*)   Major mollusk [106] group; clams and mussels. Bivalves typically have two similar hinged shells that are each asymmetrical across the midline.

bony vertebrates (*Osteichthyes*) [34]   Vertebrates [32] in which the skeleton is usually ossified to form bone. Includes the ray-finned fishes [126], coelacanths [127], lungfishes [128], and tetrapods [36].

brachiopods (*Brachiopoda*) [103]   Lophotrochozoans [26] with two similar hinged shells that are each symmetrical across the midline. Superficially resemble bivalve mollusks, except for the shell symmetry.

brittle stars (*Ophiuroidea*)   Echinoderms [119] with five long, whip-like arms radiating from a distinct central disk that contains the reproductive and digestive organs.

brown algae (*Phaeophyta*) [49]   Multicellular, almost exclusively marine stramenopiles [6] generally containing the pigment fucoxanthin as well as chlorophylls *a* and *c* in their chloroplasts.

bryozoans (*Ectoprocta* or *Bryozoa*) [99]   A group of marine and freshwater lophotrochozoans [26] that live in colonies attached to substrata; also known as ectoprocts or moss animals.

– C –

caecilians (*Gymnophiona*)   A group of burrowing or aquatic amphibians [129]. They are elongate, legless, with a short tail (or none at all), reduced eyes covered with skin or bone, and a pair of sensory tentacles on the head.

calcareous sponges (*Calcarea*) [95]   Filter-feeding marine sponges with spicules composed of calcium carbonate.

cellular slime molds (*Dictyostelida*)   Amoebozoans [85] in which individual amoebas aggregate under stress to form a multicellular pseudoplasmodium.

cephalochordates (*Cephalochordata*) [121]   A group of weakly swimming, eel-like benthic marine chordates [31]; also called lancelets.

cephalopods (*Cephalopoda*)   Active, predatory mollusks [106] in which the molluscan foot has been modified into muscular hydrostatic arms or tentacles. Includes octopuses, squids, and nautiluses.

cercozoans (*Cercozoa*) [83]   Unicellular eukaryotes [4] that feed by means of threadlike pseudopods. Group together with foraminiferans [82] and radiolarians [84] to comprise the rhizaria [16].

Charales [72]   Multicellular green algae with branching, apical growth and plasmodesmata between adjacent cells. The closest living relatives of the land plants [10], they retain the egg in the parent organism.

chelicerates (*Chelicerata*) [115]   A major group of arthropods [28] with pointed appendages (chelicerae) used to grasp food (as opposed to the chewing mandibles of most other arthropods). Includes the arachnids, horseshoe crabs, pycnogonids, and extinct sea scorpions.

chimaeras (*Holocephali*)   A group of bottom-dwelling, marine, scaleless chondrichthyan fishes [125] with large, permanent, grinding tooth plates (rather than the replaceable teeth found in other chondrichthyans).

chitons (*Polyplacophora*)   Flattened, slow-moving mollusks [106] with a dorsal protective calcareous covering made up of eight articulating plates.

chlamydias (*Chlamydiae*) [41]   A group of very small Gram-negative bacteria; they live as intracellular parasites of other organisms.

chlorophytes (*Chlorophyta*) [74]   The most abundant and diverse group of green algae, including freshwater, marine, and terrestrial forms; some are unicellular, others colonial, and still others multicellular. Chlorophytes use chlorophylls *a* and *c* in their photosynthesis.

choanoflagellates (*Choanozoa*) [92]   Unicellular eukaryotes [4] with a single flagellum surrounded by a collar. Most are sessile, some are colonial. The closest living relatives of the animals [21].

chondrichthyans (*Chondrichthyes*) [125]   One of the two main groups of jawed vertebrates [33]; includes sharks, rays, and chimaeras. They have cartilaginous skeletons and paired fins.

chordates (*Chordata*) [31]   One of the two major groups of deuterostomes [29], characterized by the presence (at some point in development) of a notochord, a hollow dorsal nerve cord, and a post-anal tail. Includes the cephalochordates [121], urochordates [122], and vertebrates [32].

chromalveolates (*Chromalveolata*) [5]   A contested group, said to have arisen from a common ancestor with chloroplasts derived from a red alga and supported by some molecular evidence. Major chromalveolate groups include the alveolates [7], stramenopiles [6], and haptophytes [48].

"chytrids" [90]   A convenience term used for a paraphyletic group of mostly aquatic, microscopic fungi [19] with flagellated gametes. Some exhibit alternation of generations.

ciliates (*Ciliophora*) [54]   Alveolates [7] with numerous cilia and two types of nuclei (micronuclei and macronuclei).

clitellates (*Clitellata*)   Annelids [105] with gonads contained in a swelling (called a clitellum) toward the head of the animal. Includes earthworms (oligochaetes) and leeches.

club fungi (*Basidiomycota*) [86]   Fungi [19] that, if multicellular, bear the products of meiosis on club-shaped basidia and possess a long-lasting dikaryotic stage. Some are unicellular.

club mosses (*Lycopodiophyta*) [68]   Vascular plants [11] characterized by microphylls. See lycophytes.

cnidarians (*Cnidaria*) [98]   Aquatic, mostly marine eumetazoans [23] with specialized stinging organelles (nematocysts) used for prey capture and defense, and a blind gastrovascular cavity. The closest living relatives of the ctenophores [97].

coelacanths (*Actinista*) [127]   A group of marine sarcopterygians [35] that was diverse from the Middle Devonian to the Cretaceous, but is now known from just two living species. The pectoral and anal fins are on fleshy stalks supported by skeletal elements, so they are also called lobe-finned fishes.

Coleochaetales [73]   Multicellular green algae characterized by flattened growth form composed of thin-walled cells. Thought to be the sister-group to the Charales [72] plus land plants [10].

conifers (*Pinophyta* or *Coniferophyta*) [61]   Cone-bearing, woody seed plants [12].

copepods (*Copepoda*)   Small, abundant crustaceans [117] found in marine, freshwater, or wet terrestrial habitats. They have a single eye, long antennae, and a body shaped like a teardrop.

craniates (*Craniata*)   Some biologist exclude the hagfishes [123] from the vertebrates [32], and use the term craniates to refer to the two groups combined.

crenarchaeotes (*Crenarchaeota*) [46]   A major and diverse group of archaeans [3], defined on the basis of rRNA base sequences. Many are extremophiles (inhabit extreme environments), but the group may also be the most abundant archaeans in the marine environment.

crinoids (*Crinoidea*)   Echinoderms [119] with a mouth surrounded by feeding arms, and a U-shaped gut with the mouth next to the anus. They attach to the substratum by a stalk or are free-swimming. Crinoids were abundant in the middle and late Paleozoic, but only a few hundred species have survived to the present. Includes the sea lilies and feather stars.

crocodilians (*Crocodylia*) [134]   A group of large, predatory, aquatic archosaurs [39]. The closest living relatives of birds [133]. Includes alligators, caimans, crocodiles, and gharials.

crustaceans (*Crustacea*) [117]   Major group of marine, freshwater, and terrestrial arthropods [28] with a head, thorax, and abdomen (although the head and thorax may be fused), covered with a thick exoskeleton, and with two-part appendages. Crustaceans undergo metamorphosis from a nauplius larva. Includes decapods, isopods, krill, barnacles, amphipods, copepods, and ostracods.

ctenophores (*Ctenophora*) [97]   Radially symmetrical, diploblastic marine animals [21], with a complete gut and eight rows of fused plates of cilia (called ctenes).

cyanobacteria (*Cyanobacteria*) [43]   A group of unicellular, colonial, or filamentous bacteria that conduct photosynthesis using chlorophyll *a*.

cycads (*Cycadophyta*) [64]   Palmlike gymnosperms with large, compound leaves.

cyclostomes (*Cyclostomata*)   This term refers to the possibly monophyletic group of lampreys [124] and hagfishes [123]. Molecular data support this group, but morphological data suggest that lampreys are more closely related to jawed vertebrates [33] than to hagfishes.

– D –

decapods (*Decapoda*)   A group of marine, freshwater, and semiterrestrial crustaceans [117] in which five of the eight pairs of thoracic appendages function as legs (the other three pairs, called maxillipeds, function as mouthparts). Includes crabs, lobsters, crayfishes, and shrimps.

demosponges (*Demospongiae*) [94]   The largest of the three groups of sponges [22], accounting for 90 percent of all sponge species. Demosponges have spicules made of silica, spongin fiber (a protein), or both.

deuterostomes (*Deuterostomia*) [29]   One of the two major groups of bilaterians [24], in which the mouth forms at the opposite end of the embryo from the blastopore in early development (contrast with protostomes). Includes the ambulacrarians [30] and chordates [31].

diatoms (*Bacillariophyta*) [50]   Unicellular, photosynthetic stramenopiles [6] with glassy cell walls in two parts.

dikaryotic fungi (*Dikarya*) [20]   A group of fungi [19] in which two genetically different haploid nuclei coexist and divide within the same hypha; includes club fungi [86] and sac fungi [87].

dinoflagellates (*Dinoflagellata*) [52]   A group of alveolates [7] usually possessing two flagella, one in an equatorial groove and the other in a longitudinal groove; many are photosynthetic.

diplomonads (*Diplomonadida*) [81]   A group of eukaryotes [4] lacking mitochondria; most have two nuclei, each with four associated flagella.

– E –

ecdysozoans (*Ecdysozoa*) [27]   One of the two major groups of protostomes [25], characterized by periodic molting of their exoskeletons. Nematodes [112] and arthropods [28] are the largest ecdysozoan groups.

echinoderms (*Echinodermata*) [119]   A major group of marine deuterostomes [29] with fivefold radial symmetry (at some stage of life) and an endoskeleton made of calcified plates and spines. Includes sea stars, crinoids, sea urchins, sea cucumbers, and brittle stars.

elasmobranchs (*Elasmobranchii*)   The largest group of chondrichthyan fishes [125]. Includes sharks, skates, and rays. In contrast to the other group of living chondrichthyans (the chimaeras), they have replaceable teeth.

embryophytes   See land plants [10].

eudicots (*Eudicotyledones*)[55]   A group of angiosperms [14] with pollen grains possessing three openings. Typically with two cotyledons, net-veined leaves, taproots, and floral organs typically in multiples of four or five.

euglenids (*Euglenida*) [78]   Flagellate excavates characterized by a pellicle composed of spiraling strips of protein under the plasma membrane; the mitochondria have disk-shaped cristae. Some are photosynthetic.

eukaryotes (*Eukarya*) [4]   Organisms made up of one or more complex cells in which the genetic material is contained in nuclei. Contrast with archaeans [3] and bacteria [2].

eumetazoans (*Eumetazoa*) [23]   Those animals [21] characterized by body symmetry, a gut, a nervous system, specialized types of cell junctions, and well-organized tissues in distinct cell layers (although there have been secondary losses of some of these characteristics in some eumetazoans).

euphyllophytes (*Euphyllophyta*)   The group of vascular plants [11] that is sister to the lycophytes [68] and which includes all plants with megaphylls.

euryarchaeotes (*Euryarchaeota*) [47]   A major group of archaeans [3], diagnosed on the basis of rRNA sequences. Includes many methanogens, extreme halophiles, and thermophiles.

eutherians (*Eutheria*)   A group of viviparous mammals [130], eutherians are well developed at birth (contrast to prototherians and marsupials, the other two groups of mammals). Most familiar mammals outside the Australian and South American regions are eutherians (see Table 33.1).

excavates (*Excavata*) [15]   Diverse group of unicellular, flagellate eukaryotes, many of which possess a feeding groove; some lack mitochondria.

– F –

"ferns"   Vascular plants [11] usually possessing large, frondlike leaves that unfold from a "fiddlehead." Not a monophyletic group, although most fern species are encompassed in a monophyletic clade, the leptosporangiate ferns [65].

flatworms (*Platyhelminthes*) [100]   A group of dorsoventrally flattened and generally elongate soft-bodied lophotrochozoans [26]. May be free-living or parasitic, found in marine, freshwater, or damp terrestrial environments. Major flatworm groups include the tapeworms, flukes, monogeneans, and turbellarians.

flowering plants   See angiosperms [14].

flukes (*Trematoda*)   A group of wormlike parasitic flatworms [100] with complex life cycles that involve several different host species. May be paraphyletic with respect to tapeworms.

foraminiferans (*Foraminifera*) [82]   Amoeboid organisms with fine, branched pseudopods that form a food-trapping net. Most produce external shells of calcium carbonate.

fungi (*Fungi*) [19]   Eukaryotic heterotrophs with absorptive nutrition based on extracellular digestion; cell walls contain chitin. Major fungal groups include the microsporidia [91], "chytrids" [90], "zygospore fungi" [89], arbuscular mycorrhizal fungi [88], sac fungi [87], and club fungi [86].

– G –

gastropods (*Gastropoda*)   The largest group of mollusks [106]. Gastropods possess a well-defined head with two or four sensory tentacles (often terminating in eyes) and a ventral foot. Most species have a single coiled or spiraled shell. Common in marine, freshwater, and terrestrial environments.

ginkgo (*Ginkgophyta*) [63]   A gymnosperm [13] group with only one living species. The ginkgo seed is surrounded by a fleshy tissue not derived from an ovary wall and hence not a fruit.

glass sponges (*Hexactinellida*) [93]   Sponges [22] with a skeleton composed of four- and/or six-pointed spicules made of silica.

glaucophytes (*Glaucophyta*) [76]   Unicellular freshwater algae with chloroplasts containing traces of peptidoglycan, the characteristic cell wall material of bacteria.

gnathostomes (*Gnathostomata*)   See jawed vertebrates [33].

gnetophytes (*Gnetophyta*) [62]   A gymnosperm [13] group with three very different lineages; all have wood with vessels, unlike other gymnosperms.

green plants (*Viridiplantae*) [9]   Organisms with chlorophylls *a* and *b*, cellulose-containing cell walls, starch as a carbohydrate storage product, and chloroplasts surrounded by two membranes.

gymnosperms (*Gymnospermae*) [13]   Seed plants [12] with seeds "naked" (i.e., not enclosed in carpels). Probably monophyletic, but status still in doubt. Includes the conifers [61], gnetophytes [62], ginkgo [63], and cycads [64].

– H –

hagfishes (*Myxini*) [123]   Elongate, slimy-skinned vertebrates [32] with three small accessory hearts, a partial cranium, and no stomach or paired fins. See also craniata; cyclostomes.

haptophytes (*Haptophyta*) [48]   Unicellular, photosynthetic chromalveolates [5] with two slightly unequal, smooth flagella. Abundant as phytoplankton, some form marine algal blooms.

hemichordates (*Hemichordata*) [120]   One of the two primary groups of ambulacrarians [30]; marine wormlike organisms with a three-part body plan.

heteroloboseans (*Heterolobosea*) [79]   Colorless excavates [15] that can transform among amoeboid, flagellate, and encysted stages.

hexapods (*Hexapoda*) [118]   Major group of arthropods [28] characterized by a reduction (from the ancestral arthropod condition) to six walking appendages, and the consolidation of three body segments to form a thorax. Includes insects and their relatives (see Table 32.2).

high-GC Gram-positives (*Actinobacteria*) [42]   Gram-positive bacteria with a relatively high (G+C)/(A+T) ratio of their DNA, with a filamentous growth habit.

hornworts (*Anthocerophyta*) [69]   Nonvascular plants with sporophytes that grow from the base. Cells contain a single large, platelike chloroplast.

horsehair worms (*Nematomorpha*) [111]   A group of very thin, elongate, wormlike freshwater ecdysozoans [27]. Largely nonfeeding as adults, they are parasites of insects and crayfish as larvae.

horseshoe crabs (*Xiphosura*)   Marine chelicerates [115] with a large outer shell in three parts: a carapace, an abdomen, and a tail-like telson. There are only five living species, but many additional species are known from fossils.

horsetails (*Sphenophyta* or *Equisetophyta*) [66] Vascular plants [11] with reduced megaphylls in whorls.

hydrozoans (*Hydrozoa*) A group of cnidarians [98]. Most species go through both polyp and mesuda stages, although one stage or the other is eliminated in some species.

– I –

insects (*Insecta*) The largest group within the hexapods [118]. Insects are characterized by exposed mouthparts and one pair of antennae containing a sensory receptor called a Johnston's organ. Most have two pairs of wings as adults. There are more described species of insects than all other groups of life [1] combined, and many species remain to be discovered. The major insect groups are described in Table 32.2.

"invertebrates" A convenience term that encompasses any animal [21] that is not a vertebrate [32].

isopods (*Isopoda*) Crustaceans [117] characterized by a compact head, unstalked compound eyes, and mouthparts consisting of four pairs of appendages. Isopods are abundant and widespread in salt, fresh, and brackish water, although some species (the sow bugs) are terrestrial.

– J –

jawed vertebrates (*Gnathostomata*) [33] A major group of vertebrates [32] with jawed mouths. Includes chondrichthyans [125], ray-finned fishes [126], and sarcopterygians [35].

– K –

kinetoplastids (*Kinetoplastida*) [77] Unicellular, flagellate organisms characterized by the presence in their single mitochondrion of a kinetoplast (a structure containing multiple, circular DNA molecules).

kinorhynchs (*Kinorhyncha*) [109] Small (< 1 mm) marine ecdysozoans [27] with bodies in 13 segments and a retractable proboscis.

korarchaeotes (*Korarchaeota*) A group of archaeans [3] known only by evidence from nucleic acids derived from hot springs. Its phylogenetic relationships within the Archaea are unknown.

krill (*Euphausiacea*) A group of shrimplike marine crustaceans [117] that are important components of the zooplankton.

– L –

lampreys (*Petromyzontiformes*) [124] Elongate, eel-like vertebrates [32] that often have rasping and sucking disks for mouths.

lancelets (*Cephalochordata*) See cephalochordates [121].

land plants (*Embryophyta*) [10] Plants with embryos that develop within protective structures; also called embryophytes. Sporophytes and gametophytes are multicellular. Land plants possess a cuticle. Major groups are the liverworts [71], mosses [70], hornworts [69], and vascular plants [11].

larvaceans (*Larvacea*) Solitary, planktonic urochordates [122] that retain both notochords and nerve cords throughout their lives.

lepidosaurs (*Lepidosauria*) [132] Reptiles [38] with overlapping scales. Includes tuataras and

squamates (lizards, snakes, and amphisbaenians).

leptosporangiate ferns (*Pteridopsida* or *Polypodiopsida*) [65] Vascular plants [11] usually possessing large, frondlike leaves that unfold from a "fiddlehead," and possessing thin-walled sporangia.

life (*Life*) [1] The monophyletic group that includes all known living organisms. Characterized by a nucleic-acid based genetic system (DNA or RNA), metabolism, and cellular structure. Some parasitic forms, such as viruses, have secondarily lost some of these features and rely on the cellular environment of their host.

liverworts (*Hepatophyta*) [71] Nonvascular plants lacking stomata; stalk of sporophyte elongates along its entire length.

loboseans (*Lobosea*) A group of unicellular amoebozoans [85]; includes the most familiar amoebas (e.g., *Amoeba proteus*).

"lophophorates" Not a monophyletic group. A convenience term used to describe several groups of lophotrochozoans [26] that have a feeding structure called a lophophore (a circular or U-shaped ridge around the mouth that bears one or two rows of ciliated, hollow tentacles).

lophotrochozoans (*Lophotrochozoa*) [26] One of the two main groups of protostomes [25]. This group is morphologically diverse, and is supported primarily on information from gene sequences. Includes bryozoans [99], flatworms [100], rotifers [101], ribbon worms [102], brachiopods [103], phoronids [104], annelids [105], and mollusks [106].

loriciferans (*Loricifera*) [110] Small (< 1 mm) ecdysozoans [27] with bodies in four parts, covered with six plates.

low-GC Gram-positives (*Firmicutes*) [44] A diverse group of bacteria [2] with a relatively low (G+C)/(A+T) ratio of their DNA, often but not always Gram-positive, some producing endospores.

lungfishes (*Dipnoi*) [128] A group of aquatic sarcopterygians [35] that are the closest living relatives of the tetrapods [36]. They have a modified swim bladder used to absorb oxygen from air, so some species can survive the temporary drying of their habitat.

lycophytes (*Lycopodiophyta*) [68] Vascular plants [11] characterized by microphylls; includes club mosses, spike mosses, and quillworts.

– M –

magnoliids (*Magnoliidae*) [57] A major group of angiosperms [14] possessing two cotyledons and pollen grains with a single opening. The group is defined primarily by nucleotide sequence data; it is more closely related to the eudicots and monocots than to three other small angiosperm groups.

mammals (*Mammalia*) [130] A group of tetrapods [36] with hair covering all or part of their skin; females produce milk to feed their developing young. Includes the prototherians, marsupials, and eutherians.

marsupials (*Marsupialia*) Mammals [130] in which the female typically has a marsupium (a pouch for rearing young, which are born at an

extremely early stage in development). Includes such familiar mammals as opossums, koalas, and kangaroos.

metazoans (*Metazoa*) See animals [21].

microbial eukaryotes See "protists."

microsporidia (*Microsporidia*) [91] A group of parasitic unicellular fungi [19] that lack mitochondria and have walls that contain chitin.

mollusks (*Mollusca*) [106] One of the major groups of lophotrochozoans [26], mollusks have bodies composed of a foot, a mantle (which often secretes a hard, calcareous shell), and a visceral mass. Includes monoplacophorans, chitons, bivalves, gastropods, and cephalopods.

monilophytes (*Monilophyta*) A group of vascular plants [11], sister to the seed plants [12], characterized by overtopping and possession of megaphylls; includes the ferns [65], horsetails [66], and whisk ferns [67].

monocots (*Monocotyledones*) [56] Angiosperms [14] characterized by possession of a single cotyledon, usually parallel leaf veins, a fibrous root system, pollen grains with a single opening, and floral organs usually in multiples of three.

monogeneans (*Monogenea*) A group of ectoparasitic flatworms [100].

monoplacophorans (*Monoplacophora*) Mollusks [106] with segmented body parts and a single, thin, flat, rounded, bilateral shell.

mosses (*Bryophyta*) [70] Nonvascular plants with true stomata and erect, "leafy" gametophytes; sporophytes elongate by apical cell division.

moss animals See bryozoans [99].

myriapods (*Myriapoda*) [116] Arthropods [28] characterized by an elongate, segmented trunk with many legs. Includes centipedes and millipedes.

– N –

nanoarchaeotes (*Nanoarchaeota*) A group of extremely small, thermophilic archaeans [3] with a much-reduced genome. The only described example can survive only when attached to a host organism.

nematodes (*Nematoda*) [112] A very large group of elongate, unsegmented ecdysozoans [27] with thick, multilayer cuticles. They are among the most abundant and diverse animals, although most species have not yet been described. Include free-living predators and scavengers, as well as parasites of most species of land plants [10] and animals [21].

neognaths (*Neognathae*) The main group of birds [133], including all living species except the ratites (ostrich, emu, rheas, kiwis, cassowaries) and tinamous. See palaeognaths.

– O –

oligochaetes (*Oligochaeta*) An annelid [105] group whose members lack parapodia, eyes, and anterior tentacles, and have few setae. Earthworms are the most familiar oligochaetes.

onychophorans (*Onychophora*) [114] Elongate, segmented ecdysozoans [27] with many pairs of soft, unjointed, claw-bearing legs. Also known as velvet worms.

oomycetes (*Oomycota*) [51]   Water molds and relatives; absorptive heterotrophs with nutrient-absorbing, filamentous hyphae.

opisthokonts (*Opisthokonta*) [18]   A group of unikonts [17] in which the flagellum on motile cells, if present, is posterior. The opisthokonts include the fungi [19], animals [21], and choanoflagellates [92].

ostracods (*Ostracoda*)   Marine and freshwater crustaceans [117] that are laterally compressed and protected by two clamlike calcareous or chitinous shells.

– P –

palaeognaths (*Palaeognathae*)   A group of secondarily flightless or weakly flying birds [133]. Includes the flightless ratites (ostrich, emu, rheas, kiwis, cassowaries) and the weakly flying tinamous.

parabasalids (*Parabasalia*) [80]   A group of unicellular eukaryotes [4] that lack mitochondria; they possess flagella in clusters near the anterior of the cell.

phoronids (*Phoronida*) [104]   A small group of sessile, wormlike marine lophotrochozoans [26] that secrete chitinous tubes and feed using a lophophore.

placoderms (*Placodermi*)   An extinct group of jawed vertebrates [33] that lacked teeth. Placoderms were the dominant predators in Devonian oceans.

placozoans (*Placozoa*) [96]   A poorly known group of structurally simple, asymmetrical, flattened, transparent animals found in coastal marine tropical and subtropical seas. Most evidence suggests that placozoans are the sister-group of eumetazoans [23].

Plantae [8]   The most broadly defined plant group. In most parts of this book, we use the word "plant" as synonymous with "land plant" [10], a more restrictive definition.

plasmodial slime molds (*Myxogastrida*)   Amoebozoans [85] that in their feeding stage consist of a coenocyte called a plasmodium.

pogonophorans (*Pogonophora*)   Deep-sea annelids [105] that lack a mouth or digestive tract; they feed by taking up dissolved organic matter, facilitated by endosymbiotic bacteria in a specialized organ (the trophosome).

polychaetes (*Polychaeta*)   A group of mostly marine annelids [105] with one or more pairs of eyes and one or more pairs of feeding tentacles; parapodia and setae extend from most body segments. May be paraphyletic with respect to the clitellates.

priapulids (*Priapulida*) [108]   A small group of cylindrical, unsegmented, wormlike marine ecdysozoans [27] that takes its name from its phallic appearance.

"progymnosperms"   Paraphyletic group of extinct vascular plants [11] that flourished from the Devonian through the Mississippian periods. The first truly woody plants, and the first with vascular cambium that produced both secondary xylem and secondary phloem, they reproduced by spores rather than by seeds.

"prokaryotes"   Not a monophyletic group; as commonly used, includes the bacteria [2] and archaeans [3]. A term of convenience encompassing all cellular organisms that are not eukaryotes.

proteobacteria (*Proteobacteria*) [45]   A large and extremely diverse group of Gram-negative bacteria that includes many pathogens, nitrogen fixers, and photosynthesizers. Includes the alpha, beta, gamma, delta, and epsilon proteobacteria.

"protists"   This term of convenience does not describe a monophyletic group but is used to encompass a large number of distinct and distantly related groups of eukaryotes, many but far from all of which are microbial and unicellular. Essentially a "catch-all" term for any eukaryote group not contained within the land plants [10], fungi [19], or animals [21].

protostomes (*Protostomia*) [25]   One of the two major groups of bilaterians [24]. In protostomes, the mouth typically forms from the blastopore (if present) in early development (contrast with deuterostomes). The major protostome groups are the lophotrochozoans [26] and ecdysozoans [27].

prototherians (*Prototheria*)   A mostly extinct group of mammals [130], common during the Cretaceous and early Cenozoic. The five living species—the echidnas and the duck-billed platypus—are the only extant egg-laying mammals.

pterobranchs (*Pterobranchia*)   A small group of sedentary marine hemichordates [120] that live in tubes secreted by the proboscis. They have one to nine pairs of arms, each bearing long tentacles that capture prey and function in gas exchange.

pycnogonids (*Pycnogonida*)   Treated in this book as a group of chelicerates [115], but sometimes considered an independent group of arthropods [28]. Pycnogonids have reduced bodies and very long, slender legs. Also called sea spiders.

– R –

radiolarians (*Radiolaria*) [84]   Amoeboid organisms with needlelike pseudopods supported by microtubules. Most have glassy internal skeletons.

ray-finned fishes (*Actinopterygii*) [126]   A highly diverse group of freshwater and marine bony vertebrates [34]. They have reduced swim bladders that often function as hydrostatic organs and fins supported by soft rays (lepidotrichia). Includes most familiar fishes.

red algae (*Rhodophyta*) [75]   Mostly multicellular, marine and freshwater algae characterized by the presence of phycoerythrin in their chloroplasts.

reptiles (*Reptilia*) [38]   One of the two major groups of extant amniotes [37], supported on the basis of similar skull structure and gene sequences. The term "reptiles" traditionally excluded the birds [133], but the resulting group is then clearly paraphyletic. As used in this book, the reptiles include turtles [131], lepidosaurs [132], birds [133], and crocodilians [134].

rhizaria (*Rhizaria*) [16]   Mostly amoeboid unicellular eukaryotes with pseudopods, many with external or internal shells. Includes the foraminiferans [82], cercozoans [83], and radiolarians [84].

rhyniophytes (*Rhyniophyta*)   A group of early vascular plants [11] that appeared in the Silurian and became extinct in the Devonian. Possessed dichotomously branching stems with terminal sporangia but no true leaves or roots.

ribbon worms (*Nemertea*) [102]   A group of unsegmented lophotrochozoans [26] with an eversible proboscis used to capture prey. Mostly marine, but some species live in fresh water or on land.

rotifers (*Rotifera*) [101]   Tiny (< 0.5 mm) lophotrochozoans [26] with a pseudocoelomic body cavity that functions as a hydrostatic organ and a ciliated feeding organ called the corona that surrounds the head. They live in freshwater and wet terrestrial habitats.

roundworms (*Nematoda*) [112]   *See* nematodes.

– S –

sac fungi (*Ascomycota*) [87]   Fungi that bear the products of meiosis within sacs (asci) if the organism is multicellular. Some are unicellular.

salamanders (*Caudata*)   A group of amphibians [129] with distinct tails in both larvae and adults and limbs set at right angles to the body.

salps   *See* thaliaceans.

sarcopterygians (*Sarcopterygii*) [35]   One of the two major groups of bony vertebrates [34], characterized by jointed appendages (paired fins or limbs).

scyphozoans (*Scyphozoa*)   Marine cnidarians [98] in which the medusa stage dominates the life cycle. Commonly known as jellyfish.

sea cucumbers (*Holothuroidea*)   Echinoderms [119] with an elongate, cucumber-shaped body and leathery skin. They are scavengers on the ocean floor.

sea spiders   *See* pycnogonids.

sea squirts   *See* ascidians.

sea stars (*Asteroidea*)   Echinoderms [119] with five (or more) fleshy "arms" radiating from an indistinct central disk. Also called starfishes.

sea urchins (*Echinoidea*)   Echinoderms [119] with a test (shell) that is covered in spines. Most are globular in shape, although some groups (such as the sand dollars) are flattened.

"seed ferns"   A paraphyletic group of loosely related, extinct seed plants that flourished in the Devonian and Carboniferous. Characterized by large, frondlike leaves that bore seeds.

seed plants (*Spermatophyta*) [12]   Heterosporous vascular plants [11] that produce seeds; most produce wood; branching is axillary (not dichotomous). The major seed plant groups are gymnosperms [13] and angiosperms [14].

sow bugs   *See* isopods.

spirochetes (*Spirochaetes*) [40]   Motile, Gram-negative bacteria with a helically coiled structure and characterized by axial filaments.

sponges (*Porifera*) [22]   A group of relatively asymmetric, filter-feeding animals that lack a gut or nervous system and generally lack differentiated tissues. Includes glass sponges [93], demosponges [94], and calcareous sponges [95].

springtails (*Collembola*)   Wingless hexapods [118] with springing structures on the third and fourth segments of their bodies. Springtails are extremely abundant in some environments (especially in soil, leaf litter, and vegetation).

squamates (*Squamata*)   The major group of lepidosaurs [132], characterized by the possession of movable quadrate bones (which allow the upper jaw to move independently of the rest of the skull) and hemipenes (a paired set of eversible penises, or penes) in males. Includes the lizards (a paraphyletic group), snakes, and amphisbaenians.

star anise (*Austrobaileyales*) [58]   A group of woody angiosperms [14] thought to be the sister-group of the clade of flowering plants that includes eudicots [55], monocots [56], and magnoliids [57].

starfish (*Asteroidea*)   *See* sea stars.

stramenopiles (*Heterokonta* or *Stramenopila*) [6] Organisms having, at some stage in their life cycle, two unequal flagella, the longer possessing rows of tubular hairs. Chloroplasts, when present, surrounded by four membranes. Major stramenopile groups include the brown algae [49], diatoms [50], and oomycetes [51].

– T –

tapeworms (*Cestoda*)   Parasitic flatworms [100] that live in the digestive tracts of vertebrates as adults, and usually in various other species of animals as juveniles.

tardigrades (*Tardigrada*) [113]   Small (< 0.5 mm) ecdysozoans [27] with fleshy, unjointed legs and no circulatory or gas exchange organs. They live in marine sands, in temporary freshwater pools, and on the water films of plants. Also called water bears.

tetrapods (*Tetrapoda*) [36]   The major group of sarcopterygians [35]; includes the amphibians [129] and the amniotes [37]. Named for the presence of four jointed limbs (although limbs have been secondarily reduced or lost completely in several tetrapod groups).

thaliaceans (*Thaliacea*)   A group of solitary or colonial planktonic marine urochordates [122]. Also called salps.

therians (*Theria*)   Mammals [130] characterized by viviparity (live birth). Includes eutherians and marsupials.

theropods (*Theropoda*)   Archosaurs [39] with bipedal stance, hollow bones, a furcula ("wishbone"), elongated metatarsals with three-fingered feet, and a pelvis that points backwards. Includes many well-known extinct dinosaurs (such as *Tyrannosaurus rex*), as well as the living birds [133].

tracheophytes   *See* vascular plants [11].

trilobites (*Trilobita*)   An extinct group of arthropods [28] related to the chelicerates [115]. Trilobites flourished from the Cambrian through the Permian.

tuataras (*Rhyncocephalia*)   A group of lepidosaurs [132] known mostly from fossils; there are just two living tuatara species. The quadrate bone of the upper jaw is fixed firmly to the skull. Sister group of the squamates.

tunicates   *See* ascidians.

turbellarians (*Turbellaria*)   A group of free-living, generally carnivorous flatworms [100]. Their monophyly is questionable.

turtles (*Testudines*) [131]   A group of reptiles [38] with a bony carapace (upper shell) and plastron (lower shell) that encase the body.

– U –

unikonts (*Unikonta*) [17]   A group of eukaryotes [4] whose motile cells possess a single flagellum. Major unikont groups include the amoebozoans [85], fungi [19], and animals [21].

urochordates (*Urochordata*) [122]   A group of chordates [31] that are mostly saclike filter feeders as adults, with motile larvae stages that resemble a tadpole.

– V –

vascular plants (*Tracheophyta*) [11]   Plants with xylem and phloem. Major groups include the lycophytes [68] and euphyllophytes.

vertebrates (*Vertebrata*) [32]   The largest group of chordates [31], characterized by a rigid endoskeleton supported by the vertebral column and an anterior skull encasing a brain. Includes hagfishes [123], lampreys [124], and the jawed vertebrates [33], although some biologists exclude the hagfishes from this group. *See also* craniates.

– W –

water bears   *See* tardigrades.

water lilies (*Nymphaeaceae*) [59]   A group of aquatic, freshwater angiosperms [14] that are rooted in soil in shallow water, with round floating leaves and flowers that extend above the water's surface. They are the sister-group to most of the remaining flowering plants, with the exception of the genus *Amborella* [60].

whisk ferns (*Psilotophyta*) [67]   Vascular plants [11] lacking leaves and roots.

– Y –

"yeasts"   A convenience term for several distantly related groups of unicellular fungi [19].

– Z –

"zygospore fungi" (*Zygomycota*, if monophyletic) [89]   A convenience term for a probably paraphyletic group of fungi [19] in which hyphae of differing mating types conjugate to form a zygosporangium.

Appendix B: Some Measurements Used in Biology

MEASURES OF	UNIT	EQUIVALENTS	METRIC → ENGLISH CONVERSION
Length	meter (m)	base unit	1 m = 39.37 inches = 3.28 feet
	kilometer (km)	1 km = 1000 (10^3) m	1 km = 0.62 miles
	centimeter (cm)	1 cm = 0.01 (10^{-2}) m	1 cm = 0.39 inches
	millimeter (mm)	1 mm = 0.1 cm = 10^{-3} m	1 mm = 0.039 inches
	micrometer (μm)	1 μm = 0.001 mm = 10^{-6} m	
	nanometer (nm)	1 nm = 0.001 μm = 10^{-9} m	
Area	square meter (m^2)	base unit	1 m^2 = 1.196 square yards
	hectare (ha)	1 ha = 10,000 m^2	1 ha = 2.47 acres
Volume	liter (L)	base unit	1 L = 1.06 quarts
	milliliter (mL)	1 mL = 0.001 L = 10^{-3} L	1 mL = 0.034 fluid ounces
	microliter (μL)	1 μL = 0.001 mL = 10^{-6} L	
Mass	gram (g)	base unit	1 g = 0.035 ounces
	kilogram (kg)	1 kg = 1000 g	1 kg = 2.20 pounds
	metric ton (mt)	1 mt = 1000 kg	1 mt = 2,200 pounds = 1.10 ton
	milligram (mg)	1 mg = 0.001 g = 10^{-3} g	
	microgram (μg)	1 μg = 0.001 mg = 10^{-6} g	
Temperature	degree Celsius (°C)	base unit	°C = (°F – 32)/1.8
			0°C = 32°F (water freezes)
			100°C = 212°F (water boils)
			20°C = 68°F ("room temperature")
			37°C = 98.6°F (human internal body temperature)
	Kelvin (K)*	K = °C + 273	0 K = –460°F
Energy	joule (J)		1 J ≈ 0.24 calorie = 0.00024 kilocalorie[†]

*0 K (–273°C) is "absolute zero," a temperature at which molecular oscillations approach 0—that is, the point at which motion all but stops.

[†]A *calorie* is the amount of heat necessary to raise the temperature of 1 gram of water 1°C. The *kilocalorie*, or nutritionist's calorie, is what we commonly think of as a calorie in terms of food.

Answers to Self-Quizzes

Chapter 2
1.	b	6.	a
2.	d	7.	d
3.	c	8.	a
4.	c	9.	c
5.	d	10.	b

Chapter 3
1.	e	6.	a
2.	e	7.	c
3.	c	8.	e
4.	d	9.	b
5.	b	10.	d

Chapter 4
1.	c	6.	c
2.	c	7.	e
3.	c	8.	b
4.	d	9.	c
5.	e	10.	b

Chapter 5
1.	b	6.	e
2.	d	7.	a
3.	c	8.	d
4.	e	9.	b
5.	a	10.	d

Chapter 6
1.	e	6.	c
2.	c	7.	c
3.	a	8.	b
4.	d	9.	e
5.	c	10.	c

Chapter 7
1.	d	6.	a
2.	d	7.	e
3.	c	8.	d
4.	c	9.	c
5.	d	10.	a

Chapter 8
1.	c	6.	c
2.	e	7.	d
3.	b	8.	b
4.	c	9.	d
5.	c	10.	e

Chapter 9
1.	d	6.	d
2.	d	7.	e
3.	e	8.	d
4.	e	9.	a
5.	c	10.	e

Chapter 10
1.	e	6.	d
2.	b	7.	d
3.	d	8.	d
4.	b	9.	d
5.	e	10.	b

Chapter 11
1.	d	6.	d
2.	c	7.	e
3.	b	8.	d
4.	d	9.	d
5.	c	10.	c

Chapter 12*
1.	e	6.	d
2.	a	7.	b
3.	d	8.	b
4.	d	9.	b
5.	d	10.	b

Chapter 13
1.	c	6.	b
2.	a	7.	d
3.	c	8.	d
4.	b	9.	c
5.	e	10.	c

Chapter 14
1.	c	6.	d
2.	d	7.	b
3.	e	8.	d
4.	b	9.	d
5.	a	10.	c

Chapter 15
1.	a	6.	b
2.	c	7.	d
3.	b	8.	d
4.	b	9.	d
5.	d	10.	b

Chapter 16
1.	b	6.	c
2.	e	7.	c
3.	a	8.	d
4.	e	9.	b
5.	b	10.	a

Chapter 17
1.	c	6.	e
2.	c	7.	b
3.	a	8.	b
4.	e	9.	b
5.	e	10.	a

Chapter 18
1.	b	6.	b
2.	b	7.	c
3.	a	8.	a
4.	c	9.	e
5.	e	10.	d

Chapter 19
1.	c	6.	c
2.	b	7.	e
3.	a	8.	b
4.	b	9.	a
5.	d	10.	b

Chapter 20
1.	c	6.	b
2.	a	7.	a
3.	c	8.	e
4.	b	9.	d
5.	c	10.	b

Chapter 21
1.	d	6.	d
2.	e	7.	a
3.	d	8.	e
4.	c	9.	b
5.	d	10.	c

Chapter 22
1.	b	6.	b
2.	e	7.	e
3.	a	8.	a
4.	b	9.	e
5.	e	10.	d

Chapter 23
1.	c	6.	c
2.	e	7.	a
3.	d	8.	e
4.	c	9.	c
5.	a	10.	e

Chapter 24
1.	a	6.	e
2.	a	7.	e
3.	d	8.	e
4.	a	9.	b
5.	b	10.	e

Chapter 25
1.	d	6.	a
2.	b	7.	c
3.	e	8.	b
4.	c	9.	c
5.	a	10.	d

Chapter 26
1.	e	6.	b
2.	e	7.	d
3.	a	8.	d
4.	c	9.	b
5.	e	10.	d

*Answers to Chapter 12 Genetics Problems

1. $BB \times bb$; $bb \times bb$; $Bb \times bb$; $Bb \times Bb$

2. 1/32

3a. Autosomal dominant

3b. 1/4

4a. Males (XY) contain only one allele and will show only one color, black (X^BY) or yellow (X^bY). Females can be heterozygous (X^BX^b).

4b. X^bY, yellow

5. The body color (G/g) and wing size (A/a) genes are linked; eye color (R/r) is unlinked to the other two genes. The map distance is 18.5 units.

6. Yellow, blue, and white in a 1:2:1 ratio

7. F_1 all wild-type, $PpSwsw$; F_2 will have phenotypes in the ratio 9:3:3:1. See Figure 12.7 (p. 244) for analogous genotypes.

8a. Ratio of phenotypes in F_2 is 3:1 (double dominant to double recessive).

8b. The F_1 are $PpByby$; they produce just two kinds of gametes (Pby and pBy). Combine them carefully and see the 1:2:1 phenotypic ratio fall out in the F_2.

8c. Pink-blistery

8d. See Figures 11.17 and 11.19 (pp. 224–226). Crossing over took place in the F_1 generation.

9. $Rraa$ and $RrAa$

10a. $w^+ > w^e > w$

10b. Parents are w^ew and w^+Y. Progeny are w^+w^e, w^+w, w^eY, and wY.

11a. BX^a, BY, bX^a, bY

11b. Mother bbX^AX^a, father BbX^aY, son BbX^aY, daughter bbX^aX^a

12. 75 percent

13. Because the gene is carried on mitochondrial DNA, it is passed through the mother only. Thus if the woman does not have the disease but her husband does, their child will not be affected. On the other hand, if the woman has the disease but her husband does not, their child will have the disease.

Chapter 27
1.	a	6.	b
2.	e	7.	d
3.	c	8.	b
4.	d	9.	a
5.	c	10.	d

Chapter 28
1.	d	6.	e
2.	c	7.	c
3.	e	8.	b
4.	b	9.	b
5.	b	10.	d

Chapter 29
1.	d	6.	c
2.	c	7.	a
3.	d	8.	e
4.	a	9.	c
5.	d	10.	a

Chapter 30
1.	b	6.	a
2.	d	7.	e
3.	e	8.	a
4.	c	9.	e
5.	d	10.	c

Chapter 31
1.	c	6.	b
2.	d	7.	c
3.	c	8.	d
4.	d	9.	e
5.	e	10.	d

Chapter 32
1.	b	6.	e
2.	e	7.	b
3.	c	8.	d
4.	d	9.	d
5.	a	10.	e

Chapter 33
1.	d	6.	e
2.	a	7.	a
3.	c	8.	e
4.	d	9.	c
5.	d	10.	b

Chapter 34
1.	c	6.	b
2.	b	7.	b
3.	e	8.	c
4.	e	9.	a
5.	a	10.	d

Chapter 35
1.	c	6.	d
2.	d	7.	d
3.	b	8.	e
4.	b	9.	e
5.	b	10.	a

Chapter 36
1.	d	6.	c
2.	d	7.	e
3.	c	8.	d
4.	a	9.	d
5.	c	10.	e

Chapter 37
1.	a	6.	c
2.	e	7.	b
3.	c	8.	c
4.	d	9.	a
5.	b	10.	b

Chapter 38
1.	d	6.	e
2.	b	7.	a
3.	e	8.	c
4.	b	9.	c
5.	d	10.	d

Chapter 39
1.	e	6.	a
2.	b	7.	b
3.	c	8.	c
4.	c	9.	e
5.	d	10.	a

Chapter 40
1.	c	6.	b
2.	c	7.	e
3.	a	8.	a
4.	d	9.	e
5.	b	10.	b

Chapter 41
1.	b	6.	b
2.	a	7.	d
3.	b	8.	e
4.	e	9.	c
5.	e	10.	c

Chapter 42
1.	a	6.	a
2.	b	7.	d
3.	e	8.	d
4.	e	9.	a
5.	c	10.	d

Chapter 43
1.	c	6.	d
2.	e	7.	d
3.	a	8.	d
4.	d	9.	d
5.	d	10.	a

Chapter 44
1.	a	6.	c
2.	c	7.	b
3.	e	8.	b
4.	a	9.	b
5.	d	10.	c

Chapter 45
1.	d	6.	e
2.	d	7.	c
3.	c	8.	c
4.	c	9.	d
5.	e	10.	d

Chapter 46
1.	d	6.	e
2.	d	7.	b
3.	a	8.	c
4.	e	9.	c
5.	e	10.	d

Chapter 47
1.	c	6.	c
2.	a	7.	a
3.	e	8.	c
4.	d	9.	a
5.	d	10.	c

Chapter 48
1.	e	6.	d
2.	a	7.	e
3.	b	8.	a
4.	c	9.	a
5.	b	10.	e

Chapter 49
1.	e	6.	b
2.	d	7.	c
3.	a	8.	c
4.	b	9.	a
5.	c	10.	d

Chapter 50
1.	d	6.	d
2.	a	7.	b
3.	c	8.	d
4.	d	9.	c
5.	c	10.	e

Chapter 51
1.	b	6.	d
2.	e	7.	a
3.	c	8.	b
4.	a	9.	d
5.	b	10.	d

Chapter 52
1.	d	6.	b
2.	a	7.	e
3.	d	8.	a
4.	a	9.	c
5.	d	10.	e

Chapter 53
1.	b	6.	d
2.	e	7.	e
3.	c	8.	e
4.	a	9.	a
5.	d	10.	e

Chapter 54
1.	d	6.	c
2.	a	7.	c
3.	a	8.	c
4.	d	9.	d
5.	a	10.	b

Chapter 55
1.	c	6.	b
2.	c	7.	e
3.	a	8.	c
4.	d	9.	e
5.	c	10.	d

Chapter 56
1.	a	6.	a
2.	c	7.	b
3.	b	8.	d
4.	c	9.	e
5.	d	10.	b

Chapter 57
1.	a	6.	b
2.	a	7.	e
3.	b	8.	b
4.	d	9.	d
5.	e	10.	e

Chapter 58
1.	e	6.	e
2.	d	7.	a
3.	b	8.	e
4.	c	9.	d
5.	c	10.	b

Chapter 59
1.	b	6.	a
2.	d	7.	d
3.	e	8.	c
4.	e	9.	a
5.	b	10.	c

Glossary

- A -

abiotic (a' bye ah tick) [Gk. *a*: not + *bios*: life] Nonliving. (Contrast with biotic.)

abscisic acid (ABA) (ab sighs' ik) A plant growth substance with growth-inhibiting action. Causes stomata to close; involved in a plant's response to salt and drought stress.

abscission (ab sizh' un) [L. *abscissio*: break off] The process by which leaves, petals, and fruits separate from a plant.

absorption (1) Of light: complete retention, without reflection or transmission. (2) Of water or other molecules: soaking up (taking in through pores or by diffusion).

absorption spectrum A graph of light absorption versus wavelength of light; shows how much light is absorbed at each wavelength.

absorptive heterotroph An organism (usually a fungus) that obtains its food by secreting digestive enzymes into the environment to break down large food molecules, then absorbing the breakdown products.

abyssal plain (uh biss' ul) [Gk. *abyssos*: bottomless] The deep ocean floor.

accessory pigments Pigments that absorb light and transfer energy to chlorophylls for photosynthesis.

acetylcholine (ACh) A neurotransmitter that carries information across vertebrate neuromuscular junctions and some other synapses. It is then broken down by the enzyme acetylcholinesterase (AChE).

acetyl coenzyme A (acetyl CoA) A compound that reacts with oxaloacetate to produce citrate at the beginning of the citric acid cycle; a key metabolic intermediate in the formation of many compounds.

acid [L. *acidus*: sharp, sour] A substance that can release a proton in solution. (Contrast with base.)

acid growth hypothesis The hypothesis that auxin increases proton pumping, thereby lowering the pH of the cell wall and activating enzymes that loosen polysaccharides. Proposed to explain auxin-induced cell expansion in plants.

acid precipitation Precipitation that has a lower pH than normal as a result of acid-forming precursor molecules introduced into the atmosphere by human activities.

acidic Having a pH of less than 7.0 (a hydrogen ion concentration greater than 10^{-7} molar). (Contrast with basic.)

acoelomate An animal that does not have a coelom.

acrosome (a' krow soam) [Gk. *akros*: highest + *soma*: body] The structure at the forward tip of an animal sperm which is the first to fuse with the egg membrane and enter the egg cell.

ACTH *See* corticotropin.

actin [Gk. *aktis*: ray] A protein that makes up the cytoskeletal microfilaments in eukaryotic cells and is one of the two contractile proteins in muscle.

action potential An impulse in a neuron taking the form of a wave of depolarization or hyperpolarization.

action spectrum A graph of a biological process versus light wavelength; shows which wavelengths are involved in the process.

activation energy (E_a) The energy barrier that blocks the tendency for a chemical reaction to occur.

active site The region on the surface of an enzyme or ribozyme where the substrate binds, and where catalysis occurs.

active transport The energy-dependent transport of a substance across a biological membrane against a concentration gradient—that is, from a region of low concentration (of that substance) to one of high concentration. (*See also* primary active transport, secondary active transport; contrast with facilitated diffusion, passive transport.)

adaptation (a dap tay' shun) (1) In evolutionary biology, a particular structure, physiological process, or behavior that makes an organism better able to survive and reproduce. Also, the evolutionary process that leads to the development or persistence of such a trait. (2) In sensory neurophysiology, a sensory cell's loss of sensitivity as a result of repeated stimulation.

adaptive radiation An evolutionary radiation that results in an array of related species that live in a variety of environments and differ in the characteristics they use to exploit those environments.

adenine (A) (a' den een) A nitrogen-containing base found in nucleic acids, ATP, NAD, and other compounds.

adenosine triphosphate *See* ATP.

adrenal gland (a dree' nal) [L. *ad*: toward + *renes*: kidneys] An endocrine gland located near the kidneys of vertebrates, consisting of two glandular parts, the cortex and medulla.

adrenaline *See* epinephrine.

adrenocorticotropic hormone *See* corticotropin.

adsorption Binding of a gas or a solute to the surface of a solid.

adventitious roots (ad ven ti' shus) [L. *adventitius*: arriving from outside] Roots originating from the stem at ground level or below; typical of the fibrous root system of monocots.

aerenchyma In plants, parenchymal tissue containing air spaces.

aerobic (air oh' bic) [Gk. *aer*: air + *bios*: life] In the presence of oxygen; requiring oxygen. (Contrast with anaerobic.)

afferent (af' ur unt) [L. *ad*: toward + *ferre*: to carry] Carrying to, as in a neuron that carries impulses to the central nervous system (afferent neuron), or a blood vessel that carries blood to a structure. (Contrast with efferent.)

age structure The distribution of the individuals in a population across all age groups.

AIDS Acquired immune deficiency syndrome, a condition caused by human immunodeficiency virus (HIV) in which the body's T-helper cells are reduced, leaving the victim subject to opportunistic diseases.

air sacs Structures in the respiratory system of birds that receive inhaled air; they keep fresh air flowing unidirectionally through the lungs, but are not themselves gas exchange surfaces.

aldosterone (al dohs' ter own) A steroid hormone produced in the adrenal cortex of mammals. Promotes secretion of potassium and reabsorption of sodium in the kidney.

aleurone layer In some seeds, a tissue that lies beneath the seed coat and surrounds the endosperm. Secretes digestive enzymes that break down macromolecules stored in the endosperm.

allantoic membrane In animal development, an outgrowth of extraembryonic endoderm plus adjacent mesoderm that forms the allantois, a saclike structure that stores metabolic wastes produced by the embryo.

allantois (al lun twah') [Gk. *allant*: sausage] An extraembryonic membrane enclosing a sausage-shaped sac that stores the embryo's nitrogenous wastes.

allele (a leel') [Gk. *allos*: other] The alternate form of a genetic character found at a given locus on a chromosome.

allele frequency The relative proportion of a particular allele in a specific population.

allergic reaction [Ger. *allergie*: altered] An overreaction of the immune system to amounts of an antigen that do not affect most people; often involves IgE antibodies.

allometric growth A pattern of growth in which some parts of the body of an organism grow faster than others, resulting in a change in body proportions as the organism grows.

allopatric speciation (al' lo pat' rick) [Gk. *allos*: other + *patria*: homeland] The formation of two species from one when reproductive isolation occurs because of the interposition of (or crossing of) a physical geographic barrier such as a river. Also called geographic speciation. (Contrast with sympatric speciation.)

allopolyploidy The possession of more than two chromosome sets that are derived from more than one species.

allosteric regulation (al lo steer' ik) [Gk. *allos*: other + *stereos*: structure] Regulation of the activity of a protein (usually an enzyme) by the binding of an effector molecule to a site other than the active site.

alpha diversity Species diversity within a single community or habitat. (Contrast with beta diversity, gamma diversity.)

α (alpha) helix A prevalent type of secondary protein structure; a right-handed spiral.

alternation of generations The succession of multicellular haploid and diploid phases in some sexually reproducing organisms, notably plants.

alternative splicing A process for generating different mature mRNAs from a single gene by splicing together different sets of exons during RNA processing.

altruism Pertaining to behavior that benefits other individuals at a cost to the individual who performs it.

alveolus (al ve′ o lus) (plural: alveoli) [L. *alveus*: cavity] A small, baglike cavity, especially the blind sacs of the lung.

amensalism (a men′ sul ism) Interaction in which one animal is harmed and the other is unaffected. (Contrast with commensalism, mutualism.)

amine An organic compound containing an amino group (NH_2).

amino acid An organic compound containing both NH_2 and COOH groups. Proteins are polymers of amino acids.

amino acid replacement A change in the nucleotide sequence that results in one amino acid being replaced by another.

ammonotelic (am moan′ o teel′ ic) [Gk. *telos*: end] Pertaining to an organism in which the final product of breakdown of nitrogen-containing compounds (primarily proteins) is *ammonia*. (Contrast with ureotelic, uricotelic.)

amnion (am′ nee on) The fluid-filled sac within which the embryos of reptiles (including birds) and mammals develop.

amniote egg A shelled egg surrounding four extraembryonic membranes and embryo-nourishing yolk. This evolutionary adaptation permitted mammals and reptiles to live and reproduce in drier environments than can most amphibians.

amphipathic (am′ fi path′ ic) [Gk. *amphi*: both + *pathos*: emotion] Of a molecule, having both hydrophilic and hydrophobic regions.

amplitude The magnitude of change over the course of a regular cycle.

amygdala A component of the limbic system that is involved in fear and fear memory.

amylase (am′ ill ase) An enzyme that catalyzes the hydrolysis of starch, usually to maltose or glucose.

anabolic reaction (an uh bah′ lik) [Gk. *ana*: upward + *ballein*: to throw] A synthetic reaction in which simple molecules are linked to form more complex ones; requires an input of energy and captures it in the chemical bonds that are formed. (Contrast with catabolic reaction.)

anaerobic (an ur row′ bic) [Gk. *an*: not + *aer*: air + *bios*: life] Occurring without the use of molecular oxygen, O_2. (Contrast with aerobic.)

analogy (a nal′ o jee) [Gk. *analogia*: resembling] A resemblance between two features that is due to convergent evolution rather than to common ancestry. The structures are said to be *analogous*, and each is an *analog* of the others. (Contrast with homology.)

anaphase (an′ a phase) [Gk. *ana*: upward] The stage in cell nuclear division at which the first separation of sister chromatids (or, in the first meiotic division, of paired homologs) occurs.

ancestral trait The trait originally present in the ancestor of a given group; may be retained or changed in the descendants of that ancestor.

androgen (an′ dro jen) Any of the several male sex steroids (most notably testosterone).

aneuploidy (an′ you ploy dee) A condition in which one or more chromosomes or pieces of chromosomes are either lacking or present in excess.

angiotensin (an′ jee oh ten′ sin) A peptide hormone that raises blood pressure by causing peripheral vessels to constrict. Also maintains glomerular filtration by constricting efferent vessels and stimulates thirst and the release of aldosterone.

angular gyrus A part of the human brain believed to be essential for integrating spoken and written language.

animal hemisphere The metabolically active upper portion of some animal eggs, zygotes, and embryos; does not contain the dense nutrient yolk. (Contrast with vegetal hemisphere.)

anion (an′ eye on) [Gk. *ana*: upward progress] A negatively charged ion. (Contrast with cation.)

anisogamous (an eye sog′ a muss) [Gk. *aniso*: unequal + *gamos*: marriage] Having morphologically dissimilar male and female gametes. (Contrast with isogamous.)

annual A plant whose life cycle is completed in one growing season. (Contrast with biennial, perennial.)

antagonistic interactions Interactions between two species in which one species benefits and the other is harmed. Includes predation, herbivory, and parasitism.

antenna system In photosynthesis, a group of different molecules that cooperate to absorb light energy and transfer it to a reaction center.

anterior Toward or pertaining to the tip or headward region of the body axis. (Contrast with posterior.)

anterior pituitary The portion of the vertebrate pituitary gland that derives from gut epithelium and produces tropic hormones.

anther (an′ thur) [Gk. *anthos*: flower] A pollen-bearing portion of the stamen of a flower.

antheridium (an′ thur id′ ee um) [Gk. *antheros*: blooming] The multicellular structure that produces the sperm in nonvascular land plants and ferns.

antibody One of the myriad proteins produced by the immune system that specifically binds to a foreign substance in blood or other tissue fluids and initiates its removal from the body.

anticodon The three nucleotides in transfer RNA that pair with a complementary triplet (a codon) in messenger RNA.

antidiuretic hormone (ADH) A hormone that promotes water reabsorption by the kidney. ADH is produced by neurons in the hypothalamus and released from nerve terminals in the posterior pituitary. Also called vasopressin.

antigen (an′ ti jun) Any substance that stimulates the production of an antibody or antibodies in the body of a vertebrate.

antigenic determinant The specific region of an antigen that is recognized and bound by a specific antibody. Also called an epitope.

antiparallel Pertaining to molecular orientation in which a molecule or parts of a molecule have opposing directions.

antiporter A membrane transport protein that moves one substance in one direction and another in the opposite direction. (Contrast with symporter, uniporter.)

antisense RNA A single-stranded RNA molecule complementary to, and thus targeted against, an mRNA of interest to block its translation.

anus (a′ nus) An opening through which solid digestive wastes are expelled, located at the posterior end of a tubular gut.

aorta (a or′ tah) [Gk. *aorte*: aorta] The main trunk of the arteries leading to the systemic (as opposed to the pulmonary) circulation.

aortic body A chemosensor in the aorta that senses a decrease in blood supply or a dramatic decrease in partial pressure of oxygen in the blood.

aortic valve A one-way valve between the left ventricle of the heart and the aorta that prevents backflow of blood into the ventricle when it relaxes.

apex (a′ pecks) The tip or highest point of a structure, as of a growing stem or root.

aphasia a deficit in the ability to use or understand words.

apical (a′ pi kul) Pertaining to the *apex*, or tip, usually in reference to plants.

apical dominance In plants, inhibition by the apical bud of the growth of axillary buds.

apical hook A form taken by the stems of many eudicot seedlings that protects the delicate shoot apex while the stem grows through the soil.

apical meristem The meristem at the tip of a shoot or root; responsible for a plant's primary growth.

apomixis (ap oh mix′ is) [Gk. *apo*: away from + *mixis*: sexual intercourse] The asexual production of seeds.

apoplast (ap′ oh plast) In plants, the continuous meshwork of cell walls and extracellular spaces through which material can pass without crossing a plasma membrane. (Contrast with symplast.)

apoptosis (ap uh toh′ sis) A series of genetically programmed events leading to cell death.

aposematism Warning coloration; bright colors or striking patterns of toxic or mimetic prey species that act as a warning to predators.

appendix In the human digestive system, the vestigial equivalent of the cecum, which serves no digestive function.

aquaporin A transport protein in plant and animal cell membranes through which water passes in osmosis.

aquatic (a kwa′ tic) [L. *aqua*: water] Pertaining to or living in water. (Contrast with marine, terrestrial.)

aqueous (a′ kwee us) Pertaining to water or a watery solution.

aquifer A large pool of groundwater.

archegonium (ar′ ke go′ nee um) The multicellular structure that produces eggs in nonvascular land plants, ferns, and gymnosperms.

archenteron (ark en′ ter on) [Gk. *archos*: first + *enteron*: bowel] The earliest primordial animal digestive tract.

area phylogenies Phylogenies in which the names of the taxa are replaced with the names of the places where those taxa live or lived.

arms race A series of reciprocal adaptations between species involved in antagonistic interactions, in which adaptations that increase the fitness of a consumer species exert selection pressure on its resource species to counter the consumer's adaptation, and vice versa.

arteriole A small blood vessel arising from an artery that feeds blood into a capillary bed.

artery A muscular blood vessel carrying oxygenated blood away from the heart to other parts of the body. (Contrast with vein.)

artificial selection   The selection by plant and animal breeders of individuals with certain desirable traits.

ascus (ass' cus) (plural: asci) [Gk. *askos*: bladder] In sac fungi, the club-shaped sporangium within which spores (ascospores) are produced by meiosis.

asexual reproduction   Reproduction without sex.

assisted reproductive technologies (ARTs)   Any of several procedures that remove unfertilized eggs from the ovary, combine them with sperm outside the body, and then place fertilized eggs or egg–sperm mixtures in the appropriate location in a female's reproductive tract for development.

association cortex   In the vertebrate brain, the portion of the cortex involved in higher-order information processing, so named because it integrates, or associates, information from different sensory modalities and from memory.

associative learning   A form of learning in which two unrelated stimuli become linked to the same response.

astrocyte [Gk. *astron*: star] A type of glial cell that contributes to the blood–brain barrier by surrounding the smallest, most permeable blood vessels in the brain.

atherosclerosis (ath' er oh sklair oh' sis) [Gk. *athero*: gruel, porridge + *skleros*: hard] A disease of the lining of the arteries characterized by fatty, cholesterol-rich deposits in the walls of the arteries. When fibroblasts infiltrate these deposits and calcium precipitates in them, the disease become arteriosclerosis, or "hardening of the arteries."

atom [Gk. *atomos*: indivisible] The smallest unit of a chemical element. Consists of a nucleus and one or more electrons.

atomic mass   *See* atomic weight.

atomic number   The number of protons in the nucleus of an atom; also equals the number of electrons around the neutral atom. Determines the chemical properties of the atom.

atomic weight   The average of the mass numbers of a representative sample of atoms of an element, with all the isotopes in their normally occurring proportions. Also called atomic mass.

ATP (adenosine triphosphate)   An energy-storage compound containing adenine, ribose, and three phosphate groups. When it is formed from ADP, useful energy is stored; when it is broken down (to ADP or AMP), energy is released to drive endergonic reactions.

ATP synthase   An integral membrane protein that couples the transport of protons with the formation of ATP.

atrial natriuretic peptide   A hormone released by the atrial muscle fibers of the heart when they are overly stretched, which decreases reabsorption of sodium by the kidney and thus blood volume.

atrioventricular node   A modified node of cardiac muscle that organizes the action potentials that control contraction of the ventricles.

atrium (a' tree um) [L. *atrium*: central hall] An internal chamber. In the hearts of vertebrates, the thin-walled chamber(s) entered by blood on its way to the ventricle(s). Also, the outer ear.

auditory system   A sensory system that uses mechanoreceptors to convert pressure waves into receptor potentials; includes structures that gather sound waves, direct them to a sensory organ, and amplify their effect on the mechanoreceptors.

autocatalysis [Gk. *autos*: self + *kata*: to break down] A positive feedback process in which an activated enzyme acts on other inactive molecules of the same enzyme to activate them.

autocrine   A chemical signal that binds to and affects the cell that makes it. (Contrast with paracrine.)

autoimmunity   An immune response by an organism to its own molecules or cells.

autonomic nervous system (ANS)   The portion of the peripheral nervous system that controls such involuntary functions as those of guts and glands. Also called the involuntary nervous system.

autopolyploidy   The possession of more than two entire chromosomes sets that are derived from a single species.

autosome   Any chromosome (in a eukaryote) other than a sex chromosome.

autotroph (au' tow trowf') [Gk. *autos*: self + *trophe*: food] An organism that is capable of living exclusively on inorganic materials, water, and some energy source such as sunlight (photoautotrophs) or chemically reduced matter (see chemolithotrophs). (Contrast with heterotroph.)

auxin (awk' sin) [Gk. *auxein*: to grow] In plants, a substance (the most common being indoleacetic acid) that regulates growth and various aspects of development.

avirulence (Avr) genes   Genes in a pathogen that may trigger defenses in plants. *See* gene-for-gene resistance.

Avogadro's number   The number of atoms or molecules in a mole (weighed out in grams) of a substance, calculated to be 6.022×10^{23}.

axillary bud   A bud that forms in the angle (axil) where a leaf meets a stem.

axon [Gk. *axle*] The part of a neuron that conducts action potentials away from the cell body.

axon hillock   The junction between an axon and its cell body, where action potentials are generated.

axon terminals   The endings of an axon; they form synapses and release neurotransmitter.

- B -

B cell   A type of lymphocyte involved in the humoral immune response of vertebrates. Upon recognizing an antigenic determinant, a B cell develops into a plasma cell, which secretes an antibody. (Contrast with T cell.)

bacillus (bah sil' us) [L: little rod] Any of various rod-shaped bacteria.

bacterial artificial chromosome (BAC)   A DNA cloning vector used in bacteria that can carry up to 150,000 base pairs of foreign DNA.

bacteriophage (bak teer' ee o fayj) [Gk. *bakterion*: little rod + *phagein*: to eat] Any of a group of viruses that infect bacteria. Also called phage.

bacteroids   Nitrogen-fixing organelles that develop from endosymbiotic bacteria.

bark   All tissues external to the vascular cambium of a plant.

baroreceptor [Gk. *baros*: weight] A pressure-sensing cell or organ. Sometimes called a stress receptor.

basal body   A centriole found at the base of a eukaryotic flagellum or cilium.

basal metabolic rate (BMR)   The minimum rate of energy turnover in an awake (but resting) bird or mammal that is not expending energy for thermoregulation.

base   (1) A substance that can accept a hydrogen ion in solution. (Contrast with acid.) (2) In nucleic acids, the purine or pyrimidine that is attached to each sugar in the sugar–phosphate backbone.

base pair (bp)   In double-stranded DNA, a pair of nucleotides formed by the *complementary base pairing* of a purine on one strand and a pyrimidine on the other.

basic   Having a pH greater than 7.0 (i.e., having a hydrogen ion concentration lower than 10^{-7} molar). (Contrast with acidic.)

basidiocarp   A fruiting structure produced by club fungi.

basidium (bass id' ee yum)   In club fungi, the characteristic sporangium in which four spores are formed by meiosis and then borne externally before being shed.

basilar membrane   A membrane in the human inner ear whose flexion in response to sound waves activates hair cells; flexes at different locations in response to different pitches of sound.

basophil   A type of phagocytic white blood cell that releases histamine and may promote T cell development.

Batesian mimicry   The convergence in appearance of an edible species (mimic) with an unpalatable species (model).

benefit   An improvement in survival and reproductive success resulting from performing a behavior or having a trait. (Contrast with cost.)

benthic zone [Gk. *benthos*: bottom] The bottom of the ocean.

beta diversity   Between-habitat diversity; a measure of the change in species composition from one community or habitat to another. (Contrast with alpha diveristy, gamma diversity.)

β (beta) pleated sheet   A type of protein secondary structure; results from hydrogen bonding between polypeptide regions running antiparallel to each other.

biennial   A plant whose life cycle includes vegetative growth in the first year and flowering and senescence in the second year. (Contrast with annual, perennial.)

bilateral symmetry   The condition in which only the right and left sides of an organism, divided by a single plane through the midline, are mirror images of each other.

bilayer   A structure that is two layers in thickness. In biology, most often refers to the phospholipid bilayer of membranes. (*See* phospholipid bilayer.)

bile   A secretion of the liver made up of bile salts synthesized from cholesterol, various phospholipids, and bilirubin (the breakdown product of hemoglobin). Emulsifies fats in the small intestine.

binary fission   Reproduction of a prokaryote by division of a cell into two comparable progeny cells.

binocular vision   Overlapping visual fields of an animal's two eyes; allows the animal to see in three dimensions.

binomial nomenclature   A taxonomic naming system in which each species is given two names (a genus name followed by a species name).

biofilm   A community of microorganisms embedded in a polysaccharide matrix, forming a highly resistant coating on almost any moist surface.

biogeochemical cycle   Movement of inorganic elements such as nitrogen, photsphorus, and carbon through living organisms and the physical environment.

biogeographic region   One of several defined, continental-scale regions of Earth, each of which has a biota distinct from that of the others. (Contrast with biome.)

biogeography   The scientific study of the patterns of distribution of populations, species, and ecological communities across Earth.

bioinformatics   The use of computers and/or mathematics to analyze complex biological information, such as DNA sequences.

biological species concept   The definition of a species as a group of actually or potentially interbreeding natural populations that are reproductively isolated from other such groups. (Contrast with lineage species concept; morphological species concept.)

biology [Gk. *bios*: life + *logos*: study]   The scientific study of living things.

bioluminescence   The production of light by biochemical processes in an organism.

biomass   The total weight of all the organisms, or some designated group of organisms, in a given area.

biome (bye' ome)   A major division of the ecological communities of Earth, characterized primarily by distinctive vegetation. A given biogeographic region contains many different biomes.

bioremediation   The use by humans of other organisms to remove contaminants from the environment.

biosphere (bye' oh sphere)   All regions of Earth (terrestrial and aquatic) and Earth's atmosphere in which organisms can live.

biota (bye oh' tah)   All of the organisms—animals, plants, fungi, and microorganisms—found in a given area. (Contrast with flora, fauna.)

biotechnology   The use of cells or living organisms to produce materials useful to humans.

biotic (bye ah' tick) [Gk. *bios*: life]   Alive. (Contrast with abiotic.)

biotic interchange   The dispersal of species from two different biotas into the region they had not previously inhabited, as when two formerly separated land masses fuse.

blade   The thin, flat portion of a leaf.

blastocoel (blass' toe seal) [Gk. *blastos*: sprout + *koilos*: hollow]   The central, hollow cavity of a blastula.

blastocyst (blass' toe cist)   An early embryo formed by the first divisions of the fertilized egg (zygote). In mammals, a hollow ball of cells.

blastodisc (blass' toe disk)   An embryo that forms as a disk of cells on the surface of a large yolk mass; comparable to a blastula, but occurring in animals such as birds and reptiles, in which the massive yolk restricts in incomplete cleavage.

blastomere   Any of the cells produced by the early divisions of a fertilized animal egg.

blastula (blass' chu luh)   An early stage of the animal embryo; in many species, a hollow sphere of cells surrounding a central cavity, the blastocoel. (Contrast with blastodisc.)

block to polyspermy   Any of several responses to entry of a sperm into an egg that prevent more than one sperm from entering the egg.

blood   A fluid tissue that is pumped around the body; a component of the circulatory system.

blood–brain barrier   A property of blood vessels in the brain that prevents most chemicals from diffusing from the blood into the brain.

blue-light receptors   Pigments in plants that absorb blue light (400–500 nm). These pigments mediate many plant responses including phototropism, stomatal movements, and expression of some genes.

body plan   The general structure of an animal, the arrangement of its organ systems, and the integrated functioning of its parts.

bond   *See* chemical bond.

bottleneck   *See* population bottleneck.

bone   A rigid component of vertebrate skeletal systems that contains an extracellular matrix of insoluble calcium phosphate crystals as well as collagen fibers.

Bowman's capsule   An elaboration of the renal tubule, composed of podocytes, that surrounds and collects the filtrate from the glomerulus.

brain   The centralized integrative center of a nervous system.

brainstem   The portion of the vertebrate brain between the spinal cord and the forebrain, made up of the medulla, pons, and midbrain.

brassinosteroids   Plant steroid hormones that mediate light effects promoting the elongation of stems and pollen tubes.

Broca's area   A portion of the human brain essential for speech. Located in the frontal lobe just in front of the primary motor cortex.

bronchioles   The smallest airways in a vertebrate lung, branching off the bronchi.

bronchus (plural: bronchi)   The major airway(s) branching off the trachea into the vertebrate lung.

brown fat   In mammals, fat tissue that is specialized to produce heat. It has many mitochondria and capillaries, and a protein that uncouples oxidative phosphorylation.

budding   Asexual reproduction in which a more or less complete new organism grows from the body of the parent organism, eventually detaching itself.

buffer   A substance that can transiently accept or release hydrogen ions and thereby resist changes in pH.

bulbourethral glands   Secretory structures of the human male reproductive system that produce a small volume of an alkaline, mucoid secretion that helps neutralize acidity in the urethra and lubricate it to facilitate the passage of semen.

bulk flow   The movement of a solution from a region of higher pressure potential to a region of lower pressure potential.

bundle of His   Fibers of modified cardiac muscle that conduct action potentials from the atria to the ventricular muscle mass.

bundle sheath cell   Part of a tissue that surrounds the veins of plants; contains chloroplasts in C_4 plants.

– C –

C3 plants   Plants that produce 3PG as the first stable product of carbon fixation in photosynthesis and use ribulose bisphosphate as a CO_2 receptor.

C4 plants   Plants that produce oxaloacetate as the first stable product of carbon fixation in photosynthesis and use phosphoenolpyruvate as CO_2 acceptor. C_4 plants also perform the reactions of C_3 photosynthesis.

calcitonin   Hormone produced by the thyroid gland; lowers blood calcium and promotes bone formation. (Contrast with parathyroid hormone.)

calorie [L. *calor*: heat]   The amount of heat required to raise the temperature of 1 gram of water by 1°C. Physiologists commonly use the kilocalorie (kcal) as a unit of measure (1 kcal = 1,000 calories). Nutritionists also use the kilocalorie, but refer to it as the *Calorie* (capital C).

Calvin cycle   The stage of photosynthesis in which CO_2 reacts with RuBP to form 3PG, 3PG is reduced to a sugar, and RuBP is regenerated, while other products are released to the rest of the plant. Also known as the Calvin–Benson cycle.

calyx (kay' licks) [Gk. *kalyx*: cup]   All of the sepals of a flower, collectively.

CAM   *See* crassulacean acid metabolism.

Cambrian explosion   The rapid diversification of multicellular life that took place during the Cambrian period.

cAMP (cyclic AMP)   A compound formed from ATP that acts as a second messenger.

cancellous bone   A type of bone with numerous internal cavities that make it appear spongy, although it is rigid. (Contrast with compact bone.)

canopy   The leaf-bearing part of a tree. Collectively, the aggregate of the leaves and branches of the larger woody plants of an ecological community.

capillaries [L. *capillaris*: hair]   Very small tubes, especially the smallest blood-carrying vessels of animals between the termination of the arteries and the beginnings of the veins. Capillaries are the site of exchange of materials between the blood and the interstitial fluid.

capsid   The outer shell of a virus that encloses its nucleic acid.

carbohydrates   Organic compounds containing carbon, hydrogen, and oxygen in the ratio 1:2:1 (i.e., with the general formula $C_nH_{2n}O_n$). Common examples are sugars, starch, and cellulose.

carbon skeleton   The chains or rings of carbon atoms that form the structural basis of organic molecules. Other atoms or functional groups are attached to the carbon atoms.

carboxylase   An enzyme that catalyzes the addition of carboxyl groups to a substrate.

cardiac (kar' dee ak) [Gk. *kardia*: heart]   Pertaining to the heart and its functions.

cardiac cycle   Contraction of the two atria of the heart, followed by contraction of the two ventricles and then relaxation.

cardiac muscle   A type of muscle tissue that makes up, and is responsible for the beating of, the heart. Characterized by branching cells with single nuclei and a striated (striped) appearance. (Contrast with smooth muscle, skeletal muscle.)

cardiovascular system [Gk. *kardia*: heart + L. *vasculum*: small vessel]   The heart, blood, and vessels are of a circulatory system.

carnivore [L. *carn*: flesh + *vovare*: to devour]   An organism that eats animal tissues. (Contrast with detritivore, herbivore, omnivore.)

carotenoid (ka rah' tuh noid)   A yellow, orange, or red lipid pigment commonly found as an accessory pigment in photosynthesis; also found in fungi.

carotid body   A chemosensor in the carotid artery that senses a decrease in blood supply or a dramatic decrease in partial pressure of oxygen in the blood.

carpel (kar' pel) [Gk. *karpos*: fruit]   The organ of the flower that contains one or more ovules.

carrier   (1) In facilitated diffusion, a membrane protein that binds a specific molecule and transports it through the membrane. (2) In respiratory and photosynthetic electron transport, a participating substance such as NAD that exists in both oxidized and reduced forms. (3) In genetics, a person heterozygous for a recessive trait.

carrying capacity (K)   The number of individuals in a population that the resources of its environment can support.

cartilage   In vertebrates, a tough connective tissue found in joints, the outer ear, and elsewhere. Forms the entire skeleton in some animal groups.

cartilage bone   A type of bone that begins its development as a cartilaginous structure resembling the future mature bone, then gradually hardens into mature bone. (Contrast with membranous bone.)

Casparian strip   A band of cell wall containing suberin and lignin, found in the endodermis. Restricts the movement of water across the endodermis.

caspase   One of a group of proteases that catalyze cleavage of target proteins and are active in apoptosis.

catabolic reaction (kat uh bah' lik) [Gk. *kata*: to break down + *ballein*: to throw]   A synthetic reaction in which complex molecules are broken down into simpler ones and energy is released. (Contrast with anabolic reaction.)

catabolite repression   In the presence of abundant glucose, the diminished synthesis of catabolic enzymes for other energy sources.

catalyst (kat' a list) [Gk. *kata*: to break down]   A chemical substance that accelerates a reaction without itself being consumed in the overall course of the reaction. Catalysts lower the activation energy of a reaction. Enzymes are biological catalysts.

cation (cat' eye on)   An ion with one or more positive charges. (Contrast with anion.)

caudal [L. *cauda*: tail]   Pertaining to the tail, or to the posterior part of the body.

cDNA   *See* complementary DNA.

cDNA library   A collection of complementary DNAs derived from mRNAs of a particular tissue at a particular time in the life cycle of an organism.

cecum (see' cum) [L. blind]   A blind branch off the large intestine. In many nonruminant mammals, the cecum contains a colony of microorganisms that contribute to the digestion of food.

cell   The simplest structural unit of a living organism. In multicellular organisms, the building blocks of tissues and organs.

cell adhesion molecules   Molecules on animal cell surfaces that affect the selective association of cells into tissues during development of the embryo.

cell cycle   The stages through which a cell passes between one division and the next. Includes all stages of interphase and mitosis.

cell division   The reproduction of a cell to produce two new cells. In eukaryotes, this process involves nuclear division (mitosis) and cytoplasmic division (cytokinesis).

cell fate   The type of cell that an undifferentiated cell in an embryo will become in the adult.

cell junctions   Specialized structures associated with the plasma membranes of epithelial cells. Some contribute to cell adhesion, others to intercellular communication.

cell recognition   Binding of cells to one another mediated by membrane proteins or carbohydrates.

cell theory   States that cells are the basic structural and physiological units of all living organisms, and that all cells come from preexisting cells.

cell wall   A relatively rigid structure that encloses cells of plants, fungi, many protists, and most prokaryotes, and which gives these cells their shape and limits their expansion in hypotonic media.

cellular immune response   Immune system response mediated by T cells and directed against parasites, fungi, intracellular viruses, and foreign tissues (grafts). (Contrast with humoral immune response.)

cellular respiration   The catabolic pathways by which electrons are removed from various molecules and passed through intermediate electron carriers to O_2, generating H_2O and releasing energy.

cellulose (sell' you lowss)   A straight-chain polymer of glucose molecules, used by plants as a structural supporting material.

central dogma   The premise that information flows from DNA to RNA to polypeptide.

central nervous system (CNS)   That portion of the nervous system that is the site of most information processing, storage, and retrieval; in vertebrates, the brain and spinal cord. (Contrast with peripheral nervous system.)

central vacuole   In plant cells, a large organelle that stores the waste products of metabolism and maintains turgor.

centrifuge [L. *centrum*: center + *fugere*: to flee]   A laboratory device in which a sample is spun around a central axis at high speed. Used to separate suspended materials of different densities.

centriole (sen' tree ole)   A paired organelle that helps organize the microtubules in animal and protist cells during nuclear division.

centromere (sen' tro meer) [Gk. *centron*: center + *meros*: part]   The region where sister chromatids join.

centrosome (sen' tro soam)   The major microtubule organizing center of an animal cell.

cephalization (sef ah luh zay' shun) [Gk. *kephale*: head]   The evolutionary trend toward increasing concentration of brain and sensory organs at the anterior end of the animal.

cerebellum (sair uh bell' um) [L. diminutive of *cerebrum*, brain]   The brain region that controls muscular coordination; located at the anterior end of the hindbrain.

cerebral cortex   The thin layer of gray matter (neuronal cell bodies) that overlies the cerebrum.

cerebrum (su ree' brum) [L. brain]   The dorsal anterior portion of the forebrain, making up the largest part of the brain of mammals; the chief coordination center of the nervous system; consists of two *cerebral hemispheres*.

cervix (sir' vix) [L. neck]   The opening of the uterus into the vagina.

cGMP (cyclic guanosine monophosphate)   An intracellular messenger that is part of signal transmission pathways involving G proteins. (*See* G protein.)

channel protein   An integral membrane protein that forms an aqueous passageway across the membrane in which it is inserted through which specific solutes may pass.

chaperone   A protein that guards other proteins by counteracting molecular interactions that threaten their three-dimensional structure.

character   In genetics, an observable feature, such as eye color. (Contrast with trait.)

character displacement   An evolutionary phenomenon in which species that compete for the same resources within the same territory tend to diverge in morphology and/or behavior.

chemical bond   An attractive force stably linking two atoms.

chemical evolution   The theory that life originated through the chemical transformation of inanimate substances.

chemical reaction   The change in the composition or distribution of atoms of a substance with consequent alterations in properties.

chemical synapse   Neural junction at which neurotransmitter molecules released from a presynaptic cell induce changes in a postsynaptic cell. (Contrast with electrical synapse.)

chemically gated channel   A type of gated channel that opens or closes depending on the presence or absence of a specific molecule, which binds to the channel protein or to a separate receptor that in turn alters the three-dimensional shape of channel protein.

chemiosmosis   Formation of ATP in mitochondria and chloroplasts, resulting from a pumping of protons across a membrane (against a gradient of electrical charge and of pH), followed by the return of the protons through a protein channel with ATP synthase activity.

chemoautotroph   *See* chemolithotroph.

chemoheterotroph   An organism that must obtain both carbon and energy from organic substances. (Contrast with chemolithotroph, photoautotroph, photoheterotroph.)

chemolithotroph [Gk. *lithos*: stone, rock]   An organism that uses carbon dioxide as a carbon source and obtains energy by oxidizing inorganic substances from its environment; also called chemoautotroph. (Contrast with chemoheterotroph, photoautotroph, photoheterotroph.)

chemoreceptor   A sensory receptor cell that senses specific molecules (such as odorant molecules or pheromones) in the environment.

chiasma (kie az' muh) (plural: chiasmata) [Gk. cross]   An X-shaped connection between paired homologous chromosomes in prophase I of meiosis. A chiasma is the visible manifestation of crossing over between homologous chromosomes.

chitin (kye' tin) [Gk. *kiton*: tunic]   The characteristic tough but flexible organic component of the exoskeleton of arthropods, consisting of a complex, nitrogen-containing polysaccharide. Also found in cell walls of fungi.

chlorophyll (klor' o fill) [Gk. *kloros*: green + *phyllon*: leaf]   Any of several green pigments associated with chloroplasts or with certain bacterial membranes; responsible for trapping light energy for photosynthesis.

chloroplast [Gk. *kloros*: green + *plast*: a particle] An organelle bounded by a double membrane containing the enzymes and pigments that perform photosynthesis. Chloroplasts occur only in eukaryotes.

choanocyte (ko' an uh site) The collared, flagellated feeding cells of sponges.

cholecystokinin (ko' luh sis tuh kai' nin) A hormone produced and released by the lining of the duodenum when it is stimulated by undigested fats and proteins. It stimulates the gallbladder to release bile and slows stomach activity.

chorion (kor' ee on) [Gk. *khorion*: afterbirth] The outermost of the membranes protecting mammal, bird, and reptile embryos; in mammals it forms part of the placenta.

chromatid (kro' ma tid) A newly replicated chromosome, from the time molecular duplication occurs until the time the centromeres separate (during anaphase of mitosis or of meiosis II).

chromatin The nucleic acid–protein complex that makes up eukaryotic chromosomes.

chromosomal mutation Loss of or changes in position/direction of a DNA segment on a chromosome.

chromosome (krome' o sowm) [Gk. *kroma*: color + *soma*: body] In bacteria and viruses, the DNA molecule that contains most or all of the genetic information of the cell or virus. In eukaryotes, a structure composed of DNA and proteins that bears part of the genetic information of the cell.

chylomicron (ky low my' cron) Particles of lipid coated with protein, produced in the gut from dietary fats and secreted into the extracellular fluids.

chyme (kime) [Gk. *kymus*: juice] Created in the stomach; a mixture of ingested food with the digestive juices secreted by the salivary glands and the stomach lining.

cilium (sil' ee um) (plural: cilia) [L. eyelash] Hairlike organelle used for locomotion by many unicellular organisms and for moving water and mucus by many multicellular organisms. Generally shorter than a flagellum.

circadian rhythm (sir kade' ee an) [L. *circa*: approximately + *dies*: day] A rhythm of growth or activity that recurs about every 24 hours.

circannual rhythm [L. *circa*: + *annus*: year] A rhythm of growth or activity that recurs on a yearly basis.

circulatory system A system consisting of a muscular pump (heart), a fluid (blood or hemolymph), and a series of conduits (blood vessels) that transports materials around the body.

citric acid cycle In cellular respiration, a set of chemical reactions whereby acetyl CoA is oxidized to carbon dioxide and hydrogen atoms are stored as NADH and $FADH_2$. Also called the Krebs cycle.

clade [Gk. *klados*: branch] A monophyletic group made up of an ancestor and all of its descendants.

class I MHC molecules Cell surface proteins that participate in the cellular immune response directed against virus-infected cells.

class II MHC molecules Cell surface proteins that participate in the cell–cell interactions (of T-helper cells, macrophages, and B cells) of the humoral immune response.

cleavage The first few cell divisions of an animal zygote. *See also* complete cleavage, incomplete cleavage.

climate The long-term average atmospheric conditions (temperature, precipitation, humidity, wind direction and velocity) found in a region.

climax community The final stage of succession; a community that is capable of perpetuating itself under local climatic and soil conditions and persists for a relatively long time.

clinal variation [Gk. *klinein*: to lean] Gradual change in the phenotype of a species over a geographic gradient.

clonal deletion Inactivation or destruction of lymphocyte clones that would produce immune reactions against the animal's own body.

clonal selection Mechanism by which exposure to antigen results in the activation of selected T- or B-cell clones, resulting in an immune response.

clone [Gk. *klon*: twig, shoot] (1) Genetically identical cells or organisms produced from a common ancestor by asexual means. (2) To produce many identical copies of a DNA sequence by its introduction into, and subsequent asexual reproduction of, a cell or organism.

closed circulatory system Circulatory system in which the circulating fluid is contained within a continuous system of vessels. (Contrast with open circulatory system.)

coastal zone The marine life zone that extends from the shoreline to the edge of the continental shelf. Characterized by relatively shallow, well-oxygenated water and relatively stable temperatures and salinities.

coccus (kock' us) (plural: cocci) [Gk. *kokkos*: berry, pit] Any of various spherical or spheroidal bacteria.

cochlea (kock' lee uh) [Gk. *kokhlos*: snail] A spiral tube in the inner ear of vertebrates; it contains the sensory cells involved in hearing.

codominance A condition in which two alleles at a locus produce different phenotypic effects and both effects appear in heterozygotes.

codon Three nucleotides in messenger RNA that direct the placement of a particular amino acid into a polypeptide chain. (Contrast with anticodon.)

coelom (see' loam) [Gk. *koiloma*: cavity] An animal body cavity, enclosed by muscular mesoderm and lined with a mesodermal layer called peritoneum that also surrounds the internal organs.

coenocytic (seen' a sit ik) [Gk. *koinos*: common + *kytos*: container] Referring to the condition, found in some fungal hyphae, of "cells" containing many nuclei but enclosed by a single plasma membrane. Results from nuclear division without cytokinesis.

coenzyme A nonprotein organic molecule that plays a role in catalysis by an enzyme.

coevolution Evolutionary processes in which an adaptation in one species leads to the evolution of an adaptation in a species with which it interacts; also known as reciprocal adaptation.

cofactor An inorganic ion that is weakly bound to an enzyme and required for its activity.

cohesin A protein involved in binding chromatids together.

cohesion The tendency of molecules (or any substances) to stick together.

cohort (co' hort) [L. *cohors*: company of soldiers] A group of similar-aged organisms.

coleoptile A sheath that surrounds and protects the shoot apical meristem and young primary leaves of a grass seedling as they move through the soil.

collagen [Gk. *kolla*: glue] A fibrous protein found extensively in bone and connective tissue.

collecting duct In vertebrates, a tubule that receives urine produced in the nephrons of the kidney and delivers that fluid to the ureter for excretion.

collenchyma (cull eng' kyma) [Gk. *kolla*: glue + *enchyma*: infusion] A type of plant cell, living at functional maturity, which lends flexible support by virtue of primary cell walls thickened at the corners. (Contrast with parenchyma, sclerenchyma.)

colon [Gk. *kolon*] The large intestine.

commensalism [L. *com*: together + *mensa*: table] A type of interaction between species in which one participant benefits while the other is unaffected.

communication A signal from one organism (or cell) that alters the functioning or behavior of another organism (or cell).

community Any ecologically integrated group of species of microorganisms, plants, and animals inhabiting a given area.

compact bone A type of bone with a solid, hard structure. (Contrast with cancellous bone.)

companion cell In angiosperms, a specialized cell found adjacent to a sieve tube element.

comparative experiment Experimental design in which data from various unmanipulated samples or populations are compared, but in which variables are not controlled or even necessarily identified. (Contrast with controlled experiment.)

comparative genomics Computer-aided comparison of DNA sequences between different organisms to reveal genes with related functions.

competition In ecology, use of the same resource by two or more species when the resource is present in insufficient supply for the combined needs of the species.

competitive exclusion A result of competition between species for a limiting resource in which one species completely eliminates the other.

competitive inhibitor A nonsubstrate that binds to the active site of an enzyme and thereby inhibits binding of its substrate. (Contrast with noncompetitive inhibitor.)

complement system A group of eleven proteins that play a role in some reactions of the immune system. The complement proteins are not immunoglobulins.

complementary base pairing The AT (or AU), TA (or UA), CG, and GC pairing of bases in double-stranded DNA, in transcription, and between tRNA and mRNA.

complementary DNA (cDNA) DNA formed by reverse transcriptase acting with an RNA template; essential intermediate in the reproduction of retroviruses; used as a tool in recombinant DNA technology; lacks introns.

complete cleavage Pattern of cleavage that occurs in eggs that have little yolk. Early cleavage furrows divide the egg completely and the blastomeres are of similar size. (Contrast with incomplete cleavage.)

complete metamorphosis A change of state during the life cycle of an organism in which the body is almost completely rebuilt to produce an individual with a very different body form. Characteristic of insects such as butterflies, moths, beetles, ants, wasps, and flies.

compound   (1) A substance made up of atoms of more than one element. (2) Made up of many units, as in the *compound eyes* of arthropods.

concerted evolution   The common evolution of a family of repeated genes, such that changes in one copy of the gene family are replicated in other copies of the gene family.

condensation reaction   A chemical reaction in which two molecules become connected by a covalent bond and a molecule of water is released ($AH + BOH \rightarrow AB + H_2O$.) (Contrast with hydrolysis reaction.)

conditional mutation   A mutation that results in a characteristic phenotype only under certain environmental conditions.

conduction   The transfer of heat from one object to another through direct contact.

cone   (1) In conifers, a reproductive structure consisting of spore-bearing scales extending from a central axis. (Contrast with strobilus.) (2) In the vertebrate retina, a type of photoreceptor cell responsible for color vision.

conidium   (ko nid' ee um) (plural: conidia) [Gk. *konis*: dust]   A type of haploid fungal spore borne at the tips of hyphae, not enclosed in sporangia.

conjugation   (kon ju gay' shun) [L. *conjugare*: yoke together]   (1) A process by which DNA is passed from one cell to another through a *conjugation tube*, as in bacteria. (2) A nonreproductive sexual process by which *Paramecium* and other ciliates exchange genetic material.

connective tissue   A type of tissue that connects or surrounds other tissues; its cells are embedded in a collagen-containing matrix. One of the four major tissue types in multicellular animals.

connexon   In a gap junction, a protein channel linking adjacent animal cells.

consensus sequences   Short stretches of DNA that appear, with little variation, in many different genes.

conservation biology   An applied science that carries out investigations with the aim of maintaining the diversity of life on Earth.

conserved   Pertaining to a gene or trait that has evolved very slowly and is similar or even identical in individuals of highly divergent groups.

conspecifics   Individuals of the same species.

constant region   The portion of an immunoglobulin molecule whose amino acid composition determines its class and does not vary among immunoglobulins in that class. (Contrast with variable region.)

constitutive   Always present; produced continually at a constant rate. (Contrast with inducible.)

consumer   An organism that eats the tissues of some other organism.

continental drift   The gradual movements of the world's continents that have occurred over billions of years.

contractile vacuole   (kon trak' tul)   A specialized vacuole that collects excess water taken in by osmosis, then contracts to expel the water from the cell.

controlled experiment   An experiment in which a sample is divided into groups whereby experimental groups are exposed to manipulations of an independent variable while one group serves as an untreated control. The data from the various groups are then compared to see if there are changes in a dependent variable as a result of the experimental manipulation. (Contrast with comparative experiment.)

convection   The transfer of heat to or from a surface via a moving stream of air or fluid.

convergent evolution   Independent evolution of similar features from different ancestral traits.

copulation   Reproductive behavior that results in a male depositing sperm in the reproductive tract of a female.

cork cambium   [L. *cambiare*: to exchange]   In plants, a lateral meristem that produces secondary growth, mainly in the form of waxy-walled protective cells, including some of the cells that become bark.

cornea   The clear, transparent tissue that covers the eye and allows light to pass through to the retina.

corolla   (ko role' lah) [L. *corolla*: a small crown]   All of the petals of a flower, collectively.

coronary artery   (kor' oh nair ee)   An artery that supplies blood to the heart muscle.

coronary thrombosis   A fibrous clot that blocks a coronary artery.

corpus luteum   (kor' pus loo' tee um) (plural: corpora lutea) [L. yellow body]   A structure formed from a follicle after ovulation; produces hormones important to the maintenance of pregnancy.

corridor   A connection between habitat patches through which organisms can disperse; plays a critical role in maintaining subpopulations.

cortex   [L. *cortex*: covering, rind]   (1) In plants, the tissue between the epidermis and the vascular tissue of a stem or root. (2) In animals, the outer tissue of certain organs, such as the adrenal gland (adrenal cortex) and the brain (cerebral cortex).

corticosteroids   Steroid hormones produced and released by the cortex of the adrenal gland.

corticotropin   A tropic hormone produced by the anterior pituitary hormone that stimulates cortisol release from the adrenal cortex. Also called adrenocorticotropic hormone (ACTH).

corticotropin-releasing hormone   A releasing hormone produced by the hypothalamus that controls the release of cortisol from the anterior pituitary.

cortisol   A corticosteroid that mediates stress responses.

cost–benefit analysis   An approach to evolutionary studies that assumes an animal has a limited amount of time and energy to devote to each of its activities, and that each activity has fitness costs as well as benefits. (*See also* trade-off.)

cotyledon   (kot' ul lee' dun) [Gk. *kotyledon*: hollow space]   A "seed leaf." An embryonic organ that stores and digests reserve materials; may expand when seed germinates.

countercurrent flow   An arrangement that promotes the maximum exchange of heat, or of a diffusible substance, between two fluids by having the fluids flow in opposite directions through parallel vessels close together.

countercurrent multiplier   The mechanism that increases the concentration of the interstitial fluid in the mammalian kidney through countercurrent flow in the loops of Henle and selective permeability and active transport of ions by segments of the loops of Henle.

covalent bond   Chemical bond based on the sharing of electrons between two atoms.

CpG islands   DNA regions rich in C resides adjacent to G residues. Especially abundant in promoters, these regions are where methylation of cytosine usually occurs.

crassulacean acid metabolism (CAM)   A metabolic pathway enabling the plants that possess it to store carbon dioxide at night and then perform photosynthesis during the day with stomata closed.

critical night length   In the photoperiodic flowering response of short-day plants, the length of night above which flowering occurs and below which the plant remains vegetative. (The reverse applies in the case of long-day plants.)

critical period   *See* sensitive period.

cross section   A section taken perpendicular to the longest axis of a structure. Also called a transverse section.

crossing over   The mechanism by which linked genes undergo recombination. In general, the term refers to the reciprocal exchange of corresponding segments between two homologous chromatids.

crypsis   [Gk. *kryptos*: hidden]   The resemblance of an organism to some part of its environment, which helps it to escape detection by enemies.

cryptochromes   [Gk. *kryptos*: hidden + *kroma*: color]   Photoreceptors mediating some blue-light effects in plants and animals.

ctene   (teen) [Gk. *cteis*: comb]   In ctenophores, a comblike row of cilia-bearing plates. Ctenophores move by beating the cilia on their eight ctenes.

culture   (1) A laboratory association of organisms under controlled conditions. (2) The collection of knowledge, tools, values, and rules that characterize a human society.

cuticle   (1) In plants, a waxy layer on the outer body surface that retards water loss. (2) In ecdysozoans, an outer body covering that provides protection and support and is periodically molted.

cyclic AMP   *See* cAMP.

cyclic electron transport   In photosynthetic light reactions, the flow of electrons that produces ATP but no NADPH or O_2.

cyclin   A protein that activates a cyclin-dependent kinase, bringing about transitions in the cell cycle.

cyclin-dependent kinase (Cdk)   A proetin kinase whose target proteins are involved in transitions in the cell cycle and which is active only when complexed with additional protein subunits, called cyclins.

cytokine   A regulatory protein made by immune system cells that affects other target cells in the immune system.

cytokinesis   (sy' toe kine ee' sis) [Gk. *kytos*: container + *kinein*: to move]   The division of the cytoplasm of a dividing cell. (Contrast with mitosis.)

cytokinin   (sy' toe kine' in)   A member of a class of plant growth substances that plays roles in senescence, cell division, and other phenomena.

cytoplasm   The contents of the cell, excluding the nucleus.

cytoplasmic determinants   In animal development, gene products whose spatial distribution may determine such things as embryonic axes.

cytoplasmic segregation   The asymmetrical distribution of cytoplasmic determinants in a developing animal embryo.

cytosine (C) (site' oh seen)   A nitrogen-containing base found in DNA and RNA.

cytoskeleton   The network of microtubules and microfilaments that gives a eukaryotic cell its shape and its capacity to arrange its organelles and to move.

cytosol   The fluid portion of the cytoplasm, excluding organelles and other solids.

cytotoxic T cells (T$_C$)   Cells of the cellular immune system that recognize and directly eliminate virus-infected cells. (Contrast with T-helper cells.)

- D -

DAG   *See* diacylglycerol.

daughter chromosomes   During mitosis, the separated chromatids from the beginning of anaphase onward.

dead space   The lung volume that fails to be ventilated with fresh air (because the lungs are never completely emptied during exhalation).

deciduous [L. *deciduus*: falling off]   Pertaining to a woody plant that sheds it leaves but does not die.

declarative memory   Memory of people, places, events, and things that can be consciously recalled and described. (Contrast with procedural memory.)

decomposer   An organism that metabolizes organic compounds in debris and dead organisms, releasing inorganic material; found among the bacteria, protists, and fungi. *See also* detritivore, saprobe.

defensin   A type of protein made by phagocytes that kills bacteria and enveloped viruses by insertion into their plasma membranes.

degeneracy   The situation in which a single amino acid may be represented by any of two or more different codons in messenger RNA. Most of the amino acids can be represented by more than one codon.

deletion   A mutation resulting from the loss of a continuous segment of a gene or chromosome. Such mutations almost never revert to wild type. (Contrast with duplication, point mutation.)

demethylase   An enzyme that catalyzes the removal of the methyl group from cytosine, reversing DNA methylation.

demography   The study of population structure and of the processes by which it changes.

denaturation   Loss of activity of an enzyme or nucleic acid molecule as a result of structural changes induced by heat or other means.

dendrite [Gk. *dendron*: tree]   A fiber of a neuron which often cannot carry action potentials. Usually much branched and relatively short compared with the axon, and commonly carries information to the cell body of the neuron.

denitrification   Metabolic activity by which nitrate and nitrite ions are reduced to form nitrogen gas; carried by certain soil bacteria.

denitrifiers   Bacteria that release nitrogen to the atmosphere as nitrogen gas (N_2).

density-dependent   Pertaining to a factor with an effect on population size that increases in proportion to population density.

density-independent   Pertaining to a factor with an effect on population size that acts independently of population density.

deoxyribonucleic acid   *See* DNA.

deoxyribose   A five-carbon sugar found in nucleotides and DNA.

depolarization   A change in the resting potential across a membrane so that the inside of the cell becomes less negative, or even positive, compared with the outside of the cell. (Contrast with hyperpolarization.)

derived trait   A trait that differs from the ancestral trait. (Contrast with shared derived trait.)

dermal tissue system   The outer covering of a plant, consisting of epidermis in the young plant and periderm in a plant with extensive secondary growth. (Contrast with ground tissue system and vascular tissue system.)

desmosome (dez' mo sowm) [Gk. *desmos*: bond + *soma*: body]   An adhering junction between animal cells.

desmotubule   A membrane extension connecting the endoplasmic retitulum of two plant cells that traverses the plasmodesma.

determination   In development, the process whereby the fate of an embryonic cell or group of cells (e.g., to become epidermal cells or neurons) is set.

determinate growth   A growth pattern in which the growth of an organism or organ ceases when an adult state is reached; characteristic of most animals and some plant organs. (Contrast with indeterminate growth.)

detritivore (di try' ti vore) [L. *detritus*: worn away + *vorare*: to devour]   An organism that obtains its energy from the dead bodies or waste products of other organisms.

developmental module   A functional entity in the embryo encompassing genes and signaling pathways that determine a physical structure independently of other such modules.

developmental plasticity   The capacity of an organism to alter its pattern of development in response to environmental conditions.

diacylglycerol (DAG)   In hormone action, the second messenger produced by hydrolytic removal of the head group of certain phospholipids.

diapause   A period of developmental or reproductive arrest, entered in response to day length, that enables an organism to better survive.

diaphragm (dye' uh fram) [Gk. *diaphrassein*: barricade]   (1) A sheet of muscle that separates the thoracic and abdominal cavities in mammals; responsible for breathing. (2) A method of birth control in which a sheet of rubber is fitted over the woman's cervix, blocking the entry of sperm.

diastole (dye ass' toll ee) [Gk. dilation]   The portion of the cardiac cycle when the heart muscle relaxes. (Contrast with systole.)

diencephalon   The portion of the vertebrate forebrain that develops into the thalamus and hypothalamus.

differential gene expression   The hypothesis that, given that all cells contain all genes, what makes one cell type different from another is the difference in transcription and translation of those genes.

differentiation   The process whereby originally similar cells follow different developmental pathways; the actual expression of determination.

diffuse coevolution   The evolution of similar traits in suites of species experiencing similar selection pressures imposed by other suites of species with which they interact.

diffusion   Random movement of molecules or other particles, resulting in even distribution of the particles when no barriers are present.

dihybrid cross   A mating in which the parents differ with respect to the alleles of two loci of interest.

dikaryon (di care' ee ahn) [Gk. *di*: two + *karyon*: kernel]   A cell or organism carrying two genetically distinguishable nuclei. Common in fungi.

dioecious (die eesh' us) [Gk. *di*: two + *oikos*: house]   Pertaining to organisms in which the two sexes are "housed" in two different individuals, so that eggs and sperm are not produced in the same individuals. Examples: humans, fruit flies, date palms. (Contrast with monoecious.)

diploblastic   Having two cell layers. (Contrast with triploblastic.)

diploid (dip' loid) [Gk. *diplos*: double]   Having a chromosome complement consisting of two copies (homologs) of each chromosome. Designated 2*n*.

diplontic   A type of life cycle in which gametes are the only haploid cells and mitosis occurs only in diploid cells. (Contrast with haplontic.)

direct transduction   A cell signaling mechanism in which the receptor acts as the effector in the cellular response. (Contrast with indirect transduction.)

directional selection   Selection in which phenotypes at one extreme of the population distribution are favored. (Contrast with disruptive selection, stabilizing selection.)

disaccharide   A carbohydrate made up of two monosaccharides (simple sugars).

dispersal   Movement of organisms away from a parent organism or from an existing population.

dispersion   The distribution of individuals in space within a population.

disruptive selection   Selection in which phenotypes at both extremes of the population distribution are favored. (Contrast with directional selection; stabilizing selection.)

distal   Away from the point of attachment or other reference point. (Contrast with proximal.)

distal convoluted tubule   The portion of a renal tubule from where it reaches the renal cortex, just past the loop of Henle to where it joins a collecting duct. (Compare with proximal convoluted tubule.)

disturbance   A short-term event that disrupts populations, communities, or ecosystems by changing the environment.

disulfide bridge   The covalent bond between two sulfur atoms (–S—S–) linking two molecules or remote parts of the same molecule.

DNA (deoxyribonucleic acid)   The fundamental hereditary material of all living organisms. In eukaryotes, stored primarily in the cell nucleus. A nucleic acid using deoxyribose rather than ribose.

DNA fingerprint   An individual's unique pattern of allele sequences, commonly short tandem repeats and single nucleotide polymorphisms.

DNA helicase   An enzyme that functions to unwind the double helix.

DNA ligase   Enzyme that unites broken DNA strands during replication and recombination.

DNA methylation   The addition of methyl groups to to bases in DNA, usually cytosine or guanine.

DNA methyltransferase   An enzyme that catalyzes the methylation of DNA.

DNA microarray   A small glass or plastic square onto which thousands of single-stranded DNA sequences are fixed so that hybridization of cell-derived RNA or DNA to the target sequences can be performed.

DNA polymerase   Any of a group of enzymes that catalyze the formation of DNA strands from a DNA template.

DNA topoisomerase   An enzyme that unwinds and winds coils of DNA that form during replication and transcription.

docking protein   A receptor protein that binds (docks) a ribosome to the membrane of the endoplasmic reticulum by binding the signal sequence attached to a new protein being made at the ribosome.

domain   (1) An independent structural element within a protein. Encoded by recognizable nucleotide sequences, a domain often folds separately from the rest of the protein. Similar domains can appear in a variety of different proteins across phylogenetic groups (e.g., "homeobox domain"; "calcium-binding domain"). (2) In phylogenetics, the three monophyletic branches of life (Bacteria, Archaea, and Eukarya).

dominance   In genetics, the ability of one allelic form of a gene to determine the phenotype of a heterozygous individual in which the homologous chromosomes carry both it and a different (recessive) allele. (Contrast with recessive.)

dormancy   A condition in which normal activity is suspended, as in some spores, seeds, and buds.

dorsal   [L. *dorsum*: back]   Toward or pertaining to the back or upper surface. (Contrast with ventral.)

dorsal lip   In amphibian embryos, the dorsal segment of the blastopore. Also called the "organizer," this region directs the development of nearby embryonic regions.

double fertilization   In angiosperms, a process in which the nuclei of two sperm fertilize one egg. One sperm's nucleus combines with the egg nucleus to produce a zygote, while the other combines with the same egg's two polar nuclei to produce the first cell of the triploid endosperm (the tissue that will nourish the growing plant embryo).

double helix   Refers to DNA and the (usually right-handed) coil configuration of two complementary, antiparallel strands.

downregulation   A negative feedback process in which continuous high concentrations of a hormone can decrease the number of its receptors. (Contrast with upregulation.)

duodenum   (do' uh dee' num)   The beginning portion of the vertebrate small intestine. (Contrast with ileum, jejunum.)

duplication   A mutation in which a segment of a chromosome is duplicated, often by the attachment of a segment lost from its homolog. (Contrast with deletion.)

- E -

ecdysone   [eck die' sone] [Gk. *ek*: out of + *dyo*: to clothe]   In insects, a hormone that induces molting.

ecological efficiency   The overall transfer of energy from one trophic level to the next, expressed as the ratio of consumer production to producer production.

ecology   [Gk. *oikos*: house]   The scientific study of the interaction of organisms with their living (biotic) and nonliving (abiotic) environments.

ecosystem   (eek' oh sis tum)   The organisms of a particular habitat, such as a pond or forest, together with the physical environment in which they live.

ecosystem engineer   An organism that builds structures that alter existing habitats or create new habitats.

ecosystem services   Processes by which ecosystems maintain resources that benefit human society.

ectoderm   [Gk. *ektos*: outside + *derma*: skin]   The outermost of the three embryonic germ layers first delineated during gastrulation. Gives rise to the skin, sense organs, and nervous system.

ectotherm   [Gk. *ektos*: outside + *thermos*: heat]   An animal that is dependent on external heat sources for regulating its body temperature (Contrast with endotherm.)

edema   (i dee' mah) [Gk. *oidema*: swelling]   Tissue swelling caused by the accumulation of fluid.

edge effect   The changes in ecological processes in a community caused by physical and biological factors originating in an adjacent community.

effector protein   In cell signaling, a protein responsible for the cellular reponse to a signal transduction pathway.

efferent   (ef' ur unt) [L. *ex*: out + *ferre*: to bear]   Carrying outward or away from, as in a neuron that carries impulses outward from the central nervous system (efferent neuron), or a blood vessel that carries blood away from a structure. (Contrast with afferent.)

egg   In all sexually reproducing organisms, the female gamete; in birds, reptiles, and some other vertebrates, a structure within which early embryonic development occurs. *See also* amniote egg, ovum.

electrical synapse   A type of synapse at which action potentials spread directly from presynaptic cell to postsynaptic cell. (Contrast with chemical synapse.)

electrocardiogram (ECG or EKG)   A graphic recording of electrical potentials from the heart.

electrochemical gradient   The concentration gradient of an ion across a membrane plus the voltage difference across that membrane.

electroencephalogram (EEG)   A graphic recording of electrical potentials from the brain.

electromagnetic radiation   A self-propagating wave that travels though space and has both electrical and magnetic properties.

electron   A subatomic particle outside the nucleus carrying a negative charge and very little mass.

electron shell   The region surrounding the atomic nucleus at a fixed energy level in which electrons orbit.

electron transport   The passage of electrons through a series of proteins with a release of energy which may be captured in a concentration gradient or chemical form such as NADH or ATP.

electronegativity   The tendency of an atom to attract electrons when it occurs as part of a compound.

electrophoresis   *See* gel electrophoresis.

element   A substance that cannot be converted to simpler substances by ordinary chemical means.

elongation   (1) In molecular biology, the addition of monomers to make a longer RNA or protein during transcription or translation. (2) Growth of a plant axis or cell primarily in the longitudinal direction.

embolus   (em' buh lus) [Gk. *embolos*: stopper]   A circulating blood clot. Blockage of a blood vessel by an embolus or by a bubble of gas is referred to as an *embolism*. (Contrast with thrombus.)

embryo   [Gk. *en*: within + *bryein*: to grow]   A young animal, or young plant sporophyte, while it is still contained within a protective structure such as a seed, egg, or uterus.

embryonic stem cell (ESC)   A pluripotent cell in the blastocyst.

embryo sac   In angiosperms, the female gametophyte. Found within the ovule, it consists of eight or fewer cells, membrane bounded, but without cellulose walls between them.

emergent property   A property of a complex system that is not exhibited by its individual component parts.

emigration   The deliberate and usually oriented departure of an organism from the habitat in which it has been living.

3′ end   (3 prime)   The end of a DNA or RNA strand that has a free hydroxyl group at the 3′ carbon of the sugar (deoxyribose or ribose).

5′ end   (5 prime)   The end of a DNA or RNA strand that has a free phosphate group at the 5′ carbon of the sugar (deoxyribose or ribose).

endemic   (en dem' ik) [Gk. *endemos*: native]   Confined to a particular region, thus often having a comparatively restricted distribution.

endergonic   A chemical reaction in which the products have higher free energy than the reactants, thereby requiring free energy input to occur. (Contrast with exergonic.)

endocrine gland   (en' doh krin) [Gk. *endo*: within + *krinein*: to separate]   An aggregation of secretory cells that secretes hormones into the blood. The *endocrine system* consists of all *endocrine cells* and endocrine glands in the body that produce and release hormones. (Contrast with exocrine gland.)

endocytosis   A process by which liquids or solid particles are taken up by a cell through invagination of the plasma membrane. (Contrast with exocytosis.)

endoderm   [Gk. *endo*: within + *derma*: skin]   The innermost of the three embryonic germ layers delineated during gastrulation. Gives rise to the digestive and respiratory tracts and structures associated with them.

endodermis   In plants, a specialized cell layer marking the inside of the cortex in roots and some stems. Frequently a barrier to free diffusion of solutes.

endomembrane system   A system of intracellular membranes that exchange material with one another, consisting of the Golgi apparatus, endoplasmic reticulum, and lysosomes when present.

endometrium   The epithelial lining of the uterus.

endoplasmic reticulum (ER)   [Gk. *endo*: within + L. *reticulum*: net]   A system of membranous tubes and flattened sacs found in the cytoplasm of eukaryotes. Exists in two forms: rough ER, studded with ribosomes; and smooth ER, lacking ribosomes.

endorphins   Molecules in the mammalian brain act as neurotransmitters in pathways that control pain.

endoskeleton [Gk. *endo*: within + *skleros*: hard] An internal skeleton covered by other, soft body tissues. (Contrast with exoskeleton.)

endosperm [Gk. *endo*: within + *sperma*: seed]   A specialized triploid seed tissue found only in angiosperms; contains stored nutrients for the developing embryo.

endospore [Gk. *endo*: within + *spora*: to sow] In some bacteria, a resting structure that can survive harsh environmental conditions.

endosymbiosis theory [Gk. *endo*: within + *sym*: together + *bios*: life]   The theory that the eukaryotic cell evolved via the engulfing of one prokaryotic cell by another.

endothelium   The single layer of epithelial cells lining the interior of a blood vessel.

endotherm [Gk. *endo*: within + *thermos*: heat] An animal that can control its body temperature by the expenditure of its own metabolic energy. (Contrast with ectotherm.)

endotoxin   A lipopolysaccharide that forms part of the outer membrane of certain Gram-negative bacteria that is released when the bacteria grow or lyse. (Contrast with exotoxin.)

energetic cost   The difference between the energy an animal expends in performing a behavior and the energy it would have expended had it rested.

energy   The capacity to do work or move matter against an opposing force. The capacity to accomplish change in physical and chemical systems.

energy budget   A quantitative description of all paths of energy exchange between an animal and its environment.

enkephalins   Molecules in the mammalian brain act as neurotransmitters in pathways that control pain.

enthalpy (*H*)   The total energy of a system.

entropy (*S*) (en' tro pee) [Gk. *tropein*: to change] A measure of the degree of disorder in any system. Spontaneous reactions in a closed system are always accompanied by an increase in entropy.

enveloped virus   A virus enclosed within a phospholipid membrane derived from its host cell.

environment   Whatever surrounds and interacts with or otherwise affects a population, organism, or cell. May be external or internal.

environmentalism   The use of ecological knowledge, along with economics, ethics, and many other considerations, to inform both personal decisions and public policy relating to stewardship of natural resources and ecosystems.

enzyme (en' zime) [Gk. *zyme*: to leaven (as in yeast bread)]   A catalytic protein that speeds up a biochemical reaction.

epi- [Gk. upon, over]   A prefix used to designate a structure located on top of another; for example, epidermis, epiphyte.

epiblast   The upper or overlying portion of the avian blastula which is joined to the hypoblast at the margins of the blastodisc.

epiboly   The movement of cells over the surface of the blastula toward the forming blastopore.

epitope   *See* antigenic determinant.

epidermis [Gk. *epi*: over + *derma*: skin]   In plants and animals, the outermost cell layers. (Only one cell layer thick in plants.)

epididymis (epuh did' uh mus) [Gk. *epi*: over + *didymos*: testicle]   Coiled tubules in the testes that store sperm and conduct sperm from the seminiferous tubules to the vas deferens.

epigenetics   The scientific study of changes in the expression of a gene or set of genes that occur without change in the DNA sequence.

epinephrine (ep i nef' rin) [Gk. *epi*: over + *nephros*: kidney]   The "fight or flight" hormone produced by the medulla of the adrenal gland; it also functions as a neurotransmitter. (Also known as adrenaline.)

epistasis   Interaction between genes in which the presence of a particular allele of one gene determines whether another gene will be expressed.

epithelium   A type of animal tissue made up of sheets of cells that lines or covers organs, makes up tubules, and covers the surface of the body; one of the four major tissue types in multicellular animals.

equilibrium   Any state of balanced opposing forces and no net change.

ER   *See* endoplasmic reticulum.

error signal   In regulatory systems, any difference between the set point of the system and its current condition.

erythrocyte (ur rith' row site) [Gk. *erythros*: red + *kytos*: container]   A red blood cell.

erythropoietin   A hormone produced by the kidney in response to lack of oxygen that stimulates the production of red blood cells.

esophagus (i soff' i gus) [Gk. *oisophagos*: gullet] That part of the gut between the pharynx and the stomach.

essential acids   Amino acids or fatty acids that an animal cannot synthesize for itself and must obtain from its food.

essential element   A mineral nutrient required for normal growth and reproduction in plants and animals.

ester linkage   A condensation (water-releasing) reaction in which the carboxyl group of a fatty acid reacts with the hydroxyl group of an alcohol. Lipids are formed in this way.

estivation (ess tuh vay' shun) [L. *aestivalis*: summer]   A state of dormancy and hypometabolism that occurs during the summer; usually a means of surviving drought and/or intense heat. (Contrast with hibernation.)

estrogen   Any of several steroid sex hormones; produced chiefly by the ovaries in mammals.

estrus (es' trus) [L. *oestrus*: frenzy]   The period of heat, or maximum sexual receptivity, in some female mammals. Ordinarily, the estrus is also the time of release of eggs in the female.

ethology [Gk. *ethos*: character + *logos*: study]   An approach to the study of animal behavior that focuses on studying many species in natural environments and addresses questions about the evolution of behavior.

ethylene   One of the plant growth hormones, the gas $H_2C=CH_2$. Involved in fruit ripening and other growth and developmental responses.

eukaryotes (yew car' ree oats) [Gk. *eu*: true + *karyon*: kernel or nucleus]   Organisms whose cells contain their genetic material inside a nu-

cleus. Includes all life other than the viruses, archaea, and bacteria.

eusocial   Pertaining to a social group that includes nonreproductive individuals, as in honey bees.

eutrophication (yoo trofe' ik ay' shun) [Gk. *eu*: truly + *trephein*: to flourish]   The addition of nutrient materials to a body of water, resulting in changes in ecological processes and species composition therein.

evaporation   The transition of water from the liquid to the gaseous phase.

evolution   Any gradual change. Most often refers to organic or Darwinian evolution, which is the genetic and resulting phenotypic change in populations of organisms from generation to generation. (*See* macroevolution, microevolution; contrast with speciation.)

evolutionary radiation   The proliferation of many species within a single evolutionary lineage.

evolutionary reversal   The reappearance of an ancestral trait in a group that had previously acquired a derived trait.

excision repair   A mechanism that removes damaged DNA and replaces it with the appropriate nucleotide.

excited state   The state of an atom or molecule when, after absorbing energy, it has more energy than in its normal, ground state. (Contrast with ground state.)

excretion   Release of metabolic wastes by an organism.

exergonic   A chemical reaction in which the products of the reaction have lower free energy than the reactants, resulting in a release of free energy. (Contrast with endergonic.)

exocrine gland (eks' oh krin) [Gk. *exo*: outside + *krinein*: to separate]   Any gland, such as a salivary gland, that secretes to the outside of the body or into the gut. (Contrast with endocrine gland.)

exocytosis   A process by which a vesicle within a cell fuses with the plasma membrane and releases its contents to the outside. (Contrast with endocytosis.)

exon   A portion of a DNA molecule, in eukaryotes, that codes for part of a polypeptide. (Contrast with intron.)

exoskeleton (eks' oh skel' e ton) [Gk. *exos*: outside + *skleros*: hard]   A hard covering on the outside of the body to which muscles are attached. (Contrast with endoskeleton.)

exotoxin   A highly toxic, usually soluble protein released by living, multiplying bacteria. (Contrast with endotoxin.)

expanding triplet repeat   A three-base-pair sequence in a human gene that is unstable and can be repeated a few to hundreds of times. Often, the more the repeats, the less the activity of the gene involved. Expanding triplet repeats occur in some human diseases such as Huntington's disease and fragile-X syndrome.

experiment   A testing process to support or disprove hypotheses and to answer questions. The basis of the scientific method. *See* comparative experiment, controlled experiment.

expiratory reserve volume   The amount of air that can be forcefully exhaled beyond the normal tidal expiration. (Contrast with inspiratory reserve volume, tidal volume, vital capacity.)

exploitation competition   Competition in which individuals reduce the quantities of their shared resources. (Contrast with interference competition.)

exponential growth   Growth, especially in the number of organisms in a population, which is a geometric function of the size of the growing entity: the larger the entity, the faster it grows. (Contrast with logistic growth.)

expression vector   A DNA vector, such as a plasmid, that carries a DNA sequence that includes the adjacent sequences for its expression into mRNA and protein in a host cell.

expressivity   The degree to which a genotype is expressed in the phenotype; may be affected by the environment.

extensor   A muscle that extends an appendage.

external fertilization   The release of gametes into the environment; typical of aquatic animals. Also called spawning. (Contrast with internal fertilization.)

external gills   Highly branched and folded extensions of the body surface that provide a large surface area for gas exchange with water; typical of larval amphibians and many larval insects.

extinction   The termination of a lineage of organisms.

extracellular matrix   A material of heterogeneous composition surrounding cells and performing many functions including adhesion of cells.

extraembryonic membranes   Four membranes that support but are not part of the developing embryos of reptiles, birds, and mammals, defining these groups phylogenetically as amniotes. (See amnion, allantois, chorion, and yolk sac.)

- F -

F₁   The first filial generation; the immediate progeny of a parental (P) mating.

F₂   The second filial generation; the immediate progeny of a mating between members of the F_1 generation.

facilitated diffusion   Passive movement through a membrane involving a specific carrier protein; does not proceed against a concentration gradient. (Contrast with active transport, diffusion.)

facilitation   In succession, modification of the environment by a colonizing species in a way that allows colonization by other species. (Contrast with inhibition.)

facultative anaerobe   A prokaryote that can shift its metabolism between anaerobic and aerobic operations modes on the presence or absence of O_2. (Alternatively, facultative aerobe.)

fast-twitch fibers   Skeletal muscle fibers that can generate high tension rapidly, but fatigue rapidly ("sprinter" fibers). Characterized by an abundance of enzymes of glycolysis.

fat   A triglyceride that is solid at room temperature. (Contrast with oil.)

fate map   A diagram of the blastula showing which cells (blastomeres) are "fated " to contribute to specific tissues and organs in the mature body.

fatty acid   A molecule made up of a long nonpolar hydrocarbon chain and a polar carboxyl group. Found in many lipids.

fauna (faw' nah)   All the animals found in a given area. (Contrast with flora.)

feces [L. *faeces*: dregs]   Waste excreted from the digestive system.

fecundity (m_x)   The average number of offspring produced by each female.

feedback information   In regulatory systems, information about the relationship between the set point of the system and its current state.

feedforward information   In regulatory systems, information that changes the set point of the system.

fermentation (fur men tay' shun) [L. *fermentum*: yeast]   The anaerobic degradation of a substance such as glucose to smaller molecules such as lactic acid or alcohol with the extraction of energy.

fertilization   Union of gametes. Also known as syngamy.

fetus   Medical and legal term for the stages of a developing human embryo from about the eighth week of pregnancy (the point at which all major organ systems have formed) to the moment of birth.

fiber   In angiosperms, an elongated, tapering sclerenchyma cell, usually with a thick cell wall, that serves as a support function in xylem. (See also muscle fiber.)

fibrin   A protein that polymerizes to form long threads that provide structure to a blood clot.

fibrinogen   A circulating protein that can be stimulated to fall out of solution and provide the structure for a blood clot.

fibrous root system   A root system typical of monocots composed of numerous thin adventitious roots that are all roughly equal in diameter. (Contrast with taproot system.)

Fick's law of diffusion   An equation that describes the factors that determine the rate of diffusion of a molecule from an area of higher concentration to an area of lower concentration.

fight-or-flight response   A rapid physiological response to a sudden threat mediated by the hormone epinephrine.

filter feeder   An organism that feeds on organisms much smaller than itself that are suspended in water or air by means of a straining device.

first law of thermodynamics   The principle that energy can be neither created nor destroyed.

fission   See binary fission.

fitness   The contribution of a genotype or phenotype to the genetic composition of subsequent generations, relative to the contribution of other genotypes or phenotypes. (See also inclusive fitness.)

flagellum (fla jell' um) (plural: flagella) [L. *flagellum*: whip]   Long, whiplike appendage that propels cells. Prokaryotic flagella differ sharply from those found in eukaryotes.

fixed action pattern   In ethology, a genetically determined behavior that is performed without learning, stereotypic (performed the same way each time), and not modifiable by learning.

flexor   A muscle that flexes an appendage.

flora (flore' ah)   All of the plants found in a given area. (Contrast with fauna.)

floral meristem   In angiosperms, a meristem that forms the floral organs (sepals, petals, stamens, and carpels).

floral organ identity genes   In angiosperms, genes that determine the fates of floral meristem cells; their expression is triggered by the products of meristem identity genes.

florigen   A plant hormone involved in the conversion of a vegetative shoot apex to a flower.

flower   The sexual structure of an angiosperm.

fluid feeder   An animal that feeds on fluids it extracts from the bodies of other organisms; examples include nectar-feeding birds and blood-sucking insects.

fluid mosaic model   A molecular model for the structure of biological membranes consisting of a fluid phospholipid bilayer in which suspended proteins are free to move in the plane of the bilayer.

follicle [L. *folliculus*: little bag]   In female mammals, an immature egg surrounded by nutritive cells.

follicle-stimulating hormone (FSH)   A gonadotropin produced by the anterior pituitary.

food chain   A portion of a food web, most commonly a simple sequence of prey species and the predators that consume them.

food vacuole   Membrane enclosed structure formed by phagocytosis in which engulfed food particles are digested by the action of lysosomal enzymes.

food web   The complete set of food links between species in a community; a diagram indicating which ones are the eaters and which are eaten.

forebrain   The region of the vertebrate brain that comprises the cerebrum, thalamus, and hypothalamus.

fossil   Any recognizable structure originating from an organism, or any impression from such a structure, that has been preserved over geological time.

fossil fuels   Fuels, including oil, natural gas, coal, and peat, formed over geologic time from organic material buried in anaerobic sediments.

founder effect   Random changes in allele frequencies resulting from establishment of a population by a very small number of individuals.

fovea [L. *fovea*: a small pit]   In the vertebrate retina, the area of most distinct vision.

frame-shift mutation   The addition or deletion of a single or two adjacent nucleotides in a gene's sequence. Results in the misreading of mRNA during translation and the production of a nonfunctional protein. (Contrast with missense mutation, nonsense mutation, silent mutation.)

Frank–Starling law   The stroke volume of the heart increases with increased return of blood to the heart.

free energy (G)   Energy that is available for doing useful work, after allowance has been made for the increase or decrease of disorder.

freeze-fracturing   Method of tissue preparation for transmission and scanning electron microscopy in which a tissue is frozen and a knife is then used to crack open the tissue; the fracture often occurs in the path of least resistance, within a membrane.

frequency-dependent selection   Selection that changes in intensity with the proportion of individuals in a population having the trait.

fruit   In angiosperms, a ripened and mature ovary (or group of ovaries) containing the seeds. Sometimes applied to reproductive structures of other groups of plants.

functional genomics   The assignment of functional roles to the proteins encoded by genes identified by sequencing entire genomes.

functional group   A characteristic combination of atoms that contribute specific properties when attached to larger molecules.

fundamental niche   A species' niche as defined by its physiological capabilities. (Contrast with realized niche.)

- G -

G cap   A chemically modified GTP added to the 5' end of mRNA; facilitates binding of mRNA to ribosome and prevents mRNA breakdown.

G1   In the cell cycle, the gap between the end of mitosis and the onset of the S phase.

G2   In the cell cycle, the gap between the S (synthesis) phase and the onset of mitosis.

G protein   A membrane protein involved in signal transduction; characterized by binding GDP or GTP.

gain of function mutation   A mutation that results in a protein with a new function. (Contrast with loss of function mutation.)

gallbladder   In the human digestive system, an organ in which bile is stored.

gametangium (gam uh tan' gee um) (plural: gametangia) [Gk. *gamos*: marriage + *angeion*: vessel]   Any plant or fungal structure within which a gamete is formed.

gamete (gam' eet) [Gk. *gamete/gametes*: wife, husband]   The mature sexual reproductive cell: the egg or the sperm.

gametogenesis (ga meet' oh jen' e sis)   The specialized series of cellular divisions that leads to the production of gametes. (*See also* oogenesis, spermatogenesis.)

gametophyte (ga meet' oh fyte)   In plants and photosynthetic protists with alternation of generations, the multicellular haploid phase that produces the gametes. (Contrast with sporophyte.)

gamma diversity   The regional diversity found over a range of communities or habitats in a geographic region. (Contrast with alpha diveristy, beta diversity.)

ganglion (gang' glee un) (plural: ganglia) [Gk. *tumor*]   A cluster of neurons that have similar characteristics or function.

ganglion cells   Cells at the front of the human retina that transmit information from the bipolar cells to the brain.

gap junction   A 2.7-nanometer gap between plasma membranes of two animal cells, spanned by protein channels. Gap junctions allow chemical substances or electrical signals to pass from cell to cell.

gastric pits   Deep infoldings in the walls of the stomach lined with secretory cells.

gastrin   A hormone secreted by cells in the lower region of the stomach that stimulates the secretion of digestive juices as well as movements of the stomach.

gastrovascular cavity   Serving for both digestion (gastro) and circulation (vascular); in particular, the central cavity of the body of jellyfish and other cnidarians.

gastrulation   Development of a blastula into a gastrula. In embryonic development, the process by which a blastula is transformed by massive movements of cells into a *gastrula*, an embryo with three germ layers and distinct body axes.

gated channel   A membrane protein that changes its three-dimensional shape, and therefore its ion conductance, in response to a stimulus. When open, it allows specific ions to move across the membrane.

gel electrophoresis (e lek' tro fo ree' sis) [L. *electrum*: amber + Gk. *phorein*: to bear]   A technique for separating molecules (such as DNA fragments) from one another on the basis of their electric charges and molecular weights by applying an electric field to a gel.

gene [Gk. *genes*: to produce]   A unit of heredity. Used here as the unit of genetic function which carries the information for a single polypeptide or RNA.

gene family   A set of similar genes derived from a single parent gene; need not be on the same chromosomes. The vertebrate globin genes constitute a classic example of a gene family.

gene flow   Exchange of genes between populations through migration of individuals or movements of gametes.

gene-for-gene resistance   In plants, a mechanism of resistance to pathogens in which resistance is triggered by the specific interaction of the products of a pathogen's *Avr* genes and a plant's *R* genes.

gene pool   All of the different alleles of all of the genes existing in all individuals of a population.

gene therapy   Treatment of a genetic disease by providing patients with cells containing functioning alleles of the genes that are nonfunctional in their bodies.

gene tree   A graphic representation of the evolutionary relationships of a single gene in different species or of the members of a gene family.

genetic code   The set of instructions, in the form of nucleotide triplets, that translate a linear sequence of nucleotides in mRNA into a linear sequence of amino acids in a protein.

genetic drift   Changes in gene frequencies from generation to generation as a result of random (chance) processes.

genetic map   The positions of genes along a chromosome as revealed by recombination frequencies.

genetic marker   (1) In gene cloning, a gene of identifiable phenotype that indicates the presence of another gene, DNA segment, or chromosome fragment. (2) In general, a DNA sequence such as a single nucleotide polymorphism whose presence is correlated with the presence of other linked genes on that chromosome.

genetic structure   The frequencies of different alleles at each locus and the frequencies of different genotypes in a Mendelian population.

genetic switches   Mechanisms that control how the genetic toolkit is used, such as promoters and the transcription factors that bind them. The signal cascades that converge on and operate these switches determine when and where genes will be turned on and off.

genetic toolkit   In evolutionary developmental biology, DNA sequences controlling developmental mechanisms that have been conserved over evolutionary time.

genetics   The scientific study of the structure, functioning, and inheritance of genes, the units of hereditary information.

genome (jee' nome)   The complete DNA sequence for a particular organism or individual.

genomic equivalence   The principle that no information is lost from the nuclei of cells as they pass through the early stages of embryonic development.

genomic imprinting   The form of a gene's expression is determined by parental source (i.e., whether the gene is inherited from the male or female parent).

genomic library   All of the cloned DNA fragments generated by the action of a restriction endonuclease on a genome.

genomics   The scientific study of entire sets of genes and their interactions.

genotype (jean' oh type) [Gk. *gen*: to produce + *typos*: impression]   An exact description of the genetic constitution of an individual, either with respect to a single trait or with respect to a larger set of traits. (Contrast with phenotype.)

genus (jean' us) (plural: genera) [Gk. *genos*: stock, kind]   A group of related, similar species recognized by taxonomists with a distinct name used in binomial nomenclature.

germ cell [L. *germen*: to beget]   A reproductive cell or gamete of a multicellular organism. (Contrast with somatic cell.)

germ layers   The three embryonic layers formed during gastrulation (ectoderm, mesoderm, and endoderm). Also called cell layers or tissue layers.

germ line mutation   Mutation in a cell that produces gametes (i.e., a germ line cell). (Contrast with somatic mutation.)

germination   Sprouting of a seed or spore.

gestation (jes tay' shun) [L. *gestare*: to bear]   The period during which the embryo of a mammal develops within the uterus. Also known as pregnancy.

ghrelin   A hormone produced and secreted by cells in the stomach that stimulates appetite.

gibberellin (jib er el' lin)   A class of plant growth hormones playing roles in stem elongation, seed germination, flowering of certain plants, etc.

gill   An organ specialized for gas exchange with water.

gizzard (giz' erd) [L. *gigeria*: cooked chicken parts]   A muscular port of the stomach of birds that grinds up food, sometimes with the aid of fragments of stone.

glia (glee' uh) [Gk. *glia*: glue]   Cells of the nervous system that do not conduct action potentials.

glomerular filtration rate (GFR)   The rate at which the blood is filtered in the glomeruli of the kidney.

glomerulus (glo mare' yew lus) [L. *glomus*: ball]   Sites in the kidney where blood filtration takes place. Each glomerulus consists of a knot of capillaries served by afferent and efferent arterioles.

glucagon   Hormone produced by alpha cells of the pancreatic islets of Langerhans. Glucagon stimulates the liver to break down glycogen and release glucose into the circulation.

gluconeogenesis   The biochemical synthesis of glucose from other substances, such as amino acids, lactate, and glycerol.

glucose [Gk. *gleukos*: sugar, sweet]   The most common monosaccharide; the monomer of the polysaccharides starch, glycogen, and cellulose.

glycerol (gliss' er ole)   A three-carbon alcohol with three hydroxyl groups; a component of phospholipids and triglycerides.

glycogen (gly' ko jen)   An energy storage polysaccharide found in animals and fungi; a branched-chain polymer of glucose, similar to starch.

glycolipid   A lipid to which sugars are attached.

glycolysis (gly kol' li sis) [Gk. *gleukos*: sugar + *lysis*: break apart]   The enzymatic breakdown of glucose to pyruvic acid.

glycoprotein   A protein to which sugars are attached.

glycosidic linkage   Bond between carbohydrate (sugar) molecules through an intervening oxygen atom (–O–).

glycosylation   The addition of carbohydrates to another type of molecule, such as a protein.

glyoxysome (gly ox' ee soam)   An organelle found in plants, in which stored lipids are converted to carbohydrates.

Golgi apparatus (goal' jee)   A system of concentrically folded membranes found in the cytoplasm of eukaryotic cells; functions in secretion from cell by exocytosis.

gonad (go' nad) [Gk. *gone*: seed]   An organ that produces gametes in animals: either an ovary (female gonad) or testis (male gonad).

gonadotropin   A type of trophic hormone that stimulates the gonads.

gonadotropin-releasing hormone (GnRH)   Hormone produced by the hypothalamus that stimulates the anterior pituitary to secrete ("release") gonadotropins.

Gondwana   The large southern land mass that existed from the Cambrian (540 mya) to the Jurassic (138 mya). Present-day remnants are South America, Africa, India, Australia, and Antarctica.

grafting   Artificial transplantation of tissue from one organism to another. In horticulture, the transfer of a bud or stem segment from one plant onto the root of another as a form of asexual reproduction.

Gram stain   A differential purple stain useful in characterizing bacteria. The peptidoglycan-rich cell walls of Gram-positive bacteria stain purple; cell walls of Gram-negative bacteria generally stain orange.

gravitropism [Gk. *tropos*: to turn]   A directed plant growth response to gravity.

gray matter   In the nervous system, tissue that is rich in neuronal cell bodies. (Contrast with white matter.)

greenhouse gases   Gases in the atmosphere, such as carbon dioxide and methane, that are transparent to sunlight, but trap heat radiating from Earth's surface, causing heat to build up at Earth's surface.

gross primary production   The amount of energy captured by the primary producers in a community.

gross primary productivity (GPP)   The rate at which the primary producers in a community turn solar energy into stored chemical energy via photosynthesis.

ground meristem   That part of an apical meristem that gives rise to the ground tissue system of the primary plant body.

ground tissue system   Those parts of the plant body not included in the dermal or vascular tissue systems. Ground tissues function in storage, photosynthesis, and support.

growth   An increase in the size of the body and its organs by cell division and cell expansion.

growth factor   A chemical signal that stimulates cells to divide.

growth hormone   A peptide hormone released by the anterior pituitary that stimulates many anabolic processes.

guanine (G) (gwan' een)   A nitrogen-containing base found in DNA, RNA, and GTP.

guard cells   In plants, specialized, paired epidermal cells that surround and control the opening of a stoma (pore). *See* stoma.

guild   In ecology, a group of species that exploit the same resource, but in slightly different ways.

gustation   The sense of taste.

gut   An animal's digestive tract.

- H -

habitat   The particular environment in which an organism lives. A *habitat patch* is an area of a particular habitat surrounded by other habitat types that may be less suitable for that organism.

hair cell   A type of mechanoreceptor in animals. Detects sound waves and other forms of motion in air or water.

half-life   The time required for half of a sample of a radioactive isotope to decay to its stable, nonradioactive form, or for a drug or other substance to reach half its initial dosage.

halophyte (hal' oh fyte) [Gk. *halos*: salt + *phyton*: plant]   A plant that grows in a saline (salty) environment.

Hamilton's rule   The principle that, for an apparent altruistic behavior to be adaptive, the fitness benefit of that act to the recipient times the degree of relatedness of the performer and the recipient must be greater than the cost to the performer.

haplodiploidy   A sex determination mechanism in which diploid individuals (which develop from fertilized eggs) are female and haploid individuals (which develop from unfertilized eggs) are male; typical of hymenopterans.

haploid (hap' loid) [Gk. *haploeides*: single]   Having a chromosome complement consisting of just one copy of each chromosome; designated 1*n* or *n*. (Contrast with diploid.)

haplontic   A type of life cycle in which the zygote is the only diploid cell and mitosis occurs only in haploid cells. (Contrast with diplontic.)

haplotype   Linked nucleotide sequences that are usually inherited as a unit (as a "sentence" rather than as individual "words").

Hardy–Weinberg equililbrium   In a sexually reproducing population, the allele frequency at a given locus that is not being acted on by agents of evolution; the conditions that would result in no evolution in a population.

haustorium (haw stor' ee um) (plural: haustoria)[L. *haustus*: draw up]   A specialized hypha or other structure by which fungi and some parasitic plants draw nutrients from a host plant.

Haversian systems   Units of organization in compact bone that reflect the action of intercommunicating osteoblasts.

heart   In circulatory systems, a muscular pump that moves extracellular fluid around the body.

heat of vaporization   The energy that must be supplied to convert a molecule from a liquid to a gas at its boiling point.

heat shock proteins   Chaperone proteins expressed in cells exposed to high or low temperatures or other forms of environmental stress.

helical   Shaped like a screw or spring; this shape occurs in DNA and proteins.

helper T cells   *See* T-helper cells.

hemiparasite   A parasitic plant that can photosynthesize, but derives water and mineral nutrients from the living body of another plant. (Contrast with holoparasite.)

hemizygous (hem' ee zie' gus) [Gk. *hemi*: half + *zygotos*: joined]   In a diploid organism, having only one allele for a given trait, typically the case for X-linked genes in male mammals and Z-linked genes in female birds. (Contrast with homozygous, heterozygous.)

hemoglobin (hee' mo glow bin) [Gk. *heaema*: blood + L. *globus*: globe]   Oxygen-transporting protein found in the red blood cells of vertebrates (and found in some invertebrates).

Hensen's node   In avian embryos, a structure at the anterior end of the primitive groove; determines the fates of cells passing over it during gastrulation.

hepatic (heh pat' ik) [Gk. *hepar*: liver]   Pertaining to the liver.

herbivore (ur' bi vore) [L. *herba*: plant + *vorare*: to devour]   An animal that eats plant tissues. (Contrast with carnivore, detritivore, omnivore.)

heritable trait   A trait that is at least partly determined by genes.

hermaphroditism (her maf' row dite ism)   The coexistence of both female and male sex organs in the same organism.

hetero- [Gk.: *heteros*: other, different]   A prefix indicating two or more different conditions, structures, or processes. (Contrast with homo-.)

heterochrony   Alteration in the timing of developmental events, leading to different results in the adult organism.

heterocyst   A large, thick-walled cell type in the filaments of certain cyanobacteria that performs nitrogen fixation.

heteromorphic (het' er oh more' fik) [Gk. *heteros*: different + *morphe*: form]   Having a different form or appearance, as two heteromorphic life stages of a plant. (Contrast with isomorphic.)

heterosporous (het' er os' por us)   Producing two types of spores, one of which gives rise to a female megaspore and the other to a male microspore. (Contrast with homosporous.)

heterosis   The superior fitness of heterozygous offspring as compared with that of their dissimilar homozygous parents. Also called hybrid vigor.

heterotherm   An animal that regulates its body temperature at a constant level at some times but not others, such as a hibernator.

heterotroph (het' er oh trof) [Gk. *heteros*: different + *trophe*: feed]   An organism that requires preformed organic molecules as food. (Contrast with autotroph.)

heterotrophic succession   Succession in detritus-based communities, which differs from other types of succession in taking place without the participation of plants.

heterotypic   Pertaining to adhesion of cells of different types. (Contrast with homotypic.)

heterozygous (het' er oh zie' gus) [Gk. *heteros*: different + *zygotos*: joined]   In diploid organisms, having different alleles of a given gene on the pair of homologs carrying that gene. (Contrast with homozygous.)

hexose [Gk. *hex*: six]   A sugar containing six carbon atoms.

hibernation [L. *hibernum*: winter] The state of inactivity of some animals during winter; marked by a drop in body temperature and metabolic rate.

hierarchical sequencing An approach to DNA sequencing in which genetic markers are mapped and DNA sequences are aligned by matching overlapping sites of known sequence. (Contrast with shotgun sequencing.)

high-density lipoproteins (HDLs) Lipoproteins that remove cholesterol from tissues and carry it to the liver; HDLs are the "good" lipoproteins associated with good cardiovascular health.

high-throughput sequencing Rapid DNA sequencing on a micro scale in which many fragments of DNA are sequenced in parallel.

highly repetitive sequences Short (less than 100 bp), nontranscribed DNA sequences, repeated thousands of times in tandem arrangements.

hindbrain The region of the developing vertebrate brain that gives rise to the medulla, pons, and cerebellum.

hippocampus [Gr. sea horse] A part of the forebrain that takes part in long-term memory formation.

histamine (hiss' tah meen) A substance released by damaged tissue, or by mast cells in response to allergens. Histamine increases vascular permeability, leading to edema (swelling).

histone Any one of a group of proteins forming the core of a nucleosome, the structural unit of a eukaryotic chromosome.

HIV Human immunodeficiency virus, the retrovirus that causes acquired immune deficiency syndrome (AIDS).

holoparasite A fully parasitic plant (i.e., one that does not perform photosynthesis).

homeobox 180-base-pair segment of DNA found in certain homeotic genes; regulates the expression of other genes and thus controls large-scale developmental processes.

homeostasis (home' ee o sta' sis) [Gk. *homos*: same + *stasis*: position] The maintenance of a steady state, such as a constant temperature or a stable social structure, by means of physiological or behavioral feedback responses.

homeotic genes Genes that act during development to determine the formation of an organ from a region of the embryo.

homeotic mutation Mutation in a homeotic gene that results in the formation of a different organ than that normally made by a region of the embryo.

homo- [Gk. *homos*: same] A prefix indicating two or more similar conditions, structures, or processes. (Contrast with hetero-.)

homolog (1) In cytogenetics, one of a pair (or larger set) of chromosomes having the same overall genetic composition and sequence. In diploid organisms, each chromosome inherited from one parent is matched by an identical (except for mutational changes) chromosome—its homolog—from the other parent. (2) In evolutionary biology, one of two or more features in different species that are similar by reason of descent from a common ancestor.

homology (ho mol' o jee) [Gk. *homologia*: of one mind; agreement] A similarity between two or more features that is due to inheritance from a common ancestor. The structures are said to be *homologous*, and each is a *homolog* of the others. (Contrast with analogy.)

homoplasy (home' uh play zee) [Gk. *homos*: same + *plastikos*: shape, mold] The presence in multiple groups of a trait that is not inherited from the common ancestor of those groups. Can result from convergent evolution, evolutionary reversal, or parallel evolution.

homosporous Producing a single type of spore that gives rise to a single type of gametophyte, bearing both female and male reproductive organs. (Contrast with heterosporous.)

homotypic Pertaining to adhesion of cells of the same type. (Contrast with heterotypic.)

homozygous (home' oh zie' gus) [Gk. *homos*: same + *zygotos*: joined] In diploid organisms, having identical alleles of a given gene on both homologous chromosomes. An individual may be a homozygote with respect to one gene and a heterozygote with respect to another. (Contrast with heterozygous.)

horizons The horizontal layers of a soil profile, including the topsoil (A horizon), subsoil (B horizon) and parent rock or bedrom (C horizon)

hormone (hore' mone) [Gk. *hormon*: to excite, stimulate] A chemical signal produced in minute amounts at one site in a multicellular organism and transported to another site where it acts on target cells.

host An organism that harbors a parasite or symbiont and provides it with nourishment.

Hox genes Conserved homeotic genes found in vertebrates, *Drosophila*, and other animal groups. Hox genes contain the homeobox domain and specify pattern and axis formation in these animals.

human chorionic gonadotropin (hCG) A hormone secreted by the placenta which sustains the corpus luteum and helps maintain pregnancy.

humoral immune response The response of the immune system mediated by B cells that produces circulating antibodies active against extracellular bacterial and viral infections. (Contrast with cellular immune response.)

humus (hew' mus) The partly decomposed remains of plants and animals on the surface of a soil.

hybrid (high' brid) [L. *hybrida*: mongrel] (1) The offspring of genetically dissimilar parents. (2) In molecular biology, a double helix formed of nucleic acids from different sources.

hybridize (1) In genetics, to combine the genetic material of two distinct species or of two distinguishable populations within a species. (2) In molecular biology, to form a double-stranded nucleic acid in which the two strands originate from different sources.

hybrid vigor *See* heterosis.

hybridoma A cell produced by the fusion of an antibody-producing cell with a myeloma (tumor) cell; it produces monoclonal antibodies.

hydrocarbon A compound containing only carbon and hydrogen atoms.

hydrogen bond A weak electrostatic bond which arises from the attraction between the slight positive charge on a hydrogen atom and a slight negative charge on a nearby oxygen or nitrogen atom.

hydrologic cycle The movement of water from the oceans to the atmosphere, to the soil, and back to the oceans.

hydrolysis reaction (high drol' uh sis) [Gk. *hydro*: water + *lysis*: break apart] A chemical reaction that breaks a bond by inserting the components of water (AB + H$_2$O → AH + BOH). (Contrast with condensation reaction.)

hydrophilic (high dro fill' ik) [Gk. *hydro*: water + *philia*: love] Having an affinity for water. (Contrast with hydrophobic.)

hydrophobic (high dro foe' bik) [Gk. *hydro*: water + *phobia*: fear] Having no affinity for water. Uncharged and nonpolar groups of atoms are hydrophobic. (Contrast with hydrophilic.)

hydroponic Pertaining to a method of growing plants with their roots suspended in nutrient solutions instead of soil.

hydrostatic pressure Pressure generated by compression of liquid in a confined space. Generated in plants, fungi, and some protists with cell walls by the osmotic uptake of water. Generated in animals with closed circulatory systems by the beating of a heart.

hydrostatic skeleton A fluid-filled body cavity that transfers forces from one part of the body to another when acted on by surrounding muscles.

hydroxyl group The —OH group found on alcohols and sugars.

hyper- [Gk. *hyper*: above, over] Prefix indicating above, higher, more. (Contrast with hypo-.)

hyperpolarization A change in the resting potential across a membrane so that the inside of a cell becomes more negative compared with the outside of the cell. (Contrast with depolarization.)

hypersensitive response A defensive response of plants to microbial infection in which phytoalexins and pathogenesis-related proteins are produced and the infected tissue undergoes apoptosis to isolate the pathogen from the rest of the plant.

hypertonic Having a greater solute concentration. Said of one solution compared with another. (Contrast with hypotonic, isotonic.)

hypha (high' fuh) (plural: hyphae) [Gk. *hyphe*: web] In the fungi and oomycetes, any single filament.

hypo- [Gk. *hypo*: beneath, under] Prefix indicating underneath, below, less. (Contrast with hyper-.)

hypoblast The lower tissue portion of the avian blastula which is joined to the epiblast at the margins of the blastodisc.

hypothalamus The part of the brain lying below the thalamus; it coordinates water balance, reproduction, temperature regulation, and metabolism.

hypothermia Below-normal body temperature.

hypothesis A tentative answer to a question, from which testable predictions can be generated. (Contrast with theory.)

hypotonic Having a lesser solute concentration. Said of one solution in comparing it to another. (Contrast with hypertonic, isotonic.)

hypoxia A deficiency of oxygen.

- I -

ileum The final segment of the small intestine.

imbibition Water uptake by a seed; first step in germination.

immediate hypersensitivity A rapid, extensive overreaction of the immune system against an allergen, resuting in the release of large amounts of histamine. (Contrast with delayed hypersensitivity.)

immediate memory A form of memory for events happening in the present that is almost perfectly photographic, but lasts only seconds. (Contrast with short-term memory, long-term memory.)

immune system [L. *immunis*: exempt from] A system in an animal that recognizes and attempts to eliminate or neutralize foreign substances such as bacteria, viruses, and pollutants.

immunoassay The use of antibodies to measure the concentration of an antigen in a sample.

immunoglobulins A class of proteins containing a tetramer consisting of four polypeptide chains—two identical light chains and two identical heavy chains—held together by disulfide bonds; active as receptors and effectors in the immune system.

immunological memory The capacity to more rapidly and massively respond to a second exposure to an antigen than occurred on first exposure.

imperfect flower A flower lacking either functional stamens or functional carpels. (Contrast with perfect flower.)

implantation The process by which the early mammalian embryo becomes attached to and embedded in the lining of the uterus.

imprinting In animal behavior, a rapid form of learning in which an animal learns, during a brief critical period, to make a particular response, which is maintained for life, to some object or other organism. *See also* genomic imprinting.

in vitro [L. in glass] A biological process occurring outside of the organism, in the laboratory. (Contrast with in vivo.)

in vivo [L. alive] A biological process occurring within a living organism or cell. (Contrast with in vitro.)

inbreeding Breeding among close relatives.

inclusive fitness The sum of an individual's genetic contribution to subsequent generations both via production of its own offspring and via its influence on the survival of relatives who are not direct descendants.

incomplete cleavage A pattern of cleavage that occurs in many eggs that have a lot of yolk, in which the cleavage furrows do not penetrate all of it. (*See also* discoidal cleavage, superficial cleavage; contrast with complete cleavage.)

incomplete dominance Condition in which the heterozygous phenotype is intermediate between the two homozygous phenotypes.

incomplete metamorphosis Insect development in which changes between instars are gradual.

independent assortment During meiosis, the random separation of genes carried on nonhomologous chromosomes into gametes so that inheritance of these genes is random. This principle was articulated by Mendel as his second law.

indeterminate growth A open-ended growth pattern in which an organism or organ continues to grow as long as it lives; characteristic of some animals and of plant shoots and roots. (Contrast with determinate growth.)

indirect transduction Cell signaling mechanism in which a second messenger mediates the interaction between receptor binding and cellular response. (Contrast with direct transduction.)

individual fitness That component of inclusive fitness resulting from an organism producing its own offspring. (Contrast with kin selection.)

induced fit A change in the shape of an enzyme caused by binding to its substrate that exposes the active site of the enzyme.

induced mutation A mutation resulting from exposure to a mutagen from outside the cell. (Contrast with spontaneous mutation.)

induced pluripotent stem cells (iPS cells) Multipotent or pluripotent animal stem cells produced from differentiated cells in vitro by the addition of several genes that are expressed.

inducer (1) A compound that stimulates the synthesis of a protein. (2) In embryonic development, a substance that causes a group of target cells to differentiate in a particular way.

inducible Produced only in the presence of a particular compound or under particular circumstances. (Contrast with constitutive.)

induction In embryonic development, the process by which a factor produced and secreted by certain cells determines the fates other cells.

inflammation A nonspecific defense against pathogens; characterized by redness, swelling, pain, and increased temperature.

inflorescence A structure composed of several to many flowers.

inflorescence meristem A meristem that produces floral meristems as well as other small leafy structures (bracts).

ingroup In a phylogenetic study, the group of organisms of primary interest. (Contrast with outgroup.)

inhibitor A substance that blocks a biological process.

initials Cells that perpetuate plant meristems, comparable to animal stem cells. When an initial divides, one daughter cell develops into another initial, while the other differentiates into a more specialized cell.

initiation In molecular biology, the beginning of transcription or translation.

initiation complex In protein translation, a combination of a small ribosomal subunit, an mRNA molecule, and the tRNA charged with the first amino acid coded for by the mRNA; formed at the onset of translation.

initiation site The part of a promoter where transcription begins.

inner cell mass Derived from the mammalian blastula (bastocyst), the inner cell mass will give rise to the yolk sac (via hypoblast) and embryo (via epiblast).

inositol trisphosphate (IP₃) An intracellular second messenger derived from membrane phospholipids.

inspiratory reserve volume The amount of air that can be inhaled above the normal tidal inspiration. (Contrast with expiratory reserve volume, tidal volume, vital capacity.)

instar (in' star) An immature stage of an insect between molts.

insulin (in' su lin) [L. *insula*: island] A hormone synthesized in islet cells of the pancreas that promotes the conversion of glucose into the storage material, glycogen.

integrin In animals, a transmembrane protein that mediates the attachment of epithelial cells to the extracellular matrix.

integument [L. *integumentum*: covering] A protective surface structure. In gymnosperms and angiosperms, a layer of tissue around the ovule which will become the seed coat.

intercostal muscles Muscles between the ribs that can augment breathing movements by elevating and suppressing the rib cage.

interference competition Competition in which individuals actively interfere with one another's access to resources. (Contrast with exploitation competition.)

interference RNA (RNAi) *See* RNA interference.

interferon A glycoprotein produced by virus-infected animal cells; increases the resistance of neighboring cells to the virus.

intermediate filaments Components of the cytoskeleton whose diameters fall between those of the larger microtubules and those of the smaller microfilaments.

internal environment In multicellular organisms, the extracellular fluid surrounding the cells.

internal fertilization The release of sperm into the female reproductive tract; typical of most terrestrial animals. (Contrast with external fertilization.)

internal gills Gills enclosed in protective body cavities; typical of mollusks, arthropods, and fishes.

interneuron A neuron that communicates information between two other neurons.

internode The region between two nodes of a plant stem.

interphase In the cell cycle, the period between successive nuclear divisions during which the chromosomes are diffuse and the nuclear envelope is intact. During interphase the cell is most active in transcribing and translating genetic information.

interspecific competition Competition between members of two or more species. (Contrast with intraspecific competition.)

interstitial fluid Extracellular fluid that is not contained in the vessels of a circulatory system.

intestine The portion of the gut following the stomach, in which most digestion and absorption occurs.

intraspecific competition Competition among members of the same species. (Contrast with interspecific competition.)

intrinsic rate of increase (r) The rate at which a population can grow when its density is low and environmental conditions are highly favorable.

intron Portion of a of a gene within the coding region that is transcribed into pre-mRNA but is spliced out prior to translation. (Contrast with exon.)

invasive species An exotic species that reproduces rapidly, spreads widely, and has negative effects on the native species of the region to which it has been introduced.

invasiveness The ability of a pathogen to multiply in a host's body. (Contrast with toxigenicity.)

inversion A rare 180° reversal of the order of genes within a segment of a chromosome.

ion (eye' on) [Gk. *ion*: wanderer] An electrically charged particle that forms when an atom gains or loses one or more electrons.

ion channel An integral membrane protein that allows ions to diffuse across the membrane in which it is embedded.

ionic bond An electrostatic attraction between positively and negatively charged ions.

ionotropic receptors   A receptor that that directly alters membrane permeability to a type of ion when it combines with its ligand.

iris (eye' ris) [Gk. *iris*: rainbow]   The round, pigmented membrane that surrounds the pupil of the eye and adjusts its aperture to regulate the amount of light entering the eye.

island biogeography   A theory proposing that the number of species on an island (or in another geographically defined and isolated area) represents a balance, or equilibrium, between the rate at which species immigrate to the island and the rate at which resident species go extinct.

islets of Langerhans   Clusters of hormone-producing cells in the pancreas.

iso- [Gk. *iso*: equal]   Prefix used for two separate entities that share some element of identity.

isogamous   Having male and female gametes that are morphologically identical. (Contrast with anisogamous.)

isomers   Molecules consisting of the same numbers and kinds of atoms, but differing in the bonding patterns by which the atoms are held together.

isomorphic (eye so more' fik) [Gk. *isos*: equal + *morphe*: form]   Having the same form or appearance, as when the haploid and diploid life stages of an organism appear identical. (Contrast with heteromorphic.)

isotonic   Having the same solute concentration; said of two solutions. (Contrast with hypertonic, hypotonic.)

isotope (eye' so tope) [Gk. *isos*: equal + *topos*: place]   Isotopes of a given chemical element have the same number of protons in their nuclei (and thus are in the same position on the periodic table), but differ in the number of neutrons.

isozymes   Enzymes of an organism that have somewhat different amino acid sequences but catalyze the same reaction.

iteroparous [L. *itero*, to repeat + *pario*, to beget]   Reproducing multiple times in a lifetime. (Contrast with semelparous.)

- J -

jejunum (jih jew' num)   The middle division of the small intestine, where most absorption of nutrients occurs. (*See* duodenum, ileum.)

jelly coat   The outer protective layer of a sea urchin egg, which triggers an acrosomal reaction in sperm.

joint   In skeletal systems, a junction between two or more bones.

juvenile hormone   In insects, a hormone maintaining larval growth and preventing maturation or pupation.

- K -

K-strategist   A species whose life history strategy allows it to persist at or near the carrying capacity (*K*) of its environment. (Contrast with *r*-strategist.)

karyogamy   The fusion of nuclei of two cells. (Contrast with plasmogamy.)

karyotype   The number, forms, and types of chromosomes in a cell.

keystone species   Species that have a dominant influence on the composition of a community.

kidneys   A pair of excretory organs in vertebrates.

kin selection   That component of inclusive fitness resulting from helping the survival of rela-

tives containing the same alleles by descent from a common ancestor. (Contrast with individual fitness.)

kinase   *See* protein kinase.

kinetic energy (kuh-net' ik) [Gk. *kinetos*: moving]   The energy associated with movement. (Contrast with potential energy.)

kinetochore (kuh net' oh core)   Specialized structure on a centromere to which microtubules attach.

knockout   A molecular genetic method in which a single gene of an organism is permanently inactivated.

Koch's postulates   A set of rules for establishing that a particular microorganism causes a particular disease.

Krebs cycle   *See* citric acid cycle.

- L -

lagging strand   In DNA replication, the daughter strand that is synthesized in discontinuous stretches. (*See* Okazaki fragments.)

larva (plural: larvae) [L. *lares*: guiding spirits]   An immature stage of any animal that differs dramatically in appearance from the adult.

lateral [L. *latus*: side]   Pertaining to the side.

lateral gene transfer   The transfer of genes from one species to another, common among bacteria and archaea.

lateral line   A sensory system in fishes consisting of a canal filled with water and hair cells running down each side under the surface of the skin, which senses disturbances in the surrounding water.

lateral meristem   Either of the two meristems, the vascular cambium and the cork cambium, that give rise to a plant's secondary growth.

lateral root   A root extending outward from the taproot in a taproot system; typical of eudicots.

laticifers (luh tiss' uh furs)   In some plants, elongated cells containing secondary plant products such as latex.

Laurasia   The northernmost of the two large continents produced by the breakup of Pangaea.

laws of thermodynamics [Gk. *thermos*: heat + *dynamis*: power]   Laws derived from studies of the physical properties of energy and the ways energy interacts with matter. (*See also* first law of thermodynamics, second law of thermodynamics.)

leaching   In soils, a process by which mineral nutrients in upper soil horizons are dissolved in water and carried to deeper horizons, where they are unavailable to plant roots.

leading strand   In DNA replication, the daughter strand that is synthesized continuously. (Contrast with lagging strand.)

leaf (plural: leaves)   In plants, the chief organ of photosynthesis.

leaf primordium (plural: primordia)   An outgrowth on the side of the shoot apical meristem that will eventually develop into a leaf.

leghemoglobin   In nitrogen-fixing plants, an oxygen-carrying protein in the cytoplasm of nodule cells that transports enough oxygen to the nitrogen-fixing bacteria to support their respiration, while keeping free oxygen concentrations low enough to protect nitrogenase.

lek   A display ground within which male animals compete for and defend small display areas as a means of demonstrating their territorial prowess and winning opportunities to mate.

lens   In the vertebrate eye, a crystalline protein structure that makes fine adjustments in the focus of images falling on the retina.

lenticel (len' ti sill)   In plants, a spongy region in the periderm that allows gas exchange.

leptin   A hormone produced by fat cells that is believed to provide feedback information to the brain about the status of the body's fat reserves.

leukocyte   *See* white blood cell.

lichen (lie' kun)   An organism resulting from the symbiotic association of a fungus and either a cyanobacterium or a unicellular alga.

life cycle   The entire span of the life of an organism from the moment of fertilization (or asexual generation) to the time it reproduces in turn.

life history strategy   The way in which an organism partitions its time and energy among growth, maintenance, and reproduction.

ligament   A band of connective tissue linking two bones in a joint.

ligand (lig' and)   Any molecule that binds to a receptor site of another (usually larger) molecule.

light reactions   The initial phase of photosynthesis, in which light energy is converted into chemical energy.

light-independent reactions   The phase of photosynthesis in which chemical energy captured in the light reactions is used to drive the reduction of CO_2 to form carbohydrates.

lignin   A complex, hydrophobic polyphenolic polymer in plant cell walls that crosslinks other wall polymers, strengthening the walls, especially in wood.

limbic system   A group of evolutionarily primitive structures in the vertebrate telencephalon that are involved in emotions, drives, instinctive behaviors, learning, and memory.

liming   Application of compounds such as calcium carbonate, calcium hydroxide, or magnesium carbonate—commonly known as *lime*— to soil to reverse its acidification and increase the availability of calcium to plants.

limiting resource   The required resource whose supply most strongly influences the size of a population.

lineage species concept   The definition of a species as a branch on the tree of life, which has a history that starts at a speciation event and ends either at extinction or at another speciation event. (Contrast with biological species concept; morphological species concept.)

linkage   Association between genes on the same chromosome such that they do not show random assortment and seldom recombine; the closer the genes, the lower the frequency of recombination.

lipase (lip' ase; lye' pase)   An enzyme that digests fats.

lipid (lip' id) [Gk. *lipos*: fat]   Nonpolar, hydrophobic molecules that include fats, oils, waxes, steroids, and the phospholipids that make up biological membranes.

lipid bilayer   *See* phospholipid bilayer.

liver   A large digestive gland. In vertebrates, it secretes bile and is involved in the formation of blood.

loam   A type of soil consisting of a mixture of sand, silt, clay, and organic matter. One of the best soil types for agriculture.

locus (low' kus) (plural: loci, low' sigh)   In genetics, a specific location on a chromosome. May be considered synonymous with *gene*.

logistic growth   Growth, especially in the size of an organism or in the number of organisms in a population, that slows steadily as the entity approaches its maximum size. (Contrast with exponential growth.)

long-day plant (LDP)   A plant that requires long days (actually, short nights) in order to flower.

long-term depression (LTD)   A long-lasting decrease in the responsiveness resulting from continuous, repetitive, low-level stimulation. (Contrast with long-term potentiation.)

long-term potentiation (LTP)   A long-lasting increase in the responsiveness of a neuron resulting from a period of intense stimulation. (Contrast with long-term depression.)

loop of Henle (hen' lee)   Long, hairpin loop of the mammalian renal tubule that runs from the cortex down into the medulla and back to the cortex; creates a concentration gradient in the interstitial fluids in the medulla.

lophophore   A U-shaped fold of the body wall with hollow, ciliated tentacles that encircles the mouth of animals in several different groups. Used for filtering prey from the surrounding water.

loss of function mutation   A mutation that results in the loss of a functional protein. (Contrast with gain of function mutation.)

low-density lipoproteins (LDLs)   Lipoproteins that transport cholesterol around the body for use in biosynthesis and for storage; LDLs are the "bad" lipoproteins associated with a high risk of cardiovascular disease.

lumen (loo' men) [L. *lumen*: light]   The open cavity inside any tubular organ or structure, such as the gut or a renal tubule.

lung   An internal organ specialized for respiratory gas exchange with air.

luteinizing hormone (LH)   A gonadotropin produced by the anterior pituitary that stimulates the gonads to produce sex hormones.

lymph [L. *lympha*: liquid]   A fluid derived from blood and other tissues that accumulates in intercellular spaces throughout the body and is returned to the blood by the lymphatic system.

lymph node   A specialized structure in the vessels of the lymphatic system. Lymph nodes contain lymphocytes, which encounter and respond to foreign cells and molecules in the lymph as it passes through the vessels.

lymphatic system   A system of vessels that returns interstitial fluid to the blood.

lymphocyte   One of the two major classes of white blood cells; includes T cells, B cells, and other cell types important in the immune system.

lymphoid tissues   Tissues of the immune system that are dispersed throughout the body, consisting of the thymus, spleen, bone marrow, and lymph nodes.

lysis (lie' sis) [Gk. *lysis*: break apart]   Bursting of a cell.

lysogeny   A form of viral replication in which the virus becomes incorporated into the host chromosome and remains inactive. Also called a lysogenic cycle. (Contrast with lytic cycle.)

lysosome (lie' so soam) [Gk. *lysis*: break away + *soma*: body]   A membrane-enclosed organelle originating from the Golgi apparatus and containing hydrolytic enzymes. (Contrast with secondary lysosome.)

lysozyme (lie' so zyme)   An enzyme in saliva, tears, and nasal secretions that hydrolyzes bacterial cell walls.

lytic cycle   A viral reproductive cycle in which the virus takes over a host cell's synthetic machinery to replicate itself, then bursts (lyses) the host cell, releasing the new viruses. (Contrast with lysogeny.)

- M -

M phase   The portion of the cell cycle in which mitosis takes place.

macroevolution [Gk. *makros*: large]   Evolutionary changes occurring over long time spans and usually involving changes in many traits. (Contrast with microevolution.)

macromolecule   A giant (molecular weight > 1,000) polymeric molecule. The macromolecules are the proteins, polysaccharides, and nucleic acids.

macronutrient   In plants, a mineral element required in concentrations of at least 1 milligram per gram of plant dry matter; in animals, a mineral element required in large amounts. (Contrast with micronutrient.)

macrophage (mac' roh faj)   Phagocyte that engulfs pathogens by endocytosis.

MADS box   DNA-binding domain in many plant transcription factors that is active in development.

maintenance methylase   An enzyme that catalyzes the methylation of the new DNA strand when DNA is replicated.

major histocompatibility complex (MHC)   A complex of linked genes, with multiple alleles, that control a number of cell surface antigens that identify self and can lead to graft rejection.

malignant   Pertaining to a tumor that can grow indefinitely and/or spread from the original site of growth to other locations in the body. (Contrast with benign.)

Malpighian tubule (mal pee' gy un)   A type of protonephridium found in insects.

map unit   The distance between two genes as calculated from genetic crosses; a recombination frequency.

marine [L. *mare*: sea, ocean]   Pertaining to or living in the ocean. (Contrast with aquatic, terrestrial.)

mark–recapture method   A method of estimating population sizes of mobile organisms by capturing, marking, and releasing a sample of individuals, then capturing another sample at a later time.

mass extinction   A period of evolutionary history during which rates of extinction are much higher than during intervening times.

mass number   The sum of the number of protons and neutrons in an atom's nucleus.

mast cells   Cells, typically found in connective tissue, that release histamine in response to tissue damage.

maternal effect genes   Genes coding for morphogens that determine the polarity of the egg and larva in fruit flies.

mating type   A particular strain of a species that is incapable of sexual reproduction with another member of the same strain but capable of sexual reproduction with members of other strains of the same species.

maximum likelihood   A statistical method of determining which of two or more hypotheses (such as phylogenetic trees) best fit the observed data, given an explicit model of how the data were generated.

mechanically gated channel   A molecular channel that opens or closes in response to mechanical force applied to the plasma membrane in which it is inserted.

mechanoreceptor   A cell that is sensitive to physical movement and generates action potentials in response.

medulla (meh dull' luh)   (1) The inner, core region of an organ, as in the adrenal medulla (adrenal gland) or the renal medulla (kidneys). (2) The portion of the brainstem that connects to the spinal cord.

medusa (plural: medusae)   In cnidarians, a free-swimming, sexual life cycle stage shaped like a bell or an umbrella.

megaphyll   The generally large leaf of a fern, horsetail, or seed plant, with several to many veins. (Contrast with microphyll.)

megaspore [Gk. *megas*: large + *spora*: to sow]   In plants, a haploid spore that produces a female gametophyte.

megastrobilus   In conifers, the female (seed-bearing) cone. (Contrast with microstrobilus.)

meiosis (my oh' sis) [Gk. *meiosis*: diminution]   Division of a diploid nucleus to produce four haploid daughter cells. The process consists of two successive nuclear divisions with only one cycle of chromosome replication. In *meiosis I*, homologous chromosomes separate but retain their chromatids. The second division *meiosis II*, is similar to mitosis, in which chromatids separate.

melatonin   A hormone released by the pineal gland. Involved in photoperiodicity and circadian rhythms.

membrane potential   The difference in electrical charge between the inside and the outside of a cell, caused by a difference in the distribution of ions.

membranous bone   A type of bone that develops by forming on a scaffold of connective tissue. (Contrast with cartilage bone.)

memory cells   Long-lived lymphocytes produced by exposure to antigen. They persist in the body and are able to mount a rapid response to subsequent exposures to the antigen.

Mendel's laws   See independent assortment; segregation.

meristem [Gk. *meristos*: divided]   Plant tissue made up of undifferentiated actively dividing cells.

meristem identity genes   In angiosperms, a group of genes whose expression initiates flower formation, probably by switching meristem cells from a vegetative to a reproductive fate.

mesenchyme (mez' en kyme) [Gk. *mesos*: middle + *enchyma*: infusion]   Embryonic or unspecialized cells derived from the mesoderm.

mesoderm [Gk. *mesos*: middle + *derma*: skin]   The middle of the three embryonic germ layers first delineated during gastrulation. Gives rise to the skeleton, circulatory system, muscles, excretory system, and most of the reproductive system.

mesoglea (mez' uh glee uh) [Gk. *mesos*: middle + *gloia*, glue]   A thick, gelatinous noncellular layer that separates the two cellular tissue layers of ctenophores, cnidarians, and scyphozoans.

mesophyll (mez' uh fill) [Gk. *mesos*: middle + *phyllon*: leaf]   Chloroplast-containing, photosynthetic cells in the interior of leaves.

messenger RNA (mRNA)   Transcript of a region of one of the strands of DNA; carries information (as a sequence of codons) for the synthesis of one or more proteins.

meta- [Gk.: between, along with, beyond]   Prefix denoting a change or a shift to a new form or level; for example, as used in metamorphosis.

metabolic pathway   A series of enzyme-catalyzed reactions so arranged that the product of one reaction is the substrate of the next.

metabolism (meh tab' a lizm) [Gk. *metabole*: change]   The sum total of the chemical reactions that occur in an organism, or some subset of that total (as in respiratory metabolism).

metabolome   The quantitative description of all the small molecules in a cell or organism.

metabotropic receptor   A receptor that that indirectly alters membrane permeability to a type of ion when it combines with its ligand.

metagenomics   The practice of analyzing DNA from environmental samples without isolating intact organisms.

metamorphosis (met' a mor' fo sis) [Gk. *meta*: between + *morphe*: form, shape]   A change occurring between one developmental stage and another, as for example from a tadpole to a frog. (*See* complete metamorphosis, incomplete metamorphosis.)

metanephridia   The paired excretory organs of annelids.

metaphase (met' a phase)   The stage in nuclear division at which the centromeres of the highly supercoiled chromosomes are all lying on a plane (the metaphase plane or plate) perpendicular to a line connecting the division poles.

metapopulation   A population divided into subpopulations, among which there are occasional exchanges of individuals.

metastasis (meh tass' tuh sis)   The spread of cancer cells from their original site to other parts of the body.

methylation   The addition of a methyl group (—CH_3) to a molecule.

MHC   *See* major histocompatibility complex.

micelle   A particle of lipid covered with bile salts that is produced in the duodenum and facilitates digestion and absorption of lipids.

microclimate   A subset of climatic conditions in a small specific area, which generally differ from those in the environment at large, as in an animal's underground burrow.

microevolution   Evolutionary changes below the species level, affecting allele frequencies. (Contrast with macroevolution.)

microfibril   Crosslinked cellulose polymers, forming strong aggregates in the plant cell wall.

microfilament   In eukaryotic cells, a fibrous structure made up of actin monomers. Microfilaments play roles in the cytoskeleton, in cell movement, and in muscle contraction.

microglia   Glial cells that act as macrophages and mediators of inflammatory responses in the central nervous system.

micronutrient   In plants, a mineral element required in concentrations of less than 100 micrograms per gram of plant dry matter; in animals, a mineral element required in concentrations of less than 100 micrograms per day. (Contrast with macronutrient.)

microphyll   A small leaf with a single vein, found in club mosses and their relatives. (Contrast with megaphyll.)

micropyle (mike' roh pile) [Gk. *mikros*: small + *pylon*: gate]   Opening in the integument(s) of a seed plant ovule through which pollen grows to reach the female gametophyte within.

microRNA   A small, noncoding RNA molecule, typically about 21 bases long, that binds to mRNA to inhibit its translation.

microspore [Gk. *mikros*: small + *spora*: to sow]   In plants, a haploid spore that produces a male gametophyte.

microstrobilus   In conifers, male pollen-bearing cone. (Contrast with megastrobilus.)

microtubules   Tubular structures found in centrioles, spindle apparatus, cilia, flagella, and cytoskeleton of eukaryotic cells. These tubules play roles in the motion and maintenance of shape of eukaryotic cells.

microvilli (sing.: microvillus)   Projections of epithelial cells, such as the cells lining the small intestine, that increase their surface area.

midbrain   One of the three regions of the vertebrate brain. Part of the brainstem, it serves as a relay station for sensory signals sent to the cerebral hemispheres.

middle lamella (la mell' ah) [L. *lamina*: thin sheet]   A layer of polysaccharides that separates plant cells; a shared middle lamella lies outside the primary walls of the two cells.

mineral nutrients   Inorganic ions required by organisms for normal growth and reproduction.

mismatch repair   A mechanism that scans DNA after it has been replicated and corrects any base-pairing mismatches.

missense mutation   A change in a gene's sequence that results in a change in the sequence of the amino acid specified by the corresponding codon. (Contrast with frame-shift mutation, nonsense mutation, silent mutation.)

mitochondrial matrix   The fluid interior of the mitochondrion, enclosed by the inner mitochondrial membrane.

mitochondrion (my' toe kon' dree un) (plural: mitochondria) [Gk. *mitos*: thread + *chondros*: grain]   An organelle in eukaryotic cells that contains the enzymes of the citric acid cycle, the respiratory chain, and oxidative phosphorylation.

mitosis (my toe' sis) [Gk. *mitos*: thread]   Nuclear division in eukaryotes leading to the formation of two daughter nuclei, each with a chromosome complement identical to that of the original nucleus.

model systems   Also known as model organisms, these include the small group of species that are the subject of extensive research. They are organisms that adapt well to laboratory situations and findings from experiments on them can apply across a broad range of species. Classic examples include white rats and the fruit fly *Drosophila*.

moderately repetitive sequences   DNA sequences repeated 10–1,000 times in the eukaryotic genome. They include the genes that code for rRNAs and tRNAs, as well as the DNA in telomeres.

Modern Synthesis   An understanding of evolutionary biology that emerged in the early twentieth century as the principles of evolution were integrated with the principles of modern genetics.

modularity   In evolutionary developmental biology, the principle that the molecular pathways that determine different developmental processes operate independently from one another. *See also* developmental module.

mole   A quantity of a compound whose weight in grams is numerically equal to its molecular weight expressed in atomic mass units. Avogadro's number of molecules: 6.023×10^{23} molecules.

molecular clock   The approximately constant rate of divergence of macromolecules from one another over evolutionary time; used to date past events in evolutionary history.

molecular evolution   The scientific study of the mechanisms and consequences of the evolution of macromolecules.

molecular tool kit   A set of developmental genes and proteins that is common to most animals and is hypothesized to be responsible for the evolution of their differing developmental pathways.

molecular weight   The sum of the atomic weights of the atoms in a molecule.

molecule   A chemical substance made up of two or more atoms joined by covalent bonds or ionic attractions.

molting   The process of shedding part or all of an outer covering, as the shedding of feathers by birds or of the entire exoskeleton by arthropods.

monoclonal antibody   Antibody produced in the laboratory from a clone of hybridoma cells, each of which produces the same specific antibody.

monoculture   In agriculture, a large-scale planting of a single species of domesticated crop plant.

monoecious (mo nee' shus) [Gk. *mono*: one + *oikos*: house]   Pertaining to organisms in which both sexes are "housed" in a single individual that produces both eggs and sperm. (In some plants, these are found in different flowers within the same plant.) Examples include corn, peas, earthworms, hydras. (Contrast with dioecious.)

monohybrid cross   A mating in which the parents differ with respect to the alleles of only one locus of interest.

monomer [Gk. *mono*: one + *meros*: unit]   A small molecule, two or more of which can be combined to form oligomers (consisting of a few monomers) or polymers (consisting of many monomers).

monophyletic (mon' oh fih leht' ik) [Gk. *mono*: one + *phylon*: tribe]   Pertaining to a group that consists of an ancestor and all of its descendants. (Contrast with paraphyletic, polyphyletic.)

monosaccharide   A simple sugar. Oligosaccharides and polysaccharides are made up of monosaccharides.

monosomic   Pertaining to an organism with one less than the normal diploid number of chromosomes.

monosynaptic reflex   A neural reflex that begins in a sensory neuron and makes a single synapse before activating a motor neuron.

morphogen   A diffusible substance whose concentration gradient determines a developmental pattern in animals and plants.

morphogenesis (more' fo jen' e sis) [Gk. *morphe*: form + *genesis*: origin]   The development of form; the overall consequence of determination, differentiation, and growth.

morphological species concept   The definition of a species as a group of individuals that look alike. (Contrast with biological species concept; lineage species concept.)

morphology (more fol' o jee) [Gk. *morphe*: form + *logos*: study, discourse]   The scientific study of organic form, including both its development and function.

mosaic development   Pattern of animal embryonic development in which each blastomere contributes a specific part of the adult body. (Contrast with regulative development.)

motif   *See* structural motif.

motile (mo' tul)   Able to move from one place to another. (Contrast with sessile.)

motor cortex   The region of the cerebral cortex that contains motor neurons that directly stimulate specific muscle fibers to contract.

motor neuron   A neuron carrying information from the central nervous system to a cell that produces movement.

motor proteins   Specialized proteins that use energy to change shape and move cells or structures within cells. *See* dynein, kinesin.

motor unit   A motor neuron and the muscle fibers it controls.

mouth   An opening through which food is taken in, located at the anterior end of a tubular gut.

mRNA   *See* messenger RNA.

mucosal epithelium   An epithelial cell layer containing cells that secrete mucus; found in the digestive and respiratory tracts. Also called mucosa.

Müllerian mimicry   Convergence in appearance of two or more unpalatable species.

multipotent   Having the ability to differentiate into a limited number of cell types. (Contrast with pluripotent, totipotent.)

muscle fiber   A single muscle cell. In the case of skeletal muscle, a syncitial, multinucleate cell.

muscle tissue   Excitable tissue that can contract through the interactions of actin and myosin; one of the four major tissue types in multicellular animals. There are three types of muscle tissue: skeletal, smooth, and cardiac.

mutagen (mute' ah jen) [L. *mutare*: change + Gk. *genesis*: source]   Any agent (e.g., chemicals, radiation) that increases the mutation rate.

mutation   A change in the genetic material not caused by recombination.

mutualism   A type of interaction between species that benefits both species.

mycelium (my seel' ee yum) [Gk. *mykes*: fungus]   In the fungi, a mass of hyphae.

mycorrhiza (my' ko rye' za) (plural: mycorrhizae) [Gk. *mykes*: fungus + *rhiza*: root]   An association of the root of a plant with the mycelium of a fungus.

myelin (my' a lin)   Concentric layers of plasma membrane that form a sheath around some axons; myelin provides the axon with electrical insulation and increases the rate of transmission of action potentials.

myocardial infarction   Blockage of an artery that carries blood to the heart muscle.

myofibril (my' oh fy' bril) [Gk. *mys*: muscle + L. *fibrilla*: small fiber]   A polymeric unit of actin or myosin in a muscle.

myoglobin (my' oh globe' in) [Gk. *mys*: muscle + L. *globus*: sphere]   An oxygen-binding molecule found in muscle. Consists of a heme unit and a single globin chain; carries less oxygen than hemoglobin.

myosin   One of the two contractile proteins of muscle.

- N -

natural killer cell   A type of lymphocyte that attacks virus-infected cells and some tumor cells as well as antibody-labeled target cells.

natural selection   The differential contribution of offspring to the next generation by various genetic types belonging to the same population. The mechanism of evolution proposed by Charles Darwin.

nauplius (naw' plee us) [Gk. *nauplios*: shellfish]   A bilaterally symmetrical larval form typical of crustaceans.

necrosis (nec roh' sis) [Gk. *nekros*: death]   Premature cell death caused by external agents such as toxins.

negative feedback   In regulatory systems, information that decreases a regulatory response, returning the system to the set point. (Contrast with positive feedback.)

negative regulation   A type of gene regulation in which a gene is normally transcribed, and the binding of a repressor protein to the promoter prevents transcription. (Contrast with positive regulation.)

nematocyst (ne mat' o sist) [Gk. *nema*: thread + *kystis*: cell]   An elaborate, threadlike structure produced by cells of jellyfishes and other cnidarians, used chiefly to paralyze and capture prey.

nephron (nef' ron) [Gk. *nephros*: kidney]   The functional unit of the kidney, consisting of a structure for receiving a filtrate of blood and a tubule that reabsorbs selected parts of the filtrate.

Nernst equation   A mathematical statement that calculates the potential across a membrane permeable to a single type of ion that differs in concentration on the two sides of the membrane.

nerve   A structure consisting of many neuronal axons and connective tissue.

nervous tissue   Tissue specialized for processing and communicating information; one of the four major tissue types in multicellular animals.

net primary productivity (NPP)   The rate at which energy captured by photosynthesis is incorporated into the bodies of primary producers bodies through growth and reproduction.

net primary production   The amount of primary producer biomass made available for consumption by heterotrophs.

neural network   An organized group of neurons that contains three functional categories of neurons—afferent neurons, interneurons, and efferent neurons—and is capable of processing information.

neural tube   An early stage in the development of the vertebrate nervous system consisting of a hollow tube created by two opposing folds of the dorsal ectoderm along the anterior–posterior body axis.

neurohormone   A chemical signal produced and released by neurons that subsequently acts as a hormone.

neuromuscular junction   Synapse (point of contact) where a motor neuron axon stimulates a muscle fiber cell.

neuron (noor' on) [Gk. *neuron*: nerve]   A nervous system cell that can generate and conduct action potentials along an axon to a synapse with another cell.

neurotransmitter   A substance produced in and released by a neuron (the presynaptic cell) that diffuses across a synapse and excites or inhibits another cell (the postsynaptic cell).

neurulation   Stage in vertebrate development during which the nervous system begins to form.

neutral allele   An allele that does not alter the functioning of the proteins for which it codes.

neutral theory   A view of molecular evolution that postulates that most mutations do not affect the amino acid being coded for, and that such mutations accumulate in a population at rates driven by genetic drift and mutation rates.

neutron (new' tron)   One of the three fundamental particles of matter (along with protons and electrons), with mass approximately 1 amu and no electrical charge.

niche (nitch) [L. *nidus*: nest]   The set of physical and biological conditions a species requires to survive, grow, and reproduce. (*See also* fundamental niche, realized niche.)

nitrate reduction   The process by which nitrate (NO_3^-) is reduced to ammonia (NH_3).

nitric oxide (NO)   An unstable molecule (a gas) that serves as a second messenger causing smooth muscle to relax. In the nervous system it operates as a neurotransmitter.

nitrifiers   Chemolithotrophic bacteria that oxidize ammonia to nitrate in soil and in seawater.

nitrogen fixation   Conversion of atmospheric nitrogen gas (N_2) into a more reactive and biologically useful form (ammonia), which makes nitrogen available to living things. Carried out by nitrogen-fixing bacteria, some of them free-living and others living within plant roots.

nitrogenase   An enzyme complex found in nitrogen-fixing bacteria that mediates the stepwise reduction of atmospheric N_2 to ammonia and which is strongly inhibited by oxygen.

node [L. *nodus*: knob, knot]   In plants, a (sometimes enlarged) point on a stem where a leaf is or was attached.

node of Ranvier   A gap in the myelin sheath covering an axon; the point where the axonal membrane can fire action potentials.

nodule   A specialized structure in the roots of nitrogen-fixing plants that houses nitrogen-fixing bacteria, in which oxygen is maintained at a low level by leghemoglobin.

noncompetitive inhibitor   A nonsubstrate that inhibits the activity of an enzyme by binding to a site other than its active site. (Contrast with competitive inhibitor.)

noncyclic electron transport In photosynthesis, the flow of electrons that forms ATP, NADPH, and O_2.

nondisjunction   Failure of sister chromatids to separate in meiosis II or mitosis, or failure of homologous chromosomes to separate in meiosis I. Results in aneuploidy.

nonpolar   Having electric charges that are evenly balanced from one end to the other. (Contrast with polar.)

nonrandom mating   Selection of mates on the basis of a particular trait or group of traits.

nonsense mutation Change in a gene's sequence that prematurely terminates translation by changing one of its codons to a stop codon.

nonsynonymous substitution A change in a gene from one nucleotide to another that changes the amino acid specified by the corresponding codon (i.e., AGC →AGA, or serine → arginine). (Contrast with synonymous substitution.)

norepinephrine A neurotransmitter found in the central nervous system and also at the post-ganglionic nerve endings of the sympathetic nervous system. Also called noradrenaline.

normal flora Microorganisms that normally live and reproduce on or in the body without causing disease, and which form a nonspecific defense against pathogens by competing with them for space and nutrients.

notochord (no' tow kord) [Gk. notos: back + chorde: string] A flexible rod of gelatinous material serving as a support in the embryos of all chordates and in the adults of tunicates and lancelets.

nucleic acid (new klay' ik) A polymer made up of nucleotides, specialized for the storage, transmission, and expression of genetic information. DNA and RNA are nucleic acids.

nucleic acid hybridization A technique in which a single-stranded nucleic acid probe is made that is complementary to, and binds to, a target sequence, either DNA or RNA. The resulting double-stranded molecule is a hybrid.

nucleoid (new' klee oid) The region that harbors the chromosomes of a prokaryotic cell. Unlike the eukaryotic nucleus, it is not bounded by a membrane.

nucleolus (new klee' oh lus) A small, generally spherical body found within the nucleus of eukaryotic cells. The site of synthesis of ribosomal RNA.

nucleoside A nucleotide without the phosphate group; a nitrogenous base attached to a sugar.

nucleosome A portion of a eukaryotic chromosome, consisting of part of the DNA molecule wrapped around a group of histone molecules, and held together by another type of histone molecule. The chromosome is made up of many nucleosomes.

nucleotide The basic chemical unit in nucleic acids, consisting of a pentose sugar, a phosphate group, and a nitrogen-containing base.

nucleotide substitution A change of one base pair to another in a DNA sequence.

nucleus (new' klee us) [L. nux: kernel or nut] (1) In cells, the centrally located compartment of eukaryotic cells that is bounded by a double membrane and contains the chromosomes. (2) In the brain, an identifiable group of neurons that share common characteristics or functions.

nutrient A food substance; or, in the case of mineral nutrients, an inorganic element required for completion of the life cycle of an organism.

- O -

obligate anaerobe An anaerobic prokaryote that cannot survive exposure to O_2.

odorant A molecule that can bind to an olfactory receptor.

oil A triglyceride that is liquid at room temperature. (Contrast with fat.)

Okazaki fragments Newly formed DNA making up the lagging strand in DNA replication. DNA ligase links Okazaki fragments together to give a continuous strand.

olfactory [L. olfacere: to smell] Pertaining to the sense of smell (olfaction).

oligodendrocyte A type of glial cell that myelinates axons in the central nervous system.

oligosaccharide A polymer containing a small number of monosaccharides.

ommatidia [Gk. omma: eye] The units that make up the compound eye of some arthropods.

omnivore [L. omnis: everything + vorare: to devour] An organism that eats both animal and plant material. (Contrast with carnivore, detritivore, herbivore.)

oncogene [Gk. onkos: mass, tumor + genes: born] A gene that codes for a protein product that stimulates cell proliferation. Mutations in oncogenes that result in excessive cell proliferation can give rise to cancer.

oocyte See primary oocyte, secondary oocyte.

oogenesis (oh' eh jen e sis) [Gk. oon: egg + genesis: source] Gametogenesis leading to production of an ovum.

oogonium (oh' eh go' nee um) (plural: oogonia) (1) In some algae and fungi, a cell in which an egg is produced. (2) In animals, the diploid progeny of a germ cell in females.

operator The region of an operon that acts as the binding site for the repressor.

open circulatory system Circulatory system in which extracellular fluid leaves the vessels of the circulatory system, percolates between cells and through tissues, and then flows back into the circulatory system to be pumped out again. (Contrast with closed circulatory system.)

operon A genetic unit of transcription, typically consisting of several structural genes that are transcribed together; the operon contains at least two control regions: the promoter and the operator.

opportunity cost The sum of the benefits an animal forfeits by not being able to perform some other behavior during the time when it is performing a given behavior.

opsin (op' sin) [Gk. opsis: sight] The protein portion of the visual pigment rhodopsin. (See rhodopsin.)

optic chiasm [Gk. chiasma: cross] Structure on the lower surface of the vertebrate brain where the two optic nerves come together.

optical isomers Two isomers that are mirror images of each other.

optimal foraging theory The application of a cost–benefit approach to feeding behavior to identify the fitness value of feeding choices.

orbital A region in space surrounding the atomic nucleus in which an electron is most likely to be found.

organ [Gk. organon: tool] A body part, such as the heart, liver, brain, root, or leaf. Organs are composed of different tissues integrated to perform a distinct function. Organs, in turn, are integrated into organ systems.

organ identity genes In angiosperms, genes that specify the different organs of the flower. (Compare with homeotic genes.)

organ of Corti Structure in the inner ear that transforms mechanical forces produced from pressure waves ("sound waves") into action potentials that are sensed as sound.

organ system An interrelated and integrated group of tissues and organs that work together in a physiological function.

organelle (or gan el') Any of the membrane-enclosed structures within a eukaryotic cell. Examples include the nucleus, endoplasmic reticulum, and mitochondria.

organic (1) Pertaining to any chemical compound that contains carbon. (2) Pertaining to any aspect of living matter, e.g., to its evolution, structure, or chemistry.

organism Any living entity.

organizer Region of the early amphibian embryo that directs early embryonic development. Also known as the primary embryonic organizer.

organogenesis The formation of organs and organ systems during development.

origin of replication (ori) DNA sequence at which helicase unwinds the DNA double helix and DNA polymerase binds to initiate DNA replication.

orthology (or thol' o jee) Type of homology in which the divergence of homologous genes can be traced to speciation events. (Contrast with paralogy.)

osmoconformer An aquatic animal that equilibrates the osmolarity of its extracellular fluid that is the same as with that of the external environment.

osmolarity The concentration of osmotically active particles in a solution.

osmoregulation Regulation of the chemical composition of the body fluids of an organism.

osmosis (oz mo' sis) [Gk. osmos: to push] Movement of water across a differentially permeable membrane, from one region to another region where the water potential is more negative.

ossicle (oss' ick ul) [L. os: bone] The calcified construction unit of echinoderm skeletons.

osteoblast (oss' tee oh blast) [Gk. osteon: bone + blastos: sprout] A cell that lay down the protein matrix of bone.

osteoclast (oss' tee oh clast) [Gk. osteon: bone + klastos: broken] A cell that dissolves bone.

osteocyte An osteoblast that has become enclosed in lacunae within the bone it has built.

outgroup In phylogenetics, a group of organisms used as a point of reference for comparison with the groups of primary interest (the ingroup).

oval window The flexible membrane that, when moved by the bones of the middle ear, produces pressure waves in the inner ear.

ovarian cycle In human females, the monthly cycle of events by which eggs and hormones are produced. (Contrast with uterine cycle).

ovary (oh' var ee) [L. ovum: egg] Any female organ, in plants or animals, that produces an egg.

overtopping Plant growth pattern in which one branch differentiates from and grows beyond the others.

oviduct In mammals, the tube serving to transport eggs to the uterus or to outside of the body.

oviparity Reproduction in which eggs are released by the female and development is external to the mother's body. (Contrast with viviparity.)

ovoviviparity Pertaining to reproduction in which fertilized eggs develop and hatch within the mother's body but are not attached to the mother by means of a placenta.

ovulation Release of an egg from an ovary.

ovule (oh' vule)   In plants, a structure comprising the megasporangium and the integument, which develops into a seed after fertilization.

ovum (oh' vum) (plural: ova) [L. egg]   The female gamete.

oxidation (ox i day' shun)   Relative loss of electrons in a chemical reaction; either outright removal to form an ion, or the sharing of electrons with substances having a greater affinity for them, such as oxygen. Most oxidations, including biological ones, are associated with the liberation of energy. (Contrast with reduction.)

oxidative phosphorylation   ATP formation in the mitochondrion, associated with flow of electrons through the respiratory chain.

oxygenase   An enzyme that catalyzes the addition of oxygen to a substrate from O_2.

oxytocin   A hormone released by the posterior pituitary that promotes social bonding.

- P -

pancreas (pan' cree us)   A gland located near the stomach of vertebrates that secretes digestive enzymes into the small intestine and releases insulin into the bloodstream.

Pangaea (pan jee' uh) [Gk. pan: all, every]   The single land mass formed when all the continents came together in the Permian period.

para- [Gk. para: akin to, beside]   Prefix indicating association in being along side or accessory to.

parabronchi   Passages in the lungs of birds through which air flows.

paracrine [Gk. para: near]   Pertaining to a chemical signal, such as a hormone, that acts locally, near the site of its secretion. (Contrast with autocrine.)

paralogy (par al' o jee)   Type of homology in which the divergence of homologous genes can be traced to gene duplication events. (Contrast with orthology.)

paraphyletic (par' a fih leht' ik) [Gk. para: beside + phylon: tribe]   Pertaining to a group that consists of an ancestor and some, but not all, of its descendants. (Contrast with monophyletic, polyphyletic.)

parasite   An organism that consumes parts of an organism much larger than itself (known as its host). Parasites sometimes, but not always, kill their host.

parasympathetic nervous system   The division of the autonomic nervous system that works in opposition to the sympathetic nervous system. (Contrast with sympathetic nervous system.)

parathyroid glands   Four glands on the posterior surface of the thyroid gland that produce and release parathyroid hormone.

parathyroid hormone (PTH)   A hormone secreted by the parathyroid glands that stimulates osteoclast activity and raises blood calcium levels. Also called parathormone.

parenchyma (pair eng' kyma)   A plant tissue composed of relatively unspecialized cells without secondary walls.

parent rock   The soil horizon consisting of the rock that is breaking down to form the soil. Also called bedrock, or the C horizon.

parental (P) generation   The individuals that mate in a genetic cross. Their offspring are the first filial (F_1) generation.

parsimony   Preferring the simplest among a set of plausible explanations of any phenomenon.

parthenocarpy   Formation of fruit from a flower without fertilization.

parthenogenesis [Gk. parthenos: virgin]   Production of an organism from an unfertilized egg.

particulate theory   In genetics, the theory that genes are physical entities that retain their identities after fertilization.

passive transport   Diffusion across a membrane; may or may not require a channel or carrier protein. (Contrast with active transport.)

patch clamping   A technique for isolating a tiny patch of membrane to allow the study of ion movement through a particular channel.

pathogen (path' o jen) [Gk. pathos: suffering + genesis: source]   An organism that causes disease.

pattern formation   In animal embryonic development, the organization of differentiated tissues into specific structures such as wings.

pedigree   The pattern of transmission of a genetic trait within a family.

pelagic zone [Gk. pelagos: sea]   The open ocean.

penetrance   The proportion of individuals with a particular genotype that show the expected phenotype.

penis   An accessory sex organ of male animals that enables the male to deposit sperm in the female's reproductive tract.

pentose [Gk. penta: five]   A sugar containing five carbon atoms.

PEP carboxylase   The enzyme that combines carbon dioxide with PEP to form a 4-carbon dicarboxylic acid at the start of C_4 photosynthesis or of crassulacean acid metabolism (CAM).

pepsin [Gk. pepsis: digestion]   An enzyme in gastric juice that digests protein.

pepsinogen   Inactive secretory product that is converted into pepsin by low pH or by enzymatic action.

peptide linkage   The bond between amino acids in a protein; formed between a carboxyl group and amino group (CO—NH^-) with the loss of water molecules.

peptidoglycan   The cell wall material of many bacteria, consisting of a single enormous molecule that surrounds the entire cell.

perennial (per ren' ee al) [L. per: throughout + annus: year]   A plant that survives from year to year. (Contrast with annual, biennial.)

perfect flower   A flower with both stamens and carpels; a hermaphroditic flower. (Contrast with imperfect flower.)

pericycle [Gk. peri: around + kyklos: ring or circle]   In plant roots, tissue just within the endodermis, but outside of the root vascular tissue. Meristematic activity of pericycle cells produces lateral root primordia.

periderm   The outer tissue of the secondary plant body, consisting primarily of cork.

period   (1) A category in the geologic time scale. (2) The duration of a single cycle in a cyclical event, such as a circadian rhythm.

peripheral nervous system (PNS)   The portion of the nervous system that transmits information to and from the central nervous system, consisting of neurons that extend or reside outside the brain or spinal cord and their supporting cells. (Contrast with central nervous system.)

peristalsis (pair' i stall' sis)   Wavelike muscular contractions proceeding along a tubular organ, propelling the contents along the tube.

peritoneum   The mesodermal lining of the body cavity in coelomate animals.

peroxisome   An organelle that houses reactions in which toxic peroxides are formed and then converted to water.

petal [Gk. petalon: spread out]   In an angiosperm flower, a sterile modified leaf, nonphotosynthetic, frequently brightly colored, and often serving to attract pollinating insects.

petiole (pet' ee ole) [L. petiolus: small foot]   The stalk of a leaf.

pH   The negative logarithm of the hydrogen ion concentration; a measure of the acidity of a solution. A solution with pH = 7 is said to be neutral; pH values higher than 7 characterize basic solutions, while acidic solutions have pH values less than 7.

phage (fayj)   See bacteriophage.

phagocyte [Gk. phagein: to eat + kystos: sac]   One of two major classes of white blood cells; one of the nonspecific defenses of animals; ingests invading microorganisms by phagocytosis.

phagocytosis   Endocytosis by a cell of another cell or large particle.

pharming   The use of genetically modified animals to produce medically useful products in their milk.

pharynx [Gk. throat]   The part of the gut between the mouth and the esophagus.

phenotype (fee' no type) [Gk. phanein: to show]   The observable properties of an individual resulting from both genetic and environmental factors. (Contrast with genotype.)

phenotypic plasticity   See developmental plasticity.

pheromone (feer' o mone) [Gk. pheros: carry + hormon: excite, arouse]   A chemical substance used in communication between organisms of the same species.

phloem (flo' um) [Gk. phloos: bark]   In vascular plants, the vascular tissue that transports sugars and other solutes from sources to sinks.

phosphate group   The functional group —OPO_3H_2.

phosphodiester linkage   The connection in a nucleic acid strand, formed by linking two nucleotides.

phospholipid   A lipid containing a phosphate group; an important constituent of cellular membranes. (See lipid.)

phospholipid bilayer   The basic structural unit of biological membranes; a sheet of phospholipids two molecules thick in which the phospholipids are lined up with their hydrophobic "tails" packed tightly together and their hydrophilic, phosphate-containing "heads" facing outward. Also called lipid bilayer.

phosphorylation   Addition of a phosphate group.

photoautotroph   An organism that obtains energy from light and carbon from carbon dioxide. (Contrast with chemolithotroph, chemoheterotroph, photoheterotroph.)

photoheterotroph   An organism that obtains energy from light but must obtain its carbon from organic compounds. (Contrast with chemolithotroph, chemoheterotroph, photoautotroph.)

photon (foe' ton) [Gk. photos: light]   A quantum of visible radiation; a "packet" of light energy.

photoperiodicity Control of an organism's physiological or behavioral responses by the length of the day or night.

photoreceptor (1) In plants, a pigment that triggers a physiological response when it absorbs a photon. (2) In animals, a sensory receptor cell that senses and responds to light energy.

photorespiration Light-driven uptake of oxygen and release of carbon dioxide, the carbon being derived from the early reactions of photosynthesis.

photosynthesis (foe tow sin' the sis) [literally, "synthesis from light"] Metabolic processes, carried out by green plants, by which visible light is trapped and the energy used to synthesize compounds such as ATP and glucose.

photosystem [Gk. *phos*: light + *systema*: assembly] A light-harvesting complex in the chloroplast thylakoid composed of pigments and proteins.

photosystem I In photosynthesis, the reactions that absorb light at 700 nm, passing electrons to ferrodoxin and thence to NADPH. Rich in chlorophyll *a*.

photosystem II In photosynthesis, the reactions that absorb light at 660 nm, passing electrons to the electron transport chain in the chloroplast. Rich in chlorophyll *b*.

phototropins A class of blue light receptors that mediate phototropism and other plant responses.

phototropism [Gk. *photos*: light + *trope*: turning] A directed plant growth response to light.

phycobilin Photosynthetic pigment that absorbs red, yellow, orange, and green light and is found in cyanobacteria and some red algae.

phylogenetic tree A graphic representation of lines of descent among organisms or their genes.

phylogeny (fy loj' e nee) [Gk. *phylon*: tribe, race + *genesis*: source] The evolutionary history of a particular group of organisms or their genes.

physiology (fiz' ee ol' o jee) [Gk. *physis*: natural form] The scientific study of the functions of living organisms and the individual organs, tissues, and cells of which they are composed.

phytoalexins Substances toxic to pathogens, produced by plants in response to fungal or bacterial infection.

phytochrome (fy' tow krome) [Gk. *phyton*: plant + *chroma*: color] A plant pigment regulating a large number of developmental and other phenomena in plants.

phytomers In plants, the repeating modules that compose a shoot, each consisting of one or more leaves, attached to the stem at a node; an internode; and one or more axillary buds.

phytoplankton Photosynthetic plankton.

phytoremediation A form of bioremediation that uses plants to clean up environmental pollution.

pigment A substance that absorbs visible light.

pineal gland Gland located between the cerebral hemispheres that secretes melatonin.

pinocytosis Endocytosis by a cell of liquid containing dissolved substances.

pistil [L. *pistillum*: pestle] The structure of an angiosperm flower within which the ovules are borne. May consist of a single carpel, or of several carpels fused into a single structure. Usually differentiated into ovary, style, and stigma.

pith In plants, relatively unspecialized tissue found within a cylinder of vascular tissue.

pituitary gland A small gland attached to the base of the brain in vertebrates. Its hormones control the activities of other glands. Also known as the hypophysis.

placenta (pla sen' ta) The organ in female mammals that provides for the nourishment of the fetus and elimination of the fetal waste products.

plankton Free-floating small aquatic organisms. Photosynthetic members of the plankton are referred to as phytoplankton.

planula (plan' yew la) [L. *planum*: flat] A free-swimming, ciliated larval form typical of the cnidarians.

plaque (plack) [Fr.: a metal plate or coin] (1) A circular clearing in a layer (lawn) of bacteria growing on the surface of a nutrient agar gel. (2) An accumulation of prokaryotic organisms on tooth enamel. Acids produced by these microorganisms cause tooth decay. (3) A region of arterial wall invaded by fibroblasts and fatty deposits. (*See* atherosclerosis.)

plasma (plaz' muh) The liquid portion of blood, in which blood cells and other particulates are suspended.

plasma cell An antibody-secreting cell that develops from a B cell; the effector cell of the humoral immune system.

plasma membrane The membrane that surrounds the cell, regulating the entry and exit of molecules and ions. Every cell has a plasma membrane.

plasmid A DNA molecule distinct from the chromosome(s); that is, an extrachromosomal element; found in many bacteria. May replicate independently of the chromosome.

plasmodesma (plural: plasmodesmata) [Gk. *plassein*: to mold + *desmos*: band] A cytoplasmic strand connecting two adjacent plant cells.

plasmogamy The fusion of the cytoplasm of two cells. (Contrast with karyogamy.)

plastid Any of the plant cell organelles that house biochemical pathways for photosynthesis.

plate tectonics [Gk. *tekton*: builder] The scientific study of the structure and movements of Earth's lithospheric plates, which are the cause of continental drift.

platelet A membrane-bounded body without a nucleus, arising as a fragment of a cell in the bone marrow of mammals. Important to blood-clotting action.

pleiotropy (plee' a tro pee) [Gk. *pleion*: more] The determination of more than one character by a single gene.

pleural membrane [Gk. *pleuras*: rib, side] The membrane lining the outside of the lungs and the walls of the thoracic cavity. Inflammation of these membranes is a condition known as pleurisy.

pluripotent [L. *pluri*: many + *potens*: powerful] Having the ability to form all of the cells in the body. (Contrast with multipotent, totipotent.)

podocytes Cells of Bowman's capsule of the nephron that cover the capillaries of the glomerulus, forming filtration slits.

point mutation A mutation that results from the gain, loss, or substitution of a single nucleotide.

polar Having separate and opposite electric charges at two ends, or poles. (Contrast with nonpolar.)

polar body A nonfunctional nucleus produced by meiosis during oogenesis.

polar nuclei In angiosperms, the two nuclei in the central cell of the megagametophyte; following fertilization they give rise to the endosperm.

polarity (1) In chemistry, the property of unequal electron sharing in a covalent bond that defines a polar molecule. (2) In development, the difference between one end of an organism or structure and the other.

pollen [L. *pollin*: fine flour] In seed plants, microscopic grains that contain the male gametophyte (microgametophyte) and gamete (microspore).

pollen tube A structure that develops from a pollen grain through which sperm are released into the megagametophyte.

pollination The process of transferring pollen from an anther to the stigma of a pistil in an angiosperm or from a strobilus to an ovule in a gymnosperm.

poly- [Gk. *poly*: many] A prefix denoting multiple entities.

poly A tail A long sequence of adenine nucleotides (50–250) added after transcription to the 3' end of most eukaryotic mRNAs.

polyandry Mating system in which one female mates with multiple males.

polygyny Mating system in which one male mates with multiple females.

polymer [Gk. *poly*: many + *meros*: unit] A large molecule made up of similar or identical subunits called monomers. (Contrast with monomer, oligomer.)

polymerase chain reaction (PCR) An enzymatic technique for the rapid production of millions of copies of a particular stretch of DNA where only a small amount of the parent molecule is available.

polymorphic (pol' lee mor' fik) [Gk. *poly*: many + *morphe*: form, shape] Coexistence in a population of two or more distinct traits.

polyp (pah' lip) [Gk. *poly*: many + *pous*: foot] In cnidarians, a sessile, asexual life cycle stage.

polypeptide A large molecule made up of many amino acids joined by peptide linkages. Large polypeptides are called proteins.

polyphyletic (pol' lee fih leht' ik) [Gk. *poly*: many + *phylon*: tribe] Pertaining to a group that consists of multiple distantly related organisms, and does not include the common ancestor of the group. (Contrast with monophyletic, paraphyletic.)

polyploidy (pol' lee ploid ee) The possession of more than two entire sets of chromosomes.

polyribosome (polysome) A complex consisting of a threadlike molecule of messenger RNA and several (or many) ribosomes. The ribosomes move along the mRNA, synthesizing polypeptide chains as they proceed.

polysaccharide A macromolecule composed of many monosaccharides (simple sugars). Common examples are cellulose and starch.

pons [L. *pons*: bridge] Region of the brainstem anterior to the medulla.

population Any group of organisms coexisting at the same time and in the same place and capable of interbreeding with one another.

population bottleneck A period during which only a few individuals of a normally large population survive.

population density The number of individuals of a population per unit of area or volume.

population dynamics   The patterns and processes of change in populations.

population genetics   The study of genetic variation and its causes within populations.

portal blood vessels   Blood vessels that begin and end in capillary beds.

positive cooperativity   Occurs when a molecule can bind several ligands and each one that binds alters the conformation of the molecule so that it can bind the next ligand more easily. The binding of four molecules of O_2 by hemoglobin is an example of positive cooperativity.

positive feedback   In regulatory systems, information that amplifies a regulatory response, increasing the deviation of the system from the set point. (Contrast with negative feedback.)

positive regulation   A form of gene regulation in which a regulatory macromolecule is needed to turn on the transcription of a structural gene; in its absence, transcription will not occur. (Contrast with negative regulation.)

post- [L. *postere*: behind, following after]   Prefix denoting something that comes after.

postabsorptive state   State in which no food remains in the gut and thus no nutrients are being absorbed. (Contrast with absorptive state.)

posterior   Toward or pertaining to the rear. (Contrast with anterior.)

postsynaptic cell   The cell that receives information from a neuron at a synapse. (Contrast with presynaptic neuron.)

postzygotic reproductive barriers   Barriers to the reproductive process that occur after the union of the nuclei of two gametes. (Contrast with prezygotic reproductive barriers.)

potential energy   Energy not doing work, such as the energy stored in chemical bonds. (Contrast with kinetic energy.)

precapillary sphincter   A cuff of smooth muscle that can shut off the blood flow to a capillary bed.

pre-mRNA (precursor mRNA)   Initial gene transcript before it is modified to produce functional mRNA. Also known as the primary transcript.

predator   An organism that kills and eats other organisms.

prereplication complex   In eukaryotes, a complex of proteins that binds to DNA at the initiation of DNA replication.

pressure flow model   An effective model for phloem transport in angiosperms. It holds that sieve element transport is driven by an osmotically driven pressure gradient between source and sink.

pressure potential   The hydrostatic pressure of an enclosed solution in excess of the surrounding atmospheric pressure. (Contrast with solute potential, water potential.)

presynaptic neuron   The neuron that transmits information to another cell at a synapse. (Contrast with postsynaptic cell.)

prey [L. *praeda*: booty]   An organism consumed by a predator as an energy source.

prezygotic reproductive barriers   Barriers to the reproductive process that occur before the union of the nuclei of two gametes (Contrast with postzygotic reproductive barriers.)

primary active transport   Active transport in which ATP is hydrolyzed, yielding the energy required to transport an ion or molecule against its concentration gradient. (Contrast with secondary active transport.)

primary cell wall   In plant cells, a structure that forms at the middle lamella after cytokinesis, made up of cellulose microfibrils, hemicelluloses, and pectins. (Contrast with secondary cell wall.)

primary consumer   An organism (herbivore) that eats plant tissues.

primary growth   In plants, growth that is characterized by the lengthening of roots and shoots and by the proliferation of new roots and shoots through branching. (Contrast with secondary growth.)

primary immune response   The first response of the immune system to an antigen, involving recognition by lymphocytes and the production of effector cells and memory cells. (Contrast with secondary immune response.)

primary lysosome   *See* lysosome.

primary meristem   Meristem that produces the tissues of the primary plant body.

primary oocyte   (oh' eh site) [Gk. *oon*: egg + *kytos*: container]   The diploid progeny of an oogonium. In many species, a primary oocyte enters prophase of the first meiotic division, then remains in developmental arrest for a long time before resuming meiosis to form a secondary oocyte and a polar body.

primary plant body   That part of a plant produced by primary growth. Consists of all the *nonwoody* parts of a plant; many herbaceous plants consist entirely of a primary plant body. (Contrast with secondary plant body.)

primary producer   A photosynthetic or chemosynthetic organism that synthesizes complex organic molecules from simple inorganic ones.

primary sex determination   Genetic determination of gametic sex, male or female. (Contrast with secondary sex determination.)

primary spermatocyte   The diploid progeny of a spermatogonium; undergoes the first meiotic division to form secondary spermatocytes.

primary succession   Succession that begins in an area initially devoid of life, such as on recently exposed glacial till or lava flows. (Contrast with secondary succession.)

primary structure   The specific sequence of amino acids in a protein.

primase   An enzyme that catalyzes the synthesis of a primer for DNA replication.

primer   Strand of nucleic acid, usually RNA, that is the necessary starting material for the synthesis of a new DNA strand, which is synthesized from the 3′ end of the primer.

primordium (plural: primordia) [L. origin]   The most rudimentary stage of an organ or other part.

prion   An infectious protein that can proliferate by converting the inactive form of a particular protein into an active protein.

pro- [L.: first, before, favoring]   A prefix often used in biology to denote a developmental stage that comes first or an evolutionary form that appeared earlier than another. For example, prokaryote, prophase.

probe   A segment of single stranded nucleic acid used to identify DNA molecules containing the complementary sequence.

procambium   Primary meristem that produces the vascular tissue.

procedural memory   Memory of motor tasks. Cannot be consciously recalled and described. (Contrast with declarative memory.)

processive   Pertaining to an enzyme that catalyzes many reactions each time it binds to a substrate, as DNA polymerase does during DNA replication.

progesterone [L. *pro*: favoring + *gestare*: to bear]   A female sex hormone that maintains pregnancy.

prolactin   A hormone released by the anterior pituitary, one of whose functions is the stimulation of milk production in female mammals.

proliferating cell nuclear antigen (PCNA)   A protein complex that ensures processivity of DNA replication in eukaryotes.

prometaphase   The phase of nuclear division that begins with the disintegration of the nuclear envelope.

promoter   A DNA sequence to which RNA polymerase binds to initiate transcription.

prophage (pro' fayj)   The noninfectious units that are linked with the chromosomes of the host bacteria and multiply with them but do not cause dissolution of the cell. Prophage can later enter into the lytic phase to complete the virus life cycle.

prophase (pro' phase)   The first stage of nuclear division, during which chromosomes condense from diffuse, threadlike material to discrete, compact bodies.

prostaglandin   Any one of a group of specialized lipids with hormone-like functions. It is not clear that they act at any considerable distance from the site of their production.

prostate gland   In male humans, surrounds the urethra at its junction with the vas deferens; supplies an acid-neutralizing fluid to the semen.

prosthetic group   Any nonprotein portion of an enzyme.

proteasome   In the eukaryotic cytoplasm, a huge protein structure that binds to and digests cellular proteins that have been tagged by ubiquitin.

protein (pro' teen) [Gk. *protos*: first]   Long-chain polymer of amino acids with twenty different common side chains. Occurs with its polymer chain extended in fibrous proteins, or coiled into a compact macromolecule in enzymes and other globular proteins.

protein kinase (kye' nase)   An enzyme that catalyzes the addition of a phosphate group from ATP to a target protein.

protein kinase cascade   A series of reactions in response to a molecular signal, in which a series of protein kinases activates one another in sequence, amplifying the signal at each step.

proteoglycan   A glycoprotein containing a protein core with attached long, linear carbohydrate chains.

proteolysis [protein + Gk. *lysis*: break apart]   An enzymatic digestion of a protein or polypeptide.

proteome   The set of proteins that can be made by an organism. Because of alternative splicing of pre-mRNA, the number of proteins that can be made is usually much larger than the number of protein-coding genes present in the organism's genome.

protoderm   Primary meristem that gives rise to the plant epidermis.

proton (pro' ton) [Gk. *protos*: first, before]   (1) A subatomic particle with a single positive charge.

The number of protons in the nucleus of an atom determine its element. (2) A hydrogen ion, H^+.

proton pump   An active transport system that uses ATP energy to move hydrogen ions across a membrane, generating an electric potential.

proton-motive force   Force generated across a membrane having two components: a chemical potential (difference in proton concentration) plus an electrical potential due to the electrostatic charge on the proton.

protonephridium   The excretory organ of flatworms, made up of a tubule and a flame cell.

protoplast   The living contents of a plant cell; the plasma membrane and everything contained within it.

provirus   Double-stranded DNA made by a virus that is integrated into the host's chromosome and contains promoters that are recognized by the host cell's transcription apparatus.

proximal   Near the point of attachment or other reference point. (Contrast with distal.)

proximal convoluted tubule   The initial segment of a renal tubule, closest to the glomerulus. (Compare with distal convoluted tubule.)

proximate cause   The immediate genetic, physiological, neurological, and developmental mechanisms responsible for a behavior or morphology. (Contrast with ultimate cause.)

pseudocoelomate (soo' do see' low mate) [Gk. *pseudes*: false + *koiloma*: cavity]   Having a body cavity, called a pseudocoel, consisting of a fluid-filled space in which many of the internal organs are suspended, but which is enclosed by mesoderm only on its outside.

pseudogene [Gk. *pseudes*: false]   A DNA segment that is homologous to a functional gene but is not expressed because of changes to its sequence or changes to its location in the genome.

pseudopod (soo' do pod) [Gk. *pseudes*: false + *podos*: foot]   A temporary, soft extension of the cell body that is used in location, attachment to surfaces, or engulfing particles.

pulmonary [L. *pulmo*: lung]   Pertaining to the lungs.

pulmonary circuit   The portion of the circulatory system by which blood is pumped from the heart to the lungs or gills for oxygenation and back to the heart for distribution. (Contrast with systemic circuit.)

pulmonary valve   A one-way valve between the right ventricle of the heart and the pulmonary artery that prevents backflow of blood into the ventricle when it relaxes.

Punnett square   Method of predicting the results of a genetic cross by arranging the gametes of each parent at the edges of a square.

pupa (pew' pa) [L. *pupa*: doll, puppet]   In certain insects (the Holometabola), the encased developmental stage between the larva and the adult.

pupil   The opening in the vertebrate eye through which light passes.

purine (pure' een)   One of the two types of nitrogenous bases in nucleic acids. Each of the purines—adenine and guanine—pairs with a specific pyrimidine.

Purkinje fibers   Specialized heart muscle cells that conduct excitation throughout the ventricular muscle.

pyrimidine (per im' a deen)   One of the two types of nitrogenous bases in nucleic acids. Each

of the pyrimidines—cytosine, thymine, and uracil—pairs with a specific purine.

pyrogen   Molecule that produces a rise in body temperature (fever); may be produced by an invading pathogen or by cells of the immune system in response to infection.

pyruvate   A three-carbon acid; the end product of glycolysis and the raw material for the citric acid cycle.

pyruvate oxidation   Conversion of pyruvate to acetyl CoA and CO_2 that occurs in the mitochondrial matrix in the presence of O_2.

- Q -

Q_{10}   A value that compares the rate of a biochemical process or reaction over 10°C temperature ranges. A process that is not temperature-sensitive has a Q_{10} of 1; values of 2 or 3 mean the reaction speeds up as temperature increases.

quantitative trait loci   A set of genes that determines a complex character that exhibits quantitative variation.

quaternary structure   The specific three-dimensional arrangement of protein subunits.

- R -

R group   The distinguishing group of atoms of a particular amino acid; also known as a side chain.

r-strategist   A species whose life history strategy allows for a high intrinsic rate of population increase (*r*). (Contrast with *K*-strategist.)

radial symmetry   The condition in which any two halves of a body are mirror images of each other, providing the cut passes through the center; a cylinder cut lengthwise down its center displays this form of symmetry.

radiation   The transfer of heat from warmer objects to cooler ones via the exchange of infrared radiation. *See also* electromagnetic radiation; evolutionary radiation.

radicle   An embryonic root.

radioisotope   A radioactive isotope of an element. Examples are carbon-14 (^{14}C) and hydrogen-3, or tritium (^3H).

reactant   A chemical substance that enters into a chemical reaction with another substance.

reaction center   A group of electron transfer proteins that receive energy from light-absorbing pigments and convert it to chemical energy by redox reactions.

realized niche   A species' niche as defined by its interactions with other species. (Contrast with fundamental niche.)

receptive field   The area of visual space that activates a particular cell in the visual system.

receptor   *See* receptor protein, sensory receptor cell.

receptor-mediated endocytosis   Endocytosis initiated by macromolecular binding to a specific membrane receptor.

receptor potential   The change in the resting potential of a sensory cell when it is stimulated.

receptor protein   A protein that can bind to a specific molecule, or detect a specific stimulus, within the cell or in the cell's external environment.

recessive   In genetics, an allele that does not determine phenotype in the presence of a dominant allele. (Contrast with dominance.)

reciprocal adaptation   *See* coevolution.

reciprocal crosses   A pair of matings in one of which a female of genotype A mates with a male of genotype B and in the other of which a female of genotype B mates with a male of genotype A.

recognition sequence   *See* restriction site.

recombinant   Pertaining to an individual, meiotic product, or chromosome in which genetic materials originally present in two individuals end up in the same haploid complement of genes.

recombinant DNA   A DNA molecule made in the laboratory that is derived from two or more genetic sources.

recombinant frequency   The proportion of offspring of a genetic cross that have phenotypes different from the parental phenotypes due to crossing over between linked genes during gamete formation.

reconciliation ecology   The practice of making exploited lands more biodiversity-friendly.

rectum   The terminal portion of the gut, ending at the anus.

redox reaction   A chemical reaction in which one reactant becomes oxidized and the other becomes reduced.

reduction   Gain of electrons by a chemical reactant; any reduction is accompanied by an oxidation. (Contrast with oxidation.)

refractory period   The time interval after an action potential during which another action potential cannot be elicited from an excitable membrane.

regeneration   The development of a complete individual from a fragment of an organism.

regulative development   A pattern of animal embryonic development in which the fates of the first blastomeres are not absolutely fixed. (Contrast with mosaic development.)

regulatory sequence   A DNA sequence to which the protein product of a regulatory gene binds.

regulatory gene   A gene that codes for a protein (or RNA) that in turn controls the expression of another gene.

regulatory system   A system that uses feedback information to maintain a physiological function or parameter at an optimal level. (Contrast with controlled system.)

regulatory T cells (T_{reg})   Class of T cells that mediates tolerance to self antigens.

reinforcement   The evolution of enhanced reproductive isolation between populations due to natural selection for greater isolation.

releaser   Sensory stimulus that triggers performance of a stereotyped behavior pattern.

REM (rapid-eye-movement) sleep   A sleep state characterized by vivid dreams, skeletal muscle relaxation, and rapid eye movements. (Contrast with slow-wave sleep.)

renal [L. *renes*: kidneys]   Relating to the kidneys.

renal tubule   A structural unit of the kidney that collects filtrate from the blood, reabsorbs specific ions, nutrients, and water and returns them to the blood, and concentrates excess ions and waste products such as urea for excretion from the body.

replication   The duplication of genetic material.

replication complex   The close association of several proteins operating in the replication of DNA.

replication fork   A point at which a DNA molecule is replicating. The fork forms by the unwinding of the parent molecule.

replicon   A region of DNA replicated from a single origin of replication.

reporter gene   A genetic marker included in recombinant DNA to indicate the presence of the recombinant DNA in a host cell.

repressor   A protein encoded by a regulatory gene that can bind to a specific operator and prevent transcription of the operon. (Contrast with activator.)

reproductive isolation   Condition in which two divergent populations are no longer exchanging genes. Can lead to speciation.

rescue effect   The process by which individuals moving between subpopulations of a metapopulation may prevent declining subpopulations from becoming extinct.

resistance (R) genes   Plant genes that confer resistance to specific strains of pathogens.

resource   Something in the environment required by an organism for its maintenance and growth that is consumed in the process of being used.

resource partitioning   A situation in which selection pressures resulting from interspecific competition cause changes in the ways in which the competing species use the limiting resource, thereby allowing them to coexist.

respiration (res pi ra' shun) [L. *spirare*: to breathe] (1) Cellular respiration. (2) Breathing.

respiratory chain   The terminal reactions of cellular respiration, in which electrons are passed from NAD or FAD, through a series of intermediate carriers, to molecular oxygen, with the concomitant production of ATP.

respiratory gases   Oxygen (O_2) and carbon dioxide (CO_2); the gases that an animal must exchange between its internal body fluids and the outside medium (air or water).

resting potential   The membrane potential of a living cell at rest. In cells at rest, the interior is negative to the exterior. (Contrast with action potential, electrotonic potential.)

restoration ecology   The science and practice of restoring damaged or degraded ecosystems.

restriction enzyme   Any of a type of enzyme that cleaves double-stranded DNA at specific sites; extensively used in recombinant DNA technology. Also called a restriction endonuclease.

restriction fragment length polymorphism   See RFLP.

restriction point (R)   The specific time during G1 of the cell cycle at which the cell becomes committed to undergo the rest of the cell cycle.

restriction site   A specific DNA base sequence that is recognized and acted on by a restriction endonuclease.

reticular system   A central region of the vertebrate brainstem that includes complex fiber tracts conveying neural signals between the forebrain and the spinal cord, with collateral fibers to a variety of nuclei that are involved in autonomic functions, including arousal from sleep.

retina (rett' in uh) [L. *rete*: net]   The light-sensitive layer of cells in the vertebrate or cephalopod eye.

retinoblastoma protein   A protein that inhibits an animal cell from passing through the restriction point; inactivation of this protein is necessary for the cell cycle to proceed.

retrovirus   An RNA virus that contains reverse transcriptase. Its RNA serves as a template for cDNA production, and the cDNA is integrated into a chromosome of the host cell.

reverse genetics   Method of genetic analysis in which a phenotype is first related to a DNA variation, then the protein involved is identified.

reverse transcriptase   An enzyme that catalyzes the production of DNA (cDNA), using RNA as a template; essential to the reproduction of retroviruses.

RFLP   Restriction fragment length polymorphism, the coexistence of two or more patterns of restriction fragments resulting from underlying differences in DNA sequence.

rhizoids (rye' zoids) [Gk. root]   Hairlike extensions of cells in mosses, liverworts, and a few vascular plants that serve the same function as roots and root hairs in vascular plants. The term is also applied to branched, rootlike extensions of some fungi and algae.

rhizome (rye' zome)   An underground stem (as opposed to a root) that runs horizontally beneath the ground.

rhodopsin   A photopigment used in the visual process of transducing photons of light into changes in the membrane potential of photoreceptor cells.

ribonucleic acid   See RNA.

ribose   A five-carbon sugar in nucleotides and RNA.

ribosomal RNA (rRNA)   Several species of RNA that are incorporated into the ribosome. Involved in peptide bond formation.

ribosome   A small particle in the cell that is the site of protein synthesis.

ribozyme   An RNA molecule with catalytic activity.

risk cost   The increased chance of being injured or killed as a result of performing a behavior, compared to resting.

RNA (ribonucleic acid)   An often single stranded nucleic acid whose nucleotides use ribose rather than deoxyribose and in which the base uracil replaces thymine found in DNA. Serves as genome from some viruses. (See rRNA, tRNA, mRMA, and ribozyme.)

RNA interference (RNAi)   A mechanism for reducing mRNA translation whereby a double-stranded RNA, made by the cell or synthetically, is processed into a small, single-stranded RNA, whose binding to a target mRNA results in the latter's breakdown.

RNA polymerase   An enzyme that catalyzes the formation of RNA from a DNA template.

RNA splicing   The last stage of RNA processing in eukaryotes, in which the transcripts of introns are excised through the action of small nuclear ribonucleoprotein particles (snRNP).

rod cells   Light-sensitive cell in the vertebrate retina; these sensory receptor cells are sensitive in extremely dim light and are responsible for dim light, black and white vision.

root   The organ responsible for anchoring the plant in the soil, absorbing water and minerals, and producing certain hormones. Some roots are storage organs.

root cap   A thimble-shaped mass of cells, produced by the root apical meristem, that protects the meristem; the organ that perceives the gravitational stimulus in root gravitropism.

root hair   A long, thin process from a root epidermal cell that absorbs water and minerals from the soil solution.

root system   The organ system that anchors a plant in place, absorbs water and dissolved minerals, and may store products of photosynthesis from the shoot system.

rough endoplasmic reticulum (RER)   The portion of the endoplasmic reticulum whose outer surface has attached ribosomes. (Contrast with smooth endoplasmic reticulum.)

rRNA   See ribosomal RNA.

rubisco   Contraction of ribulose bisphosphate carboxylase/oxygenase, the enzyme that combines carbon dioxide or oxygen with ribulose bisphosphate to catalyze the first step of photosynthetic carbon fixation or photorespiration, respectively.

ruminant   Herbivorous, cud-chewing mammals such as cows or sheep, characterized by a stomach that consists of four compartments: the rumen, reticulum, omasum, and abomasum.

- S -

S phase   In the cell cycle, the stage of interphase during which DNA is replicated. (Contrast with G_1 phase, G_2 phase, M phase.)

saltatory conduction [L. *saltare*: to jump]   The rapid conduction of action potentials in myelinated axons; so called because action potentials appear to "jump" between nodes of Ranvier along the axon.

saprobe [Gk. *sapros*: rotten]   An organism (usually a bacterium or fungus) that obtains its carbon and energy by absorbing nutrients from dead organic matter.

sarcomere (sark' o meer) [Gk. *sark*: flesh + *meros*: unit]   The contractile unit of a skeletal muscle.

sarcoplasm   The cytoplasm of a muscle cell.

sarcoplasmic reticulum   The endoplasmic reticulum of a muscle cell.

saturated fatty acid   A fatty acid in which all the bonds between carbon atoms in the hydrocarbon chain are single bonds—that is, all the bonds are saturated with hydrogen atoms. (Contrast with unsaturated fatty acid.)

scientific method   A means of gaining knowledge about the natural world by making observations, posing hypotheses, and conducting experiments to test those hypotheses.

Schwann cell   A type of glial cell that myelinates axons in the peripheral nervous system.

scion   In horticulture, the bud or stem from one plant that is grafted to a root or root-bearing stem of another plant (the stock).

sclerenchyma (skler eng' kyma) [Gk. *skleros*: hard + *kymus*: juice]   A plant tissue composed of cells with heavily thickened cell walls. The cells are dead at functional maturity. The principal types of sclerenchyma cells are fibers and sclereids.

second law of thermodynamics   The principle that when energy is converted from one form to another, some of that energy becomes unavailable for doing work.

second messenger   A compound, such as cAMP, that is released within a target cell after a hormone (the first messenger) has bound to a surface receptor on a cell; the second messenger triggers further reactions within the cell.

secondary active transport   A form of active transport that does not use ATP as an energy source; rather, transport is coupled to ion diffusion down a concentration gradient established by primary active transport.

secondary cell wall   A thick, cellulosic structure internal to the primary cell wall formed in some plant cells after cell expansion stops (Contrast with primary cell wall.)

secondary consumer   An organism that eat primary consumers.

secondary growth   In plants, growth that contributes to an increase in girth. (Contrast with primary growth.)

secondary immune response   A rapid and intense response to a second or subsequent exposure to an antigen, initiated by memory cells. (Contrast with primary immune response.)

secondary lysosome   Membrane-enclosed organelle formed by the fusion of a primary lysosome with a phagosome, in which macromolecules taken up by phagocytosis are hydrolyzed into their monomers. (Contrast with lysosome.)

secondary metabolite   A compound synthesized by a plant that is not needed for basic cellular metabolism. Typically has an antiherbivore or antiparasite function.

secondary plant body   That part of a plant produced by secondary growth; consists of woody tissues. (Contrast with primary plant body.)

secondary sex determination   Formation of secondary sexual characteristics (i.e., those other than gonads), such as external sex organs and body hair. (Contrast with primary sex determination.)

secondary spermatocyte   One of the products of the first meiotic division of a primary spermatocyte.

secondary structure   Of a protein, localized regularities of structure, such as the α helix and the β pleated sheet.

secondary succession   Succession after a disturbance that did not eliminate all the organisms originally living on the site. (Contrast with primary succession.)

secretin (si kreet′ in)   A peptide hormone secreted by the upper region of the small intestine when acidic chyme is present. Stimulates the pancreatic duct to secrete bicarbonate ions.

sedimentary rock   Rock formed by the accumulation of sediment grains on the bottom of a body of water.

seed   A fertilized, ripened ovule of a gymnosperm or angiosperm. Consists of the embryo, nutritive tissue, and a seed coat.

segmentation   Division of an animal body into segments.

segmentation genes   Genes that determine the number and polarity of body segments.

segregation   In genetics, the separation of alleles, or of homologous chromosomes, from each other during meiosis so that each of the haploid daughter nuclei produced contains one or the other member of the pair found in the diploid parent cell, but never both. This principle was articulated by Mendel as his first law.

selective permeability   Allowing certain substances to pass through while other substances are excluded; a characteristic of membranes.

self-incompatability   In plants, the possession of mechanisms that prevent self-fertilization.

semelparous   [L. *semel*: once + *pario*: to beget] Reproducing only once in a lifetime. (Contrast with iteroparous.)

semen (see′ men) [L. *semin*: seed]   The thick, whitish liquid produced by the male reproductive system in mammals, containing the sperm.

semiconservative replication   The way in which DNA is synthesized. Each of the two partner strands in a double helix acts as a template for a new partner strand. Hence, after replication, each double helix consists of one old and one new strand.

seminiferous tubules   The tubules within the testes within which sperm production occurs.

senescence [L. *senescere*: to grow old]   Aging; deteriorative changes with aging; the increased probability of dying with increasing age.

sensitive period   The life stage during which some particular type of learning must take place, or during which it occurs much more easily than at other times. Typical of song learning among birds.

sensor   *See* sensory receptor cell.

sensory neuron   A specialized neuron that transduces a particular type of sensory stimulus into action potentials.

sensory receptor cell   Cell that is responsive to a particular type of physical or chemical stimulation.

sensory transduction   The transformation of environmental stimuli or information into neural signals.

sepal (see′ pul) [L. *sepalum*: covering]   One of the outermost structures of the flower, usually protective in function and enclosing the rest of the flower in the bud stage.

septum (plural: septa) [L. *wall*]   (1) A partition or cross-wall appearing in the hyphae of some fungi. (2) The bony structure dividing the nasal passages.

sequence alignment   A method of identifying homologous positions in DNA or amino acid sequences by pinpointing the locations of deletions and insertions that have occurred since two (or more) organisms diverged from a common ancestor.

Sertoli cells   Cells in the seminiferous tubules that nurture the developing sperm.

sessile (sess′ ul) [L. *sedere*: to sit]   Permanently attached; not able to move from one place to another. (Contrast with motile.)

set point   In a regulatory system, the threshold sensitivity to the feedback stimulus.

sex chromosome   In organisms with a chromosomal mechanism of sex determination, one of the chromosomes involved in sex determination.

sex linkage   The pattern of inheritance characteristic of genes located on the sex chromosomes of organisms having a chromosomal mechanism for sex determination.

sexual reproduction   Reproduction involving the union of gametes.

sexual selection   Selection by one sex of characteristics in individuals of the opposite sex. Also, the favoring of characteristics in one sex as a result of competition among individuals of that sex for mates.

Shannon diversity index   A formula for quantifying diversity that takes both species richness and species evenness into account; based on a mathematical expression of the certainty with which the next item sampled in a series can be predicted.

shared derived trait   *See* synapomorphy.

shoot system   In plants, the organ system consisting of the leaves, stem(s), and flowers.

short tandem repeat (STR)   A short (1–5 base pairs), moderately repetitive sequence of DNA. The number of copies of an STR at a particular location varies between individuals and is inherited.

shotgun sequencing   A relatively rapid method of DNA sequencing in which a DNA molecule is broken up into overlapping fragments, each fragment is sequenced, and high-speed computers analyze and realign the fragments. (Contrast with hierarchical sequencing.)

side chain   *See* R group.

sieve tube element   The characteristic cell of the phloem in angiosperms, which contains cytoplasm but relatively few organelles, and whose end walls (*sieve plates*) contain pores that form connections with neighboring cells.

signal sequence   The sequence of a protein that directs the protein protein to a particular organelle.

signal transduction pathway   The series of biochemical steps whereby a stimulus to a cell (such as a hormone or neurotransmitter binding to a receptor) is translated into a response of the cell.

silent mutation   A change in a gene's sequence that has no effect on the amino acid sequence of a protein because it occurs in noncoding DNA or because it does not change the amino acid specified by the corresponding codon. (Contrast with frame-shift mutation, missense mutation, nonsense mutation.)

similarity matrix   A matrix used to compare the degree of divergence among pairs of objects. For molecular sequences, constructed by summing the number or percentage of nucleotides or amino acids that are identical in each pair of sequences.

single nucleotide polymorphisms (SNPs)   Inherited variations in a single nucleotide base in DNA that differ between individuals.

single-strand binding protein   In DNA replication, a protein that binds to single strands of DNA after they have been separated from each other, keeping the two strands separate for replication.

sink   In plants, any organ that imports the products of photosynthesis, such as roots, developing fruits, and immature leaves. (Contrast with source.)

sinoatrial node (sigh′ no ay′ tree al) [L. *sinus*: curve + *atrium*: chamber]   The pacemaker of the mammalian heart.

siRNAs (small interfering RNAs)   Short, double-stranded RNA molecules used in RNA interference.

sister chromatid   Each of a pair of newly replicated chromatids.

sister groups   Two phylogenetic groups that are each other's closest relatives.

skeletal muscle   A type of muscle tissue characterized by multinucleated cells containing highly ordered arrangements of actin and myosin microfilaments. Also called striated muscle. (Contrast with cardiac muscle, smooth muscle.)

skeletal systems   Organ systems that provide rigid supports against which muscles can pull to create directed movements.

sliding DNA clamp   Protein complex that keeps DNA polymerase bound to DNA during replication.

sliding filament theory   Mechanism of muscle contraction based on the formation and breaking of crossbridges between actin and myosin filaments, causing the filaments to slide together.

slow-wave sleep   A state of deep, restorative sleep characterized by high-amplitude slow waves in the EEG. (Contrast with REM sleep.)

small intestine   The portion of the gut between the stomach and the colon; consists of the duodenum, the jejunum, and the ileum.

small nuclear ribonucleoprotein particle (snRNP)   A complex of an enzyme and a small nuclear RNA molecule, functioning in RNA splicing.

smooth endoplasmic reticulum (SER)   Portion of the endoplasmic reticulum that lacks ribosomes and has a tubular appearance. (Contrast with rough endoplasmic reticulum.)

smooth muscle   Muscle tissue consisting of sheets of mononucleated cells innervated by the autonomic nervous system. (Contrast with cardiac muscle, skeletal muscle.)

sodium–potassium (Na+–K+) pump   Antiporter responsible for primary active transport; it pumps sodium ions out of the cell and potassium ions into the cell, both against their concentration gradients. Also called a sodium–potassium ATPase.

soil horizon   See horizon.

solute   A substance that is dissolved in a liquid (solvent) to form a solution.

solute potential   A property of any solution, resulting from its solute contents; it may be zero or have a negative value. The more negative the solute potential, the greater the tendency of the solution to take up water through a differentially permeable membrane. (Contrast with pressure potential, water potential.)

solution   A liquid (the solvent) and its dissolved solutes.

solvent   Liquid in which a substance (solute) is dissolved to form a solution.

somatic cell   [Gk. *soma*: body]   All the cells of the body that are not specialized for reproduction. (Contrast with germ cell.)

somatic mutation   Permanent genetic change in a somatic cell. These mutations affect the individual only; they are not passed on to offspring. (Contrast with germ line mutation.)

somatosensory cortex   The region of the cerebral cortex that receives input from mechanosensors distributed throughout the body.

somatostatin   Peptide hormone made in the hypothalamus that inhibits the release of other hormones from the pituitary and intestine.

somite   (so' might)   One of the segments into which an embryo becomes divided longitudinally, leading to the eventual segmentation of the animal as illustrated by the spinal column, ribs, and associated muscles.

Sorenson's index   Mathematical formula that measures beta diversity.

source   In plants, any organ that exports the products of photosynthesis in excess of its own needs, such as a mature leaf or storage organ. (Contrast with sink.)

spatial summation   In the production or inhibition of action potentials in a postsynaptic cell, the interaction of depolarizations and hyperpolarizations produced at different sites on the postsynaptic cell. (Contrast with temporal summation.)

spawning   See external fertilization.

speciation   (spee' see ay' shun)   The process of splitting one population into two populations that are reproductively isolated from one another.

species   (spee' sees) [L. kind]   The base unit of taxonomic classification, consisting of an ancestor–descendant group of populations of evolutionarily closely related, similar organisms. The more narrowly defined "biological species" consists of individuals capable of interbreeding with each other but not with members of other species.

species–area relationship   The relationship between the size of an area and the numbers of species it supports.

species richness   The total number of species living in a region.

specific defenses   Defensive reactions of the vertebrate immune system that are based on the reaction of an antibody to a specific antigen. (Contrast with nonspecific defenses.)

specific heat   The amount of energy that must be absorbed by a gram of a substance to raise its temperature by one degree centigrade. By convention, water is assigned a specific heat of one.

sperm   [Gk. *sperma*: seed]   The male gamete.

spermatid   One of the products of the second meiotic division of a primary spermatocyte; four haploid spermatids, which remain connected by cytoplasmic bridges, are produced for each primary spermatocyte that enters meiosis.

spermatogenesis   (spur mat' oh jen' e sis) [Gk. *sperma*: seed + *genesis*: source]   Gametogenesis leading to the production of sperm.

spermatogonia   In animals, the diploid progeny of a germ cell in males.

spherical symmetry   The simplest form of symmetry, in which body parts radiate out from a central point such that an infinite number of planes passing through that central point can divide the organism into similar halves.

spicule   [L. arrowhead]   A hard, calcareous skeletal element typical of sponges.

spinal reflex   The conversion of afferent to efferent information in the spinal cord without participation of the brain.

sphincter   (sfink' ter) [Gk. *sphinkter*: something that binds tightly]   A ring of muscle that can close an orifice, for example, at the anus.

spindle   Array of microtubules emanating from both poles of a dividing cell during mitosis and playing a role in the movement of chromosomes at nuclear division. Named for its shape.

spleen   Organ that serves as a reservoir for venous blood and eliminates old, damaged red blood cells from the circulation.

spliceosome   RNA–protein complex that splices out introns from eukaryotic pre-mRNAs.

splicing   See RNA splicing.

spontaneous mutation   A genetic change caused by internal cellular mechanisms, such as an error in DNA replication. (Contrast with induced mutation.)

sporangiophore   A stalked reproductive structure produced by zygospore fungi that extends from a hypha and bears one or many sporangia.

sporangium   (spor an' gee um) (plural: sporangia) [Gk. *spora*: seed + *angeion*: vessel or reservoir]   In plants and fungi, any specialized stucture within which one or more spores are formed.

spore   [Gk. *spora*: seed]   (1) Any asexual reproductive cell capable of developing into an adult organism without gametic fusion. In plants, haploid spores develop into gametophytes, diploid spores into sporophytes. (2) In prokaryotes, a resistant cell capable of surviving unfavorable periods.

sporocyte   Specialized cells of the diploid sporophyte that will divide by meiosis to produce four haploid spores. Germination of these spores produces the haploid gametophyte.

sporophyte   (spor' o fyte) [Gk. *spora*: seed + *phyton*: plant]   In plants and protists with alternation of generations, the diploid phase that produces the spores. (Contrast with gametophyte.)

stabilizing selection   Selection against the extreme phenotypes in a population, so that the intermediate types are favored. (Contrast with disruptive selection.)

stamen   (stay' men) [L. *stamen*: thread]   A male (pollen-producing) unit of a flower, usually composed of an anther, which bears the pollen, and a filament, which is a stalk supporting the anther.

starch   [O.E. *stearc*: stiff]   A polymer of glucose; used by plants to store energy.

Starling's forces   The two opposing forces responsible for water movement across capillary walls: blood pressure, which squeezes water and small solutes out of the capillaries, and osmotic pressure, which pulls water back into the capillaries.

start codon   The mRNA triplet (AUG) that acts as a signal for the beginning of translation at the ribosome. (Contrast with stop codon.)

stele   (steel) [Gk. *stylos*: pillar]   The central cylinder of vascular tissue in a plant stem.

stem   In plants, the organ that holds leaves and/or flowers and transports and distributes materials among the other organs of the plant.

stem cell   In animals, an undifferentiated cell that is capable of continuous proliferation. A stem cell generates more stem cells and a large clone of differentiated progeny cells. (See also embryonic stem cell.)

steroid   Any of a family of lipids whose multiple rings share carbons. The steroid cholesterol is an important constituent of membranes; other steroids function as hormones.

sticky ends   On a piece of two-stranded DNA, short, complementary, one-stranded regions produced by the action of a restriction endonuclease. Sticky ends facilitate the joining of segments of DNA from different sources.

stigma   [L. *stigma*: mark, brand]   The part of the pistil at the apex of the style that is receptive to pollen, and on which pollen germinates.

stimulus   [L. *stimulare*: to goad]   Something causing a response; something in the environment detected by a receptor.

stock   In horticulture, the root or root-bearing stem to which a bud or piece of stem from another plant (the scion) is grafted.

stoma   (plural: stomata) [Gk. *stoma*: mouth, opening]   Small opening in the plant epidermis that permits gas exchange; bounded by a pair of guard cells whose osmotic status regulates the size of the opening.

stomatal crypt   In plants, a sunken cavity below the leaf surface in which a stoma is sheltered from the drying effects of air currents.

stop codon   Any of the three mRNA codons that signal the end of protein translation at the ribosome: UAG, UGA, UAA.

stratosphere   The upper part of Earth's atmosphere, above the troposphere; extends from approximately 18 kilometers upward to approximately 50 kilometers above the surface.

stratum (plural strata) [L. *stratos*: layer]   A layer of sedimentary rock laid down at a particular time in the past.

stretch receptor   A modified muscle cell embedded in the connective tissue of a muscle that acts as a mechanoreceptor in response to stretching of that muscle.

striated muscle   *See* skeletal muscle.

strobilus (plural: strobili)   One of several cone-like structures in various groups of plants (including club mosses, horsetails, and conifers) associated with the production and dispersal of reproductive products.

stroma   The fluid contents of an organelle such as a chloroplast or mitochondrion.

structural gene   A gene that encodes the primary structure of a protein not involved in the regulation of gene expression.

structural isomers   Molecules made up of the same kinds and numbers of atoms, in which the atoms are bonded differently.

structural motif   A three-dimensional structural element that is part of a larger molecule. For example, there are four common motifs in DNA-binding proteins: helix-turn-helix, zinc finger, leucine zipper, and helix-loop-helix.

style [Gk. *stylos*: pillar or column]   In the angiosperm flower, a column of tissue extending from the tip of the ovary, and bearing the stigma or receptive surface for pollen at its apex.

sub- [L. under]   A prefix used to designate a structure that lies beneath another or is less than another. For example, subcutaneous (beneath the skin); subspecies.

suberin   A waxlike lipid that is a barrier to water and solute movement across the Casparian strip of the endodermis.

submucosa (sub mew koe' sah)   The tissue layer just under the epithelial lining of the lumen of the digestive tract.

subsoil   The soil horizon lying below the topsoil and above the parent rock (bedrock); the zone of infiltration and accumulation of materials leached from the topsoil. Also called the B horizon.

substrate (sub' strayte)   The molecule or molecules on which an enzyme exerts catalytic action.

substratum (plural: substrata)   The base material on which a sessile organism lives.

succession   The gradual, sequential series of changes in the species composition of a community following a disturbance.

succulence   In plants, possession of fleshy, water-storing leaves or stems; an adaptation to dry environments.

superficial cleavage   A variation of incomplete cleavage in which cycles of mitosis occur without cell division, producing a syncytium (a single cell with many nuclei).

suprachiasmatic nuclei (SCN)   In mammals, two clusters of neurons just above the optic chiasm that act as the master circadian clock.

surface area-to-volume ratio   For any cell, organism, or geometrical solid, the ratio of surface area to volume; this is an important factor in setting an upper limit on the size a cell or organism can attain.

surface tension   The attractive intermolecular forces at the surface of liquid; especially important in water.

surfactant   A substance that decreases the surface tension of a liquid. Lung surfactant, secreted by cells of the alveoli, is mostly phospholipid and decreases the amount of work necessary to inflate the lungs.

survivorship (l_x)   In life tables, the proportion of individuals in a cohort that are alive at age *x*. A graph of this data is a survivorship curve.

suspensor   In the embryos of seed plants, the stalk of cells that pushes the embryo into the endosperm and is a source of nutrient transport to the embryo.

sustainable   Pertaining to the use and management of ecosystems in such a way that humans benefit over the long term from specific ecosystem goods and services without compromising others.

symbiosis (sim' bee oh' sis) [Gk. *sym*: together + *bios*: living]   The living together of two or more species in a prolonged and intimate relationship.

symmetry   Pertaining to an attribute of an animal body in which at least one plane can divide the body into similar, mirror-image halves. (*See* bilateral symmetry, radial symmetry.)

sympathetic nervous system   The division of the autonomic nervous system that works in opposition to the parasympathetic nervous system. (Contrast with parasympathetic nervous system.)

sympatric speciation (sim pat' rik) [Gk. *sym*: same + *patria*: homeland]   Speciation due to reproductive isolation without any physical separation of the subpopulation. (Contrast with allopatric speciation.)

symplast   The continuous meshwork of the interiors of living cells in the plant body, resulting from the presence of plasmodesmata. (Contrast with apoplast.)

symporter   A membrane transport protein that carries two substances in the same direction. (Contrast with antiporter, uniporter.)

synapomorphy   A trait that arose in the ancestor of a phylogenetic group and is present (sometimes in modified form) in all of its members, thus helping to delimit and identify that group. Also called a shared derived trait.

synapse (sin' aps) [Gk. *syn*: together + *haptein*: to fasten]   A specialized type of junction where a neuron meets its target cell (which can be another neuron or some other type of cell) and information in the form of neurotransmitter molecules is exchanged across a synaptic cleft.

synapsis (sin ap' sis)   The highly specific parallel alignment (pairing) of homologous chromosomes during the first division of meiosis.

synergids [Gk. *syn*: together + *ergos*: work]   In angiosperms, the two cells accompanying the egg cell at one end of the megagametophyte.

syngamy   *See* fertilization.

synonymous (silent) substitution   A change of one nucleotide in a sequence to another when that change does not affect the amino acid specified (i.e., UUA → UUG, both specifying leucine). (Contrast with nonsynonymous substitution, missense substitution, nonsense substitution.)

systematics   The scientific study of the diversity and relationships among organisms.

systemic acquired resistance   A general resistance to many plant pathogens following infection by a single agent.

systemic circuit   Portion of the circulatory system by which oxygenated blood from the lungs or gills is distributed throughout the rest of the body and returned to the heart. (Contrast with pulmonary circuit.)

systems biology   The scientific study of an organism as an integrated and interacting system of genes, proteins, and biochemical reactions.

systole (sis' tuh lee) [Gk. *systole*: contraction]   Contraction of a chamber of the heart, driving blood forward in the circulatory system. (Contrast with diastole.)

- T -

T cell   A type of lymphocyte involved in the cellular immune response. The final stages of its development occur in the thymus gland. (Contrast with B cell; *see also* cytotoxic T cell, T-helper cell.)

T cell receptor   A protein on the surface of a T cell that recognizes the antigenic determinant for which the cell is specific.

T-helper cell (T_H)   Type of T cell that stimulates events in both the cellular and humoral immune responses by binding to the antigen on an antigen-presenting cell; target of the HIV-I virus, the agent of AIDS. (Contrast with cytotoxic T cells.)

T tubules   A system of tubules that runs throughout the cytoplasm of a muscle fiber, through which action potentials spread.

taproot system   A root system typical of eudicots consisting of a primary root (*taproot*) that extends downward by tip growth and outward by initiating lateral roots. (Contrast with fibrous root system.)

target cell   A cell with the appropriate receptors to bind and respond to a particular hormone or other chemical mediator.

taste bud   A structure in the epithelium of the tongue that includes a cluster of chemoreceptors innervated by sensory neurons.

TATA box   An eight-base-pair sequence, found about 25 base pairs before the starting point for transcription in many eukaryotic promoters, that binds a transcription factor and thus helps initiate transcription.

taxon (plural: taxa) [Gk. *taxis*: arrange, put in order]   A biological group (typically a species or a clade) that is given a name.

telencephalon   The outer, surrounding structure of the embryonic vertebrate forebrain, which develops into the cerebrum.

telomerase   An enzyme that catalyzes the addition of telomeric sequences lost from chromosomes during DNA replication.

telomeres (tee' lo merz) [Gk. *telos*: end + *meros*: units, segments]   Repeated DNA sequences at the ends of eukaryotic chromosomes.

telophase (tee' lo phase) [Gk. *telos*: end]   The final phase of mitosis or meiosis during which chromosomes became diffuse, nuclear envelopes reform, and nucleoli begin to reappear in the daughter nuclei.

template   A molecule or surface on which another molecule is synthesized in complementary fashion, as in the replication of DNA.

template strand   In double-stranded DNA, the strand that is transcribed to create an RNA transcript that will be processed into a protein. Also refers to a strand of RNA that is used to create a complementary RNA.

temporal summation   In the production or inhibition of action potentials in a postsynaptic cell, the interaction of depolarizations or hyperpolarizations produced by rapidly repeated stimulation of a single point on the postsynaptic cell. (Contrast with spatial summation.)

tendon   A collagen-containing band of tissue that connects a muscle with a bone.

tepal   A sterile, modified, nonphotosynthetic leaf of an angiosperm flower that cannot be distinguished as a petal or a sepal.

termination   In molecular biology, the end of transcription or translation.

terminator   A sequence at the 3′ end of mRNA that causes the RNA strand to be released from the transcription complex.

terrestrial (ter res′ tree al) [L. *terra*: earth]   Pertaining to or living on land. (Contrast with aquatic, marine.)

tertiary structure   In reference to a protein, the relative locations in three-dimensional space of all the atoms in the molecule. The overall shape of a protein. (Contrast with primary, secondary, and quaternary structures.)

test cross   Mating of a dominant-phenotype individual (who may be either heterozygous or homozygous) with a homozygous-recessive individual.

testis (tes′ tis) (plural: testes) [L. *testis*: witness]   The male gonad; the organ that produces the male gametes.

tetanus [Gk. *tetanos*: stretched]   (1) A state of sustained maximal muscular contraction caused by rapidly repeated stimulation. (2) In medicine, an often fatal disease ("lockjaw") caused by the bacterium *Clostridium tetani*.

tetrad [Gk. *tettares*: four]   During prophase I of meiosis, the association of a pair of homologous chromosomes or four chromatids.

thalamus [Gk. *thalamos*: chamber]   A region of the vertebrate forebrain; involved in integration of sensory input.

theory [Gk. *theoria*: analysis of facts]   A far-reaching explanation of observed facts that is supported by such a wide body of evidence, with no significant contradictory evidence, that it is scientifically accepted as a factual framework. Examples are Newton's theory of gravity and Darwin's theory of evolution. (Contrast with hypothesis.)

thermoneutral zone [Gk. *thermos*: temperature]   The range of temperatures over which an endotherm does not have to expend extra energy to thermoregulate.

thermophile (ther′ muh fyle)[Gk. *thermos*: temperature + *philos*: loving]   An organism that lives exclusively in hot environments.

thoracic cavity [Gk. *thorax*: breastplate]   The portion of the mammalian body cavity bounded by the ribs, shoulders, and diaphragm. Contains the heart and the lungs.

thoracic duct   The connection between the lymphatic system and the circulatory system.

threshold   The level of depolarization that causes an electrically excitable membrane to fire an action potential.

thrombus (throm′ bus) [Gk. *thrombos*: clot]   A blood clot that forms within a blood vessel and remains attached to the wall of the vessel. (Contrast with embolus.)

thylakoid (thigh la koid) [Gk. *thylakos*: sack or pouch]   A flattened sac within a chloroplast. Thylakoid membranes contain all of the chlorophyll in a plant, in addition to the electron carriers of photophosphorylation. Thylakoids stack to form grana.

thymine (T)   Nitrogen-containing base found in DNA.

thymus [Gk. *thymos*: warty]   A ductless, glandular lymphoid tissue, involved in development of the immune system of vertebrates. In humans, the thymus degenerates during puberty.

thyroid gland [Gk. *thyreos*: door-shaped]   A two-lobed gland in vertebrates. Produces the hormone thyroxin.

thyrotropin   Hormone produced by the anterior pituitary that stimulates the thyroid gland to produce and release thyroxine. Also called thyroid-stimulating hormone (TSH).

thyrotropin-releasing hormone (TRH)   Hormone produced by the hypothalamus that stimulates the anterior pituitary to release thyrotropin.

thyroxine   Hormone produced by the thyroid gland; controls many metabolic processes.

tidal volume   The amount of air that is exchanged during each breath when a person is at rest.

tight junction   A junction between epithelial cells in which there is no gap between adjacent cells.

tissue   A group of similar cells organized into a functional unit; usually integrated with other tissues to form part of an organ.

tissue system   In plants, any of three organized groups of tissues—dermal tissue, vascular tissue, and ground tissue—that are established during embryogenesis and have distinct functions.

titin   A protein that holds bundles of myosin filaments in a centered position within the sarcomeres of muscle cells. The largest protein in the human body.

tonoplast   The membrane of the plant central vacuole.

topsoil   The uppermost soil horizon; contains most of the organic matter of soil, but may be depleted of most mineral nutrients by leaching. Also called the A horizon.

totipotent [L. *toto*: whole, entire + *potens*: powerful]   Possessing all the genetic information and other capacities necessary to form an entire individual. (Contrast with multipotent, pluripotent.)

trachea (tray′ kee ah) [Gk. *trakhoia*: tube]   A tube that carries air to the bronchi of the lungs of vertebrates. When plural (*tracheae*), refers to the major airways of insects.

tracheary element   Either of two types of xylem cells—tracheids and vessel elements—that undergo apoptosis before assuming their transport function.

tracheid (tray′ kee id)   A type of tracheary element found in the xylem of nearly all vascular plants, characterized by tapering ends and walls that are pitted but not perforated. (Contrast with vessel element.)

trade-off   The relationship between the fitness benefits conferred by an adaptation and the fitness costs it imposes. For an adaptation to be favored by natural selection, the benefits must exceed the costs.

trait   In genetics, a specific form of a character: eye color is a character; brown eyes and blue eyes are traits. (Contrast with character.)

transcription   The synthesis of RNA using one strand of DNA as a template.

transcription factors   Proteins that assemble on a eukaryotic chromosome, allowing RNA polymerase II to perform transcription.

transduction   (1) Transfer of genes from one bacterium to another by a bacteriophage. (2) In sensory cells, the transformation of a stimulus (e.g., light energy, sound pressure waves, chemical or electrical stimulants) into action potentials.

transfection   Insertion of recombinant DNA into animal cells.

transfer RNA (tRNA)   A family of double-stranded RNA molecules. Each tRNA carries a specific amino acid and anticodon that will pair with the complementary codon in mRNA during translation.

transformation   (1) A mechanism for transfer of genetic information in bacteria in which pure DNA from a bacterium of one genotype is taken in through the cell surface of a bacterium of a different genotype and incorporated into the chromosome of the recipient cell. (2) Insertion of recombinant DNA into a host cell.

transgenic   Containing recombinant DNA incorporated into the genetic material.

translation   The synthesis of a protein (polypeptide). Takes place on ribosomes, using the information encoded in messenger RNA.

translocation   (1) In genetics, a rare mutational event that moves a portion of a chromosome to a new location, generally on a nonhomologous chromosome. (2) In vascular plants, movement of solutes in the phloem.

transmembrane protein   An integral membrane protein that spans the phospholipid bilayer.

transpiration [L. *spirare*: to breathe]   The evaporation of water from plant leaves and stem, driven by heat from the sun, and providing the motive force to raise water (plus mineral nutrients) from the roots.

transpiration–cohesion–tension mechanism   Theoretical basis for water movement in plants: evaporation of water from cells within leaves (transpiration) causes an increase ion surface tension, pulling water up through the xylem. Cohesion of water occurs because of hydrogen bonding.

transposable element   A segment of DNA that can move to, or give rise to copies at, another locus on the same or a different chromosome.

transposon   Mobile DNA segment that can insert into a chromosome and cause genetic change.

triglyceride   A simple lipid in which three fatty acids are combined with one molecule of glycerol.

triploblastic   Having three cell layers.

trisomic   Containing three rather than two members of a chromosome pair.

tRNA   *See* transfer RNA.

trochophore (troke′ o fore) [Gk. *trochos*: wheel + *phoreus*: bearer]   A radially symmetrical larval form typical of annelids and mollusks, distinguished by a wheel-like band of cilia around the middle.

trophic cascade   The progression over successively lower trophic levels of the indirect effects of a predator.

trophic level [Gk *trophes*: nourishment]   A group of organisms united by obtaining their energy from the same part of the food web of a biological community.

trophoblast [Gk *trophes*: nourishment + *blastos*: sprout]   At the 32-cell stage of mammalian development, the outer group of cells that will become part of the placenta and thus nourish the growing embryo. (Contrast with inner cell mass.)

tropic hormones   Hormones produced by the anterior pituitary that control the secretion of hormones by other endocrine glands.

tropomyosin [troe poe my' oh sin]   One of the three protein components of an actin filament; controls the interactions of actin and myosin necessary for muscle contraction.

troponin   One of the three components of an actin filament; binds to actin, tropomyosin, and Ca^{2+}.

troposphere   The lowest atmospheric zone, reaching upward from the Earth's surface approximately 10–17 km. Zone in which virtually all water vapor is located.

true-breeding   A genetic cross in which the same result occurs every time with respect to the trait(s) under consideration, due to homozygous parents.

trypsin   A protein-digesting enzyme. Secreted by the pancreas in its inactive form (trypsinogen), it becomes active in the duodenum of the small intestine.

tubulin   A protein that polymerizes to form microtubules.

tumor [L. *tumor*: a swollen mass]   A disorganized mass of cells. Malignant tumors spread to other parts of the body.

tumor necrosis factor   A family of cytokines (growth factors) that causes cell death and is involved in inflammation.

tumor suppressor   A gene that codes for a protein product that inhibits cell proliferation; inactive in cancer cells. (Contrast with oncogene.)

turgor pressure [L. *turgidus*: swollen]   *See* pressure potential.

turnover   In freshwater ecosystems, vertical movements of water that bring nutrients and dissolved CO_2 to the surface and O_2 to deeper water.

tympanic membrane [Gk. *tympanum*: drum]   The eardrum.

- U -

ubiquinone (yoo bic' kwi known) [L. *ubique*: everywhere]   A mobile electron carrier of the mitochondrial respiratory chain. Similar to plastoquinone found in chloroplasts.

ubiquitin   A small protein that is covalently linked to other cellular proteins identified for breakdown by the proteosome.

ultimate cause   In ethology, the evolutionary processes that produced an animal's capacity and tendency to behave in particular ways. (Contrast with proximate cause.)

uniporter [L. *unus*: one + *portal*: doorway]   A membrane transport protein that carries a single substance in one direction. (Contrast with antiporter, symporter.)

unsaturated fatty acid   A fatty acid whose hydrocarbon chain contains one or more double bonds. (Contrast with saturated fatty acid.)

upregulation   A process by which the abundance of receptors for a hormone increases when

hormone secretion is suppressed. (Contrast with downregulation.)

upwelling zones   Areas of the ocean where cool, nutrient-rich water from deeper layers rises to the surface.

uracil (U)   A pyrimidine base found in nucleotides of RNA.

urea   A compound that is the main excreted form of nitrogen by many animals, including mammals.

ureotelic   Pertaining to an organism in which the final product of the breakdown of nitrogen-containing compounds (primarily proteins) is urea. (Contrast with ammonotelic, uricotelic.)

ureter (your' uh tur)   Long duct leading from the vertebrate kidney to the urinary bladder or the cloaca.

urethra (you ree' thra)   In most mammals, the canal through which urine is discharged from the bladder and which serves as the genital duct in males.

uric acid   A compound that serves as the main excreted form of nitrogen in some animals, particularly those which must conserve water, such as birds, insects, and reptiles.

uricotelic   Pertaining to an organism in which the final product of the breakdown of nitrogen-containing compounds (primarily proteins) is uric acid. (Contrast with ammonotelic, ureotelic.)

urinary bladder   A structure in which urine is stored until it can be excreted to the outside of the body.

urine (you' rin)   In vertebrates, the fluid waste product containing the toxic nitrogenous by-products of protein and amino acid metabolism.

uterine cycle   In human females, the monthly cycle of events by which the endometrium is prepared for the arrival of a blastocyst. (Contrast with ovarian cycle).

uterus (yoo' ter us) [L. *utero*: womb]   A specialized portion of the female reproductive tract in mammals that receives the fertilized egg and nurtures the embryo in its early development. Also called the womb.

- V -

vaccination   Injection of virus or bacteria or their proteins into the body, to induce immunization. The injected material is usually attenuated (weakened) before injection.

vacuole (vac' yew ole)   Membrane-enclosed organelle in plant cells that can function for storage, water concentration for turgor, or hydrolysis of stored macromolecules.

vagina (vuh jine' uh) [L. *sheath*]   In female animals, the entry to the reproductive tract.

van der Waals forces   Weak attractions between atoms resulting from the interaction of the electrons of one atom with the nucleus of another. This type of attraction is about one-fourth as strong as a hydrogen bond.

variable region   The portion of an immunoglobulin molecule or T-cell receptor that includes the antigen-binding site and is responsible for its specificity. (Contrast with constant region.)

vas deferens (plural: vasa deferentia)   Duct that transfers sperm from the epididymis to the urethra.

vasa recta   Blood vessels that parallel the loops of Henle and the collecting ducts in the renal medulla of the kidney.

vascular (vas' kew lar) [L. *vasculum*: a small vessel]   Pertaining to organs and tissues that conduct fluid, such as blood vessels in animals and xylem and phloem in plants.

vascular bundle   In vascular plants, a strand of vascular tissue, including xylem and phloem as well as thick-walled fibers.

vascular cambium (kam' bee um) [L. *cambiare*: to exchange]   In plants, a lateral meristem that gives rise to secondary xylem and phloem.

vascular tissue system   The transport system of a vascular plant, consisting primarily of xylem and phloem.

vasopressin   *See* antidiuretic hormone.

vector   (1) An agent, such as an insect, that carries a pathogen affecting another species. (2) A plasmid or virus that carries an inserted piece of DNA into a bacterium for cloning purposes in recombinant DNA technology.

vegetal hemisphere   The lower portion of some animal eggs, zygotes, and embryos, in which the dense nutrient yolk settles. The *vegetal pole* is to the very bottom of the egg or embryo. (Contrast with animal hemisphere.)

vegetative   Nonreproductive, nonflowering, or asexual.

vegetative meristem   An apical meristem that produces leaves.

vegetative reproduction   Asexual reproduction through the modification of stems, leaves, or roots.

vein [L. *vena*: channel]   A blood vessel that returns blood to the heart. (Contrast with artery.)

ventral [L. *venter*: belly, womb]   Toward or pertaining to the belly or lower side. (Contrast with dorsal.)

ventricle   A muscular heart chamber that pumps blood through the lungs or through the body.

venule   A small blood vessel draining a capillary bed that joins others of its kind to form a vein. (Contrast with arteriole.)

vernalization [L. *vernalis*: spring]   Events occurring during a required chilling period, leading eventually to flowering.

vertebral column [L. *vertere*: to turn]   The jointed, dorsal column that is the primary support structure of vertebrates.

very low-density lipoproteins (VLDLs)   Lipoproteins that consist mainly of triglyceride fats, which they transport to fat cells in adipose tissues throughout the body; associated with excessive fat deposition and high risk for cardiovascular disease.

vesicle   Within the cytoplasm, a membrane-enclosed compartment that is associated with other organelles; the Golgi complex is one example.

vessel element   A type of tracheary element with perforated end walls; found only in angiosperms. (Contrast with tracheid.)

vestibular system (ves tib' yew lar) [L. *vestibulum*: an enclosed passage]   Structures within the inner ear that sense changes in position or momentum of the head, affecting balance and motor skills.

vicariant event (vye care' ee unt) [L. *vicus*: change]   The splitting of a taxon's range by the imposition of some barrier to dispersal.

villus (vil' lus) (plural: villi) [L. *villus*: shaggy hair or beard]   A hairlike projection from a membrane; for example, from many gut walls.

virion (veer' e on)   The virus particle, the minimum unit capable of infecting a cell.

virulence [L. *virus*: poison, slimy liquid]   The ability of a pathogen to cause disease and death.

virus   Any of a group of ultramicroscopic particles constructed of nucleic acid and protein (and, sometimes, lipid) that require living cells in order to reproduce. Viruses evolved multiple times from different cellular species.

vital capacity   The maximum capacity for air exchange in one breath; the sum of the tidal volume and the inspiratory and expiratory reserve volumes.

vitamin [L. *vita*: life]   An organic compound that an organism cannot synthesize, but nevertheless requires in small quantities for normal growth and metabolism.

vitelline envelope   The inner, proteinaceous protective layer of a sea urchin egg.

viviparity (vye vi par' uh tee)   Reproduction in which fertilization of the egg and development of the embryo occur inside the mother's body. (Contrast with oviparity.)

vivipary   Premature germination in plants.

voltage-gated channel   A type of gated channel that opens or closes when a certain voltage exists across the membrane in which it is inserted.

vomeronasal organ (VNO)   Chemosensory structure embedded in the nasal epithelium of amphibians, reptiles, and many mammals. Often specialized for detecting pheromones.

- W -

water potential   In osmosis, the tendency for a system (a cell or solution) to take up water from pure water through a differentially permeable membrane. Water flows toward the system with a more negative water potential. (Contrast with solute potential, pressure potential.)

water vascular system   In echinoderms, a network of water-filled canals that functions in gas exchange, locomotion, and feeding.

wavelength   The distance between successive peaks of a wave train, such as electromagnetic radiation.

weathering   The mechanical and chemical processes by which rocks are broken down into soil particles.

Wernicke's area   A region in the temporal lobe of the human brain that is involved with the sensory aspects of language.

white blood cells   Cells in the blood plasma that play defensive roles in the immune system. Also called leukocytes.

white matter   In the central nervous system, tissue that is rich in axons. (Contrast with gray matter.)

wild-type   Geneticists' term for standard or reference type. Deviants from this standard, even if the deviants are found in the wild, are usually referred to as mutant. (Note that this terminology is not usually applied to human genes.)

wood   Secondary xylem tissue.

- X -

xerophyte (zee' row fyte) [Gk. *xerox*: dry + *phyton*: plant]   A plant adapted to an environment with limited water supply.

xylem (zy' lum) [Gk. *xylon*: wood]   In vascular plants, the tissue that conducts water and minerals; xylem consists, in various plants, of tracheids, vessel elements, fibers, and other highly specialized cells.

- Y -

yolk [M.E. *yolke*: yellow]   The stored food material in animal eggs, rich in protein and lipids.

yolk sac   In reptiles, birds, and mammals, the extraembryonic membrane that forms from the endoderm of the hypoblast; it encloses and digests the yolk.

- Z -

zeaxanthin   A blue-light receptor involved in the opening of plant stomata.

zona pellucida   A jellylike substance that surrounds the mammalian ovum when it is released from the ovary.

zoospore (zoe' o spore) [Gk. *zoon*: animal + *spora*: seed]   In algae and fungi, any swimming spore. May be diploid or haploid.

zygote (zye' gote) [Gk. *zygotos*: yoked]   The cell created by the union of two gametes, in which the gamete nuclei are also fused. The earliest stage of the diploid generation.

zymogen   The inactive precursor of a digestive enzyme; secreted into the lumen of the gut, where a protease cleaves it to form the active enzyme.

Illustration Credits

Cover: © Dr. Merlin D. Tuttle/Photo Researchers, Inc.
Frontispiece: © Art Wolfe/www.artwolfe.com.

Photographs appearing behind chapter numbers:

Part 1 *HMS Beagle*: Painting by Ronald Dean, reproduced by permission of the artist and Richard Johnson, Esquire.
Part 2 *Plant cells*: © Ed Reschke/Peter Arnold Inc.
Part 3 *Stoma*: © Andrew Syred/SPL/Photo Researchers, Inc.
Part 4 *Anaphase*: © Nasser Rusan.
Part 5 *Sheep*: © Roddy Field, the Roslin Institute.
Part 6 *Trilobite*: Courtesy of the Amherst College Museum of Natural History and the Trustees of Amherst College.
Part 7 *Diatoms*: © Scenics & Science/Alamy.
Part 8 *Flowers*: © Ed Reschke/Peter Arnold Inc.
Part 9 *Leopard*: Courtesy of Andrew D. Sinauer.
Part 10 *Hummingbird*: © Yufeng Zhou/istockphoto.com.

Table of Contents

Chapter 1 *Frogs*: © Pete Oxford/Minden Pictures. *T. Hayes*: © Pamela S. Turner. 1.1A: © Eye of Science/SPL/Photo Researchers, Inc. 1.1B: Science Photo Library/Photolibrary.com. 1.1C: © Steve Gschmeissner/Photo Researchers, Inc. 1.1D: © Frans Lanting. 1.1E: © Glen Threlfo/Auscape/Minden Pictures. 1.1F: © Piotr Naskrecki/Minden Pictures. 1.1G: © Tui De Roy/Minden Pictures. 1.2A: From R. Hooke, 1664. *Micrographia*. 1.2B: © John Durham/SPL/Photo Researchers, Inc. 1.2C: © Biophoto Associates/Photo Researchers, Inc. 1.3 *Maple*: © Simon Colmer & Abby Rex/Alamy. 1.3 *Spruce*: David McIntyre. 1.3 *Lily pad*: © Pete Oxford/Naturepl.com. 1.3 *Cucumber*: David McIntyre. 1.3 *Pitcher plant*: © Nick Garbutt/Naturepl.com. 1.5A: © Frans Lanting. 1.5B: © Stefan Huwiler/Rolfnp/Alamy. 1.6 *Organism*: © Nico Smit/istockphoto.com. 1.6 *Population*: © blickwinkel/Alamy. 1.6 *Community*: © Georgie Holland/AGE Fotostock. 1.6 *Biosphere*: Courtesy of NASA. 1.7: © P&R Fotos/AGE Fotostock. 1.9: © Michael Abbey/Visuals Unlimited, Inc. 1.11: Courtesy of Wayne Whippen. 1.13: From T. Hayes et al., 2003. *Environ. Health Perspect.* 111: 568.

Chapter 2 *Hair*: © Steve Gschmeissner/Photo Researchers, Inc. *Barber*: © Digital Vision/Alamy. 2.3: From N. D. Volkow et al., 2001. *Am. J. Psychiatry* 158: 377. 2.14: © Pablo H Caridad/Shutterstock. 2.15: © Denis Miraniuk/Shutterstock.

Chapter 3 *T. Rex*: © The Natural History Museum, London. *Chicken*: David McIntyre. 3.8: Data from PDB 1IVM. T. Obita, T. Ueda, & T. Imoto, 2003. *Cell. Mol. Life Sci.* 60: 176. 3.10A: Data from PDB 2HHB. G. Fermi et al., 1984. *J. Mol. Biol.* 175: 159. 3.16C *left*: © Biophoto Associates/Photo Researchers, Inc. 3.16C *middle*: © Ken Wagner/Visuals Unlimited. 3.16C *right*: © CNRI/SPL/Photo Researchers, Inc. 3.17 *Cartilage*: © Robert Brons/Biological Photo Service. 3.17 *Beetle*: © Scott Bauer/USDA.

Chapter 4 *Phoenix*: Courtesy of NASA/JPL/UA/Lockheed Martin. *Ice*: Courtesy of NASA/JPL-Caltech/University of Arizona/Texas A&M University. 4.4: Data from S. Arnott & D. W. Hukins, 1972. *Biochem. Biophys. Res. Commun.* 47(6): 1504. 4.8: Courtesy of the Argonne National Laboratory. 4.13B: Courtesy of Janet Iwasa, Szostak group, MGH/Harvard. 4.14: © Stanley M. Awramik/Biological Photo Service. 4.14 *inset*: © Dennis Kunkel Microscopy, Inc.

Chapter 5 *Heart cell*: © Roger J. Bick & Brian J. Poindexter/UT-Houston Medical School/Photo Researchers, Inc. *Surgery*: © The Stock Asylum, LLC/Alamy. 5.1: After N. Campbell, 1990. *Biology*, 2nd Ed., Benjamin Cummings. 5.1 *Protein*: Data from PDB 1IVM. T. Obita, T. Ueda, & T. Imoto, 2003. *Cell. Mol. Life Sci.* 60: 176. 5.1 *T4*: © Dept. of Microbiology, Biozentrum/SPL/Photo Researchers, Inc. 5.1 *Bacterium*: © Jim Biddle/Centers for Disease Control. 5.1 *Plant cells*: © Michael Eichelberger/Visuals Unlimited, Inc. 5.1 *Frog egg*: David McIntyre. 5.1 *Bird*: © Steve Byland/Shutterstock. 5.1 *Baby*: Courtesy of Sebastian Grey Miller. 5.3 *Light microscope*: © Radu Razvan/Shutterstock. 5.3 *Bright-field*: Courtesy of the IST Cell Bank, Genoa. 5.3 *Phase-contrast*: © Michael W. Davidson, Florida State U. 5.3 *Stained*: © Richard J. Green/SPL/Photo Researchers, Inc. 5.3 *Confocal*: © Dr. Gopal Murti/SPL/Photo Researchers, Inc. 5.3 *Electron microscope*: © Sinclair Stammers/Photo Researchers, Inc. 5.3 *TEM*: © Dr. Gopal Murti/Visuals Unlimited. 5.3 *SEM*: © K. R. Porter/SPL/Photo Researchers, Inc. 5.3 *Freeze-fracture*: © D. W. Fawcett/Photo Researchers, Inc. 5.4: © J. J. Cardamone Jr. & B. K. Pugashetti/Biological Photo Service. 5.5A: © Dennis Kunkel Microscopy, Inc. 5.5B: Courtesy of David DeRosier, Brandeis U. 5.7 *Mitochondrion*: © K. Porter, D. Fawcett/Visuals Unlimited. 5.7 *Cytoskeleton*: © Don Fawcett, John Heuser/Photo Researchers, Inc. 5.7 *Nucleolus*: © Richard Rodewald/Biological Photo Service. 5.7 *Peroxisome*: © E. H. Newcomb & S. E. Frederick/Biological Photo Service. 5.7 *Cell wall*: © Biophoto Associates/Photo Researchers, Inc. 5.7 *Ribosome*: From M. Boublik et al., 1990. *The Ribosome*, p. 177. Courtesy of American Society for Microbiology. 5.7 *Centrioles*: © Barry F. King/Biological Photo Service. 5.7 *Plasma membrane*: Courtesy of J. David Robertson, Duke U. Medical Center. 5.7 *Rough ER*: © Don Fawcett/Science Source/Photo Researchers, Inc. 5.7 *Smooth ER*: © Don Fawcett, D. Friend/Science Source/Photo Researchers, Inc. 5.7 *Chloroplast*: © W. P. Wergin, E. H. Newcomb/Biological Photo Service. 5.7 *Golgi apparatus*: Courtesy of L. Andrew Staehelin, U. Colorado. 5.8: © D. W. Fawcett/Photo Researchers, Inc. 5.9A: © Barry King, U. California, Davis/Biological Photo Service. 5.9B: © Biophoto Associates/Science Source/Photo Researchers, Inc. 5.10: © B. Bowers/Photo Researchers, Inc. 5.11: © Sanders/Biological Photo Service. 5.12: © K. Porter, D. Fawcett/Visuals Unlimited. 5.13 *left*: © W. P. Wergin, E. H. Newcomb/Biological Photo Service. 5.13 *right*: © W. P. Wergin/Biological Photo Service. 5.14A: © Michael Eichelberger/Visuals Unlimited, Inc. 5.14B: © Ed Reschke/Peter Arnold Inc. 5.14C: © Gerald & Buff Corsi/Visuals Unlimited. 5.15A: David McIntyre. 5.15A *inset*: © Richard Green/Photo Researchers, Inc. 5.15B: David McIntyre. 5.15B *inset*: Courtesy of R. R. Dute. 5.16: Courtesy of M. C. Ledbetter, Brookhaven National Laboratory. 5.17: Courtesy of Vic Small, Austrian Academy of Sciences, Salzburg, Austria. 5.19: Courtesy of N. Hirokawa. 5.20A *upper*: © SPL/Photo Researchers, Inc. 5.20A *lower*, 5.20B: © W. L. Dentler/Biological Photo Service. 5.22: From N. Pollock et al., 1999. *J. Cell Biol.* 147: 493. Courtesy of R. D. Vale. 5.23: © Michael Abbey/Visuals Unlimited. 5.24: © Biophoto Associates/Photo Researchers, Inc. 5.25 *left*: Courtesy of David Sadava. 5.25 *upper right*: From J. A. Buckwalter & L. Rosenberg, 1983. *Coll. Rel. Res.* 3: 489. Courtesy of L. Rosenberg. 5.25 *lower right*: © J. Gross, Biozentrum/SPL/Photo Researchers, Inc. 5.27: Courtesy of Noriko Okamoto and Isao Inouye.

Chapter 6 *Patient*: From Alzheimer, A. 1906. Über einen eigenartigen schweren Erkrankungsprozess der Hirnrinde. *Neurologisches Centralblatt* 23: 1129. *Plaques*: © G. W. Willis/Photolibrary.com. 6.2: After L. Stryer, 1981. *Biochemistry*, 2nd Ed., W. H. Freeman. 6.4: © D. W. Fawcett/Photo Researchers, Inc. 6.7A: Courtesy of D. S. Friend, U. California, San Francisco. 6.7B: Courtesy of Darcy E. Kelly, U. Washington. 6.7C: Courtesy of C. Peracchia. 6.10A *left*: © Stanley Flegler/Visuals Unlimited, Inc. 6.10A *center*: © David M. Phillips/Photo Researchers, Inc. 6.10B *left*: © Ed Reschke/Peter Arnold Inc. 6.13: From G. M. Preston et al., 1992. *Science* 256: 385. 6.19: From M. M. Perry, 1979. *J. Cell Sci.* 39: 26.

Chapter 7 *Voles*: Courtesy of Todd Ahern. *Oxytocin*: Data from PDB 1NPO. J. P. Rose et al., 1996. *Nat. Struct .Biol.* 3: 163. 7.3: © Biophoto Associates/Photo Researchers, Inc. 7.4: Data from PDB 3EML. V. P. Jaakola et al., 2008. *Science* 322: 1211. 7.16: © Stephen A. Stricker, courtesy of Molecular Probes, Inc.

Chapter 8 *Dairy*: © Bob Randall/ istockphoto.com. *Maasai*: ©blickwinkel/Alamy. 8.1: Courtesy of Violet Bedell-McIntyre. 8.5B: © Satoshi Kuribayashi/OSF/Photolibrary.com. 8.11A: Data from PDB 1AL6. B. Schwartz et al., 1997. 8.11B: Data from PDB 1BB6. V. B. Vollan et al., 1999. *Acta Crystallogr. D. Biol. Crystallogr.* 55: 60. 8.11C: Data from PDB 1AB9. N. H. Yennawar, H. P. Yennawar, & G. K. Farber, 1994. *Biochemistry* 33: 7326. 8.13: Data from PDB 1IG8 (P. R. Kuser et al., 2000. *J. Biol. Chem.* 275: 20814) and 1BDG (A. M. Mulichak et al., 1998 *Nat. Struct. Biol.* 5: 555).

Chapter 9 *Marathoners*: © Chuck Franklin/ Alamy. *Mouse*: © Royalty-Free/Corbis. 9.9: From Y. H. Ko et al., 2003. *J. Biol. Chem.* 278: 12305. Courtesy of P. Pedersen.

Chapter 10 *Rainforest*: © Jon Arnold Images/Photolibrary.com. *FACE*: Courtesy of David F. Karnosky. 10.1: © Andrew Syred/ SPL/Photo Researchers, Inc. 10.6: © Martin Shields/Alamy. 10.13: Courtesy of Lawrence Berkeley National Laboratory. 10.17A: © E. H. Newcomb & S. E. Frederick/Biological Photo Service. 10.21: © Aflo Foto Agency/Alamy.

Chapter 11 *Cells*: © Parviz M. Pour/Photo Researchers, Inc. *Model*: Data from PDB 1GUX. J. Lee et al., 1998. *Nature* 391: 859. 11.1A: © SPL/Photo Researchers, Inc. 11.1B: © Biodisc/ Visuals Unlimited/Alamy. 11.1C: © Robert Valentic/Naturepl.com. 11.2B: © John J. Cardamone Jr./Biological Photo Service. 11.8 *Chromosome*: Courtesy of G. F. Bahr. 11.8 *Nucleus*: © D. W. Fawcett/Photo Researchers, Inc. 11.9 *inset*: © Biophoto Associates/Science Source/Photo Researchers, Inc. 11.10B: © Conly L. Rieder/Biological Photo Service. 11.11: © Nasser Rusan. 11.13A: © T. E. Schroeder/Biological Photo Service. 11.13B: © B. A. Palevitz, E. H. Newcomb/Biological Photo Service. 11.14A: © Dr. John Cunningham/ Visuals Unlimited. 11.14B: © Steve Gschmeissner/Photo Researchers, Inc. 11.15 *left*: © Andrew Syred/SPL/Photo Researchers, Inc. 11.15 *center*: David McIntyre. 11.15 *right*: © Gerry Ellis, DigitalVision/PictureQuest. 11.16: Courtesy of Dr. Thomas Ried and Dr. Evelin Schröck, NIH. 11.17: © C. A. Hasenkampf/Biological Photo Service. 11.18: Courtesy of J. Kezer. 11.22A: © Gopal Murti/ Photo Researchers, Inc. 11.23: © Dennis Kunkel Microscopy, Inc.

Chapter 12 *Vavilov*: Courtesy of the Library of Congress/New York World-Telegram & Sun Collection. *Harvest*: © Paul Nevin/ Photolibrary.com. 12.1: © the Mendelianum. 12.11: Courtesy the American Netherland Dwarf Rabbit Club. 12.14: Courtesy of Madison, Hannah, and Walnut. 12.15: Courtesy of the Plant and Soil Sciences eLibrary (http:// plantandsoil.unl.edu); used with permission from the Institute of Agriculture and Natural Resources at the University of Nebraska. 12.16: © Grant Heilman Photography/Alamy. 12.17:

© Peter Morenus/U. of Connecticut. 12.26A: Courtesy of L. Caro and R. Curtiss.

Chapter 13 *Velociraptor*: © Joe Tucciarone/ Photo Researchers, Inc. *DNA art*: © Simon Fraser/Karen Rann/SPL/Photo Researchers, Inc. 13.3: © Biozentrum, U. Basel/SPL/Photo Researchers, Inc. 13.6 *X-ray crystallograph*: Courtesy of Prof. M. H. F. Wilkins, Dept. of Biophysics, King's College, U. London. 13.6B: © CSHL Archives/Peter Arnold, Inc. 13.8A: © A. Barrington Brown/Photo Researchers, Inc. 13.8B: Data from S. Arnott & D. W. Hukins, 1972. *Biochem. Biophys. Res. Commun.* 47(6): 1504. 13.14A: Data from PDB 1SKW. Y. Li et al., 2001. *Nat. Struct. Mol. Biol.* 11: 784. 13.20C: © Dr. Peter Lansdorp/Visuals Unlimited.

Chapter 14 *Mycobacterium*: © Dennis Kunkel Microscopy, Inc. *Ethiopia*: © Jenny Matthews/ Alamy. 14.3: Data from PDB 1I3Q. P. Cramer et al., 2001. *Science* 292: 1863. 14.9: From D. C. Tiemeier et al., 1978. *Cell* 14: 237. 14.12: Data from PDB 1EHZ. H. Shi & P. B. Moore, 2000. *RNA* 6: 1091. 14.14: Data from PDB 1GIX and 1G1Y. M. M. Yusupov et al., 2001. *Science* 292: 883. 14.18B: Courtesy of J. E. Edström and *EMBO J.*

Chapter 15 *Destruction*: © Suzanne Plunkett/ AP Images. *Baby 81*: © Gemunu Amarasinghe/ AP Images. 15.3: © Stanley Flegler/Visuals Unlimited, Inc. 15.8: © Philippe Plailly/Photo Researchers, Inc. 15.10: Bettmann/CORBIS. 15.11 *Butterfly*: © Bershadsky Yuri/ Shutterstock. 15.11 *Bacteria*: Courtesy of Janice Haney Carr/CDC. 15.11 *Fungus*: © Warwick Lister-Kaye/istockphoto.com. 15.17: From C. Harrison et al., 1983. *J. Med. Genet.* 20: 280. 15.19: © Simon Fraser/Photo Researchers, Inc.

Chapter 16 *Alcoholic*: © LJSphotography/ Alamy. *CREB*: Data from PDB 1T2K. D. Panne et al., 2004. *EMBO J.* 23: 4384. 16.2A: © Dennis Kunkel Microscopy, Inc. 16.2B: © Lee D. Simon/ Photo Researchers, Inc. 16.6: © NIBSC/Photo Researchers, Inc. 16.21A: Courtesy of the Centers for Disease Control.

Chapter 17 *Mastiff and chihuahua*: © moodboard RF/Photolibrary.com. *Whippets*: © Stuart Isett/Anzenberger/Eyevine. 17.3: © Sam Ogden/Photo Researchers, Inc. 17.5: Based on an illustration by Anthony R. Kerlavage, Institute for Genomic Research. *Science* 269: 449 (1995). 17.12: Courtesy of O. L. Miller, Jr. 17.17: From P. H. O'Farrell, 1975. *J. Biol. Chem.* 250: 4007. Courtesy of Patrick H. O'Farrell.

Chapter 18 *Kuwait*: © K. Bry—UNEP/Peter Arnold Inc. *A. Chakrabarty*: © Ted Spiegel/ Corbis. 18.5: © Martin Shields/Alamy. 18.15: Courtesy of the Golden Rice Humanitarian Board, www.goldenrice.org. 18.16: Courtesy of Eduardo Blumwald.

Chapter 19 *Horse*: © Benoit Photo. *Centrifuge*: Courtesy of Cytori Therapeutics. 19.4: © Roddy Field, the Roslin Institute. 19.11A: From J. E. Sulston & H. R. Horvitz, 1977. *Dev. Bio.* 56: 100. 19.14C: Courtesy of J. Bowman. 19.17: Courtesy of W. Driever and C. Nüsslein-Vollhard. 19.18B: Courtesy of C. Rushlow and M. Levine. 19.18C: Courtesy of T. Karr. 19.18D: Courtesy of S. Carroll and S. Paddock. 19.20: Courtesy of F. R. Turner, Indiana U.

Chapter 20 *Fly head*: © Science Photo Library RF/Photolibrary.com. *Mutant leg*: From G. Halder et al., 1995. *Science* 267: 1788. Courtesy of W. J. Gehring and G. Halder. 20.1 *Mouse*: © orionmystery@flickr/Shutterstock. 20.1 *Fly*: David McIntyre. 20.1 *Shark*: © Kristian Sekulic/ Shutterstock. 20.1 *Squid*: © Gergo Orban/ Shutterstock. 20.3: © David M. Phillips/Photo Researchers, Inc. 20.5: © Bone Clones, www.boneclones.com. 20.6: Courtesy of J. Hurle and E. Laufer. 20.7: Courtesy of J. Hurle. 20.8 *Cladogram*: After R. Galant & S. Carroll, 2002. *Nature* 415: 910. 20.8 *Insect*: © Stockbyte/ PictureQuest. 20.8 *Centipede*: © Burke/Triolo/ Brand X Pictures/PictureQuest. 20.9: From M. Kmita and D. Duboule, 2003. *Science* 301: 331. 20.10: © Neil Hardwick/Alamy. 20.11: © Rob Valentic/ANTPhoto.com. 20.12A: © Erick Greene. 20.12B: Courtesy of John Gruber. 20.13: © Nigel Cattlin, Holt Studios International/ Photo Researchers, Inc. 20.15: Courtesy of Mike Shapiro and David Kingsley.

Chapter 21 *Pandemic*: Courtesy of the Naval Historical Foundation. *Researcher*: Courtesy of James Gathany/Centers for Disease Control. 21.1A: Painting by Ronald Dean, reproduced by permission of the artist and Richard Johnson, Esquire. 21.2A: © Luis César Tejo/ Shutterstock. 21.2B: © Duncan Usher/Alamy. 21.2C: © PetStockBoys/Alamy. 21.2D: © Arco Images GmbH/Alamy. 21.9A: © Tom Ulrich/ OSF/Photolibrary.com. 21.9B: © Kim Karpeles/ Alamy. 21.11: David McIntyre. 21.14: Courtesy of David Hillis. 21.16: © Jason Gallier/Alamy. 21.17: David McIntyre. 21.21A: © Reinhard Dirscherl/WaterFrame - Underwater Images/ Photolibrary.com. 21.21B: © Marevision/AGE Fotostock. 21.22 *Snake*: © Joseph T. Collins/ Photo Researchers, Inc. 21.22 *Newt*: © Robert Clay/Visuals Unlimited.

Chapter 22 *HIV*: © James Cavallini/Photo Researchers, Inc. *Chimpanzees*: © John Cancalosi/Naturepl.com. 22.6A: Courtesy of William Jeffery. 22.6B: © Jurgen Freund/ Naturepl.com. 22.6C: © Michael & Patricia Fogden/Minden Pictures. 22.6D: © Mark Kostich/Shutterstock. 22.8: © Larry Jon Friesen. 22.10: © Alexandra Basolo. 22.13A: © Krieger, C./AGE Fotostock. 22.13B: © Krieger, C./ AGE Fotostock. 22.13C: © Ed Reschke/Peter Arnold Inc.

Chapter 23 *Cuatro Cienegas*: © George Grall/National Geographic. *Fly*: © Dr. David Phillips/Visuals Unlimited, Inc. 23.1A *left*: © R. L. Hambley/Shutterstock. 23.1A *right*: © Frank Leung/istockphoto.com. 23.1B: © Norman Bateman/istockphoto.com. 23.10: © OSF/ Photolibrary.com. 23.11 *G. olivacea*: © Charles Melton/Visuals Unlimited, Inc. 23.11 *G. carolinensis*: © Michael Redmer/Visuals Unlimited, Inc. 23.12A: © Yufeng Zhou/istockphoto.com. 23.12B: © P01017 Desmette Frede/AGE Fotostock. 23.12C: © J. S. Sira/ Photolibrary.com. 23.12D: © Daniel L. Geiger/SNAP/Alamy. 23.13: Courtesy of Donald A. Levin. 23.14: © Boris I. Timofeev/ Pensoft. 23.16A: © Tony Tilford/ Photolibrary.com. 23.16B: © W. Peckover/ VIREO. 23.17 *Madia*: © Peter K. Ziminsky/ Visuals Unlimited. 23.17 *Argyroxiphium*: © Ron Dahlquist/Pacific Stock/Photolibrary.com. 23.17 *Wilkesia*: © Gerald D. Carr. 23.17 *Dubautia*: © Noble Proctor/The National

Audubon Society Collection/Photo Researchers, Inc.

Chapter 24 *Electric fish:* © Jane Burton/Naturepl.com. *Ray:* © David Fleetham/Alamy. 24.3: *Rice:* data from PDB 1CCR. H. Ochi et al., 1983. *J. Mol. Biol.* 166: 407. *Tuna:* data from PDB 5CYT. T. Takano, 1984. 24.4: From P. B. Rainey & M. Travisano, 1998. *Nature* 394: 69. © Macmillan Publishers Ltd. 24.7A *Langur:* © blickwinkel/Alamy. 24.7A *Longhorn:* Courtesy of David Hillis. 24.7B: © M. Graybill/J. Hodder/Biological Photo Service.

Chapter 25 *Meganeura:* © Graham Cripps/NHMPL. *Modern dragonfly:* © Natasha Litova/istockphoto.com. *Grand Canyon:* © Tim Fitzharris/Minden Pictures. 25.4A: Photos courtesy of P.F. Hoffman, Geological Survey of Canada. 25.4B: © Robin Smith/Photolibrary.com. 25.8: © Martin Bond/SPL/Photo Researchers, Inc. 25.9: David McIntyre. 25.11 *left:* © Ken Lucas/Visuals Unlimited. 25.11 *center:* © The Natural History Museum, London. 25.11 *right:* Courtesy of Martin Smith. 25.12 *Cambrian:* © John Sibbick/NHMPL. 25.12 *Marella:* Courtesy of the Amherst College Museum of Natural History, The Trustees of Amherst College. 25.12 *Ottoia:* © Alan Sirulnikoff/Photo Researchers, Inc. 25.12 *Anomalocaris:* © Kevin Schafer/Alamy. 25.12 *Devonian:* © Tom McHugh/Field Museum, Chicago/Photo Researchers, Inc. 25.12 *Codiacrinus:* Courtesy of the Amherst College Museum of Natural History, The Trustees of Amherst College. 25.12 *Phacops:* Courtesy of the Amherst College Museum of Natural History, The Trustees of Amherst College. 25.12 *Eridophyllum:* © Mark A. Schneider/Photo Researchers, Inc. 25.12 *Nautiloid:* © The Natural History Museum, London. 25.12 *Permian:* © John Sibbick/NHMPL. 25.12 *Cacops:* © Albert Copley/Visuals Unlimited, Inc. 25.12 *Walchia:* © The Natural History Museum, London. 25.12 *Triassic:* © OSF/Photolibrary.com. 25.12 *Ferns:* © Ken Lucas/Visuals Unlimited. 25.12 *Coelophysis:* © Ken Lucas/Visuals Unlimited. 25.12 *Cretaceous:* © OSF/Photolibrary.com. 25.12 *Gryposaurus:* Courtesy of the Amherst College Museum of Natural History, The Trustees of Amherst College. 25.12 *Magnolia:* © The Natural History Museum, London. 25.12 *Tertiary:* © Publiphoto/Photo Researchers, Inc. 25.12 *Hyracotherium:* Courtesy of the Amherst College Museum of Natural History, The Trustees of Amherst College. 25.12 *Plesiadapis:* © The Natural History Museum, London. 25.13: Courtesy of Conrad C. Labandeira, Department of Paleobiology, National Museum of Natural History, Smithsonian Institution.

Chapter 26 *Antarctica:* Courtesy of Emily Gercke. *Yellowstone:* Courtesy of Jim Peaco/National Park Service. 26.2: © Dennis Kunkel Microscopy, Inc. 26.3B: © Steve Gschmeissner/Photo Researchers, Inc. 26.4: From F. Balagaddé et al., 2005. *Science* 309: 137. Courtesy of Frederick Balagaddé. 26.5A: © David M. Phillips/Visuals Unlimited. 26.5B: Courtesy of the CDC. 26.6A: © J. A. Breznak & H. S. Pankratz/Biological Photo Service. 26.6B: © J. Robert Waaland/Biological Photo Service. 26.7: © USDA/Visuals Unlimited. 26.8: © Steven Haddock and Steven Miller. 26.12: Courtesy of David Cox/CDC. 26.13: Courtesy of Randall

C. Cutlip. 26.14: © David Phillips/Visuals Unlimited. 26.15A: © Paul W. Johnson/Biological Photo Service. 26.15B: © H. S. Pankratz/Biological Photo Service. 26.15C: © Bill Kamin/Visuals Unlimited. 26.16: © Dr. Kari Lounatmaa/Photo Researchers, Inc. 26.17: © Dr. Gary Gaugler/Visuals Unlimited. 26.18: © Michael Gabridge/Visuals Unlimited. 26.20: © Geoff Kidd/Photo Researchers, Inc. 26.21: From K. Kashefi & D. R. Lovley, 2003. *Science* 301: 934. Courtesy of Kazem Kashefi. 26.23: © David Sanger Photography/Alamy. 26.24: From H. Huber et al., 2002. *Nature* 417: 63. © Macmillan Publishers Ltd. Courtesy of Karl O. Stetter. 26.25A: © Science Photo Library RF/Photolibrary.com. 26.25B, C: © Russel Kightley/SPL/Photo Researchers, Inc. 26.25D: © Science Photo Library RF/Photolibrary.com. 26.25E: animate4.com ltd./Photo Researchers, Inc. 26.25F: © Russell Kightley/Photo Researchers, Inc. 26.26: © Nigel Cattlin/Alamy.

Chapter 27 *Bug:* © Martin Dohrn/Photo Researchers, Inc. *Leishmania:* © Dennis Kunkel Microscopy, Inc. 27.4: © Michael Abbey/Photo Researchers, Inc. 27.7A: © Wim van Egmond/Visuals Unlimited. 27.7B: © Science Photo Library RF/Photolibrary.com. 27.8A: © Georgette Douwma/Naturepl.com. 27.8B: © David Patterson, Linda Amaral Zettler, Mike Peglar, & Tom Nerad/micro*scope. 27.9: © London School of Hygiene/SPL/Photo Researchers, Inc. 27.10A: © Bill Bachman/Photo Researchers, Inc. 27.10B: © Markus Geisen/NHMPL. 27.11: © Science Photo Library RF/Photolibrary.com. 27.16: © Dennis Kunkel Microscopy, Inc. 27.17A: © SPL/Photo Researchers, Inc. 27.17B: © Dennis Kunkel Microscopy, Inc. 27.17C: © Paul W. Johnson/Biological Photo Service. 27.17D: © Steve Gschmeissner/Photo Researchers, Inc. 27.19: © Scenics & Science/Alamy. 27.20A: © Marevision/AGE Fotostock. 27.20B: © Larry Jon Friesen. 27.20C: © J. N. A. Lott/Biological Photo Service. 27.20D: © Gerald & Buff Corsi/Visuals Unlimited, Inc. 27.21: © James W. Richardson/Visuals Unlimited. 27.22A: © Wim van Egmond/Visuals Unlimited, Inc. 27.22B: © Doug Sokell/Visuals Unlimited, Inc. 27.23A: © Carolina Biological/Visuals Unlimited. 27.23B: © Marevision/AGE Fotostock. 27.24A: © J. Paulin/Visuals Unlimited. 27.24B: © Dr. David M. Phillips/Visuals Unlimited. 27.26: © Andrew Syred/SPL/Photo Researchers, Inc. 27.27A: © William Bourland/micro*scope. 27.27B: © David Patterson & Aimlee Laderman/micro*scope. 27.28A: © Eye of Science/Photo Researchers, Inc. 27.29: Courtesy of R. Blanton and M. Grimson.

Chapter 28 *Silurian:* © Richard Bizley/Photo Researchers, Inc. *Rainforest:* © Photo Resource Hawaii/Alamy. 28.2A: © Bob Gibbons/OSF/Photolibrary.com. 28.2B: © Larry Mellichamp/Visuals Unlimited. 28.4: © J. Robert Waaland/Biological Photo Service. 28.6: © U. Michigan Exhibit Museum. 28.8A: Courtesy of the Biology Department Greenhouses, U. Massachusetts, Amherst. 28.9: After C. P. Osborne et al., 2004. *PNAS* 101: 10360. 28.11A: David McIntyre. 28.11C: © Harold Taylor/Photolibrary.com. 28.12A: © mediacolor's/Alamy. 28.12B: © Ed Reschke/Peter Arnold, Inc. 28.13A: © Dr. John D. Cunningham/Visuals Unlimited, Inc. 28.13B: © Danilo Donadoni/AGE Fotostock. 28.14: © Daniel

Vega/AGE Fotostock. 28.15A: David McIntyre. 28.15B: © Carolina Biological/Visuals Unlimited. 28.16A: © Bjorn Svensson/SPL/Photo Researchers, Inc. 28.16B: © J. N. A. Lott/Biological Photo Service. 28.17: Courtesy of the Biology Department Greenhouses, U. Massachusetts, Amherst. 28.18A: © Michael & Patricia Fogden/Minden Pictures. 28.18B: © E.A. Janes/AGE Fotostock. 28.18C: Courtesy of the Talcott Greenhouse, Mount Holyoke College. 28.19 *inset:* David McIntyre.

Chapter 29 *Masada:* © Eddie Gerald/Alamy. *Coconut:* © Ben Osborne/Naturepl.com. 29.1 *Cycad:* David McIntyre. 29.1 *Ginkgo:* © hypnotype/Shutterstock. 29.1 *Conifer:* © Irina Tischenko/istockphoto.com. 29.1 *Magnolia:* © Dole/Shutterstock. 29.4: © Natural Visions/Alamy. 29.5A: © Susumu Nishinaga/Photo Researchers, Inc. 29.6A: © Victoria Field/Shutterstock. 29.6B: © Topic Photo Agency IN/AGE Fotostock. 29.6C: © Pichugin Dmitry/Shutterstock. 29.6D: © Frans Lanting. 29.7A *left:* David McIntyre. 29.7A *right:* © Stan W. Elems/Visuals Unlimited. 29.7B *left:* David McIntyre. 29.7B *right:* © Dr. John D. Cunningham/Visuals Unlimited. 29.9: Courtesy of Jim Peaco/National Park Service. 29.10A: David McIntyre. 29.10B: © koer/Shutterstock. 29.10C: © Marion Nickig/Picture Press/Photolibrary.com. 29.11A: © Plantography/Alamy. 29.11B: © Thomas Photography LLC/Alamy. 29.16A: © bamby/Shutterstock. 29.16B: © blickwinkel/Alamy. 29.16C: David McIntyre. 29.16D: © Michel de Nijs/istockphoto.com. 29.16E: © Yuri Vainshtein/istockphoto.com. 29.16F: © Denis Pogostin/istockphoto.com. 29.18A: Photo by David McIntyre, courtesy of the University of Massachusetts Biology Department Greenhouses. 29.18B: David McIntyre. 29.18C: © WILDLIFE GmbH/Alamy. 29.18D: © blickwinkel/Alamy. 29.18E: David McIntyre. 29.18F: © Mike Donenfeld/Shutterstock. 29.19A: © amit erez/istockphoto.com. 29.19B: © Tootles/Shutterstock. 29.19C: © Charlotte Erpenbeck/Shutterstock. 29.20A, B: Courtesy of David Hillis. 29.20C: © Susan Law Cain/istockphoto.com.

Chapter 30 *Fusarium:* © Dr. Gary Gaugler/Visuals Unlimited. *Ant:* © L. E. Gilbert/Biological Photo Service. 30.3: © David M. Phillips/Visuals Unlimited. 30.4B: © Dr. Jeremy Burgess/Photo Researchers, Inc. 30.6: © Richard Packwood/Photolibrary.com. 30.7A: © G. T. Cole/Biological Photo Service. 30.8: © N. Allin & G. L. Barron/Biological Photo Service. 30.9A: Courtesy of David Hillis. 30.9B: David McIntyre. 30.9C: Courtesy of David Hillis. 30.11A: © R. L. Peterson/Biological Photo Service. 30.11B: © Ken Wagner/Visuals Unlimited. 30.12A: © J. Robert Waaland/Biological Photo Service. 30.12B: © M. F. Brown/Visuals Unlimited. 30.12C: © Dr. John D. Cunningham/Visuals Unlimited. 30.12D: © Biophoto Associates/Photo Researchers, Inc. 30.13: © Eye of Science/Photo Researchers, Inc. 30.14: © John Taylor/Visuals Unlimited. 30.15: © G. L. Barron/Biological Photo Service. 30.16A: © Photoshot Holdings Ltd/Alamy. 30.16B: © Biosphoto/Le Moigne Jean-Louis/Peter Arnold Inc. 30.17: © Andrew Syred/SPL/Photo Researchers, Inc. 30.18A: © Jämsen Jorma/AGE Fotostock. 30.18B: © Matt Meadows/Peter Arnold Inc.

Chapter 31 *Placozoan:* © Ana Yuri Signorovitch. *Sponge:* © Fred Bavendam/Minden Pictures. *Ctenophore:* © Norbert Wu/Minden Pictures. 31.2A: Courtesy of J. B. Morrill. 31.2B: From G. N. Cherr et al., 1992. *Microsc. Res. Tech.* 22: 11. Courtesy of J. B. Morrill. 31.4A: © Ed Robinson/Photolibrary.com. 31.4B: © Steve Gschmeissner/Photo Researchers, Inc. 31.4C: © DEA/Christian Ricci/Photolibrary.com. 31.5A: © Jurgen Freund/Naturepl.com. 31.5B: © John Bell/istockphoto.com. 31.6A: © John A. Anderson/istockphoto.com. 31.6B: © Kevin Schafer/DigitalVision/Photolibrary.com. 31.6B inset: © Mike Rogal/Shutterstock. 31.7B: © David Patterson & Aimlee Laderman/micro*scope. 31.8A: © IntraClique LLC/Shutterstock. 31.8B: © Stockbyte/PictureQuest. 31.9: Adapted from F. M. Bayerand & H. B. Owre, 1968. *The Free-Living Lower Invertebrates,* Macmillan Publishing Co. 31.10A: © Cathy Keifer/Shutterstock. 31.10B: David McIntyre. 31.12A: © First Light/Alamy. 31.12B: © Charlie Bishop/istockphoto.com. 31.13A: © Dave Watts/Naturepl.com. 31.13B: © Larry Jon Friesen. 31.14 inset: © Scott Camazine/Phototake/Alamy. 31.15: © Larry Jon Friesen. 31.16A: © Jurgen Freund/Naturepl.com. 31.16B: David McIntyre. 31.16C: © David Wrobel/Visuals Unlimited. 31.18B: © Larry Jon Friesen. 31.19: Adapted from F. M. Bayerand & H. B. Owre, 1968. *The Free-Living Lower Invertebrates,* Macmillan Publishing Co. 31.20A: © Larry Jon Friesen. 31.20B: © Georgette Douwma/Naturepl.com. 31.20C: © Larry Jon Friesen. 31.21A: © Jurgen Freund/Naturepl.com. 31.21B: © Stephan Kerkhofs/Shutterstock. 31.22: Adapted from F. M. Bayerand & H. B. Owre, 1968. *The Free-Living Lower Invertebrates,* Macmillan Publishing Co.

Chapter 32 *Strepsipterans:* Courtesy of Dr. Hans Pohl. *Wasp:* © Sean McCann. 32.2: © blickwinkel/Alamy. 32.3A: From D. C. García-Bellido & D. H. Collins, 2004. *Nature* 429: 40. Courtesy of Diego García-Bellido Capdevila. 32.3B: © Piotr Naskrecki/Minden Pictures. 32.6: © Alexis Rosenfeld/Photo Researchers, Inc. 32.7A: © Larry Jon Friesen. 32.8B: © Robert Brons/Biological Photo Service. 32.9B: © Larry Jon Friesen. 32.10A: © Fred Bavendam/Minden Pictures. 32.11: © David Wrobel/Visuals Unlimited. 32.13A: © Larry Jon Friesen. 32.13B: Courtesy of Cindy Lee Van Dover. 32.13C: © Pakhnyushcha/Shutterstock. 32.13D: © Larry Jon Friesen. 32.15A: © Larry Jon Friesen. 32.15B: © Dave Fleetham/Photolibrary.com. 32.15C: © Larry Jon Friesen. 32.15F: © Reinhard Dirscherl/Alamy. 32.16A: Courtesy of Jen Grenier and Sean Carroll, U.Wisconsin. 32.16B: Courtesy of Graham Budd. 32.16C: Courtesy of Reinhardt Møbjerg Kristensen. 32.17B: © Grave/Photo Researchers, Inc. 32.17C: © Steve Gschmeissner/Photo Researchers, Inc. 32.18: © Pascal Goetgheluck/Photo Researchers, Inc. 32.19A: © Michael Fogden/Photolibrary.com. 32.19B: © Steve Gschmeissner/Photo Researchers, Inc. 32.20: © Kevin Schafer/Alamy. 32.21A: © David M Dennis/OSF/Photolibrary.com. 32.21B: © John R. MacGregor/Peter Arnold, Inc. 32.22A: © David Shale/Naturepl.com. 32.22B: © Frans Lanting. 32.23A: © Kelly Swift, www.swiftinverts.com. 32.23B: © Larry Jon Friesen. 32.23C: © Nigel Cattlin/Alamy. 32.23D: Photo by Eric Erbe; colorization by

Chris Pooley/USDA ARS. 32.24A, B: © Larry Jon Friesen. 32.24C: © Solvin Zankl/Naturepl.com. 32.24D: © Larry Jon Friesen. 32.24E: © Norbert Wu/Minden Pictures. 32.27: © Oxford Scientific Films/Photolibrary.com. 32.28: © Mark Moffett/Minden Pictures. 32.29A: © John C. Abbott/Abbott Nature Photography. 32.29B: © Dr. Torsten Heydenreich/imagebroker/Alamy. 32.29C: © Larry Jon Friesen. 32.29D: David McIntyre. 32.29E: © Papilio/Alamy. 32.29F: © CorbisRF/Photolibrary.com. 32.29G: © Larry Jon Friesen. 32.29H: © Juniors Bildarchiv/Alamy.

Chapter 33 *Gastric-brooding frog:* © Michael Tyler/ANTPhoto.com. *Marsupial frog:* © Michael Fogden/OSF/Photolibrary.com. 33.2: From S. Bengtson, 2000. Teasing fossils out of shales with cameras and computers. *Palaeontologia Electronica* 3(1). 33.4A: © Hal Beral/Visuals Unlimited. 33.4B: © Jose B. Ruiz/Naturepl.com. 33.4C: © WaterFrame/Alamy. 33.4D: © Peter Scoones/Photo Researchers, Inc. 33.4E: © Larry Jon Friesen. 33.5A: © C. R. Wyttenbach/Biological Photo Service. 33.6A: © Stan Elems/Visuals Unlimited, Inc. 33.6B: © Larry Jon Friesen. 33.7A: © Marevision/AGE Fotostock. 33.7B: © David Wrobel/Visuals Unlimited. 33.9A: © Brandon Cole Marine Photography/Alamy. 33.9B left: © Marevision/AGE Fotostock. 33.9B right: © anne de Haas/istockphoto.com. 33.11B: © Roger Klocek/Visuals Unlimited. 33.12A: © David B. Fleetham/OSF/Photolibrary.com. 33.12B: © Kelvin AitkenAGE Fotostock. 33.12C: © Norbert Wu/Minden Pictures. 33.13A: © Peter Pinnock/ImageState/Alamy. 33.13B, C: © Larry Jon Friesen. 33.13D: © Norbert Wu/Minden Pictures. 33.14A: © Peter Scoones/Photo Researchers, Inc. 33.14B: © Tom McHugh/Photo Researchers, Inc. 33.14C: © Ted Daeschler/Academy of Natural Sciences/VIREO. 33.16A: © Morley Read/Naturepl.com. 33.16B: © Michael & Patricia Fogden/Minden Pictures. 33.16C: © Jack Goldfarb/Design Pics, Inc./Photolibrary.com. 33.16D: Courtesy of David Hillis. 33.19A: © Dave B. Fleetham/OSF/Photolibrary.com. 33.19B: © C. Alan Morgan/Peter Arnold, Inc. 33.19C: © Cathy Keifer/Shutterstock. 33.19D: © Larry Jon Friesen. 33.20A: © Susan Flashman/istockphoto.com. 33.20B: © Gerry Ellis, DigitalVision/PictureQuest. 33.21A: From X. Xu et al., 2003. *Nature* 421: 335. © Macmillan Publishers Ltd. 33.21B: © Tom & Therisa Stack/Painet. 33.22: © Melinda Fawver/istockphoto.com. 33.23A: © Tim Zurowski/All Canada Photos/Photolibrary.com. 33.23B: © Roger Wilmhurst/Foto Natura/Minden Pictures. 33.23C: Courtesy of Andrew D. Sinauer. 33.23D: © Tom Vezo/Minden Pictures. 33.24A: © imagebroker.net/Photolibrary.com. 33.24B: © Dave Watts/Alamy. 33.25A: © Ingo Arndt/Naturepl.com. 33.25B: © JTB Photo Communications/Photolibrary.com. 33.25C: © Michael & Patricia Fogden/Minden Pictures. 33.26A: © Design Pics Inc/Photolibrary.com. 33.26B: © Barry Mansell/Naturepl.com. 33.26C: © Michael S. Nolan/AGE Fotostock. 33.26D: © John E Marriott/All Canada Photos/AGE Fotostock. 33.28: © Pete Oxford/Minden Pictures. 33.29A: © mike lane/Alamy. 33.29B: © Eric Isselée/Shutterstock. 33.30A: © Steve Bloom Images/Alamy. 33.30B: © Anup Shah/AGE Fotostock. 33.30C: © Lars Christensen/

istockphoto.com. 33.30D: © Anup Shah/Minden Pictures.

Chapter 34 *Svalbard:* Courtesy of the Global Crop Diversity Trust. *Seed:* © Dr. Richard Kessel & Dr. Gene Shih/Visuals Unlimited, Inc. 34.2A: David McIntyre. 34.2C: © Biosphoto/Thiriet Claudius/Peter Arnold Inc. 34.3A: David McIntyre. 34.3B: © Steven Wooster/Garden Picture Library/Photolibrary.com. 34.3C: © Renee Lynn/Photo Researchers, Inc. 34.6: © Biophoto Associates/Photo Researchers, Inc. 34.9A: © Biodisc/Visuals Unlimited. 34.9B: © P. Gates/Biological Photo Service. 34.9C: © Biophoto Associates/Photo Researchers, Inc. 34.9D: © Jack M. Bostrack/Visuals Unlimited. 34.9E: © John D. Cunningham/Visuals Unlimited. 34.9F: © J. Robert Waaland/Biological Photo Service. 34.9G: © Randy Moore/Visuals Unlimited. 34.10 upper: © Larry Jon Friesen. 34.10 lower: © Biodisc/Visuals Unlimited. 34.11B: © Microfield Scientific LTD/Photo Researchers, Inc. 34.13A: © Larry Jon Friesen. 34.13B: © Ed Reschke/Peter Arnold, Inc. 34.13C: © Dr. James W. Richardson/Visuals Unlimited. 34.14A left: © Ed Reschke/Peter Arnold, Inc. 34.14A right: © Biodisc/Visuals Unlimited. 34.14B: © Ed Reschke/Peter Arnold, Inc. 34.15B: Courtesy of Thomas Eisner, Cornell U. 34.15C: © Susumu Nishinaga/Photo Researchers, Inc. 34.18: © Biodisc/Visuals Unlimited. 34.19: © David M Dennis/OSF/Photolibrary.com.

Chapter 35 *Planting:* © Robert Harding Picture Library Ltd/Alamy. *Drought:* Courtesy of the International Rice Research Institute/IRRI file photo. 35.4: © Nigel Cattlin/Alamy. 35.9A: © David M. Phillips/Visuals Unlimited. 35.10: After G. D. Humble & K. Raschke, 1971. *Plant Physiology* 48: 447. 35.12: © R. Kessel & G. Shih/Visuals Unlimited. 35.13: © M. H. Zimmermann.

Chapter 36 *Family:* Dorothea Lange/Library of Congress, Prints & Photographs Division, FSA/OWI Collection. *Haiti:* Courtesy of the NASA/Goddard Space Flight Center Scientific Visualization Studio. 36.1: David McIntyre. 36.5: © Kathleen Blanchard/Visuals Unlimited. 36.9 left: © E. H. Newcomb & S. R. Tandon/Biological Photo Service. 36.9 right: © Dr. Jeremy Burgess/SPL/Photo Researchers, Inc. 36.12A: © J. H. Robinson/The National Audubon Society Collection/Photo Researchers, Inc. 36.12B: © Kim Taylor/Naturepl.com. 36.13: Courtesy of Susan and Edwin McGlew.

Chapter 37 *N. Borlaug:* © Micheline Pelletier/Sygma/Corbis. *Rice:* Courtesy of Drs. Matsuoka and Ashikari. 37.2: © Visions of America/Joe Sohm/Photolibrary.com. 37.3: From J. M. Alonso and J. R. Ecker, 2006. *Nature Reviews Genetics* 7: 524. 37.4: Courtesy of J. A. D. Zeevaart, Michigan State U. 37.5: © Sylvan Wittwer/Visuals Unlimited, Inc. 37.12: © Ed Reschke/Peter Arnold, Inc. 37.16: David McIntyre.

Chapter 38 *Girl:* © Image Source/ZUMA Press. *Postcards:* © Amoret Tanner/Alamy. 38.1A: David McIntyre. 38.1B1: © Tish1/Shutterstock. 38.1B2: © Pierre BRYE/Alamy. 38.1C: David McIntyre. 38.3A: © Biosphoto/Hazan Muriel/Peter Arnold Inc.

Index

Sodium channels action potentials and, 953–954, 955